AF256352

Joachim Gottsmann · Jürgen Neuberg
Bettina Scheu

Editors

Volcanic Unrest

From Science to Society

Editors
Joachim Gottsmann
School of Earth Sciences
University of Bristol
Bristol, Bath and North East Somerset
UK

Jürgen Neuberg
Faculty of Environment
University of Leeds
Leeds, UK

Bettina Scheu
Ludwig Maximilian University of
  Munich
Munich, Bayern
Germany

https://doi.org/10.1007/978-3-319-58412-6

Volcanic unrest is a manifestation of complex subsurface processes which might or might not herald an imminent volcanic eruption. Short-term (hours to few weeks) unrest activity before an eruption may in hindsight be interpreted as "pre-eruptive". Protracted periods or waning of unrest without a clear relationship between unrest and future eruptive activity may be deemed "non-eruptive".

Periods of unrest challenge both scientists and decision-makers to interact effectively and efficiently to minimise societal risk during the unrest and in anticipation of a possible imminent eruption.

The scientific challenge is to identify whether unrest phenomena are pre-eruptive or non-eruptive early on in a developing unrest crisis. The challenge for decision-makers is to respond with the "right" decision during an emerging crisis.

Expectations from many different stakeholders including the media and the general public need to be managed amid epistemic (the unknown subsurface processes hidden from the observer) and aleatoric uncertainty (their stochastic variability).

Numerous examples in the recent past have shown that periods of unrest are hazardous and costly even without leading to imminent eruption (e.g. at La Soufriere, Guadeloupe, in 1976; at Campi Flegrei, Italy, in 1982–1984; at Long Valley Caldera/Mammoth Mountain, USA, in the 1980s and 1990s).

Tensions between different stakeholders can arise particularly during protracted periods (years to decades) of unrest at volcanoes with significant and growing populations, when trust in scientific knowledge is challenged amid the need for economic and societal development and growth.

Volcanic Unrest: From Science to Society—highlights the complexities of volcanic unrest from both scientific and societal perspectives and provides a summary of findings from the project entitled "Volcanic Unrest in Europe and Latin America: Phenomenology, eruption precursors, hazard forecast, and risk mitigation (VUELCO)". The multi-disciplinary and cross-boundary research project was funded by the European Commission's 7th framework programme for research, technological development and demonstration under grant agreement no. 282759 and brought together an international team of academics from the social and natural sciences, as well as personnel from volcano observatory and civil protection agencies.

The project was designed in response to a call by the EC for a "collaborative project (small- or medium-scale focused research project) for specific cooperation actions (SICA) dedicated for international cooperation partner countries with focus on Latin America".

The work was conducted at ten partner institutions across Europe and Latin America during 2011–2015. Most of the findings have been published in dedicated scientific journals over the past few years, and others will be published in the years to come.

The purpose of this open-access book is to summarise and publicise the key findings of the project for a broad audience. Seventeen and one appendix chapters focus on four broad topics relevant to the understanding of volcanic unrest in the context of scientific and societal challenges:

1. The significance of volcanic unrest at the natural hazard and risk interface,
2. Geophysical and geochemical fingerprints of unrest and precursory activity,
3. Subsurface dynamics leading to unrest phenomena,
4. Stakeholder interaction and volcanic risk governance.

This book aims to make our research accessible to both scientific and non-scientific audiences with interest in the different aspects of volcanic unrest, its impact and consequences. The chapters have been written with the intention to make the findings accessible to a broad audience. This entails a balancing act between keeping the scientific jargon at bay, whilst also satisfying the curiosity of scientific readers from different disciplines. In addition, most chapters are accompanied by Spanish-language abstracts.

As a consequence, the style of the chapters is different in several ways from the general peer-reviewed scientific literature. Most chapters have a summary/review character of findings from the project, which were originally published in dedicated journals. The chapters hence provide syntheses and articulations of concepts rather than comprehensive compilation of data and the available literature. However, all chapters provide the reader with references to original publications that will permit a wider reading and study of the findings behind the chapters.

The book is a joint effort between editors, authors, reviewers and publishers.

All chapters have undergone peer review, and we are indebted to all reviewers from the VUELCO community as well as the following external reviewers who provided their expert opinions and comments:

S. de Angelis, F. Arzilli, O. Bachmann, F. Costa, N. Deligne, J. Gardner, H. Gonnermann, A. Hicks, S. Hurwitz, S. Jenkins, P. Lesage, C. Newhall, C. Pritchard, G. Woo.

We thank J. Schwarz at Springer for her patience and expert handling of all things related to publishing this book.

Munich, Germany
Bristol, UK
Leeds, UK

Bettina Scheu
Joachim Gottsmann
Jürgen Neuberg

Contents

Volcanic Unrest and Pre-eruptive Processes: A Hazard and Risk Perspective

J. Gottsmann, J.-C. Komorowski and J. Barclay

Abstract

Volcanic unrest is complex and capable of producing multiple hazards that can be triggered by a number of different subsurface processes. Scientific interpretations of unrest data aim to better understand (i) the processes behind unrest and their associated surface signals, (ii) their future spatio-temporal evolution and (iii) their significance as precursors for future eruptive phenomena. In a societal context, additional preparatory or contingency actions might be needed because relationships between and among individuals and social groups will be perturbed and even changed in the presence of significant uncertainty. Here we analyse some key examples from three international and multidisciplinary projects (VUELCO, CASAVA and STREVA) where issues around the limits of volcanic knowledge impact on volcanic risk governance. We provide an overview of the regional and global context of volcanic unrest and highlight scientific and societal challenges with a geographical emphasis on the Caribbean and Latin America. We investigate why the forecasting of volcanic unrest evolution and the exploitability of unrest signals to forecast future eruptive behaviour and framing of response protocols is

J. Gottsmann (✉)
School of Earth Sciences, University of Bristol,
Bristol, UK
e-mail: j.gottsmann@bristol.ac.uk

J. Gottsmann
The Cabot Institute, University of Bristol, Bristol,
UK

J.-C. Komorowski
Institut de Physique du Globe de Paris et Université
Paris Diderot, Université Sorbonne Paris Cité, CNRS
UMR 7154, Paris, France

J. Barclay
School of Environmental Sciences, University of
East Anglia, Norwich, UK

Advs in Volcanology (2019) 1–21
DOI 10.1007/11157_2017_19
© The Author(s) 2017
Published Online: 17 August 2017

challenging, especially during protracted unrest. We explore limitations of current approaches to decision-making and provide suggestions for how future improvements can be made in the framework of holistic volcanic unrest risk governance. We investigate potential benefits arising from improved communication, and framing of warnings around decision-making timescales and hazard levels.

Resumen

La agitación volcánica es compleja y capaz de generar múltiples peligros que pueden ser desencadenados por un número diferente de procesos subsuperficiales. Las interpretaciones científicas sobre datos de agitación volcánica tienen como objetivo el mejor entendimiento de (i) los procesos detrás de la agitación volcánica y sus señales superficiales asociadas, (ii) su evolución espacial-temporal y (iii) su significado como precursores de fenómenos eruptivos a futuro. Dentro de un contexto social, acciones adicionales preparatorias o de contingencia podrían ser requeridas debido a que las relaciones entre individuos y dentro de grupos sociales serán perturbadas e inclusive modificadas ante la presencia de incertidumbre significativa. Aquí nosotros analizamos algunos ejemplos clave a partir de tres proyectos internacionales y multidisciplinarios (VUELCO, CASAVA y STREVA) en los cuales las cuestiones alrededor de los límites del conocimiento volcánico tienen impacto en la gestión pública del riesgo volcánico. Proveemos una perspectiva general del contexto regional y global de la agitación volcánica y sobresaltamos retos científicos y sociales con énfasis geográfico en el Caribe y América Latina. Investigamos porqué el pronóstico de la evolución en la agitación volcánica y el aprovechamiento de señales de agitación volcánica para el pronóstico de comportamiento eruptivo a futuro y el enmarque de protocolos de respuesta es un reto, especialmente durante periodos de agitación prolongada (años a décadas) en los que algunos retos surgen desde la utilización de señales de agitación para pronosticar la evolución de agitación a largo plazo y sus eventuales consecuencias. Exploramos las limitantes de actuales enfoques para la toma de decisiones y proveemos sugerencias acerca de cómo pueden hacerse reformas a futuro dentro del marco holístico de gobernabilidad ante el riesgo de agitación volcánica. Investigamos los potenciales beneficios que surgen por comunicación mejorada, y delimitando alertas alrededor de escalas de tiempo para la toma de decisiones y los niveles de alerta. Proponemos la necesidad de la cooperación a través de las fronteras científicas tradicionales, una valoración más amplia del riesgo natural y una mayor interacción de los sectores interesados.

1   Introduction

Volcanic unrest is a complex multi-hazard phenomenon of volcanism. Although it is fair to assume that probably all volcanic eruptions are preceded by some form of unrest, the cause and effect relationship between subsurface processes and resulting unrest signals (geophysical or geochemical data recorded at the ground surface, phenomenological observations) is unclear and surrounded by uncertainty (e.g., Wright and Pierson 1992). Unrest may, or may not lead to eruption in the short-term (days to months). If an eruption were to ensue it may involve the eruption of magma or may be non-magmatic and mainly driven by expanding steam and hot water (hydrothermal fluids) (Table 1). These conundrums contribute significant uncertainty to short-term hazard assessment and forecasting of volcanic activity and have profound impact on the management of unrest crises (e.g., Marzocchi and Woo 2007).

While institutional and individual decision-making in response to this unrest should promote the efficient and effective mitigation or management of risk, informed decision-making

Table 1 Summary of processes contributing to unrest signals in space and time, possible outcomes and hazards/impact of unrest

Nature of processes	Processes	Signals	Hazards/Impact	Unrest Outcome
Magmatic	Magma and/or melt and/or volatile migration (input, loss or ascent from reservoir), chemical differentiation, thermal convection, thermal perturbation (heating or cooling), pore fluid migration reservoir rejuvenation, crystallization and other phase changes	Seismicity, ground deformation, changes in potential fields, changes in gas and/or ground water chemistry, changes in heat flux, changes in volatile flux	Ground deformation, shaking and rupture and associated infrastructure damage; water table level changes; toxic gas emissions, contamination of ground water, atmosphere and crops; edifice destabilization; toxic gas emissions	Waning and return to background activity; eruptive activity (magmatic and/or phreatic)
Tectonic/gravitational	Faulting, changes in local/regional stress fields, edifice gravitational spreading, crustal loading, pore fluid migration			Waning and return to background activity; eruptive activity (magmatic and/or phreatic)
Hydrothermal	Fluid migration, phase changes, changes in temperature and/or pressure, chemical changes, pore pressure variations, porosity and permeability changes (sealing), host-rock alteration			Waning and return to background activity; phreatic eruptive activity

Processes can act individually, in unison or in any combination is fundamentally dependent on the early and reliable identification of changes in the subsurface dynamics of a volcano and their "correct" assessment as precursors to an impending eruption. However, uncertainties in identifying the causative processes of unrest impact significantly on the ability to "correctly" forecast the short-term evolution of unrest.

When a volcano evolves from dormancy through a phase of unrest, scientific interpretations of data generated by this unrest relate to (i) the processes behind unrest and their associated surface signals, (ii) their potential future spatio-temporal evolution (i.e., hydrothermal vs. phreatic vs. magmatic processes and their intensity) and (iii) their significance as precursors for future eruptive phenomena. Scientific interpretations framed towards the governance of and social responses to the risk implicit in the potential onset of an eruption focus on: (i) understanding the epistemic (relating to the limits of existing knowledge) and aleatoric (relating to the intrinsic variability of natural processes) uncertainties surrounding these data and their impact on decision making and emergency management, (ii) the communication of these uncertainties to emergency managers and the citizens at risk, and (iii) understanding how best to manage evolving crises through the use of forecasted scenarios.

2 Motivation

The analysis presented in this chapter synthesises wider results and experiences gained in three major research consortia with focus on volcanic hazards and risks: (1) The VUELCO project, (2) the CASAVA project, and (3) the STREVA project.

The European Commission funded VUELCO project (2011–2015; "Volcanic unrest in Europe and Latin America: Phenomenology, eruption precursors, hazard forecast, and risk mitigation; www.vuelco.net) focused on multi-disciplinary research on the origin, nature and significance of

volcanic unrest and pre-eruptive processes from the scientific contributions generated by collaboration of ten partners in Europe and Latin America. Dissecting the science of monitoring data from unrest periods at six target volcanoes in Italy (Campi Flegrei caldera), Spain (Tenerife), the West Indies (Montserrat), Mexico (Popocatepetl) and Ecuador (Cotopaxi) the consortium created strategies for (1) enhanced monitoring capacity and value, (2) mechanistic data interpretation and (3) identification of eruption precursors and (4) crises stakeholder interaction during unrest.

The CASAVA project (2010–2014; Agence nationale de la recherche, France; *Understanding and assessing volcanic hazards, scenarios, and risks in the Lesser Antilles—implications for decision-making, crisis management, and pragmatic development;* https://sites.google.com/site/casavaanr/, last accessed 11-10-2016) implemented an original strategy of multi-disciplinary fundamental research on the quantitative assessment of volcanic risk for the Lesser Antilles region with emphasis on Guadeloupe and Martinique. The aim of the project was to improve the capacity to anticipate and manage volcanic risks in order to reduce reactive 'repairing' post-crisis solutions and promote the emergence of a society of proactive volcanic risk prevention in case of a future eruption. Part of this was achieved via a forensic analysis of past crises, described here.

The STREVA Project (2012–2018 funded by the UK Natural Environment and Economic and Social Research Councils; www.streva.ac.uk) was designed as a large interdisciplinary project to develop new means to understand how volcanic risk should be assessed and framed. It uses the 'forensic' interdisciplinary analysis of past volcanic eruptions in four settings to understand the key drivers of volcanic risk. The aim is to use this analysis to generate future plans that will reduce the negative consequences of future eruptions on populations and their assets. STREVA works closely with partners in the Caribbean, Ecuador and Colombia, focussing the forensic analysis on long-lived eruptions of Soufrière Hills Volcano (Montserrat) and Tungurahua (Ecuador) and shorter duration eruptions of La Soufrière (St. Vincent) and Nevado del Ruiz (Colombia). The focus of the 'forensic analysis' process in the STREVA project has been to understand the key drivers of risk and resilience during long-lived volcanic crises. Nonetheless the analysis of the initial phases of activity from these eruptions provide some insights into the acute uncertainties of unrest and the social, political and scientific consequences of that uncertainty.

3 Volcanic Unrest: Scientific and Social Context

Volcanic unrest can be defined in a scientific context: "The deviation from the background or baseline behaviour of a volcano towards a behaviour or state which is a cause for concern in the short-term (hours to few months) because it might prelude an eruption" (Phillipson et al. 2013). The term "eruption" in the context of a possible unrest outcome could either relate to a magmatic or non-magmatic (phreatic or hydrothermal) origin including the possible evolution from phreatic to magmatic activity or an alternation or mix between the two (e.g., Rouwet et al. 2014). In a social context, these concerns might necessitate additional preparatory or contingency actions in response to the unrest phenomena or the preparation for an eruption given that the organisation and preparedness of communities and those who manage them will be perturbed and even changed in the context of significant uncertainty (Barclay et al. 2008 and next section).

4 Challenges and Key Questions Relating to Volcanic Unrest

4.1 Wider Perspective

Whether or not unrest results in eruption, either of magmatic or non-magmatic origin, and whether (in hindsight) "correct" or "false" forecasts are issued to suggest there could be an imminent eruption are among the central questions that need answering as soon as unrest is detected.

The cost of scientific uncertainty regarding the causes and outcome of volcanic unrest may be substantial not only in terms of direct or indirect financial implications such as explored in Sect. 5, but also regarding knock-on (secondary) effects such as public trust in the accuracy or inaccuracy of scientific knowledge, public perception of the relationships between signals of unrest and volcanic risk and future public compliance with orders to evacuate or improve preparedness in the medium to long term.

A multitude of subsurface processes may contribute to unrest signals and some are summarised in Table 1. Not all processes are pre-eruptive and the challenge lies in deciphering the causes of unrest with a view to establish early on in a developing crises whether a volcanic system develops towards a state where an eruption may ensue. Whether or not unrest leads to eruption depends on many parameters. In general the main concern during volcanic unrest lies with the potential for a magmatic eruption. For this to occur magma must rise from depth and break through the surface. The dilemma for scientists is that magma movement does not create uniquely attributable unrest signals and does not necessarily lead to eruption (Table 1). For example, seismicity and ground uplift, both common indicator of unrest, may be induced by the replenishment of a magma reservoir, the ascent of magma towards the surface or the redistribution of aqueous fluids and fluid phase changes (see Salvage et al. 2017; Hickey et al. 2017; Mothes et al. 2017 for examples from VUELCO volcanoes). Similarly, an increase in the gas and heat flux (Christopher et al. 2015) at the surface may be induced by magmatic or hydrothermal processes and even tectonic stress changes have also been shown to trigger such behaviour (e.g., Hill et al. 1995). In fact, non-magmatic eruptions are associated with significant hazards and have or could have caused fatalities in the past such as for example Bandai in 1888 (Sekiya and Kikuchi

1890), Te Maari Tongariro in 2012 (e.g., Jolly et al. 2014) and recently at Ontake in 2014 (e.g., Maeno et al. 2016). Many unrest processes contribute to non-eruptive secondary hazards such as flank instability and collapse (e.g. Reid 2004).

4.2 Uncertain Causes and Uncertain Effects

Substantial uncertainties surround both the interpretation of the drivers of unrest and the assessment of the potential evolution and outcome of unrest. Critical questions include: Will an eruption ensue? If so, will it occur in the short-term (days to months) or long-term (years to decades)? What will be the nature and intensity of the eruption (magmatic vs. phreatic)?

In the case of magmatic unrest, magma ascent towards the surface can lead to a magmatic eruption with potential for the formation of lava flows, pyroclastic flows, lahars, ash-fall and ballistics. These processes impact the proximal (few tens to hundreds of meters), medial (kilometers) and distal (tens of kilometres or more) areas around the volcano. Conversely unrest driven by sub-surface hydrothermal activity may peak in a phreatic eruption and while impacted areas are rather proximal to the volcano, associated ballistics and dilute pyroclastic density currents triggered by laterally-directed explosions and emplacement of a debris avalanche from a partial edifice collapse can lead to an anomalously high loss of lives as recently shown by the September 27, 2014 Mount Ontake eruption, the deadliest eruption in more than 100 years in Japan (e.g. Maeno et al. 2016).

The challenge, however, is to identify and discriminate signals that are indicative of reactivation leading towards a major expulsion of magmatic material from those associated with a slight deviation from background levels and potential waning of unrest phenomena (Table 1).

The fundamental limitation for volcanologists is that it is not possible to directly observe causative processes at depth. Thus interpretations of these drivers rely on the secondary interpretation of observable signals associated with those processes (Salvage et al. 2017) or the reproduction of interpreted processes via laboratory experiments (Wadsworth et al. 2016). In addition, many volcanic processes are intrinsically non-linear and characterized by a chain-link reaction such that minor variations of some uncertain parameters might have ultimately significant consequences on the eruptive outcome. Such non-linear processes coupled with epistemic and aleatoric uncertainties are complex to understand and model. This chapter analyses some key examples across the three aforementioned projects where issues around the limits of volcanic knowledge exacerbated risk and makes suggestion for how future improvements can be made.

4.3 The Hazard and Risk Interface

Scientific Challenges

In the light of the above, from a scientific point of view the early identification of the cause of unrest and its likely outcome and evolution is pivotal for effective and efficient risk assessment, risk management and the design of mitigation efforts. In order to address the key scientific question of whether unrest is a prelude to imminent eruption or whether it will wane after some time without eruption several questions require answering first (note, that the list is not exhaustive):

- Is the anomalous behaviour unambiguously indicative for a change in the volcano's behaviour and for a deviation from its background state?
- How reliable is the assessment of unrest as a prelude to eruption, particularly in the absence of data on past events?

- What are the mechanistic processes at depth leading to observed unrest signals?
- Are monitoring signals indicative of magmatic, hydrothermal or tectonic unrest?
- Can the unrest be caused by perturbations and changes in the host-rock properties (e.g. porosity, permeability, mechanical properties) rather than by distinct endogenic processes of hydrothermal or magmatic origin?
- What are the uncertainties surrounding monitoring signals and inferred sub-surface processes (see Hickey et al. 2017 and Salvage et al. 2017)?
- Do secondary processes (e.g. hydrothermal system perturbation, meteorological forcing) modify primary signals from deeper-seated magmatic processes?
- What are the consequences of signal modification for the assessment of the process-to-signal-to-outcome causal link?
- Does one follow a deterministic or probabilistic approach for observations and forecasting (e.g., Hincks et al. 2014; Aspinall and Woo 2014; Rouwet et al. 2017)?
- What is the likelihood of a specific eruptive or non-eruptive scenario to manifest (e.g., Bartolini et al. 2017)?
- Which types of eruptions did the volcano produce in the past?
- If an eruption is to occur, what is its likely nature: magmatic, or phreatic or a mix?
- How much lead-time before eruption is there based on previous experience; how much lead-time is there in the absence of previous experience?
- Which eruptive or non-eruptive unrest episodes at analogue volcanoes can provide clues for the interpretation of signals and forecasting of unrest evolution and outcome (e.g., Sheldrake et al. 2016)?
- What is the likely size of the eruption and the associated hazards and risks and impacted area?
- What is the temporal evolution of eruptive intensity once the eruption has started? i.e.,

what is the likelihood that the eruption (a) will have its paroxysmal phase in the first 24 h of eruption (42% of eruptions do, according to Siebert et al. (2015)); or (b) will have a more progressive escalation over several months that will culminate in a paroxysm; or (c) will be characterised by peaks in activity separated by more or less long-lasting pauses or strong decline of activity preceding another rapid increase and peak of activity?

Societal Challenges

At the same time, the political, sociological, cultural and economic (grouped here under the term 'societal') implications from unrest need addressing in order to respond appropriately to the emerging natural hazard (Wynne 1992) Here we provide a (non-exhaustive) list of questions for risk managers and/or politicians in the context of risk governance during volcanic unrest:

- What is the best-practice to provide maximum response time, while minimizing vulnerability and optimizing the cost/benefit ratio (see Fig. 1) of mitigation actions in a developing unrest crises?
- What is the best practice to issue or raise an alert?
- When and how to decide to raise an alert and to take action?
- What are the potential (legal) consequences of a false positive or false negative (see Table 2 and Bretton et al. 2015)?
- What are the consequences of a true positive (Table 2)?
- What is the basis for raising an alarm: the outcomes of unrest (e.g., instability of buildings due to ground deformation or seismicity; toxic degassing and environmental contamination) or the potential for eruption?
- How to best disseminate what information on unrest and its potential consequences, when, and via which communication vehicle(s) to the public?

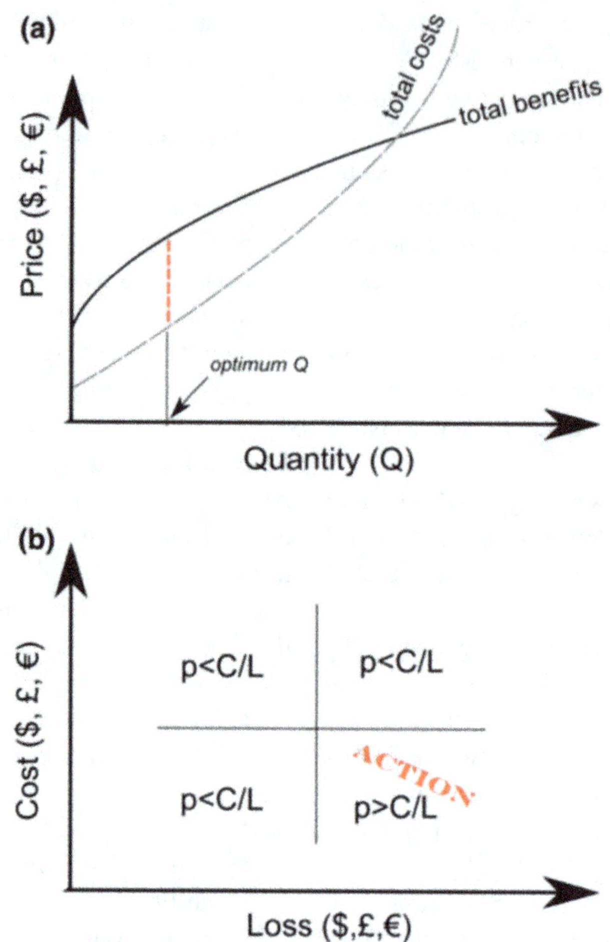

Fig. 1 Cost-benefit relationship as a tool for decision-making. **a** In the context of volcanic unrest risk management, actions of given quantity Q (for example, number of shelters or evacuees) are associated with costs in relation to their expected benefits (expressed by a financial value). An optimal relationship between costs of mitigation efforts and resultant benefits can be achieved when the difference between investment and benefit is greatest (shown by *stippled red line*). The example is based on concepts of capital management theory presented in Brealey et al. (2011). **b** Cost (C) versus loss (L) model for volcanic risk management (after Marzocchi and Woo 2007). If, in this decision-making framework, the expected expense (cost) for mitigation action is to be minimised, then action is required if the probability (p) of an adverse event to occur exceeds the ratio between the cost of the action and the expected loss (L/C). See discussion for a wider appraisal of the challenges arsing from such an analysis

Table 2 Concept of successful and unsuccessful forecasting

	Event forecast	Event not forecast
Event occurs	True positive	False negative (Type II error)
Event does not occur	False positive (Type I error)	True negative

- How to account for uncertainty and the diversity of expert opinions in deciding the alert level?

- In what context does this occur such as political pressures, concurrent natural or other hazards (pandemic, famine, cyclone, etc.)

4.4 Cost-Benefit Analysis (CBA)

In the previous paragraphs, several questions related to how to get both the scientific analysis ('what is going on?') and the societal response ('how to respond?') 'right'. One measure employed to quantify the economic consequences of action or no-action under imminent threat and a tool for informed institutional decision-making is the cost-benefit analysis (Marzocchi and Woo 2007) whereby one aims to find a good answer to the question: "Given an assessment of costs and benefits related to risk mitigation efforts, which actions should be recommended?" Figure 1 shows the concept of evaluating the optimum ratio between the cost and benefit of mitigation efforts and provides a cost-benefit matrix for the design of action plans in response to a future [short-term in context of this chapter) adverse event of given probability (p) (Brealey et al. 2011; Marzocchi and Woo 2007)]. A critical issue in CBA is the 'minimum value of a human life', which we will not discuss further here. The interested reader is referred to, for example, Woo (2015) for further details on this quantification. Another interesting point relates to what might be regarded as a 'cost' and a 'benefit' in a response to an unfolding unrest crisis with an uncertain outcome (see also Sect. 6).

5 Global and Regional Context of Volcanic Unrest

5.1 Unrest Durations and Characteristics

Phillipson et al. (2013) reviewed global unrest reports of the Smithsonian Institution Global Volcanism Program (GVP) between January 2000 and July 2011 to establish the nature and length of unrest activity, to test whether there are common temporal patterns in unrest indicators and to test whether there is a link between the length of inter-eruptive periods and unrest duration across different volcano types.

Using available formation on unrest at 228 volcanoes they defined unrest timelines to demonstrate how unrest evolved over time and highlight different classes of unrest including reawakening, pulsatory, prolonged, sporadic and intra-eruptive unrest (see Fig. 2 for an example from Cotopaxi volcano). Statistical analyses of the data indicate that pre-eruptive unrest (where there is a causal link between unrest and an eruption within the observation period) duration was different across different volcano types with 50% of stratovolcanoes erupting within one month of reported unrest. The median average

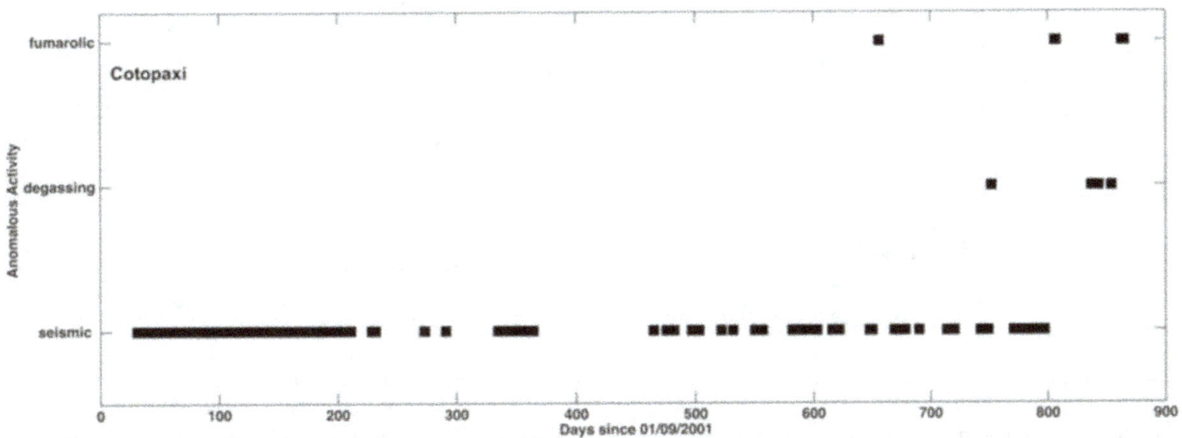

Fig. 2 Timeline of reported anomalous activity at Cotopaxi volcano (Ecuador) in 2001/2002. This period of pulsatory unrest lasted for more than 3 years with a heightened level of activity in 2001 and 2002. The unrest did not lead to an eruption in the short-term (weeks to months), but Cotopaxi entered an eruptive phase in August 2015 after a short-period of renewed unrest activity starting in April 2015 (see Mothes et al. 2017 for details). The data shown in the graph are from Phillipson et al. (2013)

duration of pre-eruptive unrest at large calderas was about two months, while at shield volcanoes a median average five months of unrest was reported before eruptive activity. The shortest median average duration is reported for complex volcanoes where eruptive unrest was short at only two days. Overall there appears to be only a very weak correlation between the length of the inter-eruptive period and pre-eruptive unrest duration. This may indicate that volcanoes with long periods of quiescence between eruptions will not necessarily undergo prolonged periods of unrest before their next eruption (Fig. 3). Phillipson et al. (2013) found statistically relevant information only from reports of anomalous seismic behaviour, most other monitoring signals are either not recorded or not reported as unrest criteria. The authors reported a noteworthy lack of geodetic data/information and in particular

satellite remote sensing data in the available reports. Recently Biggs et al. (2014), addressed the latter and systematically analysed 198 volcanoes with more than 18 years of satellite remote sensing deformation data for their deformation behaviour. 54 volcanoes that showed deformation also erupted during the observation period. Their analysis does not imply any causal link, or even a temporal relationship between any specific eruptions and episodes of deformation and is hence not directly comparable to the causal and predictive analysis by Phillipson et al. (2013). However, given that 46% of deforming volcanoes erupted while 94% of non-deforming volcanoes did not erupt provides "strong evidential worth of using deformation data as a trait association with eruption" (Biggs et al. 2014).

It is important to note that exploitable records on volcanic unrest are limited and the available

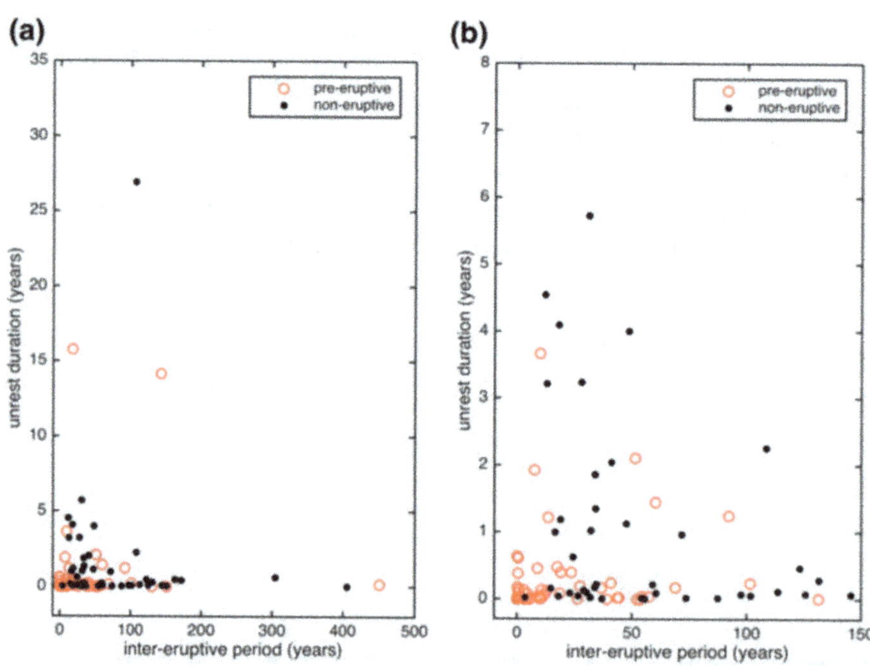

Fig. 3 Comparison between the inter-eruptive period (IEP) and unrest duration (UD) from the data set presented in Phillipson et al. (2013). **a** shows entire data set (n = 118) **b** shows a subset of the data for clarity of inter-eruptive periods <150 years. The p-values of the Pearson's correlation test are $p = 0.93$ for the entire data set, $p = 0.60$ for the subset of non-eruptive unrest (n = 58) and $p = 0.20$ for the subset of eruptive unrest (n = 60). The null hypothesis ("the UD is independent of the IEP") is hence statistically acceptable when considering the entire data set. Considering the subset of pre-eruptive unrest, however, the statistical tests do not provide enough evidence to fully accept the null hypothesis since the associated p-value of 0.20 might hint towards some weak correlation between the two variables

data sets are far from complete. Key issues are the lack of or poor instrumentation at most volcanoes, the lack of reporting by observers particularly if an unrest turns out to be minor and without immediate consequences, and the lack of integrating unrest data from satellite remote sensing. The GVP generally lacks the post-facto integration of unrest indicators from satellite-remote sensing data (e.g., Fournier et al. (2010) and Biggs et al. (2014) for deformation and Carn et al. (2011) for degassing). In this respect, it is vitally important to recognise and support initiatives to collate and exploit worldwide volcano monitoring data such as for example the WOVOdat project (Venezky and Newhall 2007). Only by significantly increasing the knowledge-base on the spatial and temporal evolution of the unrest-eruption relationship can we embark on statistically sound exploitations of the data with a potential to improve forecasting capabilities early on in developing unrest crises.

5.2 Socio-Economic Contexts

The Wider Perspective

Nowadays, about 800 million people live on or in direct vicinity of active volcanoes (Brown et al. 2015). The overwhelming majority of this population lives in low and middle income countries (countries with an annual gross national income per capita of less than US$12,700) including the focus area of the VUELCO, STREVA and CASAVA projects: the wider Latin American (LA) region extending from Mexico, through Central America and the Caribbean to South America. This region hosts about 330 Holocene volcanic centres compared to 84 in Europe and one quarter of the reported global fatalities attributed to volcanic events occurred there (Global Volcanism Program 2013).

Volcanic disasters are among the least audited of all natural disasters and therefore our knowledge on the impact of volcanic activity beyond claiming lives is largely incomplete (Benson 2006; Auker et al. 2013). Huge uncertainty surround estimates for indirect losses from for example disease or starvation as a result of volcanic activity. Beyond increased human vulnerability, the direct and indirect financial impacts from volcanic activity can be immense as demonstrated by the relatively small-scale eruption of Iceland's Eyjafjallajökull volcano in April 2010 and the associated air travel disruption. This eruption demonstrated the vulnerability of modern infrastructure to volcanic hazards on an unprecedented scale with losses to the aviation industry alone at a minimum of US$2.5 Billion (European Commission 2010).

Equally there are social, political and financial implications for "false positives" related to volcanic unrest. In these instances actions are taken in response to an imminent threat, which then did not manifest. In the case of volcanic unrest the imminent threat is generally defined as a volcanic eruption, although the multi-hazard nature of volcanic unrest (e.g., ground shaking, ground uplift or subsidence, ground rupture, ground instability, toxic gas emissions, contaminated water supplies) and possibly ensuing eruptive activity (magmatic vs. phreatomagmatic vs. phreatic) makes the definition of 'imminent threat' rather complex.

Although there is little systematic gathering and synthesis of data relating to financial or social losses associated with these episodes there are some well-documented analyses. Examples include:

(1) On Guadeloupe in the French West Indies a major evacuation over a period of 4 months in excess of 70,000 individuals was initiated in 1976, as a result of abnormal levels of volcanic seismicity and degassing (see also next section). The estimated cost of the unrest was about US$340 Million at the 1976 exchange rate (data compiled using Lepointe 1999; Tazieff 1980; Blérald 1986; Baunay 1998; Kokelaar 2002; Annen and Wagner 2003), which translates to more than US$ 1.2 Billion at the time of this writing (July 2016). At the time the cost equaled to ca. 60% of the Gross National Product of the Guadeloupe economy (Blérald 1986). 90% of these costs were incurred by the costs of the evacuation,

and the costs associated with the rehabilitation and salvage of the economy in Guadeloupe after the evacuation.

(2) Unrest at Rabaul volcano in Papua New Guinea (an LDC) between 1983 and 1985, had significant adverse implications for both the private and public sectors. Considerable economic costs were incurred, estimated at over US$22.2 Million at the 1984 rate of exchange although an eruption did not occur until 10 years later (Benson 2006).

(3) Evacuation and rehousing of 40,000 inhabitants of the Pozzuoli area in the Campi Flegrei volcanic area of Italy resulted as a response to intense seismicity and ground uplift in the early 1980s. Although decision-makers did not release notice that this was in part due to the threat from an imminent eruption (see also Sect. 4.3.2), it is true that the re-location of these inhabitants moved them from the area of highest threat in the event of an eruption. At the time there was no agreement amongst scientists as to the cause of the unrest (Barberi et al. 1984) and the scientific discussion as to the cause of these events is still ongoing more than 30 years after the crisis.

The following paragraphs focus on two examples of short-term and long-term volcanic unrest crises response and provide more detailed insights into the volcanic risk governance in two different jurisdictions.

Short-Term Crisis Example: The 1976–1977 La Soufrière of Guadeloupe Unrest

The unrest on Guadeloupe culminated in a series of explosive eruptions of hot gas, mud and rock (termed phreatic eruption) without the direct eruption of magma before waning in 1977 (Feuillard et al. 1983; Komorowski et al. 2005; Hincks et al. 2014). Fortunately no fatalities were caused by the activity. Had the unrest on Guadeloupe led to a magmatic eruption, then the cost of the unrest would have likely been negligible. Although the precautionary evacuation caused a substantial economic loss with severe social consequences, it is acknowledged that the

"proportion of evacuees who would have owed their lives to the evacuation, had there been a major eruption, was substantial" (Woo 2008). The CASAVA project undertook an exhaustive hindsight analysis of the process of scientific decision-making for the unrest and eruptive crisis of 1976–1977 at La Soufrière de Guadeloupe. The crisis caused significant hardships and loss of livelihood for the evacuated population and the whole society in Guadeloupe as a result of controversial crisis management associated with a forecast of a major magmatic eruption that did not occur (false positive) (Feuillard et al. 1983; Fiske 1984; Komorowski et al. 2005; Hincks et al. 2014). Given the evidence of continued escalating pressurisation and the uncertain transition to a devastating magmatic eruption, authorities declared a 4-month evacuation of ca. 70,000 people on August 15, 1976 that provoked severe socio-economical consequences for months to years thereafter. This evacuation is still perceived as unnecessary and reflecting an exaggerated use of the "principle of precaution" on behalf of the government.

However, some level of risk governance (i.e. evacuation of the most exposed area) was justified in hindsight given the persistent ashfalls and environmental contamination from acid degassing as well as the hazards from a series of non-magmatic eruptions (e.g., pyroclastic flows from laterally directed explosions, partial edifice collapse, mudflows) (Komorowski et al. 2005; Hincks et al. 2014).

The (in hindsight) erroneous identification of the presence of *'fresh glass'* in the ejecta and its interpretation as evidence of the magmatic origin of the unrest and thus of its possible outcome, led to a major controversy amongst scientists that was widely echoed in the media. Lack of a comprehensive monitoring network prior to the crisis, limited knowledge of the eruptive history, and living memory of past devastating eruptions in the Lesser Antilles contributed to a high degree of scientific uncertainty and a publically-expressed lack of consensus and trust in available expertise. Consequently analysis, forecast, and crisis response were highly challenging for scientists and authorities in the context of escalating and fluctuating activity and societal pressure. The high uncertainty about a so-called "unequivocal" impending disaster fostered a binary zero-sum strongly opinionated approach in the scientific discourse. The public debate thus became polarized on issues of opposing "truths" served by contrasted scientific expertise rather than on how science could help constrain epistemic and aleatoric uncertainty and foster improved decision-making in the context of uncertainty (Komorowski et al. 2017). This situation acted as an ideal crucible to fuel a media-hyped controversy on the crisis and its management. A recent retrospective Bayesian Belief Network analysis of this crisis (Hincks et al. 2014) demonstrates that a formal evidential case could have been made to support the authorities' concerns about public safety and decision to evacuate in 1976.

As part of the CASAVA project we conducted focus group interviews, issued questionnaires, and ran role playing games with the population currently living in areas potentially threatened by renewed unrest and eruptive activity from La Soufrière, (be it magmatic or non-magmatic). We found that the current population's risk perception increases to a level of preparing to evacuate chiefly on the basis of the timing and nature of scientific information issued publically by the volcano observatory. This implies that the population is prone to self-evacuate ahead of any official evacuation order given by the authorities in charge of civil protection and crisis response.

Long-Term Crises Examples: Soufrière Hills (Montserrat) and Tungurahua (Ecuador)

The forensic analyses of the STREVA project have focussed on the integration of new social-science based understandings of population response and recovery with the scientific insights prompted by these long-lived eruptions. This has similarities with the 'FORIN' approach advocated by the International Program on Integrated Risk for Disaster Reduction (Burton 2010). In this description we focus particularly on the initial stages of the eruptions.

The long-lived volcanic crisis of the Soufrière Hills Volcano is probably one of the most written about volcanic eruptions, encompassing a wide variety of perspectives, scientific, social-scientific and personal, in that writing. As a consequence of the activity on the island of Montserrat a population of over 10,500 was reduced to just 2850 (the population has since risen to 4922 [2011 census], Hicks and Few 2015). At the onset of eruption (1995) an assessment of risk existed (Wadge and Isaacs 1988) but was not acted on or acknowledged by the authorities, and so preparedness was low, exacerbated by the recent passage of Hurricane Hugo (1989) which had caused 11 fatalities and rendered 3000 homeless. Governance on Montserrat was reforming in the wake of the economic and social crisis induced by the hurricane (Wilkinson 2015). The protracted uncertainty in the early stages of the eruption coupled with a lack of coherence in governance between the UK and local governments lead to the protracted evacuation of 1300 people in temporary public shelters, which suffered from overcrowding, lack of privacy, poor sanitation and lack of access to good nutrition. Ultimately, this led to a partial disregard for evacuation advice and a strong pulse of outwards migration. In the longer term, the long-lived volcanic eruption has acted to exaggerate pre-existing vulnerabilities in the local population (Hicks and Few 2015).

The early stages of the current Tungurahua (Ecuador) eruptive episode that started in 1999 typify a further challenge for the management of unrest prior to or between surface activity at the early stages of a volcanic crisis. Initially the local population were evacuated by a compulsory evacuation order but when the initial phases proceeded more slowly than had been expected by local authorities and communities, civil unrest and disturbance happened with the re-occupation by force and ultimately abandonment of the evacuation order. These arose from the acute economic and social pressures visited on the population by the evacuation (Mothes et al. 2015). Subsequently, the response of the monitoring organisation to these pressures represents a new archetype for collaborative monitoring and management of restive volcanoes (Mothes et al. 2015; Stone et al. 2014). The growth of trust, and attempts to maximise resilience in the face of repeated unrest episodes provides strong evidence for collaborative approaches to risk management (Few et al. 2017). Nonetheless tensions still exist, largely arising from our current incapacity to predict the intensity or magnitude of eruptions from signals relating to new unrest. There can be problems in this risk system implicit in anticipating the 'maximum expected' outcome from unrest.

6 Discussion

6.1 The Caveats of Volcanic Unrest Response

Managing volcanic unrest episodes is extremely complex and challenging due to the multi-hazard nature of unrest. The risks to be assessed and mitigated include both those associated with the unrest itself as well as those from the potential future eruptive activity. Whilst ground deformation, seismicity, thermal flux or anomalous degassing are indicators of possible future activity these phenomena also pose significant immediate threats to population, infrastructure and other assets in affected areas during the unrest.

From a scientific point of view, hazard assessment relating to eruptive activity has made considerable progress in recent years partly through the deployment of increasingly powerful computational models and simulation capabilities (e.g., Esposti Ongaro et al. 2007; Manville et al. 2013) as well as through advances in the development of probabilistic eruption forecasting tools (e.g., Marzocchi et al. 2008; Aspinall 2006; Aspinall and Woo 2014) and improvements to fundamental understandings of the root drivers of changing activity (e.g., Cashman and Sparks 2013).

Despite these crucial advances for short-term eruption forecasting, the knowledge-base on volcanic unrest, its significance as an eruption precursor, its exploitability regarding forecasting of potential eruptive behaviours and framing of response protocols (e.g., CBA) remains weak for a number of reasons:

(1) The scientific interpretation of volcanic unrest is surrounded by substantial uncertainty, ambiguity and ignorance (Stirling 2010) regarding causes and eventual outcome. Since the contributing subsurface processes cannot be directly observed, volcanic unrest is likely among the least understood phenomena in volcanology for a variety of reasons:

(i) Incomplete knowledge of the mechanistic processes and their dynamic behaviour over time within a magma reservoir and its surroundings (host-rock, hydrothermal system, meteoric recharge, local and regional structural context) that trigger the geophysical, geochemical and geodetic signals recorded at the surface during unrest periods (Table 1).

(ii) Consequently, the interpretation, of the departure of monitoring signals from a long-term baseline level or in the absence of baseline data a crescendo or decrescendo of signals collected during periods of unrest are often ambiguous or non-unique. While this can in practice be addressed in models through epistemic and aleatoric uncertainties, ambiguities in the interpretation will remain.

(2) Ambiguity, uncertainty and ignorance (Stirling 2010) have impact on probabilistic forecasting of duration, spatio-temporal evolution, causal relationship between sequential events and outcomes of unrest episodes (see Sandri et al. 2017) and on remedial actions to mitigate current and future adverse effects. Uncertainties in the decision-making process may give rise to "false alerts" (i.e., false positives; see Table 2) and actions by civil protection with adverse impacts on the compliance of affected communities in future unrest events.

(3) Lack of globally accepted and standardised approach for the terminology, methodology,

criteria, protocols and best practice employed to evaluate and respond to volcanic unrest by different stakeholders such as academia, volcano observatories and the Civil Protection Agencies. This absence of commonly recognised standards can result in the critical issue of managerial risk vulnerability (i.e., standard equivocality' after Bretton et al. 2015). It also often impacts negatively on the effectiveness of communication between stakeholders, hinders or delays effective and efficient decision-making processes and hampers the dialogue among members of scientific, governmental and civil communities (De la Cruz-Reyna and Tilling 2008). However, it is important to note that internationally-defined standards should not be rigidly imposed irrespective of local, cultural, political and social practices (e.g., Bretton et al. 2015; IAVCEI 2016).

(4) Globally, there is no commonly accepted and standardised denominator between those that provide and those that receive scientific advice regarding the level of appropriate scientific complexity to be considered. This may hamper a wider discourse on scientific and technological advances in the quantification of unrest phenomena and resultant uncertainties with other stakeholders. From the scientist's perspective this may generate the notion that the public, administrators, mass media and governmental entities do not appreciate the "excellence of the science" behind unrest characterisation and use the inherent uncertainty as a rationale to go into denial over the hazards posed during unrest. From a sociological point of view, however, decision-making apparently prompted solely by the present or likely volcanic hazards, does not account for local context and can result in a lack of trust in either scientific expertise or government representatives, (Johnson 1987; Haynes et al. 2008; Christie et al. 2015; Komorowski et al. 2017).

6.2 Some Ways Forward

The issues identified above can contribute less optimal unrest response and risk mitigation actions. Although there are other strong contributors to societal vulnerability, we have shown that scientific uncertainty combined with a lack of social awareness and preparedness does act to increase the vulnerability of a society to hazardous unrest phenomena with possibly adverse outcomes. Here, we propose future avenues which can form part of a Risk Governance Framework (IGRC 2017; Fig. 4) including research that could gather critical evidence for some of the key drivers of decisions that result in adverse outcomes for affected populations in the face of an unrest crisis. Such research could also contribute to the analysis of and identification of key targets for future research in volcanology and the social sciences.

(a) Cost-Benefit Analysis

CBA, where the economic impacts of different decisions are quantified, can be difficult at the unrest hazard and risk interface. The case studies presented here demonstrate that intangible assets such as social and cultural cohesion and capital as well as trust (in the context of CBA analysis this would be intentional trust in the sense of Dasgupta (1988); i.e., the subjective probability assigned to compassionate action by an individual or a group of individuals) between different stakeholders can have a strong impact on individual and institutional vulnerabilities during crises.

Analyses that include a wider range of definitions and types of 'costs' and 'benefits' of mitigation efforts (e.g. loss of empowerment, a loss of cultural identity or cultural references), informed by past experiences would facilitate a discourse between different stakeholders. This would entail the need to attribute a financial value to, for example, mental well-being, social networks and cohesions and would necessarily trigger a wider discourse of the impacts of decision-making beyond the avoidance of 'cost

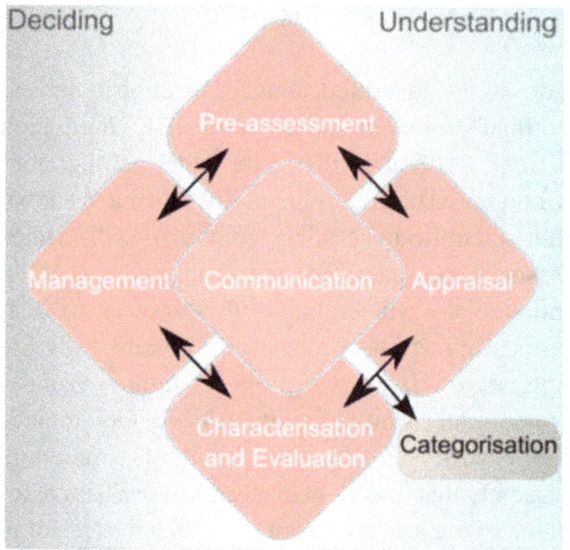

Fig. 4 The *International Risk Governance Council* (IRGC 2017) Risk Governance Framework adapted for the specific case of volcanic unrest. *Hazard and Risk Pre-assessment*—"peacetime framing" the hazard and risk in order to provide a structured definition of the baseline behaviour of the volcano and its consequences, of how the hazard and risk are framed by different stakeholders, of how the risk may best be handled, and of the thresholds to be met or exceeded to declare a state of unrest. *Hazard and Risk Appraisal*—combining a scientific risk assessment of the current unrest hazards (using for example a rating scheme of unrest intensity, e.g., Potter et al. 2015) and its probability with a systematic concern assessment (of public concerns and perceptions) to provide the knowledge base for subsequent decisions in an emerging unrest crises. *Risk Characterisation and Evaluation*—in which the scientific data and a thorough understanding of societal values affected by the risk are used to evaluate the risk as acceptable, tolerable (requiring mitigation), or intolerable (unacceptable). *Risk Management*—the actions and remedies needed to avoid, reduce transfer or retain the unrest risk and risks from probable unrest outcomes. *Risk Communication*—how stakeholders and civil society understand the unrest risk and participate in the risk governance process. *Risk Categorisation and Evaluation*—categorising the knowledge about the cause-effect relationships as either simple, complex, uncertain or ambiguous. In the context of volcanic unrest this may include the categorisation of the outcome of unrest and probable future eruptive activity

to lives'. Ideally this discourse would be framed during 'peace-time' (i.e. not in response to an unfolding unrest crisis) and involve participation from a wide spectrum of scientific and societal stakeholders. The necessity to move beyond circumscribed appraisal methods such as the CBA is also evident from the response to

protracted (several years or decades) unrest requiring an above back-ground level of long-term vigilance (e.g. yellow/vigilance level for La Soufrière of Guadeloupe since 1999, Komorowski et al. 2005; OVSG-IPGP 1999–2016 or at the Campi Flegrei caldera since 1969; Ricci et al. 2013). In such cases there are obvious long-term strategies that could be developed to improve social well-being and economic development (e.g. developing resilient critical infrastructures such as roads, bridges, public electrical water and sewer systems and communications networks) that would significantly enhance the quality of life for years of "peace time" from the volcano while ensuring a more efficient crisis response and recovery should the volcano erupt and impact the society.

(b) Improved communication

Open and multi-directional communication processes are of paramount importance in fostering the development of a shared representation and understanding among all stakeholders of the nature, magnitude, dynamics, and societal and environmental consequences of unrest and its potential eruptive outcome on multiple spatio-temporal scales (e.g. Barclay et al. 2008, 2015; Komorowski et al. 2017). While communicating this information in a timely and comprehensible format is challenging, the evidence presented here suggests that a continuous discourse is needed between different stakeholders ideally both before, during and after an unrest situation. The case studies presented here demonstrate that part of this discourse should involve a discussion about the appropriate scientific complexity in the communication between scientific and non-scientific stakeholders is essential, so that the information exchange is 'useful, usable and used' (Aitsi-Selmi et al. 2016) and fit for the decision-making purpose to which it is intended (Fischhoff 2013). Wider discourse could for example include regular information bulletins from monitoring agents to the civil society and authorities, the development of scenario-based approaches in simulation exercises involving the civil society and the wider appraisal of less tangible 'assets' (i.e., live stock or cultural capital) in risk governance efforts.

Dialogues between those responsible for monitoring hazards and those responsible for managing risk, as well as the communities at risk cannot only help to understand the most important aspects of scientific information to convey but could also lead to an improved understanding of the context into which emergency response actions must be made (e.g. Christie et al. 2015), and encourage citizens at risk to act on advice. In particular more systematic studies that analyse the effectiveness of different techniques and strategies in achieving these goals would be very useful (see Fearnley et al. 2017 for a recent compilation). These efforts should help address reluctance by the public to follow emergency-response advice in an emerging unrest crises.

(c) Wider natural risk appraisal

In a similar vein, the implementation of advice on volcanic risk could be more effective if it is considered in the context of other natural risks and social challenges (e.g., Wilkinson et al. 2016). By definition the onset of a volcanic eruption involves the anticipation of impacts from multiple hazards but the risk associated with volcanic hazards are often considered in isolation, and as a low probability, high consequence hazard, ignored in advance of an unrest crisis. This lack of dialogue and preparation has been identified above as a strong contributor to tensions during unrest crises. Volcanic regions only very rarely suffer solely from the impacts of a single natural hazard (e.g. volcanic small-island developing states discussed in Wilkinson et al. 2016; Komorowski et al. 2017). Therefore methods that consider the multi-hazard context more clearly may ultimately help communities at risk cope with uncertainty in face of volcanic hazards. This may be particularly the case, if they are able to identify 'co-benefits' during volcano "peace time" where preparedness or mitigation measures yield benefits for more than one hazard scenario (Wilkinson et al. 2016). This improvement of social well-being is likely to allow the society to take better decisions when times of impeding adversity arise.

(d) Framing of warnings around decision-making timescales and hazard level

Typically changes in alert levels are strongly tied to pre-determined changes in geophysical and geochemical signals or phenomenological observations and have carefully worked out associated actions. In our case studies, difficulties have arisen when the time-scale over which mitigating actions can be taken is much shorter than needed to implement mitigating actions such as evacuation or much longer than the timescale over which unrest or new eruptive activity impacts on the population at risk. In the case of the former, lives or assets may be put at risk and in the case of the latter, possessions and livelihoods can be negatively impacted with repercussions on trust and political stability. Managing decade or longer periods of protracted moderate-level unrest amid significant epistemic and aleatoric uncertainty on its outcome constitutes major challenges for scientists, authorities, the population, and the media.

The development of novel probabilistic formalism for decision-making could help reduce scientific uncertainty and better assist public officials in making urgent evacuation decisions and policy choices should the current and ongoing unrest lead to renewed eruptive activity. To improve decision-making around changing alert or hazard levels, improved modelling efforts of the time-scales and pathways of population mobilisation or actions (both as forward modelling and as analysis of past events) and better understanding of the consequences of protracted unrest or eruptive activity on the vulnerabilities of affected populations (e.g. Few et al. 2017) could improve choices to be made in responding to changing or escalating activity as well as chain-link scenarios.

Further, focussing on the time-scales associated with the *responses* to unrest (from the time taken to mobilise populations in an acute emergency, to the time-limits of tolerability of evacuation processes and finally the time-scales over which services and livelihoods deteriorate in response to protracted unrest) could provide important indicators for the time-scales over which alert levels (and attendant actions) need attention. In turn this perspective could inform scientific targets for improved forecasting, with strong effort expended to reduce uncertainty over time intervals that match those most critical to effective societal action.

7 Conclusions

We have identified a number of scientific and sociological problems surrounding volcanic unrest and have highlighted key aspects of risk governance at the interface between scientists, emergency managers and wider societal stakeholders. We have in particular focussed on the issue of scientific uncertainty and its impact on preparatory or contingency actions that might be needed because relationships between and among individuals and social groups will be perturbed or even changed. Especially, during periods of protracted unrest (years to decades) challenges arise from the exploitability of unrest signals to forecast long-term unrest evolution and its eventual outcome. This impacts directly on establishing the probability for and the timing and type of future eruptive behaviour as well as the definition of appropriate response protocols. To improve communication and trust between stakeholders as well as the framing of warnings around decision-making timescales and hazard levels of unrest, we propose that bridging across traditional scientific boundaries, wider natural risk appraisal and broader stakeholder interaction is needed.

Acknowledgements The VUELCO project received funding by the EC-FP7 program under grant agreement number 282759, the CASAVA project (ANR-09-RISK-002) was funded by the Agence Nationale de la Recherche (France) and the STREVA Project (NE/J020052/1) was funded by the UK NERC and ESRC. We acknowledge the useful insights and comments from two anonymous reviewers.

References

Aitsi-Selmi A, Blanchard K, Murray V (2016) Ensuring science is useful, usable and used in global disaster risk reduction and sustainable development: a view through the Sendai Framework lens. Palgrave Commun (2) doi:10.1057/palcomms.2016.16

Annen C, Wagner J-J (2003) The impact of volcanic eruptions during the 1990s. Nat Hazards Rev 4:169–175

Aspinall WP (2006) Structured elicitation of expert judgment for probabilistic hazard and risk assessment in volcanic eruptions. In: Mader HM, Coles SG, Connor CB, Connor LJ (eds) Statistics in volcanology. The geological society for IAVCEI, 15–30

Aspinall WP, Woo G (2014) Santorini unrest 2011–2012: an immediate Bayesian belief network analysis of eruption scenario probabilities for urgent decision support under uncertainty. J Appl Volcanol 3:12

Auker MR, Sparks RSJ, Siebert L, Crossweller S, Ewert J (2013) A statistical analysis of the global historical volcanic fatalities record. J Appl Volcanol 2:2

Barclay J, Haynes K, Mitchell T, Solana C, Teeuw R, Darnell A, Crosweller HS, Cole P, Pyle DM, Lowe C, Fearnley C, Kelman I (2008) Framing volcanic risk within disaster risk reduction: finding ways for the social and physical sciences to work together. In: Liverman DGE, Pereria CPG, Marker B (eds) Communicating environmental geoscience. Geological Society Special Publication, 305

Barclay J, Haynes K, Houghton BF, Johnston DM (2015) Social processes in volcanic risk reduction. In: Sigurdsson H et al (eds) Encyclopaedia of volcanology, 2nd edn. Elsevier, Amsterdam.

Barberi F, Corrado G, Innocenti F, Luongo G (1984) Phlegraean Fields 1982–1984: brief chronicle of a volcano emergency in a densely populated area. Bull Volcanol 47:175–185

Bartolini S, Marti J, Sobradelo R, Becerril L (2017) Probabilistic e-tools for hazard assessment and risk management. Adv Volcanol 1–15

Baunay Y (1998) Gestion des risques et des crises: l'exemple de deux crises d'origines volcaniques dans les Caraïbes: Montserrat 1995–1998 et la Guadeloupe 1976–1977. Mémoire de Maîtrise de Géographie, Université de Paris 1, Panthéon-Sorbonne, Paris, unpublished, 164 pp.

Biggs J, Ebmeier SK, Aspinall WP, Lu Z, Pritchard ME, Sparks RSJ, Mather TA (2014) Global link between deformation and volcanic eruption quantified by satellite imagery: Nat Commun 5

Blérald APh (1986) Histoire éruptive de la Guadeloupe et de la Martinique du XVIIème siècle à nos jours, Editions Karthala, Paris, 336 p

Benson C (2006) Volcanoes and the economy. In: Marti J, Ernst GGJ (eds) Volcanoes and the environment. Cambridge University Press, pp 440–467.

Brealey R, Myers S, Allen F (2011) Principles of corporate finance. McGraw, 872 p

Bretton R, Gottsmann J, Aspinall WP, Christie R (2015) Implications of legal scrutiny processes (including the L'Aquila trial and other recent court cases) for future volcanic risk governance. J Appl Volcanol 4:18

Brown SK, Auker MR, Sparks RSJ (2015) Populations around Holocene volcanoes and development of a population exposure index. In: Loughlin SC et al (eds) Global volcanic hazards and risk 'populations around volcanoes and the development of a population exposure index. Cambridge

Burton I (2010) Forensic disaster investigations in depth: a new case study model. Environ Mag 52(5):36–41

Cashman KV, Sparks RSJ (2013) How volcanoes work: a 25 year perspective. Geol Soc Am Bull. doi:10.1130/b30720.1

Carn SA, Froyd KD, Anderson BE, Wennberg P, Crounse J, Spencer K, Dibb JE, Krotkov NA, Browell EV, Hair JW, Diskin G, Sachse G, Vay SA (2011) In situ measurements of tropospheric volcanic plumes in Ecuador and Colombia during TC4. J Geophys Res Atmos, 116

Christie R, Cooke O, Gottsmann J (2015) Fearing the knock on the door: critical security studies insights into limited cooperation with disaster management regimes. J Appl Volcanol 4:19

Christopher TE, Blundy J, Cashman K, Cole P, Edmonds M, Smith PJ, Sparks RSJ, Stinton A (2015) Crustal-scale degassing due to magma system destabilization and magma-gas decoupling at Soufrière Hills Volcano, Montserrat: Geochemistry. Geophys Geosyst 16(9):2797–2811

Dasgupta P (1988) Trust as a commodity. In: Gambetta D (ed) Trust: making and breaking of cooperative relations. Blackwell, Oxford, 49–72

De la Cruz-Reyna S, Tilling RI (2008) Scientific and public responses to the ongoing volcanic crisis at Popocatépetl Volcano, Mexico: importance of an effective hazards-warning system. J Volcanol Geotherm Res 170:121–134

Esposti Ongaro T, Cavazzoni C, Erbacci G, Neri A, Salvetti MV (2007) A parallel multiphase flow code for the 3D simulation of volcanic explosive eruptions. Parallel Comput 33:541–560

European Commission (2010) Press Release 27. April, 2010). http://ec.europa.eu/commission_2010-2014/kallas/headlines/news/2010/04/doc/information_note_volcano_crisis.pdf

Fearnley C, Bird D, Haynes K, Jolly G, McGuire B (eds) (2017) Observing the volcano world: volcano crisis communication. Advances in volcanology, IAVCEI. Springer, Berlin, 350 p

Feuillard M, Allegre CJ, Brandeis G, Gaulon R, Le Mouel JL, Mercier JC, Pozzi JP, Semet MP (1983) The 1975–1977 crisis of La Soufrière de Guadeloupe (FWI): a still-born magmatic eruption. J Volcanol Geotherm Res 16:317–334

Few R, Armijos T, Barclay J (2017) Living with Volcan Tungurahua: the dynamics of vulnerability during prolonged volcanic activity. Geoforum 80:72–81

Fischhoff B (2013) The sciences of science communication. In: Proceedings of the National Academy of Sciences of the United States of America, 110 (Supplement 3), 14033–14039. doi:10.1073/pnas.1213273110

Fiske R (1984) Volcanologists, journalists, and the concerned local public: a tale of two crises in the eastern Caribbean. In: Boyd F (ed) Explosive volcanism: inception, evolution and hazards. Studies in geophysics. National Academy Press, Washington DC, pp 170–176

Fournier TJ, Pritchard ME, Riddick SN (2010) Duration, magnitude, and frequency of subaerial volcano deformation events: new results from Latin America using InSAR and a global synthesis. Geochem Geophys Geosyst 11:Q01003. doi:10.01029/02009GC002558

Global Volcanism Program (2013) Volcanoes of the World, v. 4.5.5. In: Venzke E (ed) Smithsonian Institution. Downloaded 04 May 2017. http://dx.doi.org/10.5479/si.GVP.VOTW4-2013

Haynes K, Barclay J, Pidgeon NF (2008) The issue of trust and its influence on risk communication during a volcanic crisis. Bull Volcanol 70:605–621

Hicks A, Few R (2015) Trajectories of social vulnerabiltiy during the Soufriere Hills volcanic crisis. J Appl Volcanol 4:10

Hill DP, Johnston MJS, Langbein JO, Bilham R (1995) Response of Long Valley caldera to the Mw = 7.3 Landers, California, Earthquake. J Geophys Res 100 (B7):12985–13005

Hickey J, Gottsmann J, Mothes P, Odbert HM, Prutkin I, Vajda P (2017) The ups and downs of volcanic unrest: insights from integrated geodesy and numerical modelling. Adv Volcanol 1–17

Hincks T, Komorowski J-C, Sparks RSJ, Aspinall WP (2014) Retrospective analysis of uncertain eruption precursors at La Soufrière volcano, Guadeloupe, 1975–77: volcanic hazard assessment using a Bayesian Belief Network approach. J Appl Volcanol 3:3

IAVCEI task force on crises protocols (2016) Toward IAVCEI guidelines on the roles and responsibilities of scientists involved in volcanic hazard evaluation, risk mitigation, and crisis response. Bull Volcanol 8(31). doi:10.1007/s00445-00016-01021-00448

IRGC (2017) https://www.irgc.org/risk-governance/irgc-risk-governance-framework/. Last accessed 05/02/2017.

Johnson BB (1987) Accounting for the social context of risk communication. Sci Technol Stud, 103–111

Jolly AD, Jousset P, Lyons JJ, Carniel R, Fournier N, Fry B, Miller C (2014) Seismo-acoustic evidence for an avalanche driven phreatic eruption through a beheaded hydrothermal system: an example from the 2012 Tongariro eruption. J Volcanol Geotherm Res 286:331–347

Kokelaar BP (2002) Setting, chronology and consequences of the eruption of Soufrière Hills Volcano, Montserrat (1995–1999). In: Druitt TH, Kokelaar BP (eds) The eruption of Soufrière Hills Volcano, Montserrat, from 1995 to 1999, vol 21. Geological Society, London, Memoirs, pp 1–44

Komorowski J-C, Boudon G, Semet M, Beauducel F, Anténor-Habazac C, Bazin S, Hammouya G (2005) Guadeloupe. In: Lindsay JM, Robertson REA, Shepherd JB, Ali S (eds) Volcanic Atlas of the Lesser Antilles, Seismic Research Unit, The University of the West Indies, Trinidad and Tobago, WI, 65-102

Komorowski J-C, Morin J, Jenkins S, Kelman I (2017) Challenges of volcanic crises on small islands states. In: Fearnley C, Bird D, Haynes K, Jolly G, McGuire B (eds) Observing the volcano world: volcano crisis communication', Advances in Volcanology, IAVCEI, Springer, Berlin, pp 1–18. doi:10.1007/11157_2015_15

Lepointe E (1999) Le réveil du volcan de la Soufrière en 1976: la population guadeloupéenne à l'épreuve du danger. In: Yacou A (ed) Les catastrophes naturelles aux Antilles – D'une Soufrière à l'autre. CERC Université Antilles et de la Guyane, Editions Karthala, Paris, pp 15–71

Maeno F, Nakada S, Oikawa T, Yoshimoto M, Komori J, Ishizuka Y (2016) Reconstruction of a phreatic eruption on 27 September 2014 at Ontake volcano, central Japan, based on proximal pyroclastic density current and fallout deposits. Earth, Planets and Space 68:82

Manville V, Major JJ, Fagents SA (2013) Modeling lahar behavior and hazards. In: Fagents S, Gregg T, Lopes R (eds) Modeling volcanic processes: the physics and mathematics of volcanism. Cambridge University Press, pp 300–330

Marzocchi W, Woo G (2007) Probabilistic eruption forecasting and the call for an evacuation. Geophys Res. Lett 34:L22310. doi:10.21029/22007GL031922

Marzocchi W, Sandri L, Selva J (2008) BET_EF: a probabilistic tool for long- and short-term eruption forecasting. Bull Volc 70:623–632

Mothes P, Yepes HA, Hall ML, Ramon PA, Steele AL, Ruiz MC (2015) The scientific-community interface over the 15 year eruptive episode of the Tungurahua Volcano, Ecuador. J Appl Volcanol 4:9

Mothes P, Ruiz MC, Viracucha EG, Ramón PA, Hernández S, Hidalgo S, Bernard B, Gaunt EH, Jarrín P, Yépez MA, Espín PA (2017) Geophysical footprints of Cotopaxi's unrest and minor eruptions in 2015: an opportunity to test scientific and community preparedness. Adv Volcanol 1–30

OVSG-IPGP (1999–2013) Bulletin mensuel de l'activité volcanique et sismique de Guadeloupe. Monthly public report, Observatoire Volcanologique et Sismologique de Guadeloupe - Institut de Physique du Globe de Paris, Gourbeyre (ISSN 1622-4523). http://www.ipgp.fr/fr/ovsg/bulletins-mensuels-de-lovsg. Last accessed 11-10-2016

Phillipson G, Sobradelo R, Gottsmann J (2013) Global volcanic unrest in the 21st century: an analysis of the first decade. J Volcanol Geotherm Res 264:183–196

Potter S, Scott B, Jolly G, Neall V, Johnston D (2015) Introducing the volcanic unrest index (VUI): a tool to quantify and communicate the intensity of volcanic unrest. Bull Volcanol 77:1–15

Reid ME (2004) Massive collapse of volcano edifices triggered by hydrothermal pressurization. Geology 32:373–376

Ricci T, Barberi F, Davis MS, Isaia R, Nave R (2013) Volcanic risk perception in the Campi Flegrei area. J Volcanol Geoth Res 254:118–130

Rouwet D, Sandri L, Marzocchi W, Gottsmann J, Selva J, Tonini R, Papale P (2014) Recognizing and tracking volcanic hazards related to non-magmatic unrest: a review. J Appl Volcanol 3:17

Rouwet D, Constantinescu R, Sandri L (2017) Deterministic versus probabilistic volcano monitoring: not "or" but "and". Adv Volcanol. doi:10.1007/11157_2017_8

Sandri L, Tonini R, Rouwet D, Constantinescu R, Mendoza-Rosas AT, Andrade D, Bernard B (2017) The need to quantify hazard related to non-magmatic unrest: from BET_EF to BET_UNREST. Adv Volcanol. doi:10.1007/11157_12017_11159

Salvage RO, Karl S, Neuberg J (2017) Volcano seismology: detecting unrest in wiggly lines. Adv Volcanol 1–17

Sekiya S, Kikuchi Y (1890) The eruption of Bandai-san. Tokyo Imp Univ Coll Sci J 3:91–172

Sheldrake TE, Sparks RSJ, Cashman KV, Wadge G, Aspinall WP (2016) Similarities and differences in the historical records of lava dome-building volcanoes: Implications for understanding magmatic processes and eruption forecasting. Earth Sci Rev 160:240–263

Siebert L, Cottrell E, Venzke E, Andrews B (2015) Earth's volcanoes and their eruptions: an overview. In: Sigurdsson H et al (eds) The encyclopedia of volcanoes, 2nd edn. Elsevier, Amsterdam. http://dx.doi.org/10.1016/B978-0-12-385938-9.00012-2

Stone J, Barclay J, Simmons P, Cole PD, Loughlin SC, Ramon P, Mothes P (2014) Risk reduction through community-based monitoring: the vigías of Tungurahua. Ecuador. J. Appl. Volcanol. 3:11

Stirling A (2010) Keep it complex. Nature 468:1029–1031

Tazieff H (1980) About the Soufrière of Guadeloupe. J Volcanol Geotherm Res 8:3–6

Venezky D, Newhall C (2007) "WOVOdat Design Document; The Schema, Table Descriptions, and Create Table Statements for the Database of Worldwide Volcanic Unrest (WOVOdat Version 1.0)." U.S. Geological Survey Open-File Report, 184.

Woo G (2008) Probabilistic criteria for volcano evacuation decisions. Nat Hazards 45(1):87–97

Woo G (2015) Cost-Benefit analysis is volcanic risk. In: Papale P (ed) Volcanic Hazards, Risks and Disasters. Springer, pp 289–300. https://www.elsevier.com/books/volcanic-hazards-risks-and-disasters/papale/978-0-12-396453-3

Wadge G, Isaacs MC (1988) Mapping the volcanic hazards from Soufriere Hills Volcano, Montserrat, West Indies using an image processor. J Geol Soc 145(4):541–551

Wadsworth FB, Vasseur J, Scheu B, Kendrick JE, Lavallée Y, Dingwell DB (2016) Universal scaling of fluid permeability during volcanic welding and sediment diagenesis. Geology 44(3):219–222

Wilkinson E (2015) Beyond the volcanic crisis: co-governance of risk in Montserrat. J Appl Volcanol 4:3

Wilkinson E, Lovell E, Carby B, Barclay J, Robertson REA (2016) The dilemmas of risk sensitive development on a small volcanic island. Resources 5 (2):21. doi:10.3390/resources5020021

Wright TL, Pierson T (1992) Living with volcanoes. No. 1073. US Geologic Survey

Wynne B (1992) Uncertainty and environmental learning: reconceiving science and policy in the preventive paradigm. Glob Environ Change 2:111–127

The Role of Laws Within the Governance of Volcanic Risks

R. J. Bretton, J. Gottsmann and R. Christie

Abstract

The governance of volcanic risks does not take place in a vacuum. In many cultures, volcanic risks are perceived to be susceptible to governance with the objective of achieving their effective mitigation, and have become the responsibility of the institutions and stakeholders of relevant social communities. An array of international, national and local laws dictate governance infrastructures, the roles of duty holders and beneficiaries and the relationships between them (the stakeholders), duties and rights (the stakes) and acceptable standards of safety and wellbeing (the ultimate rewards). Many regional, national and local stakeholders (individuals and entities) have a range of different, yet complementary, roles, duties, rights and powers. Much of this chapter, which has two main sections, represents a summary of a longer paper (Bretton et al. 2015) that addresses legal aspects of the future governance of volcanic risks. After a general introduction to relevant terminology in the first section, the second section describes the significant threat posed by periods of volcanic unrest.

R. J. Bretton (✉) · J. Gottsmann
School of Earth Sciences, University of Bristol,
Wills Memorial Building, Queens Road, BS8 1RJ
Bristol, UK

R. J. Bretton · J. Gottsmann · R. Christie
The Cabot Institute, University of Bristol, Wills
Memorial Building, Queens Road, BS8 1RJ Bristol,
UK

R. Christie
School of Sociology, Politics and International
Studies, University of Bristol, 4 Priory Road, BS8
1TU Bristol, UK

Advs in Volcanology (2019) 23–34
DOI 10.1007/11157_2017_29
© The Author(s) 2017
Published Online: 28 November 2017

The third section contains a general introduction to the critical concept of risk which lies at the heart of governance and provides a more detailed description of the many roles that national laws play. Reference is also made to international law which has an increasingly important role in the absence of relevant national laws, or when national laws are inadequate, ineffective or unenforced.

Keywords

Hazard · Risk · Risk governance · Legal duties

1 Introduction

This chapter describes the ways in which laws create the administrative and functional infrastructures that facilitate the effective mitigation of volcanic risks.

For the sake of brevity and clarity, we draw upon the existing rich discourse on relevant terminology (e.g. Fournier d'Albe 1979; Luhmann 1998, 1992; Power 2007, 2009; UN/ISDR 2009; Smith and Petley 2009; MIAVITA 2012) and adopt a number of brief working definitions.

A *'hazard'* is an event (defined by risk-related temporal, spatial and other parameters) that may cause adverse effects. It is a complex function being the "probability of any particular area being affected by a destructive volcanic manifestation within a given period of time" (Fournier d'Albe 1979, 321). A *'volcanic hazard'* is a volcanic scenario (defined by risk-related temporal, spatial and physical parameters) that may cause adverse consequences to people and/or valued assets.

In the early 1970s, definitions of risk identified the product of three separate and distinct elements—'vulnerability' on 'exposure' to a defined 'hazard' (UNESCO 1972). Bankoff et al. (2004) refers to volcanic hazards being one of three variables (hazard, exposure and vulnerability) that are convolved to produce volcanic risks. For the purposes of governance, it may be helpful to quantify each identified exposure in units of both number (i.e. the number of people exposed) and time (given that, by way of

example, non-resident workers may be less exposed than full-time residents and 'vulnerability' may be related to length of exposure).[1]

'Governance' is a complex concept attracting a multitude of definitions (Walker et al. 2010). Although it "encompasses an array of organisations, practices and ideas" (Rothstein et al. 2012) that change over time, for sake of clarity, we embrace following definition.

> Governance is the sum of the many ways individuals and institutions, public and private, manage their common affairs. It is a continuing process through which conflicting or diverse interests may be accommodated and co-operative action may be taken. It includes formal institutions and regimes empowered to enforce compliance, as well as informal arrangements that people and institutions either have agreed to or perceive to be in their interest (Commission on Global Governance 1995, 4).

'Risk governance' includes all attempts to manage the three constituent variables of risk including steps to mitigate volcanic hazards (there are very few successful examples of this), reduce the exposure of people, assets etc. and reduce their vulnerability when exposed. This expression is adopted not only as an analytic term to describe stakeholders undertaking mitigation activities but also for 'normative' (i.e. evaluative standard) purposes. Risk governance

---

[1]The 18th report of the Scientific Advisory Committee on Montserrat Volcanic Activity contains a good example of a risk assessment which adopts this approach. It differentiates between the risks faced by residents in Zone A and those of workers involved in the shipment of sand from Plymouth Jetty.

has a set of definable 'good' qualities that provide for the effective integration of the key components of how risks are handled by risk stakeholders (Walker et al. 2010; IRGC 2009).

2 Geological Background

Active volcanism can involve complex multi-hazard phenomena. Precursory unrest provides, by means of its signals, the monitoring data upon which evidence-based short-term hazard analysis is grounded. However, periods of mild unrest, even if they may not lead to an eruption, can themselves present a range of hazards including earthquakes, ground deformation, hydrothermal changes/eruptions and gas/water chemistry changes. These precursory hazards can create societal risks that can escalate unnecessarily and therefore require very careful management. Unrest periods create, not only uncertainty about what is happening and resulting public alarm, anxiety and speculation, but also demands for information and advice (Johnston et al. 2002).

The evolution of an unrest period will depend upon its underlying causative processes, which can lead to different outcomes in different locations and with different spatial and physical properties (Rouwet et al. 2014; Sobradelo and Marti 2015).

Volcanic hazard communications and risk mitigation decisions rely upon the suitable and sufficient collection, and the correct analysis and interpretation of monitoring data, and the geological record (Newhall and Hoblitt 2002; Sparks et al. 2012; Rouwet et al. 2014). The analysis of monitoring data, which will often be limited in both quantity and quality, is challenging and there are many uncertainties in identifying causes and thereafter anticipating the evolution of unrest and imminent eruption (Sparks et al. 2012; Phillipson et al. 2013; Sobradelo and Marti 2015).

Hazard analysis is difficult and the risk governance stakes are high. Poorly handled unrest periods cause social, economic and political problems, even without an eruption. "Adverse response may take the form of the release of inappropriate advice, media speculation, unwarranted emergency declarations and premature cessation of economic activity and community services" (Johnston et al. 2002, 228).

3 Risk Governance and Roles of Law

The concept of risk as something that can be managed through human intervention is a relatively new one and important because it has become an increasingly pervasive concept in many societies. Risk is also associated with notions of choice, responsibility and blame (OECD 2015).

Risk evolved from its modest origins in the seventeenth century and became in the nineteenth century a principle for the objectification of possible experience—not only of the hazards of personal life and private venture, but also of the common venture of society (Gordon 1991).

By the late nineteenth century, risk had "become central to the rhetoric of regulation". State regulation of risk emerged as the means by which the state controlled economic activities in Western societies. The traditional objects of state regulation were manufactured risks, most particularly those resulting from scientific and technological innovation within manufacturing processes. The usual style of state regulation was "command and control" by imposing formal, structured and active risk management duties. The state exercised control through the promulgation of primary (i.e. enabling) and secondary (i.e. detailed implementing) laws and policing through specialist inspectorates.

In the twenty first century, regulation is no longer confined to non-natural, human-made risks. Many risks are, in whole or in part, recurring social manifestations (i.e. human-made phenomena) with negative consequences (Lauta 2014). In many cultures, particularly western cultures, they are no longer perceived as the consequences of external forces occurring independently of society and insusceptible to mitigation by society. Accordingly, they are now positioned within, and have become the responsibility of, the institutions and stakeholders of relevant social communities (Lauta 2014). These human-made risks are perceived to be susceptible to regulation with the objective of achieving their effective mitigation. By way of illustration, the population of Naples has greatly increased since 1944 and many would argue that the resulting increase in volcanic risk exposure is human-made and capable of regulation.

Low probability-high impact risks pose a particular challenge for legislators. In fact there are three related challenges, namely scientific uncertainty, a low likelihood of occurrence, and significant societal consequences. Whilst the elevated consequences of these risks call for some level of regulation, the intrinsic uncertainty and low probability of their occurrence make it difficult to review the evidentiary scientific justification, to assess costs and benefits, and to identify means by which chosen regulatory goals can be pursued (Simoncini 2013).

In the absence of a tragedy, it is difficult to measure the performance of law-backed societal risk governance by the usual measures of: (1) economy (e.g. value for money) for input and process; (2) efficiency (e.g. quality delivered on time) for process and output; and (3) effectiveness for output and outcome. The indicators of outcome (the intended and unintended results) of the integrated governance system will be related to the impacts on, and the consequences for, public good, safety, security, health and welfare but it will be a challenge for any related targets (e.g. benchmarks and performance standards) to be SMART—Specific, Measurable, Achievable, Relevant and Timed (OECD 2002).

By contrast, in a fact-finding process of scrutiny after a tragedy, the use of SMART targets may become more practicable. It may be possible to measure hazard characterisation outputs against planned targets for timely delivery, user-friendliness, outcome-focussed, and temporal/spatial/intensity forecast accuracy. Based upon findings of fact, it may be feasible to quantify the resulting risk-mitigation impact measured in lives and assets saved.

Notwithstanding these challenges, many jurisdictions have national laws that attempt to regulate the management of risks arising from natural hazards. Many reflect the shift in paradigm, at both international and national levels, from focussing on ex-post, reactive response (the phases of emergency response and post-disaster longer term recovery) to ex-ante, pro-active risk management and mitigation (the phase of planning and preparedness) (UN SC-DRR 2009).

As illustrated in Fig. 1, national laws create governance infrastructures, duties of care and duty holders, rights and rights holders, enabling powers, regulators, enforcement powers, and lastly scrutiny venues. Each will now to considered in turn.

3.1 The Creation of National Risk Governance Infrastructures

National laws tend to identify, authorise and fund risk governance bodies (e.g. government departments and agencies, and public corporations) and public officials (e.g. individuals such governors, mayors, prefects and village heads) within a coherent legal and administrative framework, in other words, a risk governance infrastructure. These laws often use and build upon existing entities within existing administrative frameworks that have multi-level national, regional, district, municipal etc. political divisions and subdivisions.

In some jurisdictions, formal legal infrastructures anticipate and rely upon less formal structures and relationships at local levels nearer at-risk communities. For example, in Ecuador, the risk governance infrastructure relies upon the engagement and commitment of local representatives (e.g. chiefs and elders) and volunteers, such as hazard wardens/monitors, for both hazard data gathering and risk mitigation.

In some jurisdictions, such as Italy, the laws favour the imposition of duties upon individuals, rather than impersonal legal entities such as government departments/agencies and public companies. Legal duties may be founded upon an

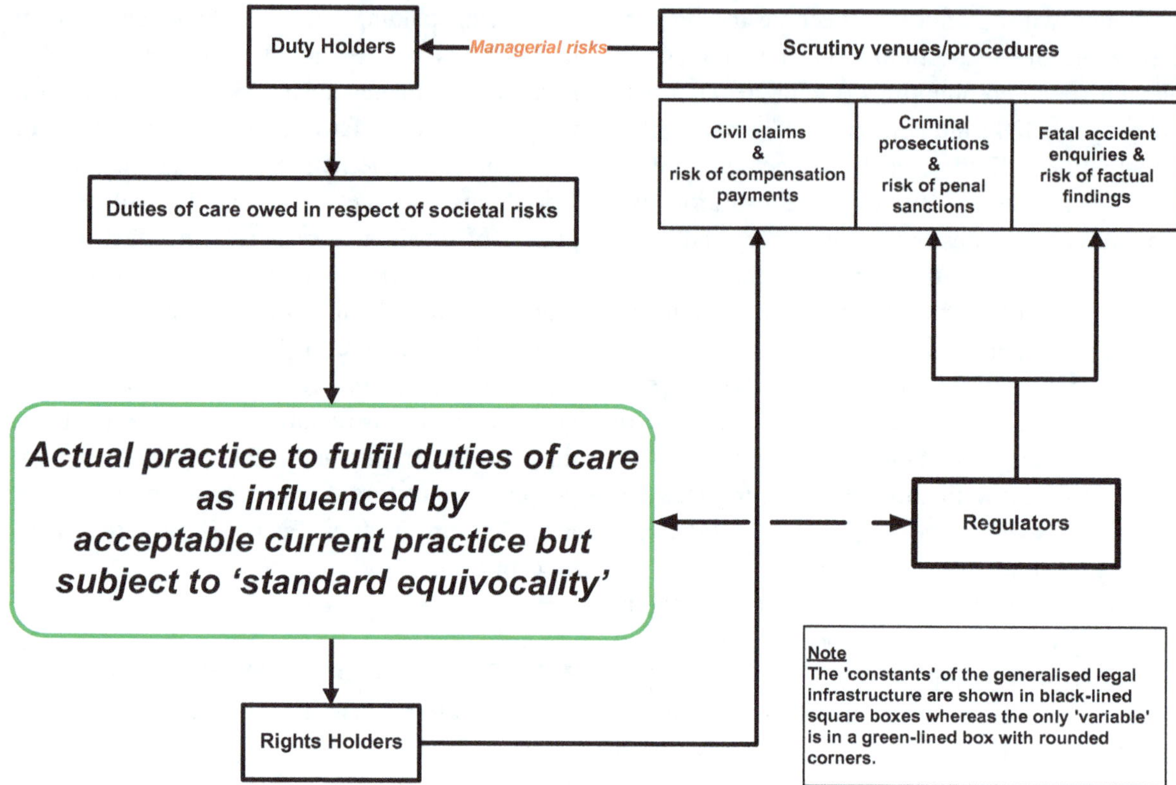

Fig. 1 The many roles of law in the governance of natural hazards set out in a generalised legal framework

individual having effective decision-making powers and control over financial resources rather than upon an individual holding a particular title or occupying a particular post (Bergman et al. 2007).

These infrastructures can be complex, confusing, fragmented and multi-level. They are often the creations of multiple sets of national primary (enabling) and secondary (detailed implementing) legislation supplemented as necessary by further provisions at ministerial, inter-ministerial, regional, provincial and local levels of government.

Occasionally additional specialised bodies are established (e.g. emergency management agencies, research/monitoring institutes and volcano observatories) with the creation of statutory roles to be filled by appointed individuals.

In a few known jurisdictions (e.g. USA, Canada and the Philippines) laws also regulate to varying degrees the qualification, licensing and registration of geologists and the practice of geology per se.

3.2 The Creation of Duty and Rights

National laws allocate to bodies and individuals (duty holders such as volcano observatories and civil protection authorities) high level management functions with responsibilities (duties of care), which are owed to the particular classes of people for whose benefit the duties were created (rights holders).

Since modest beginnings in the 1840's near Vesuvius Italy, the role of over 100 volcano observatories around the world has evolved. Observatories have at least two overlapping roles which involve a synergy of observation and theory. They have been described as 'critical in the volcanic risk reduction cycle'[2] (Jolly 2015, 302), and employ and/or engage scientists who practice at the hazard-risk interface. The World Organisation of Volcano Observatories (WOVO), a Commission of the International

---

[2]This cycle includes periods before, during and after periods of volcanic unrest that may or may not lead to an eruption (Jolly 2015, 302).

Table 1.1 Volcano observatories—three contrasting regimes

Montserrat—One individual with no powers of delegation

Section 8 of the Montserrat Volcano Observatory Act 2002 states that the Director of the Observatory shall be responsible for "reporting on the status of the volcanic activity in a regular and timely manner to the appropriate authorities" and "assisting in the dissemination to the public of information concerning the status of volcanic activity"

Alaska, USA—A group of institutions

"The Alaska Volcano observatory is a joint programme of the United States Geological Survey (USGS), the Geophysical Institute of the University of Alaska Fairbanks (UAFGI) and the State of Alaska Division of Geological and Geophysical Surveys (ADGGS)" (Jolly 2015, 299)

New Zealand—One institution under powers of delegation given to one individual

The Institute of Geological and Nuclear Sciences Limited (or GNS Science) is a Crown research institute created in 1992. The GNS has "sole responsibility for providing volcanic activity warnings and hence provides the function of a volcano observatory" (Jolly 2015, 299) under wide powers of delegation given to the Director of Civil Defence Emergency Management in the Civil Defence Emergency Management Act 2002

Association of Volcanology and Chemistry of the Earth's Interior (IAVCEI), is an organization of and for volcano observatories of the world. WOVO's website notes that its "members are institutions that are engaged in volcano surveillance and, in most cases, are responsible for warning authorities and the public about hazardous volcanic unrest".

Examples of three contrasting regimes are given in Table 1.1.

In some jurisdictions, general disaster management obligations are also imposed on 'the community' (i.e. members of the public) and non-government business entities. At-risk individuals and communities, businesses (such as aeroplane makers and operators within the aviation industry) and insurers have active and critical roles to play in the governance of volcanic risks, however, their roles are not the principal focus of either this chapter or Bretton et al. (2015). The needs and sentiments of all duty and rights holders, who depend upon and use geoscientific knowledge of volcanic hazards, must be identified and reflected carefully and clearly in the roles and interface practices of volcanologists.

During an emerging period of volcanic unrest, the relevant duty holders may change as the defined duties are transferred from one duty holder to another—sometimes as a result of changing hazard or risk characterisations. These duties of care can be framed in a wide variety of ways. They may relate to general health and safety, not specifying any risk creator (a particular hazard, natural or otherwise) or they may be more specific, identifying a particular hazard (ground uplift, earthquakes etc.).

Rights holders may be given the right to a safe and healthy environment, and to be represented, consulted or engaged in risk decisions and/or given information. Additional rights may be given to certain categories of persons due to special vulnerabilities and/or the influence of social structures and practices. These categories may include women, the very poor, older persons, children and people with disabilities (IFRC 2015).

There are two main types of duty of care, which are called here, respectively, 'functional' and 'goal-setting'. Functional duties dictate the fulfilment of a particular role (e.g. a duty to undertake monitoring, to prepare plans and programmes for emergency preparedness or to provide emergency preparedness communications and warnings). Goal-setting duties require the achievement of an outcome (e.g. a duty to ensure the safety and wellbeing of identified rights holders). Not even within the highly regulated field of occupational health are these safety goals absolute (i.e. unqualified). The imposition of an unrealistic absolute duty would give rights holders a theoretical guarantee of health and safety within a risk-free environment.

As a general rule, 'qualified' duties of care are therefore laid down. These duties represent democratic statements or mandates of 'acceptable' or 'optimal' risk after mitigation, and express a rational trade-off between safety and risk (Hood and Rothstein 2001; Rothstein 2014). Rothstein (2014) notes "After all, what is an acceptable risk other than a euphemistic boundary between an *acceptable* adverse outcome and an *unacceptable* failure".

Compliance with qualified duties inevitably requires duties holders to perform a risk-focussed cost/benefit analysis and a test of proportionality. A duty holder wishing to establish that societal risks have been reduced 'as low as reasonably practicable' has to show that the costs (the sacrifice) of further feasible safety measures would be grossly disproportionate to the additional safety benefits of those measures (based upon UK Office for Nuclear Regulation 2013).

Laws rarely, if ever, attempt to dictate, in either general or more detailed terms, the societal risk management arrangements that will be required to either fulfil a functional role or achieve a stated safety outcome. In practice, an assessment of societal vulnerability has two main stages. Firstly, the nature and scope of duties of care must be identified and delineated. This is essentially a matter of law and involves the legal interpretation of primary and secondary legislation and, if relevant, case law. Secondly, it is necessary to identify the actions that the duty holder should take to fulfil those duties. This is far more difficult. Competent lawyers can describe the safety function or outcome required in law—in football terms the dimensions and position of the goal. However, they can offer very limited guidance regarding the practical measures that the competent societal risk manager will need to take to achieve legal compliance (i.e. how to actually get the ball over the goal line).

In the case of food standards and occupational health and safety standards, it is common for national laws to set up government agencies to carry out research, to set performance standards and offer approved codes of practice or authoritative guidance. By contrast, in respect of volcanic hazards, at both the international and national levels, there appear to be neither:

(1) law-based performance standards offering guidance to societal risk managers; or (2) law endorsed self-regulatory regimes such as those that frequented the early stages of food regulation.

Within the 'goal-setting' legislative approach, referred to above, it is implicit that there is an obligation on duties holders to establish the nature and suitability (i.e. the legal adequacy) of their societal risk management arrangements. This difficult justification will usually be done post-facto, in other words, only after the risk outcome (perhaps a disaster properly so-called) is known and legal consequences are already being considered (Simoncini 2011). The justification will cover, but not be limited to, the arrangements that were necessary for the planning, organisation, control, monitoring and review of societal risk mitigation measures. To complete the authors' footballing analogy, post-facto legal processes are analogous to slow motion TV replays, in full view of partisan onlookers and experts with hindsight. They determine what has happened, whether or not a goal has been scored and, if not, why not and what effect any missed goal had on the final score (i.e. whether legal compliance has actually been achieved and, if not, why not and what the consequences should have been if compliance had been achieved).

'Standard equivocality', which is the absence of commonly recognised standards (norms), is likely to exist in the absence of clear 'legal' requirements, approved codes of practice or guidance. The resulting challenges faced by duty holders are: (1) to find or design authoritative standards or benchmarks to steer their societal risk management arrangements; and thereby (2) to increase their chances of fulfilling their societal risk duties of care and achieving legal compliance; and thereby (3) to minimise their vulnerability to managerial risks.

Rothstein (2002) and Hood (1986) have noted that, in the absence of commonly agreed and practical principles or methodologies by which compliance can be measured ('standard-unequivocality'), process compliance is difficult to monitor and enforce. Other obstacles to monitoring, surveillance and enforcement include inherent scientific uncertainty, a dynamic state of scientific knowledge, a lack of expertise within regulatory agencies, and often complex and fragmented multi-level infrastructures. Donovan and Oppenheimer (2014) note complexities in governmental structures presented major challenges to managing volcanic eruptions in Montserrat. Recent crises including the 2010 Icelandic Eyjafjallajökull eruption have highlighted the difficulty of co-ordinating and synthesising scientific input from many different disciplines and institutions and translating these into useful policy advice at very short notice (OECD 2015).

3.3 The Creation of Powers

National laws have traditionally granted defined 'authorities' to government duty holders backed by administration, protection and intervention powers (ordinary, extraordinary and emergency). Governance, with an emphasis more on control than protection, has often been achieved by the exercise of authority using linear "coercion and enforcement" (Walker et al. 2010).

3.4 The Creation of Regulators, Enforcement Powers and Scrutiny Venues

Laws often establish, resource and empower regulators to monitor the performance of duty holders and to take enforcement actions, including prosecutions, against them if necessary. Examples of regulators include the Labour Standards Agency in Japan, the Department of Labour Health and Safety Service in New Zealand and the Occupational Safety and Health Agency in the United States of America. These regulators often have very wide powers similar to, and sometimes exceeding, those of police forces. They include the power to enter premises, to investigate and inspect, to acquire and preserve evidence and to serve notices.

Laws provide the formal scrutiny venues (courts, tribunals etc.) and related procedures for: (1) the ex-ante pro-active enforcement of duties

of care by regulators, generally health and safety agencies; and (2) the ex-post facto reactive scrutiny of events, the identification of duty holders, the assessment of what happened and what should have happened and, if appropriate, the imposition of sanctions and/or the granting of remedies. The latter procedures are required at a national level to comply with the international law ex-post facto obligations which are now considered.

3.5 The Role of International Law

In the absence of relevant national laws, or when national laws are inadequate, ineffective or unenforced, there is room for the intervention of international law. The European Court of Human Rights (ECHR) has taken the lead and it is suggested here that in time the Inter-American Court of Human Rights will follow. The European Convention of Human Rights (EConHR) lays down a positive obligation on States to take appropriate steps to safeguard the lives of citizens within their jurisdiction. Article 2(1) EConHR provides that "Everyone's right to life shall be protected by law. No one shall be deprived of his life intentionally save in the execution of a sentence of a court following his conviction of a crime for which this penalty is provided by law."

In the context of the management of natural hazards, the most important case involving Article 2 arose in 2008. Budayeva and others v Russia (2008) ECHR 15339/02 concerned a mudslide. In this case, the EHCR considered principles that had been applied in Oneryildiz v Turkey (2004) to a human-made hazard—an industrial risk or dangerous activity such as the operation of a waste site. They were subsequently adopted in Kolyadenko and others v Russia (2012) in respect of natural flash floods.

Budayeva and others v Russia
The town of T was situated in a mountain district. Two tributaries passed through it and were known to have associated mudslides. A mud collector and a dam were constructed in order to protect T. The dam was seriously damaged by a mud and debris

flow in August 1999, so funds were requested to construct observation posts to warn of mudslides until it could be repaired, and to carry out certain emergency works to the dam.

Those measures were never implemented. A number of mudslides occurred in July 2000, killing eight residents, including the first applicant's husband, and destroying the applicants' homes.

It was decided to dispense with a criminal investigation into the circumstances of the death of the first applicant's husband, and claims of compensation by the first applicant and others were refused on the basis that a mudslide of such exceptional force could neither have been predicted nor stopped. However, the applicants were granted substitute housing and a lump-sum emergency allowance.

The applicants complained to the ECHR, inter alia, that the authorities had violated the substantive limb of Article 2 of the EConHR. The first applicant asserted that the authorities were responsible for the death of her husband and she and the other applicants asserted that the authorities had failed to take appropriate measures to mitigate the risks to their lives posed by natural hazards.

The Court concluded that the relevant authorities were aware of the mudslides (the hazards) and their capacity to cause devastating consequences (the risks). There was no ambiguity about the scope and timing of the work that needed to be performed (the risk mitigation actions). After 1999, risk mitigation was not given proper consideration by the decision makers and budgetary bodies (the duty holders) and there was no functioning early warning system. State responsibility for the deaths had never been investigated. Each applicant was awarded compensation.

The EHCR determined that the obligation in Article 2 entails, above all, a primary duty on the State to put in place a clear legislative and administrative framework designed to provide effective deterrence against threats to the right to life. This applies in the context of any activity, whether public or not, in which the right to life may be at stake and extends not only to industrial risks and dangerous activities but also to actions and omissions to control natural hazards.

In the cases of Oneryildiz v Turkey (2004), Budayeva v Russia (2008), and Kolyadenko v Russia (2012), the EHCR determined that there is a positive obligation: (1) ex-ante to take substantive regulatory measures to manage risks; and (2) ex-post facto to ensure that any risk eventuated fatalities are followed by a public investigation. In relation to the latter, procedures must exist for identifying not only shortcomings in the ex-ante regulatory measures but also any errors committed by those responsible (i.e. duty holders). If there are any shortcomings and the infringement of the right to life was not intentional, it is not necessary for criminal proceedings to be brought in every case. It may satisfactory to make available to the victims civil law remedies (either alone or in conjunction with a criminal law remedy), enabling any responsibility of the parties concerned to be established and any appropriate civil redress, such as an order for the payment of damages, to be obtained.

The positive obligations of EConHR State duty holders under the ECHR are summarised in Fig. 2.

3.6 The Role of International Institutions and Agencies

In March 2015, the International Federation of Red Cross and Red Crescent Societies (IFRC) and the United Nations (UN) Development Programme issued the pilot version of "The checklist on law and disaster risk reduction". It encourages accountability mechanisms within legislation to address failures to fulfil risk governance responsibilities. In particular it advocates laws: (1) to establish public reporting or parliamentary oversight mechanisms and transparency requirements for government entities tasked with risk governance responsibilities; (2) to give a mandated role to the judiciary in enhancing accountability; (3) to provide enforceable incentives for compliance and disincentives for non-compliance; and (4) to establish legal and/or administrative sanctions (as appropriate) for public officials individuals and businesses for a gross ("particularly egregious") failure to fulfil their duties (IFRC 2015, 16).

The prioritisation of mitigation before response and recovery was recognised within the Sendai Framework for Disaster Risk Reduction 2015–2020 (the Sendai Framework) which

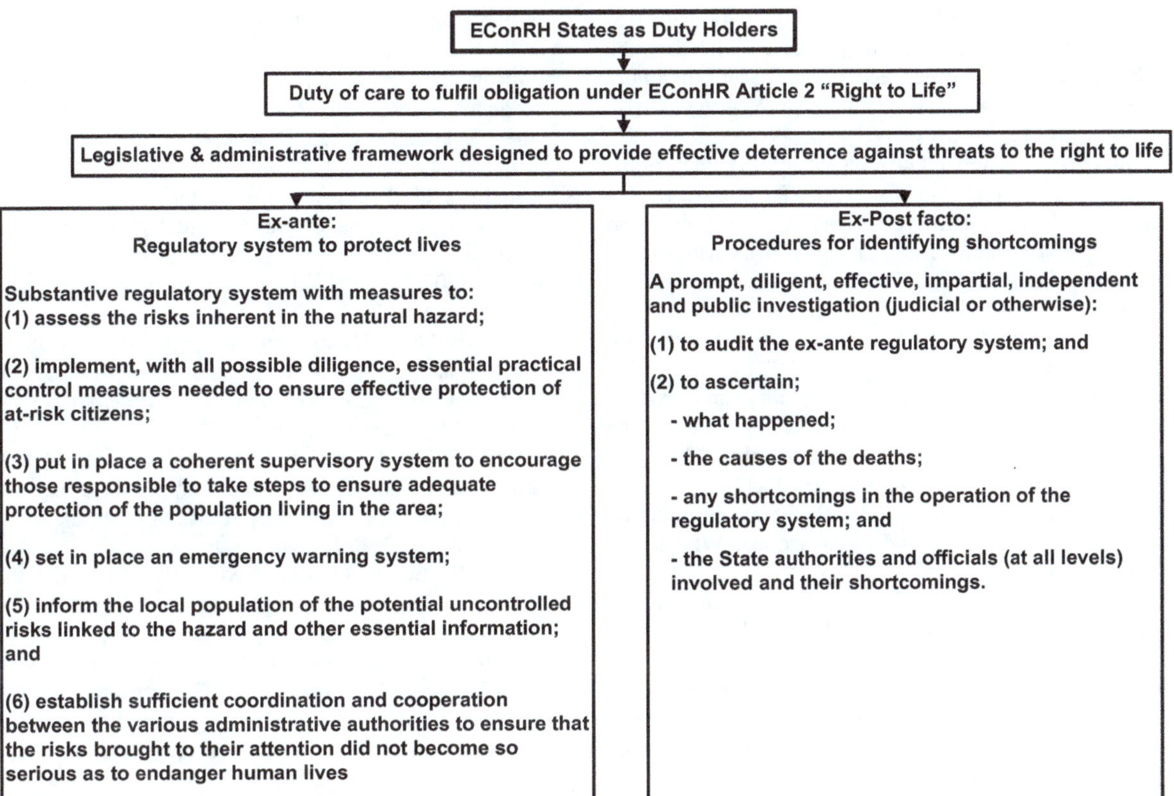

Fig. 2 The obligations of EConHR 'States' to manage natural hazards

emerged from the United Nations 3rd World Conference on Disaster Risk Reduction (UN/ISDR 2015). It is suggested here that one inevitable effect of the Sendai Framework will be to enhance the importance of not only the collection and interpretation of monitoring data but also the better characterisation of unrest periods. For the reasons stated in Sect. 2, periods of volcanic unrest, even if they do not lead to an eruption, present multiple hazards and risks which require very careful assessment and mitigation.

national laws, supplemented by international laws and initiatives, few countries would have the complex administrative infrastructures necessary for the mitigation of volcanic risks.

Although emergency response may still dominate thinking and funding in some jurisdictions, national laws are unlikely to diminish in number and/or reach in the light of the emerging international law governance norms, the IFRC/UN law checklist and the Sendai Framework.

4 Conclusions

In many cultures, volcanic risks are perceived to be susceptible to governance and have become the responsibility of the institutions and stakeholders of relevant social communities.

Laws create the stakeholders, the stakes and the standards of risk governance. Without

References

Bankoff G, Frerks G, Hilhorst D (2004) Mapping vulnerability – disasters, development & people. Earthscan, London and New York

Bergman D, Davis C, Rigby B (2007) International comparison of health and safety responsibilities of company directors, HSE research report RR 53, 207. http://www.hse.gov.uk/leadership/international.pdf

Bretton RJ, Gottsmann J, Aspinall WP, Christie R (2015) Implications of legal scrutiny processes (including the L'Aquila trial and other recent court cases) for future volcanic risk governance. J Appl Volcanol 4:18. https://doi.org/10.1186/s13617-015-0034-x

Commission on Global Governance (1995) Our global neighbourhood. Oxford University Press, Oxford

Donovan A, Oppenheimer C (2014) Science, policy and place in volcanic disasters: insights from Montserrat. Environ Sci Policy 38:150–161 (May)

Fournier d'lbe EM (1979) Objectives of volcanic monitoring and prediction. J Geol Soc136:321–326. doi: https://doi.org/10:1144/gsjgs.136.3.0321

Gordon C (1991) Governmental rationality: an introduction. In Burchell G, Gordon C, Miller P (eds) The foucault effect: studies in Governmentality. Harvester Wheatsheaf, Hemel Hempstead, pp 1–52

Hood C (1986) Administrative analysis: an introduction to rules, enforcement and organisations. Wheatsheaf Books, Sussex

Hood C, Rothstein H (2001) Risk regulation under pressure: problem solving or blame shifting? Adm Soc 33(1):21–53

International Federation of Red Cross, Red Crescent Societies (IFRC) and United Nations Development Programme (2015) The checklist on law and disaster risk reduction, Pilot version, March 2015. IFRC, Geneva, Switzerland. http://www.ifrc.org/PageFiles/115542/The-checklist-on-law-and-drr.pdf

International Risk Governance Council (IRGC) (2009) What is risk governance? www.irgc.org

Jolly G (2015) The role of volcano observatories in risk reduction. In: Loughlin SC, Sparks RSJ, Brown SK, Jenkins SF, Vye-Brown C (eds) Global volcanic hazards and risk. Cambridge University Press, Cambridge

Johnston D, Scott B, Houghton B, Paton D, Dowrick D, Villamor P, Savage J (2002) Social and economic consequences of historic caldera unrest at the Taupo volcano, New Zealand and the management of future episodes of unrest. Bull New Zealand Soc Earthquake Eng 35(4):215–229

Lauta KC (2014) Disaster law. Routledge, Oxford and New York

Luhmann N (1998) Observations on modernity. Stanford University Press, Stanford, CA

Luhmann N (1992) Risk: a sociology theory. de Gruyter, Berlin

MIAVITA (2012) Handbook for volcanic risk management: prevention, crisis management and resilience. MIAVITA team, Orleans

Newhall CG, Hoblitt RP (2002) Constructing event trees for volcanic crises. Bull Volcanol 64:3–20

OECD (Organisation for Economic Co-operation and Development) (2002) Outcome focussed management in the United Kingdom by Ellis K, Mitchell S. OECD Publications, Paris, France

OECD (2015) "Scientific advice for policy making: the role and responsibility of expert bodies and individual scientists", OECD science, technology and industry policy papers, no 21, OECD Publishing, Paris. http://dx.doi.org/10.1787/5js33l1jcpwb-en

Phillipson G, Sobradelo R, Gottsmann J (2013) Global volcanic unrest in the 21st century: an analysis of the first decade. J Volcanol Geoth Res 264(2013):183–196

Power M (2007) Organised uncertainty: designing a world of risk management. Oxford University Press, Oxford

Power M (2009) The risk management of nothing. Acc, Organ Soc 34(2009):849–855

Rothstein H (2002) Neglected risk regulation: the institutional attenuation problem, centre for analysis of risk and regulation. London School of Economics and Political Science, London

Rothstein H (2014) Exploring national cultures of risk governance accessed via http://www.lse.ac.uk/researchAndExpertise/units/CARR/publications/CARRmagR&R25-Rothstein.pdf

Rothstein H, Borraz O, Huber M (2012) Risk and the limits of governance: exploring varied patterns of risk-based governance across. Europe Regul Gov 1–21 http://doi.org/10.111/j.1748-5991.2012.01153.x

Rouwet D, Sandri L, Marzocchi W, Gottsmann J, Selva J, Tonini R, Papale P (2014) Recognizing and tracking volcanic hazards related to non-magmatic unrest: a review. J Appl Volcanol 3(1):17. https://doi.org/10.1186/s13617-014-0017-3

Simoncini M (2011) Regulating catastrophic risks by standards. EJRR (1):37–50

Simoncini M (2013) Governing air traffic management in the single European sky: the search for possible solutions to safety issues. Eur Law Rev Issue 2 (2013):209–228

Smith K, Petley DN (2009) Environmental hazards: assessing risk and reducing disaster. Routledge, London and New York

Sobradelo R, Martí J (2015) Short-term volcanic hazard assessment through Bayesian inference: retrospective application to the Pinatubo 1991 volcanic unrest. J Volcanol Geoth Res 290(2015):1–11

Sparks RSJ, Aspinall WP, Auker M, Crosweller S, Hincks T, Mahony S, Nadim F, Polley J, Syre E (2012) Mapping and characterising volcanic risk, magmatic rifting and active volcanism conference, Addis Ababa, 12 Jan 2012

United Kingdom Office for Nuclear Regulation (2013) Guidance for the demonstration of ALARP (As Low As Reasonably Practicable) NS-TAST-GD-005 Revision 6

United Nations Educational, Scientific and Cultural Organization (UNESCO) (1972) Report of consultative meeting of experts on the statistical study of natural hazards and their consequences Document SC/WS/500 11pp, Paris, France

United Nations Development Programme Safer Communities through Disaster Risk Reduction in Development Programme SC-DRR (2009) Lessons learned: disaster management legal reform—The Indonesian experience UND-, Jakarta, Indonesia

UN/ISDR (2009) Terminology on disaster risk reduction. Switzerland, Geneva

UN/ISDR (2015) Sendai framework for disaster risk reduction 2015-30

Walker G, Whittle R, Medd W, Watson N (2010) Risk governance and natural hazards, Cap Haz-Net WP2 Report, Lancaster Environment Centre, Lancaster University, Lancaster. http://caphaz-net.org/outcomess-results

Legal Authorities and Case Law

Budayeva, others v Russia (2008) Applications 15339/02, 21166/02, 20058/02, 11673 & 15343/02, judgment of 20 March 2008, ECHR 15339/02

Kolyadenko, others v Russia (2012) Applications 17423/05, 20534/05, 20678/05, 23263/05, 24283/05, 35673/05, ECHR 17423/05

Oneryildiz v Turkey (2004) Application 48939/99 18 BHRC 145, 41 EHRR 20

Deterministic Versus Probabilistic Volcano Monitoring: Not "or" But "and"

D. Rouwet, R. Constantinescu and L. Sandri

Abstract

Volcanic eruption forecasting and hazard assessment are multi-disciplinary processes with scientific and social implications. Our limited knowledge and the randomness of the processes behind a volcanic eruption yield the need to quantify uncertainties on volcano dynamics. With deterministic and probabilistic methods for volcanic hazard assessment not always being in agreement, we propose a combined approach that bridges the two schools of thoughts in order to improve future volcano monitoring. Expert elicitation has proven to be an effective way to bind deterministic research within a probabilistic framework aiming to reduce the uncertainties related to any hazard forecast; yet, numerous exercises based on expert elicitation have revealed that the attempt to reduce uncertainties led to the creation of new ones, often unquantifiable, created by human nature and reasoning during stressful situations. Such reasoning ignores the complexity of volcanic processes and the fact that every scenario has a probability to occur. The recent probabilistic methods and tools marry probabilistic and deterministic approaches and lead to unprecedented models. Nevertheless, probabilistic hazard assessment is often misunderstood as not all of the researchers involved have backgrounds in such matters. A probabilistic method cannot stand-alone

D. Rouwet (✉) · L. Sandri
Istituto Nazionale di Geofisica e Vulcanologia,
Sezioni di Bologna, Bologna, Italy
e-mail: dmitrirouwet@gmail.com; dmitri.
rouwet@ingv.it

R. Constantinescu
Seismic Research Centre, University of West Indies,
Saint Augustine, Trinidad and Tobago

Present Address:
R. Constantinescu
School of Geosciences, University of South Florida,
4202 E. Fowler Avenue, NES 107, Tampa, FL
33620-5550, USA

Advs in Volcanology (2019) 35–46
DOI 10.1007/11157_2017_8

Published Online: 21 June 2017

as it depends on data input obtained by deterministic approaches. We propose that, given the symbiotic relationship between the two methods, a probabilistic framework can play a role of moderator between various deterministic disciplines, thus creating a coherent environment for discussion and debate among seismologists, geodesists, geochemists. This can be achieved by training all scientists involved in hazard assessment, probability theory and data interpretation, while at least one group member objectively uses the information provided to produce the probabilities. Hence, numerical outcomes can be interpreted transparently as they represent the quantification of experts' knowledge and related uncertainties. A probabilistic method that incorporates the joint opinions of a group of multi-disciplinary researchers facilitates a more straightforward way of communicating scientific information to decision-makers.

Resumen extendid

La previsión de erupciones volcánicas y la evaluación del peligro son procesos multidisciplinarios, con implicaciones tanto científicas como sociales. Nuestro conocimiento limitado de los procesos detrás de una erupción volcánica y su aleatoriedad genera la necesidad de cuantificar las incertidumbres sobre las dinámicas del volcán y de mejorar la política de la toma de decisiones durante una crisis volcánica. Sabiendo que existe un desacuerdo sobre el uso de métodos determinísticos o probabilísticos durante la evaluación de la peligrosidad volcánica, revisamos ambos métodos y proponemos un enfoque que sirve como puente entre las dos escuelas de pensamiento y que pueda mejorar las capacidades de monitoreo volcánico en el futuro, hacia el reconocimiento en tiempo de manifestaciones volcánicas y amenazas relacionadas. La elicitación de expertos resulta ser una manera efectiva para relacionar la investigación determinística con el marco probabilístico para poder reducir la incertidumbre relacionada a cualquier intento de previsión de erupción; sin embargo, numerosos ejercicios basados en elicitaciones de expertos revelaron el hecho que este intento de reducir la incertidumbre creó nuevas incertidumbres, a menudo imposible de cuantificar, siendo generada por la naturaleza del pensamiento humano durante situaciones de estrés. El proceso general es sujeto a la personalidad de un/a investigador/a o un grupo de investigadores y sus ideas basadas en su experiencia. Esta manera de pensar interfiere con la complejidad intrínseca de los procesos volcánicos y con el hecho que cada escenario tiene una probabilidad de ocurrencia. Los métodos e instrumentos probabilísticos recientes juntaron los investigadores probabilísticos y determinísticos lo que resultó en modelos e interpretaciones de información sobre volcanes sin precedentes. Sin embargo, la novedad de la evaluación probabilística de peligrosidad es, a menudo, incomprendida debido al hecho que no todos los investigadores involucrados tienen una formación en estas materias teóricas. El método probabilístico no puede existir autónomamente ya que depende de datos de entrada obtenido a través de los estudios determinísticos. Proponemos que, dada la relación simbiótica entre ambos métodos, un marco probabilístico puede jugar el papel como moderador entre las varias disciplinas deterministas, creando un ambiente coherente para discusiones y debates entre científicos (e.g., sismólogos, geodetas, geoquímicos). Este se puede obtener por medio de entrenamiento de todos los científicos involucrados en el monitoreo volcánico en la teoría probabilística y la interpretación de datos, mientras que al menos un miembro del grupo utiliza de manera objetiva la información disponible para producir las probabilidades numéricas. Así, los resultados numéricos pueden ser interpretados sin duda alguna, ya que representan la cuantificación del conocimiento de los expertos y las incertidumbres relacionadas. Un método probabilístico que incorpora las opiniones conjuntas de un grupo de investigadores multidisciplinarios facilitará una manera más transparente de comunicación de la información científica hacia las autoridades civiles, así mejorando (1) el proceso de toma de decisiones durante una crisis y la mitigación a largo plazo, y (2) el estado de medidas de preparación que incorpora los varios aspectos sociales. En el futuro, los reportes de previsiones emitidos por los científicos deberían incluir los resultados numéricos de los modelos probabilísticos; la arquitectura de monitoreo se debería expandir más allá del arreglo clásico "sismo-deformación-gas" hacia un arreglo "sismo-deformación-gas-probabilidad". Este capítulo de opinión pretende proponer una ideología posible, con el máximo respeto para el volcán, la sociedad, los científicos individuales o los grupos de científicos, y para las autoridades que toman las decisiones, con un objetivo común: mejorar la previsión de amenazas relacionadas a los volcanes para proteger la sociedad.

Keywords

Volcano monitoring · Probabilistic hazard assessment · Deterministic research · Bridging and symbiosis · Best practice scheme

Palabras clave

Monitoreo volcánico · Evaluación probabilística de amenaza · Investigación determinística · Puente y simbiosis · Esquema de mejor práctica

1 Introduction

Volcanoes are intrinsically complex and unpredictable systems manifesting non-linear behaviour in space and time, on the long- and short-term. Understanding how volcanoes evolve with time through the various stages of activity—from quiescence through unrest to eruption—is highly challenging. Awareness of these facts is a basic requisite when working with/on volcanoes. A major goal in volcanology is quantifying uncertainties on volcano dynamics, and learning how to translate these to decision makers and, occasionally, to the population. The combination of social implications and forecasting future behaviour of complex natural systems makes volcanology a rather unique but "tricky business". Besides the need to quantify uncertainties and better understand our limited knowledge on volcanoes it is necessary to legally protect volcanologists when exporting their knowledge outside their protected professional community (Bretton et al. 2015).

During the past 40–50 years, volcano monitoring and eruption forecasting during volcanic crises has been largely dominated by the deterministic approach (Sparks 2003, i.e. most likely scenarios). Experts with different research backgrounds track changes in "their" parameters related to volcanic activity, afterwards discussed in "protected" round tables, generally behind locked doors, to eventually come up with a single voice. This single voice does not and should not reflect possible internal conflicts or disagreements behind the closed door. Such disagreement is an often-unstated expression of the uncertainty due to our lack of knowledge on volcanic processes, and the possible unavailability of data (i.e. epistemic uncertainty) and due to the intrinsic randomness of the volcanic process studied (i.e. aleatory uncertainty). The power of the single voice from the group of experts often misleads the receivers of the message (decision makers, authorities or lay public), believing the experts are "sure" on the evolution of volcanic activity. This is one of the reasons why volcanologists are often, correctly or incorrectly, highly trusted professionals by the public (Haynes et al. 2008; Donovan et al. 2011).

During the last decade, this "untouchable aura" around volcano monitoring based on deterministic research has vanished with the introduction of probabilistic hazard and eruption forecasting (e.g., Newhall and Hoblitt 2002; Sparks 2003; Marzocchi et al. 2004, 2008; Sparks and Aspinall 2004; Marzocchi and Bebbington 2012; Sobradelo et al. 2014; Sobradelo and Marti 2015). A recent chapter by Newhall and Pallister (2015) starts from the same false dichotomy, aiming to spouse the deterministic and probabilistic points of view in the highly applicable method of "Multiple Data Sets".

Marzocchi and Woo (2007) propose a rational on decision-making based on the hazard/risk separation principal, using a cost-benefit analyses as the guiding tool.

Among the methods for probabilistic hazard assessment and eruption forecasting, many are based on a Bayesian approach (e.g., Marzocchi et al. 2004, 2008; Sobradelo et al. 2014; Sobradelo and Marti 2015), that allows describing the probability of interest not as a single numerical value, but as a probability distribution. In this view, the probability of an eruption occurring, or of a given hazardous event hitting a target area, is described both by a best-evaluation value (for example the mean or the median of the probability distribution), and by a dispersion around such value (represented by standard deviations or by a confidence interval). These two quantities can be directly related to two different sources of uncertainty: the aleatory one and the epistemic one. In this way, Bayesian approaches allows quantifying also to what extent our probabilistic assessments are constrained by data and knowledge. In other words, now we know, as a group of volcano-experts, to which degree we can be wrong in our forecasts behind the closed doors. This black-on-white awareness brought to light by probability density functions has led to some key questions, from the in- and outside worlds: (1) are we, as volcano-experts, replaceable by a numerical approach?, and (2) we thought you, volcanologists, knew what was happening, but it seems you don't know.

This chapter critically reviews both "philosophies" of eruption forecasting and tracking of volcanic unrest and related hazards, in search of a combined approach that could become a guideline for future volcanic surveillance architectures. But we still need bridges between two schools (deterministic and probabilistic) apparently speaking a different language. Remember that both methodologies aim for the same goal: the timely recognition when volcanoes become hazardous in their various ways of expressions. This is our common professional and social responsibility as volcanologists.

2 Forecasts based on Deterministic Research

The goal of volcano monitoring based on deterministic research is to link temporal variations of physical-chemical parameters with variations in the state of unrest of the physical object volcano (i.e. unrest, magmatic unrest, non-magmatic unrest, eruption, hazard; Rouwet et al. 2014). Every volcanic eruption is intrinsically preceded by magma rise towards the surface. The major aim in eruption forecasting is the quick recognition of such magma rise by changes in the physical parameters (deformation, seismicity) and chemical parameters. The most direct way to do so is to determine how, where, when and why the physical object "volcano" responds to magma rise.

Despite the straight-to-goal approach, large uncertainties exist: (1) a volcano remains a complex system (aleatory uncertainty), and (2) our knowledge on the volcano remains limited (epistemic uncertainty). How do we know if we have detected all the signals the volcano eventually releases? Which of these signals are we considering in our forecast framework, and why? Sometimes we may dismiss some signals as being not pertinent, or simply because we are unable to correlate them either with our 'understanding', or with the rest of the signals. Some 'signals' are considered as stand-alone, at the moment they occur, others are considered within a time evolution.

The quality of the forecast largely depends on the interpretation of the signals and the hypothesis/model of future activity developed as a result of this assumed scientific stringency. The latter is related to the experience and expertise of the deterministic researcher, or better, on how the experience and expertise is perceived by individual researchers or groups, decision makers and the researcher her/himself. It is known that the most informative and valid opinion may not always be that of the most respected or distinguished professional (Selva et al. 2012).

A big advantage in volcano monitoring based on deterministic research is the fact that, if independent monitoring approaches (e.g., geochemistry vs. geophysics) point toward a similar hypothesis on future hazard in time and space, the future scenario will become more likely. Finding a larger number of arguments in favour of certain scenarios is surely an efficient way to decipher volcanic unrest.

The timescale of the forecast is highly ambiguous and based on the limited knowledge on how the volcano (or analogue volcanoes) behaved in the past within the desired time-scale. Sometimes all the 'signals' converge towards an obvious conclusion, yet, there are numerous cases in which activity stopped or pulled back, sometimes for years before an actual eruption. So, the deterministic approach, which is based mostly on a recurrence interval and 'experience' of the volcanologist, is limited in providing a sound time scale for the evolution of the 'activity', and hence, for a forecast. In deterministic monitoring the "time" concept is not unambiguously defined, can be case dependent or even change within the evolution of an unrest phase. Fortunately, with converging signals through time, the monitoring time window often becomes narrower when building up towards increased unrest or eruption, although the exact time window cannot be rigorously chosen.

Besides the instrumental accuracies (detection limits, analytical errors, data quality), the uncertainty of the forecast cannot be quantified before an eruption. The only way to decrease this "unquantifiable" uncertainty is by increasing our knowledge on volcanoes, be it the specific volcano in unrest or any volcano that has shown similar behaviour in the past. The current development of methods to increase the quality (e.g., novel approaches, numerical modelling) and quantity of data (increase frequency, e.g., by remote sensing and real-time transmission) will undoubtedly help to achieve better insights into volcanic systems.

3 Probabilistic Forecasts

Probabilistic methods and tools for both short- and long-term time windows are more and more in the spotlight (Marzocchi et al. 2008; Sandri et al. 2009, 2012, 2014; Lindsay et al. 2010; Selva et al. 2010, 2011, 2012; Sobradelo et al. 2014; Sobradelo and Marti 2015; Bartolini et al. 2013; Becerril et al. 2014). A key review on probabilistic volcano monitoring can be found in Marzocchi and Bebbington (2012).

Within this opinion chapter, we highlight some critical aspects of the probabilistic forecasting method, without entering in the technical and operational details (see Tonini et al. 2016 and Sandri et al. 2017 for further reading).

A probabilistic forecast can provide a global but clear, numerical view of the opinion of, generally, a group of people, based on the volcanic history and knowledge of the volcano. Lately, probabilistic forecasts are more and more applied in real crisis situations. Thus far, the efficiency or accuracy have hardly been evaluated, probably due to the fact that only recently we are reaching statistically relevant numbers of cases to test this critical issue (Newhall and Pallister 2015). Once high numbers of applications are reached, the numerical outcomes of probabilistic methods can even be considered to support long-term hazard analyses, by becoming input information itself.

In practice, probabilistic hazard assessment and eruption forecasting frameworks constructed on the Event-Tree methodology (Newhall and Hoblitt 2002; Newhall and Pallister 2015) rely on a dataset of information about the past activity of a volcano (i.e. past data), theoretical/mathematical models (i.e. prior data) and a series of monitoring signals (i.e. parameters) that allow us to track the changes in the system with time (Marzocchi et al. 2008; Sobradelo et al. 2014). This information allows us to compute the probabilities of a specific hazardous outcome. As any such application reveals, the quality of the numerical output depends on the quality and quantity of the input. The risk exists that using a dataset for long-term probabilistic assessment will introduce an uncertainty, since the operators are often biased by the hypothesis or model coming forth of the dataset. Data should hence be considered "just" facts. In both deterministic and probabilistic hazard assessment, volcanologists rely on information about past eruptions: eruptive behaviour, eruption frequency, and eruption style. Such catalogues of information are inevitably incomplete. For instance, traces of smaller scale events could have been literally eroded away from the geological record, buried or masked by larger events and, hence, relics of precursory activity cannot be deduced. Consequently, one should limit the part of the catalogue used, for the period and specific kind of event you desire to forecast, for which it is reasonably complete. As such, the foundation of a probabilistic framework represents a source of intrinsic uncertainty, that can somehow be overcome by quite robust methods for estimating completeness of sections of catalogues (Moran et al. 2011). The most intuitive solution, simply choosing a smaller dataset, usually representing the most recent years/centuries of a volcano's activity reduces this uncertainty. But this choice alone will be reflected in the quality of the probabilistic assessment. The unknowns of the data catalogue represent the uncertainty in a probabilistic framework, thus forcing volcanologists to 'select' how far to track back in time, and which information to use. In other words, we select e.g., only the last 300 years of activity of a volcano just simply because we believe to be more certain on what happened, instead of choosing the last 2000 years. What is the real scientific control of these choices? We cannot say with an acceptable certainty that a volcano will behave like it did in the last 300 or 2000 years. Indeed, with the recent probabilistic methods we are able to quantify the uncertainty of such choices but we are yet to find a sound scientific mechanism that allows us to make objective decisions regarding the data set. After all, the end goal of eruption forecasting is to give a prediction by analysing signals from an extremely complex system governed by a large degree of freedom.

Another aspect to tackle is the use of monitoring information, especially for short-term forecasts. Asking for numerical thresholds for monitoring parameters at the various nodes of event tree structures is intrinsically wrong, as a numerical threshold is an expression of certainty on something we cannot be certain about. For this, volcanologists use monitoring parameters in order to detect anomalies with respect to the volcano's background activity to be able to track their evolution with time. Moreover, from the beginning, we rely on a subjective choice when we define the unrest, unrest being commonly agreed upon as a state of elevated activity above background that causes concern (Phillipson et al. 2013). This cause of concern, expressed in numerical thresholds is a subjective choice: experts involved in volcano monitoring usually decide thresholds above/below which the volcano is considered in unrest. But is a volcano really in unrest simply because we observe one day an anomaly in one of the parameters? And if so, how is the choice of a threshold scientifically sound, since most of the time it is based on the "expert's experience"? Of course, expert's experience should not be dismissed and never replaced by computer codes, but indeed, such a choice is associated with a large uncertainty that is extremely difficult to quantify. Even the act of reducing the uncertainty of such a choice relies on corroboration with information from other sources (e.g., analogue volcanoes) that is again subjective itself.

The "fuzzy-threshold approach" (upper and lower thresholds) somehow resolves this problem, as it tracks the degree of anomaly, emphasising from a state in which the volcano is 'not causing concern' to one 'causing a degree of concern'. Thresholds can be case-dependent and are therefore in most cases themselves biased. Especially in the case of poorly monitored volcanoes, or of volcanoes without a monitored stage of unrest (e.g., towards higher nodes in the event tree), Boolean (Y/N) parameters are highly preferable.

During an unrest crisis it might be tempting to adapt the numerical values of the thresholds, when e.g., the previous threshold is exceeded while the volcano does not "react on this parameter" as we thought it would have. Nevertheless, once thresholds for parameters are fixed, they should not be modified, in order to track the time evolution of probabilities (and related uncertainties).

4 Recommendations: Not "or" But "and"

4.1 Expert Elicitation: A Solution?

In general, any choice made by an expert panel regarding when a volcano enters a phase of unrest, what information is pertinent for hazard analyses and how to interpret the precursory signals is done by a discussion-based elicitation process. Each expert in a specific volcanological sub-domain will exercise their opinions, based on the experience they have, on every parameter within a monitoring setup, with the goal to reach consensus about the most likely scenario/ threshold. All data and interpretations should be heard and evaluated. However, this process is still unduly influenced by the "stronger voice" of the group. We may be certain on something until someone else makes us doubt it. At the moment we start doubting our opinions we will be easily influenced by other, stronger opinions. On the other hand, volcanologists are often forced to make such decisions and be liable for their choices (in court of law, Bretton et al. 2015). The pressure of a volcanic 'crisis' and the feeling of liability increases scientists' reservations when faced with such choices.

A suggestion to improve the expert elicitation process is to introduce a person to act as a "Devil's advocate". This will mean that one of the experts is supposed to do exactly the opposite to the group's decision. If most of the experts agree on a scenario, it is the duty of the latter to completely disagree and evaluate the opposite scenario. This can be a good way to "account" for surprise scenarios. Other ways are to weight final results according to anonymous calibration tests (Cooke method; Cooke 1991; Aspinall et al. 2003; Aspinall 2006, 2010) and/or anonymous estimation of the most reliable members of the group, not necessarily the loudest.

The introduction of probabilistic hazard assessment methods in the multi-disciplinary volcanological community has first led to a discredit of the purely deterministic approach. After the usefulness of the probabilistic approach has been demonstrated, and confusion on the different philosophies has disappeared, or at least decreased, the awareness on framing the various "niche" research branches in a bigger picture resulted into constructive discussions among the various research groups and individuals involved. This results in coherent group thinking and a more collaborative atmosphere among volcanologists. Expert elicitation on its turn has obliged researchers with various backgrounds to absorb new data and ideas from one another. This is definitely a positive side-effect of the probabilistic approaches and expert elicitations.

4.2 How to Interpret Uncertainties?

Probabilistic volcanic hazard assessment and eruption forecasting is a relatively new concept in modern volcanology, and often reserved for those with a background in statistics. But, as reality showed (Constantinescu et al. 2016), most of the volcanologists are not fully aware of the probability theory and how its results should be interpreted. A panel of volcano experts usually comprises seismologists, geochemists, geodesists, geologists, petrologists, and not all of them necessarily have a background in probabilistic approaches, especially when such approach is based on the integration of opinions of all members of such a group. One idea to cope with this problem is to train the group members in how the probabilistic methodology works and how results should be interpreted, while another member of the group (the so-called "PROB-runner") objectively uses the information provided by the expert panel to produce the probabilistic forecast. In this way, the members of the group can interpret the numerical outcomes accounting for the associated uncertainty without questioning and second-guessing the output because they are already aware of the process (Newhall and Pallister 2015; Constantinescu et al. 2016).

The whole idea of the elicitation approach is to allow your mind to explore each possibility without influencing one of the possible outcomes just because one expert believes more in one outcome than the other. It is some sort of letting go. People don't like to admit they might be wrong, so the Event Tree and Cooke elicitation approaches allow them to anonymously change their views upon elicitation. If one is capable of admitting fallibility and look at the big picture with an open mind, allowing all possibilities to unfold, then full discussion can occur and resulting estimates of probabilities will eventually have lower uncertainties. In the end, the idea of probabilities is that all scenarios are possible to happen, some with a larger probability others with a lower one; all have probabilities (and related uncertainties) and nothing should be dismissed simply because 'I strongly believe it can't be, so I don't agree'.

4.3 Trust in Scientists?

Even the best scientists can make mistakes. If volcanic unrest or activity is badly forecasted, initial trust in scientists may dissolve in legal proceeding. Ideally, scientists who act to the best of their knowledge should be protected rather than being blamed if they make a bad forecast (Bretton et al. 2015). Trust in scientists depends on how these scientists are portrayed to society and actually how scientific aspects of volcanology are presented to the public. This should be a system with two-way feedback (Christie et al. 2015). Many countries lack volcano-education among communities, but people know that there are some scientists that 'know what they are doing'. People feel protected, but this is a false feeling of safety, propped up by ignorance of the real situation. When disaster strikes, scientists are often the first to blame. If the community will be involved fully in the mitigation and preparedness process (e.g., Gregg et al. 2004; Rouwet et al. 2013; Dohaney et al. 2015), they may be guided by 'compassion' and will understand that volcanologists cannot stop an eruption and protect people, and eventually the blame or trust too often falls on the elected authorities. Elected authorities should be the liaison between science and the general public. Trust is inherent when you are aware of the problem and the person dealing with it. Trust in scientists may grow because they successfully predict an eruption, but sustained growth in trust is due to multi-yearly exposure of the scientific staff to the public (Christie et al. 2015). This involves long years of planning, investing and engagement in educational campaigns. Trust is something that comes in time and involves a feedback loop between people and the scientists.

Within the current scope of this opinion chapter, the probabilistic method often seems to serve as a more transparent way of bridging between the scientists and the elected authorities (decision makers). First, the use of probabilities inherently implies some uncertainty, which in scientifically literate society is essential for public trust. Second, elicitation tools reflect a joint-opinion of a group of experts rather than of one. There is a need for training and a full understanding of probabilistic results by the officials. The role of probabilistic tools should, in our opinion, not intervene in communication protocols with the lay public, but should be rather "restricted" to transmit information to decision-makers, representing part of the voice of the group of volcanologists. The efforts in communicating towards the lay public, in order to build "trust" amongst the population, should be decoupled from the background of the involved scientist (deterministic or probabilistic). Communication protocols are independent of the applied scientific method, and researchers should become more skilled to transmit their information openly towards the public, with the awareness of the uncertainty their information contains (Hicks and Few 2015).

4.4 Towards Collaborative Volcano Monitoring

A major accomplishment of the probabilistic method is to have increased harmony among the various deterministic research environments. This new dynamic favours the refining of previous conceptual models that originate from deterministic research, as the reference frame has become more complete.

Nevertheless, the probabilistic research approach is not yet fully accepted by the deterministic community due to criticism and anxiety to be "replaced" by the probabilistic method (VUELCO simulations Colima, Campi Flegrei, Cotopaxi and Dominica). This concern is unnecessary, since the first requisite for the probabilistic method to function is the availability of data, information, a priori believes and models, originating from deterministic research. More input information for the probabilistic method means significant decreases in the epistemic uncertainty of probabilistic outcomes.

Moreover, during volcanic crisis situations, deterministic researchers still stick to the "round table" approach and the lack of time inhibits to

efficiently interact with the researchers that run the probabilistic models (e.g., VUELCO simulation exercises). The latter need more detailed feedback and input information for single parameters at the various nodes of the event-tree. Despite refining the probabilistic model during pre-crisis by expert elicitations, in the heat of the moment of the crisis, a wrong interpretation of a numerical value provided by determinists will sometimes lead to disastrous numerical outcomes in the probabilistic models. The PROB-runner cannot be blamed for not being an expert in all fields in volcanology, incorporated through parameters in e.g., BET or HASSET.

Three solutions to this crucial issue are proposed: (1) probabilistic model runners should actively take part in the "closed-door" discussions by the experts from the various fields, before incorporating numerical values in probabilistic models, (2) a separate team of experts that profoundly know the needs and functioning of probabilistic models should flank the PROB-runners during crisis. Both realities are not yet accomplished, and/or (3) an event tree structure, put forward by a facilitator between the deterministic and probabilistic research teams should serve as the base to guide scientific discussion and get fast to the nucleus of the crisis (Newhall and Pallister 2015). Future simulation exercises on volcanic crisis situations should focus on this training approach, in order to be prepared when real the crisis strikes.

Moreover, the outcomes of probabilistic models have to be included in the final reports transmitted to authorities, and be respected as one of the many monitoring tools, without decreasing or increasing their weight and value within the still deterministic-dominated general opinion. Since a probabilistic framework offers a measure of the uncertainty, any interpretation should not be taken for granted, neither decision makers should make decisions based solely on a probability. Probabilities should be considered as an addition to the information upon which decisions are made, and not as a decisive factor.

5   Take Home Ideas

Probabilistic forecasting has become an inherent part within a multi-facet view of research and volcano monitoring; neither deterministic, nor probabilistic methods, can or should stand alone. More than being a means to transmit information between volcanologists and decision making authorities, probabilistic models should also be based on, and promote, deterministic research that can be written up after "round table" discussions (Fig. 1).

Incorporation of probabilistic models in volcano monitoring has many advantages: (1) protecting against oversimplified, over-confident forecasts. Even though decision makers may initially have difficulties to understand uncertainties and prefer black-on-white numbers, Y/N forecasts, they will soon come to appreciate probabilities if they are represented in an understandable way; (2) creating harmony amongst the volcanological community because probabilities will reflect the general view of the monitoring team, and (3) legally protecting the entire monitoring team by probabilities and their uncertainties as forecasts are perfectly traceable and reproducible, if disaster strikes after "erroneous forecasts". However, to

date, probabilistic forecasts have not been rigorously evaluated to know whether they are an improvement over traditional, non-probabilistic forecast methods. For sure, they do better than traditional methods at estimating uncertainty. We still need tests on whether they are more accurate, and more useful for decision makers than older methods (Newhall and Pallister 2015).

In the future, reports should include the forecast of probabilistic models; a monitoring architecture should expand beyond the classical "seismo-deformation-gas" setup and become "seismo-deformation-gas-probability" setup (in random order of importance) (Fig. 1). Probabilistic models cannot stand alone, as they need the input and feedback from deterministic research. "Probabilists" should not communicate their numerical outcomes directly to the decision-making authorities: it is better to convey the opinion of an entire group. Probabilistic methods can serve as "moderator" among the various disciplines, while expert elicitations are the "glue" between the deterministic and probabilistic approaches (Fig. 1). Probabilistic methods should knock down walls and stimulate discussion and coherence amongst the various research branches (seismologists, geodesists,

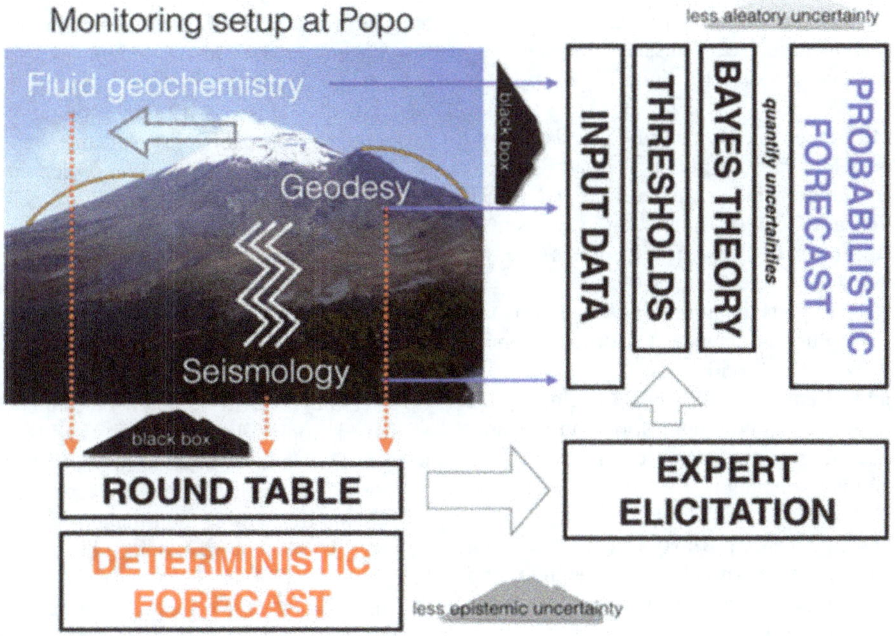

Fig. 1 From dichotomic monitoring setups (deterministic vs. probabilistic forecast, "or" setups) to an "and" strategy. The picture shows VUELCO target volcano Popocatépetl, Mexico (November 2011, D.R.). Before interpretation of the data (monitoring data, eruptive history or any a priori model) the volcano is considered a *"black box"*

geochemists, petrologists, geologists). This requires time, effort and an open-mind by all involved parties/volcanologists in volcano monitoring.

References

Aspinall WP (2006) Structured elicitation of expert judgement for probabilistic hazard and risk assessment in volcanic eruptions. In: Mader HM, Coles SG, Connor CB, Connor LJ (eds). Statistics in volcanology. IAVCEI Publications. ISBN: 978-1-86239-208-3. pp 15–30

Aspinall WP (2010) A route to more tractable expert advice. Nature 463(21):294–295

Aspinall WP, Woo G, Voight B, Baxter PJ (2003) Evidence-based volcanology: application to vol canic crises. J Volcanol Geotherm Res 128:273–285

Bartolini S, Cappello A, Martí J, Del Negro C (2013) QVAST: a new quantum GIS plugin for estimating volcanic susceptibility. Nat Hazards Earth Syst Sci 13:3031–3042. doi:10.5194/nhess-13-3031-2013

Becerril L, Bartolini S, Sobradelo R, Marti J, Morales JM, Galiendo I (2014) Long-term volcanic hazard assessment on El Hierro (Canary Islands). Nat Hazards Earth Syst Sci 14:1853–1870. doi:10.5194/nhess-14-1853-2014

Bretton RJ, Gottsmann J, Aspinall WP, Christie R (2015) Implications of legal scrutiny processes (including the L'Aquila trial and other recent court cases) for future volcanic risk governance. J Appl Volcanol 4:18. doi:10.1186/s13617-015-0034-x

Christie R, Cooke O, Gottsmann J (2015) Fearing the knock on the door: critical security studies insights into limited cooperation with disaster management regimes. J Appl Volcanol 4:19. doi:10.1186/s13617-13015-10037-13617

Constantinescu R, Robertson R, Lindsay JM, Tonini R, Sandri L, Rouwet D, Patrick Smith P, Stewart R (2016) Application of the probabilistic model BET_UNREST during a volcanic unrest simulation exercise in Dominica, Lesser Antilles. Geochem Geophys Geosyst 17:4438–4456. doi:10.1002/2016GC006485

Cooke RM (1991) Experts in uncertainty: opinion and subjective probability in science. Oxford University Press. ISBN: 9780195064650

Dohaney J, Brogt E, Kennedy B, Wilson TM, Lindsay JM (2015) Training in crisis communication and volcanic eruption forecasting: design and evaluation of an authentic role-play simulation. J Appl Volcanol 4:12. doi:10.1186/s13617-015-0030-1

Donovan A, Oppenheimer C, Bravo M (2011) Social studies of volcanology: knowledge generation and expert advice on active volcanoes. Bull Volcanol. doi:10.1007/s00445-011-0547-z

Gregg CE, Houghton BF, Johnston DM, Paton D, Swanson DA (2004) The perception of volcanic risk in Kona communities from Mauna Loa and Hualālai volcanoes, Hawai'i. J Volcanol Geotherm Res 130:179–196

Haynes K, Barclay J, Pidgeon N (2008) The issue of trust and its influence on risk communication during a volcanic crisis. Bull Volcanol 70:605–621. doi:10.1007/s00445-007-0156-z

Hicks A, Few R (2015) Trajectories of social vulnerability during the Soufrière Hills volcanic crisis. J Appl Volcanol 4:10. doi:10.1186/s13617-015-0029-7

Lindsay J, Marzocchi W, Jolly G, Constantinescu R, Selva J, Sandri L (2010) Towards real-time eruption forecasting in the Auckland Volcanic Field: application of BET_EF during the New Zealand National Disaster Exercise 'Ruaumoko'. Bull Volcanol 72:185–204. doi:10.1007/s00445-009-0311-9

Marzocchi W, Woo G (2007) Probabilistic eruption forecasting and the call for an evacuation. Geophys Res Lett 34:L22310. doi:10.21029/22007GL031922

Marzocchi W, Bebbington M (2012) Probabilistic eruption forecasting at short and long time scales. Bull Volcanol. doi:10.1007/s00445-012-0633-x

Marzocchi W, Sandri L, Gasparini P, Newhall C, Boschi E (2004) Quantifying probabilities of volcanic events: The example of volcanic hazard at Mount Vesuvius. J Geophys Res 109:B11201. doi:10.1029/2004JB003155

Marzocchi W, Sandri L, Selva J (2008) BET_EF: a probabilistic tool for long- and short-term erup tion forecasting. Bull Volcanol 70:623–632. doi:10.1007/s00445-007-0157-y

Moran SC, Newhall C, Roman DC (2011) Failed magmatic eruptions: late-stage cessation of mag ma ascent. Bull Volcanol 73:115–122. doi:10.1007/s00445-010-0444-x

Newhall C, Hoblitt RP (2002) Constructing event trees for volcanic crises. Bull Volcanol 64:3–20. doi:10.1007/s004450100173

Newhall C, Pallister J (2015) Using multi data sets to populate probabilistic volcanic event trees. In: Schroder JF, Papale P (eds) Volcanic hazards, risks, and disasters. Hazards and disasters series. Elsevier, Amsterdam, pp 203–232

Phillipson G, Sobradelo R, Gottsmann J (2013) Global volcanic unrest in the 21st century: an ana lysis of the first decade. J Volcanol Geotherm Res 264:183–196

Rouwet D, Iorio M, Polgovsky D (2013) A science & arts sensitization program in Chapultenango, 25 years after the 1982 El Chichón eruptions (Chiapas, Mexico). J Appl Volcanol 2:6. doi:10.1186/2191-5040-2-6

Rouwet D, Sandri L, Marzocchi W, Gottsmann J, Selva J, Tonini R, Papale P (2014) Recognizing and tracking hazards related to non-magmatic unrest: a review. J Appl Volcanol 3:17. doi:10.1186/s13617-014-0017-3

Sandri L, Guidoboni E, Marzocchi W, Selva J (2009) Bayesian event tree for eruption forecasting (BET_EF) at Vesusvius, Italy: a retrospective forward application to the 1631 eruption. Bull Volcanol 71:729–745. doi:10.1007/s00445-008-0261-7

Sandri L, Jolly G, Lindsay J, Howe T, Marzocchi W (2012) Combining long-term and short-term probabilistic volcanic hazard assessment with cost-benefit analysis to support decision making in a volcanic crisis from the Auckland Volcanic Field, New Zealand. Bull Volcanol 74:705–723. doi:10.1007/s00445-011-0556-y

Sandri L, Thouret J-C, Constantinescu R, Biass S, Tonini R (2014) Long-term multi-hazard assessment for El Misti volcano (Peru). Bull Volcanol 76:771. doi:10.1007/s00445-013-0771-9

Sandri L, Tonini R, Rouwet D, Constantinescu R, Mendoza-Rosas AT, Andrade D, Bernard B (2017) The need to quantify hazard related to non-magmatic unrest: from BET_EF to BET_UNREST. In: Gottsmann J, Neuberg, J, Scheu B (eds) Volcanic unrest: from science to society—IAVCEI advances in volcanology. Springer

Selva J, Costa A, Marzocchi W, Sandri L (2010) BET_VH: exploring the influence of natural uncertainties on long-term hazard from tephra fallout at Campi Flegrei (Italy). Bull Volcanol 72:717–733. doi:10.1007/s00445-010-0358-7

Selva J, Marzocchi W, Papale P, Sandri L (2012) Operational eruption forecasting at high-risk volcanoes: the case of Campi Flegrei, Naples. J Appl Volcanol 1:5. http://www.appliedvolc.com/content/1/1/5

Selva J, Orsi G, Di Vito MA, Marzocchi W, Sandri L (2011) Probability hazard map for future vent opening at Campi Felgrei caldera. Bull Volcanol, Italy. doi:10.1007/s00445-011-0528-2

Sobradelo R, Bartolini S, Marti J (2014) HASSET: a probability event tree tool to evaluate future volcanic scenarios using Bayesian inference. Bull Volcanol 76:770. doi:10.1007/s00445-013-0770-x

Sobradelo R, Marti J (2015) Short-term volcanic hazard assessment through Bayesian inference: retrospective application to the Pinatubo 1991 volcanic crisis. J Volcanol Geotherm Res 290:1–11. doi:10.1016/j.jvolgeores.2014.11.011

Sparks RSJ (2003) Forecasting volcanic eruptions. Earth Planet Sci Lett 210:1–15. doi:10.1016/S0012-821X(03)00124-9

Sparks RSJ, Aspinall WP (2004) Volcanic activity: frontiers and challenges in forecasting, pre diction and risk assessment. The State of the Planet: Frontiers and Challenges in Geo physics. Geophysical Monograph 150, IUGG Volume 19. doi:10.1029/150GM28

Tonini R, Sandri L, Rouwet D, Caudron C, Marzocchi W, Suparjan P (2016) A new Bayesian Event Tree tool to track and quantify unrest and its application to Kawah Ijen volcano. Geochem Geophys Geosyst 17:2539–2555. doi:10.1002/2016GC006327

Probabilistic E-tools for Hazard Assessment and Risk Management

Stefania Bartolini, Joan Martí, Rosa Sobradelo and Laura Becerril

Abstract

The impact of a natural event can significantly affect human life and the environment. Although fascinating, a volcanic eruption creates similar or even greater problems than more frequent natural events due to its multi-hazard nature and the intensity and extent of its potential impact. It is possible to live near a volcanic area and take advantage of the benefits that volcanoes offer, but it is also important to be aware of the existing threats and to know how to minimise risks. In this chapter, we present an integrated approach using e-tools for assessing volcanic hazard and risk management. These tools have been especially designed to assess and manage volcanic risk, to evaluate long- and short-term volcanic hazards, to conduct vulnerability analysis, and to assist decision-makers during the management of a volcanic crisis. The methodology proposed here can be implemented before an emergency in order to identify optimum mitigating actions and how these may have to be adapted as new information is obtained. These tools also allow us identifying the most appropriate probabilistic and statistical techniques for volcanological data analysis and treatment in the context of quantitative hazard and risk assessments. Understanding volcanic unrest, forecasting volcanic eruptions, and predicting the most probable scenarios, all imply a high degree of inherent uncertainty, which needs to be quantified and clearly explained when transmitting scientific information to decision-makers.

Resumen

El impacto de un evento natural puede afectar significativamente la vida humana y el medio ambiente. Aunque fascinante, una erupción volcánica

S. Bartolini (✉) · J. Martí · L. Becerril
Group of Volcanology, SIMGEO (UB-CSIC),
Institute of Earth Sciences Jaume Almera,
ICTJA-CSIC, Lluís Solé I Sabarís S/N, Barcelona
08028, Spain
e-mail: sbartolini.1984@gmail.com

J. Martí
e-mail: joan.marti@ictja.csic.es

L. Becerril
e-mail: lbecerril@ictja.csic.es

R. Sobradelo
WRN Earth Hub Leader, Analytics Technology &
Willis Research Network, London, UK
e-mail: rosa.sobradelo@willistowerswatson.com

Advs in Volcanology (2019) 47–61
DOI 10.1007/11157_2017_14
© The Author(s) 2017
Published Online: 15 July 2017

puede generar problemas parecidos o incluso mayores que otros eventos naturales más frecuentes, ya que se trata de un fenómeno multipeligro y con un impacto potencial de gran intensidad y magnitud. Vivir cerca de una zona volcánica es posible considerando sus innumerables beneficios, pero sin embargo debemos ser conscientes de la amenaza existente y tomar las medidas oportunas para minimizar el riesgo. En este capítulo, se presenta un enfoque integrado que utiliza herramientas para la evaluación de la peligrosidad y del riesgo volcánico, el análisis de vulnerabilidad, y para ayudar en la gestión de una crisis volcánica. La metodología aquí propuesta puede ser implementada antes de una emergencia con el fin de determinar las medidas óptimas de mitigación y como pueden adaptarse a medida que se obtiene nueva información. Además, estas herramientas se basan en las técnicas probabilísticas y estadísticas más adecuadas para el análisis de datos vulcanológicos y su implementación en el contexto de las evaluaciones cuantitativas del peligro y riesgo. La interpretación del *unrest*, la predicción de las erupciones volcánicas, y la identificación de los escenarios más probables, implican un alto grado de incertidumbre, que debe ser cuantificado y claramente explicado para transmitir correctamente la información científica a los gestores de las crisis volcánicas.

Keywords

Volcanic risk · Volcanic hazard · Vulnerability · Risk management · Decision-making · E-tools

Introduction

One of the most important tasks in modern volcanology is to manage volcanic risk and, consequently, to minimise it. Forecasting volcanic eruptions and predicting the most probable scenarios are tasks that are subject to high degrees of uncertainty but which need to be quantified and clearly explained when transmitting scientific information to decision-makers. Assessing eruption risk scenarios in probabilistic ways has become one of the main challenges tackled by modern volcanology (Newhall and Hoblitt 2002; Marzocchi et al. 2004; Aspinall 2006; Neri et al. 2008; Sobradelo et al. 2014).

The volcanic management cycle consists of four phases (Sobradelo et al. 2015): (i) the pre-unrest phase that includes long-term assessment, hazard and risk mapping, and estimation regarding expected scenarios, volcano monitoring, and emergency planning; (ii) the unrest phase, which includes short-term assessment, alert, communication, and information procedures, the implementation of emergency measures, and the interpretation of eruption precursors; (iii) the volcanic event itself, represented by a major change in the state of the volcano; decisions can be revised as new information is obtained and the evolution of the volcanic event is updated; (iv) the post-event phase characterized by rescue and recovery.

Previous studies, focusing attention on the first phase of volcanic crisis management, have developed different methodologies for evaluating hazards. Most of these studies are based on the use of simulation models and Geographic Information Systems (GIS) that enable volcanic hazards such as lava flows, pyroclastic density currents (PDCs), and ash fallout to be modeled and visualised (Felpeto et al. 2007; Toyos et al. 2007; Cappello et al. 2012; Martí et al. 2012; Becerril et al. 2014; Bartolini et al. 2014a, b). The use of spatio-temporal data and GIS has now become an essential part of integrated approaches to disaster risk management in light of the development of new analysis/modelling techniques. These spatial information systems are used for storage, situation analysis, modelling, and visualisation (Twigg 2004). Normally, these studies present a systematic approach based on the estimation of spatial probability, that is, a susceptibility analysis (Martí and Felpeto 2010), a temporal analysis based on Bayesian inference (Marzocchi et al. 2004; Sobradelo et al. 2014), or the evaluation of hazards (Felpeto et al. 2007; Bartolini et al. 2014a, b; Becerril et al. 2014) and vulnerability (Marti et al. 2008; Scaini et al. 2014). These studies have been applied in a number of volcanic areas such as Etna, Sicily (Cappello et al. 2013), Tenerife, Spain (Martí et al. 2012), Peru (Sandri et al. 2014), the island of El Hierro, Spain (Becerril et al. 2014), and Deception Island, Antarctica (Bartolini et al. 2014a). Other procedures have been employed to assess volcanic hazards in Campi Flegrei, Italy (Lirer et al. 2001), Furnas (São Miguel, Azores), Vesuvius, Italy (Chester et al. 2002), and Auckland, New Zealand (Sandri et al. 2012).

These studies underline the fact that scientists are aware of the relationship between volcanic hazards and socio-economic impacts and are in the process of developing new approaches and models to assess its importance. One positive aspect is that, as a result, there is now a choice of freely available models; on the other hand, these models are not integrated into a single platform and have been developed in a variety of different programming languages.

Despite the fact that these tools have been created for application in real situations and have been successfully tested in different volcanic areas and retrospectively for a number of volcanic crisis (Martí et al. 2012; Sobradelo et al. 2014; Bartolini et al. 2014a, b; Becerril et al. 2014; Scaini et al. 2014), to date they only exist as academic exercises. To convert them into practical tools ready to be used by Civil Protection managers and decision-makers, they must be checked and tested, and then adapted to the real needs of end users.

Probabilities are still the best outcome of scientific forecasting. However, they are not easily understood. Understanding the potential evolution of a volcanic crisis is crucial for designing effective mitigation strategies. One of the main issues when managing a volcanic crisis is how to make scientific information understandable for decision-makers and Civil Protection managers. Thus, we need quantitative risk-based methods for decision-making under conditions of uncertainty that can be developed and applied to volcanology. In order to resolve this problem and to take a step forward in minimising risks, we have defined an integrated approach using user-friendly e-tools, which can be run on personal computers. They are specifically useful for long- and short-term hazard assessment, vulnerability analysis, decision-making, and volcanic risk management. In this chapter, we describe the e-tools designed to manage and to minimise volcanic risk.

Volcanic Risk: Hazard, Vulnerability, and Value

In general, risk is defined as the probability or likely magnitude of a loss (Blong 2000). In volcanic risk assessment, risk depends on the adverse effects of volcanic hazards and can be defined as the product of three main factors: volcanic hazard, vulnerability to those hazards, and the value of what is at risk.

Volcanic hazard is defined as the probability of any particular area being affected by a destructive volcanic event within a given period of time (Blong 2000). The quantification and evaluation of the volcanic hazard allow us determining which areas will be affected by a given volcanic event, and to design appropriate emergency plans and territorial planning.

A vulnerability assessment uses indicators to quantify the physical vulnerability of the elements of each sub-system (buildings, transportation system, urban services, and population) and is a measurement of the proportion of the value likely to be lost as a result of a given event (Blong 2000). In fact, the values of elements-at-risk that can be directly or indirectly damaged by a given hazard will vary. Each hazardous phenomenon affects elements and infrastructures in different ways in terms of their specific physical vulnerability; this in turn will be depend on their number (of buildings, people, etc.), monetary value, surface area, and the importance of the elements-at-risk. Indicators are used to determine the specific physical vulnerability of each hazardous phenomenon and sub-system (Scaini et al. 2014).

The value is the number of human lives at stake, together with the capital value (land, buildings, etc.) and productive capacity (factories, power plants, highways, etc.) exposed to the destructive events (Blong 2000).

Risk management is a complex process (Fig. 1) since different steps are necessary for evaluating and minimising risk. It can be thought of as the sum of risk assessment—which includes risk analysis and risk evaluation—and risk control. Risk analysis aims to improve prevention tools through the collection and acquisition of data on hazards and risks, and then to disseminate it in the form of maps and scenarios (Thierry et al. 2015). This phase is characterised by five different steps: hazard identification, hazard assessment, elements-at-risk/exposure analysis, vulnerability assessment, and risk estimation (Van Westen 2013). In particular, it is important to distinguish between long- and short-term hazard assessments, which will vary according to the expected period of time over which the process will display significant variations.

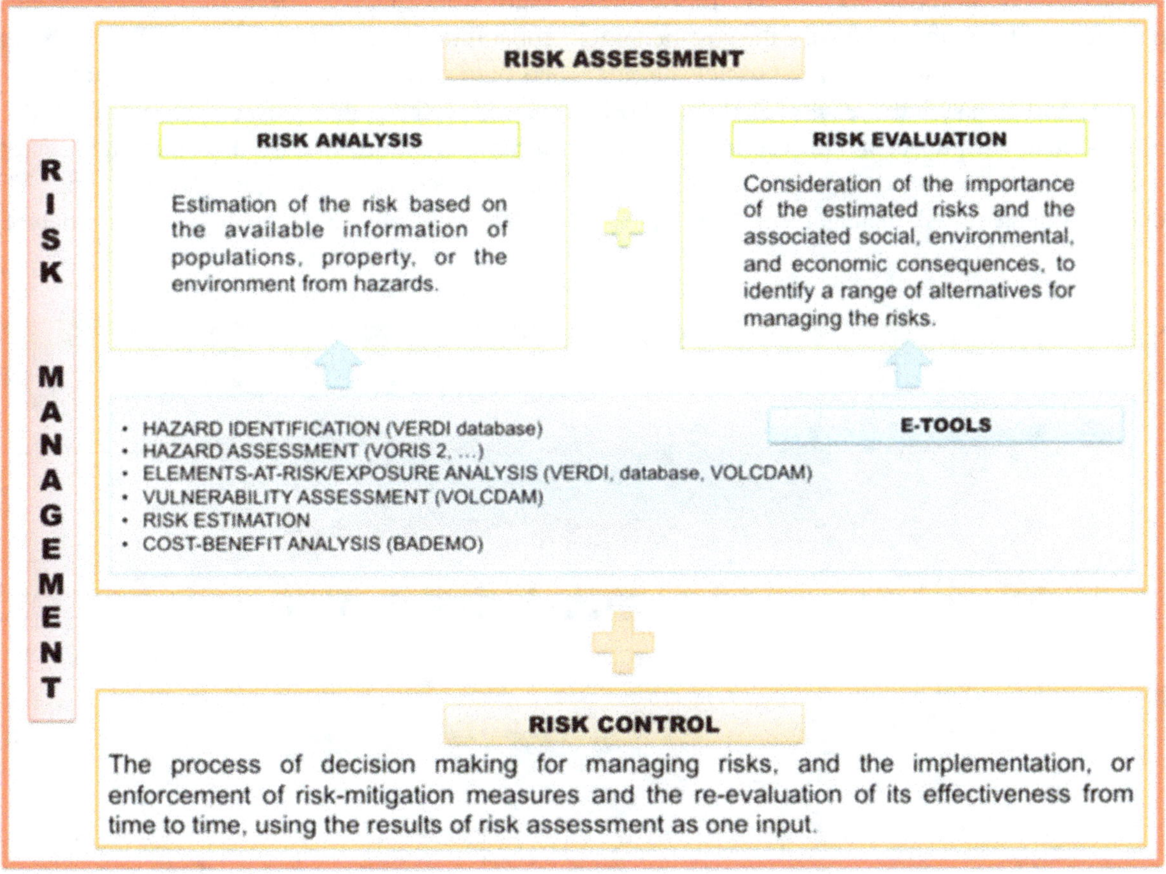

Fig. 1 Risk management schematic: steps for evaluating and minimising risk

Long-term assessment is based on historical and geological data, as well as theoretical models, and refers to the time window available—during which time the volcanic system shows no signs of unrest—before an unrest episode occurs. On the other hand, short-term assessment refers to the unrest phase, and complementary information derived from the combination of a long-term analysis with real-time monitoring data is needed to update the status of the volcanic hazard (Blong 2000). Short-term evaluation helps forecast where and when the eruption may take place and the most likely eruptive scenarios.

Once the long-term risk analysis is computed, it is then possible to adopt mitigation measures such as land-use planning and emergency preparations to reduce the risk. In addition, this long-term analysis will help manage the volcanic crisis as it will constitute the basis for the short-term analysis and, combined with a cost-benefit analysis, will assist in correct decision-making (Sobradelo et al. 2015). To evaluate the total risk related to a particular volcanic eruption we have to repeat the evaluation of the vulnerability and the cost-benefit analysis (risk evaluation) for each possible hazard scenario and then sum the results. This will allow us to estimate the impact and the economic losses that will affect society and the environment, and to identify a range of risk management alternatives.

Finally, the second part of risk management is risk control, which consists of the decision-making process involved in managing risks whose aim is to improve crisis management capabilities and implement risk-mitigation measures using the results of risk assessment as an input (Western 2013). During this phase measures must be adopted for reducing vulnerability (people and infrastructure) and developing recovery and resilience capacities after an event has taken place.

E-tools for Volcanic Hazard and Risk Management

In this section, we present different e-tools that have been specifically designed to assess and manage volcanic risk (Fig. 2). The objective is to combine freely available models to produce a new approach for minimising and managing volcanic risk. These e-tools are based on the assumption that the best way to show how probabilities work is to use the possible scenarios and outcomes of volcanic unrest (an increase in volcanic activity that may or may not precede a volcanic eruption) to design an integrated model that can act as a descriptor of scenarios. The effectiveness of these e-tools has been analysed

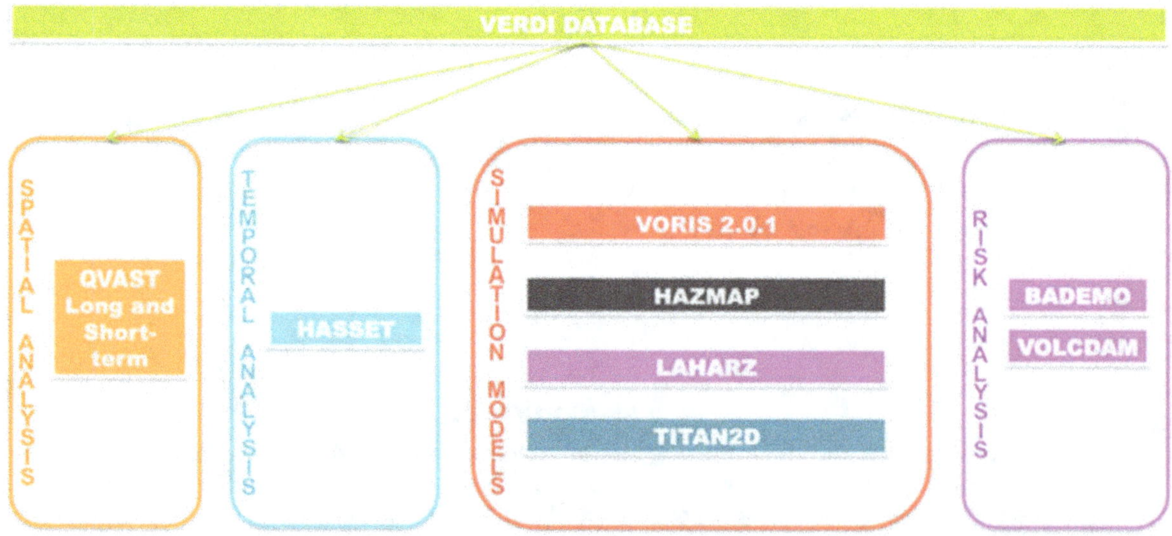

Fig. 2 E-tools for assessing and managing volcanic risk that allow to evaluate the possible hazards that could affect a volcanic area and develop appropriate hazard and risk maps

in different volcanic areas showing pros and contras of these approaches, such as in the case of El Hierro Island in Canary Islands (Becerril et al. 2014; Sobradelo et al. 2015), Tenerife in Canary Islands (Scaini et al. 2014), Deception Island in Antarctica (Bartolini et al. 2014a), and La Garrotxa Volcanic Field in Spain (Bartolini et al. 2014b). User manuals are available at https://volcanbox.wordpress.com/ and http://www.vetools.eu.

These e-tools are designed to be implemented before an emergency in order to identify (i) the optimum mitigating actions and (ii) the most appropriate probabilistic and statistical techniques for volcanological data analysis and treatment in the context of quantitative hazard and risk assessment, and (iii) how appropriate responses may change as new information is obtained. They constitute different steps, as shown in Fig. 3, in hazard assessment and risk management.

Storing Data

The results of our analyses are as important as the input parameters that are used. Given the nature, variety, and availability of the data we need to handle, one of the most important aspects of risk assessment is the management and exchange of information (De la Cruz-Reyna 1996). The quality of the data will determine the evaluation of the volcanic risk, which is an essential part of risk-based decision-making in land-use planning and emergency management. Some of the most relevant issues include how and where to store the data, in which format should they be made available, and how to facilitate its use and exchange. Thus, it is essential to have an appropriate database that is specifically adapted to the task of evaluating and managing volcanic risk. For that reason, we designed VERDI (Volcanic managEment Risk Database desIgn) (Bartolini et al. 2014c), a

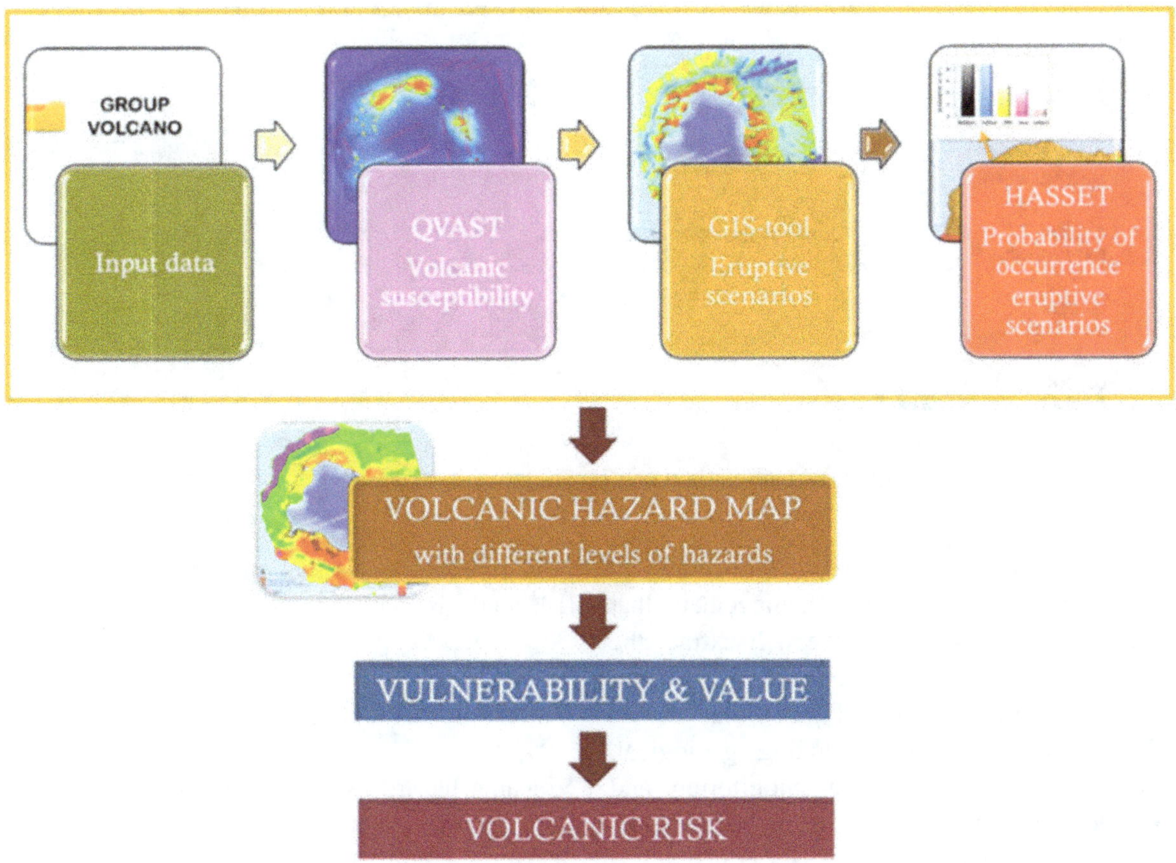

Fig. 3 Steps in volcanic risk assessment applying e-tools to be implemented before an emergency

Fig. 4 The design of the VERDI database (after Bartolini et al. 2014c)

geodatabase with an appropriate architecture for volcanic risk assessment and management that stores the data from which we will extract the input parameters to run our e-tools. VERDI is a spatial database structure (Fig. 4) that allows different types of data, including geological, volcanological, meteorological, monitoring, and socio-economic information, to be manipulated, organised, and managed. The data contained in this database are the basis for applying the probabilistic models that are the first step in our risk analysis.

Hazard E-tools

Volcanic hazard assessment consists of simulating eruptive scenarios for use in risk-based decision-making, land-use planning, and emergency management. They must necessarily be based on a good knowledge of the past eruptive history of the volcano or volcanic area, and will reveal how volcanoes erupted in the past, thereby providing clues to *how* they will erupt in the future. The first step in the quantitative assessment of volcanic hazards is the spatial probability of occurrence of a hazard, i.e. *where* the next eruption may take place and its extent. This analysis is based on the development of susceptibility maps (Martí and Felpeto 2010), that is, the spatial probability of a future vent opening given the past eruptive activity of a volcano and the simulation of possible eruptive scenarios. Another important task is to investigate the temporal probability, in other words, *when* the next eruption will occur in the future and the type of scenarios that are most likely to be involved. Thus, an evaluation of volcanic hazard enables us to infer where and when the next eruption may take place and its magnitude.

Spatial Analysis

Susceptibility analysis is the evaluation of the spatial distribution of future vent openings (Fig. 5a). This challenging issue is generally tackled using probabilistic methods that use the calculation of a kernel function at each data location to estimate probability density functions (PDFs). Commonly, a Gaussian kernel, describing a normal distribution, is used to estimate local event densities in volcanic fields, which will give the intensity of a new vent opening. This method is based on the distance from nearby volcanic structures and a smoothing parameter, also known as bandwidth. This factor is the most important parameter in the kernel function and represents the degree of randomness in the distribution of past events.

QVAST (QGIS for VolcAnic SuscepTibility), developed by Bartolini et al. (2013), is a new tool designed to generate user-friendly quantitative assessments of volcanic susceptibility (e.g. the probability of hosting a new eruptive vent). QVAST allows an appropriate method for evaluating the bandwidth for the kernel function to be selected on the basis of input parameters and the shapefile geometry, and can also evaluate the PDF with the Gaussian kernel. When different input data sets are available for the area, the total susceptibility map is obtained by assigning different weights to each of the PDFs, which are then combined via a weighted sum and modeled in a non-homogeneous Poisson process. This e-tool has been used to evaluate susceptibility on the island of El Hierro (Canary Islands) (Becerril et al. 2014) and on Deception Island (Antarctica) (Bartolini et al. 2014a).

When monitoring data generated during an unrest phase are available, the QVAST e-tool can also be used to update the susceptibility map. In fact, seismicity and surface deformation are good indicators of magma movement and during volcanic unrest variations in shallow volcano-tectonic and long-period seismicity, as well as ground deformation, are observed as the magma migrates within the volcanic system (Martí et al. 2013). Thus, QVAST is a highly useful tool that can be applied to both long- and short-term evaluations.

Temporal Analysis

HASSET (Hazard Assessment Event Tree), developed by Sobradelo et al. (2014), is a probability tool built on an event tree structure that uses Bayesian inference to estimate the probability of occurrence of a future volcanic scenario. It also evaluates the most relevant sources of uncertainty in the corresponding volcanic system. Event tree structures (Newhall and Hoblitt 2002) constitute one of the most useful and necessary tools in modern volcanology for assessing the volcanic hazard of future eruptive scenarios. An event tree is a graphic representation of events in the form of nodes and branches. It evaluates the most relevant sources of uncertainty when estimating the probability of occurrence of a future volcanic event.

The objective of this e-tool is to outline all relevant possible outcomes of volcanic unrest at progressively greater detail and to assess the hazard of each scenario by estimating its probability of occurrence within a future time interval. Each node of the event tree represents a step and contains a set of possible branches (the outcomes for that particular category). The nodes are

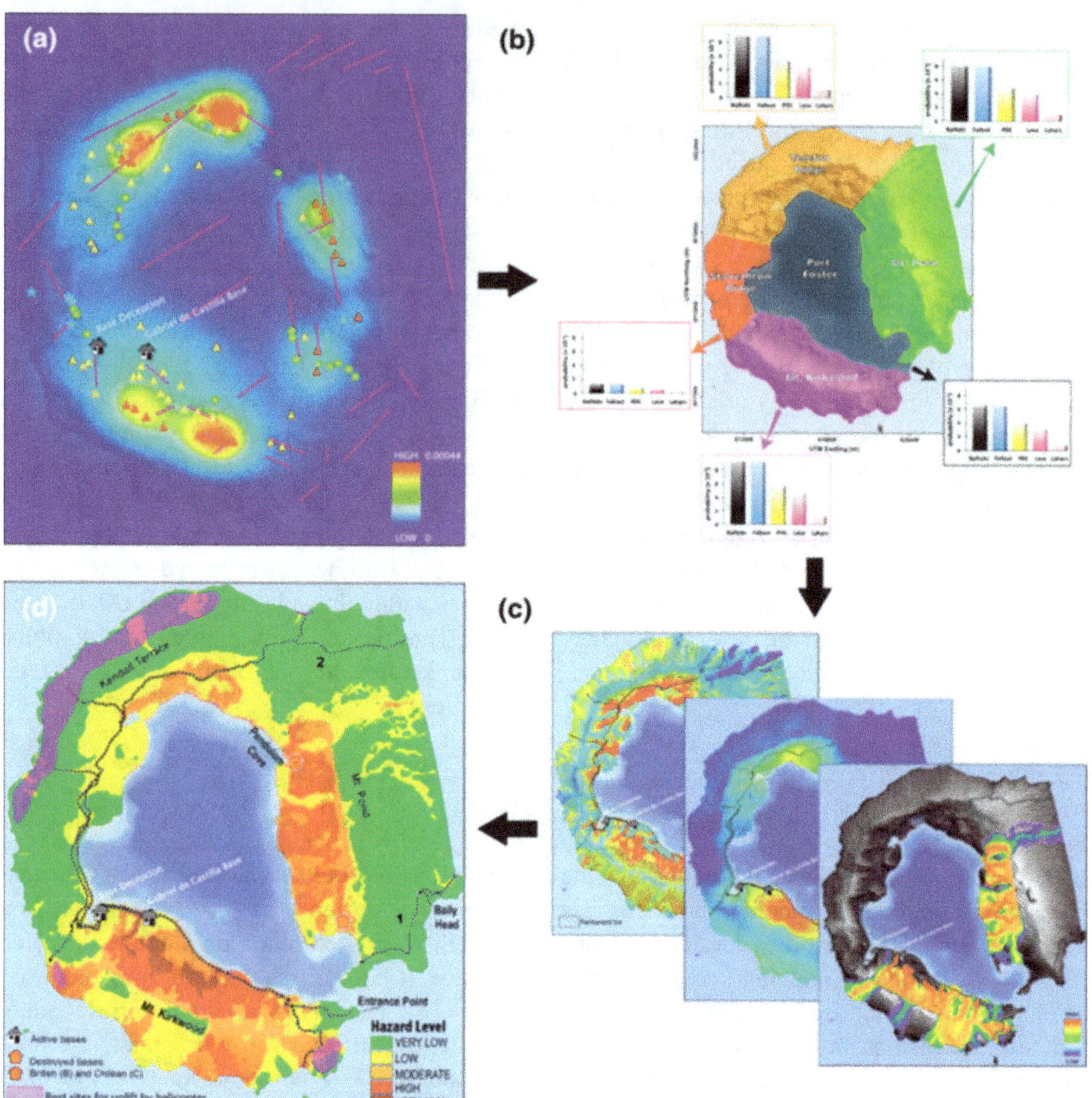

Fig. 5 Methodological approach for obtaining a qualitative hazard map: its application to Deception Island (after Bartolini et al. 2014a)

alternative steps from a general prior event, state, or condition that move towards increasingly specific subsequent events and a final outcome. HASSET (Fig. 5b) uses this event tree structure to make estimations of the probabilities for each possibility (branches and nodes) using a statistical methodology known as Bayesian Inference (Newhall and Hoblitt 2002; Marzocchi et al. 2004; Sobradelo et al. 2014). In particular, and based on comparisons with previous event trees for volcanic eruptions, HASSET accounts for the possibility of (i) flank eruptions (as opposed to only central eruptions), (ii) geothermal or tectonic unrest (as opposed to only magmatic unrest), and (iii) felsic or mafic lava composition, as well as (iv) certain volcanic hazards as possible outcomes of an eruption, and (v) the distance reached by each hazard.

A user-friendly interface guides the user through all steps and helps

– enter all the data needed for the analysis;
– compute the estimated probability for each branch in the event tree;

– compute the total estimated probability and compare up to five different scenarios.

This tool has been used for determining the eruption probability on El Teide (Tenerife, Spain) (Sobradelo and Martí 2010), El Hierro (Canary Islands) (Becerril et al. 2014), and Deception Island (Antarctica) (Bartolini et al. 2014a).

In both long-term and short-term evaluations HASSET can be useful for determining the occurrence probabilities of eruptive scenarios and can assist decision-makers assess the required mitigation actions associated with each scenario and estimate the corresponding potential risk.

Simulation Models

Simulating eruptive scenarios caused directly (e.g. lava flows, fallout, surges) and indirectly (earthquakes, landslides) by an eruption requires a detailed analysis of the past activity of the volcano or volcanic area, and must take into account all the possible hazards associated with the eruptive activity. Volcanic hazard can be assessed via two different types of approaches: deterministic, which defines a maximum extension area affected by an eruptive episode based on deposits generated by past activity, or probabilistic, based on the probability that a certain area will be affected by an eruptive process. In order to generate hazard maps (Fig. 5d), it is important to understand past eruptive behaviour and to employ physical simulation models that will permit the behaviour of future volcanic activity to be foreseen. In this type of approach, accurate and detailed geographic and cartographic data are required for high-quality analysis with a GIS.

Here, we describe some of the e-tools freely available for download that allow volcanic hazards to be evaluated (Fig. 5c):

– VORIS 2.0.1 is a GIS-based tool, developed by Felpeto et al. (2007) that allows users to simulate lava flows, fallout, and pyroclastic density current scenarios. Lava-flow simulations are based on a probabilistic model that assumes that topography is the most important factor determining the path of a lava flow. The determination of the probability of each point being invaded by lava is performed by computing several random paths with a Monte Carlo algorithm. Fallout simulation models are advection diffusion models that assume that away from the vent the transport of the particles from a Plinian column is controlled by the advective effect of the wind, by diffusion due to atmospheric turbulence, and by the settling velocity of the particles. The model for simulating pyroclastic density currents is the energy cone model proposed by Sheridan and Malin (1983). The input parameters are the topography, the collapse equivalent height (H), and the collapse equivalent angle (θ). The intersection of the energy cone, originating at the eruptive source, with the ground surface defines the distal limits of the flow.

– HAZMAP is a free program for simulating the sedimentation of volcanic particles at discrete point sources that predicts the corresponding ground deposits (deposit mode) (Macedonio et al. 2005). HAZMAP is also able to evaluate the probability of overcoming a given loading threshold in ground deposits by using a set of different wind profiles recorded on different days (probability mode). Using a statistical set of recorded wind profiles (and/or other input parameters), it can also be used to draw hazard maps for ashfall deposits. In HAZMAP, settling velocities can be calculated using several models as a direct function of particle diameters, densities, and shapes. The advantage of HAZMAP is that it is a simple tool able to predict ashfall during hypothetical or real eruptions of a given magnitude and wind profile.

– LAHARZ is a semi-empirical code for creating hazard-zonation maps that depict estimates of the location and extent of areas inundated by lahars (Schilling 1998). The input parameters for this model are the Digital Elevation Model and the lahar volume, which provide an automated method for mapping areas of potential lahar inundation. These hazard zones can be displayed in a GIS with other types of volcano hazard information such as the proximal hazard zone, infrastructure, hydrology, and population, as well as contours and shaded relief, to produce volcano hazard-zonation maps. Such maps show the proximity and intersection of potential hazard zones to people and infrastructures.

– TITAN2D is a computer program model developed by the University at Buffalo (Patra et al. 2005) that simulates granular flows over digital elevation models, based on a "thin layer model". The input for the computer code includes simulation time, minimum thickness of the final deposit, internal and bed friction angles, starting coordinates, and the initial speed and direction of the flow. Additionally, this program allows users to define the specific starting pile dimensions or a dynamic flow source. The outputs from the program (represented dynamically) are flow depth and momentum, which yield the deposit limit, run-out path, average flow velocity, inferred deposit thickness, and travel time.

Vulnerability E-tool

Once we have obtained hazard maps, the next step consists of adding population, infrastructures, and land-use data to evaluate the vulnerability associated with the impact of a determined hazard. The data required for generating vulnerability maps are very complex and varied, and depend on the observation scale. Vulnerability is directly dependent on the type of phenomena in question and on the socio-economic characteristics of the environment. VOLCANDAM is a new e-tool based on the methodology developed by Scaini et al. (2014) that generates maps estimating the expected damage caused by volcanic eruptions. VOLCANDAM (Fig. 6) consists of three main parts: exposure analysis, vulnerability assessment, and the estimation of expected damages. The exposure analysis identifies the elements exposed to the potential hazard and focuses on the relevant assets of the study area (population distribution, social and economic conditions, and productive activities and their role in the regional economy). The vulnerability analysis defines a physical vulnerability indicator for all exposed elements, as well as a corresponding qualitative vulnerability index. Systemic vulnerability considers the possible relevance of each element in the system and their interdependencies by taking into account all exposed and non-exposed elements (people, buildings, transportation network, urban services, and productive activities). Damage assessment is performed by associating a qualitative damage rating to each combination of hazard and vulnerability, bearing in mind their specific contexts and roles in the system. The way one element can be damaged—and thus lose its functionality—depends in fact on the type of hazardous event and the characteristics of the element. The result is damage maps that can be displayed at different levels of detail, depending on user preferences. This tool aims to facilitate territorial planning and risk management in active volcanic areas.

Decision-Making

The evaluation of the "direct costs" and "factors" (indirect costs) that have an impact on the economic growth of an area affected by a volcanic event needs to take into account a number of elements. A cost-benefit analysis may assist the decision-making process by evaluating the economic impact of the different scenarios. The approach used by Sobradelo et al. (2015) (Fig. 7)

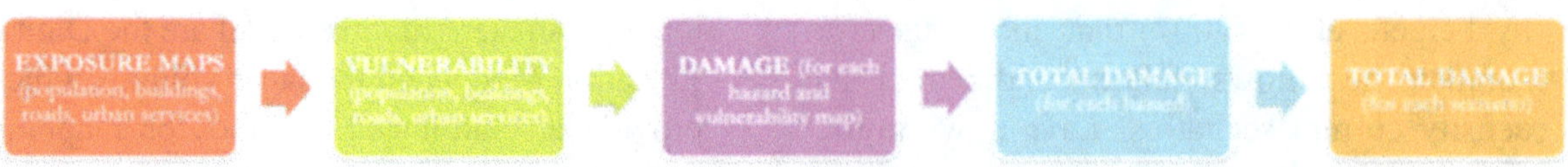

Fig. 6 Steps in the vulnerability analysis in the VOLCANDAM approach (after Scaini et al. 2014). See text for details

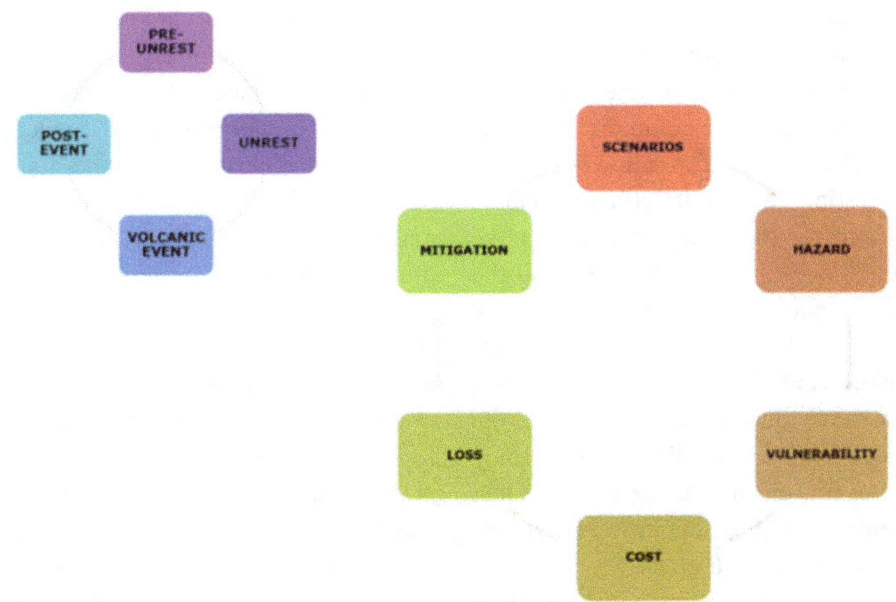

Fig. 7 The volcanic crisis management cycle: stages and phases (after Sobradelo et al. 2015). See text for details

is a Bayesian decision model that applies a general, flexible, probabilistic approach to the management of volcanic crises by combining the hazard and risk factors that decision-makers need for a holistic analysis of a volcanic crisis. These factors include eruption scenarios and their probabilities of occurrence, the vulnerability of populations and their activities, and the costs of false alarms and erroneous forecasts. This model can be implemented before an emergency to (i) pinpoint actions for reducing the vulnerability of a particular area, (ii) identify the optimum mitigating actions during an emergency and how these may change as new information is obtained, (iii) assess after an emergency the effectiveness of a mitigating response and, in light of results, (iv) how to improve strategies before another crisis occurs. BADEMO (BAyesian DEcision MOdel) is part of this integrated approach, and enables the previous analysis of the distribution of local susceptibility and vulnerability to eruptions to be combined with specific costs and potential losses. Indeed, BADEMO should be seen as a tool for improving communication between the monitoring scientists who provide volcanological information and those responsible for deciding which action plans and mitigating strategies should be put into practice.

Discussion and Conclusions

Modern volcanology is a scientific discipline with social and practical applications that derive from its direct involvement in risk reduction. Today, a multidisciplinary approach is required to risk assessment since new methodologies are constantly being developed and explored, and, for example, geo-spatial technologies are becoming extremely helpful in the disaster-risk management. A defined and solid methodology for assessing volcanic hazard and managing risk is fundamental in both long-term evaluations and in unrest phases when monitoring information is available. Furthermore, e-tools can be progressively modified and implemented in light of the outcomes obtained in order to integrate as many models as possible into the assessment and management of volcanic risk.

The possibility of foreseeing an event such as a volcanic eruption is more effective—and guarantees greater economic savings—than acting after a disaster. The potential of simple e-tools such as those presented in this chapter lies in the fact that they are freely available and have been developed on accessible and dynamic graphical user interfaces. The main advantage of using these e-tools is that new data or new model results can be easily incorporated into the procedures for updating the hazard assessment. By contrast, the use of expensive commercial e-tools hampers the exchange of information and complicates their testing in situ in volcanic fields. Furthermore, non-free tools may hinder effective risk assessment since they are often beyond the means of advisory and management groups with limited financial, technological, and manpower resources (Leidig and Teeuw 2015). However, this does not mean that anybody can use the e-tools discussed here for, on the contrary, they require expertise in volcanic hazards and related issues and their use implies scientific knowledge that will avoid incorrect outcomes.

In this chapter we have presented different statistical methodologies and e-tools for interpreting volcanic data and assessing long- and short-term volcanic hazards and vulnerability, and for carrying out cost-benefit analyses. Statistical analysis enables us to extract information about the future behaviour of a volcano by looking at the geological and historical activity of the volcanic system (VERDI database, Bartolini et al. 2014c). The QVAST tool (Bartolini et al. 2013) can be used to analyse past activity and to calculate the possibility that new vents will open (volcanic susceptibility), while the Bayesian event tree statistical method HASSET (Sobradelo et al. 2014) can be applied to calculate eruption recurrence. Using these calculations, we can identify a number of significant scenarios using GIS-based e-tools (i.e. VORIS 2.0.1, HAZMAP, …) and evaluate the potential extent of the main volcanic hazards expected to occur in volcanic areas. The results obtained allow us to generate volcanic hazard maps for different levels of hazards, evaluate vulnerability (VOLCANDAM e-tools, Scaini et al. 2014), conduct cost-benefit analysis (BADEMO e-tools, Sobradelo et al. 2015), and, finally, manage volcanic risk.

During a volcanic crisis, emergency plans must be put into practice and so different government departments need to be prepared in advance. Therefore, it is important to have conducted previous long-term hazard assessment —among many other tasks—to properly manage a volcanic crisis. They allow scientists and managers to understand the characteristics of the volcano or volcanic area and its past eruptive history, and to infer the possible eruptive scenarios that may occur in the future. With this previous information, short-term hazard assessment can be conducted when volcanic unrest starts and hazard maps can be drawn up and alert levels be defined; nevertheless, it is important to always bear in mind that the best form of protection is the evacuation of the population at risk. Volcanic monitoring is an essential part of short-term assessment and so should be performed by experts and based on a good understanding of volcanic processes. Despite the possibility of conducting cost-benefit analysis, which can help maintain the economic order, the security and health of the population should always be the main concern.

One of the purposes in the near future is to create a new software platform (VolcanBox, see VETOOLS European Project—www.vetools.eu) with a user-friendly interface. This platform will contain different e-tools that, via a homogeneous and systematic methodology, will help minimise risk. However, the feasibility and applicability of each tool will have to be analysed by different groups of experts with experience in regions possessing different volcanological and socio-economic scenarios. This evaluation must also bear in mind potential end users (i.e. Civil Protection agencies), not only to test the ability of existing tools but also to understand decision-makers' needs and requirements, and train them in the use of these tools. If we are to reduce volcanic risk we must ensure that scientists, managers, and decision-makers are all fully prepared to confront this phenomenon since the best way to guarantee risk reduction is to possess good knowledge of its causes.

Acknowledgements This work was supported by the European Commission (FP7 Theme: ENV.2011.1.3.3-1; Grant 282759: VUELCO). The English text was corrected by Michael Lockwood.

References

Aspinall WP (2006) Structured elicitation of expert judgment for probabilistic hazard and risk assessment in volcanic eruptions. In: Mader HM et al (eds) Statistics in volcanology. Special Publication of IAVCEI, Geological Society of London

Bartolini S, Cappello A, Martí J, Del Negro C (2013) QVAST: a new Quantum GIS plugin for estimating volcanic susceptibility. Nat Hazards Earth Syst Sci 13 (11):3031–3042

Bartolini S, Geyer A, Martí J, Pedrazzi D, Aguirre-Díaz G (2014a) Volcanic hazard on deception Island (South Shetland Islands. J Volcanol Geotherm Res, Antarctica). doi:10.1016/j.jvolgeores.2014.08.009

Bartolini S, Bolós X, Martí J, Riera E, Planagumà LL (2014b) Hazard assessment at the Quaternary La Garrotxa Volcanic Field (NE Iberia). Nat Hazards. doi:10.1007/s11069-015-1774-y

Bartolini S, Becerril L, Martí J (2014c) VERDI: a new volcanic management risk database design. J Volcanol Geotherm Res 288:132–143

Becerril L, Bartolini S, Sobradelo R, Martí J, Morales JM, Galindo I (2014) Long-term volcanic hazard assessment on El Hierro (Canary Islands). Nat Hazards Earth Syst Sci 14:1853–1870

Blong R (2000) Volcanic hazards and risk management. In: Sigurdsson H et al (eds) Encyclopedia of volcanoes. Academic, San Diego, pp 1215–1227

Cappello A, Neri M, Acocella V, Gallo G, Vicari A, Del Negro C (2012) Spatial vent opening probability map of Etna volcano (Sicily, Italy). Bull Volcanol 74:2083–2094. doi:10.1007/s00445-012-0647-4

Cappello A, Bilotta G, Neri M, Del Negro C (2013) Probabilistic modelling of future volcanic eruptions at Mount Etna. J Geophys Res Solid Earth 118:1925–1935. doi:10.1002/jgrb.50190

Chester DK, Dibben CJL, Duncan AM (2002) Volcanic hazard assessment in Western Europe. J Volcanol Geotherm Res 115:411–435

De La Cruz-Reyna S (1996) Long term probabilistic analysis of future explosive eruptions. In: Scarpa R, Tilling RI (eds) Monitoring and mitigation of volcanic hazards. Springer, Berlin, pp 599–629 (IAVCEI/UNESCO Volume)

Felpeto A, Martí J, Ortiz R (2007) Automatic GIS-based system for volcanic hazard assessment. J Volcanol Geotherm Res 166:106116

Leidig M, Teeuw R (2015) Free software: a review, in the context of disaster management. Inter J Appl Earth Observ Geoinfo 42:49–56

Lirer L, Petrosino P, Alberico I (2001) Volcanic hazard assessment at volcanic fields: the Campi Flegrei case history. J Volcanol Geotherm Res 101(1–4):55–75

Macedonio G, Costa A, Longo A (2005) A computer model for volcanic ash fallout and assessment of susbsequent hazard. Comput Geosci 31:837–845

Martí J, Felpeto A (2010) Methodology for the computation of volcanic susceptibility. An example for mafic and felsic eruptions on Tenerife (Canary Islands). J Volcanol Geotherm Res 195:69–77

Marti J, Spence R, Calogero E, Ordoñez A, Felpeto A, Baxter P (2008) Estimating building exposure and impact to volcanic hazards in Icod de los Vinos, Tenerife (Canary Islands). J Volcanol Geotherm Res 178:553–561

Martí J, Sobradelo R, Felpeto A, García O (2012) Eruptive scenarios of phonolitic volcanism at Teide-Pico Viejo volcanic complex (Tenerife, Canary Islands). Bull Volcanol 74:767–782. doi:10.1007/s00445-011-0569-6

Martí J, Pinel V, López C, Geyer A, Abella R, Tárraga M, Blanco MJ, Castro A, Rodríguez C (2013) Causes and mechanisms of the 2011–2012 El Hierro (Canary Islands) submarine eruption. J Geophys Res Solid Earth 118(3):823–839

Marzocchi W, Sandri L, Gasparini P, Newhall C, Boschi E (2004) Quantifying probabilities of volcanic events: the example of volcanic hazard at Mount Vesuvius. J Geophys Res 109. doi:10.1029/2004JB003155

Neri A, Aspinall W, Cioni R, Bertagnini A, Baxter P, Zuccaro G, Andronico D, Barsotti S, Cole P, Ongaro TE, Hincks T, Macedonio G, Papale P, Rosi M, Santacroce R, Woo G (2008) Developing an event tree for probabilistic hazard and risk assessment at Vesuvius. J Volcanol Geotherm Res 178 (3):397–415

Newhall CG, Hoblitt RP (2002) Constructing event trees for volcanic crisis. Bull Volcanol 64:3–20

Sandri L, Jolly G, Lindsay J, Howe T, Marzocchi W (2012) Combining long-and short-term probabilistic volcanic hazard assessment with cost-benefit analysis to support decision making in a volcanic crisis from the Auckland Volcanic Field, New Zealand. Bull Volcanol 74(3):705–723

Sandri L, Thouret JC, Constantinescu R, Biass S, Tonini R (2014) Long-term multi-hazard assessment for El Misti volcano (Peru). Bull Volcanol 76(2):1–26

Scaini C, Felpeto A, Martí J, Carniel R (2014) A GIS-based methodology for the estimation of potential volcanic damage and its application to Tenerife Island, Spain. J Volcanol Geotherm Res 278–279:40–58

Schilling SP (1998) LAHARZ: GIS program for automated mapping of lahar-inundation hazard zones. US Geological Survey Open-File Report, pp 98–638

Sheridan MF, Malin MC (1983) Application of computer-assisted mapping to volcanic hazard evaluation of surge eruptions: Vulcano, Lipari. J Volcanol Geotherm Res 17:187–202

Sobradelo R, Martí J (2010) Bayesian event tree for long-term volcanic hazard assessment: application to Teide-Pico Viejo stratovolcanoes, Tenerife, Canary islands. J Geophys Res 115. doi:10.1029/2009JB006566

Sobradelo R, Bartolini S, Martí J (2014) HASSET: a probability event tree tool to valuate future volcanic scenarios using Bayesian inference presented as a plugin for QGIS. Bull Volcanol 76:770

Sobradelo R, Martí J, Kilburn C, López C (2015) Probabilistic approach to decision-making under uncertainty during volcanic crises: retrospective application to the El Hierro (Spain) 2011 volcanic crisis. Nat Hazards 76(2):979–998

Thierry P, Neri M, Le Cozannet G, Jousset P, Costa A (2015) Preface: approaches and methods to improve risk management in volcanic areas. Nat Hazards Earth Syst Sci 15:197–201

Toyos GP, Cole PD, Felpeto A, Martí J (2007) A GIS-based methodology for hazard mapping of small pyroclastic density currents. Nat Hazards 41:99–112

Twigg J (2004) Disaster risk reduction: mitigation and preparedness in development and emergency programming. Overseas Development Institute, Humanitarian Practice Network, London, Good Practice Review no. 9

Van Westen CJ (2013) Remote sensing and GIS for natural hazards assessment and disaster risk management. In: Shroder J (editor in Chief), Bishop MP (ed) Treatise on geomorphology. vol 3, Remote Sensing and GIScience in Geomorphology. Academic Press, San Diego, pp 259–298

The Need to Quantify Hazard Related to Non-magmatic Unrest: From BET_EF to BET_UNREST

Laura Sandri, Roberto Tonini, Dmitri Rouwet, Robert Constantinescu, Ana Teresa Mendoza-Rosas, Daniel Andrade and Benjamin Bernard

Abstract

Most volcanic hazard studies focus on magmatic eruptions and their accompanying phenomena. However, hazardous volcanic events can also occur during non-magmatic unrest, defined as a state of volcanic unrest in which no migration of magma is recognised. Examples include tectonic unrest, and hydrothermal unrest that may lead to phreatic eruptions. Recent events (e.g. Ontake eruption, September 2014) have demonstrated that the successful forecasting of phreatic eruptions is still very difficult. It is therefore of paramount importance to identify indicators that define the state of non-magmatic unrest. Often, this type of unrest is driven by fluids-on-the-move, requiring alternative monitoring setups, beyond the classical seismic-geodetic-geochemical architectures. Here we present a new version of the probabilistic model BET (Bayesian Event Tree), called BET_UNREST, specifically developed to include the forecasting of non-magmatic unrest and related hazards. The structure of BET_UNREST differs from the previous BET_EF (BET for Eruption Forecasting) by adding a dedicated branch to detail non-magmatic unrest outcomes. Probabilities are calculated at each node by merging prior models and past

L. Sandri (✉) · D. Rouwet
Sezione di Bologna, Istituto Nazionale di Geofisica e Vulcanologia, Via Donato Creti 12, Bologna, Italy
e-mail: laura.sandri@ingv.it

R. Tonini
Sezione di Roma 1, Istituto Nazionale di Geofisica e Vulcanologia, Via di Vigna Murata 605, Rome, Italy

R. Constantinescu
Seismic Research Centre, University of West Indies, Gordon street, Saint Augustine, Trinidad & Tobago

Present Address:
R. Constantinescu
School of Geosciences, University of South Florida, 4202 E. Fowler Avenue, NES 107, Tampa, FL 33620-5550, USA

A.T. Mendoza-Rosas
Instituto de Geofísica, Universidad Nacional Autónoma de México. C. Universitaria, Coyoacán 04510, México, D.F., Mexico

Present Address:
A.T. Mendoza-Rosas
CONACYT-Centro de Ingeniería y, Desarrollo Industrial, Av. Playa pie de la cuesta 702, Desarrollo San Pablo, Querétaro, Qro. 76125, Mexico, C.P., Mexico

D. Andrade · B. Bernard
Escuela Politécnica Nacional, Instituto Geofísico, Quito, Ecuador

Advs in Volcanology (2019) 63–82
DOI 10.1007/11157_2017_9
© The Author(s) 2017
Published Online: 12 August 2017

data with new incoming monitoring data, and the results can be updated any time new data has been collected. Monitoring data are weighted through pre-defined thresholds of anomaly, as in BET_EF. The BET_UNREST model is introduced here, together with its software implementation PyBetUnrest, with the aim of creating a user-friendly, open-access, and straightforward tool to support short-term volcanic forecasting (already available on the VHub platform). The BET_UNREST model and PyBetUnrest tool are tested through three case studies in the frame of the EU VUELCO project.

Resumen extendido

La mayoría de los estudios sobre amenazas volcánicas están enfocados en las erupciones magmáticas y fenómenos relacionados. Sin embargo, fenómenos volcánicos peligrosos pueden también ocurrir durante una fase de "unrest" no-magmático, definido por el estado de unrest volcánico en el cual no se reconoce la migración de un magma. Ejemplos de esto son unrest tectónico (capaz de causar preocupación independientemente del resultado posterior) y unrest hidrotermal, que pueden resultar en erupciones freáticas. Eventos recientes (e.g. la erupción de Ontake en septiembre 2014) han demostrado que las erupciones freáticas siguen siendo difícilmente previsibles. Por estas razones, es de extrema importancia identificar señales que permitan definir un estado de unrest no-magmático. Muchas veces, este tipo de unrest es provocado por fluidos en movimiento, y requiere la instalación de un sistema de monitoreo alternativo, más allá de la clásica arquitectura sismo-geodético-química. En este capítulo, presentamos la nueva versión del modelo probabilístico BET (Arbol de Eventos Bayesiano, por sus siglas en inglés), llamado BET_UNREST, específicamente desarrollado para incluir la previsión de unrest no-magmático y sus peligros relacionados. La estructura de BET_UNREST difiere de la versión anterior BET_EF (BET para Previsión de Erupciones, por sus siglas en inglés), añadiendo una rama dedicada para detallar los resultados potenciales de unrest no-magmático. Las probabilidades están calculadas para cada nodo juntando modelos a priori y datos pasados con los datos nuevos, provenientes del monitoreo. Los datos de monitoreo están ponderados mediante umbrales predefinidos de anomalía, como es el caso en BET_EF. Este capítulo ilustra el modelo, y su herramienta, con tres casos de estudio, en el marco del proyecto EU VUELCO:

(i)   un análisis retrospectivo para el volcán Popocatépetl, en donde no hay necesidad de la rama hidrotermal, debido al carácter magmático; Popocatépetl ha permanecido en estado de unrest desde diciembre 1994 hasta el presente. Para esta aplicación, BET_UNREST fue corrido usando la base de datos de la UNAM (1997–2012), con una aplicación retrospectiva para prever erupciones mayores (columnas eruptivas >8 km) durante el período abril-junio 2013.

(ii) una aplicación basada en un ejercicio de simulacro en el Cotopaxi; en este caso se probó con BET_UNREST de manera retrospectiva, pero, esta vez, usando datos creados específicamente para el simulacro, junto a datos de entrada basados en la historia real del volcán preparados antes del simulacro. Presentamos la previsión de erupciones magmáticas resultantes del simulacro mismo.

(iii) el simulacro en tiempo casi-real organizado en el marco de VUELCO en Dominica (mayo 2015). El sistema volcánico de Dominica es el prototipo para BET_UNREST debido a su carácter hidrotermal. Actividad freática/freatomagmática ocurrió realmente durante el simulacro, lo cual de hecho era bastante probable según BET_UNREST (la probabilidad media de unrest hidrotermal fue de 0.73, mientras la probabilidad media de una erupción hidrotermal fue de 0.32). También se produjo un mapa de probabilidades para la apertura de ventos eruptivos en caso de erupciones magmáticas y freáticas.

Con estos ejercicios, estamos convencidos de haber llevado BET un paso más cerca hacia una implementación completa en situaciones de crisis. Al final, BET_UNREST funcionó como se esperaba. Sin embargo, es importante ser consiente de algunos puntos críticos que han resultado de estas aplicaciones, incluso realizar más pruebas para mejorar su diseño y comprobar su utilidad en casos reales en el futuro. BET_UNREST se introdujo junto a su implementación digital PyBetUnrest con el objetivo de crear un instrumento de fácil uso, libre y de acceso directo (disponible en el sitio web Vhub) para ayudar en la evaluación de la amenaza volcánica a corto plazo.

Keywords

Volcanic unrest · Forecasting · Hydrothermal · Magmatic · Bayesian inference

Palabras clave

Unrest volcánico · Previsión · Hidrotermal · Magmático · Inferencia Bayesiana

1　Introduction

Monitoring activities represent the main source of information to understand the behaviour of volcanic systems on short time-scales and, possibly, during emergency crises. In this framework, one of the main challenges of volcano monitoring is the identification and characterisation of the phase defined as "unrest", which consists of a relevant physical or chemical change in the volcanic system with respect to its background behaviour, leading to cause for concern. Unrest can be due to several factors and depends on the local characteristics of each volcanic system, making it very difficult to find general features or patterns (Phillipson et al. 2013). Unrest may be followed by volcanic eruptions due to the movement of magma, but can also be associated with other dangerous phenomena: indeed, in addition to magma-related hazards (e.g., tephra fallout, lava flows, ballistics), hydrothermal and tectonic activities, without evidence for "magma-on-the-move", can also lead to dangerous outcomes (i.e., flank collapses, gas emissions, phreatic explosions, lahars). Such hazardous events related to non-magmatic unrest are not easy to track and, in volcanic hazard evaluations, are sometimes underestimated (Rouwet et al. 2014). For instance, many volcanoes pass through a phase of hydrothermal unrest for years, decades or even centuries. Due to this long-term behavioural similarity, it is often difficult to recognise how hydrothermal unrest can lead to related hazards in the short-term. Where the driving agent and the main eruptive product is not magma, but water (liquid or vapour) and occasionally liquid sulphur, or gas, this type of unrest can lead to non-magmatic eruptions. On the other hand, non-eruptive hydrothermal unrest can also promote volcanic hazards after prolonged gas emissions, acidic fluid infiltration into aquifers, soils and the hydrologic network, or deformation induced by a rising fluid front (see Rouwet et al. 2014).

In this light, although most volcanic hazard assessments focus only on magmatic eruptions as potential hazard sources, hazardous events can also occur during non-magmatic unrest, which in this chapter is defined as a state of volcanic unrest in which no migration of magma is recognised. Examples of non-magmatic unrest include the tectonic (which causes concern independently on how it evolves and eventually ends), and hydrothermal unrest types; the latter may eventually lead to phreatic eruptions. Recent occurrences of phreatic eruptions (e.g. Ontake eruption, September 2014, Japan) have demonstrated that they are still very hard to anticipate from classical observations based on seismic-geodetic-geochemical monitoring architectures. For these reasons, it is of paramount importance to identify indicators that define the state of non-magmatic unrest. Often, this type of unrest is driven by "fluids-on-the-move", requiring alternative and innovative monitoring setups, beyond the classical ones.

In the last decade it has become crucial to provide forecasts of the possible outcomes of volcanic unrest, to give quantitative support and scientific advice to decision makers (e.g., Woo 2008; Marzocchi and Woo 2007, 2009). Because of this, event tree schemes have been proposed (e.g., Newhall and Hoblitt 2002; Marzocchi et al. 2004), and a few probabilistic tools based on event trees and Bayesian inference have been developed (e.g., BET_EF, Marzocchi et al. 2008; HASSET, Sobradelo et al. 2013) with the ability to quantify the probability of different possible outcomes related to magmatic unrest. However, the need for recognising and tracking the evolution of *any* type of volcanic unrest, and to quantify the probability linked to non-magmatic unrest as well, have led us, within the VUELCO project, to the development of a new probabilistic model, able to forecast both magmatic and non-magmatic hazardous events related to volcanic unrest: BET_UNREST. The BET_UNREST model is based on an event tree, whose structure is extended with respect to the previous schemes such as BET_EF (see the generalisation from BET_EF to BET_UNREST in Fig. 1, highlighted in red) by adding a specific branch to detail the track and outcome of non-magmatic unrest. Nonetheless, BET_UNREST adopts from BET_EF the Bayesian inferential paradigm and the ability to account both for long-term data (typically from the geological record) and short-term information from monitoring networks.

In this chapter, we briefly present the BET_UNREST model and its implementation in the PyBetUnrest software tool (Tonini et al. 2016), made with the aim of providing a user-friendly, open-access, and straightforward tool to handle probabilistic forecasts and visualise results, and that has already been included on the Vhub platform (https://vhub.org/resources/betunrest). The new event tree and tool are applied here as illustrative examples to the VUELCO target volcanoes Popocatépetl (Mexico), Cotopaxi (Ecuador) and Dominica (West Indies).

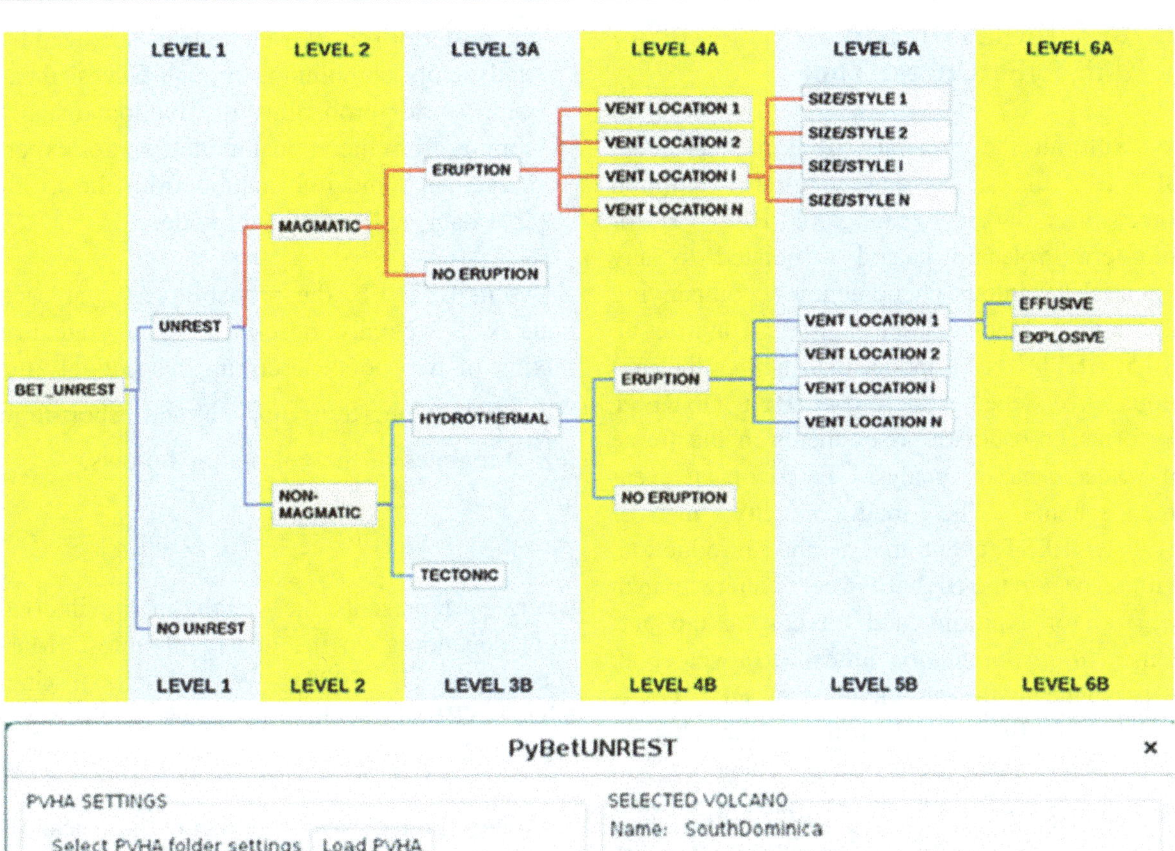

Fig. 1 The new event tree as defined for the BET_UNREST model (*on top*) and its visual implementation in the software PyBetUnrest (*on bottom*). The *red branch* corresponds to the previous BET_EF model

2 BET_UNREST Model and PyBetUnrest Tool

As with all the previous BET models (e.g., BET_EF, for short- and long-term eruption forecasting, Marzocchi et al. 2008; BET_VH, for long-term volcanic hazard associated to any potential hazardous phenomenon accompanying an eruption, Marzocchi et al. 2010; Tonini et al. 2015; BET_VHst, a model that merges the previous two, Selva et al. 2014), BET_UNREST performs probabilistic assessments in the frame of volcanic hazard analysis, based on an event tree scheme. The main novelty in the BET_UNREST event tree is the introduction, with respect to the BET_EF tree, of a new branch (Fig. 1) for exploring and forecasting the outcomes of *non-magmatic* unrest (Rouwet et al. 2014). Due to the resemblance of BET_UNREST to other BET models from a methodological and computational point of view, here we will only give a brief overview. The papers by Marzocchi et al. (2004, 2008) provide a more detailed description.

BET_UNREST probabilities are evaluated by a Bayesian inferential procedure, in order to quantify both the aleatory and epistemic uncertainty characterising the impact of volcanic eruptions in terms of eruption forecasting and/or hazard assessment. Such a procedure allows merging all the available information, such as models, a priori beliefs, past data from volcanic records and, when available, real-time monitoring data in order to include, in principle, all the knowledge about the considered volcanic system.

In general, the Bayesian inference procedure at the basis of BET_UNREST assigns a probability to each node, providing a framework where:

- probabilities are expressed through a probability density function (pdf), and not as a single number, to account for a best-evaluation value (for example the mean of the probability density function, representing a degree of aleatory uncertainty) and for a measure of the epistemic uncertainty (the dispersion of the pdf);

- the posterior pdf, at each node, is achieved by statistically combining, through Bayes' theorem, a prior probability distribution (usually coming from theoretical models and/or expert judgement) and information from the available data relevant for that node.

As in BET_EF, the probability $[\theta_k]$ at each node k is actually described by a statistical mixing of two pdfs, describing respectively the "so-called" long-term $[\theta_k^{\{\bar{M}\}}]$ and short-term $[\theta_k^{\{M\}}]$ regimes of the volcano as follows:

$$[\theta_k] = \gamma_k[\theta_k^{\{M\}}] + (1 - \gamma_k)[\theta_k^{\{\bar{M}\}}]$$

where γ_k represents the weight in the interval [0,1] depending on the degree of unrest (Marzocchi et al. 2008). With such mixing, BET_UNREST switches between the two "regimes". In practice:

- When anomalies with respect to the volcano's background activity are not observed at time $t = t_0$, BET_UNREST relies on the so-called long-term information to assign the probabilities (hereinafter also referred to as background probabilities) at the various branches. Such background probabilities (i.e., $[\theta_k^{\{\bar{M}\}}]$) are based on theoretical models and information from the geological and eruptive record of the volcano studied, or of similar volcanoes, and describe the long-term frequencies of magmatic or non-magmatic unrest, and subsequent outcomes at these volcanic systems.

- When a clear state of unrest of whatever nature is detected at $t = t_0$ by BET_UNREST, the probabilistic assignment at all the successive nodes is based mainly on the monitoring information. In practice, monitoring data are transformed into subjective pdfs (i.e., $[\theta_k^{\{M\}}]$) relative to the occurrence of magmatic or non-magmatic unrest and the following branches. Actually, at some nodes, monitoring data are not considered as relevant (for example, in forecasting the size of an eruption, magmatic or not), and here BET_UNREST continues to rely on theoretical models and long-term frequencies.

- When, at time $t = t_0$, BET_UNREST observes a "degree of unrest" (of whatever nature) without it being completely clear, the statistical mixing provides a resulting pdf which accounts for both the regimes, giving the short-term regime a weight equal to the degree of unrest, and to the long-term regime its complement.

In this way, during a phase of unrest, the past data have less (null, in the case of complete unrest) importance. The short term hazard/eruption forecasting depends exclusively on the translation of observed anomalies into pdfs describing all the branches of the event tree. This is done, separately at each node, by weighting monitoring data through pre-defined thresholds of anomaly (Marzocchi et al. 2008) and converting the resulting "degree of anomaly" into a best-evaluation probability, to which a degree of variance is associated (Fig. 2). This is a very simple and intuitive procedure, in which the basic assumptions are:

1. the first anomaly detected is the most informative
2. subsequent anomalies contribute less and less to the increase of the degree of anomaly
3. strong non-linear coupling among anomalies are neglected.

At each node, BET_UNREST evaluates the following probabilities (see also Fig. 1) by means of Bayesian inference (we give the acronyms used throughout the chapter to indicate the probability at each node in brackets):

- *Unrest*: probability ($P(U)$) of unrest in the time period $[t_0; t_0 +]$, given the monitoring observations at time $t = t_0$; the time window $|$ is defined by the user;
- *Magmatic unrest*: probability ($P(MU)$) that the unrest is due to "magma-on-the-move", given the unrest;

- *Magmatic eruption*: probability ($P(MEr)$) of a magmatic eruption, given magmatic unrest; the following sub-branches mirror the BET_EF structure, so we point the reader to Marzocchi et al. (Marzocchi et al. 2008) for them;
- *Non-magmatic unrest*: this is the complementary of the *Magmatic unrest* branch, so by definition is the probability of non-magmatic unrest, given an unrest;
- *Hydrothermal unrest*: probability ($P(HU)$) of hydrothermal unrest, given a non-magmatic unrest;
- *Tectonic unrest*: this is the complementary of the *Hydrothermal unrest* branch, so it describes the probability ($P(TU)$) of a tectonic unrest, given a non-magmatic unrest;
- *Hydrothermal eruption*: probability ($P(HEr)$) of a hydrothermal eruption, given a hydrothermal unrest;
- *Vent of hydrothermal eruption*: here we explore the spatial probability of vent opening in a hydrothermal eruption, given a hydrothermal eruption occurring; this node is an extension with respect to the event tree proposed in Rouwet et al. (2014);
- *Size of hydrothermal eruption*: probability of an explosive hydrothermal eruption, given a hydrothermal eruption occurring from a specific vent; its complementary branch is the effusive hydrothermal eruption.

In order to keep the structure of BET_UNREST as simple as possible, an effort has been made to maintain, where possible, a dichotomic branching into complementary (i.e., exhaustive and mutually exclusive) events. This is why the *Unrest* node does not branch directly into magmatic, hydrothermal and tectonic, but first it branches into magmatic-or-not. This allows a simplification in the evaluation of short-term probabilities. In particular, with this type of ramification, the user defines which monitoring measurements (plus thresholds and weight) affect the pdf of one of the two branches; the pdf of the complementary branch then comes automatically.

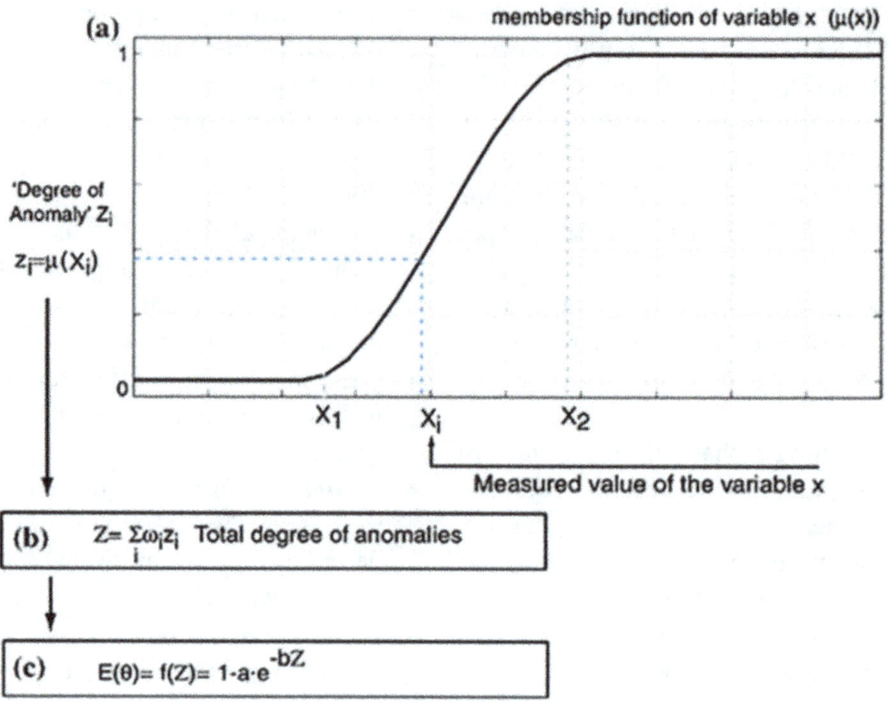

Fig. 2 This figure explains how monitoring measures are transformed into a best-evaluation probability at a given node of the event tree. First, a monitoring measure x_i is translated in a degree of anomaly z_i according to a selected anomaly function $\mu(\cdot)$ (**a**). In the above example, a measure below x_1 is considered background, above x_2 is anomalous, and in between it has a certain degree of anomaly. After collecting the degree of anomaly for all parameters considered at the node, we combine them using a weighted average (ωi is the weight of the i-th parameter) in order to obtain the total degree of anomaly (**b**). Then the total degree of anomaly is transformed into an average probability using a predefined function, in BET_UNREST, we use the function in (**c**). The parameters, weights, and thresholds are selected by the user, possibly through expert opinions' elicitation. Figure modified from (Marzocchi and Bebbington 2012)

The new BET_UNREST model is applied here with its software implementation PyBetUnrest presented in Tonini et al. (2016), which aims to provide an open and usable tool to bridge between the scientific community and decision makers, with a graphical user interface which allows the exploration of the event tree and the visualisation of the results (see Fig. 1). This solution was also implemented in the VHub cyber-infrastructure (http://vhub.org/resources/betunrest). In the present PyBetUnrest tool only one file needs to be adapted when new monitoring information is gathered. This structure makes PyBetUnrest extremely fast and user-friendly during crisis situations. More on the technical background of the BET_UNREST model and PyBetUnrest tool can be found in the VUELCO Deliverable 7.3 (at http://vhub.org) and in Tonini et al. (2016).

So far BET_UNREST and PyBetUnrest have not yet been blindly tested in real-time during an actual volcanic crisis, but only retrospectively (Tonini et al. 2016) at Kawah Ijen (Indonesia), for the time period 2010–2012 (after a learning period based on the observations from 2000 to 2010). The term "blindly" signifies that the rules of BET_UNREST (the long-term pdfs, and the monitoring parameters, thresholds and weight at the different nodes) are set *before* the beginning of the application, on different data (the *learning* dataset), and then the model is applied untouched to new data (the *voting* dataset), typically covering a different time period (as in the case of Tonini et al. 2016).

In the next section of this chapter, results and performances of the new model and tool will be discussed and validated by analysing the unrest crises for VUELCO target volcanoes Popocatépetl, Cotopaxi and Dominica through blind applications of BET_UNREST. The latter two applications show the results of the VUELCO crisis simulation exercises held in Quito (November 2014) and Dominica (May 2015).

3 BET_UNREST Applications

3.1 Popocatépetl, Mexico: A Retrospective Application Based on the Popo-DataBase

Here we apply the BET_UNREST model to Popocatépetl Volcano (Mexico), based on a catalog of monitored parameters of the 1994-ongoing eruptive period. Popocatépetl volcano awakened in December 1994, after almost 48 years of volcanic quiescence. Since 1994, Popocatépetl volcano has been one of the most active volcanoes in the world, and magmatic activity has been nearly constant. This fact raises the need to first redefine the concept of volcanic unrest for Popocatépetl, as BET_UNREST, at the *Unrest* node, requires indicative parameters to verify if the given volcano is in a state of unrest, or not. In *stricto* sensu, Popocatépetl has remained at least in a state of unrest, or even magmatic or eruptive unrest, since 1994, as its common manifestations are dome growth and vulcanian eruptive phases. The continuous state of unrest is reflected by the decision to never decrease the level of alert from orange to green (traffic light, De la Cruz-Reyna and Tilling 2008). Nevertheless, many of these eruptions are of no cause of concern (so, no unrest in *lato* sensu), neither for volcanologists nor for population. On the other hand, a practical scope of the BET_UNREST application at Popocatépetl is to forecast *major* eruptions, which can be considered a deviation from its current background activity. During the past 23,000 years, nine Plinian eruptions occurred at Popocatépetl (Mendoza-Rosas and De la Cruz-Reyna 2008), while, since 1994,

three eruptions with an eruption column >8 km have occurred. No Plinian eruptions have occurred during the 1994-ongoing eruption cycle, and thus none of the past Plinian eruptions have been monitored. For practical purposes, we thus define a *major* eruption for Popocatépetl as an eruption *with an eruption column >8 km*, as they are recorded during the current monitoring period. These eruptions have caused ash fall in the Puebla-Mexico City metropolitan area, thus having an impact on human activity. We aim at finding precursory signals for major eruptions (>8 km, VEI 3) for the period 1997–2012 (the learning period), and test the BET_UNREST retrospectively, using monitoring data of the volcanic activity observed during 2013 (the voting period). The time window, |, is defined as 1 month.

In Table 1 we report the activity carried out 24/7 with regards to monitoring at Popocatépetl, available as short-term information for unrest, origin of unrest and eruption. However, for the time period 1994–2012, the available data (as listed in Mendoza-Rosas, *VUELCO deliverable 5.1*), are restricted mainly to seismicity (VT, tremor, number of events) and visual observations (i.e. number of eruptions, column height). No real-time SO_2 flux is available for our purpose, and deformation data would need further processing. Regarding past data (long-term information for unrest, origin of unrest, and eruption), there have been 13 unrest episodes, and constant unrest since December 1994 (so, a priori probability to be in a state of unrest for the next month is about 85%). Out of the 13 unrest episodes, 6 were due to magma-on-the-move (magmatic unrest), of which 3 lead into a magmatic *major* eruption. The monitoring parameters listed in Table 2, along with respective thresholds and weight, have been identified in the UNAM (Universidad Nacional Autónoma de México) database for the period 1997–2012, and used to set BET_UNREST for Popocatépetl. The volcano is a stratocone with a higher probability of an eruption to occur from the central vent. For the period of observation (1997–2012) all eruptions were magmatic and occurred at the central crater. The a priori spatial distribution of vent opening is assigned as in Table 3. As a prior

Table 1 Activity carried out 24/7 as regards monitoring at Popocatépetl

Observations	4 cameras for visual observations
	5 three-component seismic stations
	5 BB seismic stations
	1 video camera + microwaves
	1 doppler radar
	3 biaxial inclinometers
	Geochemical observations (3 sites)
Routine actions	Automatic alarm for anomalies in seismicity
	Cell phone messages to personnel
	Comité Técnico Cientifico Asesor UNAM/CENAPRED
	Reports by SMS to population

model to define the size/style of magmatic eruptions we take the power law from Simkin and Siebert (1994). As past data we take the Mendoza-Rosas and De la Cruz-Reyna (2008) catalog for the past 23,000 years, and assume it to be complete for VEI >= 2 (Table 3).

We retrospectively applied BET_UNREST for the voting period April–June 2013, in which respectively 10, 11 and 2 eruptions of 2, 3 and 4 km-high columns were observed. No *major* eruption occurred. Observed anomalies include ash eruptions up to 130/day (all with columns <4 km), seismic tremor, incandescence in the crater/dome, and VT events (but no shallow event with depth <5 km). There was no anomalous deformation, no dome growth, and no SO_2 data available. Results of $P(MEr)$ for the retrospective application period (weekly updated) are presented in Fig. 3. For the whole period, $P(MEr)$ of a *major* eruption (>8 km eruption column) was <1% per month.

3.2 Cotopaxi, Ecuador: Retrospective Application Inspired by the VUELCO Simulation Exercise in Quito

A volcanic unrest simulation exercise for Cotopaxi volcano (5897 m.a.s.l.) was performed on November 13th, 2014 in Quito, Ecuador. The ice-capped stratovolcano, with an andesitic to

Table 2 Monitoring parameters set for BET_UNREST at Popocatépetl

	Parameters
Unrest	– # exhalations with ash (<4 km) > 20/day – Tremor Y/N – Increase VT Y/N
Magmatic unrest	– Incandescence dome Y/N, weight 2 – Duration tremor >6000 s, weight 1 – SO_2 >2000 t/d, weight 1
Magmatic eruption	– Dome growth Y/N – SO_2 >9000 t/d – Tectonic EQ > M5.5 (along the coast/arc Michoacán-Chiapas) Y/N – Incandescent debris Y/N – Change in # tremor Y/N – VT depth <5 km – Increase # VT >M2 Y/N – Duration tremor >30,000 s (inertia 2 months) Y/N – Increase # ash eruptions >2000–4000 (inertia 2 months) – Deformation Y/N

Table 3 *Left Part:* Spatial probability of vent opening for magmatic eruptions assigned for BET_UNREST at Popocatépetl: best guess a priori values. No past data are used. *Right Part:* Parameters of the magmatic eruption size distribution assigned for BET_UNREST at Popocatépetl: best guess a priori values and past data

Spatial probability of vent opening in magmatic eruptions		Size of magmatic eruption		
Vent location	A priori probability (best guess values; equivalent number of data = 1)	Size	A priori (best guess values; equivalent number of data = 1)	Past data
Central vent	0.99	VEI 1	0.83	975
North flank	0.0025	VEI 2	0.14	13
East flank	0.0025	VEI 3	0.023	3
South flank	0.0025	VEI 4	0.0038	7
West flank	0.0025	VEI $\geq$ 5	0.0008	2

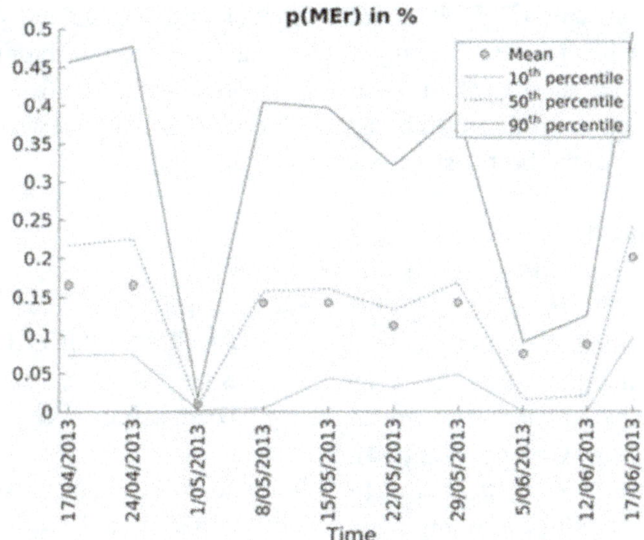

Fig. 3 Time history of probability (expressed in percentage) to have a magmatic eruption in the retrospective analysis at Popocatépetl

rhyolitic composition, is one of the most active and hazardous volcanoes in Ecuador. Historic eruptions at Cotopaxi produced large lithic-rich pyroclastic flows, ash flows, lava flows as well as large lahars (Barberi et al. 1995; Hall and Mothes 2008; Biass and Bonadonna 2011). Some lahars reached the Pacific Ocean at >200 km distance (Aguilera et al. 2004; Pistolesi et al. 2013). Recent unrest periods at Cotopaxi occurred in 1975–1976 and 2001–2002 and were characterised by increased fumarolic activity, elevated seismicity and edifice deformation (Molina et al. 2008). Fumarolic activity is a concern due to the heat transfer that may affect the ice cover resulting in non-eruptive debris flows or lahars.

A still unstable version of PyBetUnrest was set up (along with parameters and thresholds at each node for Cotopaxi volcano derived from monitoring information) before the simulation exercise, based on the available data in the literature up to the beginning of the simulation (the learning period stopped with the beginning of the exercise), in order to preliminarily test its value in decision support by providing near-real time probabilities of (i) the occurrence of unrest, (ii) the origin and nature of unrest and (iii) eruptive activity. However, during the simulation, the reports from the "volcano team" did not reflect the real eruptive and unrest history of Cotopaxi, as the past activity for the simulation was "invented". A different setting of BET_UNREST (and consequently of PyBetUnrest) on site was not possible due to the lack of time and the still premature customisability of the tool. This obliged us to set up and run the old BET_EF tool during the exercise (Constantinescu et al. 2015). Obviously, this prevented us from providing probabilistic assessment of non-magmatic events during the exercise at Cotopaxi: this would have been possible with BET_UNREST, enabling the calculation of probabilities for hydrothermal unrest and hydrothermal eruptions ($P(HU)$ and $P(HEr)$). Nevertheless, the unrest scenario proposed by the "volcano team" (Bulletins 1–5) did not emphasise a significant state of hydrothermal unrest, which, on the one hand, made our output less biased in not providing an evaluation for $P(HU)$ and $P(HEr)$; but on the other hand this simulation was probably not the best case to test BET_UNREST.

Here, we will re-run BET_UNREST and PyBetUnrest at Cotopaxi retrospectively for the unrest phases described in the five bulletins provided by the "volcano team" during the simulation exercise and using the BET_UNREST setup prepared prior to the simulation based on the *real* past activity of the volcano (Table 4). The time window | was set to 1 month. In Table 5 we show the probabilities resulting from the run of the code, after each bulletin:

(1) *Phase 0: The background activity of Cotopaxi (NO anomalies)*: results are based on the past activity of Cotopaxi, with all observation within background limits.

(2) *Phase 1 (Bulletin 1)*: the observed anomalies in this phase were limited to an increase in seismic activity compared to background level. Such an increase is indicative, according to pre-set parameters, of magma-on-the-move ($P(MU) = 0.68$). The considerable uncertainty is summarised by the 10th to 90th percentiles confidence interval.

(3) *Phase 2 (Bulletin 2)*: the observed anomalies in this phase were: a drastic increase in seismicity, an increase in SO_2 emission (5 times background levels), and a crater thermal anomaly. As a consequence, the mean P

(MU) increases, along with a decrease in the associated uncertainty.

(4) *Phase 3 (Bulletin 3)*: the observed anomalies in this phase were: an increase in VT and LP events, occurrence of tremor, appearance of new fumaroles, an increase in SO_2 emission, and an increase in the crater thermal anomaly. As a consequence, the $P(MU)$ is similar to Bulletin 2, but the $P(HU)$ increases slightly, due to the new fumaroles.

(5) *Phases 4 and 5 (Bulletins 4 and 5)*: the observed anomalies in these phases were similar, and included: intense fumarolic activity, occurrence of hybrid seismic events, an increase in SO_2 emission, and an increase in the crater thermal anomaly. As a consequence, $P(MEr)$ increases from 0.21 (phase 3) to 0.57, combined with a lower uncertainty.

3.3 Dominica, West Indies, Lesser Antilles: VUELCO Simulation Exercise, Dominica, May 2015

Dominica is characterised by hydrothermal activity manifested as thermal springs (up to boiling temperature), boiling-temperature fumarolic emissions (e.g. Valley of Desolation) and a crater lake, known as 'Boiling Lake', with a particular hydrodynamic behaviour (Fournier et al. 2009; Joseph et al. 2011; Rouwet et al. 2017). No high-temperature manifestations occur on the island, so no clear evidence of active magmatic degassing exists at the present time.

The simulation exercise, and consequently the BET_UNREST application, for the VUELCO target island of Dominica mainly focused on an unrest scenario for the southern part of the island. The purpose of the exercise was to test the tracking/assessment of an unrest period, and the decision making process undertaken by the scientific advisory group and local authorities.

Due to the hydrothermal character of Dominica, the application of BET_UNREST is highly suited. *Before* the simulation exercise, the

Table 4 Monitoring parameters set for BET_UNREST at Cotopaxi

Node-parameter#	Parameter and threshold(s) (Y/N indicates a Boolean observation)
Unrest-parameter 1	LP/month (205–335) (Garcia-Aristazabal 2010)
Unrest-parameter 2	VT/month (24–32) (Garcia-Aristazabal 2010)
Unrest-parameter 3	M Tectonic EQ (3–4)
Unrest-parameter 4	SO_2 (Y/N)
Magmatic unrest-parameter 1	EQ depth (>4.5–5.5 km)
Magmatic unrest-parameter 2	Deep VLP (Y/N)
Magmatic unrest-parameter 3	T fumarole (>119 °C)
Magmatic unrest-parameter 4	Appearance of acidic gas (Y/N)
Magmatic unrest-parameter 5	VT/month (>32)
Magmatic unrest-parameter 6	Increased deformation (Y/N)
Magmatic unrest-parameter 7	VLP + LP together (Y/N)
Magmatic unrest-parameter 8	Harmonic LP tremor (Y/N)
Magmatic unrest-parameter 9	SO_2 flux (t/d) (>100–350)
Magmatic eruption-parameter 1	sudden stop (Y/N)
Magmatic eruption-parameter 2	SO_2 flux (t/d) (>2000–2500)
Magmatic eruption-parameter 3	Tornillos (Y/N)
Hydrothermal unrest-parameter 1	New fumarole (Y/N)
Hydrothermal unrest-parameter 2	Anomalous glacier volume decrease (defrosting) (Y/N)
Hydrothermal unrest-parameter 3	LP/month (>205–335) (Garcia-Aristazabal 2010)
Hydrothermal eruption-parameter 1	Increase in T of fumarole (>120–200 °C)
Hydrothermal eruption-parameter 2	Increase in extension of fumarolic field (Y/N)
Hydrothermal eruption-parameter 3	Inflation of fumarolic field (Y/N)
Hydrothermal eruption-parameter 4	Landslides in hydrothermal areas (Y/N)
Hydrothermal eruption-parameter 5	New/extension of alteration areas (Y/N)

PyBetUnrest tool was set for Dominica, based on (1) existing literature of the past volcanic activity; (2) insights on the current hydrothermal activity; (3) discussion-based expert elicitation sessions (4 sessions at SRC and 1 at INGV-Bologna); and (4) exchanges with local experts in order to fine-tune the code with the monitoring parameters. We remark that all of this was done *prior* to the start of the simulation exercise (the learning period stopped at the beginning of the simulation exercise, as for Cotopaxi), and again no hindsight tuning was made. The long-term setup of PyBetUnrest is done by filling up a configuration file that includes the a priori and past data specifically for Dominica, whose main information is summarised in Table 6. The short-term information is listed in Table 7 (parameters and thresholds identified prior to the exercise onset, see above). Further details on the Dominica simulation exercise and on the BET_UNREST application are given in Constantinescu et al. (2016).

During the simulation exercise (May 14–15, 2015) three phases of changes in volcanic activity, each with a duration of six months, were distributed by the "volcano team" to the operators of the unrest crisis. The reports included four types of observations: (1) seismic bulletin, (2) GPS, (3) geothermal monitoring data, and (4) other observations.

The translation of the reported bulletins into the values for the selected parameters in the BET_UNREST for Dominica setup were reported back to the team of experts in real-time

Table 5 Resulting probabilities from retrospective application of BET_UNREST at Cotopaxi

		P(U)	P(MU)	P(MEr)	P(HU)	P(HEr)
Phase 0 (Background)	Mean	0.005	0.002	0.0005	0.001	0.0006
	10th prctile	0.0013	0.0002	0	0	0
	50th prctile	0.004	0.001	0.0002	0.0006	0.0002
	90th prctile	0.009	0.004	0.001	0.003	0.001
Phase 1	Mean	1	0.68	0.18	0.08	0.02
	10th prctile	1	0.07	0	0	0
	50th prctile	1	0.84	0.02	0.001	0
	90th prctile	1	1	0.69	0.30	0.04
Phase 2	Mean	1	0.83	0.22	0.05	0.013
	10th prctile	1	0.27	0	0	0
	50th prctile	1	1	0.04	0	0
	90th prctile	1	1	0.75	0.13	0.008
Phase 3	Mean	1	0.80	0.21	0.13	0.07
	10th prctile	1	0.14	0	0	0
	50th prctile	1	1	0.04	0.002	0.0003
	90th prctile	1	1	0.72	0.54	0.22
Phase 4 and 5	Mean	1	0.81	0.57	0.12	0.07
	10th prctile	1	0.23	0.02	0	0
	50th prctile	1	1	0.65	0.0004	0.0002
	90th prctile	1	1	1	0.49	0.28

Table 6 Set up of BET_UNREST at Dominica in terms of long-term information

	A priori mean (equivalent n data in brackets)	Past data
Unrest	0.5 (1)	Past data (successes) = 14 Past data (total) = 608
Magmatic	0.5 (1)	Past data (successes) = 13 past data (total) = 14
Magmatic eruption	0.58 from Phillipson et al. (2013) (1)	Past data (successes) = 0 Past data (total) = 13
Magmatic vent location	file	file
Hydrothermal vent location	file	file
Size distribution (Magmatic)	Dome extrusion: 0.83 Small explosive: 0.14 Large explosive: 0.03 (1)	Dome extrusion: 0 Small explosive: 5 Large explosive: 2

Some of the data are too many to be listed (this is indicated by the label "file" in the table). They can be provided in the form of files on request

during the simulation. In Table 8 we provide the probabilities resulting from the run of the code after each bulletin. In Fig. 4 we also provide the time evolution of some of the most relevant probability distributions, across all the time periods spanned by the simulation exercise in Dominica. For each bulletin, among the output information from PyBetUnrest, there were two

Table 7 Monitoring parameters set for BET_UNREST at Dominica

Node–parameter#	Parameter and threshold(s) (Y/N indicates a boolean observation)
Unrest-parameter 1	Increased CO_2 flux above background (Y/N)
Unrest-parameter 2	Increase in T of hot springs and/or fumaroles (Y/N)
Unrest-parameter 3	Changes in H_2O/CO_2 (Y/N)
Unrest-parameter 4	Appearance of new fumaroles and/or hot springs (Y/N)
Unrest-parameter 5	Vegetation die back (Y/N)
Unrest-parameter 6	Appearance of LPs and hybrid EQs (Y/N)
Unrest-parameter 7	Large regional tectonic event (M > 7) (Y/N)
Unrest-parameter 8	Number of VTs [if >10/day for two weeks]
Unrest-parameter 9	Detectable ground deformation (Y/N)
Magmatic unrest-parameter 1	Increase in C/S, or decrease after increase (Y/N)
Magmatic unrest-parameter 2	Detectable SO_2, HCl, HF (Y/N)
Magmatic unrest-parameter 3	Extreme increase in T [>300 °C]
Magmatic unrest-parameter 4	Any VLPs (Y/N)
Magmatic unrest-parameter 5	No. of LPs after significant VT swarms (#/day) (>5–10)
Magmatic unrest-parameter 6	Consistent increase in No. of VTs for 1 month (Y/N)
Magmatic unrest-parameter 7	Deep VTs [>8 km] (#/week) (4–5)
Magmatic unrest-parameter 8	Detectable radial deformation (localized-coherent signal) (Y/N)
Magmatic unrest-parameter 9	Surface deformation (island wide, >6 cm in over 6 months) (Y/N)
Magmatic eruption-parameter 1	Decreasing C/S after increase (Y/N)
Magmatic eruption-parameter 2	Increase in Cl, Br, F content in hot springs/pools (Y/N)
Magmatic eruption-parameter 3	Decrease in H_2O/CO_2 and/or H_2S/SO_2 and/or SO_2/HCl (Y/N)
Magmatic eruption-parameter 4	Phreatic activity (Y/N)
Magmatic eruption-parameter 5	Large thermal anomaly [incandescense] (Y/N)
Magmatic eruption-parameter 6	Landslides in hydrothermal areas (Y/N)
Magmatic eruption-parameter 7	Acceleration of VTs, LPs, hybrids [weekly] (Y/N)
Magmatic eruption-parameter 8	Presence of harmonic tremor (Y/N)
Magmatic eruption-parameter 9	Shallowing of VTs hypocenters in the ediffice or shallow depths [<3 km] (Y/N)
Magmatic eruption-parameter 10	Sudden reversal of activity (Y/N)
Hydrothermal unrest-parameter 1	Anomalous behavior of Boiling Lake [overflow, lower or higher T than usual, no return of lake, etc.] (Y/N)
Hydrothermal unrest-parameter 2	Changes in hydrothermal features (Y/N)
Hydrothermal unrest-parameter 3	Increase in B and/or NH_4 concentration in waters (Y/N)
Hydrothermal unrest-parameter 4	Increase in CH_4/CO_2 (fumaroles) (Y/N)
Hydrothermal unrest-parameter 5	Increase in T of fumaroles (Y/N)
Hydrothermal eruption-parameter 1	Increase in T of fumaroles (fuzzy 120–200 °C)
Hydrothermal eruption-parameter 2	riSe of water level in pools/overflow of BL (Y/N)
Hydrothermal eruption-parameter 3	Increase in extension of fumarolic field (Y/N)
Hydrothermal eruption-parameter 4	Muddy pools (Y/N)

(continued)

Table 7 (continued)

Node–parameter#	Parameter and threshold(s) (Y/N indicates a boolean observation)
Hydrothermal eruption-parameter 5	Boiling/bubbling of pools that previously didn't (Y/N)
Hydrothermal eruption-parameter 6	Inflation of fumarolic field (Y/N)
Hydrothermal eruption-parameter 7	Landslides in hydrothermal areas (Y/N)
Hydrothermal eruption-parameter 8	New/extension of alteration areas (Y/N)

Table 8 Resulting probabilities from real-time application of BET_UNREST at Dominica during VUELCO simulation exercise

		P(U)	P(MU)	P(MEr)	P(HU)	P(HEr)	P(TU)
Phase 1	mean	1	0.26	0.06	0.62	0.42	0.12
	10th prctile	1	0	0	0.05	0.01	0
	50th prctile	1	0.06	0	0.73	0.32	0
	90th prctile	1	0.85	0.22	1	0.95	0.5
Phase 2	mean	1	0.82	0.53	0.13	0.03	0.05
	10th prctile	1	0.29	0.01	0	0	0
	50th prctile	1	1	0.56	0.001	0	0
	90th prctile	1	1	1	0.54	0.06	0.06
Phase 3	mean	1	0.70 (0.24)	0.17 (0.07)	0.08	0.02	0.22
	10th prctile	1	0.09	0	0	0	0
	50th prctile	1	0.87	0.02	0	0	0.08
	90th prctile	1	1	0.68	0.27	0.03	0.80

In bracket estimates of mean values without including HCl anomaly in Phase 3

maps of the spatial probability of vent opening: one for the case of magmatic eruption, and one for hydrothermal eruption (Fig. 4). We believe this could be particularly useful, for example in a volcanic system like Dominica, where there are numerous areas showing hydrothermal activity, thus increasing the uncertainty on the position of a possible phreatic event.

The parameter "detectable SO_2, HCl, HF" created confusion and opened up a scientific discussion. For the sake of transparency, we provide the mean values of $P(MU)$ and $P(MEr)$ including, or not, the HCl anomaly (Table 8). Beyond the scientific implications of this issue, this concern reflected the sensitivity of BET_UNREST to the interpretation of some parameters. When relatively few monitoring parameters are provided, the weight of a single anomaly can be high: this is somehow a measure of the epistemic uncertainty.

4 Discussion and Implications for Unrest Tracking

This chapter presents the need for an updated BET model and tool that is able to account for the non-magmatic nature of some volcanic unrest episodes, which can often go under-estimated, if not totally neglected. The new model (BET_UNREST) and tool (PyBetUnrest) allow the tracking of unrest phases at volcanic systems and enables short-term volcanic forecasts. It has been fully developed within the VUELCO project, during which time it has been applied to some of the project's target volcanoes. In general, when we are able to distinguish magma-on-the-move (Rouwet et al. 2014) from the monitoring observations the new model basically "collapses" to BET_EF (or, better, the assessment of the probabilities related to

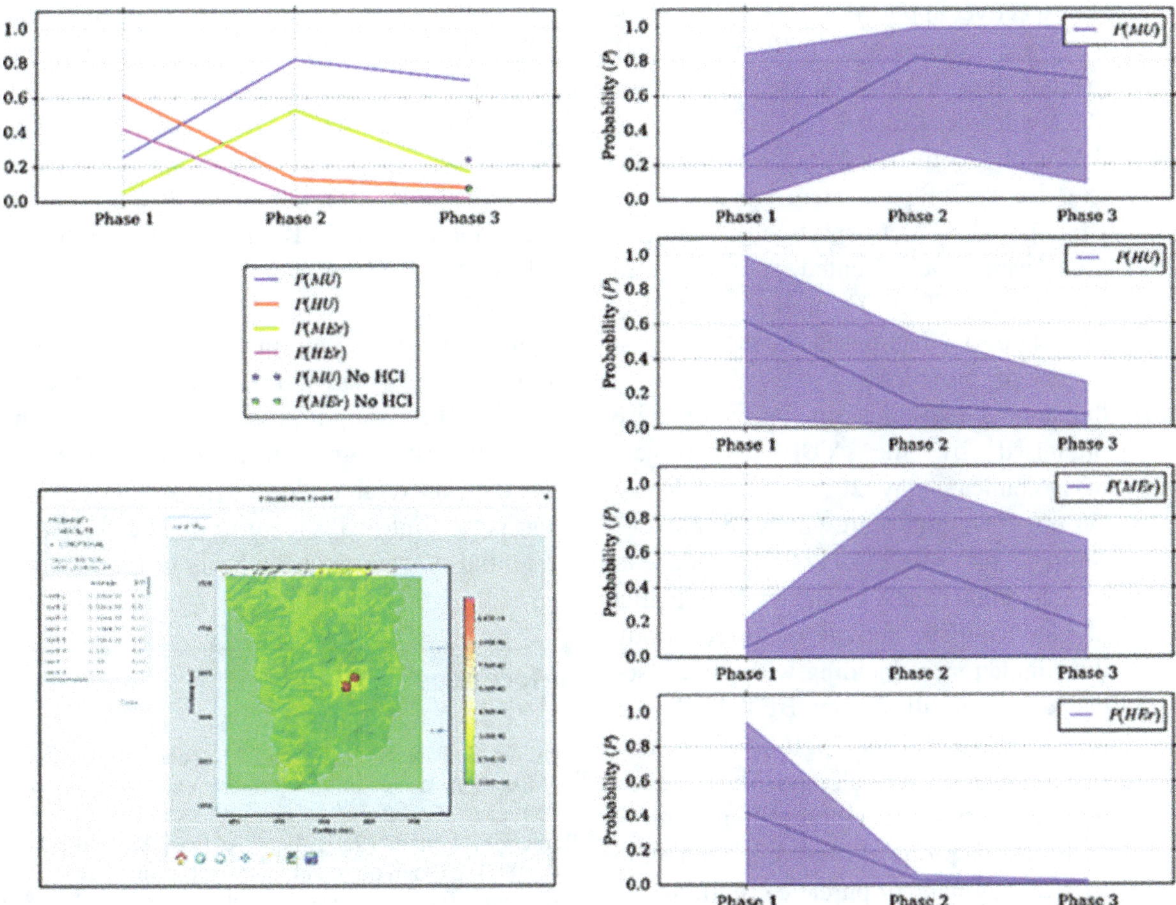

Fig. 4 Average values (*top left*) obtained by BET_UNREST during the three phases of Dominica exercise for *P*(*MU*), *P*(*HU*), *P*(*MEr*) and *P*(*HEr*). Asterisk points are the alternative average values for *P*(*MU*) *and P*(*MEr*) without considering HCl as detectable. On the *right column* the same probabilities are shown together with their confidence interval between 10th and 90th percentiles. On *bottom left*, a snapshot of PyBetUnrest tool shows the spatial probability of vent opening during Phase 1, localising the most probable position of the phreatic eruption

magmatic outcomes provided by the two models coincide). On the other hand, if we are not able to identify a magmatic "active role" in the unrest (from the available monitoring observations), BET_UNREST is still able to provide the probabilities of hazardous events that accompany non-magmatic volcanic unrest, rather than neglecting them. As discussed in Rouwet et al. (2014), a very difficult case is presented by phreatomagmatic eruptions that, sometimes, can occur without any precursors indicating magma movement. This is surely an important limit to overcome which requires further efforts to detect subtle changes in the very short-term (hours to minutes) by improving monitoring techniques.

The chapter illustrates the development and implementation of BET_UNREST model and PyBetUnrest tool through three different applications:

(i) the pure retrospective analysis at Popocatépetl volcano, where there is no compelling need for a hydrothermal branch due to the current magmatic nature of the unrest episodes. Popocatépetl has remained in unrest from December 1994 to present and, for this application, BET_UNREST and PyBetUnrest were run using the UNAM Data Base for the learning period 1997–2012, with a retrospective application aiming to forecast major eruptions (column heights greater than 8 km) for the April–June 2013 volcanic activity.

(ii) the application based on a simulation exercise at Cotopaxi. Here we tested the BET_UNREST retrospectively, but, this time, using the invented data provided during the VUELCO simulation exercise, in addition to data based on the real past history of the volcano.

(iii) the almost real-time simulation exercise organised by the VUELCO project in Dominica (May 2015). The volcanic system of Dominica presents a "prototype" setting for BET_UNREST due to its hydrothermal character. Phreatic/phreatomagmatic activity occurred during the simulation, coinciding with high associated probabilities from BET_UNREST (the average values $P(HU) = 0.73$ and $P(HEr) = 0.32$). We also positively tested the feasibility of providing different maps of the spatial probability of vent opening in case of magmatic or phreatic eruption.

As mentioned in previous sections, we implemented the BET_UNREST model into PyBetUnrest software tool using a graphical user interface aiming to provide a fast, open and user-friendly tool, which extends the usage of BET_UNREST to volcanologists with different expertise. The PyBetUnrest tool reached a mature and usable version during the Dominica simulation and its first stable release has been uploaded to Vhub cyber-infrastructure.

With these exercises we strongly believe we have brought BET a step closer to a full and proper implementation during a crisis situation. The PyBetUnrest tool eventually worked as expected, but it is important to take advantage of the lessons learned during these applications and pursue more tests that will improve its design and prove its usefulness in real-case scenarios.

As a final comment, we would like to remark that, as with any other event tree model (e.g. BET models by Marzocchi et al. 2004, 2008, 2010; HASSET model by Sobradelo et al. 2013),

one can always apply and "populate" the BET_UNREST model in any "volcanic" circumstance. The uncertainty on the results provided by BET_UNREST, and consequently their practical use, will however be strongly dependent on the available information and data used to set up the models rules. If only a few pieces of evidence are available, the models results will be characterised by a large uncertainty, and thus might be not very helpful for decision-makers. As more and more knowledge is gathered, BET_UNREST output probabilities will become more attractive from a practical point of view, since their uncertainty will be increasingly small. This is an intrinsic feature of the Bayesian inferential procedure at the basis of the model.

References

Aguilera E, Pareschi MT, Rosi M, Zanchetta G (2004) Risk from Lahars in the Northern Valleys of Cotopaxi Volcano (Ecuador). Nat Hazards 33:161–189

Barberi F, Coltelli M, Frullani A, Rosi M, Almeida E (1995) Chronology and dispersal characteristics of recently (last 5000 years) erupted tephra of Cotopaxi (Ecuador): implications for long-term eruptive forecasting. J Volcanol Geotherm Res 69:217–239

Biass S, Bonadonna C (2011) A quantitative uncertainty assessment of eruptive parameters derived from tephra deposits: the example of two large eruptions of Cotopaxi volcano, Ecuador. Bull Volcanol 73:73–90. doi:10.1007/s00445-010-0404-5

Constantinescu R, Rouwet D, Gottsmann J, Sandri L, Tonini R (2015) Tracking volcanic unrest at Cotopaxi, Ecuador: the use of BET_EF tool during an unrest simulation exercise. Geophys Res Abs, 17-EGU 2015–2251

Constantinescu R, Robertson R, Lindsay JM, Tonini R, Sandri L, Rouwet D, Patrick Smith P, Stewart R (2016) Application of the probabilistic model BET_UNREST during a volcanic unrest simulation exercise in Dominica, Lesser Antilles, Geochem Geophys Geosyst 17:4438–4456, doi:10.1002/2016GC006485

De la Cruz-Reyna S, Tilling RI (2008) Scientific and public responses to the ongoing volcanic crisis at Popocatépetl Volcano, Mexico: importance of an effective hazards-warning system. J Volcanol Geotherm Res 170:121–134

Fournier N, Witham F, Moureau-Fournier M, Bardou L (2009) Boiling Lake of Dominica, West Indies: high-temperature volcanic crater lake dynamics. J Geophys Res 114(B02203). doi:10.1029/2008JB005773

Garcia-Aristazabal A (2010) Analysis of eruptive and seismic sequences to improve the short- and long-term eruption forecasting. PhD Università deli Studi di Bologna, pp 167

Hall M, Mothes P (2008) The rhyolitic–andesitic eruptive history of Cotopaxi volcano. Ecuador Bull Volcanol 70(6):675–702

Joseph EP, Fournier N, Lindsay JM, Fischer TP (2011) Gas and water geochemistry of geothermal systems in Dominica, Lesser Antilles island arc. J Volcanol Geotherm Res 206:1–14. doi:10.1016/j.jvolgeores.2011.06.007

Marzocchi W, Bebbington M (2012) Probabilistic eruption forecasting at short and long time scales. Bull Volcanol 74:1777–1805. doi:10.1007/s00445-012-0633-x

Marzocchi W, Woo G (2007) Probabilistic eruption forecasting and the call for an evacuation. Geophys Res Lett 34:L22310. doi:10.1029/2007GL031922

Marzocchi W, Woo G (2009) Principles of volcanic risk metrics: theory and the case study of Mount Vesuvius and Campi Flegrei. Italy J Geophys Res 114:B03213. doi:10.1029/2008JB005908

Marzocchi W, Sandri L, Gasparini P, Newhall CG, Boschi E (2004) Quantifying probabilities of volcanic events: the example of volcanic hazard at Mount Vesuvius. J Geophys Res 109:B11201 doi:10.1029/2004JB003155

Marzocchi W, Sandri L, Selva J (2008) BET_EF: a probabilistic tool for long- and short-term eruption forecasting. Bull Volcanol 70:623–632

Marzocchi W, Sandri L, Selva J (2010) BET_VH: a probabilistic tool for long-term volcanic hazard assessment. Bull Volcanol 72:705–716

Mendoza-Rosas AT, De la Cruz-Reyna S (2008) A statistical method linking geological and historical eruption time series for volcanic hazard estimations: applications to active Polygenetic volcanoes. J Volcanol Geotherm Res. doi:10.1016/j.jvolgeores.2008.04.005

Molina I, Kumagai H, García-Aristizábal A, Nakano M, Mothes P (2008) Source process of very-long-period events accompanying long-period signals at Cotopaxi Volcano, Ecuador. J Volcanol Geotherm Res 176:119–133

Newhall CG, Hoblitt RP (2002) Constructing event trees for volcanic crises. Bull Volcanol 64:3–20. doi:10.1007/s004450100173

Phillipson G, Sobradelo R, Gottsmann J (2013) Global volcanic unrest in the 21st century: an analysis of the first decade. J Volcanol Geotherm Res 264:183–196

Pistolesi M, Cioni R, Rosi M, Cashman KV, Rossotti A, Aguilera E (2013) Evidence for lahar-triggering mechanisms in complex stratigraphic sequences: the post-twelfth century eruptive activity of Cotopaxi Volcano. Ecuador Bull Volcanol 75:698. doi:10.1007/s00445-013-0698-1

Rouwet D, Sandri L, Marzocchi W, Gottsmann J, Selva J, Tonini R, Papale P (2014) Recognizing and tracking hazards related to non-magmatic unrest: a review. J Appl Volcanol 3:17. doi:10.1186/s13617-014-0017-3

Rouwet D, Hidalgo S, Joseph EP, González-Ilama G (2017) Fluid geochemistry and volcanic unrest: dissolving the haze in time and space. In: Gottsmann J, Neuberg, J, Scheu B (eds) Volcanic Unrest: from Science to Society—IAVCEI Advances in Volcanology, Springer, Berlin

Selva J, Costa A, Sandri L, Macedonio G, Marzocchi W (2014) Probabilistic short-term volcanic hazard in phases of unrest: a case study for tephra fallout. J Geophys Res 119:8805–8826

Simkin T, Siebert L (1994) Volcanoes of the world, 2nd edn. Geoscience Press for the Smithsonian Institution, Tucson, p 349

Sobradelo R, Bartolini S, Marti J (2013) HASSET: a probability event tree tool to evaluate future volcanic scenarios using Bayesian inference. Bull Volcanol 76:770. doi:10.1007/s00445-013-0770-x

Tonini R, Sandri L, Thompson MA (2015) PyBetVH: a Python tool for probabilistic volcanic hazard assessment and for generation of Bayesian hazard curves and maps. Comput Geosci 79:38–46

Tonini R, Sandri L, Rouwet D, Caudron C, Marzocchi W, Suparjan (2016) A new Bayesian Event Tree tool to track and quantify unrest and its application to Kawah Ijen volcano. Geochem Ge-ophys Geosyst 17:2539–2555, doi:10.1002/2016GC006327

Woo G (2008) Probabilistic criteria for volcano evacuation decisions. Nat Hazards 87–97. doi:10.1007/s11069-007-9171-9

Groundwater flow and volcanic unrest

Alia Jasim, Brioch Hemmings, Klaus Mayer and Bettina Scheu

Abstract

Hydrology around active volcanoes is strongly controlled by the interaction between groundwater, and the fluids, dissolved elements and heat associated with magmatic intrusion. The chemical and mechanical processes associated with magmatic unrest can result in observable changes in the hydrothermal system. Consequently, observations of chemical and physical hydrothermal variations may provide insights into the state of volcanic activity. Additionally, the interaction between hydrological and volcanic systems leads to the presence of high-temperature, pressurised, and often acidic fluids, which add to, and intensify, the volcanic hazard. In the following chapter we present the major components of, and controls on, magmatic hydrothermal systems focusing on the mutual perturbation between the groundwater flow system and the volcanic system. We explore how these conditions can be modified by volcanic unrest and we identify feedbacks between dynamic hydrothermal behaviour and on-going unrest. The interaction between these systems, and therefore the associated monitoring signals, are the result of complex groundwater-volcano coupling within multi-phase flow system in evolving lithologies. Nonetheless, detailed monitoring of hydrothermal and hydrological behaviour can provide insights into unrest and the evolution of hazards at restless volcanoes.

Keywords

Hydrothermal system · Groundwater Fluid flow · Permeability · Unrest-monitoring

1 Resumen

Brevemente resumimos nuestra comprensión de los sistemas magmáticos hidrotermales y discutimos las mayores incógnitas y sus implicaciones en el vigilancia volcánica. También proveemos directrices adicionales para la recolección de datos a usarse en la calibración de la variabilidad del sistema de aguas subterráneas, alrededor de volcanes activos, como un paso crucial para desacoplar las señales magmáticas de las puramente hidrotermales.

A. Jasim (✉)
University of Bristol, Earth Science School, Wills Memorial Building, Queen's Road, Clifton BS8 1RJ, UK

K. Mayer · B. Scheu
Department of Earth and Environmental Sciences, Ludwig-Maximilians-Universität München (LMU), Theresienstrasse 41/III, 80333 Munich, Germany

B. Hemmings
GNS Science, 1 Fairway Drive, Lower Hutt, New Zealand

Advs in Volcanology (2019) 83–99
DOI 10.1007/11157_2018_33

Published Online: 23 May 2018

La interacción entre los sistemas hidrológico y volcánico es un elemento importante durante reactivación volcánica. Los cambios en el comportamiento hidrológico de un volcán activo, como la elevación del nivel del agua subterránea, la descarga de manantiales, los cambios de temperatura y de la química, pueden ser indicadores preliminares de evolución de la actividad volcánica. Las interacciones hidrológicas pueden también alterar y aumentar el peligro volcánico existente. Las interacciones físicas y químicas entre la roca encajante y los diferentes tipos de fluido pueden modificar los caminos de desgasificación, generando distribuciones de presión dinámicas dentro del edificio volcánico. Aún los procesos lentos, como el desarrollo creciente de zonas de alteración permanentes, pueden manifestarse como un peligro dinámico asociado con una reactivación continua o futura, ya que las rocas altamente cristalinas son hidrotérmicamente alteradas produciendo arcillas débiles secundarias. Discutimos los principales parámetros que controlan las reacciones y sus efectos en la distribución de la alteración en ambientes volcánicos.

Debido a la introducción del calor de la fuente en el sistema del agua saturada, se presentan peligros adicionales. Esto frecuentemente conlleva a explosiones freáticas y freato-magmáticas. La presencia de paquetes de gases bajo la superficie además incrementa este peligro. El balance entre el ingreso de agua freca fría, la desgasificación y la disipación de calor, está críticamente relacionado con la abilidad del sistema para transmitir fluidos, el mismo que evoluciona en función tanto de los procesos químicos (ej., las reacciones de dilución/precipitación mineral) como físicos (ej., fracturamiento y compactación de la roca), produciendo así propiedades hidrológicas de la roca fuertemente dependientes de la escala (ej., porosidad, permeabilidad y conductividad térmica).

En resumen, las señales físicas y químicas, o la perturbación hidrológica asociada con la reactivación magmática, son complejas y dependientes del sitio. Las contribuciones de los diferentes componentes del fluido, y sus interacciones con caminos de flujo existentes, pueden determinar cómo un sistema evoluciona en períodos de calma. Esta evolución controla la posible respuesta a la perturbación termodinámica y química asociada con la iniciación de la reactivación volcánica. Dadas las intrincadas retroalimentaciones entre el magma, la hidrología, y los cambios repentinos de los sistemas involucrados, únicamente mediciones de alta frecuencia (de horas a semanas) de la temperatura, pH, conductividad eléctrica del agua, profundidad del nivel del agua subterránea, del contenido de REE (Elementos de Tierras Raras), RFEs (Elementos de Formación de Rocas) y gas disuelto, conjuntamente con mediciones geofísicas, pueden aclarar la evolución del sistema magmático, la apertura/cierre de fracturas y la dinámica estacional del agua subterránea.

2 Introduction

Much of the research relating to the interaction between hydrological and volcanic systems has focused on the role of hydrothermal systems in the development of economic mineral deposits. hydrothermal systems have formed vast ore-deposits around the world, most of them clearly result from the interaction between magmatic and meteoric fluids (Hedenquist and Lowenstern 1994).

The interaction between hydrological and volcanic systems is an important element in volcanic unrest. Changes in hydrological behaviour, such as water table elevation, spring discharge, temperature and chemistry, at an active volcano can provide early indications of changes in volcanic activity. Hydrological interactions can also alter and augment the existing volcanic hazard. Chemical and physical interactions between host rocks and different fluid types can modify fluid degassing pathways, generating dynamic pressure distributions within a volcanic edifice. Additional hazards are also presented by the introduction of a heat source into a water saturated system, this frequently results in dangerous phreatic and phreatomagmatic explosions. Understanding the controls on hydrological and hydrothermal behaviour in volcanic settings is essential for understanding the array of hazards presented by volcanic unrest. Continued development of this understanding is also providing new volcano monitoring opportunities. Despite the clear relevance and importance of hydrological and volcanic interactions in relation to volcanic unrest, the dynamics of this interaction remain poorly constrained.

3 Hydrothermal System

Although insulated or distal, cool groundwater aquifers can respond to volcanic perturbation, the clearest manifestation of volcanic and hydrological interactions is a hydrothermal system. Fumaroles, often visible on active volcanoes, represent the surface expression of this hydrothermal system.

A magmatic hydrothermal system is composed of three main elements: a **host rock** (or reservoir), which contains a circulating **fluid**, set in motion by an igneous **heat source** (Fig. 1). While the difference in relief between stratovolcanoes and calderas can lead to contrasting hydrological systems, the lower limit of any hydrological system is commonly defined as the brittle-ductile transition zone. Within this zone, fluid pressures transition from hydrostatic to lithostatic as rock permeability becomes severely reduced (Fournier 1999). When this region is subjected to high strain rates, fracturing may occur, leading to episodic influxes of mass and heat to the hydrothermal system (Bodnar et al. 2007). While agreement exists on the definition of lower limit of a volcanic hydrological system, the same is not true for the upper limit.

We consider the water table as the upper limit of the hydrological system. However, the earth's surface could equally be considered part of the system. This adds the further complexity of flow within the unsaturated (vadose) zone (Hemmings et al. 2015a). Furthermore, the upper limit of the hydrological system closely depends on the precipitation regime. Precipitation is a function of geography, including both latitude and elevation. Together with surface processes it determines the recharge dynamics of the aquifer and

the depth and fluctuation of the water table. A number of studies suggest a correlation between the fluctuation of the water table and elevated seismicity. Both the reduction in effective stress due to a seasonal increase in hydrostatic pore pressure, and the snow unloading, lead to a seasonal peak of seismicity (Saar and Manga 2003; Christiansen et al. 2005). Furthermore, Mason et al. (2004) identify seasonal peaks in the eruption rate of volcanoes, which may be due to the load/unload seasonal stress cycle imposed by the hydrological cycle.

The definition of water table implies a water saturated medium below it. However, crater lakes (Fournier et al. 2009) and caldera settings (Bruno et al. 2007; Jasim et al. 2015) often have portion of the water table sustained by a two phases system (liquid and gas). Similarly, the condensation of magmatic gases (primarily vapour) often feeds the groundwater reservoir (Chiodini et al. 2001). While the location of the water table is important, the conventional definition does not really apply in volcanic settings, especially in high-relief stratovolcanoes, in which it is often unclear to what extent the edifice is water saturated. Specifically, the local water table may differ from the regional water table, with highly dynamic elevation changes controlled by both meteoric and volcanic processes.

Many studies suggest the presence of high elevation springs, however it is not clear whether they are fed by the regional water table or by perched saturated layers high up on the cone (Cabrera and Custodio 2004; Custodio 2007; Cruz and Oliveira Silva 2001; Hemmings et al. 2015a; Ingebritsen and Scholl 1993; Join et al. 2005; Peterson 1972). Access to wells on the flank of volcanoes and geophysical imaging methods can help resolve the hydrogeology behind such springs (Finn et al. 1987, 2001; Aizawa et al. 2008). However, in either scenario, the development of saturated flow units at high elevation has a particular relevance to forecasting volcanic hazards such as lahar, landslides and flank collapses, which can be sudden, unpredicted and deadly. Such mass wasting events all involve the displacement of material from the

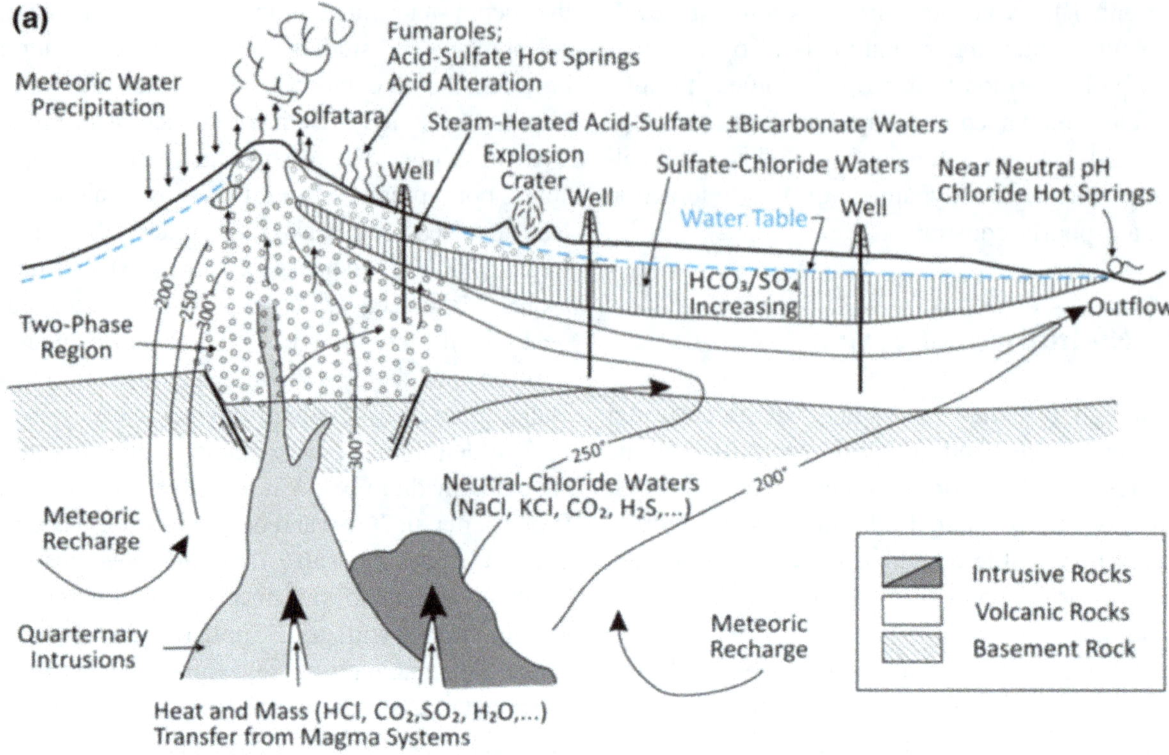

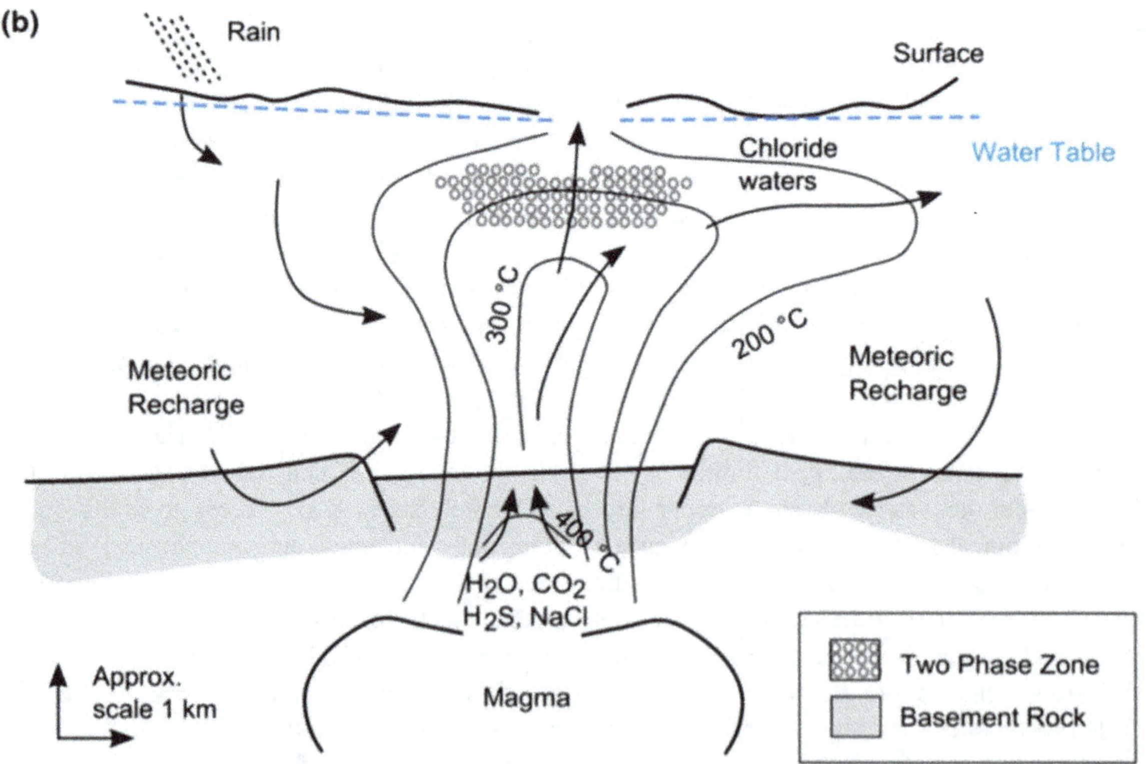

Fig. 1 Conceptual model of a hydrothermal system in, **a** high-relief volcano (modified after Goff and Janik 2000) and **b** caldera (modified after Kuhn 2004)

upper part of the volcano to the surrounding valleys. They can be triggered by gravitational instability (often in response to heavy rainfall), increase in pore pressure, reduction of rock strength, volcano-tectonic earthquake or intrusion of magma. The depressurisation induced by mass movement on the volcanic edifice may also result in the sudden reactivation of the magmatic system.

The waters circulating within volcanic systems are often high-temperature, sometimes supercritical fluids. They interact with the host rock through chemical reactions that result in hydrothermal alteration. The evolution and dynamics of hydrological and volcanic interactions are strongly controlled by fluid flow. This, in-turn, is a function of the pressure, temperature, fluid composition, and, critically, the system's ability to transmit fluids.

3.1 Fluid flow

Laminar flow through saturated porous media is described by Darcy's Law (Eq. 1),

$$q = -\frac{k}{\mu}\frac{dp}{dl} \qquad (1)$$

where q is specific discharge (m/s), k is permeability of the porous media (m^2), μ is fluid viscosity (Pa s), and dP/dl is the hydraulic head, or pressure, gradient (Pa/m) along length l (m). The fluid viscosity is usually approximated as that of water (liquid or vapour). However, the combined effect of topography and the presence of a deep source of heat and fluids likely produce vast unsaturated (or two-phase) portions within the volcano. Dissolved air and gas near the surface, the phase transition from water (liquid) to vapour due to the temperature gradient, and the decompression of upwelling fluids are some of the processes that produce two-phase (liquid and gas) fluid flow regions within a volcanic system. To extend the Darcy's equation to two phase flow we define the liquid saturation (S_w) as the fraction of a representative bulk volume of the porous medium filled by water and, similarly, the

gas saturation (S_g) as the fraction of a representative bulk volume of the porous medium filled by the gas phases, such that (Eq. 2)

$$S_g + S_w = 1, \qquad (2)$$

The capillary pressure P_c (in Pa) is due to the pressure difference between the two phases (Eq. 3), hence

$$P_c = P_g - P_w \qquad (3)$$

and is a unique function of water saturation (S_w). Finally, due to the competing flow of the two phases, the relative permeability (kr) is less than or equal to the single phase (usually water) permeability, k (m^2), of the medium. We define the relative permeability of the gas phase (kr_g) and the relative permeability of the liquid phase (kr_w) as (Eqs. 4 and 5)

$$kr_g = \frac{k_g}{k} \qquad (4)$$

$$kr_w = \frac{k_w}{k} \qquad (5)$$

where k_g and k_w are the effective permeabilities for each of the two fluids. Again the relative permeabilities are assumed to be a unique function of water saturation (S_w). Hence, we define the flow velocity vector (m/s) of the gas phase along a direction D (m) as (Eq. 6)

$$v_g = -k\frac{kr_g}{\mu_g}\left(\nabla P_g - \rho_g g \nabla D\right) \qquad (6)$$

and similarly the velocity (m/s) of the water (Eq. 7)

$$v_w = -k\frac{kr_w}{\mu_w}\left(\nabla P_w - \rho_w g \nabla D\right) \qquad (7)$$

where μ and ρ are respectively the viscosity (Pa s) and the density (kg/m^3) of the gas (g) and liquid (w) and g is the acceleration (m/s^2) due to gravity. Similarly, for non-laminar flow further terms (e.g., Forchheimer term) can be added to Darcy's equation of fluid flow to consider the inertial effects due to turbulence.

3.1.1 Permeability and Porosity

Permeability and porosity are the primary parameters controlling flow and storage of fluids in the subsurface (Manning and Ingebritsen 1999). The permeability also controls the heat regime of the hydrothermal system: high permeabilities ($\geq 10^{-14}$ m^2) favour advective transfer of heat away from the igneous source, resulting in low-temperature vapour-dominated systems. Contrastingly, low-permeabilities ($<10^{-16}$ m^2) favour slow heat conduction, also producing low-temperature systems. The hottest hydrothermal plumes reside in intermediate permeabilities, around 10^{-15} m^2 (Hayba and Ingebritsen 1997). Whilst permeability (k) is an important parameter, it is often one of the least well constrained. It can vary over 17 orders of magnitude, from $\sim 10^{-20}$ m^2 in intact crystalline rocks to $\sim 10^{-9}$ m^2 in porous and fractured basalt (Table 1).

Porosity (ϕ), the ratio of voids over total volume (voids and solid) in a given material, defines the storage capacity of the rocks; it also affects permeability and effective thermal conductivity. In many porous and fractured media, there is a positive correlation between ϕ and k, as permeability is simply the interconnected pore network. Clays and volcanic tuff are unusual in that they can have high porosity values but low permeabilities, at least in part due to their very small particle sizes, propensity to bridge pores and form aggregates (Neuzil 1994). Permeability in such lithology is greatly enhanced by the presence of discontinuities such as fissures, joints, shears and faults that can connect otherwise isolated pores.

Porosity and permeability of volcanic units are primarily controlled by the type of volcanic product (e.g., lava, pyroclastic density current, ash fall) and depositional environment. Inherent heterogeneities between deposits can be enhanced by subsequent compaction, fracturing and chemical alteration. The resulting hydrogeology is complex; hydrological rock properties (porosity, permeability and thermal conductivity) are strongly scale dependent, particularly when fluid flow is focussed along high permeability channels or fractures, as is common in volcanic settings. Such flow pathways themselves are an active component of a dynamic system. Their ability to transmit fluids and therefore the role they play in hydrothermal circulation can be modified by physical changes - the opening or closing of fractures in response to stress-field changes - and chemical alteration, which can both enhance and obstruct fluid flow, as discussed in the next section.

3.2 Chemical Reactions

Chemical alteration is an important process within a hydrothermal system. Dissolution can reduce cohesion and weaken a volcanic edifice. This can lead to catastrophic flank collapses and debris avalanches (Reid 2004). Such events can depressurise the magmatic system and trigger an eruption. Conversely, chemical precipitation and deposition processes can cause plugging in areas of intense mineralisation, this can promote pressurisation, which can also lead to flank collapse, as pore pressures increase within the edifice. Such pressurisation can also generate violent steam-driven explosions (Ingebritsen et al. 2010). Chemical dissolution and precipitation processes also alter the host-rock permeability. These processes are sensitive to the thermodynamic conditions and the proportions of different fluid components (Pirajno 2010). Feedbacks between physical and chemical flow behaviour within a hydrothermal system can modify the physical and chemical characteristics of its surface expression - hydrothermal fluid and fumarolic discharge. Therefore, monitoring the hydrothermal discharge can provide clues about the state of volcanic unrest and can even be used to help predict volcanic eruptions (Cronan et al. 1997).

To maximise the value of chemical analysis of hydrothermal discharge as a volcanic monitoring tool, and to fully understand the hazard presented by hydrological and magmatic interactions it is necessary to quantify the rates, spatial distribution and physical effects of chemical alteration within a hydrothermal system.

Table 1 Measured permeability and porosity ranges for various rock types

Rock type	Permeability (m^2)		Porosity (%)
	Min	Max	Range
Unconsolidated rocks			
Gravel	10^{-10}	10^{-7}	25–40
Clean sand	10^{-13}	10^{-9}	5–50
Silty sand	10^{-14}	10^{-10}	
Silt, loess	10^{-16}	10^{-12}	35–50
Unweathered clay[a]	10^{-20}	10^{-15}	40–80
Consolidated rocks			
Shale	10^{-20}	10^{-16}	0–10
Unfractured metamorphic and igneous	10^{-20}	10^{-17}	0–5
Sandstone	10^{-17}	10^{-13}	5–35
Limestone and dolomite	10^{-16}	10^{-13}	0–20
Fractured igneous and metamorphic	10^{-15}	10^{-13}	0–10
Permeable basalt	10^{-14}	10^{-9}	0–25
Karst limestone	10^{-13}	10^{-9}	5–50
Fractured basalt			5–50
Basalt near surface[b]	10^{-14}	10^{-12}	
Basalt at 1 km depth[b]	10^{-18}	10^{-10}	
Andesite[c]	10^{-20}	10^{-18}	0.2–0.3
Thermometamorphic[d]	10^{-18}	10^{-14}	2–17
Campi Flegrei trachy-phonolite			
Tuff, surface[e]	10^{-16}	10^{-15}	48–52
Tuff, depth[e]	10^{-17}	10^{-15}	19–52
Chaotic tuff/tuffites < 1 km depth[d]	10^{-18}	10^{-14}	6–40
Chaotic tuff/tuffites > 1 km depth[d]	10^{-18}	10^{-14}	5–36
Tuffites (HT altered)[f]	10^{-16}	10^{-16}	0.05–0.07
Lava[d]	10^{-18}	10^{-14}	7–25
Montserrat andesite			
Pyroclastic flow deposit[g]	10^{-18}	10^{-13}	
Lava[g]	10^{-17}	10^{-13}	
Lahar deposit[g]	10^{-14}	10^{-13}	

Sedimentary rocks are given for comparison. Hydrothermal systems in limestone usually develop as skarn deposit. Data from Freeze and Cherry (1979). [a]Neuzil (1994), [b]Ingebritsen et al. (2006), [c]Petrov et al. (2005), [d]Piochi et al. (2014), [e]Peluso and Arienzo (2007), [f]Giberti et al. (2006), [g]Hemmings et al. (2015a)

3.2.1 Reaction Controlling Parameters

Many factors may influence hydrothermal alteration, including temperature, pressure, rock type, fluid flux, fluid composition, and time. The relative importance of each of these has been much discussed in the literature (Gifkins et al. 2005; Pirajno 2010) and appears to vary between different case studies.

3.2.2 Fluid Composition

The dominant form of chemical alteration (dissolution and/or precipitation) is principally a function of the composition of the circulating hydrothermal fluids. Meteoric water and seawater are the main sources of fluid in hydrothermal systems with an additional and dynamic contribution of magmatic fluids. The composition of hydrothermal fluid critically affects the mineral-fluid equilibria and therefore the concentration of rock forming elements (RFEs) such as Silica (SiO_2), Sodium (Na), Potassium (K), Calcium (Ca) and Magnesium (Mg). The mineral-fluid equilibria and solubility of RFEs, as well as sulphate (SO_4^{2-}), chloride (Cl^-) and bicarbonate (HCO_{3-}), is affected by temperature, pressure and water/rock (W/R) ratio as well as the composition of the host rock. These factors also affect kinetic rate of chemical alteration. In addition, the relative mobility of elements depends on the characteristics of fluid flow, the number of phases (liquid and gas) and chemical condition along the flow path including pH, redox condition, sulfidation state, availability of ligands.

The chloride-sulphate-bicarbonate ternary diagram by Giggenbach and Soto (1992) provides a tool to identify water end-members (Fig. 2). It represents graphically the classification of thermal water suggested by Ellis and Mahon (1977) based on major ions, which identifies (i) neutral alkali-chloride waters which

result from extensive interaction with the reservoir rocks and may cause silica or carbonate supersaturation at surface condition; (ii) acid-sulphate waters which result from the condensation of volcanic gases into the shallower groundwater system and are often depleted in alkali and Cl but enriched in metals; (iii) bicarbonate waters which usually show thermodynamic equilibrium with the reservoir rock and are common at the edge of magmatic-hydrothermal systems (Ellis and Mahon 1977; Giggenbach and Soto 1992; Goff and Janik 2000).

3.2.3 Acidity of Hydrothermal Fluids

Upper regions of hydrothermal systems are often characterized by steam-heated fumarolic alteration due to the presence of acidic, sulphate-rich fluids (Rye 2005). These fluids may cause leaching of the host rocks, resulting in an increase in both rock porosity and permeability. Eventually extreme acidic fluids (pH < 2) generate the development of vuggy silica and thereby facilitate faster gas escape in the shallow zone (Mayer et al. 2016 and references therein). In the presence of abundant sulphate ions and Al-rich host rocks, within a lesser acidic environment (pH > 2), the formation of alunite dominates (alunitic alteration, Pirajno 2010).

Fig. 2 Classification ternary diagram of thermal waters from Giggenbach and Soto (1992) based on the content of major solutes: Cl^-, SO_4^{2-} and HCO_{3-}. The vertexes represent the alkali-chloride, sulphate and bicarbonate water type end members, whilst the arrows show major differentiation processes

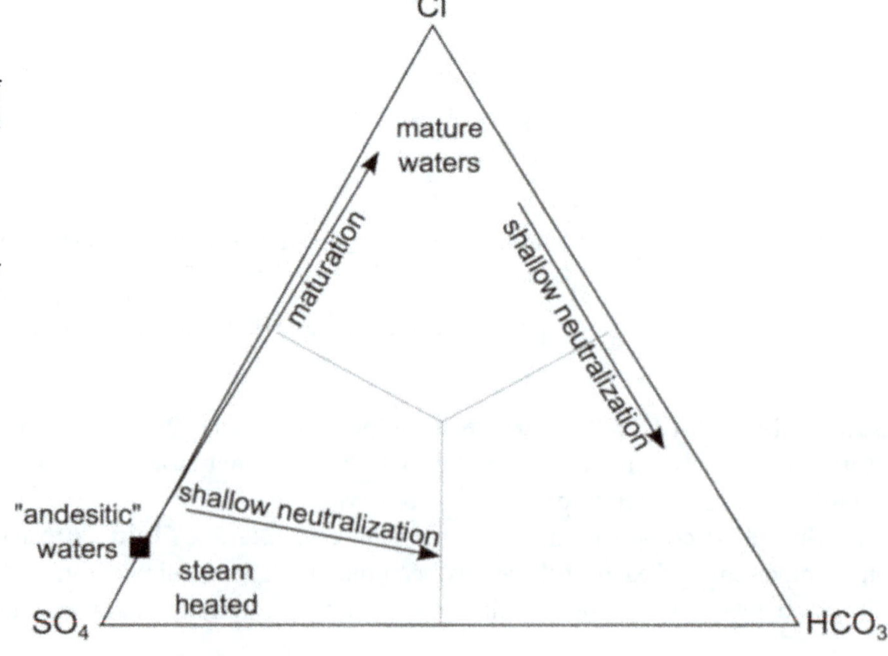

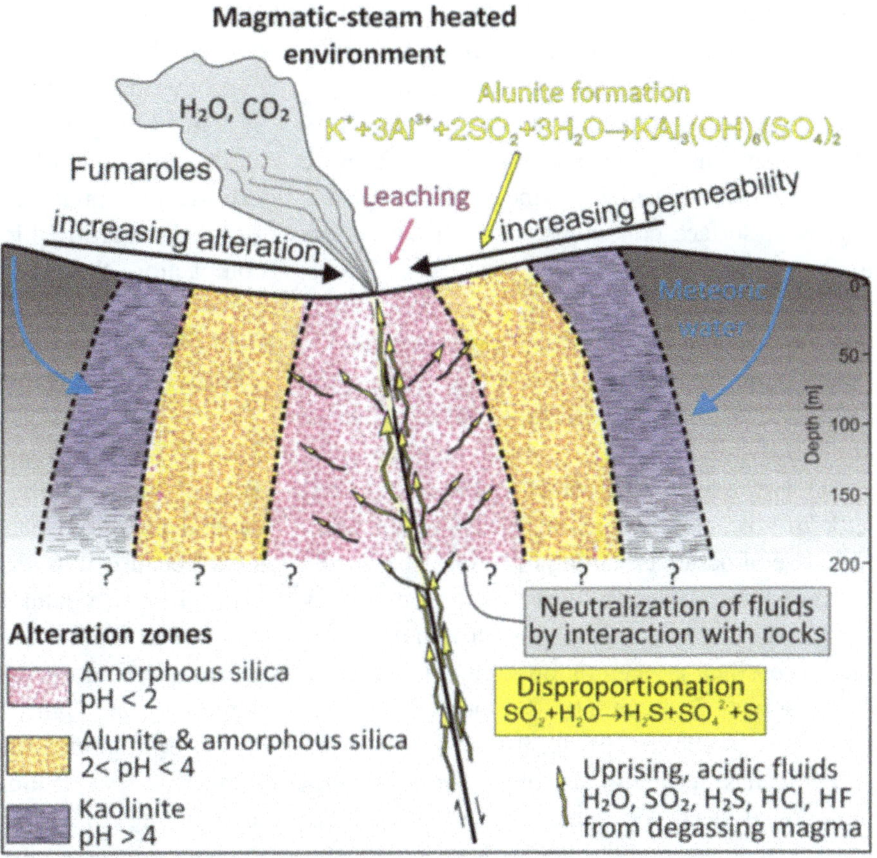

Fig. 3 Conceptual model for the formation of near surface high-sulfidation alteration. pH and composition control the development of alteration zones with increasing distance to the main fumarolic conduit. A highly permeable, acidic core characterized by amorphous silica is laterally replaced by a zone of alunite and amorphous silica. Successive neutralization of the fluids promotes the formation of kaolinite. Rock permeability as well as the degree of alteration increase toward the center of the hydrothermal activity (Mayer et al. 2016)

Often the fluids undergo progressive neutralization as they flow away from the degassing vents. This results in a sequence of alteration facies (Fig. 3) from silicic to advanced argillic to intermediate argillic (Fulignati et al. 1998).

In similar environments the distribution of kaolinite and alunite may also be affected by the presence of groundwater. Alunite preferentially forms at or above the groundwater table where atmospheric oxygen could oxidize H_2S to H_2SO_4, which is required for the formation of alunite (Mutlu et al. 2005).

3.2.4 Water/Rock Ratio

The amount of water and the rock surface area available for reactions are two of the primary controls on alteration. Therefore, the Water/Rock (W/R) ratio and rock porosity and permeability will determine the type and extent of alteration. W/R ratios range between 0.1 and 0.4 (Henley and Ellis 1983) while porosity can vary from 0 to ∼80% (Freeze and Cherry 1979; Neuzil 1994). Static systems, with low porosity and low W/R ratios, are termed "closed systems", or "rock dominated systems". In this case, the secondary minerals depend nearly entirely on temperature and are of similar chemical composition to the original rocks, although possibly in a hydrated form (typical of propylitic alteration, see below). The alteration minerals often resemble those due to metamorphism (Giggenbach 1984).

Most hydrothermal systems are, however, "open systems", or "liquid dominated systems", which are characterised by high W/R ratios. In these cases, the fluid composition has a greater importance. The minerals that precipitate in open systems are the result of alteration by mobile fluids of constantly changing composition. In these systems permeability is highly influential and systematic spatial patterns of alteration zoning are common.

3.2.5 Rock Type

Many studies support the assumptions that the chemical and mineralogical composition of the original rock will change the composition of the equilibrium solution and therefore the rate of individual dissolution/precipitation reactions (e.g. Pirajno 2010). Consequently, various investigations have attempted to assess the most easily altered minerals in the host rocks (e.g. Browne 1984). Glass, followed by olivine are the least stable phases at surface conditions, as such basalts are likely to alter more rapidly than felsic rocks. Numerous studies of basalt dissolution have shown that the presence of glass can increase dissolution rates (Wolff-Boenisch et al. 2006; Berger et al. 1994; Stefansson and Gislason 2001; Zakharova et al. 2007; Hu et al. 2010; Gudbrandsson et al. 2011). However, experimental results on the alteration of volcanic materials are biased by the dominant use of basalt as starting material.

Even though basaltic glass dissolves relatively rapidly, basalts are still low-silica magmas, and therefore silica concentrations remain higher in felsic rocks. Browne (1978) showed that, at temperatures above 280 °C, the host rock composition has a negligible effect on alteration minerals. Indeed, he gathered evidence of the same stable alteration assemblages in basalts, sandstones, rhyolites and andesites. Most accessible hydrothermal systems, however, are at temperatures below 280 °C and in such systems Browne (1978) reports high-silica zeolites in rhyolitic volcanoes, and low-silica zeolites in basaltic and andesitic systems.

3.2.6 Pressure

Pressure generally has a secondary role in hydrothermal alteration (Robb 2005). An important exception is the role of pressure in controlling boiling in hydrothermal environments. Boiling at depth leads to low-salinity vapour and high-salinity brine. This phase separation is responsible for transport and deposition of elements that are key to ore mineralization (Henley and Berger 2013). In the upper ~ 400 m, pressure is due to the weight of the hot, possibly vapour rich, water column leading to pressure gradient below hydrostatic. At the margin of the hydrothermal system mixing between cold and hot water is enhanced by the pressure difference (Henley 1985). At greater depth, pressures exceed hydrostatic, feeding the upper reservoir (Henley

1985). Sharp pressure gradients can occur between high and low permeability portions of a hydrothermal system, for example within the flow system feeding fumaroles. Major processes occur at the interface of liquid-gas phases, such as massive precipitation of minerals (Lu and Kieffer 2009). In addition, high pressures can cause rock compaction, thus reduce permeability and drive pressure solution. Conversely, rapid increase in fluid pressure can promote fracturing leading to increases in permeability.

3.2.7 Temperature

Temperature, on the other hand, controls the general alteration patterns of hydrothermal systems because it is the main control on mineral solubility (Giggenbach 1988; Oelkers et al. 2009). For example, metal chlorides and alkaline minerals are more soluble at high temperatures, while gypsum, anhydrite, calcite and dolomite show retrograde solubility below ~ 100 °C (Frazer 2014). Silica solubility increases as temperatures rise, until ~ 300 °C. With further temperature increases, silica solubility decreases (Fournier 1985). These types of thermodynamic relationships in single-phase hydrothermal systems are relatively well constrained, and are reviewed in detail by Oelkers et al. (2009).

4 Hydrothermal Systems and Unrest

Many of the conditions that control fluid flow and chemical alteration are modified by the re-activation of the magmatic system, and evolve during volcanic unrest. For example, the introduction of fresh magma into the deep portions of an active hydrothermal system can critically change the pressure and temperature conditions within the system, thus leading to the development of gas pockets in the subsurface (Jasim et al. 2015). This can rapidly lead to phreatic eruptions. Thus it is crucial to expand unrest tracking to include monitoring of non-magmatic hazards (Sandri et al. 2017—this volume).

Seismicity, gases and ground deformation are usually monitored around active and restless volcanoes (Sparks 2003). Gravity anomaly studies (Gottsmann et al. 2008; Coco et al. 2016) and acoustic waves (Ferrazzini and Aki 1987) also provide insight into subsurface processes. Measurements of water chemistry composition from hydrothermal manifestation such as boiling pools, crater lakes and thermal springs are also routinely conducted (Varekamp et al. 2001, 2009; Federico et al. 2002; Tassi et al. 2003). The increase of Rare Earth Elements (REE)/Cl, RFEs (e.g., Ca, Mg, K)/Cl and the increase in SiO_2 concentration are indicative of either intrusion of fresh magma within the hydrothermal reservoir or exposure to water/rock interaction of fresh rock due to hydrofracturing (Varekamp et al. 2008). A pH drop and temperature increase in spring water can also be indicative of an increase in magmatic activity. However the majority of springs in volcanic environment are fed by the regional groundwater reservoirs, thus representing the cooler water inflow of the hydrothermal system (Jasim 2016). Thermal waters are focused on fluid upwelling pathways such as faults and fractures (Curewitz and Karson 1997; Hemmings et al 2015b; Jasim et al. 2015).

The influx of magmatic fluids can manifest as changes in chemical compositions of hydrothermal discharges. In particular the strong field ligands Cl^- and F^-, from magmatic degassing, can mobilize metals and F^- greatly enhances the dissolution rates of aluminium silicates (Oelkers and Gislason 2001; Wolff-Boenisch et al. 2004) with detrimental effect on rock mechanical properties. However, along the upwelling flow path, mixing with surface waters and chemical reactions with the rock often occur, overprinting the magmatic signature. Magmatic gases and vapour either mix with deep circulating water (Giggenbach 1988) or they condense to form in situ thermal waters (Rye 1993). In both cases, they rapidly dissociate and form a strong acidic solution which causes cation leaching of the host rock leading to advanced argillitic alteration (Giggenbach 1988; Symonds et al. 2001). Prior to mineral precipitation, the resulting waters are enriched in Si, Na^+, K^+, Mg^{2+} and Ca^{2+}, and other metals, proportionally to their concentrations in the host rock (rock congruent dissolution) until the solution is more or less neutralised. This is therefore called the "primary neutralisation zone". Continued acid leaching in K-feldspar systems leads invariably to the formation of alunite, an important component of high-sulfidation epithermal systems, where it can replace entire masses of rocks. For example, alunite can be formed indirectly from K-feldspar through the formation of K-mica and kaolinite (Eqs. 8a–8c).

$$3KAlSi_3O_8 + 2H^+ \rightarrow KAl_3Si_3O_{10}(OH)_2 + 2K^+ + 6SiO_2 \tag{8a}$$
$$\text{K-feldspar} \qquad \text{K-mica}$$

$$KAl_3Si_3O_{10}(OH)_2 + 2H^+ + 3H_2O \rightarrow 3Al_2Si_2O_5(OH)_4 + 2K^+ \tag{8b}$$
$$\text{K-mica} \qquad \text{Kaolinite}$$

$$3Al_2Si_2O_5(OH)_4 + 2K^+ + 6H^+ + 4SO_4^{2-} \rightarrow 2KAl_3(SO_4)_2(OH)_6 + 6SiO_2 + 3H_2O \tag{8c}$$
$$\text{Kaolinite} \qquad \text{Alunite}$$

Most field and experimental observations imply that permeability in hydrothermal systems tends to decrease with time. However, most of these systems remain active for long periods of time (typically 10^3–10^6 years, Ingebritsen et al. 2010). Mechanisms must exist by which permeable pathways are maintained and or developed to allow continuing circulation of hydrothermal fluids and ongoing alteration. Such mechanisms include: (i) the periodic re-organisation of flow patterns related to spatial variations in dissolution and precipitation behaviour (Ritchie and Pritchard 2011); (ii) the dissolution of minerals because of pulses of acidic fluids (Plumlee 1999); (iii) feedback between permeability reduction, fluid pressure and rock mechanics resulting in hydrofracturing, shear dislocation, mineral dissolution and the opening of flow pathways (Barnes 2015; Weis 2015); (iv) pulsating volcanic activity causing fracturing, periodic variations in temperature, pressure and the composition of the circulating fluids (Bodnar et al. 2007); (v) Cooling of the magmatic sills and dikes may lead to thermal cracking (Cathles et al. 1997) and the thermal expansion of the rock; both of which cause increases in fracture density (Chen et al. 1999) enhancing rock permeability; and (vi) stressed induced fracturing (Tapponnier and Brace 1976). Furthermore, fluid pathways often ease the movement of magma towards the surface as shown by the 1975–1984 volcano-tectonic crisis at Krafla caldera (Iceland), which lead to the emplacement of fault-controlled pseudodikes at shallow depths (<100 m) and eruptive events (Opheim and Gudmundsson 1989).

The evolution of a hydrothermal system involves the interplay between a number of mechanisms, physical and chemical, that operate at very different timescales ranging from seconds to hundreds of years. Where volcanic unrest results in rapid modification of the hydrothermal system, hazards associated with unrest can manifest rapidly, with limited warning or precursory activity. Dynamic changes in permeability, related to rapid mineral precipitation or opening of fractures can immediately modify flow pathways. This can result in dramatic changes in heat and fluid flow, near surface pressurisation, hydrothermal outflow and phreatic explosions. Even slow processes such as the incremental development of pervasive alteration zones can manifest as a dynamic hazard in response to continued or future unrest, as strong crystalline rocks are hydrothermally altered into weak secondary clays.

5 Monitoring and Signals

The dynamic hydrological and hydrothermal response to volcanic unrest means that, boreholes, springs, fumaroles, crater lakes and geophysical imaging of the hydrological system can provide a rare window into the state of a volcano and the evolution of volcanic hazard. Hydrological monitoring itself is multi-parametric; insights can be gained from exploring physical and chemical patterns. For instance, the effect of groundwater on volcanic gases changes according to their solubility. As such, a larger proportion of SO_2, HCl and HF emitted from magma remain in solution in water compared to CO_2 and H_2S. Hence, SO_2, HCl and HF can be detected at the surface, only during intense magmatic activity or after drying of degassing pathways (Symonds et al. 2001). In the absence of active degassing, the isotopic ratio of the gases dissolved in groundwater2 such as $^3He/^4He$ (positive) and $\delta^{13}C$ (negative) can be indicative of a magmatic source (Sorey et al. 1998; Allard et al. 1997; Federico et al. 2002).

Fluctuations of the water table/spring discharge have also been frequently recorded before the onset of magmatic activity and are often interpreted as the effect of opening and closing of fractures during the intrusion of fresh magma (Tanguy 1994; Shibata and Akita 2001; Newhall et al. 2001). Alternatively, the effect of the water phase transition from liquid to gas at relatively shallow (<2 km) depth may also cause uplift of the water table (Jasim et al. 2015). Water levels in boreholes can be relatively easily monitored and have been observed to respond to tectonic and volcanic perturbations in a range of volcanic settings (e.g. Usu Volcano, Japan; Kilauea Volcano, Hawai'i; Koryajskii Volcano, Kamchatka).

Level changes have been attributed to thermal pressurisation, compression of water saturated rocks and opening of fractures in response to the intrusion of magma. However, the magnitude and even the sign of this hydrological response is a complex function of the nature of the thermal and mechanical perturbation, the orientation and connectivity of permeable pathways and even the design of the well itself. Thus, interpreting such signals in relation to magmatic unrest requires some prior understanding of the hydrological features involved.

Spring discharge fluctuations are harder to measure than well water level changes, especially on the flanks of volcanoes experiencing unrest, and are therefore less well documented. Decline in non-thermal spring discharge on Centre Hills, Montserrat, were observed prior to the onset of volcanic activity at the adjacent Soufrière Hills Volcano in 1995. This was followed by an increase after the cessation of the second eruptive phase in 2004 (Hemmings et al. 2015a). The mechanism behind such fluctuation is unclear, it may relate to fracture dynamics associated with magmatic pressurisation (and depressurisation). Regular temperature measurement and chemical analysis of spring systems and hydrological lakes in volcanic settings can provide insights into the differences and changes in flow pathways related to magmatic perturbation.

Chemical analysis of thermal springs and fumaroles are more common hydrological/hydrothermal monitoring strategies employed at active volcanoes. Changes in chemical composition and isotopic concentrations are often related to changes in the relative contribution of magmatic fluids to other groundwater species. Although there are general indicators for increased magmatic fluid contribution to discharging hydrothermal fluids, effective use of spring temperature, chemistry and discharge data as volcanic unrest monitoring tools requires a good understanding of the underlying composition of the hydrological and hydrothermal features and the likely sensitivity to different perturbation scenarios, in specific volcanic areas. For example, Taran et al. (2008) proposed that lower flowing, acidic springs at El Chichón volcano, Mexico would make a better monitoring target than near-neutral, high discharge springs. Potential chemical indicators of unrest in these springs include increase in relative concentration of Mg, and increase in Cl/B and Cl/Br ratios.

In summary, the physical and chemical signals or hydrological perturbation associated with magmatic unrest are complex and site dependant. Relative contributions of different fluid components and their interaction with existing flow pathways, can determine how a system evolves during quiescent periods. This evolution dictates the likely response to thermodynamic and chemical perturbation associated with the initiation of volcanic unrest. Given the intricate feedbacks between magma, hydrology and hazards and the sudden changes of the systems involved, only high-frequency (hour-week) monitoring of temperature, pH, electrical conductivity of water, depth of the water table, REE, rock forming elements and dissolved gas coupled with geophysical monitoring can untangle the evolution of the magmatic system, the opening/closure of fractures and the seasonal groundwater dynamic.

6 Open Questions—Important Unknowns

We have established that circulating hydrothermal fluids are highly reactive and may result in precipitation of alteration products or dissolution of the host rock, both of which may cause porosity, and permeability changes. However, the precise nature of this alteration varies with fluid chemistry, rock mineralogy and thermodynamic conditions. This uncertainty in alteration makes predicting the impact of water/rock interaction (WRI) on porosity and permeability, and therefore on fluid flow, particularly challenging. The background fluid flow regime is a critical part of the local expression of heat-flow as well as pressure distribution that surrounds a magmatic system. As such it may exercise an important control over the dynamics of the magmatic system that is currently poorly understood. Data constraining the time scales over which hydrothermal alteration occurs, related to data gathered from long term monitoring of coupled magmatic-hydrothermal systems, are thus crucial to inform ongoing interpretations and further predictions of areas experiencing magmatic-hydrothermal unrest.

Acknowledgements The authors thank Shaul Hurwitz for his constructive comments during the review of this manuscript and Pablo Palacios for carefully editing the extended abstract. This project has received funding from the European Union's Seventh Program for research, technological development, and demonstration under grant agreement no. 282759 (VUELCO). K. Mayer and B. Scheu also acknowledge the support of a PROCOPE grant (Hot Hydrothermal Volcanic Systems; project-ID 57130387), funded and implemented by the Deutscher Akademischer Austauschdienst (DAAD) in Germany, and the Ministry of Foreign and European Affairs (MAE) and the Ministry of Higher Education and Research (MESR) in France. K. Mayer and B. Scheu acknowledge the support of the ERC Advanced Investigator Grant (EVOKES—no. 247076).

Glossary

Hydrothermal system A groundwater system that has an area of recharge, an area of discharge, and a heat source. When a magma supplies the heat source and volatiles, the hydrothermal system is termed a magmatic hydrothermal system

Hydrothermal alteration hydrothermal alteration is a complex process involving chemical, mineralogical, and textural changes, due to the interaction of hot aqueous fluids and the host rocks through which they circulate

Permeability Connected pore space of a rock or lithology, controlling fluid flow within a reservoir

Porosity Ratio of voids over the total volume of the rock with respect to a reference rock-volume

Water/rock interaction (WRI) The set of chemical reactions between aqueous fluids and rocks. These reactions modify both the chemistry of the circulating fluid and the mineralogy of the host rock

Fluid generic term for either liquid or gas or both

Liquid liquid state of matter (e.g., water)

Gas gas state of matter (e.g., vapour)

Water table level below which water saturation occurs

References

Aizawa K (2008) Classification of self-potential anomalies on volcanoes and possible interpretations for their subsurface structure. J Volcanol Geoth Res 175:253–268

Allard P, Jean-Baptiste P, D'Alessandro W, Parello F, Parisi B, Flehoc C (1997) Mantlederived helium and carbon in groundwaters and gases of Mount Etna, Italy. Earth Planet Sci Lett 148:501–516

Barnes H (2015) Hydrothermal processes: the development of geochemical concepts in the latter half of the twentieth century. Geochem Perspect 4:1–93

Berger G, Claparols C, Guy C, Daux V (1994) Dissolution rate of a basalt glass in silica-rich solutions: implications for long-term alteration. Geochim Cosmochim Acta 58:4875–4886

Bodnar RJ, Cannatelli C, De Vivo B, Lima A, Belkin HE, Milia A (2007) Quantitative models for magma degassing and ground deformation (bradyseism) at Campi Flegrei, Italy: implications for future eruptions. Geology 35:791–794

Browne PRL (1978) Hydrothermal alteration in active geothermal fields. Annu Rev Earth Planet Sci 6:229–250

Browne PRL (1984) Subsurface stratigraphy and hydrothermal alteration of Eastern section of the Olkaria geothermal field, Kenya. In: Proceedings of the 6th New Zealand geothermal workshop, vol 1, pp 33–41

Bruno PPG, Ricciardi GP, Petrillo Z, Di Fiore V, Troiano A, Chiodini G (2007) Geophysical and hydrogeological experiments from a shallow hydrothermal system at Solfatara Volcano, Campi Flegrei, Italy: response to caldera unrest. J Geophys Res: Solid Earth 112:1–17

Cabrera MC, Custodio E (2004) Groundwater flow in a volcanic-sedimentary coastal aquifer: Telde area, Gran Canaria, Canary islands, Spain. Hydrogeol J 12:305–320

Cathles LM, Erendi AHJ, Barrie T (1997) How long can a hydrothermal system be sustained by a single intrusive event? Econ Geol 92:766–771

Chen Y, Xiaodong W, Fuqing Z (1999) Experiments on thermal fracture in rocks. C Sci Bull 17:1610–1612

Chiodini G, Frondini F, Cardellini C, Granieri D, Marini L, Ventura G (2001) CO_2 degassing and energy release at Solfatara volcano, Campi Flegrei, Italy. J Geophys Res 106:16213–16221

Christiansen LB, Hurwitz S, Saar MO, Ingebritsen SE, Hsieh P (2005) Seasonal seismicity at western United States volcanic centers. Earth Planet Sci Lett 240: 307–321

Coco A, Gottsmann J, Whitaker F, Rust A, Currenti G, Jasim A, Bunney S (2016) Numerical models for ground deformation and gravity changes during volcanic unrest: simulating the hydrothermal system dynamics of an active caldera. Solid Earth Discussions 7:557–577

Cronan DS, Johnson AG, Hodkinson RA (1997) Hydrothermal fluids may offer clues about impending volcanic eruptions. Eos 78:341–345

Cruz JV, Oliveira Silva M (2001) Hydrogeologic framework of Pico Island, Azores, Portugal. Hydrogeol J 9:177–189

Curewitz D, Karson JA (1997) Structural settings of hydrothermal outflow: fracture permeability maintained by fault propagation and interaction. J Volcanol Geoth Res 79:149–168

Custodio E (2007) Groundwater in volcanic hard rocks. In: Krasny and Sharp (eds) Groundwater in fractured rocks. IAH Selected Paper Series, vol 9, pp 95–108

Ellis AJ, Mahon WAJ (1977) Chemistry and geothermal systems. Academic Press (392)

Federico C, Aiuppa A, Allard P, Bellomo S, Jean-Baptiste P, Parello F, Valenza M (2002) Magma-derived gas influx and water-rock interactions in the volcanic aquifer of Mt. Vesuvius, Italy. Geochimica et Cosmochimica Acta 66:963–981

Ferrazzini V, Aki K (1987) Slow waves trapped in a fluid-filled infinite crack: implications for volcanic tremor. J Geophys Res 92:9215–9223

Finn C, Williams DL (1987) An aeromagnetic study of Mount St. Helens. J Geophys Res: Solid Earth 92:10194–10206

Finn CA, Sisson TW, Deszcz-Pan M (2001) Aerogeophysical measurements of collapse-prone hydrothermally altered zones at Mount Rainier volcano. Nature 409:600–603

Fournier RO (1985) The behavior of silica in hydrothermal solutions. In: Berger and Bethke (eds) Geology and geochemistry of epithermal systems. Reviews in Economic Geology, vol 2, pp 45–61

Fournier RO (1999) Hydrothermal processes related to movement of fluid from plastic into brittle rock in the magmatic-epithermal environment. Econ Geol 94:1193–1211

Fournier N, Witham F, Moreau-Fournier M, Bardou L (2009) Boiling Lake of Dominica, West Indies: high-temperature volcanic crater lake dynamics. J Geophys Res 114:1–17

Frazer AM (2014) Advances in understanding the evolution of diagenesis in carboniferous carbonate platforms: insights from simulations of palaeohydrology, geochemistry, and stratigraphic development. PhD Thesis, University of Bristol, UK (261)

Freeze R, Cherry J (1979) Groundwater. Prentice Hall (604)

Fulignati P, Gioncada A, Sbrana A (1998) Geologic model of the magmatic hydrothermal system of vulcano (Aeolian Islands, Italy). Mineral Petrol 62:195–222

Giberti G, Yven B, Zamora, M, Vanorio T (2006) Database on laboratory measured data on physical properties of rocks of Campi Flegrei volcanic area (Italy). In: Zollo, Capuano, Corciulo (eds) Geophysical exploration of the Campi Flegrei (Southern Italy) Caldera' interiors: data, methods and results, vol 1, pp 179–192

Gifkins CC, Herrmann W, Large RR (2005) Altered volcanic rocks: a guide to description and interpretation. Ph.D. thesis, University of Tasmania, Australia (275)

Giggenbach WF (1984) Mass transfer in hydrothermal alteration systems—A conceptual approach. Geochim Cosmochim Acta 48:2693–2711

Giggenbach WF (1988) Geothermal solute equilibria. Derivation of Na-K-Mg-Ca geoindicators. Geochim Cosmochim Acta 52:2749–2765

Giggenbach WF, Soto RC (1992) Isotopic and chemical composition of water and steam discharges from volcanic-magmatic-hydrothermal systems of the Guanacaste Geothermal Province, Costa Rica. Appl Geochem 7:309–332

Goff F, CJ Janik (2000) Geothermal systems. In: Sigurdsson, Houghton, McNutt, Rymer, Stix (eds) Encyclopedia of volcanoes, vol 1, pp 817–834

Gottsmann J, Camacho AG, Martí J, Wooller L, Fernández J, Garcia A, Rymer H (2008) Shallow structure beneath the Central Volcanic Complex of Tenerife from new gravity data: implications for its evolution and recent reactivation. Phys Earth Planet Inter 168:212–230

Gudbrandsson S, Wolff-Boenisch D, Gislason SR, Oelkers EH (2011) An experimental study of crystalline basalt dissolution from 2 < pH < 11 and temperatures from 5 to 75 C. Geochim Cosmochim Acta 75:5496–5509

Hayba D, Ingebritsen S (1997) Multiphase groundwater flow near cooling plutons. J Geophys Res 102:12235–12252

Hedenquist J, Lowenstern J (1994) The role of magmas in the formation of hydrothermal ore deposits. Nature 370:519–527

Hemmings B, Whitaker F, Gottsmann J, Hughes A (2015a) Hydrogeology of montserrat, review and new insights. J Hydrol: Reg Stud 3:1–30

Hemmings B, Gooddy D, Whitaker F, Darling GW, Jasim A, Gottsmann J (2015b) Groundwater recharge and flow on Montserrat, West Indies: insights from groundwater dating. J Hydrol: Reg Stud 4:611–622

Henley RW (1985) The geothermal framework of epithermal deposits. Rev Econ Geol 2:1–24

Henley RW, Berger BR (2013) Nature's refineries metals and metalloids in arc volcanoes. Earth Sci Rev 125:146–170

Henley R, Ellis A (1983) Geothermal Systems Ancient and Modern: a Geochemical Review. Earth Sci Rev 19:1–50

Hu SM, Zhang RH, Zhang XT, Huang WB (2010) Experimental study of water-basalt interactions in Luzong volcanic basin and its applications. Acta Petrologica Sinica 26:2681–2693

Ingebritsen SE, Scholl MA (1993) The hydrogeology of Kilauea volcano. Geothermics 22:255–270

Ingebritsen S, Ward S, Neuzil C (2006) Groundwater in geologic processes. Cambridge University Press (564)

Ingebritsen SE, Geiger S, Hurwitz S, Driesner T (2010) Numerical simulation of magmatic hydrothermal systems. Rev Geophys 48:1–33

Jasim A (2016) Exploring the complexity of groundwater flow in volcanic terrains: a combined numerical, experimental and field data approach. Ph.D. thesis, University of Bristol, UK (199)

Jasim A, Whitaker FF, Rust AC (2015) Impact of channelized flow on temperature distribution and fluid flow in restless calderas: insight from Campi Flegrei caldera, Italy. J Volcanol Geoth Res 303:157–174

Join JL, Folio JL, Robineau B (2005) Aquifers and groundwater within active shield volcanoes. Evolution of conceptual models in the Piton de la Fournaise volcano. J Volcanol Geoth Res 147:187–201

Kuhn M (2004) Reactive flow modeling of hydrothermal systems. Springer, Berlin (264)

Lu X, Kieffer S (2009) Thermodynamics and mass transport in multicomponent, multiphase H_2O systems of planetary interest. Annu Rev Earth Planet Sci 37:449–477

Manning C, Ingebritsen S (1999) Permeability of the continental crust: Implications of geothermal data and metamorphic systems. Rev Geophy 37:127–150

Mason BG, Pyle DM, Dade WB, Jupp T (2004) Seasonality of volcanic eruptions. J Geophys Res: Solid Earth 109:1–12

Mayer K, Scheu B, Montanaro C, Yilmaz TI, Isaia R, Aßbichler D, Dingwell DB (2016) Hydrothermal alteration of surficial rocks at Solfatara (Campi Flegrei): Petrophysical properties and implications for phreatic eruption processes. J Volcanol Geoth Res 320:128–143

Mutlu H, Sariiz K, Kadir S (2005) Geochemistry and origin of the Şaphane alunite deposit, Western Anatolia, Turkey. Ore Geol Rev 26:39–50

Neuzil CE (1994) How permeable are clays and shales? Water Resour 30:145–150

Newhall CG, Albano SE, Matsumoto N, Sandoval T (2001) Roles of groundwater in volcanic unrest. J Geol Soc Philippines 56:69–84

Oelkers EH, Gislason SR (2001) The mechanism, rates and consequences of basaltic glass dissolution: I. An experimental study of the dissolution rates of basaltic glass as a function of aqueous Al, Si and oxalic acid concentration at 25 C and pH = 3 and 11. Geochim Cosmochim Acta 65:3671–3681

Oelkers EH, Benezeth P, Pokrovski GS (2009) Thermodynamic databases for water-rock interaction. In: Oelkers and Schott (eds) Thermodynamics and kinetics of water-rock interaction. Reviews in Mineralogy and Geochemistry, 70, 1–37

Opheim JA, Gudmundsson A (1989) Formation and geometry of fractures, and related volcanism, of the Krafla fissure swarm, northeast Iceland. Geol Soc Am Bull 101:1608–1622

Peluso F, Arienzo I (2007) Experimental determination of permeability of Neapolitan Yellow Tuff. J Volcanol Geoth Res 160:125–136

Peterson FL (1972) Water development on tropic volcanic islands-type example: Hawaii. Ground Water 10:18–23

Petrov VA, Poluektov VV, Zharikov AV, Velichkin VI, Nasimov RM, Diaur NI, Terentiev VA, Shmonov VM, Vitovtova VM (2005) Deformation of metavolcanics in the Karachay Lake area, Southern Urals: petrophysical and mineral-chemical aspects. Geol Soc London, Spec Publ 240:307–322

Piochi M, Kilburn CRJ, Di Vito MA, Mormone A, Tramelli A, Troise C, De Natale G (2014) The volcanic and geothermally active Campi Flegrei caldera: an integrated multidisciplinary image of its buried structure. Int J Earth Sci 103:401–421

Pirajno F (2010) Hydrothermal processes and mineral systems. Springer, Berlin (1250)

Plumlee GS (1999) The environmental geology of mineral deposits. In: Plumlee and Logsdon (eds) The Environmental Geochemistry of Mineral Deposits, Part A. Processes, Techniques, and Health Issues: Society of Economic Geologists. Reviews in Economic Geology, 6, 71–116

Reid ME (2004) Massive collapse of volcano edifices triggered by hydrothermal pressurization. Geology 32:373–376

Ritchie LT, Pritchard D (2011) Natural convection and the evolution of a reactive porous medium. J Fluid Mech 673:286–317

Robb L (2005) Introduction to ore-forming processes. Blackwell Science Ltd (384)

Rye RO (1993) The evolution of magmatic fluids in the epithermal environment: the stable isotope perspective. Economica Geologica 88:733–753

Rye RO (2005) A review of the stable-isotope geochemistry of sulphate minerals in selected igneous environments and related hydrothermal systems. Chem Geol 215:5–36

Saar MO, Manga M (2003) Seismicity induced by seasonal groundwater recharge at Mt. Hood, Oregon. Earth and Planetary Science Letters 214:605–618

Sandri L, Tonini R, Rouwet D, Constantinescu R, Mendoza-Rosas AT, Andrade D, Bernard B (2017) The need to quantify hazard related to non-magmatic unrest: from BET_EF to BET_UNREST. This volume

Shibata T, Akita F (2001) Precursory changes in well water level prior to the March, 2000 eruption of Usu volcano, Japan. Geophys Res Lett 28:1799–1802

Sorey ML, Evans WC, Kennedy BM, Farrar CD, Hainsworth LJ, Hausback B (1998) Carbon dioxide and helium emissions from a reservoir of magmatic gas beneath Mammoth Mountain, California. J Geophy Res: Solid Earth 103:15303–15323

Sparks RSJ (2003) Forecasting volcanic eruptions. Earth Planet Sci Lett 210:1–15

Stefansson A, Gislason SR (2001) Chemical weathering of basalts, Southwest Iceland: effect of rock crystallinity and secondary minerals on chemical fluxes to the ocean. Am J Sci 301:513–556

Symonds RB, Gerlach TM, Reed MH (2001) Magmatic gas scrubbing: implications for volcano monitoring. J Volcanol Geoth Res 108:303–341

Tanguy JC (1994) The 1902-1905 eruptions of Montagne Pelee, Martinique: anatomy and retrospection. J Volcanol Geoth Res 60:87–107

Tapponnier P, Brace WF (1976) Development of stress-induced microcracks in Westerly granite. Int J Rock Mech Min Sci 13:103–112

Taran Y, Rouwet D, Inguaggiato S, Aiuppa A (2008) Major and trace element geochemistry of neutral and acidic thermal springs at El Chichon volcano, Mexico. J Volcanol Geoth Res 178:224–236

Tassi F, Vaselli O, Capaccioni B, Macias JL, Nencetti A, Montegrossi G, Magro G (2003) Chemical composition of fumarolic gases and spring discharges from El Chichon volcano, Mexico: Causes and implications of the changes detected over the period 1998-2000. J Volcanol Geoth Res 123:105–121

Varekamp JC (2008) The volcanic acidification of glacial Lake Caviahue, Province of Neuquen, Argentina. J Volcanol Geoth Res 178:184–196

Varekamp JC, Ouimette AP, Herman SW, Bermudez A, Delpino D (2001) Hydrothermal element fluxes from Copahue, Argentina: A "beehive" volcano in turmoil. Geology 29:1059–1062

Varekamp JC, Ouimette AP, Herman SW, Flynn KS, Bermudez A, Delpino D (2009) Naturally acid waters from Copahue volcano, Argentina. Appl Geochem 24:208–220

Weis P (2015) The dynamic interplay between saline fluid flow and rock permeability in magmatic-hydrothermal systems. Geofluids 15:350–371

Wolff-Boenisch D, Gislason SR, Oelkers EH, Putnis CV (2004) The dissolution rates of natural glasses as a function of their composition at pH 4 and 10.6, and temperatures from 25 to 74 C. Geochim Cosmochim Acta 68:4843–4858

Wolff-Boenisch D, Gislason SR, Oelkers EH (2006) The effect of crystallinity on dissolution rates and CO_2 consumption capacity of silicates. Geochim Cosmochim Acta 70:858–870

Zakharova EA, Pokrovsky OS, Dupre B, Gaillardet J, Efimova LE (2007) Chemical weathering of silicate rocks in Karelia region and Kola peninsula, NW Russia: assessing the effect of rock composition, wetlands and vegetation. Chem Geol 242:255–277

Experimental Simulations of Magma Storage and Ascent

C. Martel, R.A. Brooker, J. Andújar, M. Pichavant, B. Scaillet and J.D. Blundy

Abstract

One of the key issues in utilizing precursor signals of volcanic eruption is to reliably interpret geophysical and geochemical data in terms of magma movement towards the surface. An important first step is to identify where the magma is stored prior to ascent. This can be studied through phase-equilibrium experiments designed to replicate the phase assemblage and compositions of natural pyroclasts or by measuring volatiles in melt inclusions from previous eruptions. The second crucial step is to characterize the magmatic conditions and processes that will guide the eruption style. This may be addressed through controlled dynamic decompression or deformation experiments to examine the different rates that govern the kinetics of syn-eruptive degassing, crystallization, and strain. Comparing the compositional and textural characteristics of these experimental products with the natural samples can be used to retrieve magma ascent conditions. These experimental simulations allow interpretation of direct observations and *in situ* measurements of syn-eruptive processes leading to more accurate forecasting of future eruptive scenarios.

1 Linking Geophysical and Geochemical Warning Signals to Magmatic Processes

A key objective in volcanology is to forecast eruptions, i.e. to establish when, where, and how an eruption will occur and what magnitude it will be. The prerequisite of such forecasting is to (i) detect reliable precursory signals of magma ascent to subsurface and (ii) anticipate the eruption style in order to inform the crisis

C. Martel (✉) · J. Andújar · M. Pichavant ·
B. Scaillet
Institut Des Sciences de La Terre D'Orléans,
Université D'Orléans-CNRS-BRGM, Orléans,
France
e-mail: caroline.martel@cnrs-orleans.fr

R.A. Brooker · J.D. Blundy
School of Earth Sciences, University of Bristol,
Wills Memorial Building, Queens Road, Bristol BS8
1RJ, UK

Advs in Volcanology (2019) 101–110
DOI 10.1007/11157_2017_20
© The Author(s) 2017
Published Online: 02 August 2017

management strategy. To reach these objectives, the intensification of the geophysical and geochemical signals associated with an unrest episode has to be interpreted in terms of magma movements. This is a far from trivial task because (i) seismicity has to distinguish signals related to magma movements from those of rock fracturing and/or gas percolation (*see Chapter "Volcano Seismology: detecting unrest in wiggly lines"*), (ii) ground deformation has to precisely track magma motion towards the surface (*see Chapter "Volcano geodesy and multiparameter investigations"*), and (iii) the flux and the speciation of emitted gas at the surface has to be interpreted in terms of magma ascent and degassing (*see Chapter "Volcanic gases and low temperature volcanic fluids"*). For any of these monitoring signals, their interpretation in terms of impending eruption requires knowing at what depth beneath the volcano magma is stored (magma storage conditions) and how it progresses toward the surface (magma ascent conditions). A pertinent approach to investigate the conditions of magma storage and ascent consists of comparing petrological and textural studies of previously erupted products to the results of experimental simulations carried out under realistic magma conditions. During the last decades the development of powerful analytical and experimental tools has led to great advances in this type of investigation. Of course, this is an a posteriori approach (using previously erupted products) that relies on considering past eruptive behaviour of a given volcanic system as a guide to future activity. For this reason, it is necessary to understand the fundamental magmatic processes at any particular volcano and in the long term, build up a record that links the pre-eruptive signals with eruptive products. In this way we can successfully use the warning signals to forecast or even start to predict the timing and style of an imminent eruption.

2 Magma Storage

How and where magma is stored before an eruption are enduring and complex questions, particularly given the range of hypotheses covering single versus multiple storage regions or dyke feeder systems. Key parameters in interpreting the precursory geophysical and geochemical signals are the depth of storage and the volatile content dissolved in the magma, as the exsolution of these provides an important driving force for explosive eruptions. Magma consists principally of phenocrysts (crystals larger than 50–100 μm) coexisting with a silicate melt containing dissolved volatile species (e.g. H_2O, CO_2, S species, F, Cl, etc....). To assess the magma storage conditions, one has to determine the parameters (i.e. the pressure, temperature, redox state, volatile content) that govern equilibrium between melt and the phenocrysts. This is accomplished by comparing the phase assemblage and compositions of the natural products to those obtained by phase-equilibrium experiments where all these parameters are controlled.

2.1 Decoding Natural Pyroclasts

Phenocrysts. Mineral compositions are sensitive to various intensive (pressure, temperature, fO_2) and extensive (host liquid composition and volatile content) variables. In some cases, the compositions can be used as geothermobarometers (or hygro/oxy/chemo-meters) to retrieve first-order parameters of crystallization. For instance, we can calculate crystallization temperature from the composition of orthopyroxene-clinopyroxene pairs (e.g. Lindsley and Andersen 1983) or amphibole-plagioclase pairs (e.g. Holland and Blundy 1994), temperature and oxygen fugacity from Fe–Ti oxide pairs (e.g. Ghiorso and Evans 2008) or pressure from amphibole composition (e.g. Ridolfi and Renzulli 2012). Such calculations require two main criteria: (i) the thermobarometers must have been calibrated experimentally for conditions and compositions relevant to those of the target samples and (ii) the analysed phenocrysts must be identified as part of the phenocryst assemblage in chemical equilibrium with the melt in the reservoir under pre-eruptive conditions. This can achieved by comparison with experiments that simulate a range of variables. Establishing different 'equilibrium' events also becomes crucial as magma mixing (and associated reheating prior to eruption or further crystallization outside the storage zone) very often blur the pre-intrusion/eruption equilibrium conditions. Moreover, many eruptions involve the entrainment of crystals that did not grow from the erupted magma and are consequently out of equilibrium with the melt in which they occur. These are often termed xenocrysts or antecrysts and increasingly recognised as important aspects of magma's crystal cargo (e.g. Streck 2008; Kilgour et al 2013). The processes of mixing, reheating, and potentially decompression recorded in these crystals, may be used to provide

timescales of reservoir dynamics, such as residence times between the last magma recharge and eruption (e.g. Saunders et al. 2012). Indeed, the compositional zoning of some phenocrysts witness cycles of recharge events prior to the final eruption (e.g. Druitt et al. 2012). Diffusion chronometry provides a means to infer the crystal residence time in the reservoir prior to eruption (e.g. Costa and Morgan 2010), but again requires experimental calibration of the rates involved under various conditions.

Glass inclusions. Glass (or "melt") inclusions are aliquots of melt that are trapped in crystals, usually during stages of rapid crystal growth (Fig. 1a). If trapped in phenocrysts crystallizing in the magma chamber, these inclusions will be the only witness of the composition of the melt in equilibrium with the phenocrysts prior to eruption, provided that they remained sealed after entrapment (no volatile leak, no crystallization or post-entrapment interaction with the host mineral). Indeed, after leaving the storage region, the melt initially surrounding the phenocrysts is likely to degas and crystallize microlites upon ascent, therefore deviating significantly from its

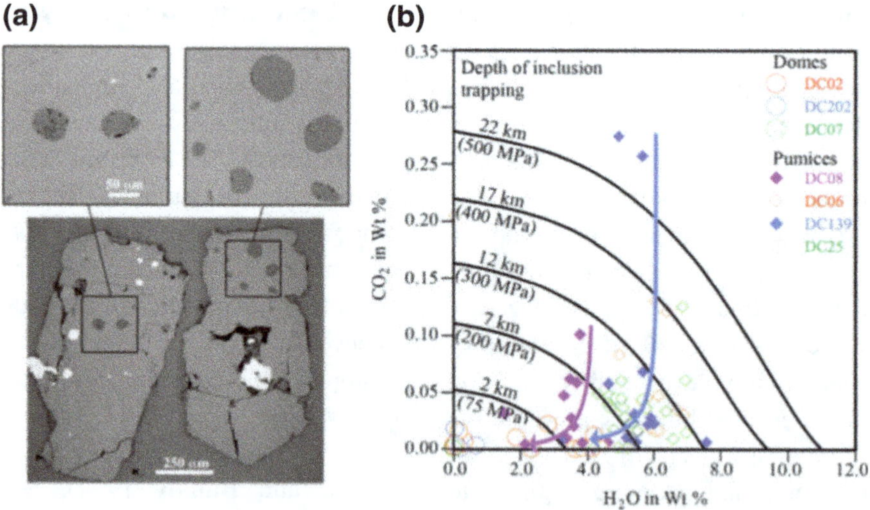

Fig. 1 Melt inclusion (*MI*) volatile contents from the Morne aux Diables complex (*MAD*), northern Dominica, as determined by ion-microprobe measurements. **a** Close up of inclusions shows they were either slightly vesiculated, during capture (fluid-saturated) or became so during ascent. The *right* hand crystal contains a very large melt inclusion around the bright oxide inclusions that has vesiculated more extensively, possibly due to cracks developing and exposing the melt to the full decompression effect during eruption. In **b** the maximum amount of H_2O and CO_2 that can be dissolved in a melt at any given depth follows pressure dependent isopleths that can be determined experimentally (in this case from Tamic et al. 2001). This represents the minimum pressure (depth of entrapment of a MI). These trends are controlled by the lower solubility of CO_2 compared to H_2O, such that there is a rapid drop in CO_2 before H_2O starts to be lost (*coloured arrows*)

pre-eruptive composition and character. The two main objectives of studying glass inclusions in phenocrysts are (i) mineral-melt thermobarometry (clinopyroxene-melt or plagioclase-melt; e.g., Putirka 2008; Water and Lange 2015) and (ii) pre-eruption volatile contents and speciation that can be converted into gas saturation pressure using experimentally-derived volatile solubility laws and which represent an end-member in the calculations of the volcanic degassing budget (dissolved vs. released gases). This latter approach requires that the magma be demonstrably volatile-saturated at the time of melt inclusion entrapment, for example by looking at relationships between dissolved volatile contents and trace elements in inclusions (Blundy and Cashman 2008). If the magma was not saturated, the calculated pressure generally represents a minimum depth prior to volatile volatile-saturated, the calculated pressure generally represents a minimum pressure estimate. As an example, the glass inclusions in phenocrysts from pyroclasts of the last eruption of Morne aux Diables, Dominica, have been analysed by ion microprobe. The data show about 6 to 8 wt% dissolved H_2O, up to 3000 ppm CO_2, together with some chlorine and fluorine. Available $H_2O–CO_2$ solubility models based on experiments for comparable compositions (e.g. Tamic et al. 2001) indicate melt entrapment during phenocryst crystallization at pressures as high as 400–500 MPa (depth of <22 km) for sample DC139 and shallower depth for sample DC08 although it is also possible this magma originally contained even more CO_2 than is recorded by any of the analysed melt inclusions. (Figure 1b). Small vesicles in the inclusions of Fig. 1a suggest there was exsolution of volatiles during ascent. The possible disequilibrium between such vesicles and melt produced during very rapid ascent is discussed in *Chapter "Magma degassing: the diffusive fractionation model and beyond"*.

Complexity of open-systems. An eruption is often triggered by the injection of new magma into the reservoir, which reheats, mingles, and mixes with the resident magma. Upon ascent to the surface, the two batches may interact to varying degrees and can further crystallize and

cool. Therefore, imprudent use of geothermobarometers may yield large pressure and temperature ranges that cannot be easily reconciled with a single/specific episode of equilibrium crystallization. In order to retrieve the storage conditions of the resident magma, i.e. prior to deep magma mixing or before possible modification within the volcanic conduit, a detailed petrological study is necessary to identify precisely the different stages of perturbation and their characteristics in terms of phase assemblage and chemical composition. Experimental petrology is one of the tools that helps to unravel the various magmatic processes at work and their relative impact on magma chemistry and magmatic evolution.

2.2 Phase-Equilibrium Experiments

Phase-equilibrium experiments use natural (or analogue) products as starting material that are subjected to high-pressure (HP) and high-temperature (HT) in various devices under controlled conditions of pressure, temperature, oxygen fugacity, and volatile content. Such experiments are powerful tools to simulate realistic magmatic conditions for the crustal reservoirs that feed volcanic systems. Experimental equipment ranges from cold-seal pressure vessels, internally-heated pressure vessels, and piston-cylinder apparatus, depending on the investigated conditions. The principle is to reproduce the natural assemblage, proportion, and chemical compositions of the phenocrysts and equilibrium coexisting melt in the magma storage region. This is then compared with the natural samples in order to retrieve the pre-eruptive crystallization conditions (Fig. 2).

The first prerequisite for such an approach is a detailed petrological and mineralogical study of the erupted samples in order to identify the magmatic processes potentially perturbing equilibrium crystallization in the reservoir. Indeed, the relevance of the experimental study relies on accurate petrological knowledge that dictates the choice of the starting material and run procedure (Pichavant et al. 2007). The second prerequisite

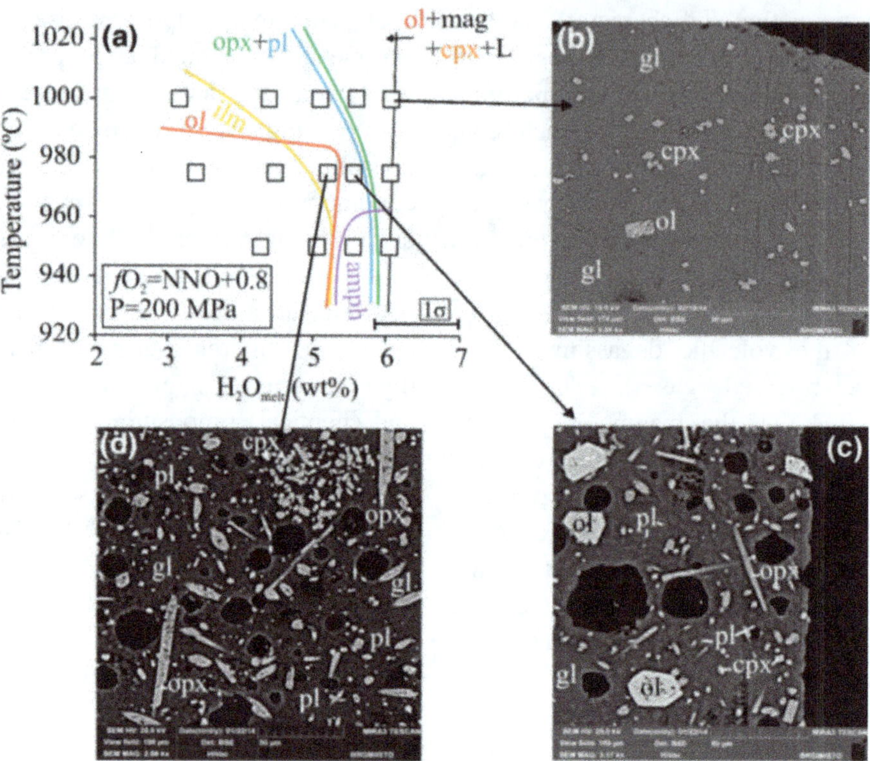

Fig. 2 Phase equilibrium experiments for the Tungurahua 2006 andesite, showing **a** mineral stability fields as a function of temperature and H_2O content at 200 MPa and oxidizing conditions (fO_2 = NNO + 0.8 log unit). SEM images of the experimental charges are shown in **b** for 1000 °C and H_2O saturation ($\sim$6.1 wt% H_2O dissolved in melt), **c** 975 °C and $\sim$5.7 wt% H_2O, and **d** 975 °C and 5.2 wt% H_2O. *Note* the drastic increase of crystal content with cooling or dehydration; *gl* for glass (*L* for silicate liquid), *ol* for olivine, *cpx* for clinopyroxene, *opx* for orthopyroxene, *pl* for plagioclase, *amph* for amphibole, *mag* for magnetite, and *ilm* for ilmenite

is the appropriate choice of the volatile species (H_2O, CO_2, sulphur, etc.) and contents to be added to the starting material. These dissolved volatiles can impact crystallization (sequence, mineral stability fields, and phase compositions; Scaillet and Pichavant 2003; Riker et al. 2015). Where volatile measurements on melt inclusions are available, these can be used, although it is possible that the melt inclusions may no longer be representative of the initial volatile content (e.g. due to possible leakage or recrystallization). Consequently, volatile species and contents become experimental parameters that have to be varied within a range that is first inferred from the study of the glass inclusions (when available), and/or based on previous work carried out on similar bulk rock compositions. It should be remembered that entrapment of melt inclusions requires crystallisation to occur and magmas may undergo substantial volatile loss (degassing) prior to any crystallisation. For instance, Blundy et al. (2010) speculate that many magmas had original CO_2 contents significantly higher than those recorded by any melt inclusions.

It is clear that an approach involving the combination of petrological study of the natural products and phase-equilibrium experiments can help to retrieve the storage conditions of magmas of a wide range of compositions. This is a prerequisite step for the interpretation of unrest signals and construction of eruptive scenarios.

3 Magma Ascent

During ascent in the volcanic conduit the silicate melt around the phenocrysts degasses by exsolving its dissolved volatiles as gas bubbles. This may lower the liquidus triggering the crystallization of microlites (i.e. crystals smaller than about 50–100 μm). The residual melt is transformed both chemically, by degassing and differentiation as microlites crystallize and physically, by an increase in melt viscosity and a change from a single liquid phase to a three-phase suspension (i.e. liquid, gas bubble, and microcrystals). Both types of transformations have drastic effects on the bulk magma flow conditions (rheology) that control ascent rate and the ductile versus brittle behaviour of the magma.

The rate of magma decompression/ascent is the key parameter that controls the kinetics of degassing and crystallization, and ultimately, the eruptive style. In silicic to intermediate systems, slow ascent rates typical of effusive eruptions such as lava flow or dome growth, i.e. cm/s to mm/s (Gardner and Rutherford 2000) yield timescales long enough for extensive degassing and crystallization. In contrast, the high ascent rates prevalent during paroxysmal Strombolian or Plinian eruptions (i.e. of the order of m/s; Gardnerand and Rutherford 2000) are able to generate physico-chemical disequilibria of both degassing and crystallization processes, driving gas overpressures that may be released explosively.

3.1 Textures of Natural Pyroclasts

Decompression-induced degassing of the magma creates gas bubbles, the number density of which has been demonstrated to correlate with the decompression rate simulated by experiments (Mourtada-Bonnefoi and Laporte 2004; Mangan et al. 2004). The growing bubbles can rapidly coalesce and form gas escape channels. In this case, the bubble number densities in the erupted/quenched pyroclasts may no longer be representative of the initial ascent rates under which nucleation was triggered. The timescale of outgassing by magma foam collapse varies from a couple of hours to about 1000 h for magmas having bulk viscosities of $\sim 10^4$ to $10^{5.5}$ Pa.s, respectively (Martel and Iacono-Marziano 2015). This restricts the use of the degassing process to simulations of rapid magma ascent rates such as those during Plinian events. To investigate longer transit times in the conduit, one requires information from magmatic processes with timescales longer than degassing. Microlite crystallization is one of those

processes, because diffusion in the melt, that controls crystal growth, occurs on timescales ranging from hours in mafic melts, to days in silicic melts. Microlite number density, volume proportion, size and shape have all been used as markers of the undercooling (liquidus temperature minus magma temperature) that drives crystallization (e.g. Hammer et al. 1999); the higher the undercooling, the more numerous, smaller, and irregularly-shaped the crystals. With decompression (and dehydration of the melt), liquidus temperature increases (as does undercooling), so that it becomes possible to infer the depth of crystallization in the conduit by relating the textural characteristics of the microlites to undercooling and pressure (Fig. 3). This approach has been used by Melnik et al. (2011) to constrain both magma flow and reservoir shape for the 1980–86 dome-forming eruptions of Mount St. Helens (USA). Such modelling requires an accurate determination of the dependence of undercooling on pressure, which can be achieved through decompression experiments (e.g. Riker et al. 2015).

3.2 Dynamic Experiments

Dynamic experiments, such as decompression, deformation or shock-wave experiments, are valuable tools to investigate degassing, crystallization, strain, mixing, or fragmentation of a magma. They provide information on the physics, chemistry, and kinetics of syn-eruptive magmatic processes, which can be used in turn to decode natural pyroclast formation in order to better identify geophysical and geochemical precursory signals of an eruption. Shock-wave experiments dedicated to the fragmentation process are covered in *Chapter "From unrest to eruption: Conditions for phreatic versus magmatic activity"*; the discussion below concentrates on decompression and deformation experiments.

Decompression experiments performed at elevated pressure (HP) and temperature (HT) have proved useful in accurately simulating magma ascent in volcanic conduits (e.g. Hammer and Rutherford 2002). In particular, experimental decompression rates can cover the most of the perceived range of magma ascent rates during

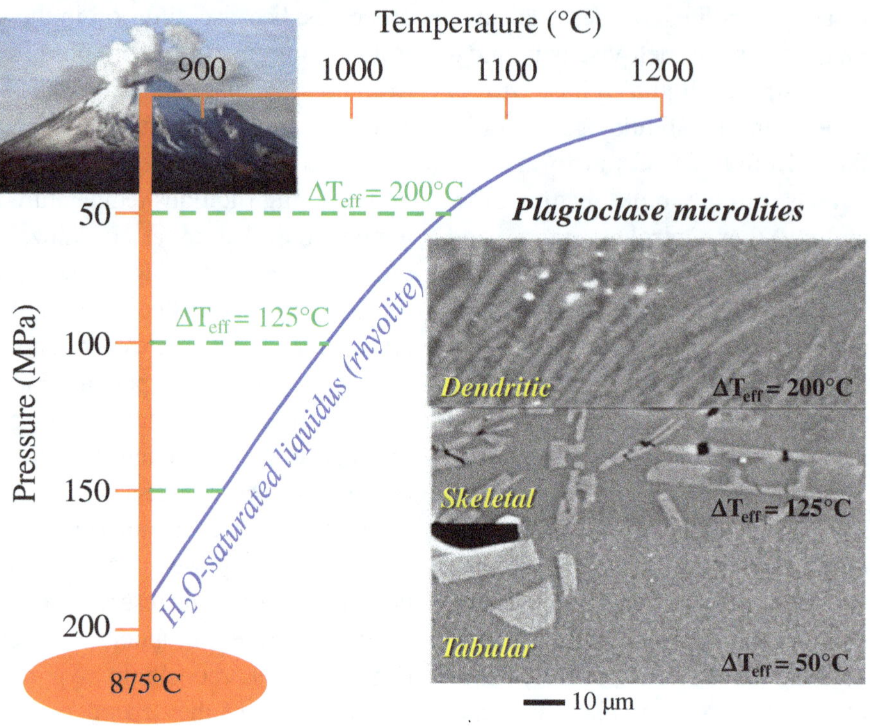

Fig. 3 Relationships between effective undercooling (ΔT_{eff}; *in green*) and microlite textural characteristics in a H_2O-saturated rhyolitic melt (modified after Mollard et al. 2012). Plagioclase microlites crystallized at 150 MPa after an isothermal quasi-instantaneous decompression from 200 MPa, i.e. $\Delta T_{eff} = 50$ °C, are represented by scarce large tabular crystals. With decreasing pressure and increasing ΔT_{eff}, microlites become more numerous, but smaller in size, and they show more complex shapes ranging from skeletal (*hollow*) to dendritic

volcanic eruptions. For instance, Plinian ascent rates of the order of m/s can be simulated experimentally by decompression durations from seconds to hours whereas the slow ascent rates recorded for dome eruptions can be reproduced by decompression durations of several days or weeks. More generally, decompression experiments can simulate different natural eruptive scenarios depending on the applied decompression rate, final pressure, and dwell time at final pressure. In basaltic H_2O- and CO_2-bearing magmas, experimental decompression in the duration range of <1–10 h has provided information on degassing processes leading to either regular or paroxysmal Strombolian eruptions (*Chapter 15 "Magma degassing: the diffusive fractionation model and beyond"*). In silicic melts, decompression pathways and durations from ten seconds to forty days have been investigated experimentally to evaluate the lifetime of rhyolitic foams as a function of bulk viscosity (Martel and Iacono-Marziano 2015). Deformation experiments performed in vessels equipped with torsion or coaxial deformation modules have shown that the lifetime of such magmatic foams is drastically reduced when a differential stress field prevails, because it enhances bubble coalescence (e.g. Okumura et al. 2009). The recent implementation of HP-HT devices that allow magma deformation at pressure coupled with in situ measurements of permeability, represents a considerable step forward for investigating the explosive-effusive transition of volcanic eruptions in the laboratory under realistic conditions (Kushnir et al. 2017).

Figure 4 illustrates how timescales of degassing and crystallization during decompression can be used to decipher eruption style. Magmas from both, Plinian and dome-forming eruptions (dome, block-and-ash flows, surges), degas during ascent. However, gases in dome-forming magmas escape from the melt (leading to dense pyroclasts)

Fig. 4 Deciphering eruption explosivity from decompression-induced degassing and crystallization experiments (see text) [*Photograph of Redoubt is from R. Clucas (USGS), Mt. Pelée is from* Lacroix (1904), *and Merapi comes from the website:* gunungmerapi. weebly.com]

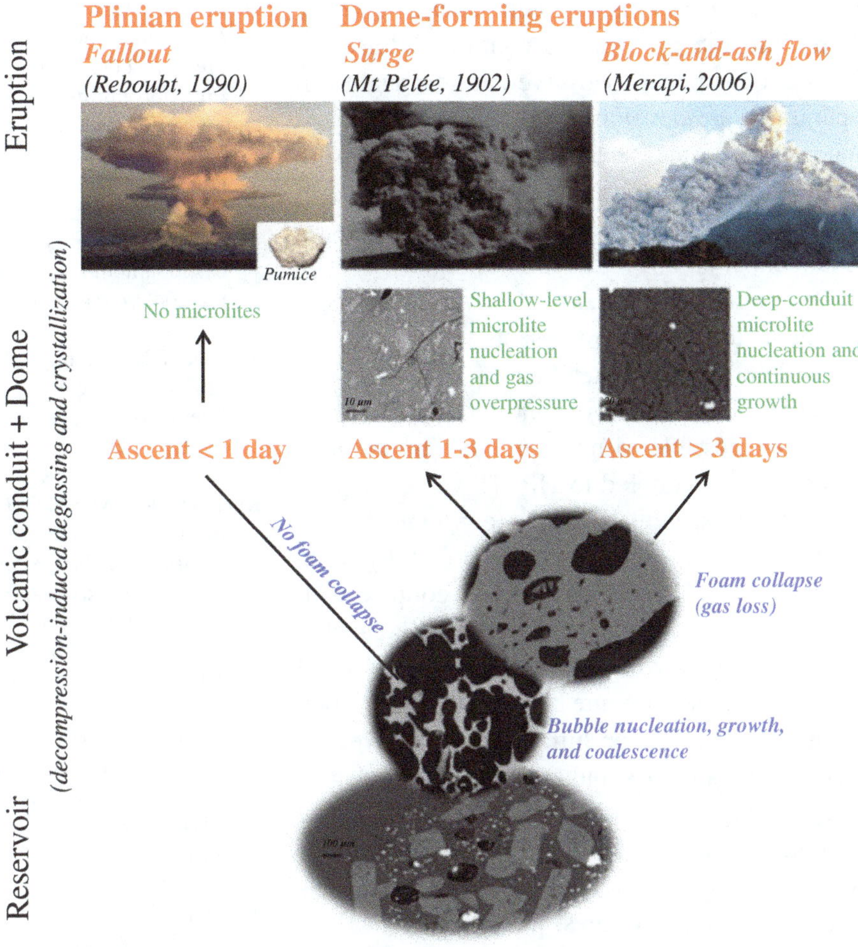

whereas Plinian foams have no time to collapse through gas escape (leading to pumiceous pyroclasts) which suggests Plinian ascent durations are limited to a couple of hours (Martel and Iacono-Marziano 2015). Furthermore, in contrast to Plinian magmas, dome-forming magmas have time to crystallize during ascent. At Mt. Pelée, the moderately-explosive block-and-ash flows in 1929–1932 may have degassed and crystallized continuously during an ascent lasting more than 3–6 days, so that little gas overpressure remains at dome level. In contrast, the devastating surges in 1902 may have resulted from rapid ascent (i.e. <3 days) that did not allow crystallization in the conduit, followed by extensive microlite crystallization at dome level (due to large effective undercooling). The exsolving gas and high overpressurization resulting from this extensive crystallization may have triggered the violent surges (Martel 2012).

This application of decompression experiments highlights the possibility of forecasting the style of an eruption provided the magma ascent rate towards the surface can be determined by some remote means (seismology, gravity, geodesy, gas discharge).

4 Future Directions

The combination of equilibrium and dynamic experiments can simulate many of the conditions relevant to magma eruption, because realistic pressures, temperatures and rates of decompression or shear are now accessible in the laboratory. The one parameter that remains impossible to simulate is an extended timescale. Experiments last typically a maximum of weeks, or occasionally months. In practical terms, this leads to small crystal sizes compared with nature, occasional difficulties in establishing equilibrium, or very short diffusion profiles for controlled 'disequilibrium' experiments. However, as new analytical techniques are developed we can start to make nanoscale measurements that allow us to measure profiles developed on laboratory timescale, providing access to ever faster natural processes (Saunders et al. 2014; Lloyd et al. 2014).

Development of *in situ* observation or measurement techniques represents a major step forward in the understanding of the magmatic processes linked to volcanic eruptions. It is becoming possible to make *in situ* observations using cameras coupled to HP-HT vessels equipped with transparent windows (e.g. Gondé et al. 2011) or 4D *in situ* X-ray tomography (e.g. Pistone et al. 2015). HP-HT vessels coupled with *in situ* analytical techniques are already capable of measuring the volatile species either dissolved in the melt under pressure by *in situ* spectroscopy techniques (Raman or Infrared) or exsolved as vapour. Laboratory simulation of magma degassing and crystallization with these new *in situ* approaches will allow the identification of potential geophysical and geochemical signals that may be used as unrest precursors.

References

Blundy JD, Cashman KV (2008) Petrologic reconstruction of magmatic system variables and processes. Miner Inclusions Volcanic Proc 69:179–239

Blundy JD, Cashman KV, Rust AC, Witham F (2010) A case for CO_2-rich arc magmas. Earth Planet Sci Lett 290(3–4):289–301

Costa F, Morgan D (2010) Time constraints from chemical equilibration in magmatic crystals. In Dosseto A, Turner SP, Van Orman JA (eds) Timescales of magmatic processes: from core to atmosphere. Wiley-Blackwell, pp 125–159

Druitt TH, Costa F, Deloule E, Dungan M, Scaillet B (2012) Decadal to monthly timescales of magma transfer and reservoir growth at a caldera volcano. Nature 482:77–80

Ghiorso MS, Evans BW (2008) Thermodynamics of rhombohedral oxide solid solutions and a revision of the Fe-Ti Two-oxide geothermometer and oxygen-barometer. Am J Sci 308:957–1039

Gondé C, Martel C, Pichavant M, Bureau H (2011) In situ bubble vesiculation in silicic magmas. American Mineralogy 96:111–124

Hammer JE, Cashman KV, Hoblitt RP, Newman S (1999) Degassing and microlite crystallization during pre-climactic events of the 1991 eruption of Mt. Pinatubo Philippines. Bull Volcanol 60:355–380

Hammer JE, Rutherford MJ (2002) An experimental study of the kinetics of decompression-induced crystallization in silicic melts. J Geophys Res 107 (ECV8):1–23

Holland T, Blundy JD (1994) Non-ideal interactions in calcic amphiboles and their bearing on amphibole-plagioclase thermometry. Contrib Miner Petrol 116(4): 433–447

Kilgour GN, Blundy JD, Cashman KV, Mader HM (2013) Small volume andesite magmas and melt-mush interactions at Ruapehu, New Zealand: evidence from melt inclusions. Contrib Minerology Petrol 166 (2):371–392

Kushnir ARL, Martel C, Champallier R, Arbaret L (2017) In situ confirmation of permeability development in shearing bubble-bearing melts and implications for volcanic outgassing. Earth Planet Sci lett 458:315–326

Lacroix A (1904) La Montagne Pelée et ses éruptions. Masson Paris, p 662

Lindsley DH, Andersen DJ (1983) A two-pyroxene thermometer. J Geophys Res 88(S02):001. doi:10. 1029/JB088iS02p0A887

Lloyd AS, Ruprecht P, Hauri EH, Rose W, Gonnerman HM, Plank T (2014) NanoSIMS results from olivine-hosted melt embayments: magma ascent rate during explosive basaltic eruptions. J Volcanol Geoth Res 283:1–18

Mangan MT, Sisson TW, Hankins WB (2004) Decompression experiments identify kinetic controls on explosive silicic eruptions. Geophys Res Lett 31: L08605. doi:10.1029/2004GL019509

Martel C (2012) Eruption dynamics inferred from microlite crystallization experiments: application to Plinian and dome-forming eruptions of Mt. Pelée (Martinique, Lesser Antilles). J Petrol 53:699–725

Martel C, Iacono-Marziano G (2015) Bubble coalescence, outgassing, and foam collapsing in decompressed rhyolitic melts. Earth Planet Sci Lett 412:173–185

Melnik OE, Blundy JD, Rust AC, Muir DD (2011) Subvolcanic plumbing systems imaged through crystal size distributions. Geology 39(4):403–406

Mollard E, Martel C, Bourdier J-L (2012) Decompression-induced crystallization in hydrated silica-rich melts: Empirical models of experimental plagioclase nucleation and growth kinetics. J Petrol 53:1743–1766

Mourtada-Bonnefoi CC, Laporte D (2004) Kinetics of bubble nucleation in a rhyolitic melt: an experimental study of the effect of ascent rate. Earth Planet Sci Lett 218:521–537

Okumura S, Nakamura M, Takeuchi S, Tsuchiyama A, Nakano T, Uesugi K (2009) Magma deformation may induce non-explosive volcanism via degassing through bubble networks. Earth Planet Sci Lett 281:267–274

Pichavant M, Costa F, Burgisser A, Scaillet B, Martel C, Poussineau S (2007) Equilibration scales in silicic to intermediate magmas—implications for experimental studies. J Petrol 48:1955–1972

Pistone M, Arzilli F, Dobson KJ, Cordonnier B, Reusser E, Ulmer P, Marone F, Whittington AG, Mancini L, Fife JL, Blundy JD (2015) Gas-driven filter pressing in magmas: insights into in-situ melt segregation from crystal mushes. Geology 43:699–702

Putirka KD (2008) Thermometers and barometers for volcanic systems. Rev Mineral Geochim 69:61–120

Ridolfi F, Renzulli A (2012) Calcic amphiboles in calc-alkaline and alkaline magmas: thermobarometric and chemometric empirical equations valid up to 1130 °C and 2.2 GPa. Contrib Mineral Petrol 163:877–895

Riker JM, Blundy JD, Rust AC, Botcharnikov RE, Humphreys MCS (2015) Experimental phase equilibria of a Mount St. Helens rhyodacite: a framework for interpreting crystallization paths in degassing silicic magmas. ContribMineral Petrol 170(6):535–560. doi:10.1007/s00410-015-1160-5

Rutherford MJ, Gardner JE (2000) Rates of magma ascent. In: Sigurdsson H (ed) Encyclopedia of volcanoes. Academic San Diego, California, pp 207–217

Saunders KE, Blundy JD, Dohmen RG, Cashman KV (2012) Linking petrology and seismology at an active volcano. Science 336(6084):1023–1027

Saunders K, Buse B, Kilburn MR, Kearns S, Blundy JD (2014) Nanoscale characterisation of crystal zoning. Chem Geol 364:20–32

Scaillet B, Pichavant M (2003) Experimental constraints on volatile abundances in arc magmas and their implications for degassing processes. In: Oppenheimer C, Pyle D, Barclay J (eds) Volcanic degassing. Geological Society, London, Special Publications 213, 23–52

Streck MJ (2008) Mineral textures and zoning as evidence for open system processes. Rev Mineral Geochem 69:595–622

Tamic N, Behrens H, Holtz F (2001) The solubility of H_2O and CO_2 in rhyolitic melts in equilibrium with a mixed CO_2-H_2O fluid phase. Chem Geol 174:333–347

Waters LE, Lange RA (2015) An updated calibration of the plagioclase-liquid hygrometer-thermometer applicable to basalts through rhyolites. Am Miner 100:2172–2184

Magma Chamber Rejuvenation: Insights from Numerical Models

C.P. Montagna, P. Papale, A. Longo and M. Bagagli

Abstract

Most volcanic systems on Earth are characterized by chemically different magmas that can be found in the erupted products throughout their history. The reasons are multiple, including variations in the mantle source and/or crustal assimilation, as well as shallower processes such as fractional crystallization or mixing and mingling. Magma chamber rejuvenation indicates the processes that happen whenever a magma intrudes from the mantle to shallower depths and encounters an already established storage zone (i.e. a magma chamber or reservoir). Magmas rising from depth are typically characterized by higher temperatures, larger volatile contents and more primitive, mantle-like compositions than those residing in the shallow crust. The interaction with magmas that have already resided at shallower depths for a while (years to thousands of years) varies the physical and chemical properties of both the involved magmatic end-members. Typically, volatile-rich magmas coming from depth are lighter than degassed shallow magma; therefore, a gravitational instability sets in as the two come into contact, which generates convection and thus intense mingling and mixing among the two. These dynamic interactions cause variations in the physical and chemical properties of the magmas themselves, as well as in the stress conditons both inside the reservoir and in the host rock. The volcanic system as a whole enters an unrest scenario, that can evolve to eruption or not depending on the specific conditions. Numerical simulations of the dynamics within magmatic systems can shed light on the features of magma chamber rejuvenation, providing the time

C.P. Montagna (✉) · P. Papale · A. Longo ·
M. Bagagli
Istituto Nazionale di Geofisica e Vulcanologia, via
U. della Faggiola 32, Pisa, Italy
e-mail: chiara.montagna@ingv.it

Present Address:
M. Bagagli
D-ERDW, ETH-Zuerich, Zürich, Switzerland

Advs in Volcanology (2019) 111–122
DOI 10.1007/11157_2017_21

Published Online: 17 August 2017

scales of mixing processes and possibly of the evolution towards eruption. Coupling with models for the visco-elastic response of the host rock allows the identification of the onset of recharge processes from the analysis of geophysical signals observed at the surface.

Keywords

Magma chamber · Magma dynamics · Magma mixing

1 Extended English/Spanish abstract

Most volcanic systems on Earth are characterized by chemically different magmas that can be found in the erupted products throughout their history, either in synchronous eruptive episodes, or in different epochs of volcanic activity. This chemical heterogeneity can have multiple reasons, and it originates from deep in the mantle, due to variations in the source and/or crustal assimilation during ascent, to shallower crustal regions, where processes such as fractional crystallization or mixing and mingling take place. Magma chamber rejuvenation comprises some of the aforementioned processes at shallow level. Whenever a magma intrudes from the mantle to shallower depths and encounters an already established storage zone (a magma chamber or reservoir), the magma already emplaced gets rejuvenated by the incoming more primitive magma. Magmas rising from the deep regions of a volcano feeding system are typically characterized by higher temperatures, larger volatile contents and more primitive, mantle-like compositions. On the other hand, magmas that have resided at shallower depths for a while (years to thousands of years) have evolved by fractional crystallization, thus they have changed their composition towards a more felsic one, and have lost most of their gaseous phase, that can escape towards the surface. The interaction between primitive and evolved magmas varies the physical and chemical properties of both the end-members involved. Typically, volatile-rich magmas coming from depth are lighter than degassed shallow magma, albeit having a higher liquid density: they are characterized by a much larger gas content. The light magma tends to rise inside the denser reservoir; a gravitational instability sets in as the two magmatic mixtures come into contact, and generates convection inside the reservoir. As a consequence, intense mingling and mixing are generated among the two end-members. These dynamic interactions cause variations in the physical and chemical properties of the magmas themselves, that loose their identity as initial end-members and become a more homogeneous mixture. The volcanic system as a whole enters an unrest scenario, that can evolve to eruption or not depending on the specific conditions. Numerical simulations of a magmatic system representing magma injection into a shallow reservoir show that mixing is very intense at the time of contact, and can be efficient on time scales of hours to day in homogeneizing the system. Depending on the geometry of the volcano feeding system, and even more on the volatile content of the incoming and resident magmas, the process can be suppressed or enhanced. Sills favour mixing, while more vertically elongated, dike-like reservoirs slow the dynamical interactions. As the presence of a gaseous phase is the engine of the gravitational instability that triggers the dynamics, a higher volatile content, which translates into a higher gas content, in the deep regions of the feeding system strongly accelerates the rejuvenation process. As mixing patterns are found almost ubiquitously in products from volcanoes around the world, comparison of the observed features to the model predictions can provide insights on the features of magma chamber rejuvenation, including the time scales over which mixing processes are efficient and possibly the timings for the evolution towards an eruption or not. Coupling to models that describe the visco-elastic response of the host rock to stress variations within the magmatic system provides hints as to how to identify recharge processes at depth from the analysis of geophysical signals observed at the surface. Characteristic features of ground deformation associated to convection and mixing is the appearance of oscillation of exremely long period, on the order of hours (Ultra-Long-Period, ULP), that can be detected by instruments such as continuous tiltmeters and dilatometers. Their records can identify the onset of the interaction among different magmas, thus provide time scales for unrest duration and evolution.

2 Introduction

Magmas evolve in many ways during their residence time within the crust, determining whether they are going to be erupted or not. Magma chamber rejuvenation takes place whenever a magma intruding from the mantle to shallower depths encounters an already established storage zone (i.e. a magma chamber or reservoir). It can involve many different processes such as reheating and melting of the residing magmas, fractional crystallization due to changes in the pressure and temperature conditions, mingling and mixing among the different components; typically, it takes place at shallow crustal depths. Magmas rising from depth are often characterized by higher temperatures, larger volatile contents and more primitive, mantle-like compositions with respect to those that have been residing at shallower levels for a while (months, years to thousands of years). This general scenario can have a variety of declinations, depending on the specific setting and physico-chemical characteristics of the magmatic mixtures involved. The shallow magma can be highly crystalline, a mush, that can be rejuvenated by the heat from the incoming component (Bachmann and Bergantz 2003, 2008; Girard and Stix 2009; Bain et al. 2013; Till et al. 2015); or, at the other end, it can still be hot and

more fluidal, especially if injection episodes are frequent (Voight et al. 2010), giving rise to mixing and mingling phenomena (Montagna et al. 2015). The interaction among the deep and shallow components changes the physical and chemical properties of both the involved magmatic end-members, triggering an unrest phase that can evolve to eruption or not depending on the specific conditions. Evidence of chamber rejuvenation both in igneous and in intrusive rocks, manifested mostly by mingling and mixing patterns, is almost ubiquitous at volcanic systems worldwide, and it is often invoked as eruption trigger.

Magma movement at depth implies mass re-distribution, pressure changes, and pressure transients which translate into variations in the gravity field, shape and slope of the volcano flanks, and seismic signals registered at the surface. Understanding the complex relationships between quantities measured by volcano monitoring networks and shallow magma processes is a crucial step for the comprehension of volcanic processes and in evaluating more realistic hazard forecast. The ability to detect the onset of magma recharge at depth is fundamental as it can provide hints to unrest duration and evolution, and possibly eruption timings.

In this work we describe a forward-modeling approach to describe magma chamber dynamics, specifically for what concerns rejuvenation episodes, and link it to the geophysical observables that are expected as a consequence. This provides a framework for the consistent interpretation of geological and geophysical records of unrest periods at active volcanoes. This methodology allows for identification of rejuvenation episodes in ground deformation records, and possibly discrimination between those episodes that lead to eruption or not.

3 Numerical Simulations of Magma Chamber Rejuvenation

3.1 Magmatic System

We refer as an archetypal case to the Phlegraean Fields magmatic system, where seismic imaging and attenuation tomographies have identified a huge (probably around 10 km wide) magma reservoir at a depth of around 8 km (Zollo et al. 2008; De Siena et al. 2010), while a variety of geophysical and geochemical evidence suggests that smaller (probably less than 1 km^3), shallower batches of magma have been forming throughout the caldera history at virtually any depth smaller than 9 km (Arienzo et al. 2010; Di Renzo et al. 2011). These shallow magma bodies have been identified as actively involved in past eruptions, which at least in some cases shortly followed the arrival of volatile-rich, less differentiated magmas from the deep feeding system (Arienzo et al. 2009; Fourmentraux et al. 2012). Chemical compositions of erupted magmas range from shoshonitic to trachytic to phonolitic; geochemical analyses on melt inclusions suggest a variety of processes contributing to this variability, such as recharge from depth, intra-chamber mixing, syn-eruptive mingling (Arienzo et al. 2010; Fourmentraux et al. 2012). The same analyses show that deep magmas are typically rich in gas, especially CO_2 (Mangiacapra et al. 2008), while shallow magmas are unusually crystal-poor, down to less than 3 wt% (Arienzo et al. 2009). To study the magmatic dynamics occurring as a consequence of a recharge event, we simplify the magmatic system retaining its most peculiar features. We model the injection of CO_2-rich shoshonitic magma coming from a deep reservoir into a shallower, much smaller chamber, containing more evolved and partially degassed phonolitic magma (see Table 1 for compositions). The two chambers are connected by a dyke. This idealized layout captures several first-order characteristics of prototype magmatic systems, including a composite structure, vertical extension, and heterogeneous composition, and it approximates systems composed by long-lived, interconnected multiple reservoirs believed to exist at many active volcanoes (Elders et al. 2011).

Table 1 Composition of the phonolite and shoshonite magma types employed in the simulations

Composition	SiO$_2$ (wt%)	TiO$_2$ (wt%)	Al$_2$O$_3$ (wt%)	Fe$_2$O$_3$ (wt%)	FeO (wt%)	MnO (wt%)	MgO (wt%)	CaO (wt%)	Na$_2$O (wt%)	K$_2$O (wt%)
Phonolite	53.5	0.6	19.8	1.6	3.2	0.1	1.8	6.8	4.7	7.9
Shoshonite	52.5	0.9	17.6	1.9	5.7	0.1	3.6	7.9	3.4	4.3

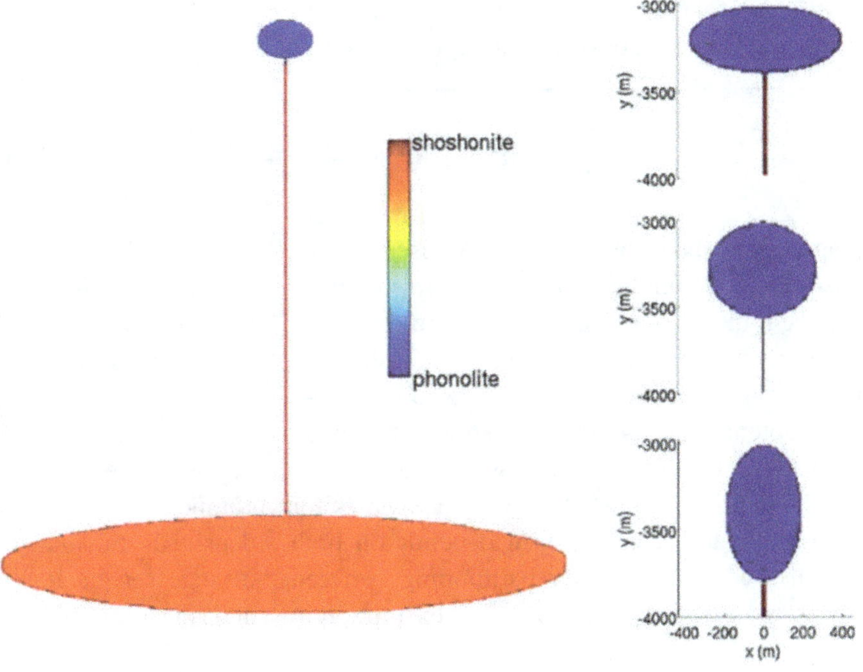

Fig. 1 Initial conditions for the numerical simulations of the magmatic system. On the *left*, the whole domain is shown, indicating the two magmatic end-members. On the *right*, the *upper* portion of the domain shows the three different geometries explored

Figure 1 shows the system domain for the numerical simulations. We assume one of the horizontal dimensions of the magmatic system to be much larger than the other, so that our domain is two-dimensional. The deep chamber is elliptical, 1 km thick and 8 km wide; its top is at 8 km depth. The geometry of the shallow chamber has been varied as shown in Fig. 1, keeping its surface area fixed. In the elliptical cases, the semi-axes measure 400 and 800 m, respectively, while the circular chamber has a radius of 283 m.

The initial conditions of the system are also shown in Fig. 1. The shallow chamber hosts a differentiated, volatile-poor phonolitic magma. Its volatile content has been varied from 0.3 wt% CO_2 and 2.5 wt% H_2O to 0.1 wt% CO_2 and 1 wt % H_2O. In the feeding dyke and deep reservoir is a less evolved, basaltic shoshonite, containing 1 wt% CO_2 and 2 wt% H_2O. Such a low water content in the phonolitic end-member derives from melt inclusion data from Phlegrean Fields (Arienzo et al. 2010). Typically, more evolved magmas are expected to have a relatively larger water content (Signorelli et al. 2001; Cannatelli et al. 2007; Pappalardo et al. 2007; Mollo et al. 2015), resulting possibly in smaller density contrasts at the interface among the two magmas thus less efficient mixing dynamics.

Volatiles partitioning between gaseous and liquid phases is computed following Papale et al. (2006) as a function of composition and pressure. Pressure at time 0 consists of a depth-dependent

turn depends on pressure, thus on the depth at which the interface is placed, which is different for each geometry of the shallow chamber (Fig. 1). Temperature differences between interacting magmas are often negligible (Sparks et al. 1977), particularly at Phlegraean Fields (Mangiacapra et al. 2008; Arienzo et al. 2010), thus the system is assumed isothermal. As a result, there is no need to speculate on the thermal status of the surrounding rock, thus reducing model uncertainties. Moreover, heat transfer effects are expected to play a minor role on the short simulated time scales (hours; Di Renzo et al. 2011).

3.2 Magma Dynamics

Interaction among the two magmas develops as a consequence of the initial gravitational instability at the interface. We solve numerically the two-dimensional space-time evolution of the system, consisting of a mixture of two different magmatic components, each of them including a liquid (silicate melt and dissolved volatiles) and a gaseous (exsolved volatiles) fractions. The equations of motion for the mixture express conservation of mass for each component k = 1, 2, and momentum for the whole mixture (Longo et al. 2012a):

$$\frac{\partial\left(\rho y_k\right)}{\partial t} + \nabla \cdot \left(\rho \mathbf{u} y_k\right) = -\nabla \cdot \left(\rho D_k \nabla y_k\right), \quad \sum_k y_k = 1$$

$$\tag{1}$$

$$\frac{\partial\left(\rho \mathbf{u}\right)}{\partial t} + \mathbf{u} \cdot \nabla\left(\rho \mathbf{u}\right) = -\nabla p + \nabla \cdot \left\{\mu\left[\nabla \mathbf{u} + \left(\nabla \mathbf{u}\right)^{\mathrm{T}} - \frac{2}{3}\nabla \cdot \mathbf{u}\right]\right\} + \rho \mathbf{g}. \tag{2}$$

magmastatic contribution superimposed to the host rock confining pressure. The interface between the two magmas, at the inlet of the shallow chamber, is gravitationally unstable, the lower magma being less dense due to its higher gas content. The dynamics is solely driven by buoyancy, without any external forcing.

The density contrast at the interface varies for each simulated scenario, as it depends on both volatiles content and their partitioning between liquid and gaseous phases; volatiles exsolution in

In the above, t is time; ρ is mixture density, y_k is mass fraction of component k, $\mathbf{u}$ is fluid velocity, D_k is the k-th coefficient of mass diffusion, p is pressure, μ is viscosity and $\mathbf{g}$ is gravity acceleration.

The magmatic mixture is considered ideal. Its density is evaluated as weighted sum of the components' densities; for each component, density is calculated using a non-ideal equation of state for the liquid phase, real gas properties and ideal mixture laws for multiphase fluids.

Mixture viscosity is computed through standard rules of mixing for one phase mixtures and with a semi-empirical relation in order to account for the effect of non-deformable gas bubbles. Liquid viscosity is modeled as in Giordano et al. (2008), and it depends on liquid composition and dissolved water content. The assumption of Newtonian rheology is justified by the very low strain rates and the crystal-free nature of the magmas. The generalized Fick's law is used to describe mass diffusion. Volatile partitioning between gaseous and liquid phases is evaluated at every point in the space-time domain as function of mixture composition and pressure as in Papale et al. (2006). All the physical properties of the two magmas are evaluated at every point in the space-time domain depending on the local conditions of pressure, velocity and mass fractions, which are the unknowns in Eqs. (1) and (2). The equations are solved numerically using GALES, a finite element C++ code specifically designed for volcanic fluid dynamics (Longo et al. 2012a).

The evolution in space and time of the system is complex and presents a number of interesting features. Figure 2 summarizes the results regarding magma dynamics, showing the evolution of composition in time in the shallow chamber for the five different simulation scenarios.

The initial inverse density contrast at the contact interface between the two magmas gives rise to convective mass transfer from the deeper parts of the system to shallower depths and vice versa. The unstable density contrast is solely due to the different volatile content of the two mixtures: the shoshonitic melt has an higher density than the phonolitic. The role played by volatiles is crucial, and it is exsolved gases that ultimately determine the buoyant dynamics. A Rayleigh-Taylor instability develops, which acts to bring the system to gravitational equilibrium by overturning it. The instability develops starting from the perturbed interface, with a first plume of light material that rises into the chamber. Depending on the initial density contrast as well as on the geometry of the shallow chamber, the initial plume starts developing at different times. The dynamics is strongly enhanced by higher density contrasts; geometry also plays an important role when density contrasts are similar, with horizontally elongated, sill-like chambers favouring convection with respect to more dyke-like setups (see also Fig. 2).

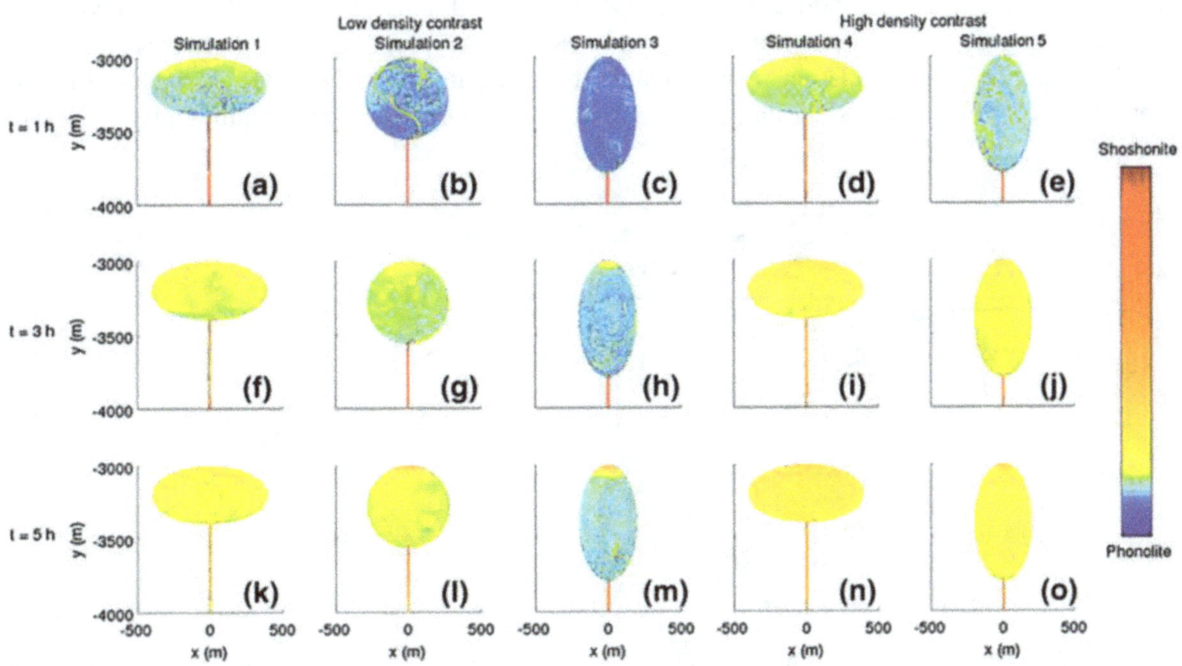

Fig. 2 Snapshots of variation of composition with time in the shallower parts of the system for the different simulations. *Columns* correspond to different simulations; *rows* correspond to different times

Plumes of light magma coming from depth keep entering the shallow reservoir as discrete filaments, following irregular trajectories and showing typical convective patterns. The lighter material tends to rise into the chamber, thereby decreasing more and more its density as volatiles exsolve in lower-pressure environments; on the other hand, the denser magmatic mixture initially residing in the chamber sinks into the feeder dyke, increasing its density by the reverse process of volatile dissolution at higher pressures. The plumes thus progressively increase their buoyancy, enhancing their expansion and acceleration. During the rise, vortexes form at the head of the plumes and subsequent plumes interact among themselves, further favouring mixing. The dynamics creates complicated patterns that maximize the interaction among the two different magmatic mixtures (Petrelli et al. 2011). Mingling is evident for all simulated conditions both within the chamber itself and even more in the feeding dyke (Fig. 2), and it is strongly intensified by the chaotic patterns that form as a consequence of deep magma injection.

Independently from system geometry or density contrast at the interface, mingling is very efficient in the feeding dyke, more than inside the upper chamber. Figure 2 shows that since the very beginning of the simulations, the magma entering the chamber is already a mixture of the two initial end-members, and not the pure shoshonitic composition.

As the dynamics proceeds, faster for higher density contrasts and sill-like setups, the gas-rich mixture tends to accumulate at the top of the chamber, thereby originating a stable density stratification that has indeed been testified at various magmatic systems (Arienzo et al. 2009). The stratification is more prominent in vertically elongated, dyke-like reservoirs (Fig. 2). The density profile along the vertical direction, evaluated averaging along horizontal planes (Fig. 3), illustrates that a quasi-stable profile is reached after some hours of simulated time.

As time proceeds, convection slows down due to smaller buoyancy of the incoming already mixed component, and the instability proceeds in time asymptotically: the more the two end-members have mingled, the less intense is convection.

The evolution of pressure in the system is highly heterogeneous in space and time. Alternating phases dominated by buoyancy and sinking at chamber inlet result in pressure fluctuations with periods of hundreds of seconds and amplitudes decreasing with time (Fig. 4). Typically pressure variations are smaller than 1 MPa; under these conditions, it is unlikely that rejuvenation can trigger eruption, as the stresses needed to create a pathway to the surface in the host rock are typically larger than that (Gudmundsson 2006).

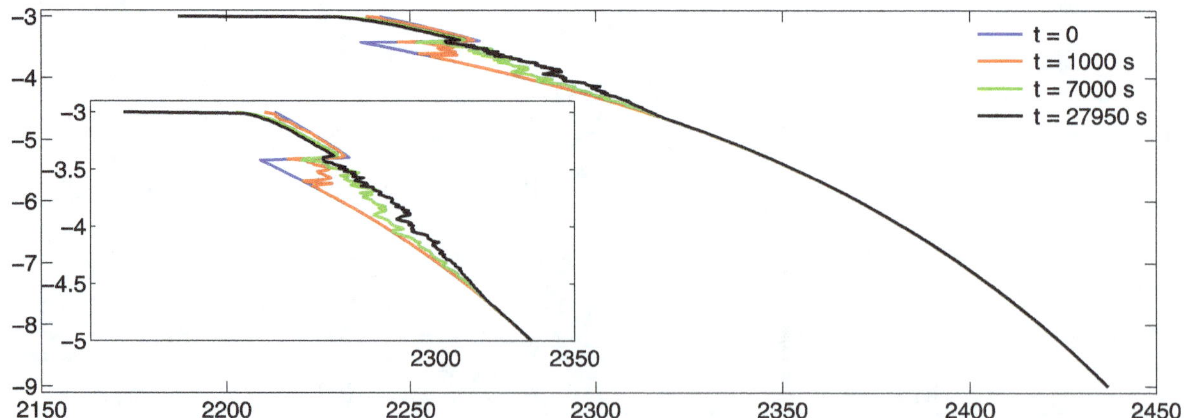

Fig. 3 Total mixture density averaged over horizontal planes as function of depth for simulation 1, at different times. The inset shows the upper 5 km of the domain; the *black line* represents the quasi-equilibrium density profile at the end of the simulation

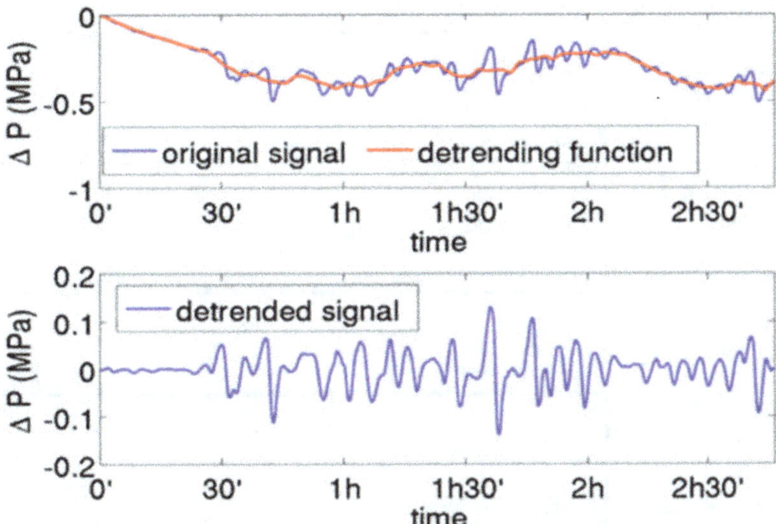

Fig. 4 Pressure variations as a function of time at a point on the boundary of the upper chamber, for simulation #1. The *upper* diagram shows the difference between the local pressure at current time and at time zero, while the *bottom* diagram shows the same quantity after subtraction of a detrending function (*red curve* in the *upper* diagram)

3.3 Ground Deformation

Determining the time–space-dependent ground displacement requires modeling the magma–rocks boundary conditions and the mechanical response of rocks, the latter depending on heterogeneous rock properties, presence and distribution of faults, interfaces, fluids, and volcano topography (e.g., O'Brien and Bean 2004). A first-order analysis performed here assumes magma–rock one-way coupling and adopts the Green's functions formulation for a homogeneous, infinite medium (Aki and Richards 2002).

We consider as point sources the fluid dynamics computational grid nodes located at the reservoir walls. As source time functions, we use the respective temporal evolutions of magmatic forces computed from pressures and stresses provided at those nodes by the numerical simulations of magma convection and mixing dynamics. Ground displacement at a series of virtual receivers is finally obtained by integrating, over all sources, the Green's functions associated with individual sources.

Continuity of pressure and stress is taken as the boundary condition along the non moving magma–rock interface. Physical properties of rocks are homogeneous averages that describe the volcanic edifices within the range of considered depths (<10 km, $v_P = 3000$ m/s; $v_P/v_S = 1/\sqrt{3}$, $\rho = 2500$ kg/m^3).

Propagation of pressure disturbances in the host rock medium reveals that the computed pressure oscillations, originated by the ingression of buoyant magma in the magma chamber, translate into Ultra Long Period ground displacement dynamics with amplitudes of millimeter to micrometer order (Fig. 5; Longo et al. 2012b). ULP ground movements like those predicted by the present modeling could not be detected by classical broadband seismometers (although more recent seismometers extend their working range up to 100–200 s periods), while they are visible in the records from other instruments, especially

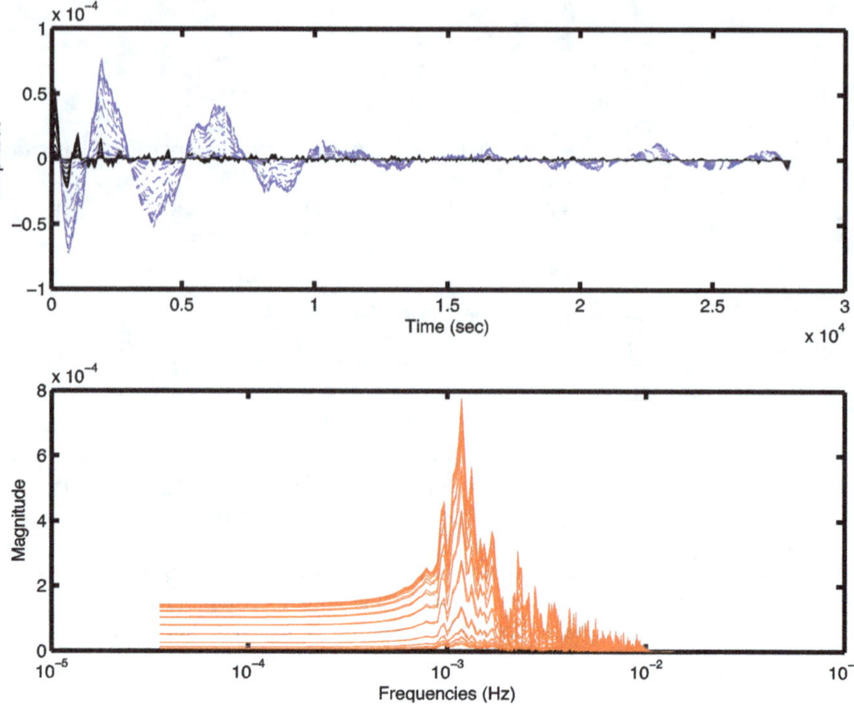

Fig. 5 Synthetic seismic signal (*blue* detrended as to represent an instrumental record, *black* filtered [0.001,0.01] Hz) and corresponding frequency spectrum for the vertical component at all synthetic stations. Example from simulation #1

borehole dilatometers characterized by high signal-to-noise ratio (Sacks et al. 1971).

4   Discussion and Conclusions

The arrival of fresh magma into an already emplaced reservoir and the consequent internal dynamics have often been invoked as possible eruption triggers, especially at Phlegraean Fields (Arienzo et al. 2010); on the other hand, footprints of magma chamber rejuvenation are often found in intrusive granites as well, testifying that it does not necessarily lead to eruption. Magma chamber rejuvenation can thus be regarded as a prototypical volcanic unrest process, that can lead to eruption or not depending on the specific system conditions (e.g. host rock compliance, volume and volatile content of injected magma).

The time scales for rejuvenation processes to be effective in magmatic reservoirs are relatively short, on the order of hours. This is consistent with what has been observed from the analyses of erupted products (Fourmentraux et al. 2012) as well as from experiments on diffusive fractionation (Perugini et al. 2010), and opens a completely new perspective in terms of unrest

duration: indications of mixing in erupted products suggest that recharge events can happen within a very short time frame from eruption, otherwise the evidence would be wiped out by the efficient mixing process (Montagna et al. 2015).

In terms of geophysical observables, convective mingling dynamics in magmatic reservoirs is associated with ultra-long-period seismic signals, characterized by frequencies in the range $10^{-2} - 10^{-4}$ Hz (Longo et al. 2012b).

The results obtained by our modeling specifically refer to the Phlegraean system. They can be extended to many other volcanoes where evidence of rejuvenation has been observed in similar magmatic settings, characterized by relatively primitive magmas that are not very different in composition and temperature, such as e.g. Mount Etna (Viccaro et al. 2006) or Stromboli (La Felice et al. 2011). For more evolved magmas, reservoirs can be dominated by crystal-rich regions (Marsh 1981; Koyaguchi and Kaneko 1999; Bachmann and Berganz 2004; Hildreth 2004; Huber et al. 2009; Cooper and Kent 2014), and they are often at a lower temperature. The approach described above must be applied with caution in such cases, as the dynamics of the incoming primitive magma is more likely to be described as flow through a porous medium (mush) than as fluid mingling and mixing. Nonetheless, there is some evidence for crystal-poor silicic magma reservoirs to be reactivated as well (Bachmann et al. 2002; Deering et al. 2011; Huber et al. 2012; Sliwinski et al. 2015; Wolff et al. 2015).

Given the short time scales over which the dynamical processes described here can be effective and lead to eruption, it would be beneficial to be able to routinely detect the signals described above for eruption forecasting and mitigation actions. This is especially true for long-dormant volcanoes such as Phlegraean Fields, one of the highest-risk volcanic areas in the world given the large population living within the caldera borders (Arienzo et al. 2010), for which there is still no widely accepted means of discriminating the precursors of an impending eruption (Druitt et al. 2012).

Acknowledgements This work has received funds from the European Union's Seventh Programme for research, technological development and demonstration under grant agreements No. 282769 VUELCO and No. 308665 MED-SUV. The manuscript has largely benefited from reviews by Fabio Arzilli and Olivier Bachmann.

Glossary

Magma chamber A storage volume for magma in the crust. Typically, during their rise from the mantle magmas accumlate in regions where there are geological discontinuities.

Rejuvenation The process by which magmas coming from depth modify chemical and physical properties of more evolved magmas already emplaced at shallower levels.

Primitive magma A magmatic melt that has composition similar to that of the mantle (smaller silica content).

Evolved magma A magmatic melt that has undergone processes within the crust that modified its chemical properties, such as fractional crystallization and crustal assimilation. Its composition is characterized by higher silica content.

Convection Exchange of mass and energy by means of cell patterns.

Green's functions Source to receiver transfer function; in this context through the volcanic rock medium.

Index

References

Aki K, Richards PG (2002) Quantitative seismology. University Science, Sausalito

Arienzo I, Civetta L, Heumann A, Woerner G, Orsi G (2009) Isotopic evidence for open system processes within the Campanian Ignimbrite (Campi Flegrei—Italy) magma chamber. Bull Volcanol 71(3):285–300

Arienzo I, Moretti R, Civetta L, Orsi G, Papale P (2010) The feeding system of Agnano—Monte Spina eruption (Campi Flegrei, Italy): Dragging the past into present activity and future scenarios. Chem Geol 270 (1–4):135–147

Bachmann O, Bergantz GW (2003) Rejuvenation of the Fish Canyon magma body: a window into the evolution of large-volume silicic magma systems. Geology 31(9):789–792

Bachmann O, Bergantz GW (2004) On the origin of crystal-poor rhyolites: Extracted from batholithic crystal mushes. J Petrol 45:1565–1582

Bachmann O, Bergantz GW (2008) The Magma Reservoirs That Feed Supereruptions. Elements 4(1):17–21

Bachmann O, Dungan MA, Lipman PW (2002) The Fish Canyon magma body, San Juan volcanic field, Colorado: Rejuvenation and eruption of an upper-crustal batholith. J Petrol 43:1469–1503

Bain AA, Jellinek AM, Wiebe RA (2013) Quantitative field constraints on the dynamics of silicic magma chamber rejuvenation and overturn. Contrib Min Petr 165(6):1275–1294

Cannatelli C, Lima A, Bodnar RJ, De Vivo B, Webster JD, Fedele L (2007) Geochemistry of melt inclusions from the Fondo Riccio and Minopoli 1 eruptions at Campi Flegrei (Italy). Chem Geol 237:418–432

Cooper KM, Kent AJR (2014) Rapid remobilization of magmatic crystals kept in cold storage. Nature 506:480–483

Deering CD, Bachmann O, Vogel TA (2011) The Ammonia Tanks Tuff: erupting a melt-rich rhyolite cap and its remobilized crystal cumulate. Earth Planet Sc Lett 310:518–525

De Siena L, Del Pezzo E, Bianco F (2010) Seismic attenuation imaging of Campi Flegrei: evidence of gas reservoirs, hydrothermal basins, and feeding systems. J Geophysi Res 115(B9):1–18

Di Renzo V, Arienzo I, Civetta L, D'Antonio M, Tonarini S, Di Vito MA, Orsi G (2011) The magmatic feeding system of the Campi Flegrei caldera: architecture and temporal evolution. Chem Geol 281(3–4):227–241

Druitt TH, Costa F, Deloule E, Dungan M, Scaillet B (2012) Decadal to monthly timescales of magma transfer and reservoir growth at a caldera volcano. Nature 482(7383):77–80

Elders WA, Friðleifsson GÓ, Zierenberg RA, Pope EC, Mortensen AK, Guðmundsson Á, Lowenstern JB, Marks NE, Owens L, Bird DK, Reed M (2011) Origin of a rhyolite that intruded a geothermal well while drilling at the Krafla volcano Iceland. Geology 39(3):231–234

Fourmentraux C, Metrich N, Bertagnini A, Rosi M (2012) Crystal fractionation, magma step ascent, and syn-eruptive mingling: the Averno 2 eruption (Phlegraean Fields, Italy). Contrib Min Petr 163(6):1121–1137

Giordano D, Russell JK, Dingwell DB (2008) Viscosity of magmatic liquids: a model. Earth Planet Sci Lett 271:123–143

Girard G, Stix J (2009) Magma recharge and crystal mush rejuvenation associated with early post-collapse Upper Basin Member rhyolites, Yellowstone caldera. Wyoming. J Petrol 50(11):2095–2125

Gudmundsson A (2006) How local stresses control magma-chamber ruptures, dyke injections, and eruptions in composite volcanoes. Earth Sci Rev 79:1–31

Hildreth WS (2004) Volcanological perspectives on Long Valley, Mammoth Mountain, and Mono Craters: several contiguous but discrete systems. J Vol Geoth Res 136:169–198

Huber C, Bachmann O, Manga M (2009) Homogenization processes in silicic magma chambers by stirring and mushification (latent heat buffering). Earth Planet Sci Lett 283:38–47

Huber C, Bachmann O, Dufek J (2012) Crystal-poor versus crystal-rich ignimbrites: a competition between stirring and reactivation. Geology 40:115–118

Koyaguchi T, Kaneko K (1999) A two-stage thermal evolution model of magmas in continental crust. J Petrol 40:241–254

La Felice S, Landi P (2011) The 2009 paroxysmal explosions at Stromboli (Italy): magma mixing and eruption dynamics. Bull Volc 73(9):1147–1154

Longo A, Barsanti M, Cassioli A, Papale P (2012a) A finite element Galerkin/least-squares method for computation of multicomponent compressible in compressible flows. Comp Fluids 67:57–71

Longo A, Papale P, Vassalli M, Saccorotti G, Montagna CP, Cassioli A, Giudice S, Boschi E (2012b) Magma convection and mixing dynamics as a source of Ultra-Long-Period oscillations. Bull Volcanol 74:873–880

Mangiacapra A, Moretti R, Rutherford MJ, Civetta L, Orsi G, Papale P (2008) The deep magmatic system of the Campi Flegrei caldera (Italy). Geophys Res Lett 35

Marsh BD (1981) On the crystallinity, probability of occurrence, and rheology of lava and magma. Contrib Min Petr 78:85–98

Mollo S, Masotta M, Forni F, Bachmann O, De Astis G, Moore G, Scarlato P (2015) A K-feldspar-liquid hygrometer specific to alkaline differentiated magmas. Chem Geol 392:1–8

Montagna CP, Papale P, Longo A (2015) Timescales of mingling in shallow magmatic reservoir. In: Caricchi L, Blundy J D (eds) Chemical, physical and temporal evolution of volcanic systems. Gelogical Society, London, Special Publications 422

O'Brien GS, Bean CJ (2004) A 3D discrete elastic lattice method for seismic wave propagation in heterogeneous media with topography. Geophys Res Lett 31, L14608

Papale P, Moretti R, Barbato D (2006) The compositional dependence of the multicomponent volatile saturation surface in silicate melts. Chem Geol 229:78–95

Pappalardo L, Ottolini L, Mastrolorenzo G (2007) The Campanian Ignimbrite (southern Italy) geochemical zoning: insight on the generation of a super-eruption from catastrophic differentiation and fast withdrawal. Contrib Min Petr 156:1–26

Perugini D, Poli G, Petrelli M, Campos CP, Dingwell DB (2010) Time-scales of recent Phlegrean Fields eruptions inferred from the application of a diffusive fractionation model of trace elements. Bull Volcanol 72(4):431–447

Petrelli M, Perugini D, Poli G (2011) Transition to chaos and implications for time-scales of magma hybridization during mixing processes in magma chambers. Lithos 125(1–2):211–220

Sacks IS, Selwyn S, Evertson DW (1971) Sacks-Evertson strainmeter, its installation in Japan and some preliminary results concerning strain steps. P Jpn Acad 47(9):707–712

Signorelli S, Vaggelli G, Carroll C, Romano MR (2001) Volatile element zonation in Campanian Ignimbrite magmas (Phlegrean Fields, Italy): evidence from the study of glass inclusions and matrix glasses. Contrib Min Petr 140:543–553

Sliwinski JT, Bachmann O, Ellis BS, Dávila-Harris P, Nelson BK, Dufek J (2015) Eruption of shallow crystal cumulates during caldera-forming events on Tenerife. Canary Islands, J Petrol 56(11):2173–2194

Sparks R, Sigurdsson H, Wilson L (1977) Magma mixing: a mechanism for triggering acid explosive eruptions. Nature 267:315–318

Till CB, Vazquez JA, Boyce JW (2015) Months between rejuvenation and volcanic eruption at Yellowstone caldera. Wyoming. Geology 43(8):695–698

Viccaro M, Ferlito C, Cortesogno L, Cristofolini R, Gaggero L (2006) Magma mixing during the 2001 event at Mount Etna (Italy): Effects on the eruptive dynamics. J Vol Geo Res 149(1–2):139–159

Voight B, Widiwijayanti C, Mattioli G S, Elsworth D, Hidayat D, Strutt M (2010) Magma-sponge hypothesis and stratovolcanoes: Case for a compressible reservoir and quasi-steady deep influx at Soufriere Hills Volcano, Montserrat. Geophys Res Lett 37(19): L00E05

Wolff JA, Ellis BS, Ramos FC, Starkel WA, Boroughs S, Olin PH, Bachmann O (2015) Remelting of cumulates as a process for producing chemical zoning in silicic tuffs: A comparison of cool, wet and hot, dry rhyolitic magma systems. Lithos 236–237:275–286

Zollo A, Maercklin N, Vassallo M, Dello Iacono D, Virieux J, Gasparini P (2008) Seismic reflections reveal a massive melt layer feeding Campi Flegrei caldera. Geophys Res Lett 35(12):L12306

Magma Mixing: History and Dynamics of an Eruption Trigger

Daniele Morgavi, Ilenia Arienzo, Chiara Montagna, Diego Perugini and Donald B. Dingwell

Abstract

The most violent and catastrophic volcanic eruptions on Earth have been triggered by the refilling of a felsic volcanic magma chamber by a hotter more mafic magma. Examples include Vesuvius 79 AD, Krakatau 1883, Pinatubo 1991, and Eyjafjallajökull 2010. Since the first hypothesis, plenty of evidence of magma mixing processes, in all tectonic environments, has accumulated in the literature allowing this natural process to be defined as fundamental petrological processes playing a role in triggering volcanic eruptions, and in the generation of the compositional variability of igneous rocks. Combined with petrographic, mineral chemistry and geochemical investigations, isotopic analyses on volcanic rocks have revealed compositional variations at different length scales pointing to a complex interplay of fractional crystallization, mixing/mingling and crustal contamination during the evolution of several magmatic feeding systems. But to fully understand the dynamics of mixing and mingling

D. Morgavi · D. Perugini
Department of Physics and Geology,
University of Perugia, Piazza Università,
06100 Perugia, Italy

D. Morgavi (✉) · D.B. Dingwell
Department Earth and Environmental Sciences,
Ludwig-Maximilian-University (LMU),
Theresienstrasse 41/III, 80333 Munich, Germany
e-mail: daniele.morgavi@unipg.it

I. Arienzo
Istituto Nazionale Di Geofisica E Vulcanologia,
Osservatorio Vesuviano, Via Diocleziano, 328,
8124 Naples, Italy

C. Montagna
Istituto Nazionale Di Geofisica E Vulcanologia,
Via Della Faggiola, 32, Pisa 56126, Italy

Advs in Volcanology (2019) 123–137
DOI 10.1007/11157_2017_30
© The Author(s) 2017
Published Online: 05 December 2017

processes, that are impossible to observe directly, at a realistically large scale, it is necessary to resort to numerical simulations of the complex interaction dynamics between chemically different magmas.

Keywords

Magma mixing · Mingling · Isotope · Modelling

1 Magma Mixing: A Brief Historical Overview

One of the first investigations on magma mixing recorded in the literature is the work of the chemist Bunsen (1851), a scholar at the University of Heidelberg who published research on the chemical variation of igneous rocks from the western region of Iceland. Through chemical analyses Bunsen highlighted that the linear correlation between pairs of chemical elements in binary plots in those Icelandic rocks was the consequence of "simple" binary mixing between two magmas with different chemical composition. Bunsen published this data and, for the first time, magma mixing was taken into account to explain the chemical variation of a suite of igneous rocks. This idea triggered a strong critical reaction from the geological community; the strongest opposition coming from Wolfgang Sartorius Freiherr von Waltershausen an expert on the Iceland and Etna volcanic areas at that time. He mostly argued against the method used by Bunsen of averaging rock analyses to calculate the starting end-members that eventually took part in the mixing process. Sartorius criticized not only the arbitrary choice of the end-members but also disliked the idea of Bunsen of an extensive layering of felsic/mafic rocks and magmas beneath Iceland.

Since the beginning of the 20th century, the experimental and thermodynamic work of Norman L. Bowen (e.g., Bowen 1928) has had a profound influence upon the way petrological processes and igneous differentiations are conceived. The conceptual model of fractional crystallization was firmly established as the most fundamental petrological process for generating the diversity of igneous rocks and remained so for many decades. Although Bowen did not explicitly deny the possibility of magma mixing, he reinterpreted field evidence of magma mixing rather as immiscibility of liquids (e.g., Bowen 1928). In 1944, Wilcox published a work on the Gardner River complex (Yellowstone, USA; Wilcox 1944) which is now considered a milestone for evidence of magma mixing, even if at that time it received strong comments from Fenner and remained one of the few papers on the topic. Only in the 1970's geoscientists started to deeply investigate magma mixing, recorded as a plethora of unequivocal evidence in both plutonic and volcanic rocks, as a major petrogenetic process (e.g., Eichelberger 1978, 1980; Blundy and Sparks 1992; Wiebe 1994; Wilcox 1999). Since the first hypotheses about the origin of mixed igneous rocks (e.g., Bunsen 1851), plenty of evidence of magma mixing processes, in all tectonic environments, throughout geological time, has accumulated in the literature allowing this natural process to be defined as a fundamental petrological process playing a key role in the generation of the compositional variability of igneous rocks and as a major process for planetary differentiation (e.g., Eichelberger 1978, 1980; Blundy and Sparks 1992; Wiebe 1994; De Campos et al. 2004; Perugini and Poli 2012; Morgavi et al. 2016).

2 Magma Mixing: Field Evidence

It is common practice in the petrological community to split magma mixing into two separate physico/chemical processes: (i) mechanical mixing (also referred to as "magma mingling"), by which two or more batches of magma mingle without chemical exchanges between them, and (ii) a chemical mixing (also referred to as "magma mixing") triggered by chemical exchanges between the interacting magmas in which elements move from one magma to the other according to compositional gradients continuously generated by the mechanical dispersion of the two magmas (e.g., Flinders and Clemens 1996). Physically, "magma mingling" is mainly controlled by the viscosity contrast between the two magmas; decreasing of the viscosity contrast results in progressively more efficient mingling dynamics (e.g., Sparks and Marshall 1986; Grasset and Albarede 1994; Bateman 1995; Poli et al. 1996; Perugini and Poli 2005). Chemically, "magma mixing" is driven by the mobility of chemical elements in the melt fractions of the two magmas (e.g., Lesher 1990; Baker 1990). Linear variations in inter-elemental plots for a set of rock samples have long been considered as the sole evidence for the occurrence of magma mixing (e.g., Fourcade and Allegre 1981).

The adoption of the above conceptual models led to the common practice of applying the term magma mingling to indicate the process acting to physically disperse (no chemical exchanges are involved) two or more magmas, whereas the term magma mixing indicates that the mingling process is also accompanied by chemical exchanges. Although such a conceptual approach may allow us to simplify the complexity of the magma mixing process and make it more tractable from the petrological point of view, unfortunately such terminology is not consistently used in the literature and this causes some misunderstanding. Although it is not always easy to clearly discriminate between the two processes, mingling is quite a rare process in nature as physical dispersion and chemical exchanges must occur in tandem during magma mixing processes (e.g., Wilcox 1999; Perugini and Poli 2012).

The most common evidence for magma mixing in igneous rocks is the occurrence of textural heterogeneity; the processes responsible for this have been discussed extensively in many works in the last decades (e.g., Eichelberger 1975; Anderson 1976; Bacon 1986; Didier and Barbarin 1991; Wada 1995; De Rosa et al. 1996; Ventura 1998; Smith 2000; Snyder 2000; De Rosa et al. 2002; Perugini et al. 2002, 2007; Perugini and Poli 2005, 2012; Pritchard et al. 2013; Morgavi et al. 2016).

In order to provide possible classification of magma mixing structures, the evidence of mechanical mixing in igneous rocks can be roughly divided into three different groups: (i) flow structures, (ii) magmatic enclaves and (iii) physico-chemical disequilibria in melts and crystals (e.g., Walker and Skelhorn 1966; Didier and Barbarin 1991; Hibbard 1995; Flinders and Clemens 1996; Wilcox 1944, 1999; Perugini et al. 2002, 2003; Streck 2008; Perugini and Poli 2012; Morgavi et al. 2016). Flow structures can be readily recognized in field outcrops as they show alternating light and dark coloured bands constituted by magmas with different compositions. Figure 1a, b shows some examples of fluid structures occurring in volcanic rocks from Grizzly Lake outcrop in Yellowstone National Park (USA) (Pritchard et al. 2013; Morgavi et al. 2016) and from Soufrière Hills volcano (Island of Montserrat, UK) (Plail et al. 2014). In particular, Fig. 1a shows flow bands of rhyolitic magma (white) intruding in a basaltic/hybrid magma (red to dark grey) whereas Fig. 1b shows an alternation of flow bands of rhyolitic (white) and hybrid magma (dark grey) across which basaltic enclaves (light grey) occur. The latter are surrounded by flow bands of hybrid filaments (blue).

Magmatic enclaves are probably the structural evidence that, according to common thinking, mostly characterize magma mixing processes. The term magmatic enclave is used to identify a discrete portion of a magma occurring within a host magma with a different composition (e.g., Wilcox 1944; Walker and Skelhorn 1966; Bacon 1986, Didier and Barbarin 1991). Generally, enclaves display quite sharp contacts with the

Fig. 1 Detailed images of the mixing features present at Grizzly Lake (Yellowstone) (**a**, **b**) and Soufrière Hills volcano (Montserrat) (**c**, **d**). Figure 1a shows the rhyolitic magma (white) has apparently intruded into the basaltic magma and the hybrid portions are present at the contact between the two end-members. Two large basaltic enclaves are visible at the bottom left and at the centre right. Figure 1b shows the stretching and folding of hybrid magma (dark grey) into a rhyolitic portion (white) with the presence of several basaltic enclaves. Figure 1c shows a basaltic enclave surrounded by andesitic magma (Soufrière Hills, from the 2010 eruption). Figure 1d exhibits at the centre a basaltic enclave in and andesitic host from Soufrière Hills volcano

host rock, although it is not rare to observe that some enclaves display engulfment and disruption of their boundaries due to infiltration of the host magma. Some examples of enclaves found in the volcanic rocks from Soufrière Hills are shown in Fig. 1c, d.

Disequilibrium textures in minerals (Fig. 2a–c) can be viewed as recorders of the thermal and compositional disequilibria operating in the magmatic system during the development of magma mixing processes. As the zoning pattern can be well preserved in minerals from both the plutonic and volcanic rocks, crystal populations from both environments can be used to reconstruct the time evolution of thermal and compositional exchanges between the two magmas during mixing. Recent studies highlighted the importance of detailed investigations of crystal compositional variability not only to reconstruct the fluid-dynamic regime governing the evolution of the igneous body, but also to understand the length-scale of the compositional variability induced by the mixing process, the latter being considered as a proxy to estimate the residence time of magmas in sub-volcanic reservoirs prior to eruption (e.g., Costa and Chakraborty 2004; Martin et al. 2008; Chamberlain et al. 2014; Perugini et al. 2003).

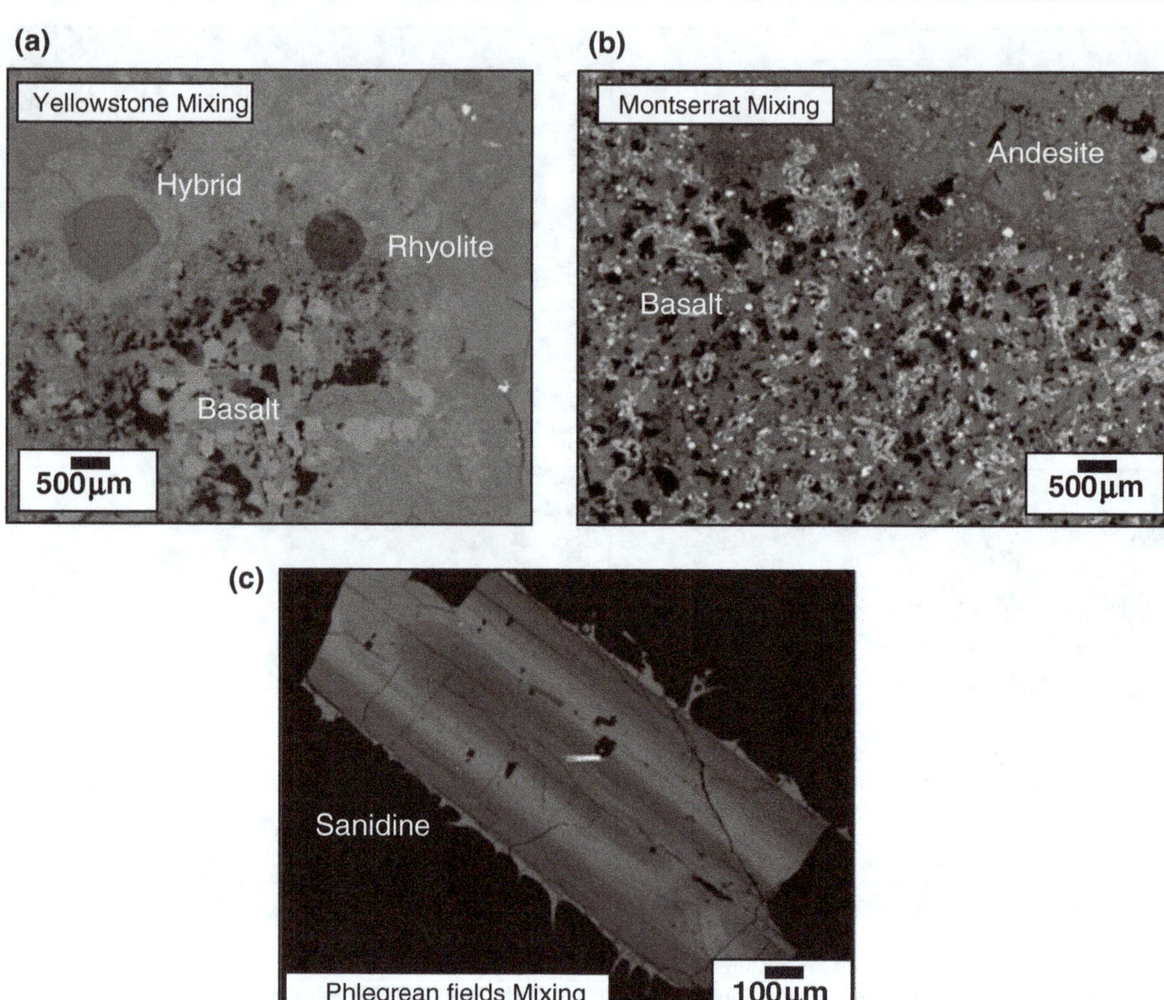

Fig. 2 Backscattered electron (BSE) image of a section of disequilibrium textures in rock and minerals from Yellowstone, Montserrat and Phlegrean Fields. **a** Mixed rock from Grizzly Lake Complex (Yellowstone) showing the interaction between the basaltic portion, the hybrid portion and the rhyolitic portion. **b** Mixed rock section from the 2010 Soufrière Hill eruption (Montserrat) showing disequilibrium texture in the basaltic and the andesitic rock. **c** Crystal from the 4.67 ± 0.09 cal ka Agnano Monte-Spina eruption (Phlegrean fields) occurred from a vent in the Agnano-San Vito area, has a darker (i.e., Ba-poorer) resorbed core and an inner rim with "swirly" zonation textures that indicate crystallization and dissolution. The outer rim is characterized by small-scale wavy oscillatory zoning that results from high-frequency growth and resorption events

3 Numerical and Experimental Studies: New Ideas for Deciphering the Complexity of Magma Mixing

Studies focused on numerical and experimental investigation of magma mixing dynamics (e.g., Perugini et al. 2003, 2008, 2015; De Campos et al. 2004, 2008, 2011; Petrelli et al. 2011; Montagna et al. 2015; Morgavi et al. 2015; Laumonier et al. 2014) can provide additional tools for a better understanding of the complexity of the mixing process, the evolution of which is governed by a continuous exchanges. One of the most striking results arising from these studies is that, during mixing, chemical elements experience a diffusive fractionation process due to the development in time of chaotic mixing dynamics (Perugini et al. 2006, 2008). This process is considered the source of strong deviations in many chemical elements from the linear variations in inter-elemental plots that would otherwise be expected, based on a conceptual model classically adopted in the geochemical modelling of magma mixing processes (e.g., Fourcade and Allegre 1981; Perugini and Poli 2012 and references therein). Recent studies on the mineralogical and geochemical features of mixed rocks (e.g., Hibbard 1981, 1995; Wallace and Bergantz 2002; Costa and Chakraborty 2004; Perugini and Poli 2005; Slaby et al. 2010), as well as those focused on quantitative analyses of morphologies related to textural heterogeneity (e.g., Wada 1995; De Rosa et al. 2002; Perugini and Poli 2005; Perugini et al. 2002, 2003) have highlighted the dominant role played by chaotic mixing dynamics in producing the substantial complexity of geochemical variations and textural patterns found in the resultant rocks (e.g., Flinders and Clemens 1996; De Campos et al. 2011; Morgavi et al. 2013a, b, c, 2016). Despite significant attention in the past, however, few works have focused on the understanding of the relationship between the morphology of the mixing patterns and the geochemical variability of the system using experimental devices (e.g., De Rosa et al. 2002).

Based upon the combination of field observations and the outcome from numerical simulations, a new experimental apparatus has been developed to perform mixing experiments using high viscosity silicate melts at high temperature (De Campos et al. 2011; Morgavi et al. 2013a, c, 2015). This device has been used to study the mixing process between natural melts, enabling the investigation of the influence of chaotic dynamics on the geochemical evolution of the system of mixing magmas (Morgavi et al. 2013a, b, c, 2015).

Preliminary results indicate that the time evolution of compositional exchanges between magmas from the experiments can be effectively modelled, leading to the prospect that the record of magma mixing processes may serve as chronometers to estimate the time interval between mixing and eruption (Perugini et al. 2010; Perugini and Poli 2012; Morgavi et al. 2013a, b, c, 2015; Perugini et al. 2015).

4 Geochemical Evidence of Magma Mixing/Mingling: An Example from the Campi Flegrei Volcanic Area

In some volcanic areas chemically and isotopically distinct magmas have been erupted, and their composition identified by analyzing the chemical composition of the erupted products (e.g., Pantelleria (Italy), Gedemsa and Fanta 'Ale (Main Ethiopian Rift), Gorely Eruptive Center (Kamchatka); Civetta et al. 1997; Giordano et al. 2014; Seligman et al. 2014). However, in other volcanic complexes the majority of the erupted products are chemically rather homogeneous, displaying a dominant composition (e.g., trachybasalt at Mt. Etna; trachyte at Campi Flegrei, Italy; basalt at Réunion Island, Indian Ocean; andesite at Popocatepetl, Mexico). Despite the roughly homogenous composition of products from these volcanic areas, their isotopic features suggest that complex open system processes occurred and superposed the main fractional crystallization trend. In fact, isotopic analyses (e.g., Sr, Nd, Pd, B) have been proven to be an important tool for discriminating between closed-system fractional crystallization and open-system magma mixing/mingling or crustal contamination (e.g., James 1982; Knesel et al. 1999; Turner and Foden 2001). Combined with petrographic, mineral chemistry and chemical investigations, isotopic analyses on volcanic rocks have revealed compositional variations at different length-scales (bulk rock, minerals, single crystals) pointing to a complex interplay of fractional crystallization, mixing/mingling and crustal contamination during the evolution of several magmatic feeding systems (e.g., Di Renzo et al. 2011 and references therein; Melluso et al. 2012; Corsaro et al. 2013 and references therein; Di Muro et al. 2014 and references therein; Brown et al. 2014). Furthermore, together with conventional isotopic analyses, current technologies permit high precision, in situ determination of Sr isotopic ratios of portions of phenocryst and glasses. In fact, microsampling by MicroMill™, coupled with isotopic measurement by Thermal

Ionization Mass Spectrometry (TIMS), allows for the performance of high precision determination of Sr isotopic composition of single crystals or portions of them. This information, unobtainable from bulk samples, has been used successfully to gather information on the time- and length-scales of the pre-eruptive magmatic processes, for identifying mantle sources and/or magmatic end-members and for tracking the time evolution of magma differentiation (e.g., Davidson et al. 1990; Davidson and Tepley 1997; Davidson et al. 1998; Knesel et al. 1999; Font et al. 2008; Kinman et al. 2009; Francalanci et al. 2012; Braschi et al. 2012; Jolis et al. 2013; Arienzo et al. 2015).

Among the active volcanic areas worldwide the volcanic hazard posed by the Campi Flegrei caldera is extremely high, due to its explosive character. Both the high volcanic hazard and the intense urbanization result in an extreme volcanic risk in this area, leading to a considerable interest in understanding which processes might contribute to triggering of eruptions and controlling? eruptive dynamics. The Campi Flegrei caldera is a nested and resurgent structure in the Campania Region, South Italy (Orsi et al. 1996), possibly formed after two large caldera forming eruptions: the Campanian Ignimbrite eruption (39 ka, Fedele et al. 2008) and the Neapolitan Yellow Tuff (15 ka, Deino et al. 2004). Its magmatic system is still active as testified by the occurrence of the last eruption in 1538 AD, as well as the present widespread fumaroles and hot springs activity, and the persistent state of unrest (Del Gaudio et al. 2010; Chiodini et al. 2003, 2012, 2015; Moretti et al. 2013). For compositionally homogenous magmas such as those extruded at the Campi Flegrei caldera (trachytes and phonolites being by far the most abundant rocks), major oxide and trace element variations cannot be used to unequivocally establish which magma evolution processes operated. Thus, together with petrographic, mineral chemistry and chemical data, isotopic investigations on volcanic rocks spanning the history of the volcano have been performed in recent decades in order to define the role of variable magmatic processes in

the evolution of its feeding system up to eruption (e.g., Civetta et al. 1997; D'Antonio et al. 1999, 2013; de Vita et al. 1999; Pappalardo et al. 2002; Fedele et al. 2008, 2009; Tonarini et al. 2004, 2009; Arienzo et al. 2010, 2011; Perugini et al. 2010; Di Vito et al. 2011; Melluso et al. 2012; Arienzo et al. 2015).

In particular, detailed investigations of the geochemical and isotopic (Sr, Nd, Pb, and B) features of the younger than 15 ka Campi Flegrei volcanic products gave understanding to how many variable magmatic components, rising from large depth and/or stagnating in middle crustal reservoir(s), recharged the shallowest reservoir(s) and interacted with magma batches left from previous eruptions (Di Renzo et al. 2011). One identified magmatic component, geochemically similar to magma from the Neapolitan ca. 0.70750, 143Nd/144Nd ratio of ca. 0.51247, 206Pb/204Pb of ca. 19.04 and d11B of ca. -7.8‰), has been the most prevalent component over the past 15 ka. A second magmatic component, having geochemical features similar to the Minopoli 2 magma (D'Antonio et al. 1999; Di Renzo et al. 2011), first erupted 10 ka ago, is shoshonitic in composition. It is the most enriched in radiogenic Sr (87Sr/86Sr of ca. 0.70850) and unradiogenic Nd and Pb (143Nd/144Nd ratio of ca. 0.51238, 206Pb/204Pb of ca. 18.90), and it is characterised by the lowest d11B value of ca. -7.4‰. The third component is trachytic in composition and is characterized by lower 206Pb/204Pb (ca. 19.08), 87Sr/86Sr (ca. 0.70720) and d11B (-9.8‰) and higher 143Nd/144Nd (ca. 0.51250), with respect to the Neapolitan Yellow Tuff component (Tonarini et al. 2009; Di Renzo et al. 2011; Arienzo et al. 2015). This third composition is known as the Astroni 6 component due to the fact that it best recognized in the Astroni 6 erupted products (Di Renzo et al. 2011). During the past 5 ka of activity, this new component has been suggested to have mixed in variable proportions with the Neapolitan Yellow Tuff and Minopoli 2 magmatic components, which dominated the Campi Flegrei volcanic activity mostly in the time span from 15 to 5 ka (Fig. 3).

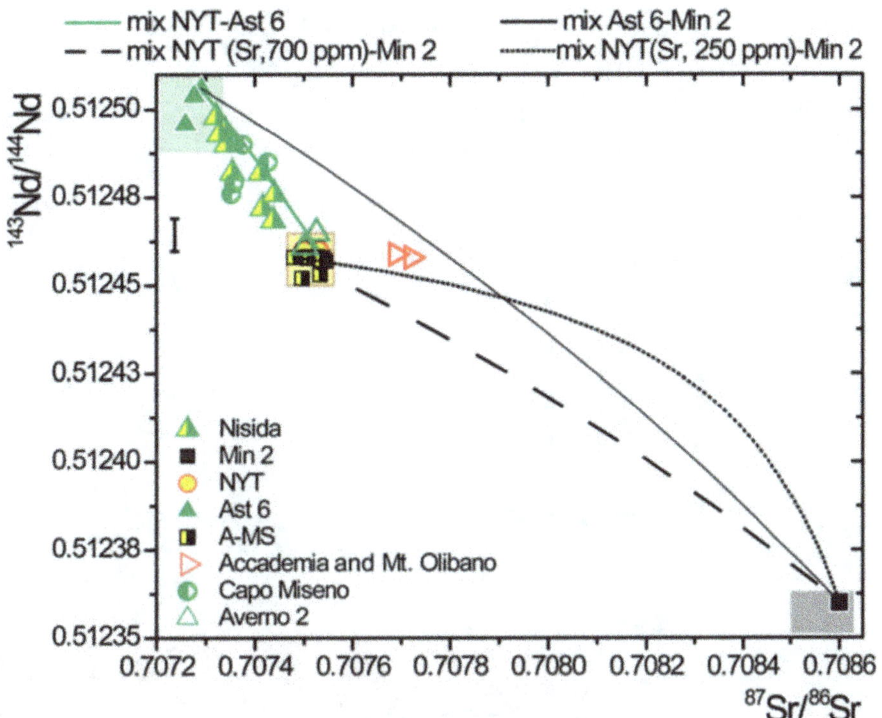

Fig. 3 Modelling of the Sr-Nd isotopic features of some of the Campi Flegrei volcanics of the past 5 ka, by assuming mixing among the Astroni 6 (Ast-6)-, Neapolitan Yellow Tuff (NYT)- and Minopoli 2 (Min 2)-like magmatic components. The green, yellow and black boxes represent the range of Sr and Nd isotopes of the products erupted during the Astroni 6, Neapolitan Yellow Tuff and the Minopoli 2 eruptions, respectively. Symbols inside the plot represent volcanic products belonging to the listed eruptions. The vertical error bar is the uncertainty in 143Nd/144Nd determination at the 2σ level of confidence; that for 87Sr/86Sr is included in the symbols. Modified after Arienzo et al. (2016)

Based on isotope investigations and melt inclusions studies, Arienzo et al. (2016) suggested that the Astroni 6 component, although undergoing differentiation during uprising, had a deep origin (larger than 8 km depth). Indeed, this magma rose not only inside and along the margins of the caldera, but also at the intersection between SE-NW and NE-SW regional fault systems mixing with the NYT-like magma component at shallower depth, and possibly entrapping crystals accumulated during older eruptions.

This detailed study of the Campi Flegrei volcanic system highlights that Sr isotopic microanalysis and, in general, more conventional isotopic analyses, coupled with petrographic, mineral chemistry and geochemical data can provide a better knowledge of the mixing/mingling processes and of the mixing end-members. In turn, they provide (i) information for evaluating the volcanic hazards and mitigating the related risks and (ii) the basic geochemical and petrologic knowledge for the numerical simulations.

5 Numerical Simulation of Magma Mingling and Mixing

To understand mixing and mingling processes at a realistically large scale, it is necessary to resort to numerical simulations of the complex interaction dynamics between chemically different magmas. Referring to the archetypal case of the Campi Flegrei magmatic system as described above, the interaction of a shoshonite and a more evolved phonolite has been investigated in detail to provide constraints on the time and length scales of the mixing dynamics. The simulated system consists of a very large, deep (8 km) reservoir connected by a dike to a shallower,

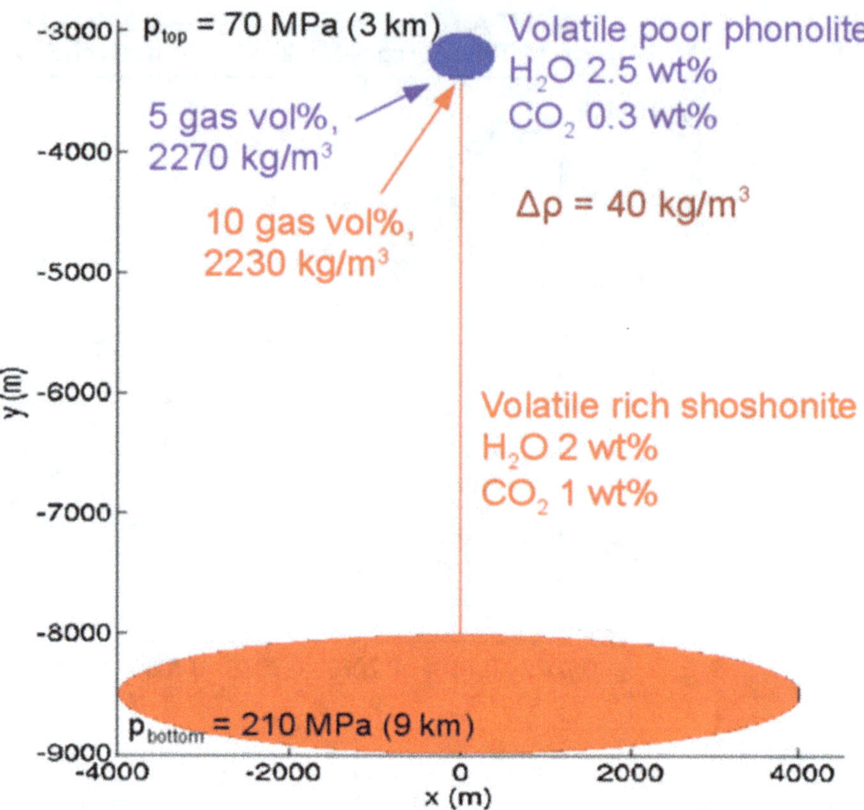

Fig. 4 Initial setup for the numerical simulations of magma chamber replenishment at Campi Flegrei caldera. The two interacting magmas are characterized by different compositions and volatile contents, thus densities and gas volume fractions

smaller chamber (Fig. 4). The chemical interactions between the two magmas cannot be resolved on the simulated large scale, as the computational costs required would be too high. The shoshonitic magma, being richer in volatiles than the resident phonolitic melt, rises into the phonolitic chamber by buoyancy, generating convection and mixing within the reservoir and in the feeding dike. Both magmatic components include a liquid (silicate melt) and a gaseous phase, that cannot decouple from the host melt. Space-time varying volatile exsolution is computed as a function of local composition, pressure and temperature following Papale et al. (2006). More details on magma chamber dynamics can be found in Chap. 8 'Magma chamber rejuvenation: insights from numerical models' of this book.

The main results show that the chaotic patterns observed in the products and in recently developed experimental setups (Morgavi et al. 2013a, b, c, 2015) are reproducible (Fig. 5), and

that the two magmas mingle very efficiently from the beginning of their interaction. As time progresses, convection slows down due to smaller buoyancy of the incoming mixed component, and the instability proceeds in time asymptotically: the more the two end-members have mingled, the less intense the convection. A time-dependent mixing efficiency η_C can be defined as:

$$\eta_C = \frac{|m_R(t) - m_R(0)|}{m_R(0)}. \tag{1}$$

In the equation above, $m_R(t)$ is the mass of the resident phonolitic magma at time t, thus $m_R(0)$ is the initial phonolite mass. The mixing efficiency η_C represents the relative variation of the mass of the initially resident magma in a certain region of the domain. Figure 6 shows the time evolution of mixing efficiency in the shallow chamber, for different simulated setups in terms of chamber geometry and volatile content. It clearly shows

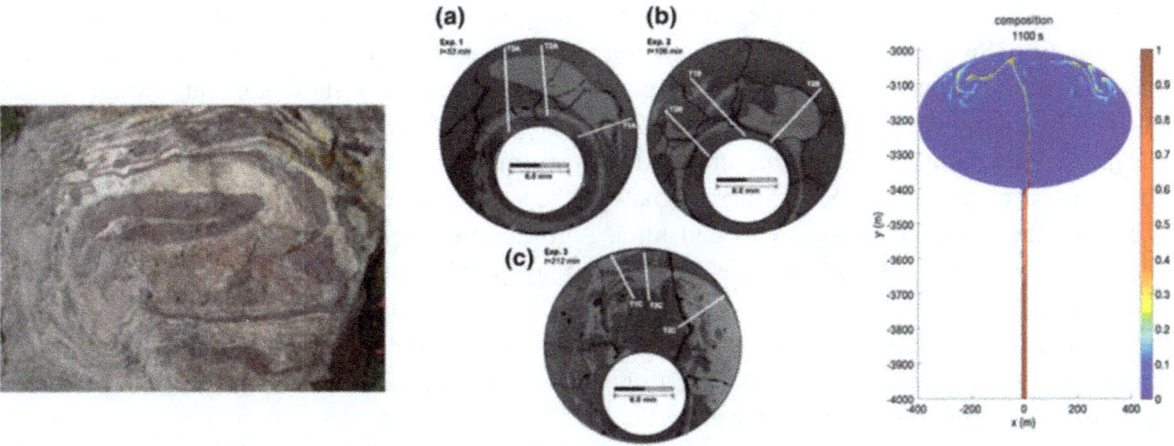

Fig. 5 Magma mixing in natural samples, experimental setups 684 and numerical modelling. Left: Lesvos (Greece) lava flow, from Perugini and Poli (2012); centre: experimental setup, from Morgavi et al. (2013b); right: simulations of magma chamber replenishment

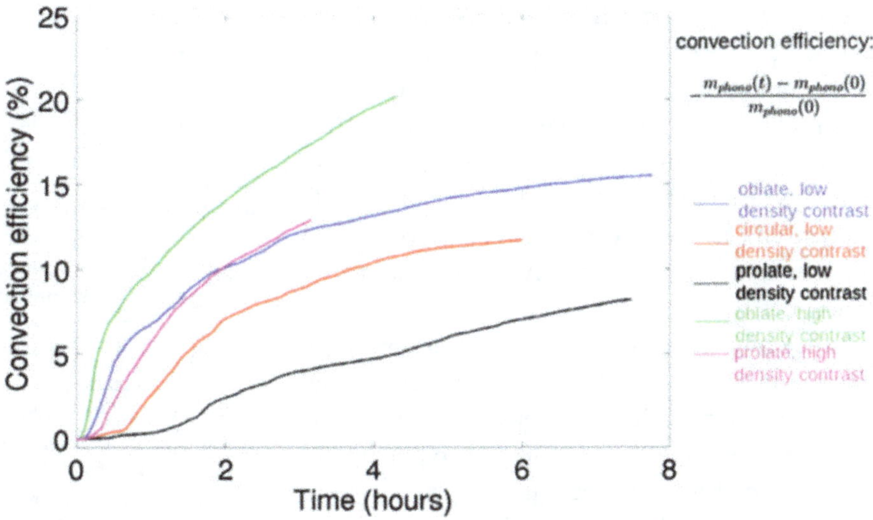

Fig. 6 Variation of convection efficiency with time in the shallow chamber for different simulated scenarios, characterized by varying geometry and total volatile content in the shallow chamber

that mingling is very effective on relatively short time scales, on the order of hours, in agreement with mixing timescales derived from the geochemical modelling of magma mixing experiments (Perugini et al. 2010). When buoyancy drives fast dynamics, more than 40% of the original end-member magmas have mingled in the feeder dyke within 4 h from the arrival of the gas-rich magma from depth. The asymptotic behaviour seems to be reached earlier for less efficient setups: after 4–5 h from the onset of the instability, the system seems to have reached a quasi-steady state. In these cases, a much smaller part of the two magmas has mingled.

6 Magma Mixing Time Scale and Eruption Trigger

The most violent and catastrophic volcanic eruptions on Earth have been triggered by the refilling of a felsic volcanic magma chamber by a hottest and more mafic magma (Kent et al. 2010; Murphy 1998). Examples include Vesuvius 79 AD (Cioni et al. 1995), Krakatau 1883 (Self 1992), Pinatubo 1991 (Kress 1997), the Campanian Ignimbrite (Arienzo et al. 2009) and Eyjafjallajökull 2010 (Sigmundsson et al. 2010). Injection of the more mafic magma into the felsic magma triggers convection dynamics and widespread mixing (Sparks et al. 1977). Vesiculation induced by convection increases magma pressure and may fracture the volcanic edifice triggering an explosive eruption. The injection and mixing process is accompanied by geophysical signals, such as earthquakes, gravity changes and ultra-long-period ground oscillations, that can now be accurately detected (Williams and Rymer 2002; Longo et al. 2012; Bagagli et al. 2017). The knowledge of the time elapsing between the beginning of mixing (and associated geophysical signals) and eruption is thus of greatest importance in forecasting the onset of a volcanic eruption.

Recent studies highlighted that in order to preserve magma mixing structures (i.e., filaments, swirls, bandings) in the rocks, the time elapsed between the beginning of mixing and the subsequent eruption must be very short; on the order of hours or days (Perugini et al. 2010). Preserved structures would indicate mixing was the last process recorded by the magmatic system and its study can unravel unprecedented information on pre-eruptive behaviour of volcanoes.

The compositional heterogeneity produced by magma mixing, and subsequently frozen in time in the volcanic products, can hence be viewed as a broken clock at a crime scene; it can potentially be used to determine the time of the incident. Following this idea and combining numerical simulations with magma mixing experiments using natural compositions and statistical analyses it was shown that for three volcanic eruptions from the Campi Flegrei volcanic system (Astroni, Averno and Agnano Monte Spina) the mixing-to-eruption timescale are of the order of a few minutes (Perugini et al. 2015). These timescales indicate that very little time elapsed from the moment mixing started until eruption. These results are in agreement with recent numerical simulations of magma mixing (Montagna et al. 2015) that highlight mixing timescales of a few hours to attain complete hybridization of magmas for the Campi Flegrei magmatic systems.

These results have implications for civil protection planning of future volcanic crisis as the high velocities of ascending magmas may imply little warning time in volcanic crises. These findings can be a starting point towards a unifying model explaining chemical exchanges in magmatic systems and supplying information on the use of chemical element mobility as geochronometers for volcanic eruptions. This may provide unparalleled clues for building an inventory of past and recent volcanic eruption timescales and could be decisive for hazard assessment in active volcanic areas.

Acknowledgements This research was funded by the European Union's Seventh Programme for research technological development and demonstration under grant agreement No 282759—VUELCO and by the ERC Consolidator Grant 612776—CHRONOS.

References

Anderson AT (1976) Magma mixing: petrological process and volcanological tool. J Volcanol Geotherm Res 1:3–33

Arienzo I, D'Antonio M, Di Renzo V, Tonarini S, Minolfi G, Orsi G, Carandente A, Belviso P, Civetta L (2015) Isotopic microanalysis sheds light on the magmatic endmembers feeding volcanic eruptions: the Astroni 6 case study (Campi Flegrei, Italy) J Volcanol Geoth Res 304:24–37

Arienzo I, Mazzeo FC, Moretti R, Cavallo A, D'Antonio M (2016) Open-system magma evolution and fluid transfer at Campi Flegrei caldera (Southern Italy) during the past 5 ka as revealed by geochemical and isotopic data: the archetype of Nisida eruption. Chem Geol 427:109–124

Bacon CR (1986) Magmatic inclusions in silicic and intermediate volcanic rocks. J Geophys Res 91:6091–6112

Bagagli M, Montagna CP, Papale P, Longo A (2017) Signature of magmatic processes in strainmeter records at Campi Flegrei (Italy). Geophys Res Lett 44(2):718–725

Baker DR (1990) Chemical Interdiffusion of Dacite and Rhyolite—Anhydrous Measurements at 1 Atm and 10 Kbar, Application of Transition-State Theory, and Diffusion in Zoned Magma Chambers. Contrib Mineral Petr 104:407–423

Bateman R (1995) The interplay between crystallization, replenishment and hybridisation in large felsic magma chambers. Earth Sci Rev 39:91–106

Blundy J, Sparks SRJ (1992) Petrogenesis of mafic enclaves in granitoids of the Adamello massif, Italy. J Petrol 33:1039–1104

Bowen N (1928) The evolution of the igneous rocks. Princeton University Press, Princeton, N.J

Braschi E, Francalanci L, Vougioukalakis GE (2012) Inverse differentiation pathway by multiple mafic magma refilling in the last magmatic activity of Nisyros Volcano, Greece. Bull Volcanol 74:1083–1100

Brown RJ, Civetta L, Arienzo I, D'Antonio M, Moretti R, Orsi G, Tomlinson EL, Albert PG, Menzies MA (2014) Geochemical and isotopic insights into the assembly, evolution and disruption of a magmatic plumbing system before and after a cataclysmic caldera Dcollapse eruption at Ischia volcano (Italy). Contrib Mineral Petrol 168, 1035. doi:https://doi.org/10.1007/s00410-014-1035-1

Bunsen W (1851) Uber die Processe der vulkanischen Gesteinsbildungen Islands. Ann Phys Chem (Dritte Reihe) 83:197–272

Chamberlain KJ, Morgan DJ, Wilson CJN (2014) Time-scales of mixing and mobilisation in the Bishop Tuff magma body: perspectives from diffusion chronometry. Contrib Mineral Petr 167:1034. https://doi.org/10.1007/s00410-014-1034-2

Chiodini G, Caliro S, De Martino P, Avino R, Gherardi F (2012) Early signals of new volcanic unrest at Campi Flegrei caldera? Insights from geochemical data and physical simulations. Geology 40:943–946

Chiodini G, Todesco M, Caliro S, Del Gaudio C, Macedonio G, Russo M (2003) Magma degassing as a trigger of bradyseismic events; the case of Phlegrean Fields (Italy). Geophys Res Lett 30(8):1434. https://doi.org/10.1029/2002GL01679

Chiodini G, Vandemeulebrouck J, Caliro S, D'Auria L, De Martino P, Mangiacapra A, Petrillo Z (2015) Evidence of thermal-driven processes triggering the 2005–2014 unrest at Campi Flegrei caldera. Earth Planet Sci Lett 414:58–67

Cioni R, Civetta L, Marianelli P, Metrich N, Santacroce R, Sbrana A (1995) Compositional layering and syn-eruptive mixing of a periodically refilled shallow magma chamber: the AD 79 Plinian eruption of Vesuvius. J Petrol 36:739–776

Civetta L, Orsi G, Pappalardo L, Fisher RV, Heiken G, Ort M (1997) Geochemical zoning, mingling, eruptive dynamics and depositional processes—the Campanian Ignimbrite, Campi Flegrei Caldera, Italy. J Volcanol Geotherm Res 75:183–219

Corsaro RA, Di Renzo V, Di Stefano S, Miraglia L, Civetta L (2013) Relationship between magmatic processes in the plumbing system of Mt. Etna and the dynamics of the eastern flank: inferences from the petrologic study of the products erupted from 1995 to 2005. J Volcanol Geotherm Res 251:75–89

Costa F, Chakraborty S (2004) Decadal time gaps between mafic intrusion and silicic eruption obtained by chemical zoning patterns in olivine. Earth Planet Sci Lett 227:517–530

D'Antonio M, Civetta L, Orsi G, Pappalardo L, Piochi M, Carandente A, de Vita S, Di Vito MA, Isaia R (1999) The present state of the magmatic system of the Campi Flegrei caldera based on a reconstruction of its behavior in the past 12 ka. J Volcanol Geotherm Res 91:247–268

D'Antonio M, Tonarini S, Arienzo I, Civetta L, Dallai L, Moretti R, Orsi G, Andria M, Trecalli A (2013) Mantle and crustal processes in the magmatism of the Campania region: inferences from mineralogy, geochemistry, and Sr–Nd–O isotopes of young hybrid volcanics of the Ischia island (South Italy). Contrib Mineral Petr 165:1173–1194. https://doi.org/10.1007/s00410-013-0853-x

Davidson JP, McMillan NJ, Moorbath S, Worner G (1990) The Nevados de Payachata volcanic region (18S, 69 W, N. Chile) II. Evidence for widespread crustal involvement in Andean magmatism. Contrib Mineral Petr 105:412–432

Davidson JP, Tepley FJ III (1997) Recharge in volcanic systems; evidence from isotopic profiles of phenocrysts. Science 275:826–829

Davidson JP, Tepley FJ III, Knesel KM (1998) Crystal isotopic stratigraphy; a method for constraining magma differentiation pathways. Eos 79:185–189

De Campos CP, Dingwell DB, Fehr KT (2004) Decoupled convection cells from mixing experiments with alkaline melts from Phlegrean Fields. Chem Geol 213:227–251

De Campos CP, Dingwell DB, Perugini D, Civetta L, Fehr TK (2008) Heterogeneities in magma chambers: insight from the behaviour of major and minor elements during mixing experiments with natural alkaline melts. Chem Geol 256:131–145

De Campos CP, Perugini D, Ertel-Ingrisch W, Dingwell DB, Poli G (2011) Enhancement of Magma mixing efficiency by Chaotic Dynamics: an experimental study. Contrib Mineral Petr 161:863–881

Deino AL, Orsi G, de Vita S, Piochi M (2004) The age of the Neapolitan Yellow Tuff caldera-forming eruption (Campi Flegrei caldera—Italy) assessed by $^{40}Ar/^{39}Ar$ dating method. J Volcanol Geotherm Res 133:157–170

De Rosa R, Donato P, Ventura G (2002) Fractal analysis of min- gled/mixed magmas: an example from the Upper Pollara eruption (Salina Island, Southern Tyrrhenian Sea, Italy). Lithos 65:299–311

De Rosa R, Mazzuoli R, Ventura G (1996) Relationships between deformation and mixing processes in lava flows: a case study from Salina (Aeolian Islands, Tyrrhenian Sea). Bull Volcanol 58:286–297

de Vita S, Orsi G, Civetta L, Carandente A, D'Antonio M, Deino A, di Cesare T, Di Vito MA, Fisher RV, Isaia R, Marotta E, Necco A, Ort M, Pappalardo L, Piochi M, Southon J (1999) The Agnano-Monte Spina eruption (4100 years BP) in the restless Campi Flegrei caldera (Italy). J Volcanol Geotherm Res 91:269–301

Del Gaudio C, Aquino I, Ricciardi GP, Ricco C, Scandone R (2010) Unrest episodes at Campi Flegrei: a reconstruction of vertical ground movements during 1905–2009. J Volcanol Geotherm Res 195:48–56

Di Muro A, Metrich N, Vergani D, Rosi M, Armienti P, Fourgeroux T, Deloule E, Arienzo I, Civetta L (2014) The shallow plumbing system of Piton de la Fournaise volcano (La Reunion island, Indian Ocean) revealed by the major 2007 caldera forming eruption. J Petrol

55(7):1287–1315. https://doi.org/10.1093/petrology/egu025

Di Renzo V, Arienzo I, Civetta L, D'Antonio M, Tonarini S, Di Vito MA, Orsi G (2011) The magmatic feeding system of the Campi Flegrei caldera: architecture and temporal evolution. Chem Geol 281: 227–241

Di Vito MA, Arienzo I, Braia G, Civetta L, D'Antonio M, Orsi G (2011) The Averno 2 fissure eruption: a recent small size explosive event at the Campi Flegrei caldera. Bull Volcanol 73:295–320. https://doi.org/10.1007/s00445-010-0417-0

Didier J, Barbarin B (1991) Enclaves and Granite Petrology, vol 13. Elsevier, Amsterdam

Eichelberger JC (1975) Origin of andesite and dacite: evidence of mixing at Glass Mountain in California and other circum-Pacific volcanoes. Geol Soc Amer Bull 86:1381–1391

Eichelberger JC (1978) Andesitic volcanism and crustal evolution. Nature 275:21–27

Eichelberger JC (1980) Vesiculation of mafic magma during replenishment of silicic magma reservoirs. Nature 288:446–450

Fedele L, Scarpati C, Lanphere M, Melluso L, Morra V, Perrotta A, Ricci G (2008) The Breccia Museo formation, Campi Flegrei, southern Italy: geochronology, chemostratigraphy and relationship with the Campanian Ignimbrite eruption. Bull Volcanol 70:1189–1219

Fedele L, Zanetti A, Morra V, Lustrino M, Melluso L, Vannucci R (2009) Clinopyroxene/liquid trace element partitioning in natural trachyte-trachyphonolite systems: insights from Campi Flegrei (southern Italy). Contrib Mineral Petr 158:337–356. https://doi.org/10.1007/s00410-009-0386-5

Flinders J, Clemens JD (1996) Non-linear dynamics, chaos, complexity and enclaves in granitoid magmas. Trans R Soc Edinburgh: Earth Sci 87:225–232

Font L, Davidson JP, Pearson DG, Nowell GM, Jerram DA, Ottley CJ (2008) Sr and Pb isotopic micro-analysis of plagioclase crystals from Skye lavas: an insight into open-system processes in a flood basalt province. J Petrol 49(8):1449–1471

Fourcade S, Allegre CJ (1981) Trace element behaviour in granite genesis: a case study the calc-alkaline plutonic association from the Querigut Complex (Pyrenees France). Contrib Mineral Petr 76:177–195

Francalanci L, Avanzinelli R, Nardini I, Tiepolo M, Davidson JP, Vannucci R (2012) Crystal recycling in the steady-state system of the active Stromboli volcano: a 2.5-ka story inferred from in situ Sr-isotopic and trace element data. Contrib Mineral Petr 163:109–131

Grasset O, Albarede F (1994) Hybridisation of mingling magmas with different densities. Earth Planet Sci Lett 121:327–332

Hibbard MJ (1981) The magma mixing origin of mantled feldspar. Contrib Mineral Petrol 76:158–170

Hibbard MJ (1995) Petrography to petrogenesis. MacMillan Publishing Company, London

Jolis EM, Freda C, Troll VR, Deegan FM, Blythe LS, McLeod CL, Davidson JP (2013) Experimental simulation of magma-carbonate interaction beneath Mt. Vesuvius, Italy. Contrib Mineral Petr 166:1335–1353

Kent AJR, Darr C, Koleszar AM, Salisbury MJ, Cooper KM (2010) Preferential eruption of andesitic magmas through recharge filtering. Nat Geosci 3: 631–636

Kinman WS, Neal CR, Davidson JP, Font L (2009) The dynamics of Kerguelen Plateau magma evolution: new insights from major element, trace element and Sr isotopic microanalysis of plagioclase hosted in Elan Bank basalts. Chem Geol 264:247–265

Knesel KM, Davidson JP, Duffield WA (1999) Evolution of silicic magma through assimilation and subsequent recharge: evidence from Sr isotopics in sanidine phenocrysts Taylor Creek Rhyolites. J Petrol 40:773–786

Kress V (1997) Magma mixing as a source for Pinatubo sulphur. Nature 389:591–593

Lesher CE (1990) Decoupling of chemical and isotopic exchange during magma mixing. Nature 344:235–237

Longo A, Papale P, Vassalli M, Saccorotti G, Montagna CP, Cassioli A, Giudice S, Boschi E (2012) Magma convection and mixing dynamics as a source of Ultra-Long-Period oscillations. Bull Volcanol 74 (4):873–880. doi:https://doi.org/10.1007/s00445-011-0570-0, https://doi.org/10.1007/s00445-011

Martin VM, Morgan DJ, Jerram DA, Caddick MJ, Prior DJ, Davidson JP (2008) Bang! Month-scale eruption triggering at Santorini Volcano. Science 321:1178

Melluso L, De' Gennaro R, Fedele L, Franciosi L, Morra V (2012) Evidence of crystallization in residual, Cl-F-rich, agpaitic, trachyphonolitic magmas and primitive Mg-rich basalt-trachyphonolite interaction in the lava domes of the Phlegrean Fields (Italy). Geol Mag 149(3):532–550

Montagna CP, Papale P, Longo A (2015) Timescales of mingling in shallow magmatic reservoir. In: Caricchi L, Blundy JD (eds) Chemical, Physical and Temporal Evolution of Volcanic Systems. Gelogical Society, London, Special Publications 422, pp 131–140

Morgavi D, Perugini D, De Campos CP, Ertl-Ingrisch W, Dingwell DB (2013a) Time evolution of chemical exchanges during mixing of rhyolitic and basaltic melts. Contrib Mineral Petr 166(2):615–638

Morgavi D, Perugini D, De Campos CP, Ertl-Ingrisch W, Dingwell DB (2013b) Morphochemistry of patterns produced by mixing of rhyolitic and basaltic melts. J Volcanol Geotherm Res 253:87–96

Morgavi D, Perugini D, De Campos CP, Ertl-Ingrisch W, Lavallee Y, Morgan L, Dingwell DB (2013c) Interactions Between Rhyolitic and Basaltic Melts Unraveled by Chaotic Magma Mixing Experiments. Chem Geol 346:119–212

Morgavi D, Petrelli M, Vetere FP, Gonzalez-Gartia D, Perugini D (2015) High temperature apparatus for chaotic mixing of natural silicate melts. Rev Sci Instrum 86(10):105108

Morgavi D, Arzilli F, Pritchard C, Perugini D, Mancini L, Larson P, Dingwell BD (2016) The Grizzly Lake complex (Yellowstone Volcano, USA): mixing between basalt and rhyolite unraveled by microanalysis and X-ray microtomography. Lithos 260:457–474

Murphy MD (1998) The role of Magma Mixing in triggering the current eruption at the Soufriere Hills Volcano, Montserrat, West Indies. Geophys Res Lett 25:3433–3436

Laumonier M, Scaillet B, Pichavant M, Champallier R, Andujar J, Arbaret L (2014) On the conditions of magma mixing and its bearing on andesite production in the crust. Nature communications, volume 5

Papale P, Moretti R, Barbato D (2006) The compositional dependence of the saturation surface of $H_2O + CO_2$ fluids in silicate melts. Chem Geol 229(1–3):78–95

Pappalardo L, Piochi M, D'Antonio M, Civetta L, Petrini R (2002) Evidence for multistage magmatic evolution during the past 60 kyr at Campi Flegrei (Italy) deduced from Sr, Nd and Pb isotopic data. J Petrol 43(8):1415–1434

Perugini D, De Campos CP, Dingwell DB, Petrelli M, Poli G (2008) Trace element mobility during magma mixing: preliminary experimental results. Chem Geol 256:146–157

Perugini D, de Campos CP, Petrelli M, Dingwell DB (2015) Concentration variance decay during magma mixing: a volcanic chronometer. Sci Rep 5:14225

Perugini D, Petrelli M, Poli G (2006) Diffusive fractionation of trace elements by chaotic mixing of magmas. Earth Planet Sci Lett 243:669–680

Perugini D, Petrelli M, Poli G, De Campos C, Dingwell DB (2010) Time-scales of recent Phlegrean fields eruptions inferred from the application of a 'Diffusive Fractionation' model of trace elements. Bull Volcanol 72:431–447

Perugini D, Poli G (2005) Viscous fingering during replenishment of felsic magma chambers by continuous inputs of mafic magmas: field evidence and fluid-mechanics experiments. Geology 33:5–8

Perugini D, Poli G (2012) The mixing of magmas in plutonic and volcanic environments: analogies and differences. Lithos. https://doi.org/10.1016/j.lithos.2012.02.002

Perugini D, Poli G, Gatta G (2002) Analysis and simulation of Magma mixing processes in 3D. Lithos 65:313–330

Perugini D, Poli G, Mazzuoli R (2003) Chaotic advection, fractals and diffusion during mixing of Magmas: evidence from Lava Flows. J Volcanol Geotherm Res 124:255–279

Perugini D, Valentini L, Poli G (2007) Insights into Magma chamber processes from the analysis of size distribution of Enclaves in Lava Flows: a case study from Vulcano Island (Southern Italy). J Volcanol Geotherm Res 166:193–203

Petrelli M, Perugini D, Poli G (2011) Transition to Chaos and implications for Time-scales of Magma Hybridization during mixing processes in Magma chambers. Lithos 125:211–220

Plail M, Barclay J, Humphreys MCS, Edmonds M, Herd RA, Christopher TE (2014) Characterization of mafic enclaves in the erupterd products of Soufriere Hills Volcano, Montserrat, 2009–2010. In Wadge G, Robertson REA, Voight B, (eds) The eruption of Soufriere Hills Volcano, Montserrat from 2000 to 2010. Geological Society, London, Memoirs, 39, 343–360

Poli G, Tommasini S, Halliday AN (1996) Trace elements and isotopic exchange during acid-basic magma interaction processes. Trans Royal Soc Edinburgh: Earth Sci 87:225–232

Pritchard CJ, Larson PB, Spell TL, Tarbert KD (2013a) Eruption-triggered mixing of extra-caldera basalt and rhyolite complexes along the East Gallatin-Washburn fault zone, Yellowstone National Park, WY, USA. Lithos 175–176:163–177. doi:https://doi.org/10.1016/j.lithos.2013.04.022

Self S (1992) Krakatau revisited: the course of events and interpretation of the 1883 eruption. Geoj Lib 28:109–121

Sigmundsson F, Hreinsdottir S, Hooper A, Arnadottir T, Pedersen R, Roberts JM, Oskarsson N, Auriac A, Decriem J, Einarsson P, Geirsson H, Hensch M, Ofeigsson BG, Sturkell E, Sveinbjornsson H, Feigl LK (2010) Intrusion triggering of the 2010 Eyjafjallajokull explosive eruption. Nature 468:426–430

Slaby E, Gotze J, Worner G, Simon K, Wrzalik R, Smigielski M (2010) K-feldspar phenocrysts in microgranular magmatic enclaves: a cathodoluminescence and geochemical study of crystal growth as a marker of magma mingling dynamics. Lithos 105:85–97

Smith JV (2000) Structures on interfaces of mingled magmas, Stewart Island, New Zealand. J Struct Geol 22:123–133

Sparks SRJ, Marshall LA (1986) Thermal and mechanical constraints on mixing between mafic and silicic magmas. J Volcanol Geotherm Res 29:99–124

SparksSRJ Sigurdsson H, Wilson L (1977) Magma mixing: a mechanism for triggering acid explosive eruptions. Nature 267:315–318

Tonarini S, D'Antonio M, Di Vito MA, Orsi G, Carandente A (2009) Geochemical and isotopical (B, Sr, Nd) evidence for mixing and mingling processes in the magmatic system feeding the Astroni volcano (4.1–3.8 ka) within the Campi Flegrei caldera (South Italy). Lithos 107:135–151

Tonarini S, Leeman WP, Civetta L, D'Antonio M, Ferrara G, Necco A (2004) B/Nb and $\delta^{11}B$ systematics in the Phlegrean Volcanic District (PVD). J Volcanol Geotherm Res 113:123–139

Ventura G (1998) Kinematic significance of mingling-rolling structures in lava flows: a case study from Porri volcano (Salina Southern Tyrrhenian Sea). Bull Volcanol 59:394–403

Wada K (1995) Fractal structure of heterogeneous ejecta from the Meakan volcano, eastern Hokkaido, Japan: implications for mixing mechanism in a volcanic conduit. J Volcanol Geotherm Res 66:69–79

Walker GPL, Skelhorn RR (1966) Some associations of acid and basic igneous rocks. Earth Sci Rev 2:9–109

Wallace G, Bergantz G (2002) Wavelet-based correlation (WBC) of crystal populations and magma mixing. Earth Planet Sci Lett 202:133–145

Wiebe RA (1994) Silicic magma chambers as traps for basaltic magmas: the Cadillac mountain intrusive complex, Mount Desert island, Maine. J Geol 102:423–427

Wilcox RE (1944) Rhyolite-basalt complex of Gardiner River, Yellowstone Park, Wyoming. Geol Soc Am Bull 55:047–1080

Wilcox RE (1999) The idea of Magma Mixing: history of a struggle for Acceptance. J Geol 107(4):421–432

Williams-Jones G, Rymer H (2002) Detecting volcanic eruption precursos: a new method using gravity and deformation measurements. J Volcanol Geotherm Res 113:379–389

Gases as Precursory Signals: Experimental Simulations, New Concepts and Models of Magma Degassing

M. Pichavant, N. Le Gall and B. Scaillet

Abstract

Volatile release during magma ascent in volcanic conduits (magma degassing) forms the basis for using volcanic gases as precursory signals. Recent high temperature high pressure experimental simulations have yielded results that challenge key assumptions related to magma degassing and are important for the interpretation of glass inclusion and gas data and for using volcanic gas as precursory signals. The experimental data show that, for ascent rates expected in natural systems, pure H_2O basaltic melts will evolve mostly close to equilibrium when decompressed from 200 to 25 MPa. In the same way, degassing of H_2O–S species evolves at near equilibrium, although this conclusion is limited by the number of S solubility data available for basaltic melts. However, degassing of CO_2 is anomalous in all studies, whether performed on basaltic or rhyolitic melts. CO_2 stays concentrated in the melt at levels far exceeding solubilities. The anomalous behaviour of CO_2, when associated with near equilibrium H_2O losses, yields post-decompression glasses with CO_2 concentrations systematically higher than equilibrium degassing curves. Therefore, there is strong experimental support for disequilibrium degassing during ascent of CO_2-bearing magmas. The existence of volatile concentration gradients around nucleated gas bubbles suggests that degassing is controlled by the respective mobilities (diffusivities) of volatiles within the melt. The recently formulated diffusive fractionation model reproduces the main characteristics, especially the volatile concentrations, of experimental glasses. The model also shows that the gas phase is more H_2O-rich than expected at equilibrium because CO_2 transfer toward the gas phase is hampered by its retention within the melt. However, only integrated gas compositions are calculated. Similarly, only bulk experimental fluid compositions are determined in recent experiments. Thus, constraints on the local gas phase are becoming necessary for the application to volcanoes. This stresses the need for the direct analysis of gas bubbles nucleated in decompression experiments. Pre-eruptive changes in volcanic CO_2/SO_2 and H_2O/CO_2 gas ratios are interpreted to reflect different pressures of gas-melt segregation in the conduit, an approach that assume gas-melt equilibrium. However, if disequilibrium magma degassing is accepted, the use of volatile saturation codes is no longer possible and caution must be exercised with the application of local equilibrium to volcanic gases. Future developments in the interpretation of gas data require progress

M. Pichavant (✉) · N. Le Gall · B. Scaillet
CNRS, Orléans, France
e-mail: michel.pichavant@cnrs-orleans.fr

Advs in Volcanology (2019) 139–154
DOI 10.1007/11157_2018_35

Published Online: 09 June 2018

from both sides, experimental and volcanological. One priority is to reduce the gap in scales between experiments and gas measurements.

Keywords

Magma ascent · Decompression experiments Disequilibrium degassing · Volatile diffusion Gas compositions

Extended Abstract

Volcanic gases are one of the main tools to monitor changes in the activity of volcanoes and forecast their eruption. Magma ascent toward the surface is associated with the exsolution of volatiles initially dissolved in the melt, a process designated as "magma degassing". Classically, the interpretation of volcanic gases relies on the assumption that degassing takes place at equilibrium. However, several observations (CO_2 contents of basaltic seafloor glasses, H_2O and CO_2 concentrations in glass inclusions, explosive basaltic volcanism) do not fit easily in such a model. Recently, decompression, ascent and degassing of magmas in volcanic conduits have been simulated by high temperature high pressure experiments. Results from these simulations stress the need to critically reconsider the whole mechanism of degassing in basaltic but also rhyolitic magmas. The new experimental data show that, at the decompression rates tested, pure H_2O basaltic melts will evolve mostly close to equilibrium when decompressed from 200 to 25 MPa. In the same way, degassing of H_2O–S species evolves at near equilibrium, although this conclusion is limited by the number of S solubility data available for basaltic melts. Degassing of CO_2 is anomalous in all studies, whether performed on basaltic or rhyolitic melts. CO_2 stays concentrated in the melt at levels far exceeding solubilities. The anomalous behaviour of CO_2, when associated with near equilibrium H_2O losses, yields post-decompression glasses with CO_2 concentrations systematically higher than equilibrium degassing curves. Therefore, there is strong experimental support for disequilibrium degassing during ascent of CO_2-bearing magmas. The existence of volatile concentration gradients around nucleated gas bubbles suggests that degassing is controlled by the respective mobilities (diffusivities) of volatiles within the melt. The contrasted diffusivities of dissolved volatile species (in particular H_2O and CO_2) selectively limit their transfer toward the gas phase for timescales typical for magma ascent. The diffusive fractionation model recently formulated reproduces the main characteristics, especially the volatile concentrations, of experimental glasses. It provides a framework to interpret the new experimental observations and the systematic deviations from equilibrium observed in CO_2-bearing systems, although coupling between volatile diffusion and vesiculation requires a more elaborate treatment. The model also shows that the gas phase is more H_2O-rich than expected at equilibrium because CO_2 transfer toward the gas phase is hampered by its retention within the melt. However, only integrated gas compositions are calculated. In the same way, only bulk experimental fluid compositions are determined in recent experiments. Since the gas phase is essential for the application to volcanoes, constraints on the local gas phase are becoming necessary. Compositions of gas bubbles in decompression experiments must be linked not only with pressure but also with volatile concentrations of local melts, specific degassing textures and mechanisms. As a way in this direction, local gas-melt equilibrium assumes that chemical equilibrium persists at the local scale, despite evidence for disequilibrium at larger scales. However, there are alternative ways to constrain the composition of nucleated gas bubbles, thus stressing the need for their direct analysis in decompression experiments. Pre-eruptive changes in CO_2/SO_2 and, in some cases, H_2O/CO_2 gas ratios observed at several basaltic volcanoes are generally interpreted to reflect different pressures of gas-melt segregation in the conduit, an approach that assume gas-melt equilibrium. However, if disequilibrium magma degassing is accepted, volatile saturation codes can no longer be directly used. Caution also must be exercised with the application of local gas-melt equilibrium to volcanic gases which are probably closer to integrated rather than to local compositions. Future developments in the interpretation of gas data require progress from both sides, experimental and volcanological. One priority is to reduce the gap in scales between experiments and gas measurements to refine interpretations of gas compositions as unrest signals.

1 Magma Degassing and Volcanic Gases as Precursory Signals

Volcanic gases are one of the main tools used to monitor changes in the activity of volcanoes and forecast their eruption. This approach is rooted in the strong pressure dependence of the solubility of volatiles (mainly H_2O, CO_2, SO_2, H_2S, Cl) in silicate melts. Accordingly, magma ascent toward the surface is associated with the exsolution of volatiles initially dissolved in the melt, a process designated as "magma degassing". The different volatiles have contrasted solubilities in silicate melts and, therefore, are expected to react differently to decompression. This forms the basis for using volcanic gas ratios to infer magma ascent and depth of gas segregation in volcanic conduits. For example, the sudden increase of gas CO_2/SO_2 ratio has been used as an indication for deep magma recharge at Stromboli (Aiuppa et al. 2010). At Soufriere Hills volcano (Montserrat), a correlation has been noted between gas HCl/SO_2 and the level of shallow activity as marked by the rate of lava extrusion and dome growth (Christopher et al. 2010; Edmonds et al. 2010).

Classically, the interpretation of volcanic gases relies on the assumption that degassing takes place at equilibrium. In the case of basaltic magmas, this assumption is supported by the high temperatures, low viscosities and high volatile diffusivities (Sparks et al. 1994). Vesiculation (i.e., the combined processes of bubble nucleation, growth and coalescence) is thought to be relatively easy in basaltic melts and degassing of basaltic magma is classically viewed as an equilibrium process. However, several observations do not fit easily in

such a model. They include (1) the existence of basaltic seafloor glasses often supersaturated in CO_2 (e.g., Aubaud et al. 2004), (2) the occurrence of glass inclusions with H_2O and CO_2 concentrations inconsistent with closed system equilibrium degassing (e.g., Metrich et al. 2010) and (3) the occurrence of explosive basaltic volcanism (e.g., Head and Wilson 2003) which implies sudden rather than gradual release of volatiles.

Recently, decompression and ascent of basaltic magmas in volcanic conduits has been simulated by high temperature high pressure petrological experiments. These simulations stress the need to critically reconsider the whole mechanism of degassing in basaltic but also more silicic magmas. In particular, the assumption of equilibrium degassing is now becoming increasingly challenged. This has major implications for the interpretation of glass inclusion and gas data and, more generally, for the use of volcanic gas as precursory signals. In this Chapter, first, the recent experimental simulations are reviewed. We show that they all demonstrate an anomalous behaviour for CO_2 which tends to stay dissolved within the melt at concentrations too high for equilibrium. Second, the diffusive fractionation model which has been proposed to account for the new experimental observations is described and critically discussed. Finally, the implications of disequilibrium degassing for experimental fluid compositions and the interpretation of volcanic gas data as precursory signals are explored.

2 Experimental Simulations

2.1 Basaltic Systems

Following early work on systems with only pure CO_2 (Lensky et al. 2006), decompression experiments on hydrous basaltic melts have been carried out recently by Pichavant et al. (2013) at 1150–1180 °C, for initial pressures of 200–250 MPa, final pressures of 100, 50 and 25 MPa and for decompression rates between $\sim$1.5 down to 0.25 m/s. Melts from Stromboli, pre-synthesized to incorporate dissolved H_2O (2.7–3.8 wt%) and CO_2 (600–1300 ppm), were used as starting materials. The experiments were of continuous decompression type, and both constant (one ramp) and variable (two ramps) decompression rates were imposed. Final melt H_2O concentrations were homogeneous and always close to equilibrium solubility values. In contrast, the rate of vesiculation was found to control the final melt CO_2 concentration. High vesicularity charges had glass CO_2 concentrations that follow theoretical equilibrium degassing paths whereas glasses from low vesicularity charges showed marked deviations from equilibrium, with CO_2 concentrations up to one order of magnitude higher than equilibrium solubilities (Fig. 1a). The experimental results were interpreted in light of the slower diffusivity of CO_2 relative to H_2O in basaltic melts.

Yoshimura (2015) decompressed a natural evolved basaltic melt containing dissolved H_2O and CO_2 at 1200 °C and between 1000 and 500 MPa. The short decompression duration of 10 min over this pressure interval simulates a very fast ascent rate (~ 32 m/s for a rock density of 2650 kg/m^3). A vesiculated glass was produced and Fourier Transform Infrared Spectroscopy (FTIR) profiles revealed large CO_2 concentration gradients in the melt adjacent to gas bubbles. In contrast, the melt H_2O content was almost constant throughout the sample. The glass volatile concentration data cover a near vertical trend in the H_2O–CO_2 diagram (Fig. 1a).

Le Gall and Pichavant (2016a) extended the decompression experiments performed by Pichavant et al. (2013), using essentially the same procedures and materials. Three starting volatile compositions were investigated: series #1 (4.91 wt% H_2O, no CO_2), series #2 (2.41 ± 0.04 wt% H_2O, 973 ± 63 ppm CO_2) and series #3 (0.98 ± 0.16 wt% H_2O, 872 ± 45 ppm CO_2). The volatile-bearing glasses were synthesized at 1200 °C and 200 MPa, then continuously decompressed at a fast decompression rate of 3 m/s in the pressure range 150–25 MPa and then rapidly quenched. Post-decompression glasses were characterized

texturally by X-ray microtomography. Volatile equilibrium was reached or approached during decompression in all series #1 melts with just water. In contrast, disequilibrium degassing occurred systematically in series #2 and #3 melts which retained elevated CO_2 concentrations (Fig. 1a). In similar experiments performed on the same three glass series but at a slower decompression rate of 1.5 m/s, Le Gall and Pichavant (2016b) found that series #1 (CO_2-free) melts followed equilibrium degassing until 100 MPa final pressure (P_{fin}). But at both 60 and 50 MPa P_{fin}, a slight H_2O-supersaturation was recognized, associated with a second bubble nucleation event that occurred at 25 MPa. In comparison, in series #2 and #3 (CO_2-bearing) melts, disequilibrium degassing was systematic, glasses retaining high non-equilibrium CO_2 concentrations (Fig. 1a).

The behavior of H_2O–, CO_2– and S-bearing basaltic melts during decompression was investigated by Le Gall et al. (2015a). Stromboli melts with 2.72 ± 0.02 wt% H_2O, 1291 ± 85 ppm CO_2 and 1535 ± 369 ppm S were synthesized at 1200 °C and 200 MPa and then decompressed to final pressures (P_{fin}) ranging from 150 to 25 MPa, followed by rapid quenching. The continuous decompressions were conducted at rates of 1.5 and 3 m/s. During decompression, S (and H_2O) were lost slightly more from the melt than expected from equilibrium degassing models, whilst significant CO_2 was retained at elevated concentrations in the melt (Fig. 1a). It was found that the degassing trend recorded by Stromboli glass inclusions could be closely reproduced by the experiments (Fig. 1b; Le Gall et al. 2015a). For andesitic melts, Fiege et al. (2014) observed that the fluid/melt partition coefficient for sulfur increases with the decompression rate. However, the influence of decompression rate on S degassing was marked only for oxidizing conditions, corresponding to sulfate as the only S species, thus making necessary to consider the different behaviour of S^{2-} and S^{6+} during degassing.

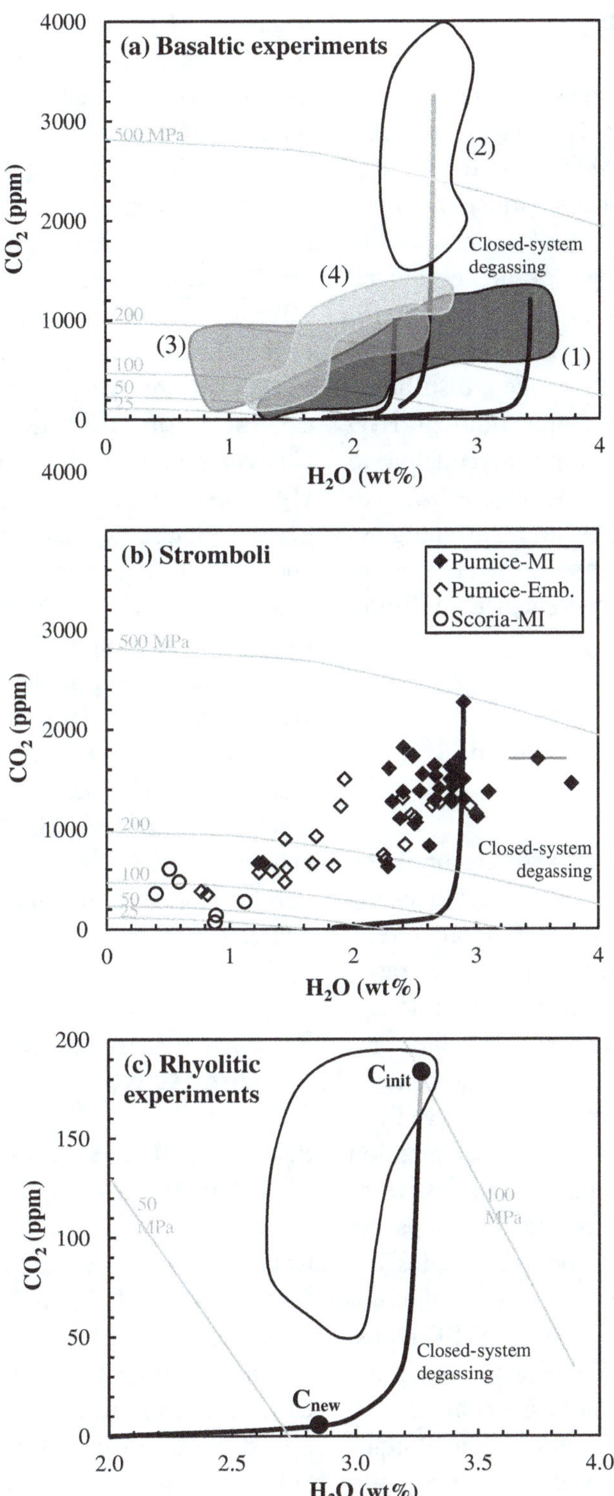

Fig. 1 H_2O–CO_2 glass concentration diagrams for **a** basaltic decompression experiments, **b** Stromboli glass inclusions and **c** rhyolitic decompression experiments. In **a**, **b** and **c**, light grey curves are isobars labelled with pressure in MPa. In **a**, fields for post-decompression glasses are distinguished with (1) referring to Pichavant et al. (2013), (2) to Yoshimura (2015), (3) to Le Gall and Pichavant (2016a, b) and (4) to Le Gall et al. (2015a). Black curves are closed-system equilibrium degassing trajectories redrawn from the original figures (Pichavant et al. 2013; Yoshimura 2015; Le Gall and Pichavant 2016a, b; Le Gall et al. 2015a). In **b**, the glass inclusion data are from Metrich et al. (2010). MI: glass inclusions, Emb: embayments. In **c**, the glass data field and the bold theoretical equilibrium closed-system degassing curve are redrawn from Yoshimura (2015). C_{init} and C_{new} are the composition of the pre-decompression melt and of the estimated post-decompression melt at the gas-melt interface, respectively (Yoshimura 2015; see also Figs. 5 and 6)

2.2 Rhyolitic Systems

Yoshimura (2015) decompressed a natural rhyolitic melt containing dissolved H_2O and CO_2 from 100 to 50 MPa at 800 °C. The duration of the decompression was 5000 s corresponding to a decompression rate of 0.38 m/s (for a density of 2650 kg/m^3). FTIR analysis of the vesiculated glass sample showed CO_2 concentration gradients in the melt away from gas bubbles. In contrast, H_2O was found to be distributed homogeneously within the sample although H_2O concentrations decreased significantly relative to the pre-decompression melt showing it had reequilibrated. On the H_2O–CO_2 diagram, the glass volatile concentrations define a near vertical array located left of the theoretical equilibrium degassing curve (Fig. 1c).

2.3 Summary of Experimental Evidence

In all experiments above, melt vesiculation is the result of decompression, in most cases single-step (constant decompression rate) and, more rarely, multi-step (variable decompression rates, Pichavant et al. 2013). Vesiculation leads to the generation of a gas (or fluid) phase. Volatiles partition between melt and gas, and volatile concentrations in post-decompression glasses evolve from those initially dissolved in pre-decompression glasses. The evaluation of equilibrium vs. disequilibrium degassing is performed by comparing volatile concentrations of post-decompression glasses with theoretical closed-system equilibrium degassing trajectories (Fig. 1). For pure H_2O melts, this equilibrium trajectory is calculated using the experimental solubility data of Lesne et al. (2011). For CO_2– and S-bearing systems, gas-melt equilibrium thermodynamic models (Newman and Lowenstern 2002; Papale et al. 2006; Burgisser et al. 2015) are used. Results show that pure H_2O basaltic systems evolve close to equilibrium when decompressed from 200 to 25 MPa with ascent rates of 1.5 and 3 m/s, although small levels of H_2O supersaturation are observed below 100 MPa (Le Gall and Pichavant 2016a, b). In the same way,

degassing of S species evolves at near equilibrium (Le Gall et al. 2015a) although the reference equilibrium model (Burgisser et al. 2015) is somewhat uncertain due to the limited number of S solubility data for calibration. Degassing of CO_2 is anomalous in all studies, whether performed on basaltic or rhyolitic melts (Pichavant et al. 2013; Yoshimura 2015; Le Gall and Pichavant 2016a, b; Le Gall et al. 2015a). CO_2 stays concentrated in the melt at concentrations far exceeding solubilities (Fig. 1). Except in the very fast basalt decompression experiment of Yoshimura (2015), the anomalous behaviour of CO_2 is associated with significant H_2O losses which results in post-decompression glasses plotting systematically left of theoretical equilibrium degassing trajectories in Fig. 1. We conclude that recent experimental studies strongly support the possibility of disequilibrium degassing, i.e., that ascending melts can keep volatile concentrations (particularly CO_2) significantly different from those expected from equilibrium modelling. The questions thus arise of (1) the mechanisms responsible for this disequilibrium behaviour and (2) of the consequences of disequilibrium degassing for the composition of the gas phase.

3 Modelling Disequilibrium Degassing

3.1 The Diffusive Fractionation Model

Both Pichavant et al. (2013) and Yoshimura (2015) observed decoupling between the behaviour of H_2O and CO_2 during experimental decompression and degassing. In both studies, CO_2 concentration gradients were found in post-decompression glasses, either around gas bubbles or near the gas-melt interface. In contrast, no such diffusion profiles were identified for H_2O, despite concentrations being lower (in most cases) than in pre-decompression glasses (Pichavant et al. 2013; Yoshimura 2015; Le Gall et al. 2015a; Le Gall and Pichavant 2016a, b).

Pichavant et al. (2013) suggested that two characteristic distances, the gas interface distance (either the distance between two bubbles in the melt or the distance to the gas-melt interface) and the volatile diffusion distance (a function of respective diffusivities of volatiles in the melt) control the degassing process. Yoshimura (2015) quantitatively formulated a diffusive fractionation model to describe the ascent and degassing of volatile-bearing magmas. The reader is referred to this work for details about the calculations. The model is based on a diffusivity of CO_2 being one log unit lower than for H_2O (e.g., Zhang and Ni 2010). Decompression trajectories computed from the model are shown on Fig. 2 for different ascent rates, from 0.1, 1, 10, 100 to ∞ m/s. Although very high ascent rates (e.g., Peslier et al. 2015) are necessary for degassing trajectories to shift significantly left to the equilibrium reference curve, the modelling results qualitatively reproduce the main characteristics of experimental post-decompression glasses, i.e., the elevated CO_2 glass concentrations, the significant H_2O losses and the melt concentration trends in H_2O–CO_2 diagrams (Fig. 1).

Yoshimura (2015) emphasized the relative simplicity of his model. For example, bubble growth was not considered as in other more elaborated theoretical treatments (e.g., Gonnermann and Manga 2005). Rather than continuously varying boundary (gas-melt) interface volatile concentrations and bubble-bubble distances as in a natural ascending magma, the calculations were performed step-by-step (i.e., at different pressures) along the decompression ramp, with fixed boundary concentrations and bubble-bubble distance (Yoshimura 2015). It is also important to note that the volatile concentrations on Fig. 2 correspond to averages computed by integrating the concentrations in the melt along the diffusion profiles (*distance integrated compositions*).

Gas phase compositions were calculated by mass balance using the initial volatile concentrations and the average volatile concentrations left in the melt after decompression and degassing (Yoshimura 2015). Results are shown on Fig. 3 and they correspond to compositions

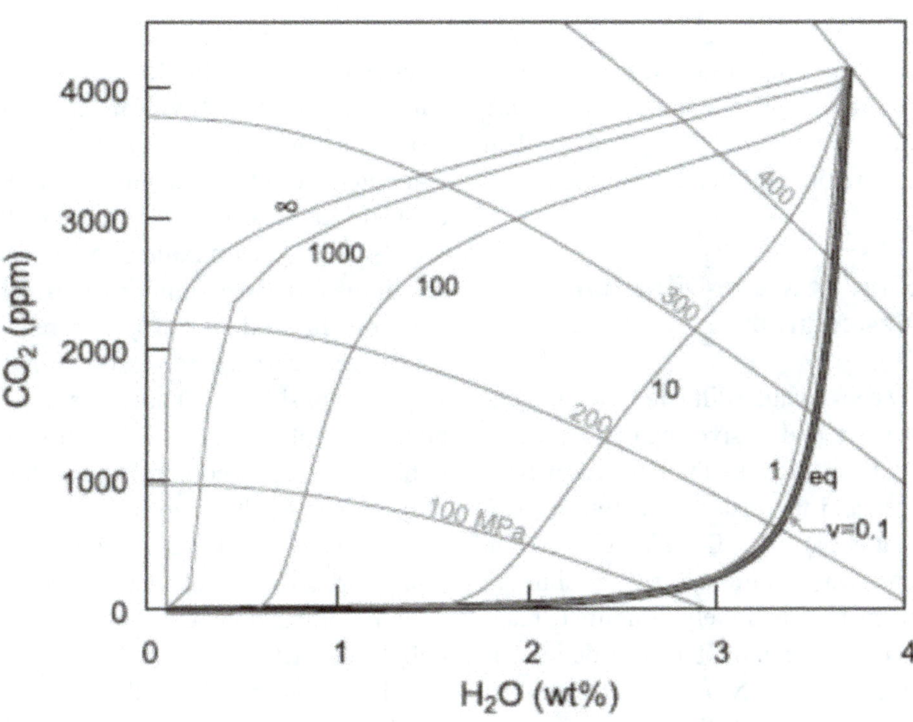

Fig. 2 H_2O and CO_2 melt volatile concentrations computed with the diffusive fractionation model for different decompression/ascent rates (v from 0.1 to ∞, in m/s). Isobars (light curves) are labelled with pressure in MPa. The heavy curve labelled "eq" is the equilibrium closed-system degassing trajectory as calculated by Yoshimura (2015). Figure redrawn from Yoshimura (2015). See text for details

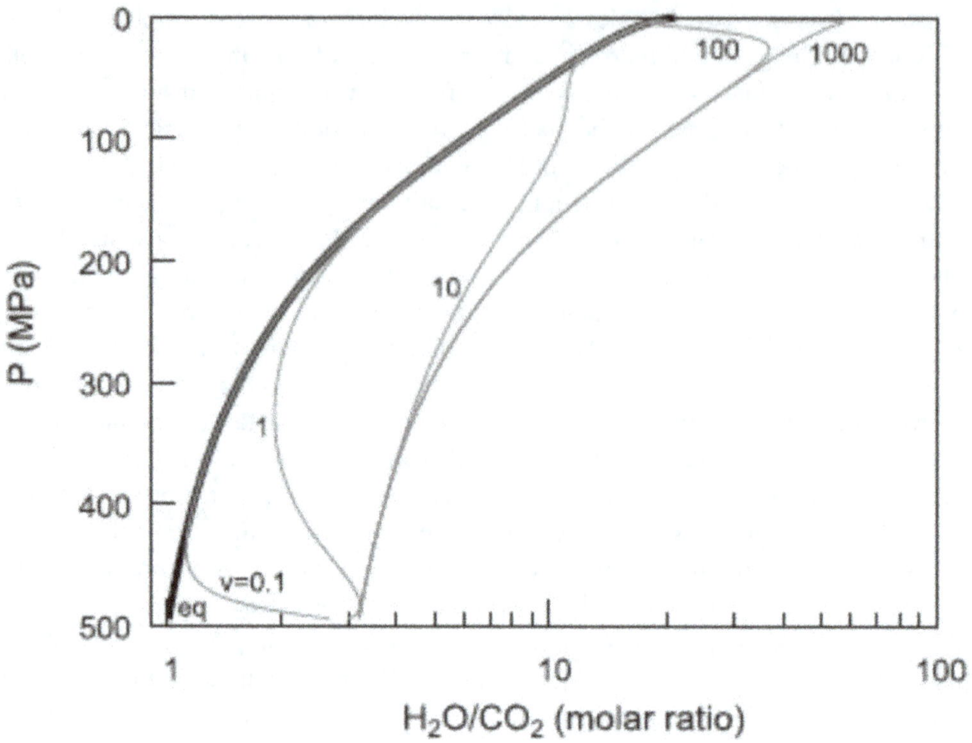

Fig. 3 Variations of the gas H_2O/CO_2 molar ratio (integrated compositions) with pressure computed from the diffusive fractionation model using different decompression/ascent rates (v from 0.1 to 1000, in m/s). The heavy curve labelled "eq" is the equilibrium closed-system degassing trajectory. Figure redrawn from Yoshimura 2015. See text for details

integrated along decompression (*pressure integrated compositions*). These compositions are more H_2O-rich (higher H_2O/CO_2 ratios) than gases generated under equilibrium degassing.

3.2 Coupling Between Diffusion and Vesiculation

Coupling between volatile diffusion and vesiculation is a necessity in diffusive degassing models because vesiculation defines the density of bubbles, their sizes and the distances between them (e.g., Pichavant et al. 2013; Le Gall et al. 2016a, b). This issue was addressed by Yoshimura (2015), although in a relatively simplified manner. The distance between bubbles was defined as being a function of only two variables, the distance between bubbles at the bottom of the decompression column (arbitrary value) and the vesicularity. Vesicularity must change along with decompression and degassing. So, the

vesicularity term should embody the textural variations associated with magma ascent. In the model of Yoshimura (2015), the vesicularity was computed from the amount of volatiles exsolved upon decompression, using an equation of state for $H_2O–CO_2$ gas mixtures to calculate the density of the gas and assuming a constant density for the melt. In so doing, it is apparent that only a vesicularity corresponding to equilibrium degassing is considered. Thus, for a given initial bubble-bubble distance, the distance between bubbles in the decompression column depends only on the equilibrium vesicularity. Degassing trajectories (Fig. 2) and integrated gas compositions (Fig. 3) were calculated on this basis.

For comparison, experimental vesicularities, bubble diameters and bubble number densities are shown on Fig. 4 for three series of basaltic melts decompressed from 200 to 25 MPa final pressure (P_{fin}) at 3 m/s (Le Gall and Pichavant 2016a). Systematic variations within the three glass series are observed depending on P_{fin}. In most cases, the vesicularity data plot intermediate between the two equilibrium vesicularity curves, which were computed in a similar way than Yoshimura (2015) but only for two end-member cases corresponding to pure H_2O and pure CO_2 gas. The vesicularity data for the series #1 melts (with pure H_2O) are in general much lower than the theoretical vesicularities calculated for pure H_2O gas. The data also show large changes in bubble sizes and bubble number densities that do not directly correlate with vesicularity. Le Gall and Pichavant (2016a) emphasized that

Fig. 4 Textural data for post-decompression experimental glasses plotted as a function of final pressure (P_{fin}) and comparison with data for natural basaltic pumices (Stromboli, Masaya). Vesicularities (**a**), bubble diameters (**b**) and bubble number densities (**c**) for three series of basaltic melts decompressed from 200 to 150, 100, 50 and to 25 MPa P_{fin} at 3 m/s (Le Gall and Pichavant 2016a). Pre-decompression melt concentrations, series #1: 4.91 wt% H_2O (no CO_2), series #2: 2.41 ± 0.04 wt% H_2O, 973 ± 63 ppm CO_2 and series #3: 0.98 ± 0.16 wt% H_2O, 872 ± 45 ppm CO_2. Note that, for charge S + S38#1 which was partially fragmented, the vesicularity (**a**) and BND (**c**) data concern the unfragmented part. The bubble diameter data (**b**) are for both the unfragmented (black symbol) and the fragmented (minimum and maximum values) parts. Figure redrawn from Le Gall and Pichavant (2016a)

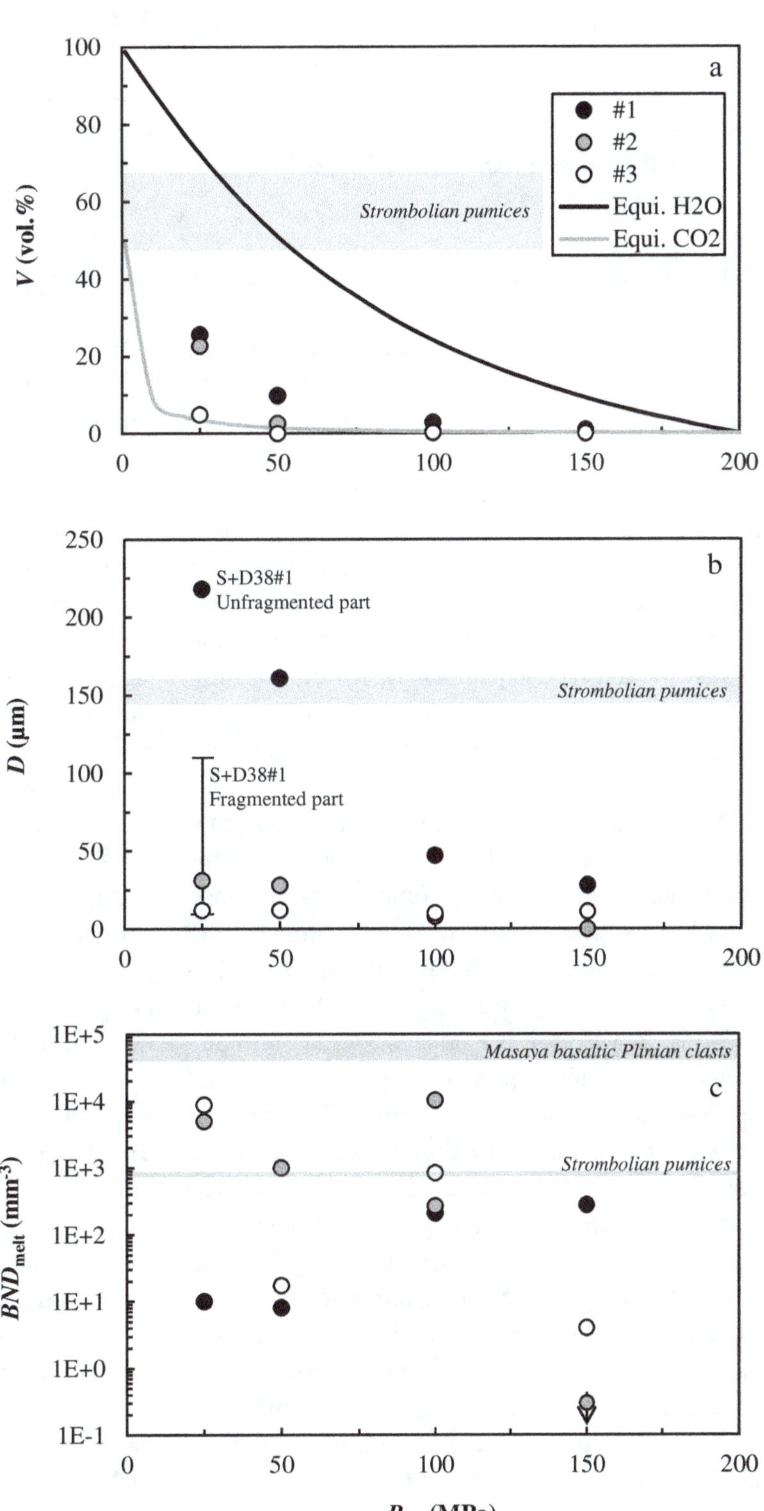

degassing textures result from several processes including bubble nucleation, growth, coalescence, plus buoyancy-driven bubble migration. We conclude that, although the diffusive fractionation model of Yoshimura (2015) provides a basis for coupling volatile diffusion calculations and vesiculation processes, more work is needed to incorporate the complex textural changes associated with ascent of volatile-bearing melts.

4 Implications for Gas Phase Compositions

4.1 Available Data and Models

Despite the limitations noted above, the diffusive fractionation model provides a framework to interpret the experimental observations and the systematic deviations from equilibrium degassing observed in CO_2-bearing systems. However, it should be emphasized that the model uses analytical data (glass volatile concentrations) and physicochemical properties (volatile diffusivities) related only to the melt phase. The question arises of the consequences of disequilibrium degassing for the gas phase composition. It is worth remembering here that the precursory signals come from gas data.

In the decompression experiments summarized above, the gas phase has not been chemically analysed although some mass balance calculations were performed to estimate the composition of the gas phase in the H_2O–, CO_2– and S-bearing experiments of Le Gall et al. (2015a). However, it is emphasized that, with this method, only bulk experimental gas compositions are provided (*charge and pressure integrated compositions*). No information is available on the composition of individual bubbles generated during decompression. The gas calculations performed by Yoshimura (2015) also use a similar mass balance approach, i.e., pressure integrated fluid compositions are given. However, the local gas at the gas-melt interface has an equilibrium composition (local gas-melt equilibrium). The differences between the disequilibrium (calculated with the model) and the equilibrium

(calculated assuming equilibrium degassing) gases (Fig. 3) is the consequence of CO_2 degassing being hampered by its retention within the melt. Therefore, disequilibrium is evidenced in the compositions of the pressure integrated fluids.

4.2 Composition of Gas Bubbles

The experiments and the diffusive fractionation model show that melt and gas both evolve under disequilibrium during magma ascent and degassing. For the melt, this conclusion is based either on volatile concentration measurements in glass at some distance of the gas/melt interface (Pichavant et al. 2013; Yoshimura 2015; Le Gall and Pichavant 2015a, 2016a, b) or on average concentrations calculated by integration along diffusion profiles (Yoshimura 2015). For the gas, constraints are available only on integrated compositions (Le Gall et al. 2015a; Yoshimura 2015). Since the gas phase is essential for the application to volcanoes, and given the interpretations proposed for the melt phase, constraints on the gas phase composition at smaller scales are becoming necessary. This requires linking compositions of gas bubbles in decompression experiments not only with pressure but also with volatile concentrations of local melts as well as with degassing textures and mechanisms.

As a way toward this direction, local gas-melt equilibrium can be assumed. This implies that chemical equilibrium persists locally between gas and melt, despite evidence for disequilibrium at larger scales. Therefore, the volatile compositions of melt and gas at the interface are defined by equilibrium partitioning of volatiles between these two phases (e.g., Dixon and Stolper 1995). To illustrate this concept, a schematic representation of the gas-melt interface for a H_2O– and CO_2-bearing melt decompressed isothermally from an initial (P_{init}) to a final (P_{fin}) pressure is shown on Fig. 5a. Initial volatile concentrations (C_{init}), together with the P_{init} and P_{fin} isobars and the equilibrium degassing trajectory are shown on the H_2O–CO_2 diagram of Fig. 5b. If local gas-melt equilibrium is assumed, the interface melt H_2O and CO_2 concentrations at P_{fin} (C_{new})

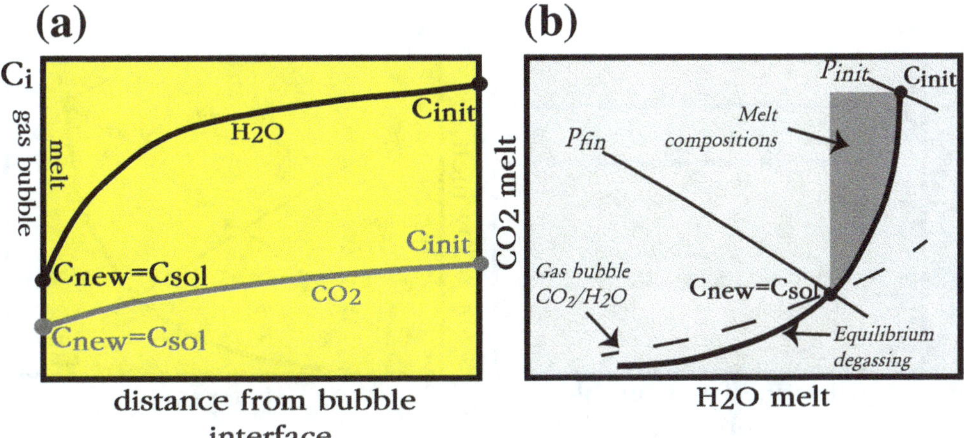

Fig. 5 Schematic illustration of local gas-melt equilibrium. **a** detail of the gas-melt interface region in a concentration (C_i) versus distance diagram where C_i refers to the volatile concentration in the melt. The gas bubble is on the left. The two curves are melt volatile concentration profiles for H_2O (black) and CO_2 (grey) respectively, generated as a result of diffusion in the melt during decompression from P_{init} to P_{fin}. C_{init} give volatile concentrations of the pre-decompression melt, C_{new} gas/melt interface volatile concentrations at P_{fin} and C_{sol} volatile solubilities at P_{fin}. Black lettering is used for H_2O and grey for CO_2. **b** H_2O–CO_2 diagram illustrating the evolution during decompression and degassing. The black bold curve is the equilibrium degassing trajectory. The two black lines are isobars labelled with initial (P_{init}) and final (P_{fin}) pressures along the decompression path. The dashed curve is the CO_2/H_2O isopleth passing through C_{sol} and it defines the composition of the gas bubble in local equilibrium with the interface melt. The shaded domain gives the range of possible melt compositions generated upon decompression from P_{init} down to P_{fin}

are equal to their solubilities (C_{sol}) at $P = P_{fin}$ (intersection of the equilibrium degassing curve with the P_{fin} isobar, Fig. 5b). Note that diffusive fractionation generates H_2O and CO_2 concentration gradients within the melt (Fig. 5a), the range of possible melt compositions during decompression being represented by the dark grey domain in Fig. 5b. The interface melt is the only melt at equilibrium with the local gas at P_{fin} which has a CO_2/H_2O corresponding to the fluid isopleth on Fig. 5b (e.g., Dixon and Stolper 1995). For a low pressure (e.g., 25 MPa), the local gas (e.g., a gas bubble nucleated at P_{fin}) is relatively H_2O-rich. In comparison, the pressure integrated gas assuming bulk *equilibrium* degassing from P_{init} to P_{fin} would be necessarily less H_2O-rich since most of the CO_2 must have been outgassed from the melt. This gas is less H_2O-rich than the pressure integrated gas produced by *disequilibrium* degassing from P_{init} to P_{fin} (Fig. 3). Thus, individual bubbles nucleated at P_{fin} can have CO_2/H_2O different from the composition of integrated gases generated continuously during decompression.

An alternative way to constrain the composition of gas bubbles is illustrated on Fig. 6. It starts from the observation that bubble nucleation is, from a kinetic point of view, an instantaneous process (e.g., Mourtada-Bonnefoi and Laporte 2002, 2004). Nucleation of a gas bubble draws volatiles from the local melt and the possibility that the initial CO_2/H_2O of the gas bubble is the same as the local melt should be considered. According to this hypothesis, represented schematically on Fig. 6, the local melt next to the nucleated bubble (C_{new}) is volatile-depleted but its CO_2/H_2O (r, Fig. 6a) is the same than the initial melt (C_{init}). Melt and gas bubble compositions are thus both located on a mixing line between C_{init} and C_{new} which passes through the origin of the H_2O–CO_2 diagram (Fig. 6b). The net result is the nucleation of individual gas bubbles more CO_2-rich than expected from local gas-melt equilibrium (Fig. 5b).

The previous discussion emphasizes the compositional variability of nucleated gas bubbles and the need for their direct analysis in decompression experiments. Comparison between the

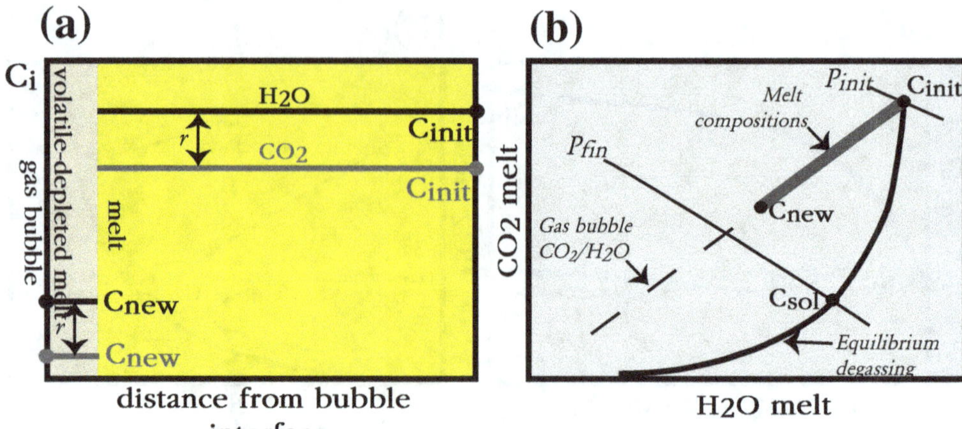

Fig. 6 Hypothetical model for the composition of a gas bubble nucleated during decompression of H_2O–, CO_2-bearing basaltic melts. **a** detail of the gas-melt interface region in a concentration (C_i) versus distance diagram where C_i refers to the volatile concentration in the melt. The gas bubble is on the left. The two horizontal lines are melt volatile concentrations for H_2O (black) and CO_2 (grey) drawn as straight lines because bubble nucleation is an instantaneous event. The narrow zone near the interface is the melt region depleted in volatiles drawn to form the bubble. C_{new} are volatile concentrations in the depleted melt region. Notice that the H_2O/CO_2 ratio (r) is identical in both the depleted and non-depleted melt regions because during nucleation volatiles are drawn from the local melt and the initial CO_2/H_2O of the gas bubble is the same as the local melt. C_{init} give volatile concentrations of the pre-decompression melt, C_{new} gas/melt interface volatile concentrations left after bubble nucleation at P_{fin}. Black lettering is used for H_2O and grey for CO_2. **b** H_2O–CO_2 diagram illustrating the evolution during decompression and degassing. The black bold curve gives the schematic location of the theoretical equilibrium degassing trajectory. The two black lines are isobars labelled with initial (P_{init}) and final (P_{fin}) pressures along the decompression path. Melts produced as a result of decompression and bubble nucleation plot on the straight line joining C_{init} and C_{new}. This line passes through the origin of the diagram because both C_{init} and C_{new} have the same CO_2/H_2O ratio. Note that the location of C_{new} along this line is arbitrary. The CO_2/H_2O ratio of the gas bubble (dashed line) is also the same as C_{init} and C_{new}. It is higher than the gas bubble controlled by local gas-melt equilibrium (Fig. 5)

CO_2/H_2O of nucleated bubbles and results of gas-melt volatile partitioning models (e.g., Dixon and Stolper 1995; Papale et al. 2006) would provide a crucial test of the local gas-melt equilibrium model (Fig. 5). If the nucleated bubbles prove to be CO_2-rich, then alternative models of control of gas composition would be supported (Fig. 6). One important aspect is that, in CO_2-bearing systems, bubble nucleation during decompression is continuous, occurring over a large pressure range (Le Gall and Pichavant 2016a, b). This is because CO_2-bearing melts are volatile-supersaturated (Fig. 1) and, so, the driving force for nucleation of new bubbles is always present. Thus, decompression of CO_2-bearing melts continuously leads to the nucleation of new bubbles, which increases the relevance of the hypothetical mechanism illustrated in Fig. 6.

5 Discussion and Perspectives for Gas Monitoring

5.1 Degassing Processes

Experimental simulations show that, for ascent rates expected in natural systems, equilibrium degassing occurs in pure H_2O melts. In contrast, results for CO_2-bearing melts conclusively demonstrate that degassing generates melt volatile concentrations out of equilibrium. The experimental database supporting this conclusion has recently expanded. It now includes basaltic and rhyolitic melts, and S-bearing as well as S-free systems. Several of those experimental decompression studies have been scaled to natural systems so that results are realistic and applicable. The decompression experiments on Stromboli basalt cover ascent rates ranging from 0.25 to 3 m/s (Pichavant et al. 2013; Le Gall and Pichavant 2016a, b; Le Gall et al. 2015a), well in the range of current estimates for basaltic magmas (e.g., Rutherford 2008; Peslier et al. 2015). An ascent rate approximately 10 times faster was used by Yoshimura (2015). Therefore, in the case of basaltic magmas, equilibrium degassing should be viewed more as a reference situation rather than as a general mechanism. This is quite a change in paradigm which has major implications for how gas signals are interpreted.

One remaining issue concerns the role of crystals on bubble nucleation, heterogeneous rather than homogeneous. All decompression studies considered in this paper were performed on very crystal-poor, if not totally crystal-free, melts and bubble nucleation appears to be mostly homogeneous. Crystals present in experimental basaltic products include Fe–Ti oxides (Le Gall and Pichavant 2015b) and rare Fe sulphides (Le Gall et al. 2015a). Le Gall and Pichavant (2015b) have documented heterogeneous nucleation of bubbles on Fe–Ti oxide crystals (and also on Fe sulphides, Le Gall et al. 2015a). Recently, Shea (2017) has stressed the importance of magnetite as a key mineral phase promoting heterogeneous bubble nucleation in natural magmas. However, Fe oxide phenocrysts and sulphides are uncommonly present in amounts exceeding a few vol.% in natural magmas. This led Le Gall and Pichavant (2015b) to conclude that heterogeneous bubble nucleation is not an important mechanism in basaltic melts if driven by Fe oxides. Yet, heterogeneous nucleation on silicate phases is still an open question. For example, olivine, clinopyroxene and plagioclase are typical phenocrysts and microlites in Stromboli basalts (e.g., Pichavant et al. 2011). On the basis of limited textural evidence, Pichavant et al. (2013) ruled out the possibility of heterogeneous nucleation of gas bubbles on clinopyroxene and olivine crystals. However, additional investigations seem warranted to guarantee full applicability of the decompression experiments above.

Disequilibrium degassing, as documented in the experiments, is the consequence of the anomalous behaviour of CO_2. CO_2-supersaturated melts are systematically generated during decompression. The interpretation suggested by Pichavant et al. (2013) and quantitatively formulated by Yoshimura (2015) is that, because of its restricted diffusive mobility within the melt, CO_2 has limited access to the gas phase for timescales typical of magma ascent. However, our knowledge of volatile diffusivities in silicate melts is still very fragmentary. There are very few diffusivity data for H_2O and CO_2 on the same melt. S is another volatile which reputedly has a slow diffusivity in silicate melts. Yet, the behaviour of S during degassing differs from that of CO_2, although we are still short of S solubility data for basaltic melts (e.g., Lesne et al. 2015). Acquisition of fundamental data (especially volatile diffusivity and solubility data for basaltic melts) is needed for the elaboration of more detailed interpretations of the decompression experiments. Future magma ascent models should also incorporate the textural complexities associated with the vesiculation process.

5.2 Gases as Unrest Signals

Pre-eruptive changes in gas ratios have been observed at several basaltic volcanoes such as Stromboli (Burton et al. 2007; Aiuppa et al. 2010), Etna (Aiuppa et al. 2007) and Villarrica (Aiuppa et al. 2017) among others. Transition from passive degassing to more explosive paroxysmal eruption regimes is marked by temporal increases of the CO_2/SO_2 gas ratio in the volcanic plume. In some cases, the CO_2/SO_2 variations are correlated with a decrease of the H_2O/CO_2 gas ratio (e.g., Aiuppa et al. 2017). These variations in volcanic gas ratios have been generally interpreted to reflect different pressures of gas-melt segregation in the conduit, high CO_2/SO_2 (and low H_2O/CO_2) indicating deep conditions and low CO_2/SO_2 (and high H_2O/CO_2) shallow conditions (e.g., Edmonds 2008; Burton et al. 2007; Allard 2010; Aiuppa et al. 2017). In this approach, the pressure-dependent evolution of the gas phase exsolved upon magma ascent and decompression is calculated by using volatile saturation codes (Newman and Lowenstern 2002; Moretti and Papale 2004; Papale et al. 2006; Burgisser et al. 2015). This implicitly assumes chemical equilibrium between gas and melt, an assumption which, as shown above, is now largely questioned. If disequilibrium magma degassing is accepted, then the consequences for the interpretation of gas signals need to be examined.

Firstly, one might argue that gas-melt equilibrium can persist at local scale, despite disequilibrium at larger scales. Thus, volatile saturation codes could still be used and applied to local gas and melt compositions, for example to model the composition of unconnected bubbles nucleated within the melt. In contrast, volcanic gases necessarily require, to be sampled, that the magma is permeable and, so, that the gas phase is connected. It is quite possible that the gases sampled are mixtures of different components, either integrated from several discrete degassing events along ascent or issued from different parts of the plumbing system. Therefore, volcanic gases are probably more representative of integrated compositions as discussed above than to compositions of local gases. We have shown previously that individual bubbles with compositions defined by local gas-melt equilibrium at a given pressure (Fig. 5) can have CO_2/H_2O different from integrated gases generated continuously during decompression (Fig. 3). We conclude to the limited applicability of local gas melt equilibrium to interpret volcanic gas ratios.

Secondly, disequilibrium gas-melt degassing due to CO_2 retention within the melt implies that CO_2/SO_2 and H_2O/CO_2 gas ratios can no longer be directly related to pressures of gas-melt segregation. Calculations using the diffusive fractionation model (Fig. 3) show that the pressure integrated gases have a higher H_2O/CO_2 (and also presumably a lower CO_2/SO_2 because CO_2 is retained within the melt) than the same gases calculated assuming equilibrium with the melt (Fig. 3). This demonstrates the possibility of changing the gas ratios depending on the degassing mechanism (equilibrium vs. disequilibrium). It is worth emphasizing that disequilibrium degassing associated with CO_2 retention produces integrated fluids that are less, not more, CO_2-rich (Fig. 3).

The CO_2-rich gases observed on basaltic volcanoes have been generally attributed to deep-seated processes such as fluxing of CO_2 or arrival of CO_2-rich magmas (e.g., Aiuppa et al. 2010, 2017; Allard 2010). In contrast, the degassing mechanism of Fig. 6 (although it needs validation from direct analysis of gas bubbles in decompression experiments) allows CO_2-rich gas bubbles to be generated at low pressures. It also provides an example of how gas ratios can be changed at constant pressure depending on the degassing mechanism. The initially CO_2-rich bubbles (Fig. 6) will probably shift rapidly with time toward lower CO_2/H_2O because of preferential diffusion of H_2O from the melt. However, nucleation is a continuous process in CO_2-bearing basaltic melts (Le Gall and Pichavant 2016a, b) and reequilibration of previously nucleated bubbles by diffusion will be accompanied by the nucleation of new CO_2-rich bubbles.

We conclude that future developments in the interpretation of gas data require progress from both sides, experimental and volcanological. Some crucial experimental information at small scale is still missing such as the composition of individual gas bubbles nucleated in the decompression experiments and the influence of crystals on bubble nucleation. In parallel, at larger scales, the representativity and the significance of the gas phase sampled on active basaltic volcanoes needs to be better demonstrated, for example by combining gas measurements with detailed textural studies of eruption products. It is expected that future work will narrow the gap in scales between experiments and gas measurements to refine interpretations of gas compositions as unrest signals.

Acknowledgements This paper has benefited from discussions with P. Allard, C. Martel, N. Metrich, A. Bertagnini, R. Moretti, P. Papale and M. Pompilio, reviews by R. Brooker and F. Wadsworth and from editorial comments by B. Scheu. Discussion with S. Yoshimura was helpful. The VUELCO consortium provided a scientifically demanding and interdisciplinary forum for the elaboration of ideas developed in this study. The Ph.D. thesis of NLG was supported by the VUELCO project.

References

Aiuppa A, Moretti R, Federico C, Giudice G, Gurrieri S, Liuzzo M, Papale P, Shinohara H, Valenza M (2007) Forecasting Etna eruptions by real-time observation of volcanic gas composition. Geology 35:1115–1118

Aiuppa A, Bertagnini A, Metrich N, Moretti R, Di Muro A (2010) A model of degassing for Stromboli volcano. Earth Planet Sci Lett 295:195–204

Aiuppa A, Bitetto M, Francofonte V, Velasquez G, Bucarey Parra C, Giudice G, Liuzzo M, Moretti R, Moussallam Y, Peters N, Tamburello G, Valderrama OA, Curtis A (2017) A CO_2-gas precursor to the March 2015 Villarrica volcano eruption. Geochem Geophys Geosyst 18:2120–2132

Allard P (2010) A CO_2-rich gas trigger of explosive paroxysms at Stromboli basaltic volcano, Italy. J Volcanol Geotherm Res 189:363–374

Aubaud C, Pineau F, Jambon A, Javoy M (2004) Kinetic disequilibrium of C, He, Ar and carbon isotopes during degassing of mid-ocean ridge basalts. Earth Planet Sci Lett 222:391–406

Burgisser A, Alletti M, Scaillet B (2015) Simulating the behavior of volatiles belonging to the C-O-H-S system in silicate melts under magmatic conditions with the software D-Compress. Comput Geosci 79:1–14

Burton M, Allard P, La Spina A, Murè F (2007) Magmatic gas composition reveals the source depth of slug-driven strombolian explosive activity. Science 317:227–230

Christopher T, Edmonds M, Humphreys MCS, Herd RA (2010) Volcanic gas emissions from Soufrière Hills Volcano, Montserrat 1995–2009, with implications for mafic magma supply and degassing. Geophys Res Lett 37: L00E04. https://doi.org/10.1029/2009gl041325

Dixon JE, Stolper EM (1995) An experimental study of water and carbon dioxide solubilities in mid-ocean ridge basaltic liquids. Part II. Applications to degassing. J Petrol 36:1633–1646

Edmonds M (2008) New geochemical insights into volcanic degassing. Philos Trans R Soc A 366:4559–4579

Edmonds M, Aiuppa A, Humphreys M, Moretti R, Giudice G, Martin RS, Herd RA, Christopher T (2010) Excess volatiles supplied by mingling of mafic magma

at an andesite arc volcano. Geochem Geophys Geosyst 11:Q04005. https://doi.org/10.1029/2009GC002781

Fiege A, Behrens H, Holtz F, Adams F (2014) Kinetic vs. thermodynamic control of degassing of H_2O– S ± Cl-bearing andesitic melts. Geochim Cosmochim Acta 125:241–264

Gonnermann HM, Manga M (2005) Non-equilibrium magma degassing: results from modelling of the ca. 1340 AD eruption of Mono craters, California. Earth Planet Sci Lett 238:1–16

Head JW III, Wilson L (2003) Deep submarine pyroclastic eruptions: theory and predicted landforms and deposits. J Volc Geotherm Res 121:155–193

Le Gall N, Pichavant M (2015b) Heterogeneous bubble nucleation on Fe–Ti oxides in H_2O- and H_2O-CO_2-bearing basaltic melts. (In preparation)

Le Gall N, Pichavant M (2016a) Homogeneous bubble nucleation in H2O- and H_2O-CO_2-bearing basaltic melts: results of high temperature decompression experiments. J Volcanol Geotherm Res 327:604–621

Le Gall N, Pichavant M (2016b) Effect of ascent rate on homogeneous bubble nucleation in the system basalt-H_2O-CO_2: Implications for Stromboli volcano. Am Mineral 101:1967–1985

Le Gall N, Pichavant M, Di Carlo I, Scaillet B (2015a) Sulfur partitioning between melt and fluid during degassing of ascending C-O-H-S-bearing basaltic magma: an experimental study. In preparation

Lensky NG, Niebo RW, Holloway JR, Lyakhovsky V, Navon O (2006) Bubble nucleation as a trigger for xenolith entrapment in mantle melts. Earth Planet Sci Lett 245:278–288

Lesne P, Scaillet B, Pichavant M, Iacono-Marziano G, Bény J-M (2011) The H_2O solubility of alkali basaltic melts: an experimental study. Contrib Mineral Petrol 162:133–151

Lesne P, Scaillet B, Pichavant M (2015) The solubility of sulphur in hydrous basaltic melts. Chem Geol 418:104–116

Metrich N, Bertagnini A, Di Muro A (2010) Conditions of magma storage, degassing and ascent at Stromboli: new insights into the volcanic plumbing system with inferences on the eruptive dynamics. J Petrol 51:603–626

Moretti R, Papale P (2004) On the oxidation state and volatile behaviour in multicomponent gas–melt equilibria. Chem Geol 213:265–280

Mourtada-Bonnefoi CC, Laporte D (2002) Homogeneous bubble nucleation in rhyolitic magmas: an experimental study of the effect of H_2O and CO_2. J Geophys Res 107(B4):2066. https://doi.org/10.1029/2001jb000290

Mourtada-Bonnefoi CC, Laporte D (2004) Kinetics of bubble nucleation in a rhyolitic melt: an experimental study of the effect of ascent rate. Earth Planet Sci Lett 218:521–537

Newman S, Lowenstern JB (2002) VolatileCalc: a silicate melt–H_2O–CO_2 solution model written in Visual Basic for Excel. Comput Geosci 28:597–604

Papale P, Moretti R, Barbato D (2006) The compositional dependence of the saturation surface of $H_2O + CO_2$ fluids in silicate melts. Chem Geol 229:78–95

Peslier AH, Bizimis M, Matney M (2015) Water disequilibrium in olivines from Hawaiian peridotites: recent metasomatism, H diffusion and magma ascent rates. Geochim Cosmochim Acta 154:98–117

Pichavant M, Pompilio M, D'Oriano C, Di Carlo I (2011) The deep feeding system of Stromboli, Italy: insights from a primitive golden pumice. Eur J Miner 23:499–517

Pichavant M, Di Carlo I, Rotolo SG, Scaillet M, Burgisser A, Le Gall N, Martel C (2013) Generation of CO_2-rich melts during basalt magma ascent and degassing. Contrib Mineral Petrol 166:545–561

Rutherford MJ (2008) Magma ascent rates. In: Putirka KD, Tepley F (eds) Minerals, inclusions and volcanic processes. Mineralogical Society of America Reviews in Mineralogy vol 69, pp 241–271

Shea T (2017) Bubble nucleation in magmas: A dominantly heterogeneous process? J Volcanol Geotherm Res 343:155–170

Sparks RSJ, Barclay J, Jaupart C, Mader HM, Phillips JC (1994) Physical aspects of magmatic degassing I. Experimental and theoretical constraints on vesiculation. In: Carroll MR, Holloway JR (eds) Volatiles in Magmas. Mineralogical Society of America Reviews in Mineralogy, vol 30, pp 413–445

Yoshimura S (2015) Diffusive fractionation of H_2O and CO_2 during magma degassing. Chem Geol 411:172–181

Zhang Y, Ni H (2010) Diffusion of H, C, and O components in silicate melts. In: Zhang Y, Cherniak DJ (eds) Diffusion in minerals and melts. Mineralogical Society of America Reviews in Mineralogy vol 72, pp 171–225

Crystals, Bubbles and Melt: Critical Conduit Processes Revealed by Numerical Models

M. E. Thomas, J. W. Neuberg and A. S. D. Collinson

Abstract

Understanding how magma moves within a conduit is an important question that is still poorly understood. In particular, estimation of the magma ascent rate is key for interpreting monitoring signals and therefore, predicting volcanic activity. This relies on understanding how strongly different magmatic processes occurring within the conduit control the ascent rate. These processes are controlled by changes in magmatic parameters such as the water content or temperature and understanding/ linking changes of such parameters to monitoring data is an essential step in the use of these data as a predictive tool. The results presented here are from a suite of conduit flow models based on Soufrière Hills Volcano, Montserrat, that assesses the influence of individual model parameters. By systematically changing these parameters, the results indicate that changes in conduit diameter and excess pressure in the magma chamber are amongst the dominant controlling variables. However, the single most important parameter controlling variations in the magma ascent rate is the volatile content. Therefore, understanding the processes controlling the volatile content within the conduit system and the outgassing of these volatiles is crucial to understanding and predicting potential unrest or eruption scenarios.

Keywords

Numerical modelling · Conduit processes Low frequency earthquakes · Magma flow Magma ascent rate

1 Introduction

A volcanic conduit provides the pathway for transport of magma and magmatic fluids within a volcano. It is possible to detect both this movement and the occurrence of conduit processes through geophysical monitoring techniques as discussed in Chap. 2 of this book. However, the extent to which changes in magma flow properties affect the data recorded on volcanoes is not well understood. Is it possible that a small change in magma temperature or water content could alter the processes or flow within the conduit enough to be recorded by geophysical monitoring instruments or simple visual observation? What effect does the size of gas bubbles within

M. E. Thomas (✉) · J. W. Neuberg
A. S. D. Collinson
School of Earth and Environment, Institute
of Geophysics and Tectonics, University of Leeds,
Leeds LS12 9JT, UK
e-mail: m.e.thomas@leeds.ac.uk

Advs in Volcanology (2019) 155–169
DOI 10.1007/11157_2018_36

Published Online: 30 August 2018

the magma have on the overall flow dynamics, and how big do these changes need to be to alter the eruption style? These types of question are addressed within this chapter in an attempt to identify the crucial parameters that cause changes in observed volcanic behaviour.

We use conduit flow models to analyse the key input parameters that control magma flow properties, such as the magma water content, crystal content and conduit geometry, to assess their relative importance to the overall magma flow dynamics. A list of all input parameters is presented in Table 1 along with the range of values studied. We focus on evolved silicic magmatic systems because of the wealth of relevant monitoring information and previous numerical modelling attempts relating to

Soufrière Hills Volcano, Montserrat—a long lived andesitic dome forming eruption (Sparks et al. 2000; Wadge et al. 2014) and excellent natural laboratory. While these initial models are based on extrusive eruptions, the results of changing the model parameters have the potential to alter the eruption style to either more violent or gentle forms and it is noted that the underlying principles discussed here are applicable to other volcanic systems, including those exhibiting signs of unrest, yet to develop into a full blown eruption. If we can develop the findings presented in this chapter into the creation of threshold levels for recorded geophysical data, it may become possible to begin to predict when a volcano will evolve from a state of unrest to eruption.

Table 1 Parameters used in the reference model and range of parameter variations

Symbol/abbreviation	Variable	"Reference" model value	Range of modelled values
–	The melt composition	Rhyolitic (>71% SiO_2) (Barclay et al. 1998)	See Table 2
b_{ni}	Bubble number density	10^{10} m^{-3} (Cluzel et al. 2008)	10^7–10^{11} m^{-3}
D_{TBL}	Thickness of thermal boundary layer over which T_{diff} is lost	0.3 m (Collier and Neuberg 2006)	0.3–0.5 m
Γ	Bubble surface tension	0.06 N m^{-1} (Lyakhovsky et al. 1996)	0.05–0.25 N m^{-1}
χ_c	Magma chamber crystal volume fraction	40% (Barclay et al. 1998)	40–50%
Ls	Slip length of brittle failure of melt	0.01	0.01–1.0 m
P_e	Excess chamber pressure above lithostatic	0 MPa	0–20 MPa
P_{top}	Pressure at conduit exit	0.09 MPa	0.09–4.5 MPa
ρ_c	Average density of crystal assemblage	2700 kg m^{-3} (Burgisser et al. 2010)	2550–3200 kg m^{-3}
ρ_m	Density of pure melt	2380 kg m^{-3} (Burgisser et al. 2010)	–
T	Magma temperature	1150 K(Devine et al. 2003)	1100–1150 K
T_{diff}	Amount of cooling at conduit wall	200 K (Collier and Neuberg, 2006)	100–200 K
τ_s	Melt shear strength	–	10^5–10^7 Pa
$W_\%$	Initial dissolved water content of magma	4.5 wt% (Barclay et al. 1998)	3–8 wt%
w, d, r	Variables that define the conduit shape and size	See Fig. 1	See Fig. 2

2   The Model

In order to assess the effect of altering the model parameters, a standard or "reference" model is defined. This reference model is based on data available in the literature that refers to Soufrière Hills Volcano, and is outlined in Fig. 1 and Table 1. The general dimensions of the modelled conduit, shown in Fig. 1 are inferred from geochemical and observational data from Soufrière Hills Volcano (Barclay et al. 1998; Sparks et al. 2000), placing minimum depth constraints of 5–6 km for the position of the magma chamber and width estimates of 30–50 m for the conduit.

2.1   Governing Equations

Conduit flow is computed with a finite element approach within the code COMSOL Multiphysics®, and modelled in an axial symmetric domain space through the compressible formulation of the Navier-Stokes equation:

$$\rho \frac{\partial \boldsymbol{u}}{\partial t} + \rho \boldsymbol{u} \cdot \nabla \boldsymbol{u} = -\nabla p + \nabla \cdot \left\{ \eta \left[\nabla \boldsymbol{u} + (\nabla \boldsymbol{u})^T \right] \right. \\ \left. - \frac{2}{3} \eta [\nabla \cdot \boldsymbol{u}] \boldsymbol{I} \right\} + \boldsymbol{F} \qquad (1)$$

and the continuity equation:

$$\frac{\partial \rho}{\partial t} + \nabla \cdot (\rho \boldsymbol{u}) = 0 \qquad (2)$$

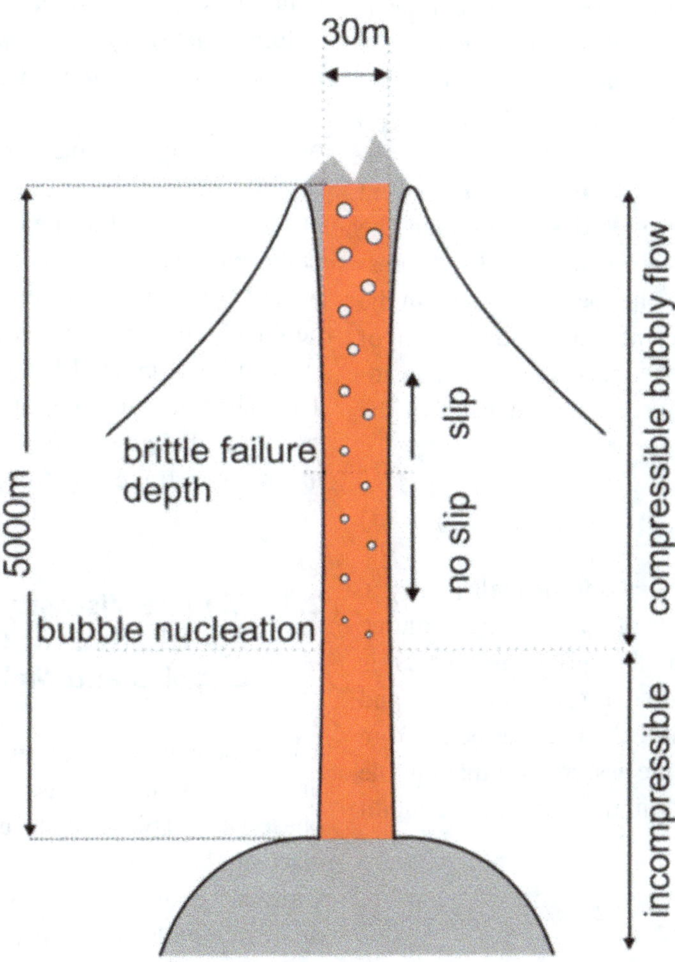

Fig. 1 Cartoon of the modelled volcanic system. Bubble nucleation and brittle failure depth vary with the model parameters considered

where ρ is density, $\boldsymbol{u}$ the velocity vector, p the pressure, η the dynamic viscosity and $\boldsymbol{F}$ the volume force vector (gravity). There is no time dependency in the model as they are solved to a steady state, so the terms $\rho \frac{\partial u}{\partial t}$ and $\frac{\partial p}{\partial t}$ in Eqs. (1) and (2) are neglected.

2.2 Magma Composition

The properties of the magma are modelled as the averaged properties of its constituents: melt, crystals and gas. For the reference model, the general composition of the melt is taken as rhyolitic, using the groundmass analysis of Montserrat dome rocks undertaken by Barclay et al. (1998). However, several melt compositions, were ultimately considered to assess the effect of melt composition on the modelled eruption dynamics (Table 2). Crystal content (χ_c) and density (ρ_c) are fixed as we assume a constant temperature and that the conduit ascent times are orders of magnitudes faster than the time required for crystal growth by decompression, meaning only the phenocrysts present in the magma chamber are accounted for and growth of microlites and microphenocrysts is not considered. The expression for the bulk density of the magma is given by:

$$\rho = \rho_m \chi_m (1 - \chi_g) + \rho_g \chi_g + \chi_c \rho_c (1 - \chi_g), \quad (3)$$

where χ_m is the initial fraction of melt ($1 - \chi_c$), ρ_m is the melt density and χ_g is the gas volume fraction (Table 1). For the gas phase, water is assumed as the only volatile species present and the gas density (ρ_g) is calculated directly from the ideal gas law with the assumption that bubble growth in is equilibrium with the conduit pressure:

$$pV = nRT, \quad (4)$$

where V is the volume of gas, R the ideal gas constant and T the temperature. The number of moles of water, n, is related to density by:

$$n = \frac{M}{m}, \quad (5)$$

where M is the molar mass of water and m is the mass of water present. Thus, combining Eqs. (7) and (8) and considering a unit volume we get:

$$\rho_g = \frac{mp}{RT} \quad (6)$$

In the reference model, a single magma temperature is used with the exception of the temperature across a thermal boundary layer (TBL) defined adjacent to the conduit wall. A linear temperature drop is applied across the TBL, to simulate the cooling of the magma abutting the country rock in a well-established conduit (Collier and Neuberg 2006). The gas volume fraction (χ_g) is calculated by determining how much water remains dissolved within the melt at a particular pressure using the solubility of H_2O in rhyolitic melts presented by Liu et al. (2005). At significant pressures, all the water is dissolved within the melt fraction and χ_g is initially zero. However, as the pressure decreases, water begins to exsolve out of the melt and forms bubbles. The absolute volume of exolved gas (V) can be calculated from the ideal gas law (4). This absolute volume of gas is then used to calculate the gas volume fraction of the bulk magma constituted by the gas phase.

2.3 Magma Viscosity—The Contribution of Crystals, Bubbles and Melt

The bulk magma viscosity (η) is determined by first calculating the viscosity of the pure melt phase (η_m). This is done using a model for the viscosity of magmatic liquids presented by Giordano et al. (2008), that predicts the viscosity of silicate melts as a function of temperature and melt composition. It is important to note that the composition used in the viscosity model is that of the pure melt phase (rhyolitic) not the overall

Table 2 Compositions of melt used in the numerical simulations

Composition[a]	SiO$_2$	Al$_2$O$_3$	TiO$_2$	FeO	MgO	MnO	CaO	Na$_2$O	K$_2$O
a	71.41	13.58	0.28	2.78	1.64	0.13	4.86	3.73	1.6
b	76.97	11.21	0.29	1.89	0.26	0.12	1.29	4.07	2.37
c	77.10	9.83	0.18	1.17	0.22	0.10	1.52	4.14	1.72
d	78.66	11.20	0.39	1.93	0.30	0.10	1.48	3.57	2.38

[a]Compositions determined through, (a) rastered electron microprobe analysis of groundmass (Barclay et al. 1998); (b) Matrix glass composition (Rutherford and Devine 2003); (c) Quartz hosted melt inclusion (Devine et al. 1998); (d) Cameca SX50 microprobe analysis of interstitial glass (Burgisser et al. 2010). All melts are rhyolitic and composition (a) is used in the defined reference model

magma composition. The whole rock composition of recent Soufrière Hills Volcano magma is andesitic (Edmonds et al. 2010), but this includes the contribution of the crystals. The viscosity model is only used to calculate the actual viscosity of the liquid component (the melt), on which the crystals (the solid) have no bearing. When the effect of crystals within the melt *is* considered, the effective viscosity of the melt (liquid) and crystal (solid) mixture (η_{mc}) increases, and can be represented by the Einstein-Roscoe equation:

$$\eta_{mc} = \eta_m \left(1 - \frac{\chi_c}{\chi_c^{\max}} \right)^{-2.5}, \qquad (7)$$

where $\chi_c^{\max}$ is the volume fraction of crystals at which the maximum packing is achieved and a commonly adopted value for this is 0.6 (*Marsh*, 1981), which is used within this study. Although this value was proposed for randomly packed spheres, and it has been shown by Marti et al. (2005) that $\chi_c^{\max}$ tends to decrease as the particle (crystal) shape becomes less isotropic. Ishibashi (2009) demonstrated that this value is a good approximation as the effect of particle shape on $\chi_c^{\max}$ is offset by effects of size heterogeneity and crystal alignment.

The presence of bubbles also affects the viscosity. If the bubbles within the magma remain un-deformed they act to increase viscosity, whilst if deformed (elongated in the direction of flow), they act to decrease *visosity* (Llewellin and Manga, 2005). Whether a bubble is in an un-deformed or deformed state is represented by the capillary number:

$$\text{Ca} = \frac{\eta_m r E}{\Gamma} \qquad (8)$$

where r is the un-deformed bubble radius, Γ, the bubble surface tension and E, a function of the strain rate within the magma flow defined below. If Ca > 1 then the bubbles can be considered deformed. The value of Ca will vary as a function of shear strain rate and elongation strain rate (Thomas and Neuberg 2012), meaning bubbles can be deformed within the model through either shear or extension. To account for strain acceleration or deceleration the dynamic capillary number (Cd) is required (Llewellin and Manga 2005). This compares the timescale over which the bubbles can respond to changes in their strain environment with the timescale over which the strain environment changes. If this value is large, the flow is termed unsteady and the bubbles are unable to deform independently in response to the flow. However, for the models considered here, conditions of unsteady flow are found only in a very small area near the exit of the conduit. Accounting for this within the models resulted in no noticeable change in the derived flow parameters, hence the computation of Cd is not considered.

Depending on the value of Ca, η is calculated using the suggested 'minimum variation' of Llewellin and Manga (2005):

$$Ca = \begin{cases} < 1 & \eta = \eta_{mc}\left(1 - \chi_g\right)^{-1} \\ > 1 & \eta = \eta_{mc}\left(1 - \chi_g\right)^{5/3} \end{cases} \quad (9)$$

By assuming the homogeneous nucleation of a number of bubbles in a unit volume of melt as a single event, which is determined experimentally through the initial bubble number density (b_{ni}) (e.g. Hurwitz and Navon 1994), the bubble radius (Lensky et al. 2002) is given by:

$$r = \left[\frac{S_0^3 \rho_m (C_0 - C_m)}{\rho_g}\right]^{1/3}, \quad (10)$$

where C_0 and C_m are the initial and remaining amount of water dissolved in the melt respectively and S_0 is the initial size of the melt shell from which each bubble grows. S_0 is related to the instantaneous bubble number density (b_n) through the expression:

$$S_0^3 = \frac{3}{4\pi b_n}. \quad (11)$$

b_n is used rather than the initial value (b_{ni}) because homogeneous nucleation is assumed. Therefore, the bubble number density must remain constant with respect to the volume of the melt fraction in Eq. 3. This also accounts for bubble coalescence and b_n is given by:

$$b_n = \frac{b_{ni}}{\chi_m}\left[\chi_m - (1 - \chi_g)\right] \quad (12)$$

2.4 Brittle Failure of Melt

It is now well established that magma, or more specifically the melt component of a magma can fail in a brittle manner (e.g. Goto 1999). This is likely to generate low-frequency (LF) earthquakes (e.g. Neuberg et al. 2006) and effect the overall flow dynamics. In order to account for these effects, it is necessary to define conditions under which the melt may fracture. Shear failure of melt occurs when the shear stress ($\eta\dot{\varepsilon}$) exceeds the

shear strength (τ_s), and has been represented as a brittle failure criterion (e.g. Tuffen et al. 2003):

$$\frac{\eta\dot{\varepsilon}}{\tau_s} > 1 \quad (13)$$

where $\dot{\varepsilon}$ is the shear strain rate. This criterion holds true under the assumption that during un-relaxed deformation the accumulation of shear stress in the melt obeys the Maxwell model:

$$\sigma_s = \frac{\eta}{\mu}\frac{\partial\sigma_s}{\partial t} = \eta\dot{\varepsilon} \quad (14)$$

where σ_s is the shear stress and μ the shear modulus.

The magma composition as discussed in Sects. 2.2 and 2.3 is considered without the effects of microlite growth, the reasons for doing so are outlined in Sect. 2.2. However, it is worth noting, that if considered, the influence of microlite growth would possibly increase the bulk viscosity significantly.

2.5 Boundary Conditions

Flow within the system is driven by a pressure gradient defined by boundary conditions at the top and bottom of the conduit. The top boundary is set to atmospheric pressure at the altitude of the conduit exit plus any overburden load from an emplaced lava dome. The bottom boundary is set to lithostatic pressure (assuming a homogeneous country rock density of 2600 kg m^{-3}) plus any imposed overpressure (P_e). Both the top and bottom pressure conditions are held constant throughout the model run. Initial boundary conditions along the length of the conduit are defined as no slip. When brittle failure of melt is considered within a model run, at the regions of the conduit wall where the brittle failure criterion was exceeded, the boundary conditions are changed to a tangential slip velocity ($\Delta\boldsymbol{u}$) defined by:

$$\Delta\boldsymbol{u} = \frac{1}{\beta}\sigma_s, \quad (15)$$

where σ_s is the tangential shear stress to the conduit wall and the coefficient β is a function of the slip length (L_s):

$$\beta = \frac{\eta}{L_s}, \qquad (16)$$

The model is then re-run to account for the effect of changing boundary conditions at the conduit walls. Where this results in an increase in the predicted failure depth, an iterative approach is used and the model is re-run with the new depth until the depth at which brittle failure of the melt stabilises. For the purposes of this study, the failure depth is considered to have converged if the depth increase between iterative runs is less than 10% of the previous observed increase.

The size and shape of the conduit is also an important factor in influencing the accent rate, so for the purpose of assessing the potential magnitude of this influence, several possible conduit shapes were modelled. While the types of boundary conditions discussed above do not change, the relative locations of the boundaries do (Fig. 2). Case (a) is the simplest geometry change and represents just a change of the conduit radius (r). Case (b) represents a narrowing of the conduit. Case (c) represents a widening of the conduit. The extent to which the geometry of the conduit is changed within the models is discussed further in the next Section.

3 Critical Conduit Processes

3.1 Using Magma Ascent Rates to Assess Model Sensitivity

The ascent rate is a key parameter in understanding volcanic hazard because it has been directly linked to eruptive behaviour (e.g. Gonnerman and Manga 2007). By gaining a better understanding of which model parameters have the greatest effect on ascent rates, we can achieve an insight into which are the most important parameters controlling explosivity, and the likely severity of the volcanic hazard. For the purpose of comparing the various models we use two velocities, defined as $\overline{V}$ and V_{2500}, where $\overline{V}$ is the average accent velocity taken along a vertical profile through the centre of the conduit, and V_{2500} is the average accent velocity taken along a horizontal profile at a depth of 2500 m within the conduit.

The ascent rate has also been linked to monitoring data such as seismicity (e.g. Thomas and Neuberg 2012) or deformation (e.g. Zobin et al. 2011), therefore, it is possible to link the changes in model parameters to recorded monitoring data. In addition, there are physically observed variations in ascent rate estimated from a variety of methods, ranging from studying mineral reaction rims around phenocrysts within erupted magma

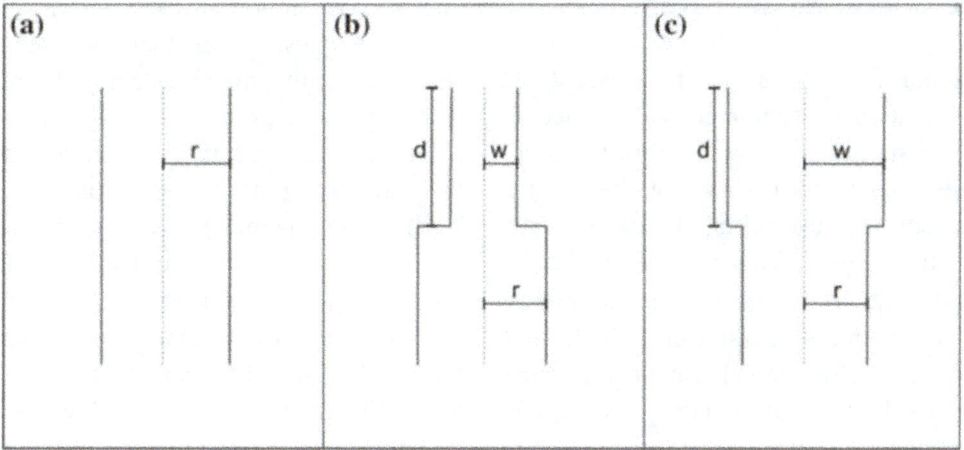

Fig. 2 Schematic diagram of the alternative conduit geometries modelled, showing **a** a constant conduit radius; **b** a narrowing conduit and **c** a widening conduit

(Rutherford and Devine 2003) to interpreting lava dome morphology (Sparks et al. 2000). This places constraints on the magnitude of changes to the modelled ascent rate engendered by altering the model input parameters we can consider realistic.

Matching the absolute values of physically observed and calculated ascent rates is currently beyond the scope of the model, however we can use the magnitude of the observed variations to provide upper and lower bounds to the extent to which the model input parameters are varied. Any changes that produce increases in ascent rate greater than two orders of magnitude over the reference model are not considered realistic in this work. This may seem at first an arbitrary discrimination, but there is a good reason that the observed or calculated ascent rates presented in the literature (e.g. Rutherford and Devine 2003; Castro and Gardner 2008) are "slow" ($<5 \times 10^{-2}$ m s^{-1}). Faster ascent rates, while likely to exist in nature, would almost certainly result in substantial fragmentation of the magma, making it very difficult to observe or calculate the actual magma ascent rate below the initial point of fragmentation. Fragmentation dynamics are not considered within the current model, hence no valid inferences or conclusions can be gained from studying the model runs that exhibit extremely fast ascent rates.

3.2 The Critical Model Parameters

Figure 3 summarises the sensitivity of ascent rate to the different model parameters presented in Table 1. The single parameter (within the modelled ranges) which has the strongest effect on the ascent velocities is the initial dissolved water content of the magma. This parameter affected both $\overline{V}$ and V_{2500} to a large degree. In contrast there are several model parameters which have little effect on the modelled ascent velocities. These include the thermal boundary layer thickness and the temperature drop across it, as well as the bubble number density and bubble surface

tension. Modifying the parameters involved in the brittle failure of the melt (magma shear strength and slip length) has a negligible effect on ascent rates and these results have not been plotted on Fig. 3. However, the contribution of the brittle failure of the melt to observed geophysical signals is considered very important, and will be discussed in Sect. 3.3.

It is unsurprising that the group of model parameters that appear to have the greatest effect on the magma ascent velocity, as seen in Fig. 3b (water content, temperature, crystal content, and chemical composition), also have the greatest effect on the magma viscosity (Sect. 2.3). Ultimately, modelling the ascent of magma is a fluid flow problem, and the properties that have the biggest effect on the fluid (magma) properties will have the biggest effect on the overall dynamics of the system. All other parameters have a much smaller direct effect on the fluid properties, and although they may be important to specific small scale magmatic processes when considered in isolation, with respect to the overall magma ascent they appear insignificant. For example, altering the properties of the bubbles within the magma, b_{ni} and Γ, the effect is to change the shape and the number of bubbles. Previous work has heavily focused on this area (e.g. Llewellin and Magna 2005) but the effect on the overall flow modelled here is minimal. The indication from this is that it is the total volatile content (water in this case) which is available that is more important to governing the overall flow dynamics, rather than how exactly it is stored in the magma. This particular observation is a key point as new estimates from Cassidy et al. (2015) suggest that basaltic South Soufrière Hills magmas (and by extension, possibly other basaltic arc magmas) have the potential to be extremely volatile-rich, containing up to >6 wt% H_2O prior to eruption. Firstly, this validates the use of high initial water contents used in the range of parameters modelled, and secondly, given the range of accent velocities generated within the models as a result of just changing the dissolved water content (the dark blue bars in

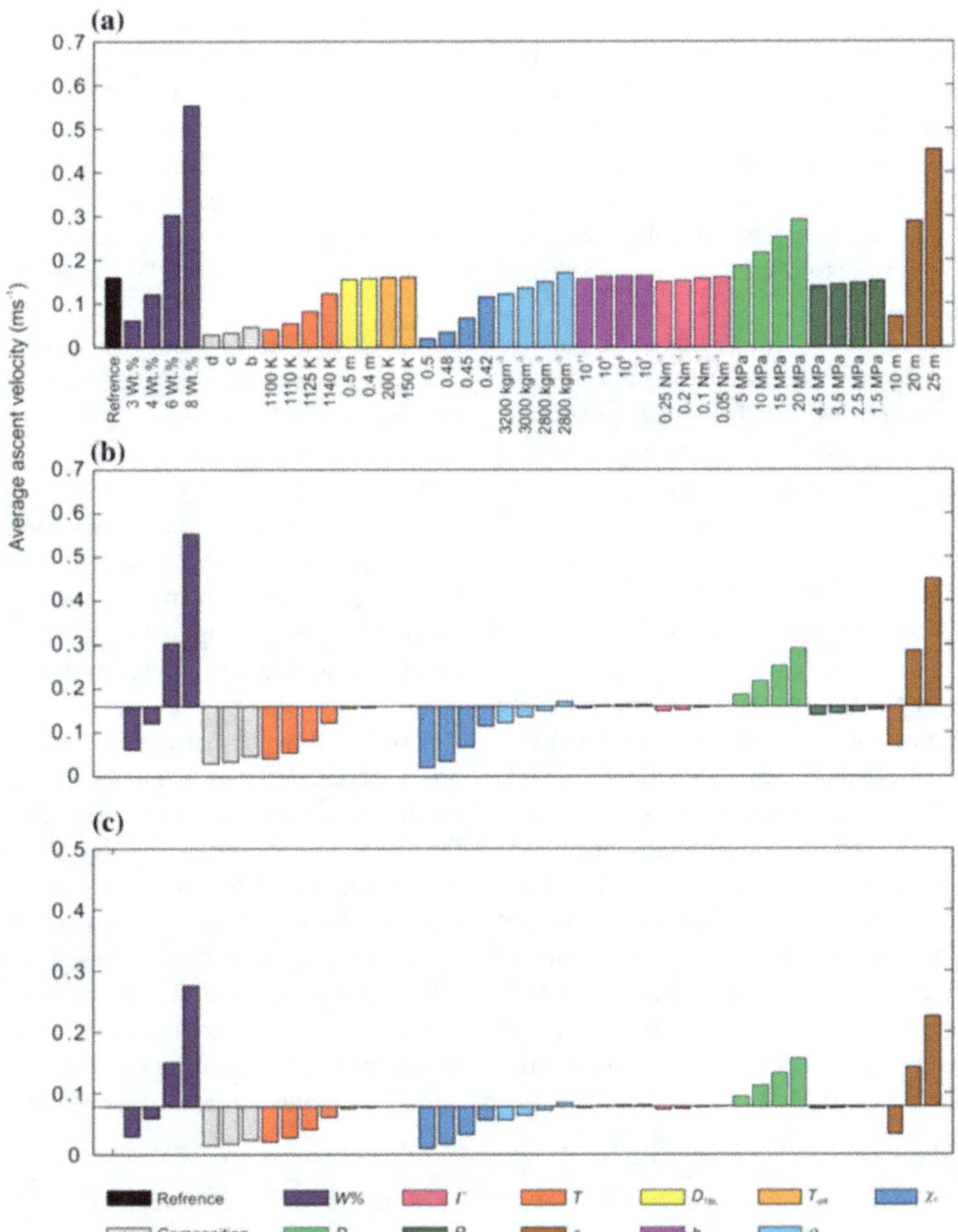

Fig. 3 a The ascent rate, $\overline{V}$ for each value of the parameters altered. The black bar represents the reference model with parameters as listed in Table 1. **b** The same data as (**a**) plotted relative to the reference model, which is represented by the horizontal black line. Parameter changes that caused increases in the ascent rates plot upward from the horizontal line, while parameter changes that caused decreases in ascent rates plot downwards. **c** V_{2500} for each value of the parameters altered, plotted relative to the reference model

Fig. 3), demonstrates that it is vitally important to obtain an accurate understanding of the magma components at the volcano of interest, rather than assuming "typical" values representative of a broad compositional category.

Outside of water content, temperature, crystal content, and chemical composition, the two parameters modelled which have the largest influence on the modelled ascent rate are the chamber overpressure and the conduit geometry. These are particularly important points when considering volcanoes entering periods of unrest following long periods of quiescence. It is problematic to achieve an accurate understanding of the magma components highlighted above at all volcanoes under these circumstances due to a likely lack of monitoring (a problem highlighted in Parts 1 and 2 of this book). Unless there has been long-term measurement of deformation occurring at the volcano now exhibiting signs unrest, it will be extremely difficult to estimate any likely overpressures in the chamber, and attempting to define the conduct geometry of a system that has not yet erupted would be almost impossible. It is therefore paramount that as much information as possible of all potentially active volcanic systems is routinely gathered before signs of unrest are detected.

3.3 Matching Observations— Explosivity and Seismicity

Although, as previously mentioned, matching the absolute values of physically observed and calculated ascent rates is currently beyond the scope of the model, key to giving the models real significance is determining whether the changes to important parameters highlighted in Sect. 3.1 can be theoretically linked to physical observations at real volcanic systems. Figure 4 shows values of V_{2500} for all of the modelled parameters in a manner similar to that presented in Fig. 3c, but in this case, the data are plotted relative to a baseline accent value of 0.02 ms^{-1}. This base line value was chosen because it has been highlighted by Rutherford and Devine (2003) as an ascent rate which may indicate a transition between effusive and explosive behaviour. This value has been obtained from quantifying the breakdown of hornblende in ascending magma, and while this technique is not an accurate barometer for defining an exact ascent velocity required for explosive eruptions, the rates calculated for non-explosive eruptive activity at Soufrière Hills volcano between the period of November 1995– September 2002 were below this value. Figure 4 shows that several model runs produced ascent rates of <0.02 ms^{-1} (by altering the melt composition or χ_c) and several other runs produced ascent rates very close to this value (by altering $W_{\%}$, T, and conduit geometry), indicating that by altering just single parameters within the system this theoretical threshold of accent rate can be crossed.

Conduit flow is treated as a closed system, so no outgassing is considered in the model, as a result the ascent velocities are overestimated (Thomas and Neuberg 2014). It is therefore predictable that if this process was included, far more of the model runs would result in ascent velocities that straddle the baseline in Fig. 4. This suggests that the ability for the ascent rate within the conduit to fluctuate, either side of values that have been linked to explosive eruptions in response to small changes in the system

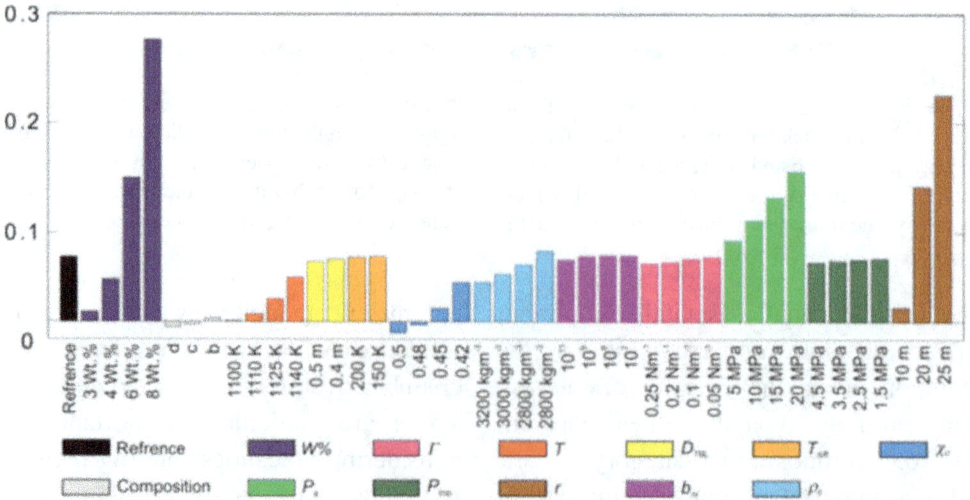

Fig. 4 The ascent rate, V_{2500} for each value of the parameters altered, plotted relative to an ascent velocity of 0.02 ms^{-1}

parameters is genuine. The requirement to accurately understand and model outgassing processes (see Sect. 4) is therefore an important capacity that is currently lacking.

One major discrepancy between physical observations and the model is the model results suggested that the brittle failure of the melt (Sect. 2.4) and the related LF seismicity would occur as a shallow process (in agreement with the work of Holland et al. (2011)).The physical observations place the location of this type of seismicity at Soufrière Hills at depths of ~ 1500 m below the conduit exit (Neuberg et al. 2006). This is because under normal conditions the shear stress required to break the melt (Eq. 14) can only be reached where the melt is extremely viscous, which occurs near the surface. In order to reach the higher shear stresses required to break the melt at greater depths, where the viscosity is lower, the shear strain rate ($\dot{\varepsilon}$) needs to increase. Since $\dot{\varepsilon}$ is equal to the lateral velocity gradient within a cylindrical conduit or dyke:

$$\dot{\varepsilon} = \frac{dv}{dx} \tag{17}$$

the simplest way to increase $\dot{\varepsilon}$ is to increase the velocity of the magma flowing within the conduit, or reduce the area through which it flows,

which since mass must be conserved also has the effect of increasing the flow velocity.

To resolve this discrepancy between model and observations we introduce a constriction within the conduit as a plausible explanation for brittle failure at greater depths. We test its effect within the reference model by including a bottleneck region at a depth of 1500 m, reducing the conduit diameter from 15 to 10 m. This bottleneck is 100 m in length, which equates to only 1/50 of the total conduit length. Figure 5 shows ascent velocity and shear strain rate profiles from the bottleneck region compared to values from the same location of the conduit in the unmodified reference model. By altering this relatively small region of the conduit, the shear strain rate increases by a factor of four. Crucially, with the exception of small changes in the magma rheology caused by the induced pressure gradients within the bottleneck, the magma viscosity has not been altered. Due to the increased value of shear strain rate the brittle failure ratio (13) will increase by the same factor. By introducing such asperities into the conduit and increasing the strain rate it is possible to drive the brittle fracture of the melt to deeper levels in the conduit that match the location of recorded LF seismicity at Soufrière Hills volcano. This further

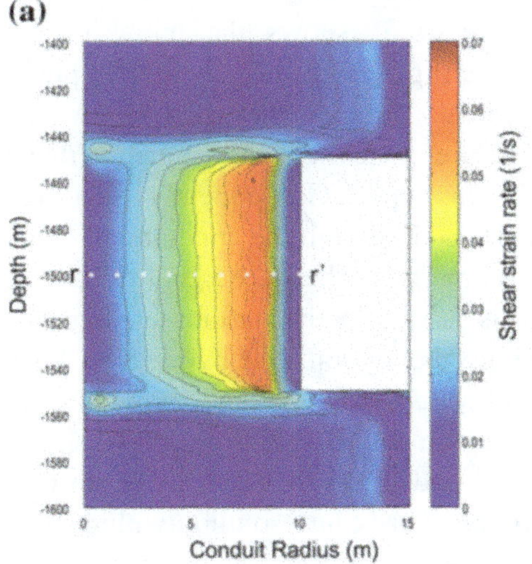

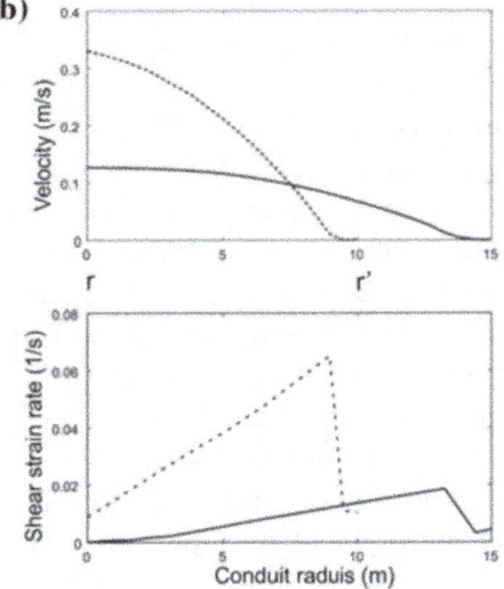

Fig. 5 a Plot of the shear strain rate within a simple bottleneck of 100 m length at intervals of 0.005. The values of shear strain rate are seen to be increases over the entire length of the bottleneck. b Cross conduit profiles

taken at the same depth for velocity and shear strain rate for the case of the unmodified reference model (solid line) and a conduit containing a bottleneck (dashed line) emphasises the importance of understanding the possible conduit geometry previously highlighted in Sect. 3.2. This is just one possible solution of generating seismicity deeper in the conduit. Any process that acts to increase the sear strain rate, such as strain localisation between crystals in a magma with a high crystal fraction (as discussed in Chap. 10 of this book) would be a possible explanation.

4   Pathways for Outgassing

The role of volatiles has been identified as the primary controlling parameter governing ascent rate within the conduit, but what was not considered was that the gas, once exsolved from the melt has the ability move independently to the melt. This is a commonly assumed process in basaltic systems, but not in silicic magmas of a relatively high viscosity as it is assumed that bubbles of gas cannot rise with significant speed within the magma. However fractures generated by brittle failure of the melt (Sect. 3.3) may provide ideal outgassing pathways for exsolved gas.

The behaviour of this exsolved gas is separated from the problem of magma ascent, and considered independent of any other parameters, through additional numerical modelling in Comsol Multiphysics (Collinson and Neuberg, 2012). In order to consider all possible outgassing pathways, including vertically through the conduit, or laterally through the walls, we model the gas response to brittle failure using a simplistic "block-style" model, with a central conduit and adjacent wall-rocks. Brittle failure is explicitly modelled as an increase in the permeability within narrow regions either side of the conduit. The problem is simplified to considering permeable flow through a static media. Consequently, the equations for Darcy's law (18), and the continuity equation (19) are amalgamated to derive a partial differential equation (20), which is solved for pressure (P):

$$\boldsymbol{u} = -\frac{k}{\eta_g}(\nabla P + \rho g \nabla z) \qquad (18)$$

$$\frac{\partial}{\partial t}(\rho \varepsilon) + \nabla \cdot (\rho \boldsymbol{u}) = 0 \qquad (19)$$

$$\frac{\partial}{\partial t}(\rho \varepsilon) + \nabla \cdot \rho \left[-\frac{k}{\eta_g}(\nabla P + \rho g \nabla z) \right] = 0 \quad (20)$$

The gas velocity (u) is then determined by using the pressure gradient within Darcy's law (18).

In contrast, with the models for ascent rate, time dependency is included in this model in order to understand the changes in the system through time, in response to a permeability (k) increase in response to brittle failure at the conduit-wall margin. Due to its abundance in volcanic systems, water is the only volatile considered here. The gas density (ρ) is calculated using the mean molar mass within the ideal gas law (6), and the gas viscosity (μ) assumed constant at 1.5×10^{-5} Pa (Collinson and Neuberg 2012).

Bulk permeabilities are set such that the conduit has a higher permeability (10^{-10} m^2) than the wallrocks (10^{-12} m^2), and initially, the conduit-wall margin is "sealed" with a very low permeability of 10^{-16} m^2. The fracturing is initiated at 1500 m, in accordance with measurements by Neuberg et al. (2006), and propagate vertically towards the surface, as an increase in permeability from 10^{-16} to 10^{-6} m^2.

Figure 6a shows the initial system, before fracturing, where the gas loss is predominantly vertical, through the conduit. In Fig. 6b, the fracture zone has propagated vertically up to 700 m depth. Consequently, the pressure has increased through the conduit and corresponding wall margins where the fractures have developed. This is due to the regions of increased permeability being confined by areas of lower permeability above. Thereby, providing a suitable environment for gas storage, which due to the increased pressurisation, may force exsolved volatiles back into solution within the melt. This change in pressurisation has resulted in a

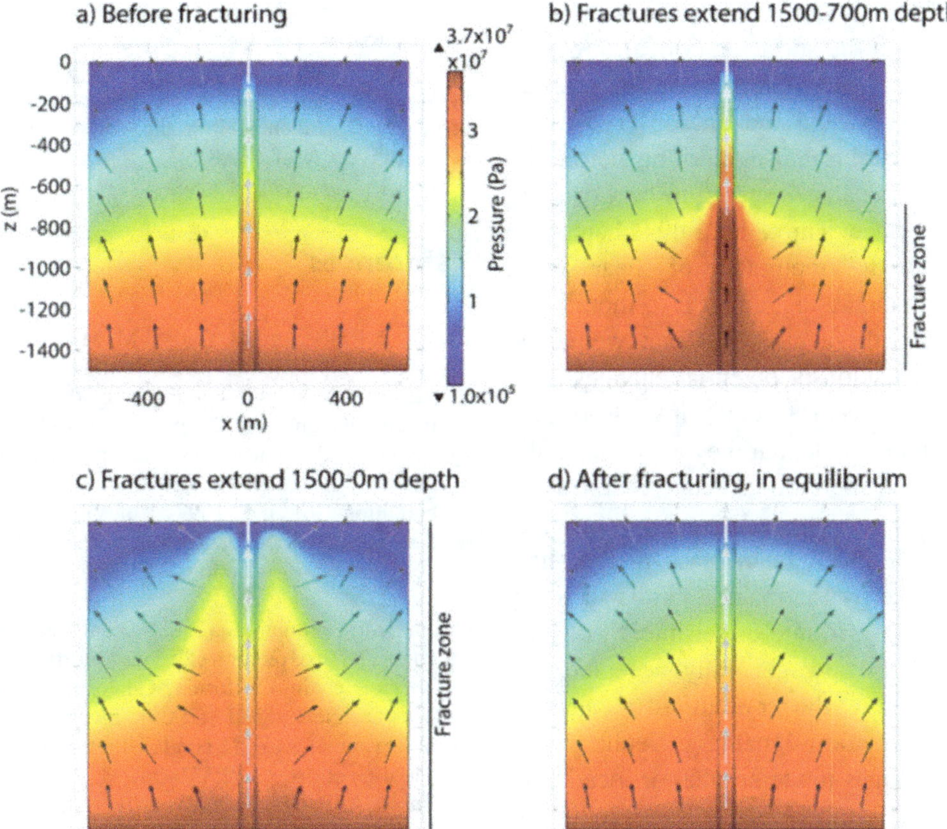

Fig. 6 a The initial system in equilibrium, before fracturing has commenced. **b** The systems of fractures has propagated upwards to a depth of 700 m. **c** The fracture zones have reached the surface. **d** The system has resumed equilibrium. Velocity range of arrows (ms⁻¹): **a** max: 0.30, min: 1.3×10^{-3}; **b** max: 0.38, min: 8.8×10^{-5}; **c** max: 0.33, min: 6.8×10^{-4}; **d** max: 0.30, min: 9.5×10^{-4}

corresponding change to the gas velocity pattern, which the lower conduit having a very low gas velocity due to the low pressure gradient. However, at shallow levels, the higher pressure gradient forces a much higher gas velocity towards the surface. On reaching the surface (Fig. 6c), the high permeability fractures result in a decrease in pressure throughout the conduit and a rapid expulsion of the stored gas. Equilibrium conditions resume approximately 4 h after the fracture zones reached the surface (Fig. 6d). This model shows a propagating fracture zone is an effective mechanism for degassing the conduit and wall margins. A key observation is the possibility for this mechanism to produce periods of cyclic activity which are observed at many silicic volcanoes (e.g. Holland et al. (2011)), which can be directly related to observed degassing patterns or through controlling the ascent rate and the

associated geophysical signals (e.g. LF seismicity) within the conduit through moderating the amount of gas stored in the system.

5 Summary and Implications

In the introduction to this chapter, we asked the question of whether it is possible that small changes in the composition of properties of the magma could cause changes significant enough to be recorded in monitoring data or even visual observations. It is clear that these small changes, particularly when considering changes in the water content or conduit diameter, can have large effects of the ascent velocity of the magma. These effects are large enough that conceivable they could be simply observed as an increased extrusion rate at the surface. At this point however it may be too late. If increased extrusions rates are being observed at the surface a critical threshold of ascent rate may have already been surpassed.

More useful would be to observe these potential changes through monitoring before the magma was physically observed to be extruding faster at the surface. Throughout this chapter we have discussed the importance of shear stress, seismicity and outgassing, and it is to monitoring data relating to these processes that we would look for an indication of potential changes.

Increases in ascent velocity may lead to a change in eruption style, but they also potentially cause changes in monitoring data. As discussed in the chapter, any increase in ascent velocity of the magma will cause an increase in the shear stress experience by the magma, and will likely lead to an increase in rate of seismicity, or alter its location. It is now generally accepted that shear stress within the conduit also causes a significant deformation signal (e.g. Neuberg et al. 2018). Any changes in the number or location of low frequency earthquakes, or changes in the near-field deformation around the conduit could therefore be inferred as changes in the magma properties. In addition, the single most important parameter identified within this study was changes in the water content. The fractures generated by the seismicity, as a result of the increases in ascent velocity, have also been shown to be important outgassing pathways. An increase in water content would lead to an increase in volatiles and these would be more easily outgassed along the created fractures. Conceptually, a clear signal of an increase in ascent velocity (and a potential early warning of a change in eruption style) caused by increase in water content of the magma, could be an increase and deepening of low frequency earthquakes, accompanied by an increase in the deformation signal followed by significant outgassing event.

In this chapter we look at a single volcanic setting, but by changing the model parameter these effects could be assessed at any volcano, and regardless of the volcano studied the relative importance of the parameters considered should remain unaltered.

Acknowledgements The authors are grateful for inspiring discussions with Geoff Kilgour (GNS New Zealand) during J. Neuberg's study leave at GNS Research Centre in Wairakei, and contributions to research ideas form former Ph.D. students within the Volcanic Studies Group at the University of Leeds.

References

Barclay J, Rutherford M, Carroll M, Murphy M, Devine J (1998) Experimental phase equilibria constraints on pre-eruptive storage conditions of the Soufrière Hills magma. Geophys Res Lett 25:3437–3440

Burgisser A, Poussineau S, Arbaret L, Druitt TH, Giachetti T, Bourdier J-L (2010) Pre-explosive conduit conditions of the 1997 Vulcanian explosions at Soufrière Hills Volcano, Montserrat: I. Pressure and vesicularity distributions. J Volcanol Geoth Res 194:27–41

Castro JM, Gardner JE (2008) Did magma ascent rate control the explosive-effusive transition at the Inyo volcanic chain, California? Geology 36:279–282

Collier L, Neuberg J (2006) Incorporating seismic observations into 2D conduit flow modelling. J Volcanol Geoth Res 152:331–346

Collinson ASD, Neuberg JW (2012) Gas storage, transport and pressure changes in an evolving permeable volcanic edifice. J Volcanol Geoth Res 243–244:1–13

Cluzel N, Laporte D, Provost A, Kannewischer I (2008) Kinetics of heterogeneous bubble nucleation in rhyolitic melts: implications for the number density of bubbles in volcanic conduits and for pumice textures. Contrib Miner Petrol 156:745–763

Devine JD, Murphy MD, Rutherford MJ, Barclay J, Sparks RSJ, Carroll MR, Young SR, Gardner JE (1998) Petrologic evidence for pre-eruptive pressure-temperature conditions, and recent reheating, of andesitic magma erupting at the Soufrière Hills Volcano, Montserrat, W.I. Geophys Res Lett 25:3669–3672

Devine JD, Rutherford MJ, Norton GE, Young SR (2003) Magma storage region processes inferred from geochemistry of Fe–Ti oxides in andesitic magma, Soufrière Hills Volcano, Montserrat, WI. J Petrol 44:1375–1400

Edmonds M, Aiuppa A, Humphreys M, Moretti R, Giudice G, Martin RS, Herd RA, Christopher T (2010) Excess volatiles supplied by mingling of mafic magma at an andesite arc volcano. Geochem Geophys Geosyst 11:Q04005

Giordano D, Russell JK, Dingwell DB (2008) Viscosity of magmatic liquids: a model. Earth Planet Sci Lett 271:123–134

Gonnermann HM, Manga M (2007) The fluid mechanics inside a volcano. Annu Rev Fluid Mech 39:321–356

Gotto A (1999) A new model for volcanic earthquake at Unzen Volcano: melt rupture model. Geophys Res Lett 26:2541–2544

Holland ASP, Watson IM, Phillips JC, Caricchi L, Dalton MP (2011) Degassing processes during lava dome growth: Insights from Santiaguito lava dome, Guatemala. J Volcanol Geoth Res 202:153–166

Hurwitz S, Navon O (1994) Bubble nucleation in rhyolitic melts: experiments at high pressure, temperature, and water content. Earth Planet Sci Lett 122:267–280

Ishibashi H (2009) Non-Newtonian behavior of plagioclase-bearing basaltic magma: subliquidus, viscosity measurement of the 1707 basalt of Fuji volcano, Japan. J Volcanol Geoth Res 181:78–88

Lensky NG, Lyakhovsky V, Navon O (2002) Expansion dynamics of volatile-supersaturated liquids and bulk viscosity of bubbly magmas. J Fluid Mech 460:39–56

Liu Y, Zhang Y, Behrens H (2005) Solubility of H_2O in rhyolitic melts at low pressures and a new empirical model for mixed H_2O-CO_2 solubility in rhyolitic melts. J Volcanol Geoth Res 143:19–235

Llewellin E, Manga M (2005) Bubble suspension rheology and implications for conduit flow. J Volcanol Geoth Res 143:205–217

Lyakhovsky V, Hurwitz S, Navon O (1996) Bubble growth in rhyolitic melts: experimental and numerical investigation. Bull Volc 58:19–32

Marsh B (1981) On the crystallinity, probability of occurrence, and rheology of lava and magma: contributions to mineralogy and petrology. Bull Volcanol 78:85–98

Marti I, Höfler O, Fischer P, Windhab EJ (2005) Rheology of concentrated suspensions containing mixtures of spheres and fibers. Rheol Acta 44:502–512

Neuberg JW, Tuffen H, Collier L, Green D, Powell T, Dingwell D (2006) The trigger mechanism of low-frequency earthquakes on Montserrat. J Volcanol Geoth Res 153:37–50

Neuberg JW, Collinson ASD, Mothes PA, Ruiz CM, Aguaiza S (2018) Understanding cyclic seismicity and ground deformation patterns at volcanoes: Intriguing lessons from Tungurahua volcano. Ecuador, Earth Planet Sci Lett 482:193–200

Rutherford MJ, Devine JD (2003) Magmatic conditions and magma ascent as indicated by hornblende phase equilibria and reactions in the 1995–2002 Soufrière hills magma. J Petrol 44:1433–1454

Sparks RSJ, Murphy MD, Lejeune AM, Watts RB, Barclay J, Young SR (2000) Control on the emplacement of the andesite lava dome of the Soufrière Hills volcano, Montserrat by degassing-induced crystallization. Terra Nova 12:14–20

Thomas ME, Neuberg JW (2012) What makes a volcano tick—A first explanation of deep multiple seismic sources in ascending magma. Geology 40:351–354

Thomas ME, Neuberg JW (2014) Understanding which parameters control shallow ascent of silicic effusive magma. Geochem Geophys Geosyst 15:4481–4506

Tuffen H, Dingwell DB, Pinkerton H (2003) Repeated fracture and healing of silicic magma generate flow banding and earthquakes? Geology 31:1089–1092

Wadge G, Robertson REA, Voight B (eds) (2014) The eruption of Soufrière Hills Volcano, Montserrat from 2000 to 2010. Geol Soc, London, Memoirs, p 39

Zobin VM, Ramirez JJ, Santiago H, Alatorre E, Navarro C (2011) Relationship between tilt changes and effusive-explosive episodes at an andesitic volcano: the 2004–2005 eruption at Volcan de Colima, Mexico. Bull Volc 73:91–99

When Does Magma Break?

Fabian B. Wadsworth, Taylor Witcher, Jérémie Vasseur, Donald B. Dingwell and Bettina Scheu

Abstract

Geophysical signals arriving at the Earth's surface originate from a source mechanism at depth but are not necessarily directly observable. Therefore, well-posed experiments can provide insights into source mechanics and, importantly, the parameters required to model aspects of the sources of unrest signals. In this Chapter we detail one such example of how experimental laboratory work has improved our understanding of unrest signals. We focus on the failure of single- and multi-phase magmas, demonstrating that the liquid viscosity, and therefore the temperature and volatile content of a magma of a given composition, is the limiting parameter in determining whether a magma will ascend viscously or whether it can fracture during ascent. This critical threshold is characterized by a Deborah number, the ratio of the timescale of relaxation to the timescale of local flow. We show that for single-phase magmatic liquids and for vigorously vesiculating magmas, a local Deborah number of 10^{-2} is the limit above which mixed viscoelastic behaviour including fracture propagation can be expected, and a Deborah number of 1 is the limit above which magma is dominantly elastic and responds in a brittle manner to applied stresses. These thresholds can be understood in terms of the onset and peak of the Debye relaxation process for viscoelastic liquids. The apparent validity of a Maxwell model permits us to predict the maximum stress that can be supported by a volcanic liquid deforming in the high Deborah number range. We use these constraints to provide a map of timescales on which we contour dominant system responses from viscous to purely brittle; valid for all magmatic liquids. Finally, we explore the scaling necessary to extend these conceptual insights to crystal- and bubble-bearing magmas valid under specific conditions.

F.B. Wadsworth (✉) · T. Witcher · J. Vasseur
D.B. Dingwell · B. Scheu
Ludwig-Maximilians-Universität, Theresienstr.41,
80333 Munich, Germany
e-mail: fabian.wadsworth@min.uni-muenchen.de

Advs in Volcanology (2019) 171–184
DOI 10.1007/11157_2017_23
© The Author(s) 2017
Published Online: 18 October 2017

The competing timescales of deformation and relaxation in magma are relevant to unrest source mechanisms that originate from magma deformation, such as long-period seismic signals that are used to predict eruption timing.

Keywords

Rheology · Strain rate · Experimental volcanology · Glass transition · Failure forecasting · Low frequency earthquakes · Viscous dissipation

Glossary

Rheology
: The study of the response of a material to an applied stress or deformation. In the volcano-sciences this typically refers to magma-rheology which is dominated by the rheology of the liquid phase (a viscous or *viscoelastic* melt) and the additional effect of suspended phases (crystals and gases).

Viscoelasticity
: A material response to an applied stress in which both elastic and viscous components of the deformation are observed. In magma, as the temperature decreases or the local strain rate increases, the elastic component becomes more dominant. When the viscous component is negligible, purely elastic behaviour and material-rupture can readily occur. See *Rheology*.

1 Introduction

There are a wide variety of observable unrest signals at active volcanoes. The key unrest signals are those that are diagnostic of a new regime of behaviour or of how the system may evolve in the future. One such family of events are the low-frequency earthquakes at volcanoes, which are thought to directly represent magma movement at moderate to shallow depths (Chouet et al. 1994; Chouet 1996; Neuberg et al. 2006; Salvage and Neuberg 2016). Such events are powerful tools particularly because they are closely associated with surface activity (Miller et al. 1998). Accelerating event rates of low frequency earthquakes can be used to retrospectively predict an eruption or dome collapse event with reasonable accuracy (Salvage and Neuberg 2016). The utility of the low-frequency event type has been solidified since physical hypotheses for the source mechanism have been put forward (Neuberg et al. 2006) and tested in scaled laboratory expeirments (Benson et al. 2008; Tuffen et al. 2008). The phenomenological observation of the use of certain signal types over others is useful and has led to accurate retrospective eruption forecasts (Voight 1988; Chouet 1996). However, knowledge of the physics of the source leads to a more diverse range of tools being deployed to understand the evolution of volcanic systems into the future. In the context of low-frequency earthquakes at volcanoes, once a source mechanism is identified (Tuffen et al. 2003; Neuberg et al. 2006), numerical models of magma ascent in the conduit beneath volcanoes can be used to reproduce the depths of the source from first principles (Thomas and Neuberg 2012) and could eventually make forecasts of other observables at the surface that would be consistent with impending eruption.

In this Chapter we explore the source mechanism of low-frequency earthquakes at volcanoes from a physical perspective using a compilation of data from scaled laboratory experiments. We use these datasets to demonstrate that the critical threshold for fracture propagation in a viscoelastic fluid such as magma is universal. We propose that this potentially simplifies the inputs required for effective modelling of source mechanics of low-frequency events. More than this, we hope that this provides a good example of how laboratory work can provide valuable insight into the physical feasibility of models proposed to explain unrest signals at volcanoes.

2 Scaling the Viscous-to-Brittle Transition in Magmas

2.1 Single-Phase Magmatic Liquids

Newtonian viscous liquids deform and flow under applied stresses such that the rate of shear strain $\dot{\gamma}$ resulting from an applied shear stress τ is proportional to the viscosity μ via $\tau = \mu\dot{\gamma}$. However, as large shear stresses are applied to viscous liquids, they can exhibit viscoelastic behaviour, which manifests as an apparent non-Newtonian relationship between τ and $\dot{\gamma}$ and can result in fracture propagation locally in the liquid. Of interest here is the transition from simple, apparently Newtonian viscous behaviour to complex viscoelastic behaviour.

Maxwell described the simplest viscoelastic model in which a liquid has a characteristic time required to relax an applied shear stress, termed the Maxwell relaxation timescale λ_r, which is given by $\lambda_r = \mu/G_\infty$ where G_∞ is the shear modulus in the purely elastic regime. The Maxwell model of viscoelasticity permits us to

predict that if a shear strain is imposed on a viscoelastic liquid, the resultant shear stress will rise rapidly to a peak value τ_i and will relax over time t according to $\tau(t) = \tau_i \exp(-t/\lambda_r)$. This simple concept has proved powerful in describing the transition between Newtonian and viscoelastic behaviour in volcanic liquids (Dingwell 1995, 1996). Indeed, this model is invoked to explain the transition between volcanic liquids and volcanic glasses on cooling (e.g. Stevenson et al. 1995), the fragmentation of fluid and bubbly magma undergoing decompression (Alidibirov and Dingwell 1996; Kameda and Kuribara 2008) and the conditions under which extensive shear fracture networks can form at or near volcanic conduit margins (Gonnermann and Manga 2003; Tuffen et al. 2003; Kendrick et al. 2014; Hornby et al. 2015). In all cases, it is useful to define a Deborah number De, which is the dimensionless ratio between the Maxwell relaxation time and the timescale of deformation λ. For simple (viscometric) shearing flow, the latter timescale is $1/\dot{\gamma}$ and thus

$$\mathrm{De} = \frac{\lambda_r}{\lambda} = \frac{\mu\dot{\gamma}}{G_\infty} \tag{1}$$

Our hypothesis is that a Deborah number of unity separates the regime between viscous ($\mathrm{De} \ll 1$) and elastic ($\mathrm{De} \gg 1$) responses to shear stresses, or equivalently, between a system that can relax shear stresses efficiently from one in which shear stresses can accumulate elastically. In two different experimental types, both Webb and Dingwell (1990a) and Cordonnier et al. (2012a, b) showed that in fact the first evidence of behaviour that is not purely viscous occurs at $\mathrm{De} \geq 10^{-2}$ and not exactly at unity. This observation will be discussed later.

The power of Eq. 1 is that simple parameters that can be measured in the laboratory can be used to predict where this transitional point of $\mathrm{De} = 1$ will be. This critical Deborah number can be termed De'. In what follows, we give examples of how μ and G_∞ have been found for a range of volcanic liquid compositions at magmatic temperatures and volatile contents before exploring the validity and applicability of the scaling in Eq. 1. The dimensionless nature of Eq. 1 means that it can be assessed for any system so long as λ can be defined.

At the core of the experimental toolkit is the determination of the fundamental quantities that are required to model volcanic processes involved in unrest. The most variable physical quantity in volcanic systems is the viscosity of the liquid phase. Using a range of techniques, this viscosity can be determined with prodigious accuracy and significant effort has been expended in mapping the full range of composition, temperature (see Giordano et al. 2008) and volatile content (Hess and Dingwell 1996) relevant to shallow magmatic settings. Multi-component models have been proposed such that the viscosity of any composition of silicate volcanic liquid on Earth can now be predicted as a function of temperature including the range of conditions between shallow magma storage and the surface (Hess and Dingwell 1996; Giordano et al. 2008). This provides parameterization of $\mu(T)$ for use in Eq. 1. The most commonly used form for $\mu(T)$ is the non-Arrhenian Vogel-Fulcher-Tammann equation of the form

$$\mu = A \exp\left(\frac{B}{T - C}\right), \qquad (2)$$

where A, B, and C are coefficients that are experimentally determined and then parameterized as a function of families of oxides in the liquid structure (Giordano et al. 2008) or as a function of the dissolved water content (for calc-alkaline rhyolites: Hess and Dingwell 1996), to give two examples.

To give examples of the range of viscosities of interest in volcanic scenarios, we give end members in Fig. 1 for a calc-alkaline rhyolite using the model from Hess and Dingwell (1996), and for the basaltic liquid composition provided in Zhang et al. (1991) using the model from Giordano et al. (2008), both contoured for a range of water contents.

Unlike viscosity, the shear modulus in the elastic regime does not vary significantly in siliciate liquids and is not strongly dependent on composition or temperature. Indeed, a short survey of the values for G_∞ in silicate glasses and liquids at a range of temperature and a huge range of composition, provided by Dingwell and Webb (1989) shows that $G_\infty = 10^{10 \pm 0.5}$ Pa. Here we use this range in order to fully parameterize De (by predicting λ_r) as a function of temperature (via Eq. 2) for any silicate liquid in the shallow crust.

2.2 Extensions to Multiphase Magmas

Except in rare circumstances such as obsidian-forming eruptions, volcanic liquids are not often erupted without some proportion of suspended pore space (either as isolated bubbles or connected networks) and rigid crystals. In either case, the utility of Eq. 1 requires additional attention. We posit that for the liquid phase between the pores or the crystals, De given by Eq. 1 holds. However, we acknowledge that estimation or measurement of the rate of shear strain locally between pores or crystals would be difficult (Deubelbeiss et al. 2011). Therefore, a more robust criterion for the viscous-to-brittle transition in crystal- or bubble-bearing magma would require that we scale the bulk strain rate $\dot{\gamma}_b$ on the system for the effect of the suspended load.

For crystals, we show some first-order scaling attempts for their effect on the critical threshold for the onset of brittle behaviour in multiphase magma. The simplest view of the local flow of liquid in crystal-bearing magma under constant shear stress is that the rate of shear strain between the crystals should scale approximately with $\dot{\gamma} = \dot{\gamma}_b(1 - \phi_x/\phi_m)^{-1}$ where ϕ_x is the suspended crystal volume fraction and ϕ_m is a jamming fraction above which no more crystals can be

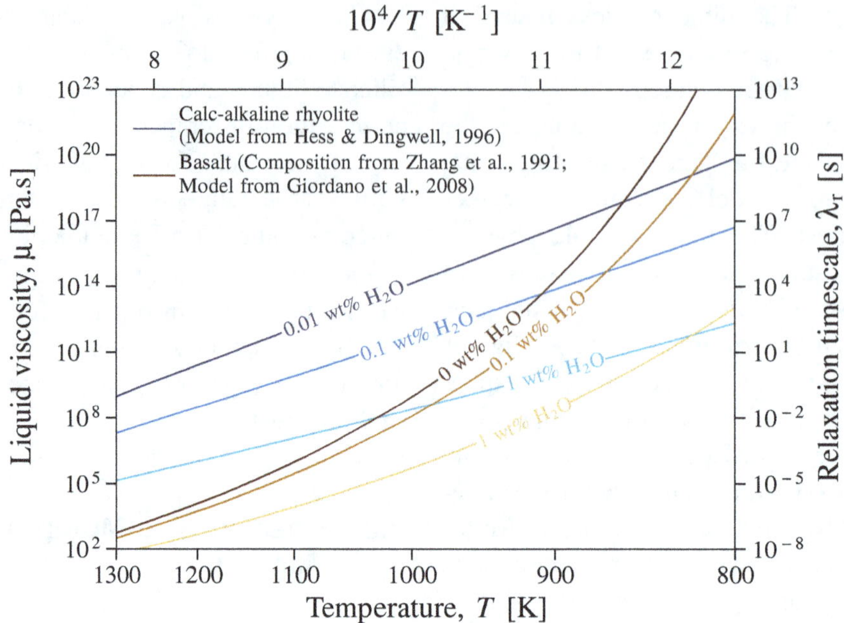

Fig. 1 The viscosity of end-member magmatic liquids with variable dissolved water concentrations. Plotted are the results for calc-alkaline rhyolite from the general viscosity model for hydrous silicic liquids (Hess and Dingwell 1996) for 0.01–1 wt% water, and a typical basaltic composition (composition from Zhang et al. 1991) calculated using the multicomponent viscosity model (Giordano et al. 2008) for 0–1 wt% water. The relaxation timescale is calculated assuming a composition independent value for G_∞ of 10^{10} Pa (Dingwell and Webb 1989)

added to the flowing system. This scaling would imply that the Deborah number for a crystal-bearing magma De$_x$ would be

$$\text{De}_x = \frac{\mu}{G_\infty}\dot{\gamma}_b\left(1 - \frac{\phi_x}{\phi_m}\right)^{-1}, \qquad (3)$$

where ϕ_m is a function of crystal shape and roughness (Mueller et al. 2010; Mader et al. 2013). A critical value of De$_x$, termed De$_x'$, is 10^{-2}, consistent with the limiting De$'$ at $\phi = 0$. A new definition of the bulk failure criterion De$'$ can now be made, which decreases as $\phi \rightarrow \phi_m$ so that De$' = De_x'(1 - \phi_x/\phi_m)$. In Fig. 2 we demonstrate how this concept predicts a linear relationship between De$_x$ and ϕ_x/ϕ_m, which in turn shows that lower bulk strain rates are required to induce brittle behaviour in crystal-bearing magmas.

This hypothesis was found to hold for simple two-phase systems by Cordonnier et al. (2012a), but where those authors required that μ be replaced by the suspension viscosity of the whole

system, which is at odds with the scaling of strain rate for local liquid effects only. Nonetheless, this approach described their data satisfactorily and the conceptual insight that crystals locally increase the liquid rate of shear strain relative to the bulk value is robust, with the implication is that the whole suspension will begin fracturing at lower bulk rates of strain relative to a single-phase liquid of the same composition as crystals are added. It may be that additional second-order effects are important at high ϕ_x, which are not accounted for in this simple analysis.

For the case of porous magmas, there are two considerations: (1) The growth of bubbles exerts a rate of shear strain in the liquid concentrated at the bubble walls, which is broadly independent of any bulk shearing deformation, and (2) like for the crystal case, the presence of bubbles changes the partitioning of the bulk rate of shear strain in the liquid between the bubbles. In the case of scenario (1), the Rayleigh-Plesset equation can be used to relate the bubble growth rate to the gas

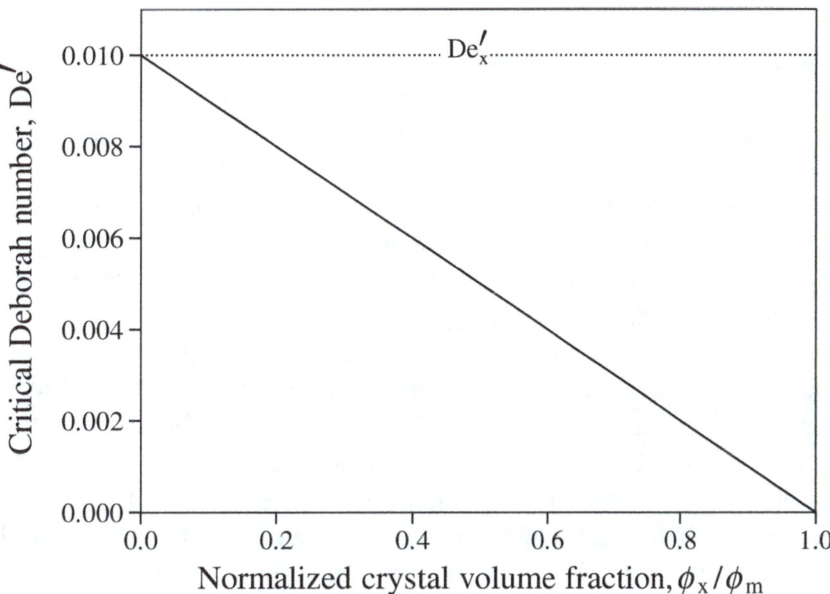

Fig. 2 The critical Deborah number required for fracturing De′ is reduced when crystals are suspended in magma. To a first-order, the De′ value is reduced proportional to $\mathrm{De}' = \mathrm{De}'_x(1 - \phi_x/\phi_m)$

pressure in the bubble, relative to the hydrostatic pressure (Sparks 1978). This can be augmented for the diffusion-controlled gradient of viscosity in the immediate liquid shell around a growing bubble (Prousevitch et al. 1993; Lensky et al. 2001). The component of the rate of shear strain tangential to the bubble wall $\dot{\gamma}_\theta$ can then be computed throughout bubble growth (Ichihara et al. 2002). If we were then to input $\dot{\gamma}_\theta$ into our computation of De, we would predict under which conditions the system would meet the criterion of $\mathrm{De}' = 10^{-2}$ locally at the bubble rim. These conditions would be best cast in terms of the critical rate of pressure or temperature change, or bulk initial volatile content that would allow the bubble to grow sufficiently fast that fractures could be propagated at the bubble wall.

In case (2), the effect of bubbles on the bulk Deborah number is less clear. We speculate that at De $\gg 1$, the stress required to fracture a bubbly system will be analogous to that required to fracture a vesicular glass. In this case, models for the effect of spherical cavities on the stress required for fracturing are valid and they predict that the stress is reduced significantly as the bulk gas volume fraction increases (Sammis and Ashby 1986). This has been confirmed in the high Deborah number regime for porous liquids analogous to volcanic systems (Vasseur et al.

2013) but remains untested in the low Deborah number regime where bubble deformation may be important.

2.3 Apparent Non-newtonian Effects

In single-phase liquids at Deborah numbers below the critical threshold at which fracturing is observed, there is evidence that during steady state shearing flow, there is a non-Newtonian relationship between applied shear stress and resultant shear rate of strain (Simmons et al. 1982; Dingwell and Webb 1989; Webb and Dingwell 1990a; Cordonnier et al. 2012b). In the onset of this non-Newtonian, the onset of this non-Newtonian behaviour appears to be well described by the value above which viscous dissipation of heat is active on the system length scale of interest (Costa and Macedonio 2005; Hess et al. 2008). This can be scaled by the Brinkman number Br

$$\mathrm{Br} = \frac{\Phi_g}{\Phi_l} = \frac{\mu\dot{\gamma}^2}{kq}, \qquad (4)$$

where Φ_g and Φ_l are the gain and loss power densities, respectively, q is the areal heat flux out of the system and k is the thermal conductivity of the material. $\Phi_g = \mu\dot{\gamma}^2$ represents the amount of energy produced in a system of a given volume by viscous dissipation of heat, while $\Phi_l = kq$ represents the energy lost due to diffusive thermal equilibration. When $Br \gg 1$, heat is efficiently produced and inefficiently lost from the system, resulting in a bulk temperature increase in the liquid. This would manifest itself as an apparent shear thinning rheology if the temperature increase where not locally accounted for, and would be most likely to be operative at high viscosities and high shear strain rates. We note that unlike the Deborah number, the Brinkman number is scale dependent (as q depends on the area available for heat transfer out of the system) and so should be assessed for each system scale separately.

---

3 The Universal Breaking Timescales of Volcanic Liquids

Laboratory data related to the viscous-to-brittle transition in magmas has been collected in a variety of geometries. Using single-phase liquids, there are two dominant geometries: (1) thin fibers of silicate liquid of basaltic, andesitic, phonolitic and rhyolitic composition were stretched under constant load in tension until the fibers snapped in a singular fracture event (Webb and Dingwell 1990b), and (2) cylinders of synthetic borosilicate liquid were compressed under constant load and the bulk temporal evolution of the rate of axial shortening was determined as viscous (relaxed) if it continuously increased to a steady or near-steady value, or brittle (unrelaxed) if the axial shortening rate jumped due to fracturing events (Cordonnier et al. 2012b). In another type of experiment, analogue vesicular liquids were decompressed at different rates from pressure (Kameda and Kuribara 2008; Kameda et al. 2013). In this type of decompression experiment, if the dominant response was viscous (relaxed), then the sample was seen to grow due to decompression-driven bubble growth, and if the dominant response was brittle (unrelaxed), then

smooth sample inflation was punctuated by visible fractures opening. In these decompression experiments there additionally was a violent rupture mode in which the fracturing was pervasive and shattered the sample in a vigorous fragmentation event. These data are selected here because the liquids used have a known relaxation time under the conditions used in the experiments and care was taken by the authors who originated the work to investigate the shear rates of strain local to the bubbles (discussed above).

In Fig. 3 we map these experimental results as a ratio of the deformation timescale to the relaxation timescale, which permits us to contour the plot for critical Deborah numbers. We colour-code the data according to the bulk mode of response of the sample to the deformation using *green* to represent purely viscous relaxed behaviour, *orange* to represent brittle unrelaxed behaviour and *red* to represent the complete violent rupture of the sample. These data suggest that the viscous relaxed behaviour transitions to unrelaxed brittle behaviour at a Deborah number of 10^{-2}, and are consistent across a huge range of experimental conditions and across both the single-phase compression and vesicular decompression experiment types. Furthermore, we see that a Deborah number of unity consistently separates the experiments for which the bulk response was unrelaxed and brittle from those for which the response was violent rupture, fragmentation or complete failure. It appears that these two thresholds, $De = 10^{-2}$ and $De = 1$ are universal even when comparing analogue room temperature liquids with high temperature silicate liquids deformed in a variety of ways. This lends power to the scaling provided by the dimensionless Deborah number and implies that we need only to define the liquid viscosity and a characteristic rate of deformation in order to predict whether a system will flow viscously or rupture violently. For example, the working viscosity of the analogue fluid used in Kameda and Kuribara (2008) is $10^0 \leq \mu \leq 10^{10}$ Pa.s, which extends to much lower used in Cordonnier et al. (2012b), and yet the scaling with the Deborah number

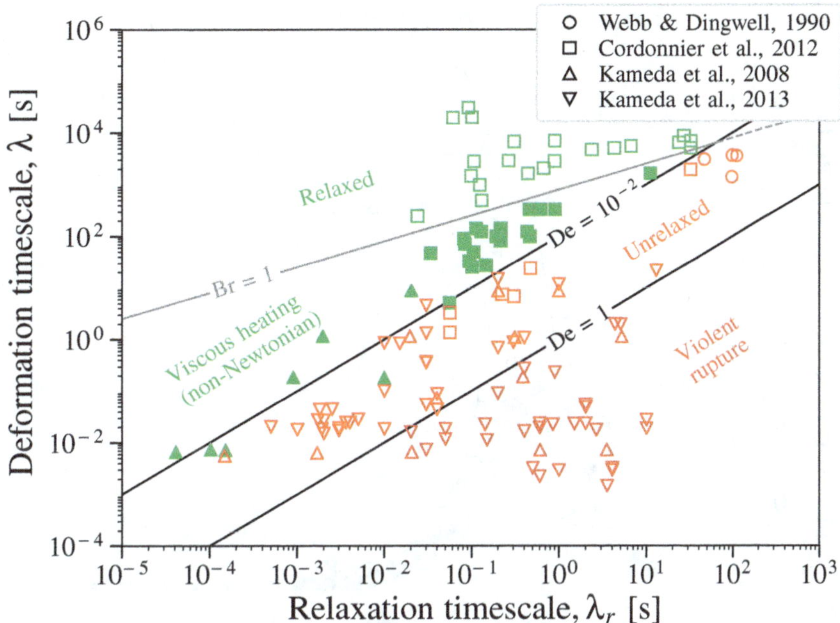

Fig. 3 A map of behaviour arising from deformation experiments on natural (Webb and Dingwell 1990b) and synthetic (Cordonnier et al. 2012b) silicate liquids and on analogues of magmatic liquids undergoing vesiculation (Kameda and Kuribara 2008; Kameda et al. 2013). The experimental results were obtained by applying a characteristic timescale of deformation λ of a liquid with a characteristic timescale of stress relaxation λ_r. Marked are ratios λ_r/λ, equivalent to Deborah numbers De, of 10^{-2} and 1, which separate experiments with a purely viscous response from those with an unrelaxed brittle response or a violent rupture response, respectively. Additionally marked is the Brinkman number Br of 1, which separates the boundary between isothermal viscous behaviour (*above the line*) and viscous behaviour in which the material heats up due to viscous dissipation as heat (*below the line*)

holds across a broad range simply because the deformation timescale was also smaller in the former example.

In the viscous field, Cordonnier et al. (2012b) additionally recorded whether samples hosted a measurable temperature increase due to viscous dissipation of heat during the experiment, or not. In Fig. 3, we plot those with a measureable temperature increase as filled symbols and those without this feature as unfilled symbols. To explain this, we plot the threshold dimensionless Brinkman number of unity $Br = 1$, using an estimation of the loss power density for the furnace and sample size used $\Phi_l = 10^{4.5} \mathrm{W.m^{-2}}$ (Cordonnier et al. 2012b). We note that the Brinkman number curve consistently divides the regimes of purely isothermal experiments and those with measurable heat gain. On this map, the position of $Br = 1$ is non-unique and depends on sample size, such that on the scale of a volcanic conduit, for example, $Br = 1$ would occur

at much higher deformation timescales for the same relaxation timescale (the curve would shift up in Fig. 3). This implies that in the natural case, viscous dissipation of heat may be far more important than shown here for the sample length scale (Costa and Macedonio 2005; Mastin 2005; Costa et al. 2007). Nevertheless, the Deborah number limits discussed appear to be universal and, importantly, are scale-independent.

To gain helpful physical insight into why $10^{-2} \leq De \leq 10^0$ is the transitional window between a purely viscous and a purely elastic response of a liquid to a deformation, we provide low-strain, high frequency oscillatory rheological measurement data for similar liquids (Fig. 4). Here, rods of single-phase liquid are subjected to a low-amplitude oscillatory strain with a forcing frequency ω at a range of temperatures similar to those used in experiments presented in Fig. 2. Here, ω is normalized with λ_r, yielding a dimensionless frequency or, equivalently, a

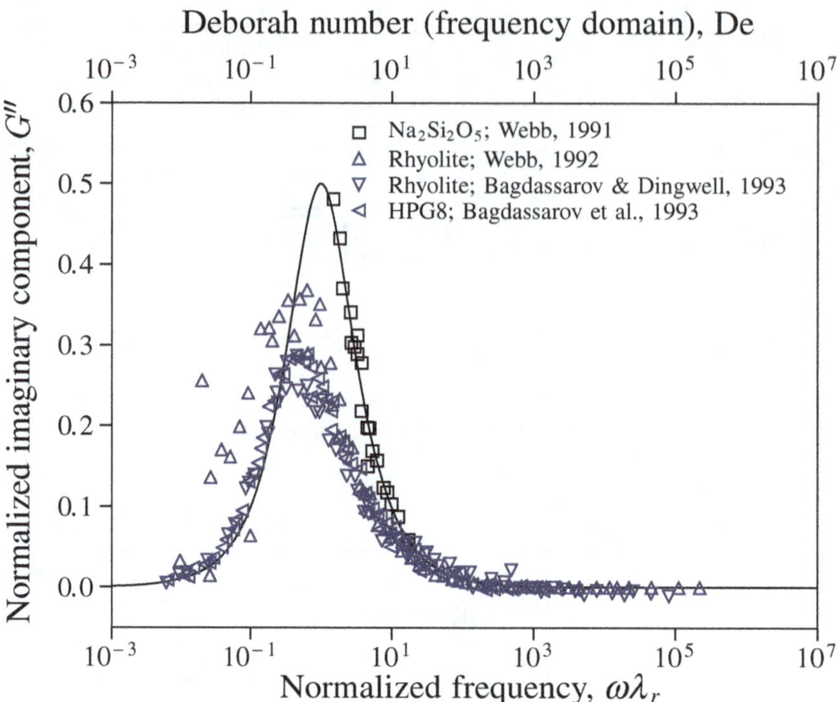

Fig. 4 The response of synthetic (Webb 1991; Bagdassarov and Dingwell 1993a) and natural (Webb 1992; Bagdassarov and Dingwell 1993b) silicate liquids to low strain, high frequency oscillatory deformation tests. This is shown as a normalized imaginary component of the shear modulus G''/G_∞ as a function of the frequency of applied oscillation normalized by the relaxation timescale $\omega\lambda_r$, which is equivalent to a Deborah number De in the frequency domain. We show the viscoelastic prediction for a Maxwell liquid for comparison (*solid curve*)

frequency-domain version of the Deborah number. The metric that we choose to track is the imaginary component of the complex elastic modulus G'' normalized by G_∞. When this value G''/G_∞ is close to zero in the low Deborah number limit, the system is dominated by liquid behaviour and when it rises from zero at increasing Deborah number, there is an increasing component of the response to the forced oscillation that is elastic. We present collated data for a sodium disilicate synthetic composition (Webb 1991), rhyolitic compositions (Webb 1992; Bagdassarov and Dingwell 1993b) and a synthetic composition used as an analogue for type calc-alkaline rhyolite systems (Bagdassarov and Dingwell 1993a). This imaginary component can be described by the generalized Debye model for viscoelasticity for systems with a single characteristic relaxation time as $G'' = \omega\lambda_r/\left[(\omega\lambda_r)^2 + 1\right]$. The data reproduce the broad shape of the Debye model, albeit with an

overprediction of G'' around $\omega\lambda_r \sim 1$ for the rhyolitic liquids, which is poorly understood. Nevertheless, we point out that the first onset of a measurable elastic component to the response of the liquid to deformation occurs at De $= \omega\lambda_r \approx 10^{-2}$. Similarly, the point above which the majority of the response to deformation is elastic (the peak of G'') occurs at De $= \omega\lambda_r \approx 1$ (Fig. 3). We use this observation to validate the two thresholds found in Fig. 3. This implies that fractures can propagate in silicate liquids when there is even a small component of elastic behaviour (De $\geq 10^{-2}$) and that those fractures can propagate vigorously when the elastic behaviour dominates over the viscous behaviour (De ≥ 1). An interpretation might also be that $10^{-2} \leq$ De ≤ 1 is the range in which fracture propagation is competing with viscous relaxation of stress and therefore the fractures are unlikely to be long, sharp-tipped or pervasive. And that at De ≥ 1, stress dissipation by fracture propagation can be localized onto longer, sharp and pervasive fracture networks.

The fact that a viscous limit to the expansion of vesicular magma also scales with our Deborah number criterion is tantalizing. Microphysically, in this case it is not fracture of a deforming homogeneous liquid, but fractures propagating at bubble walls as the local rate of strain in the liquid induced by expansion of the bubble meets the Deborah number criterion for fracture propagation (c.f. Kameda and Kuribara 2008). For complete rupture of the vesicular material—which is a fragmentation event or *violent rupture* in Fig. 3—the fractures propagating from the bubbles must interact. Presumably in the range $10^{-2} \leq \mathrm{De} < 1$, fractures propagate but the fracture tips are blunted during competing viscous relaxation such that brittle behaviour can be observed but is not catastrophic to the system. Then, as the local strain rate at the bubble wall increases further and De approaches 1, the fractures propagate in a dominantly elastic medium and can span the inter-bubble distances, interacting to produce violent rupture. Therefore, it is clear that a scaling of the Deborah number concept to bubble wall dynamics would provide a fragmentation criterion for viscoelastic vesiculating magma (c.f. Namiki and Manga 2005; Koyaguchi et al. 2008; Namiki and Manga 2008).

4 Laboratory-Scale Unrest Signals

When magma breaks in a laboratory experiment, acoustic emissions—packets of acoustic energy—are released and can be recorded (Benson et al. 2008; Lavallee et al. 2008). These signals appear to represent large total released amounts of energy when λ is short (high $\dot{\gamma}$), compared to when λ is long (low $\dot{\gamma}$) (Lavallee et al. 2008). This supports our posit that fracture networks are more likely to be pervasive and large when De is large than when De is small. For single phase silicate liquids deformed at high De, Tuffen et al. (2008) showed that the experimental acoustic event frequency range and sample fracture length

scale scaled with natural constraints of in-conduit volcanic fracture systems and natural frequencies of low-frequency volcano-seismicity, indicating that these viscoelastic fracturing events are indeed the likely source mechanism for low-frequency earthquakes at volcanoes. This confirmed the conclusion of Neuberg et al. (2006), who showed that these low-frequency events were most likely to be associated with repetitive fracturing of magma at a given depth in the conduit during ascent. Neuberg et al. (2006) further modelled magma ascent in confined geometry and showed that a De ~ 1 is met at a depth of 830 m below the surface using parameters typical for recent eruptions at Soufriere Hills volcano and the magma thereof. Thomas and Neuberg (2012) predicted a deeper source of 1500 m using the same model approach but by invoking a conduit restriction (see Chap. 9), consistent with inversions for low-frequency sources using seismic data. Therefore, viscoelastic magma fracturing in the high Deborah number regime appears to be a consistent model for the source mechanics of low-frequency volcano seismicity, often used for eruption forecasting.

Vasseur et al. (2015) showed that the forecastability of full sample rupture scales with the heterogeneity of the system—cast most simply as a porosity (Fig. 5). The implication is that the less vesicular the magma undergoing deformation, the less likely that accurate forecasts based on accelerated precursory signals can be made, with up to $\sim 120\%$ error on the timing of the rupture event for single-phase homogeneous liquids in the high De limit. It may be that the magma vesicularity, pore-network structure (Vasseur et al. 2017), crystallinity and textural anisotropy, play key roles in determining forecasting success based on low-frequency earthquakes. More experimental work is clearly required in this area, along with more rigorous scaling between acoustic and seismic events that originate from magma failure.

An implication of the models explored here, encapsulated by Eq. 1, is that the peak stress supported by a liquid σ_m can be predicted as a

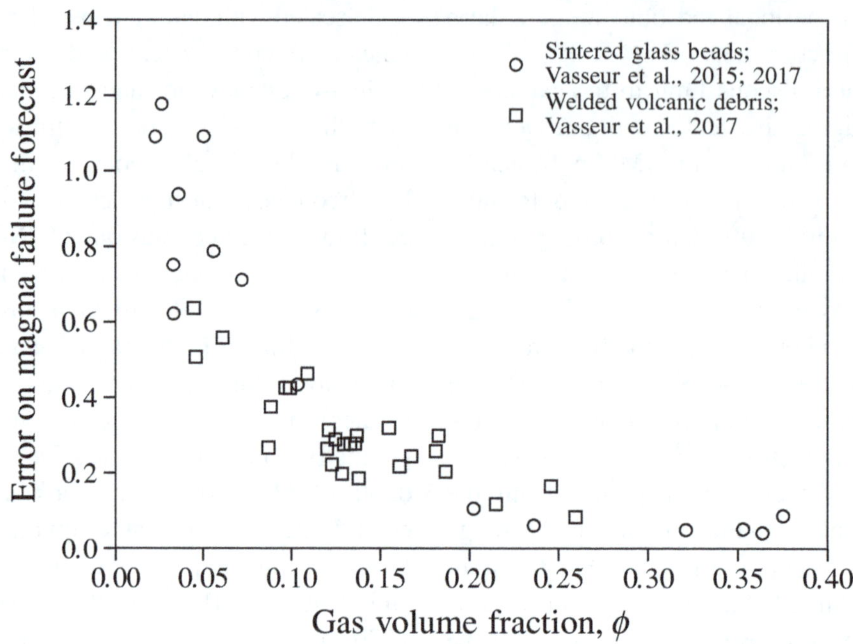

Fig. 5 In the high-De regime, the error on a prediction of failure times scales with the porosity of the material such that low porosity magmas are unpredictable and high porosity magmas are predictable (Vasseur et al. 2015). The failure times are recorded as a large stress drop during uniaxial loading, and the approach to failure is monitored using acoustic emissions generated by pre-failure micro-fracturing events

Fig. 6 The peak stress that can be supported by a shearing liquid in the low De limit is given by $\sigma_m = G_\infty De$. And in the high De limit, this value appears to asymptotically approach $\sigma_m = 10^{-2} G_\infty$ which is equivalent to the strength of glassy materials in that same regime (Simmons et al. 1982)

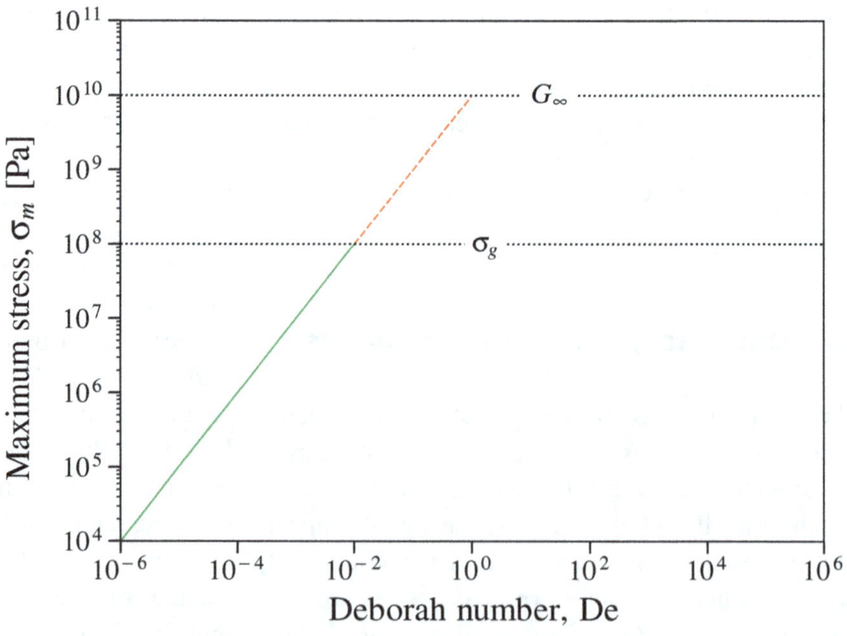

function of De by $\sigma_m = G_\infty De$. Once $De \geq 10^{-2}$, however, this linear relationship appears to be invalid as fractures can form in the liquid. It is perhaps significant that the average strength of glass in the high De regime is $\sim 10^8$ Pa (Simmons et al. 1982; Vasseur et al. 2013), consistent with a rupture threshold of $10^{-2}G_\infty$. These relationships are explored in Fig. 6.

Here we summarize just one example of how targeted laboratory experiments and dimensional analysis have demonstrated the most likely source mechanism of unrest signals commonly used to monitor magma movement and predict impending eruptions. Clearly, with scaling arguments for the applicability of an experimental set up—such as we show here with the Deborah number analysis—laboratory-based work can provide new insights into unrest mechanisms that facilitate more accurate forward-modelling of geophysical signals.

References

Alidibirov M, Dingwell DB (1996) Magma fragmentation by rapid decompression. Nature 380:146–148

Bagdassarov N, Dingwell DB, Webb SL (1993) Effect of boron, phosphorus and fluorine on shear stress relaxation in haplogranite melts. Eur J Miner 3:409–425

Bagdassarov NS, Dingwell DB (1993) Frequency dependent rheology of vesicular rhyolite. J Geophy Res: Solid Earth 98(B4):6477–6487

Benson P, Vinciguerra S, Meredith P, Young R (2008) Laboratory simulation of volcano seismicity. Science (80-) 322:249–252

Chouet BA (1996) Long-period volcano seismicity: its source and use in eruption forecasting. Nature 380:309–316. doi:10.1038/380309a0

Chouet BA, Page RA, Stephens CD et al (1994) Precursory swarms of long-period events at Redoubt Volcano (1989–1990), Alaska: their origin and use as a forecasting tool. J Volcanol Geotherm Res 62:95–135. doi:10.1016/0377-0273(94)90030-2

Cordonnier B, Caricchi L, Pistone M et al (2012a) The viscous-brittle transition of crystal-bearing silicic melt: direct observation of magma rupture and healing. Geology 40:611–614

Cordonnier B, Schmalholz SM, Hess K, Dingwell DB (2012b) Viscous heating in silicate melts: an experimental and numerical comparison. J Geophys Res 117:B02203

Costa A, Macedonio G (2005) Viscous heating effects in fluids with temperature-dependent viscosity: triggering of secondary flows. J Fluid Mech 540:21–38

Costa A, Melnik O, Vedeneeva E (2007) Thermal effects during magma ascent in conduits. J Geophys Res 112: B12205. doi:10.1029/2007JB004985

Deubelbeiss Y, Kaus BJ, Connolly JA, Caricchi L (2011) Potential causes for the non-Newtonian rheology of crystal-bearing magmas. Geochem Geophy Geosys 12(5)

Dingwell DB (1995) Relaxation in silicate melts; some applications. Rev Mineral Geochemistry 32:21–66

Dingwell DB (1996) Volcanic dilemma: flow or blow? Science (80-) 273:1054–1055. doi:10.1126/science.273.5278.1054

Dingwell DB, Webb SL (1989) Structural relaxation in silicate melts and non-Newtonian melt rheology in geologic processes. Phys Chem Miner 16:508–516

Giordano D, Russell JK, Dingwell DB (2008) Viscosity of magmatic liquids: a model. Earth Planet Sci Lett 271:123–134

Gonnermann HM, Manga M (2003) Explosive volcanism may not be an inevitable consequence of magma fragmentation. Nature 426:432–435

Hess KU, Dingwell DB (1996) Viscosities of hydrous leucogranitic melts: a non-Arrhenian model. Am Mineral 81:1297–1300

Hess K-U, Cordonnier B, Lavallee Y, Dingwell DB (2008) Viscous heating in rhyolite: an in situ experimental determination. Earth Planet Sci Lett 275:121–126. doi:10.1016/j.epsl.2008.08.014

Hornby AJ, Kendrick JE, Lamb OD, Hirose T, De Angelis S, Aulock FW, Umakoshi K, Miwa T, Henton De Angelis S, Wadsworth FB, Hess KU (2015) Spine growth and seismogenic faulting at Mt. Unzen, Japan. J Geophys Res: Solid Earth 120(6):4034–4054

Ichihara M, Rittel D, Sturtevant B (2002) Fragmentation of a porous viscoelastic material: implications to magma fragmentation. J Geophys Res: Solid Earth 107(B10)

Kameda M, Kuribara H, Ichihara M (2008) Dominant time scale for brittle fragmentation of vesicular magma by decompression. Geophy Res Lett 35(14)

Kameda M, Ichihara M, Shimanuki S, Okabe W, Shida T (2013) Delayed brittle-like fragmentation of vesicular magma analogue by decompression. J Volcanol Geoth Res 258:113–125

Kendrick JE, Lavallée, Y, Hirose T, Di Toro G, Hornby AJ, De Angelis S, Dingwell DB (2014) Volcanic drumbeat seismicity caused by stick-slip motion and magmatic frictional melting. Nat Geosci 7(6):438

Koyaguchi T, Scheu B, Mitani NK, Melnik O (2008) A fragmentation criterion for highly viscous bubbly magmas estimated from shock tube experiments. J Volcanol Geotherm Res 178:58–71

Lavallee Y, Meredith PG, Dingwell DB et al (2008) Seismogenic lavas and explosive eruption forecasting. Nature 453:507–510. doi:10.1038/nature06980

Lensky NG, Lyakhovsky V, Navon O (2001) Radial variations of melt viscosity around growing bubbles and gas overpressure in vesiculating magmas. Earth Planet Sci Lett 186:1–6

Mader HM, Llewellin EW, Mueller SP (2013) The rheology of two-phase magmas: a review and analysis. J Volcanol Geoth Res 257:135–158

Mastin LG (2005) The controlling effect of viscous dissipation on magma flow in silicic conduits. J Volcanol Geoth Res 143(1):17–28

Miller AD, Stewart RC, White R et al (1998) Seismicity associated wtih dome growth and collapse at the

Soufriere Hills Volcano, Montserrat. Geophys Res Lett 25:3401–3404

Mueller S, Llewellin EW, Mader HM (2010) The rheology of suspensions of solid particles. Proc R Soc A Math Phys Eng Sci 466:1201–1228

Namiki A, Manga M (2005) Response of a bubble bearing viscoelastic fluid to rapid decompression: implications for explosive volcanic eruptions. Earth Planet Sci Lett 236(1):269–284

Namiki A, Manga M (2008) Transition between fragmentation and permeable outgassing of low viscosity magmas. J Volcanol Geoth Res 169(1):48–60

Neuberg JW, Tuffen H, Collier L et al (2006) The trigger mechanism of low-frequency earthquakes on Montserrat. J Volcanol Geotherm Res 153:37–50. doi:10.1016/j.jvolgeores.2005.08.008

Prousevitch AA, Sahagian DL, Anderson AT (1993) Dynamics of diffusive bubble growth in magmas: isothermal case. J Geophys Res Solid Earth 98:22283–22307

Salvage R, Neuberg JW (2016) Using a cross correlation technique to refine the accuracy of the Failure Forecast Method: application to Soufrière Hills volcano, Montserrat. J Volcanol Geotherm Res 324:118–133. doi:10.1016/j.jvolgeores.2016.05.011

Sammis CG, Ashby MF (1986) The failure of brittle porous solids under compressive stress states. Acta Metall 34:511–526

Simmons JH, Mohr RK, Montrose CJ (1982) Non-Newtonian viscous flow in glass. J Appl Phys 53:4075–4080

Sparks RSJ (1978) The dynamics of bubble formation and growth in magmas: a review and analysis. J Volcanol Geotherm Res 3:1–37

Stevenson RJ, Dingwell DB, Webb SL, Bagdassarov NS (1995) The equivalence of enthalpy and shear stress relaxation in rhyolitic obsidians and quantification of the liquid-glass transition in volcanic processes. J Volcanol Geotherm Res 68:297–306

Thomas ME, Neuberg J (2012) What makes a volcano tick-a first explanation of deep multiple seismic sources in ascending magma. Geology 40:351–354. doi:10.1130/g32868.1

Tuffen H, Dingwell DB, Pinkerton H (2003) Repeated fracture and healing of silicic magma generate flow banding and earthquakes? Geology 31:1089–1092. doi:10.1130/g19777.1

Tuffen H, Smith R, Sammonds PR (2008) Evidence for seismogenic fracture of silicic magma. Nature 453:511–514

Vasseur J, Wadsworth FB, Lavallée Y et al (2013) Volcanic sintering: timescales of viscous densification and strength recovery. Geophys Res Lett 40:5658–5664

Vasseur J, Wadsworth FB, Lavallée Y, Bell AF, Main IG, Dingwell DB (2015) Heterogeneity: the key to failure forecasting. Sci Rep 5

Vasseur J, Wadsworth FB, Heap MJ, Main IG, Lavallée Y, Dingwell DB (2017) Does an inter-flaw length control the accuracy of rupture forecasting in geological materials? Earth Planet Sci Lett

Voight B (1988) A method for prediction of volcanic eruptions. Nature 332:125–130

Webb SL (1991) Shear and volume relaxation in $Na_2 Si_2 O_5$. Am Mineral 76(9–10):1449–1454

Webb SL (1992) Low-frequency shear and structural relaxation in rhyolite melt. Phys Chem Mineral 19(4):240–245

Webb SL, Dingwell DB (1990a) The onset of non-Newtonian rheology of silicate melts. Phys Chem Miner 17:125–132

Webb SL, Dingwell DB (1990b) Non-Newtonian rheology of igneous melts at high stresses and strain rates: experimental results for rhyolite, andesite, basalt, and nephelinite. J Geophys Res Solid Earth 95:15695–15701

Zhang Y, Stolper EM, Wasserburg GJ (1991) Diffusion of a multi-species component and its role in oxygen and water transport in silicates. Earth Planet Sci Lett 103:228–240

Volcano Seismology: Detecting Unrest in Wiggly Lines

R.O. Salvage, S. Karl and J.W. Neuberg

Abstract

Seismology is a useful tool to gain a better understanding of volcanic unrest in real time as it unfolds. The generation of seismic signals in a volcanic environment has been linked to a number of different physical processes occurring at depth, including fracturing of the volcanic edifice (producing high frequency seismicity) and movement of magmatic fluids (producing low frequency seismicity). Further classification of seismic signals according to their waveform similarity, in addition to their frequency content, allows greater detail in temporal and spatial changes of seismicity to be detected. At Soufrière Hills volcano, Montserrat, one of the target volcanoes of the VUELCO project, families of similar waveforms provided valuable insight into evaluating the significance of ongoing unrest. In June 1997 over 6000 more events were able to be identified over a 5 day period of interest (22 to 25 June) by using families of seismic events, rather than a standard amplitude-based detection algorithm. In total, 11 families were identified, with the events clustering into a number of swarms, suggesting a repeating and non destructive cyclic source mechanism. Since each family is believed to represent a distinct source location and mechanism, identifying 11 coexisting families

R.O. Salvage (✉)
Observatorio Vulcanológico y Sismológico
de Costa Rica, Universidad Nacional,
Apartado Postal: 2386-3000, Heredia, Costa Rica
e-mail: beckysalvage@gmail.com

S. Karl
Shell, NAM offices, Schepersmaat 2,
9405 TA Assen, The Netherlands
e-mail: sandra.karl84@gmail.com

J.W. Neuberg
School of Earth and Environment, Institute of
Geophysics and Tectonics, University of Leeds,
Leeds LS2 9JT, United Kingdom
e-mail: j.neuberg@leeds.ac.uk

Advs in Volcanology (2019) 185–201
DOI 10.1007/11157_2017_11

Published Online: 02 August 2017

reflects the complex diversity of physical processes which act simultaneous at this volcano. In July 2003, conditions at the volcano had clearly changed since only one family of seismicity was identified. The source location of this family appeared to shift with time from 8 July (when no events from the family were identified) to 12 July (where most events had a cross correlation coefficient over 0.9). In addition, the use of families appears to greatly aid hindsight forecasting attempts for the large scale dome collapses of 1997 and 2003 using the Failure Forecast Method. Knowledge of the temporal and spatial extent of seismicity during periods of unrest, its source mechanism and its relationship to physical processes at depth is essential for decision and policy makers for risk mitigation. However, the source mechanisms of such volcanic seismicity is still much debated and appears to often be misinterpreted because of compromising assumptions used in the numerical modelling of inverting such sources. Use of a spatially extended source such as a ring fault structure, rather than a single point for determining the origin of low frequency seismicity, is now thought to be more realistic for the mechanism of such events since it more accurately represents the movement of magma through a conduit. However, use of this spatially extended source instead of a simple single point results in a large underestimation of slip from P-wave amplitudes, which may lead to an underestimation in magma ascent rates, with large consequences for eruption forecasting. Additionally, the P-wave radiation patterns exhibited by these two mechanisms are remarkably similar, and can only be distinguished if the small radial radiation lobes can be determined. In a volcanic environment this is extremely difficult due to large uncertainties in earthquake source depth locations, and the implementation of small aperture seismic networks.

Español

La sismología es una herramienta geofísica valiosa que brinda información en tiempo real, permitiendo una mejor comprensión del comportamiento de sistemas volcánicos que inician un proceso de reactivación o de intensificación de la actividad. La generación de señales sísmicas en ambientes volcánicos se ha relacionado con un número diverso de procesos geofísicos que ocurren en el interior de los volcanes, incluyendo fracturamiento del edificio volcánico (produciéndose sismicidad de alta frecuencia) y movimiento de fluidos magmáticos (produciéndose sismicidad de baja frecuencia). La clasificación de señales sísmicas basada en la similitud de las formas de onda, además del contenido de frecuencias, ha permitido detectar cambios temporales y espaciales de la sismicidad con mayor detalle.

En el volcán Soufriere Hills en Monserrat, uno de los volcanes investigados como parte del Proyecto VUELCO, el reconocimiento de familias de señales sísmicas con formas de onda similares proveyó un entendimiento valioso en la evaluación de la significancia de la reactivación de su actividad volcánica. En junio de 1997, se pudieron identificar más de 6000 eventos ssmicos dentro de unperiodo particular de 4 días (entre el 22 y el 25 de junio) mediante la determinación de familias de eventos sísmicos que fueron determinados por un algoritmo estándar de detección basado en la amplitud de la señal sísmica. En total, 11 familias de eventos sísmicos fueron identificadas, con grupos de eventos conformando enjambres sísmicos, lo que sugiere un mecanismo cíclico repetitivo de una fuente sísmica no destructiva. Ya que se considera que cada familia representa una fuente con ubicación espacial y mecanismo distinto, la identificación de 11 familias refleja la compleja diversidad de los procesos geofísicos que operan simultáneamente en este volcán en particular. En julio del 2003, el régimen del volcán cambió definitivamente ya que solamente 1 familia de eventos sísmicos fue identificada. La ubicación de la fuente sísmica de esta familia parece haber migrado con el tiempo entre el 8 de julio (cuando ningún evento sísmico de esta familia fue identificado) y el 12 de julio (cuando la mayoría de eventos tuvo un coeficiente de cros-correlación superior a 0,9). Además, el establecimiento de familias de sismos parece haber sido de gran ayuda en los intentos de pronosticar los colapsos del domo en gran escala de 1997 y el 2003 utilizando el método determinístico de pronóstico de rompimiento por fatiga (Failure Forecast Method, FFM). El conocimiento de la extensión temporal y espacial de la sismicidad durante periodos de reactivación volcánica, el mecanismo de la fuente, y su relación con los procesos geofísicos a niveles profundos son aspectos esenciales para los tomadores de decisiones y ejecutores de políticas relacionadas con la mitigación de riesgos.

En general, los mecanismos de la fuente generadora de la sismicidad volcánica continúan aún bajo gran debate y parecen ser con frecuencia mal interpretados debido a simplificaciones comprometedoras usadas en los modelos numéricos de inversión de dichas fuentes. El uso de una fuente espacial extendida tal como una falla estructural tipo anular, en lugar de un solo punto para determinar el origen de sismicidad de baja frecuencia, se piensa es más realista para visualizar el mecanismo de este tipo de eventos ya que representa con más exactitud el movimiento del magma a través de un conducto. Sin embargo, el uso de este modelo resulta en la subestimación significativa del deslizamiento en la falla determinada a través de las amplitudes de las ondas P, lo que podría llevar a una subestimación de las tasas de ascenso del magma, con consecuencias cruciales para el pronóstico de erupción inminente. Por otra parte, los patrones de radiación de la onda P mostrados por estos dos mecanismos son marcadamente similares, y pueden ser solamente distinguidos si los pequeños lóbulos radiales de radiación pueden ser determinados. En un contexto volcánico, esta es una tarea extremadamente difícil debido a la gran incertidumbre inherente a la localización de la profundidad de la fuente sísmica, y a la implementación de redes sísmicas de pequeña extensión.

Keywords

Volcano seismology · Families of similar seismic events · Magmatic source mechanism · Eruption forecasting

Index Terms

Volcano seismology · Soufrière Hills · Chiles-Cerro Negro · Low frequency seismicity · Families · Cross correlation · Forecasting · Failure forecast method · Source mechanisms · Point source · CLVD · Ring fault · Spatially extended source · P-wave radiation patterns

Volcanic Unrest

The monitoring of active volcanoes for the protection of society has excelled in recent decades fuelled by rapid technological advances, which have allowed the development and deployment of more cost-effective monitoring solutions. It is now possible to detect geophysical and geochemical signals from a volcano which previously would have been below the detection threshold. The routine monitoring of volcanoes during periods of quiescence is crucial, although not always feasible, in order to assess background levels of activity at volcanoes, and thus more rapidly detect the onset of future unrest. Most simply, volcanic unrest is defined as a deviation from background levels of activity towards a level which is a cause for concern over short time scales of hours to days. Volcanic unrest does not necessarily lead to eruption, although this is the most likely outcome, with 64% of 228 volcanoes that have experienced unrest since the year 2000 culminating in eruptive events (Phillipson et al. 2013).

Current monitoring efforts at volcanoes can be grouped into three categories: measurements of surface degassing; deformation; and seismic activity, which are all thought to result from the movement of magmatic fluids at depth, and therefore may provide key insights into an impending eruption. Unrest must be detected on short timescales which is appropriate for decision-making, and therefore seismicity has remained a primary monitoring tool, as it can be remotely analysed in real-time, often by an automated system, if a number of sensors are placed around the volcano. Deviation from the background level is also often easier to determine for seismicity than for other signals.

Seismic Event Characterisation

A wide variety of seismic signals exist within volcanic settings, associated with magmatic and hydrothermal fluid movement at depth, pressurization of the volcanic edifice, and/or the surface manifestation of the interaction of these processes. The variety of signals is a reflection of the number of different processes and the great structural heterogeneities found within this contex. The characterisation of seismicity can be based upon waveform similarities, but is traditionally based upon the signals' time and frequency characteristics: different bands of frequency relate to different active source processes at depth, which can then be distinguished from one another, although the frequency bands associated with each process may overlap (Lahr et al. 1994).

The identification of seismic events in volcanic settings can assist with the detection of the onset and cessation of unrest, while the spatial and temporal patterns of occurrence may be informative of magma movement and changes in stress at depth, as well as the spatial extent of concern. An understanding of the physical processes occurring at depth is essential if accurate

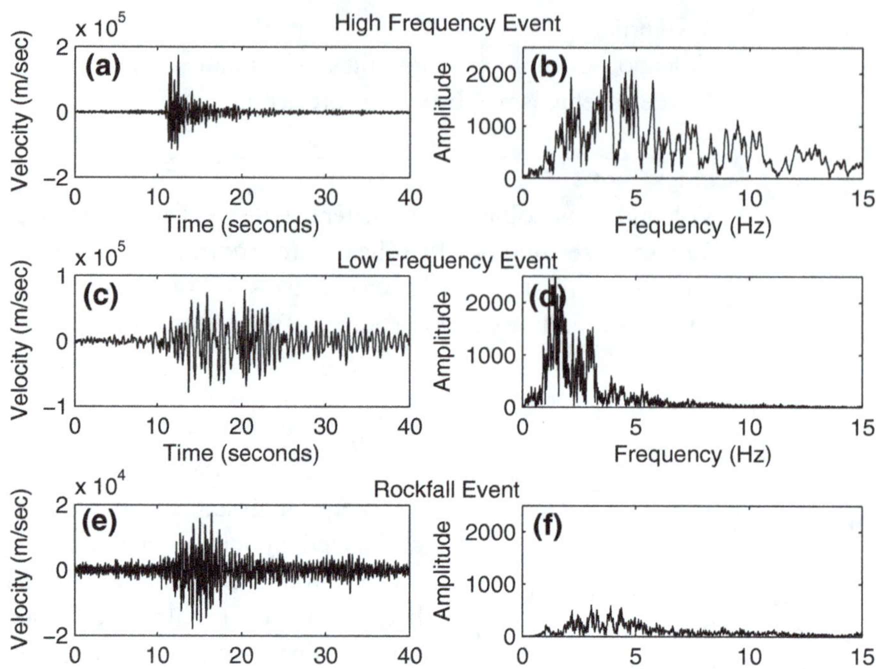

Fig. 1 Examples of waveforms and their frequency content seen in volcanic environments taken from Soufrière Hills Volcano, Montserrat in 1997. Soufrière Hills Volcano was a target volcano identified by the VUELCO project for investigation. **a, b** High frequency waveform with clear phase arrivals. **c, d** Low frequency waveforms with an emergent onset. Waveform filtered between 0.5 and 5 Hz. **e, f** Rockfall event with classic "cigar" shape

and timely forecasts of volcanic eruptions, and developing unrest scenarios, are to be made.

Classification by Frequency Content

Seismic signals originating from processes occurring at depth within the volcanic system are usually split into high- and low-frequency end-members, although in reality a continuum across the spectrum exists between the two (Chouet and Matoza 2013). High frequency seismic signals (Fig. 1a, b), also known as Volcano-Tectonic (VT) events, have energy concentrated in the frequency range of 1 to 20 Hz, generally peaking between 6 and 8 Hz (Lahr et al. 1994). They are characterized by clear, impulsive P- and S-wave arrivals, followed by a short coda. High frequency seismicity is usually attributed to brittle failure within the volcanic edifice, where magmatic processes create enough elastic strain to force the surrounding rocks into failure (Arciniega-Ceballos et al.

2003), similar to the generation of tectonic earthquakes.

Low frequency (LF) seismic signals (Fig. 1c, d) usually occupy the spectral range of 0.2 to 5 Hz (Chouet and Matoza 2013), and are frequently characterized by emergent P-wave onsets and lack of S-wave arrivals. It has been suggested that the occurrence of low frequency waveforms is linked to resonance of seismic energy trapped at a solid-fluid interface either within a crack (e.g. Chouet 1988), or a volcanic conduit (e.g. Neuberg et al. 2000). The trigger mechanism of such seismic energy is further disputed, with suggestions that it may be generated by: (1) a stick-slip motion along the conduit walls as magma ascends (e.g. Iverson et al. 2006); (2) the brittle failure of magma itself either through an increase in viscosity and strain rates (Lavallée et al. 2008), which may be due to an increase in the ascent rate of magma through the conduit (Neuberg et al. 2006), changes in the crystal and/or bubble concentration in the magma (Goto 1999), or through a change in the geometry of the conduit (Thomas and Neuberg 2012); (3) the interaction between the magmatic and hydrothermal system at depth (e.g. Nakano and Kumagai 2005); or (4) through slow rupture and failure of unconsolidated material on volcanic slopes (Bean et al. 2014).

Many volcanic seismic events fall between these two end-member categories and are termed "hybrid" events. Hybrid events are characterised by a high frequency onset with a long resonating low frequency coda, therefore distributing energy across a wider frequency spectrum (Chouet and Matoza 2013). Hybrid and LF events are often classified in the same group of volcanic seismicity, since source and path effects can result in a LF event recorded at one station being recorded as a hybrid event at another.

With the deployment of broadband sensors in many volcanic environments, it is now possible to detect seismicity within a much wider frequency band, up to 120 s periods (Chouet and Matoza 2013), known as Very Long Period (VLP) earthquakes. VLP events occupy the spectral range below 0.01 Hz. The generation of these waveforms is not yet fully understood, in particular how such long wavelengths can be generated in apparently small source volumes, although it has been linked to perturbations in the flow of fluid or gas within pressurized volcanic conduits or cracks (e.g. Dawson et al. 2011). Their large wavelength, sometimes over one hundred kilometers, means few path effects on the waveform and as such if identified, these waveforms provide an excellent choice for performing waveform inversion techniques to identify source characteristics (Chouet and Matoza 2013).

Furthermore, seismicity can be generated by surface processes, such as landslides, rockfall events, pyroclastic flows and lahars (Fig. 1e, f). These are particularly dominant during dome building eruptions and at volcanoes with glaciers during the spring and summer months due to partial melting of the ice (McNutt 2005). These signals can be exploited to determine the size and magnitude of such events, their location and their direction of travel (e.g. De Angelis et al. 2007). Typically rockfall events (small free falling rock

events) form a "cigar shaped" waveform with an emergent onset (Fig. 1e, f), whereby there is an initial increasing amplitude of the waveform as the amount of material falling down slope increases. Pyroclastic flow signals are distinguishable from rockfalls since their waveforms are at least an order of magnitude larger and they often occur over a longer duration since larger amounts of material are involved moving down slope (De Angelis et al. 2007), however the two are likely to exist on a continuum.

Classification by Waveform Similarity

Seismic waveforms can also be classified according to their similarity with other detected seismic events. The frequency content of seismic waveforms is indicative of the active processes that may be occurring within the volcanic environment and the source mechanism involved in the generation of such seismicity. The further classification of seismic events into families which all have a similar waveform shape, as well as the same frequency content, allows the depiction of temporal and spatial changes in the source mechanism and the source location on a much smaller scale (e.g. Thelen et al. 2011; Salvage and Neuberg 2016). For example, the relative relocation of families of similar seismic events at Soufrière Hills volcano, Montserrat has produced very precise source locations (e.g. De Angelis and Henton 2011). By definition, families of seismic events must be generated by the same mechanism and at the same location in order for the detected waveforms to have the same shape at the seismometer, and therefore changes in either of these parameters affect the similarity of waveforms. In many instances it is assumed that families of seismic events are generated by the same mechanism and within a similar source location, estimated at between one quarter and one tenth of the wavelength (Geller and Mueller 1980; Neuberg et al. 2006).

Waveform similarity in terms of shape and duration can be evaluated by cross correlation. Identical signals will result in a cross correlation coefficient of 1 or -1, dependent upon their

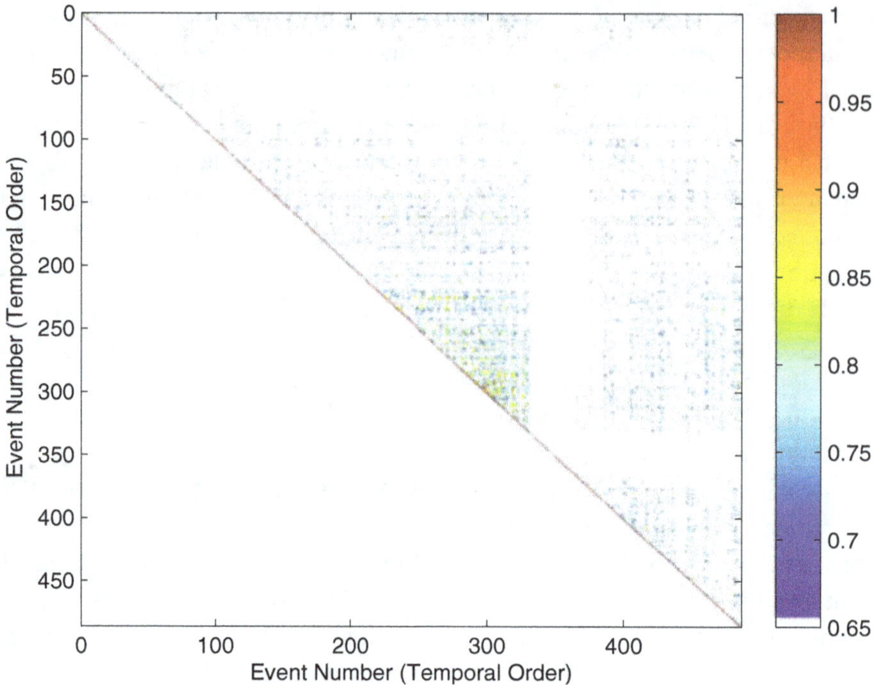

Fig. 2 Cross Correlation Matrix of events identified at Soufrière Hills volcano, Montserrat at a single station in June 1997. A total of 486 events were identified on 24 June 1997 and are shown in temporal order along the x and y axis. Events with a cross correlation coefficient of greater than 0.7 are shown on a colour scale, with those close to one being more similar. The autocorrelation of each event with itself is shown in *dark red* along the diagonal and is equal to a cross correlation coefficient of 1

relative polarity. Signals with no correlation result in a cross correlation coefficient of 0. A threshold must be chosen above which waveforms can be considered similar. Waveforms which are deemed similar can be grouped together into a family of events. The choice of similarity threshold is important: if it is too low there is a risk of placing events which are not similar into the same family; if it is too high similar events can be missed. Green and Neuberg (2006), Thelen et al. (2011) and Salvage and Neuberg (2016) suggest a cross correlation coefficient threshold of 0.7, since this is significantly above the correlation coefficient that can be produced from random correlations between noise and a waveform. Higher cross correlation coefficient thresholds can be used to identify families of almost identical waveforms, however Petersen (2007) suggests that this is probably not appropriate in volcanic settings due to additional noise in this environment.

The similarity between identified seismic events can be determined by cross correlating each individual seismic event with every other seismic event. The result are typically presented as a similarity matrix, as seen in Fig. 2, where events which are deemed to be similar are shown on the colour spectrum. However, such a matrix may include a number of families of similar events since it only determines whether each event shows similarity to any of the other earthquakes analysed. In order to identify families of similar events, events with a high cross correlation coefficient as decided by the user are grouped together and removed from the matrix. This procedure is repeated across the entire investigated time period until all events have been classified into a family, or have been removed from the matrix. A master event is then determined from each family of events as the average of the stack of similar waveforms. This is representative of the family in terms of waveform shape.

Families of similar seismicity have been identified at a number of active volcanoes around the world, including Redoubt volcano, Alaska

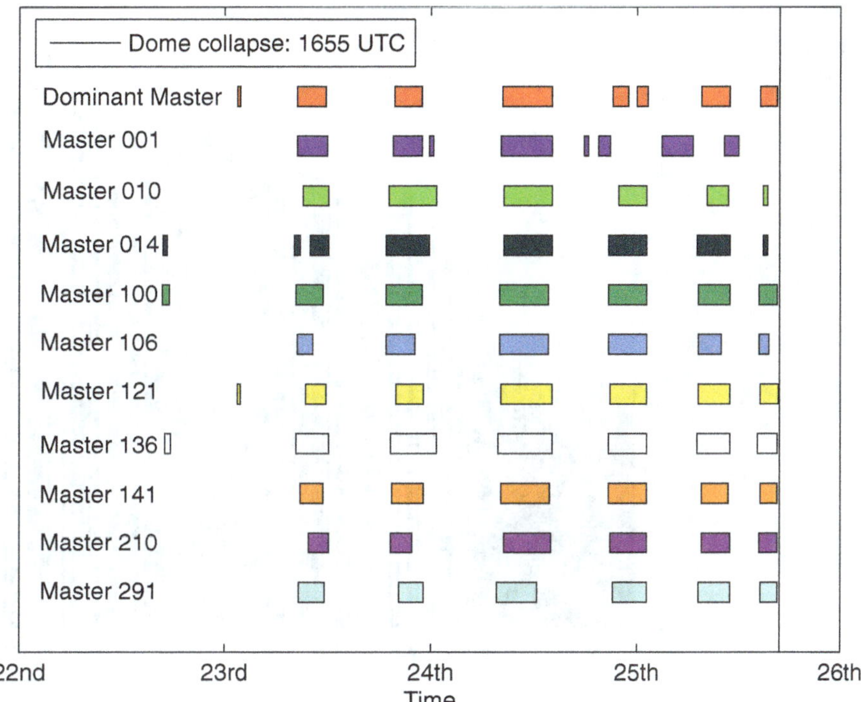

Fig. 3 Comparison of the timing and duration of swarms of families of events identified at a single station at Soufrière Hills volcano, Montserrat in June 1997. The timing of the dome collapse is represented by the vertical line on the 25 June 1997. The y axis is only an indication of each of the families present separated in space for the purpose of clarity on the plot and does not represent time or dominance; each master event is simply drawn below the last so that all can be compared. Each coloured *rectangular box* represents the times when the families were active during the 22–25 June analysis period

(e.g. Buurman et al. 2013); Mt. St. Helens, USA (Thelen et al. 2011); Colima, Mexico (Arámbula-Mendoza et al. 2011); Merapi, Indonesia (Budi-Santoso and Lesage 2016); Katla, Iceland (Sgattoni et al. 2016); and Soufrière Hills volcano, Montserrat (Green and Neuberg 2006; Ottemöller 2008; Salvage and Neuberg 2016). The identification of families rather than simply detecting seismic events and classifying them according to their frequency content is advantageous as subtle temporal and spatial patterns can be identified, allowing detailed source information to be uncovered. In addition, this technique dramatically increases the number of identified events from the continuous seismic record, since low amplitude events and closely spaced events can still be identified. For example, using a standard amplitude-based detection algorithm, 1435 seismic events were identified at Soufrière Hills volcano, Montserrat between 22 and 25 June 1997, a period of interest due to increased seismicity before a lava dome collapse. The cross correlation technique identified 7653 similar seismic events during the same time period, offering a five-fold increase in the number of detected earthquakes (Salvage and Neuberg 2016).

Low frequency families of seismicity identified during this unrest period at Soufrière Hills volcano were followed by a dome collapse on 25 June 1997. Soufrière Hills Volcano was chosen by the VUELCO project as a target volcano due to the longevity of its dome building and collapse cycles which have been ongoing since 1995, providing a wealth of associated geophysical data (Sparks and Young 2002; Wadge et al. 2014). In total, 11 distinct seismic sources (i.e. 11 families of seismicity) were identified during this period of unrest [Fig. 3; Green and Neuberg (2006); Salvage and Neuberg (2016)], which all broadly follow the same temporal pattern in the number of identified swarms present over this time period. However, the timing and duration of these

Fig. 4 The evolution of the dominant cross correlation coefficient with time at Soufrière Hills volcano, Montserrat in July 2003. A dome collapse event occurred at the time of the *vertical line* (13:30 on 12 July 2003). The temporal gaps in the data represent drops in the seismometer recordings rather than a change in the cross correlation coefficient, indicated by the white space on the x-axis

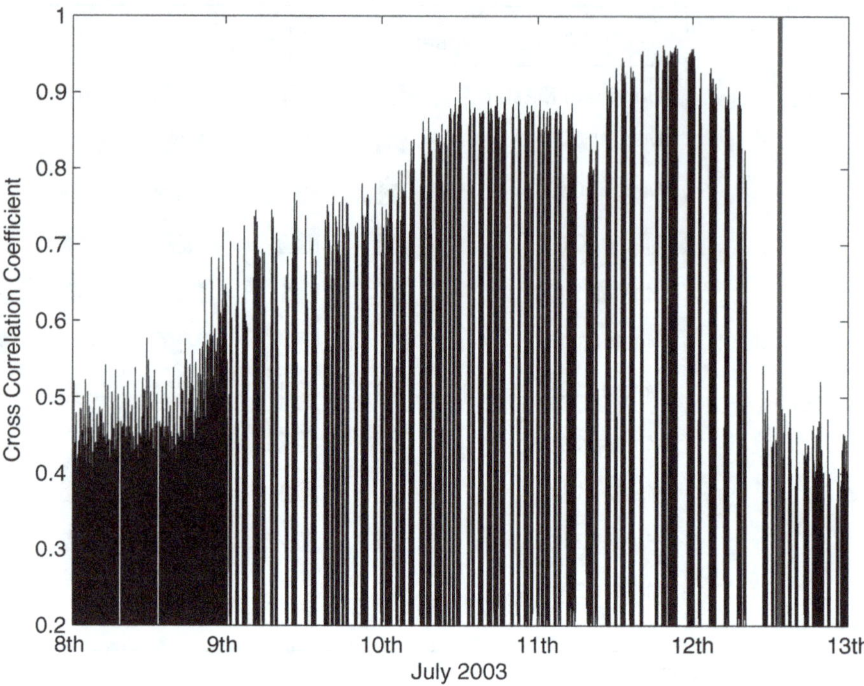

swarms can be seen to be different for each family of similar seismic events. Low frequency seismicity is associated with the movement of magmatic fluid at depth, and therefore in this case would suggest cyclic flow dynamics to generate such swarm-like behaviour. The source process for the generation of this seismicity must be stable and non-destructive in order to be repeatable (Green and Neuberg 2006; Petersen 2007), and must be able to occur at a number of different locations and/or by a number of different sources at the same time in order to generate a number of active families of events, reflecting the complex diversity of seismic sources and physical processes which act simultaneously at this volcano.

The identification of families can also be used to understand evolving seismicity with time. An evolving cross correlation coefficient with time, if not an artefact of data processing, may be indicative of a migrating source location or source mechanism. This was observed at Soufrière Hills volcano, Montserrat in July 2003 (Fig. 4; Salvage and Neuberg (2016)). The largest dome collapse to date observed at this volcano occurred on 12 July 2003, with removal of 210×10^6 m^3 of material (Herd et al. 2005), following a 4 day period from 8 to 12 July 2003

of heightened seismicity at Soufrière Hills. A migrating source mechanism can be identified from changing amplitudes in seismic events within the same family, however this characteristic cannot be identified from analysing cross correlation coefficients alone. The amplitudes of seismic events within the single family identified in July 2003 were relatively constant (Ottemöller 2008), suggesting the changing cross correlation coefficient is a consequence of a migrating source location at depth, rather than an evolving source mechanism. The generation of families ceased immediately prior to the dome collapse event, and no similar earthquakes were detected after the collapse (Fig. 4). This suggests that the physical conditions required for the generation of families were not met in the hours before, and after, the collapse event.

The analysis of families of seismicity in the time domain may also allow for the identification of spatial patterns in seismicity. Families detected at Chiles-Cerro Negro, a volcano within the Northern Andes on the border between Ecuador and Colombia in October 2014, suggests distinct temporal and spatial patterns of seismicity. The last eruption of the volcanic complex of Chiles-Cerro Negro is believed to have been

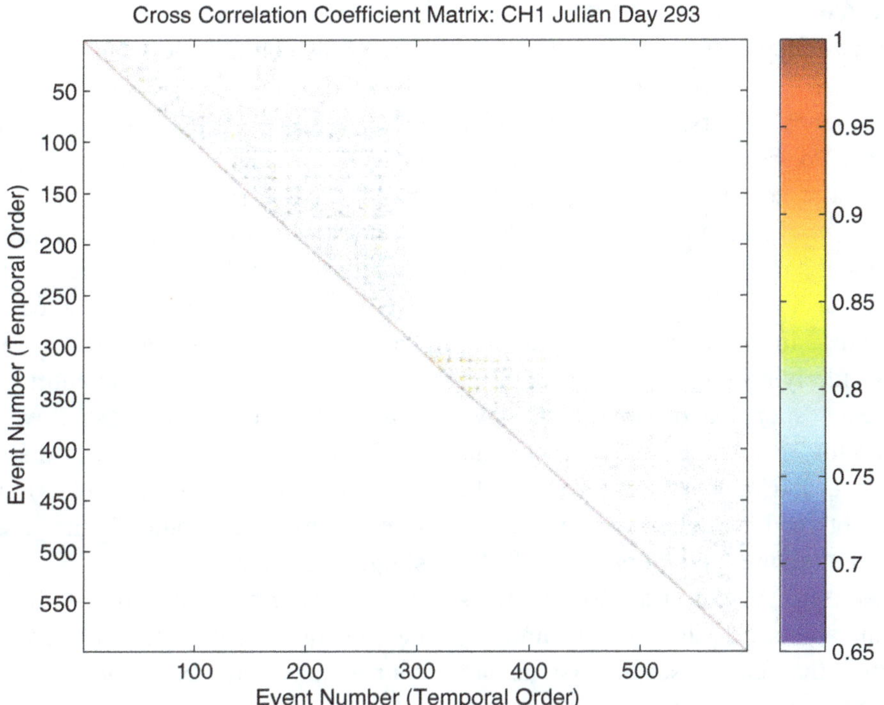

Fig. 5 Cross Correlation Matrix of events identified using a simple amplitude based detection algorithm at Chiles-Cerro Negro on 20 October 2014, at a single station. A total of 597 events were identified and are shown in temporal order along the x and y axis. Events with a cross correlation coefficient of greater than 0.7 are shown on a colour scale, with those close to one (*red*) being more similar. The autocorrelation of each event with itself is shown in dark red along the diagonal and is equal to a cross correlation coefficient of 1. Distinct clusters of similar events can be identified, thought to suggest a temporal evolution in the dominant similar seismicity

3400 years ago, although seismicity has since been detected in the area, thought to be related to an active hydrothermal system (Ruiz et al. 2013). The volcanic complex is dissected by a large fault system, believed to have been active as recently as 1868, when two large seismic events occurred (Mw 6.6 and 7.2) (Beauval et al. 2010). Seismic activity increased in 1991 and then again in July 2013. However, a dramatic increase in seismicity from less than 50 events a day to over 150 events occurred in October 2014, concentrated beneath the summit of Chiles volcano at depths of less than 10 km (Ruiz et al. 2013). Although originally not a target volcano for the VUELCO project, the volcanic complex of Chiles-Cerro Negro is an excellent example of a re-awakening volcano, having shown no signs of magmatic unrest in recent history until the events of 2014. Analysis of seismicity identified in October 2014 suggested not only the occurrence of families, but also their occurrence in distinct temporal patterns. The clustering of similar seismic events around the diagonal in a number of box-like formations within a similarity matrix suggests that a number of sources were active for discrete periods of time generating families of seismicity (Fig. 5). The distinct clusters of similar seismic waveforms may relate to a changing source location or mechanism at depth (Salvage 2015). Over this time period, no significant changes in the amplitude of events (indicative of a changing source mechanism) were identified. Since similar seismicity is thought to be generated through a similar source mechanism and a similar source location, distinct cluster of similar seismicity is most likely related to its own distinct spatial region, which generated seismicity during distinct periods of time.

The Source Mechanisms of Low Frequency Earthquakes

Since low frequency events, and in particular families, appear to be important in detecting changes in unrest at volcanoes, it is important to understand their mechanism of generation. Synthetic modelling and moment tensor inversions of low frequency seismic wavefields are powerful tools for gaining information on the source mechanisms underlying volcanic earthquakes. Once instrument response and path effects have been accounted for, real data can be compared to synthetic models, and on the basis of a best-fit approach the obtained model parameters allow insights into the nature and geometry of the source (Chouet 1996; Shuler et al. 2013). However, as the fundamental assumptions behind the commonly used moment tensor inversions are based on plane surface geometries which are believed to be too simple to explain the generation of low frequency events in a volcanic environment, the application to more complex seismic sources has so far been inconclusive. In the framework of the VUELCO project, slip along bent surfaces (a complex source) was thought of as the underlying physical motion responsible for generating seismic energy. This novel way of investigating low frequency earthquakes can explain several features of the earthquakes under investigation without introducing compromising

assumptions such as slip along a single, unbent surface, which is believed to be unrealistic.

An example of a spatially extended source generating seismicity within a volcanic environment is a volcanic conduit through which magmatic fluids move. In these instances, the generation of low frequency seismicity may be related to the brittle failure of magma itself (Neuberg et al. 2006; Lavallée et al. 2008; Thomas and Neuberg 2012) or through a stick-slip motion at the conduit edge (Iverson et al. 2006). In either case, shallow source depths (1–2 km) and short epicentral distances to seismic receivers (a few kilometers) suggest that a spatially extended source is more realistic than a single point source.

The occurrence of slip (i.e. the generation of the seismic energy itself) of spatially extended sources may either be instantaneous along two or more slip surfaces, or may occur on different slip surfaces at different times, offset by a given time increment, delta t. A ring fault structure is a numerical description of seismogenic slip of magma along all of the conduit walls within a volcanic edifice, and can be numerically modelled by considering a cylinder representing the volcanic conduit with instantaneously slipping double couple (single point) sources bounding the circumference (Fig. 6). Upward movement inside the cylinder and downward movement outside represents the movement of magma

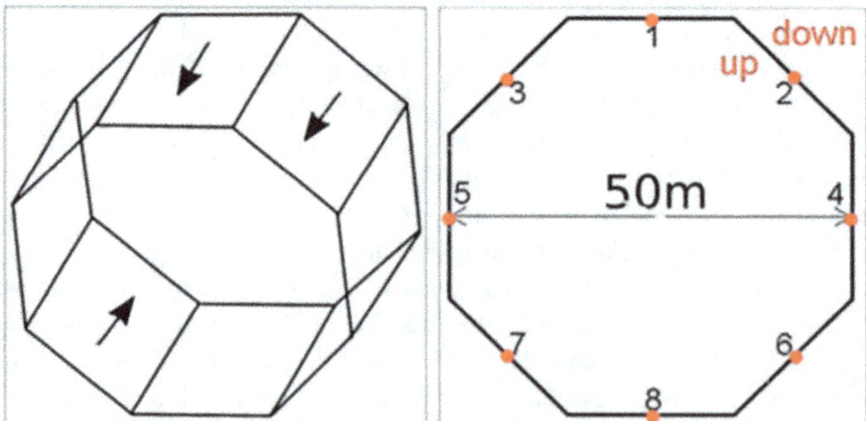

Fig. 6 Schematic representation of a ring fault structure with movement directed upwards within the cylinder to represent the flow of magma through a conduit. Each planar surface is host to a single double couple source i.e. a point source (labelled *1–8*)

through this channel. Other spatially extended sources which may evoke the generation of low frequency seismicity include: the upward movement of magma through a narrow dyke, numerically modelled by two oppositely directed double couple sources; slip along distinct segments of the volcanic conduit i.e. due to geometry changes (Thomas and Neuberg 2012), numerically modelled with double couple sources on distinct segments; or the generation of a number of seismic swarms of families of earthquakes occurring simultaneously (Salvage and Neuberg 2016), which can be numerically modelled by movement on two or more simultaneously acting ring fault structures.

Green and Neuberg (2006) suggested that accelerated magma movement at Soufrière Hills volcano can be linked to observed deformation cycles through low frequency seismic swarms, and that seismicity is only generated if significant magma movement takes place. Considering slip through extended sources brings us one step closer to estimating magma ascent rates. Once calibrated, the link between observed waveform amplitudes and the amount of seismogenic slip occurring during a seismic swarm will yield magma ascent rates and will ultimately contribute to forecasting volcanic eruptions more accurately. Point and extended source models yield great differences in observed P-wave amplitudes and waveforms, leading to remarkable differences when interpreting the amount of

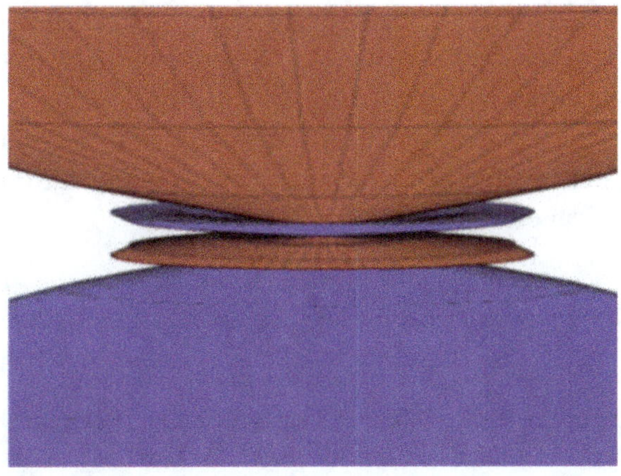

Fig. 8 Zoom of smaller radial lobes of P-wave radiation pattern for a spatially extended ring fault structure. Compressional lobes are *red*, dilatational lobes are *blue*

slip occurring and the slip rates. In the case of slip along a ring fault, P-wave amplitudes are greatly reduced due to destructive interference, in comparison to simple double couple (point) sources. As a result, observed amplitudes yield an underestimation of actual slip by more than a factor of 3 if interpreted as a point source, when in reality a spatially extended source acts at depth. This underestimation in seismic moment consequently may lead to an underestimation of magma flow rate at depth, which in turn has severe implications for eruption forecasting (Karl 2014).

Furthermore, the P-wave radiation pattern for a spatially extended ring fault structure (i.e. for modelling the movement of magma within the entire conduit made up of a number of double couple sources) shows remarkable similarity to the P-wave radiation pattern for a compensated linear vector dipole (CLVD) source, which instead is a conservation of energy (Fig. 7). The derived ring fault radiation pattern shows rotational symmetry around the depth axis, and consists of a large compressional lobe directly above the source and an inversely polarised, dilatational lobe with the same amplitude below it. Only if the small radial P-wave radiation lobes can be determined for the ring fault structure (Fig. 8) is it possible to distinguish between these two source mechanisms. Seismic networks with small apertures typical in volcanic settings and

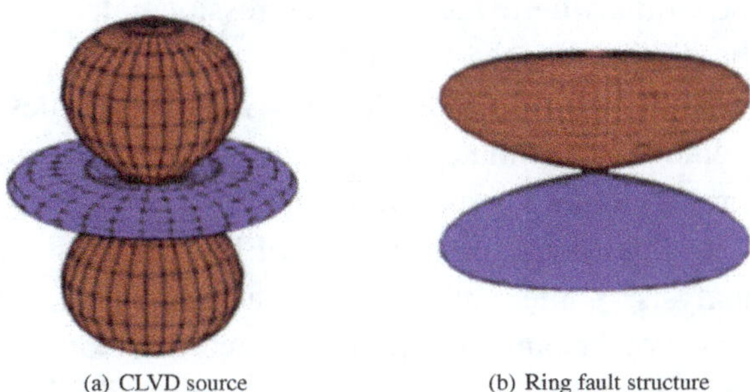

Fig. 7 P-wave radiation pattern generated for a CLVD source, and for an extended ring fault structure source. The *red* lobes are compressional, the *blue* lobes are dilatational. In both radiation patterns a large compressional lobe is found above the source, with dilatational lobes extending radially which can lead to confusion for interpretation of first motion polarity patterns

uncertainties in earthquake source depth locations will likely lead to difficulties in distinguishing between the two radiation patterns, since both can explain observed first motion polarity patterns of low frequency seismicity on volcanoes (Karl 2014).

Forecasting Eruptive Activity

The ability to forecast the timing, intensity and type of volcanic activity is one of the key issues facing volcanologists today. The most notable instances of successful volcanic forecasting use precursory activity at andesitic-dacitic volcanoes. The cataclysmic eruption of Mt. Pinatubo, Philippines on 15 June 1991 was preceded by at least two months of heightened seismicity (Harlow et al. 1996). With increases in seismicity and an alarming sudden drop in SO_2, scientists were able to successfully evacuate over 45,000 local people and 14,500 military personnel to safety by 14 June, such that less than 300 people were killed in the ensuing volcanic activity on 15 June. More recently, the 2010 eruption of Merapi, Indonesia, on 26 September was preceded by approximately 6 weeks of precursory activity (Budi-Santoso et al. 2013): rates of seismicity and SO_2 during this time were comparable to, or higher than, the highest rates observed during previous (smaller) Merapi eruptions (1992–2007), and rapid deformation was observed. Consequently, one day prior to the explosive eruption, several tens of thousands of people were evacuated from a radius extending 10 km from the volcano, resulting in a greatly lowered death toll of 35.

Volcanic eruptions are often preceded by accelerating geophysical signals, associated with the movement of magma or other fluid towards the surface. Of these precursors, seismicity is at the forefront of forecasting volcanic activity since it is frequently observed and the change from background level can be observed in real time. Since forecasting of volcanic eruptions relies on the ability to determine the timing of magma reaching the surface, low frequency seismicity may act as a forecasting tool due to its potential correlation with the movement of magmatic fluid at depth.

The Failure Forecast Method (FFM) is based on an empirical power-law relationship, which relates the acceleration of a precursor ($d^2\Omega/dt^2$) to the rate of that precursor ($d\Omega/dt$) (Voight 1988) method by:

$$\frac{d^2\Omega}{dt^2} = K\left(\frac{d\Omega}{dt}\right)^\alpha \qquad (1)$$

where K and α are empirical constants. Ω can represent a number of different geophysical precursors, for example low frequency seismic event rate (Salvage and Neuberg 2016), event rate of all recorded seismicity (Kilburn and Voight 1998), or the amplitude of seismic events (Ortiz et al. 2003). The parameter α is thought to range between 1 and 2 in volcanic environments (Voight 1988), or may even evolve from 1 towards 2 as seismicity proceeds (Kilburn 2003). α has also been calculated in hindsight as high as 3.3 for accelerating seismicity in 1991 at Mt. Pinatubo, although this extreme value appears rare and was calculated with only a small amount of seismic data (Smith and Kilburn 2010). An infinite $d\Omega/dt$ suggests an uncontrolled rate of change (a singularity) and in this environment is associated with an impending eruption. The inverse form of $d\Omega/dt$ is linear if $\alpha = 2$, and therefore in this case the solution for the timing of failure is a linear regression of inverse rate against time, with the timing of failure relating to the point where the linear regression intersects the x-axis (Voight 1988).

Although assuming that $\alpha = 2$ is the simplest method to estimate the timing of an eruption through a linear regression and therefore the most common application of the FFM in hindsight analysis, some authors have suggested that it may not be an appropriate assumption for use with the FFM (e.g. Bell et al. 2011). Additionally, some authors have argued that α may evolve with time as precursory sequences develop, which is not detailed in the FFM (Kilburn 2003). As the FFM follows a least squares regression analysis when α is equal to 2, the residual error between the observed event rate and the mean event rate of seismicity should follow a typical Gaussian distribution (Bell et al. 2011). Greenhough and Main (2008) have suggested that since earthquake occurrence is a point process, the rate uncertainties are best described by a Poisson distribution. In this instance, a generalised linear model (GLM) where $\alpha = 1$, rather than a least squares regression model ($\alpha = 2$) may be more appropriate, since it can allow for a distribution of data that is non-Gaussian (Bell et al. 2011). At Soufrière Hills volcano, however, the use a GLM to forecast the timing of eruptive events in 1997 and 2003 failed to generate an appropriate forecast (Salvage and Neuberg 2016). Hammer and Ohrnberger (2012) suggested that this may be related to the fact that a Poisson process, and therefore the GLM, is a memoryless system, meaning that past events do not influence future patterns. A memoryless system is not consistent with the fundamental assumptions of the FFM, since previous geophysical observables form the basis of such a forecast.

One of the first instances of real time forecasting using the FFM was at Redoubt volcano, when the inverse average amplitude of seismic events followed a linear regression trend for 4 days prior to a dome collapse event on 2 January 1990. Due to this trend, and the fact that the seismic intensity was far above background levels, the Alaskan Volcano Observatory issued a "formal warning" of an impending eruptive event on the morning of the 2 January, a few hours before the eruption began, although the FFM calculations suggested an eruption was likely within 0.5–2 days. A similar, if not clearer trend, that supported the forecast was found using the same precursory sequence but only using seismic events within the spectral range of 1.3–1.9 Hz (Cornelius and Voight 1994), suggesting an increased accuracy in forecasts when focusing on a single source process at depth.

Swarms of seismic events, i.e. a number of similar events within a short period of time, with typical swarm durations of hours to day, are not observed at all volcanoes, but have been commonly observed at Soufrière Hills volcano (e.g. Green and Neuberg 2006) and Redoubt volcano (e.g. Buurman et al. 2013). Using precursory seismicity and the FFM, Salvage and Neuberg (2016) forecast in hindsight the timing of a dome collapse event on 25 June 1997 at Soufrière

(a)

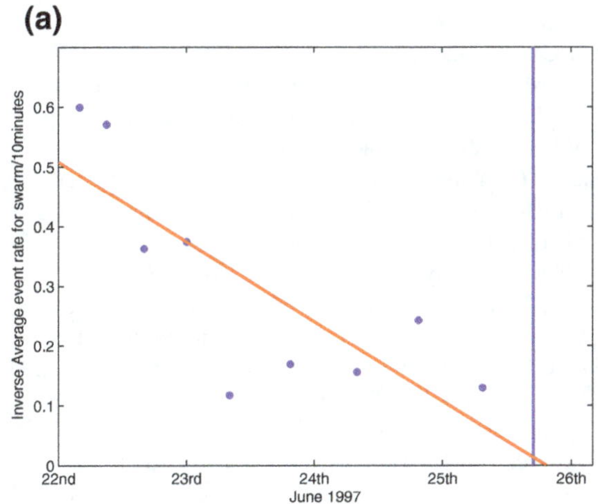

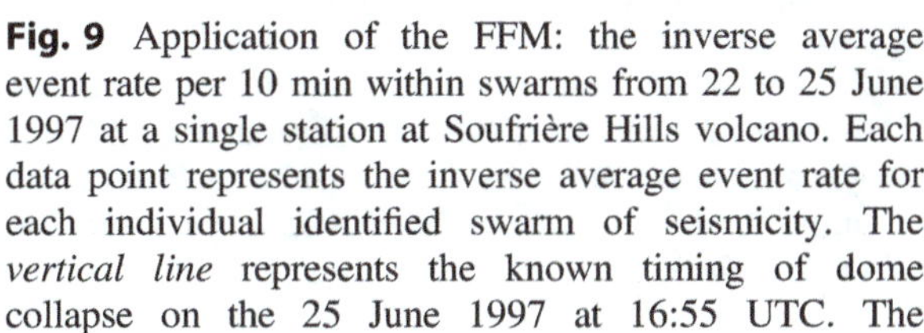

(b)

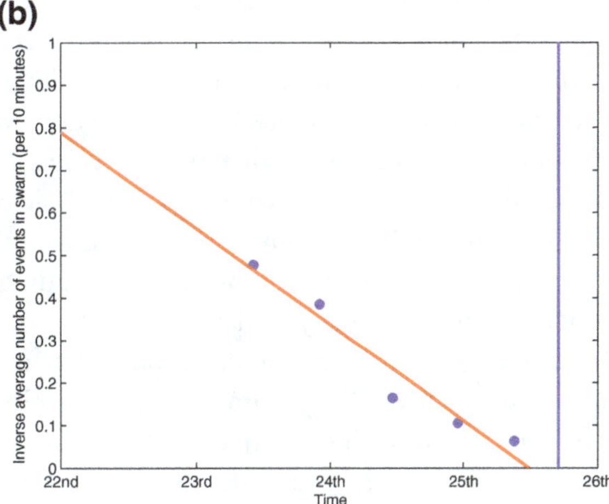

Fig. 9 Application of the FFM: the inverse average event rate per 10 min within swarms from 22 to 25 June 1997 at a single station at Soufrière Hills volcano. Each data point represents the inverse average event rate for each individual identified swarm of seismicity. The *vertical line* represents the known timing of dome collapse on the 25 June 1997 at 16:55 UTC. The graphical representation of the FFM is depicted by the linear regression (it is assumed that $\alpha = 2$ for simplicity) and the forecasted timing of failure can be read off the x-axis at the point where the linear regression crosses it. **a** All triggered low frequency seismicity. **b** Single family of similar seismicity

Hills, based upon accelerating rates of low frequency earthquakes which occurred in swarms, rather than simply the number of low frequency events over the precursory time period (Fig. 9). Using the average event rate per swarm showed a clearer accelerating pattern over the entire seismic sequence, rather than using the traditional method of binning data in units of time. More accurate forecasts were determined when using only one single family of similar events to forecast the dome collapse, rather than all low frequency seismicity mixed together. A dome collapse on 12 July 2003 at Soufrière Hills volcano was also more accurately forecast when using a single family of similar events, rather than all low frequency seismicity, which occurred during the period of unrest (Salvage and Neuberg 2016). Consequently, the use of families of seismicity, and therefore concentration upon a single active system at depth, may allow a more accurate forecast of the timing of an eruptive event in these instances.

Summary

Seismology is a powerful tool which can be used to understand processes occurring at depth, and their relationship to the surface at volcanoes, especially since seismic events are easily detected and the deviation from the background level is often notably pronounced. An increase in seismicity may be an indication of volcanic unrest, since it relates to a number of physical processes at depth including brittle failure and fracturing of the conduit or of the surrounding edifice (high frequency seismicity) and the movement of magmatic fluids at depth (low frequency seismicity). Using families of seismicity as an indicator of a single active system at depth, temporal and spatial patterns in the seismic events can be used to assess the potential migration of seismicity towards the surface, as well as to forecast the timing of volcanic eruptive events related to the acceleration of seismicity. However, a full understanding of the source mechanism of the generated seismic events is essential to ensure that the magma flow rate is estimated accurately and therefore an accurate forecast can be generated for the timing of eruption. Using seismicity in combination with other monitoring tools, we are now closer to gaining a better understanding of evolving magmatic systems at depth.

Acknowledgements The past and present staff at the Montserrat Volcano observatory are fully acknowledged for their ongoing support in the upkeep and maintenance of the seismic network, and the sharing of data. All staff at the Instituto-Geofisico in Ecuador are also fully acknowledged for providing data from Chiles-Cerro Negro volcano, for useful discussions regarding the volcanic complex, and for their continued monitoring efforts of all volcanoes in Ecuador. We thank two anonymous reviewers for their detailed comments and suggestions which greatly enhanced the quality of this manuscript. Muchas gracias también a Maria Martinez-Cruz y Javier Pacheco por su ayuda con la traducción al español.

References

Arámbula-Mendoza R, Lesage P, Valdés-González C, Varley N, Reyes-Dávila G, Navarro C (2011) Seismic activity that accompanied the effusive and explosive eruptions during the 2004–2005 period at Volcán de Colima, Mexico. J Volcanol Geoth Res 205(1):30–46

Arciniega-Ceballos A, Chouet B, Dawson P (2003) Long-period events and tremor at Popocatepetl volcano (1994–2000) and their broadband characteristics. Bull Volc 65(2):124–135

Bean CJ, De Barros L, Lokmer I, Métaxian J-P, O'Brien G, Murphy S (2014) Long-period seismicity in the shallow volcanic edifice formed from slow-rupture earthquakes. Nat Geosci 7(1):71–75

Beauval C, Yepes H, Bakun WH, Egred J, Alvarado A, Singaucho J-C (2010) Locations and magnitudes of historical earthquakes in the Sierra of Ecuador (1587–1996). Geophys J Int 181(3):1613–1633

Bell A, Naylor M, Heap M, Main I (2011) Forecasting volcanic eruptions and other material failure phenomena: an evaluation of the failure forecast method. Geophys Res Lett 38:L15304

Budi-Santoso A, Lesage P, Dwiyono S, Sumarti S, Jousset P, Metaxian J-P et al (2013) Analysis of the seismic activity associated with the 2010 eruption of Merapi Volcano, Java. J Volcanol Geoth Res 261:153–170

Budi-Santoso A, Lesage P (2016) Velocity variations associated with the large 2010 eruption of Merapi volcano, Java, retrieved from seismic multiplets and ambient noise cross-correlation. Geophys J Int 206 (1):221–240

Buurman H, West ME, Thompson G (2013) The seismicity of the 2009 Redoubt eruption. J Volcanol Geoth Res 259:16–30

Chouet B (1988) Resonance of a fluid-driven crack: radiation properties and implications for the source of long-period events and harmonic tremor. J Geophys Res Solid Earth (1978–2012) 93(B5), 4375–4400

Chouet BA (1996) New methods and future trends in seismological volcano monitoring. In: Monitoring and Mitigation of Volcano Hazards. Springer, pp. 23–97

Chouet BA, Matoza RS (2013) A multi-decadal view of seismic methods for detecting precursors of magma movement and eruption. J Volcanol Geoth Res 252:108–175

Cornelius R, Voight B (1994) Seismological aspects of the 1989–1990 eruption at Redoubt Volcano, Alaska: the materials Failure Forecast Method (FFM) with RSAM and SSAM seismic data. J Volcanol Geoth Res 62(1):469–498

Dawson PB, Chouet BA, Power J (2011) Determining the seismic source mechanism and location for an explosive eruption with limited observational data: Augustine Volcano, Alaska. Geophys Res Lett 38(3):L03302

De Angelis S, Henton S (2011) On the feasibility of magma fracture within volcanic conduits: constraints from earthquake data and empirical modelling of magma viscosity. Geophys Res Lett 38(19):L19310

De Angelis S, Bass V, Hards V, Ryan G (2007) Seismic characterization of pyroclastic flow activity at Soufrière Hills Volcano, Montserrat, 8 January 2007. Nat Hazards Earth Syst Sci 7:467–472

Geller R, Mueller C (1980) Four similar earthquakes in central California. Geophys Res Lett 7(10):821–824

Goto A (1999) A new model for volcanic earthquake at unzen volcano: melt rupture model. Geophys Res Lett 26(16):2541–2544

Green D, Neuberg J (2006) Waveform classification of volcanic low-frequency earthquake swarms and its implication at Soufrière Hills Volcano, Montserrat. J Volcanol Geoth Res 153(1):51–63

Greenhough J, Main I (2008) A poisson model for earthquake frequency uncertainties in seismic hazard analysis. Geophys Res Lett 35(19):L19313

Hammer C, Ohrnberger M (2012) Forecasting seismo-volcanic activity by using the dynamical behavior of volcanic earthquake rates. J Volcanol Geoth Res 229:34–43

Harlow DH, Power JA, Laguerta EP, Ambubuyog G, White RA, Hoblitt RP (1996) Precursory seismicity and forecasting of the June 15, 1991, eruption of Mount Pinatubo. Eruptions and lahars of Mount Pinatubo, Philippines, Fire and mud, pp 223–247

Herd RA, Edmonds M, Bass VA (2005) Catastrophic lava dome failure at Soufriere Hills volcano, Montserrat, 12–13 July 2003. J Volcanol Geoth Res 148(3):234–252

Iverson R, Dzurisin D, Gardner C, Gerlach T, LaHusen R, Lisowski M, Major J, Malone S, Messerich J, Moran S et al (2006) Dynamics of seismogenic volcanic extrusion at Mount St Helens in 2004–05. Nature 444(7118):439–443

Karl S (2014) The source mechanisms of low frequency seismic events on volcanoes. Ph.D. thesis, University of Leeds, UK, Online at: http://etheses.whiterose.ac.uk/id/eprint/8406

Kilburn C (2003) Multiscale fracturing as a key to forecasting volcanic eruptions. J Volcanol Geoth Res 125(3–4):271–289

Kilburn CR, Voight B (1998) Slow rock fracture as eruption precursor at Soufriere Hills volcano Montserrat. Geophys Res Lett 25(19):3665–3668

Lahr J, Chouet B, Stephens C, Power J, Page R (1994) Earthquake classification, location, and error analysis in a volcanic environment: implications for the magmatic system of the 1989–1990 eruptions at Redoubt Volcano, Alaska. J Volcanol Geoth Res 62(1):137–151

Lavallée Y, Meredith P, Dingwell D, Hess K-U, Wassermann J, Cordonnier B, Gerik A, Kruhl J (2008) Seismogenic lavas and explosive eruption forecasting. Nature 453(7194):507–510

McNutt SR (2005) Volcanic seismology. Annu Rev Earth Planet Sci 32:461–491

Nakano M, Kumagai H (2005) Response of a hydrothermal system to magmatic heat inferred from temporal variations in the complex frequencies of long-period events at Kusatsu-Shirane Volcano, Japan. J Volcanol Geoth Res 147(3):233–244

Neuberg J, Luckett R, Baptie B, Olsen K (2000) Models of tremor and low-frequency earthquake swarms on Montserrat. J Volcanol Geoth Res 101(1–2):83–104

Neuberg J, Tuffen H, Collier L, Green D, Powell T, Dingwell D (2006) The trigger mechanism of low-frequency earthquakes on Montserrat. J Volcanol Geoth Res 153(1):37–50

Ortiz R, Moreno H, Garcì A, Fuentealba G, Astiz M, Peña P, Sánchez N, Tárraga M (2003) Villarrica volcano (Chile): characteristics of the volcanic tremor and forecasting of small explosions by means of a material failure method. J Volcanol Geoth Res 128(1):247–259

Ottemöller L (2008) Seismic hybrid swarm precursory to a major lava dome collapse: 9–12 July 2003, Soufriere Hills Volcano, Montserrat. J Volcanol Geoth Res 177(4):903–910

Petersen T (2007) Swarms of repeating long-period earthquakes at Shishaldin Volcano, Alaska, 2001–2004. J Volcanol Geoth Res 166(3):177–192

Phillipson G, Sobradelo R, Gottsmann J (2013) Global volcanic unrest in the 21st century: an analysis of the first decade. J Volcanol Geoth Res 264:183–196

Ruiz G, Cordova A, Ruiz M, Alvarado A (2013) Informe Tecnico de los volcanes Cerro Negro y Chiles. Tech. Rep., IG-EPN, in Spanish

Salvage RO (2015) Using seismic signals to forecast volcanic processes. Ph.D. thesis, University of Leeds, UK, Online at: http://etheses.whiterose.ac.uk/12268/. Restricted until April 2018

Salvage R, Neuberg J (2016) Using a cross correlation technique to refine the accuracy of the failure forecast method: application to Soufrière Hills volcano, Montserrat. J Volcanol Geoth Res 324:118–133

Sgattoni G, Jeddi Z, Gudmundsson Ó, Einarsson P, Tryggvason A, Lund B, Lucchi F (2016) Long-period seismic events with strikingly regular temporal patterns on Katla volcano's south flank (Iceland). J Volcanol Geoth Res 324:28–40

Shuler A, Ekström G, Nettles M (2013) Physical mechanisms for vertical-CLVD earthquakes at active volcanoes. J Geophy Res Solid Earth 118(4):1569–1586

Smith R, Kilburn C (2010) Forecasting eruptions after long repose intervals from accelerating rates of rock fracture: the June 1991 eruption of Mount Pinatubo, Philippines. J Volcanol Geoth Res 191(1):129–136

Sparks R, Young S (2002) The eruption of Soufrière Hills Volcano, Montserrat (1995–1999): overview of scientific results. Geol Soc Lond Mem 21(1):45–69

Thelen W, Malone S, West M (2011) Multiplets: their behavior and utility at dacitic and andesitic volcanic centers. J Geophys Res Solid Earth (1978–2012) 116 (B8):B08210

Thomas ME, Neuberg J (2012) What makes a volcano tick—a first explanation of deep multiple seismic sources in ascending magma. Geology 40(4):351–354

Voight B (1988) A method for prediction of volcanic eruptions. Nature 332:125–130

Wadge G, Voight B, Sparks R, Cole P, Loughlin S, Robertson R (2014) An overview of the eruption of Soufriere Hills Volcano, Montserrat from 2000 to 2010. Geological Society, London, Memoirs 39(1):1–40

The Ups and Downs of Volcanic Unrest: Insights from Integrated Geodesy and Numerical Modelling

J. Hickey, J. Gottsmann, P. Mothes, H. Odbert, I. Prutkin and P. Vajda

Abstract

Volcanic eruptions are often preceded by small changes in the shape of the volcano. Such volcanic deformation may be measured using precise surveying techniques and analysed to better understand volcanic processes. Complicating the matter is the fact that deformation events (e.g., inflation or deflation) may result from magmatic, non-magmatic or mixed/hybrid sources. Using spatial and temporal patterns in volcanic deformation data and mathematical models it is possible to infer the location and strength of the subsurface driving mechanism. This can provide essential information to inform hazard assessment, risk mitigation and eruption forecasting. However, most generic models over-simplify their representation of the crustal conditions in which the deformation source resides. We present work from a selection of studies that employ advanced numerical models to interpret deformation and gravity data. These incorporate crustal heterogeneity, topography, viscoelastic rheology and the influence of temperature, to constrain unrest source parameters at Uturuncu (Bolivia), Cotopaxi (Ecuador), Soufrière Hills (Montserrat), and Teide (Tenerife) volcanoes. Such model complexities are justified by

J. Hickey and H. Odbert, previously at School of Earth Sciences, University of Bristol, UK.

J. Hickey (✉)
Camborne School of Mines, University of Exeter, Exeter, UK
e-mail: J.Hickey@exeter.ac.uk

J. Gottsmann
School of Earth Sciences, University of Bristol, Bristol, UK

J. Gottsmann
Cabot Institute, University of Bristol, Bristol, UK

P. Mothes
Instituto Geofísico, Escuela Politecnica Nacional, Quito, Ecuador

H. Odbert
Met Office, Exeter, UK

I. Prutkin
Institute of Geosciences, Jena University, Jena, Germany

P. Vajda
Earth Science Institute, Slovak Academy of Sciences, Bratislava, Slovakia

Advs in Volcanology (2019) 203–219
DOI 10.1007/11157_2017_13
© The Author(s) 2017
Published Online: 15 July 2017

geophysical, geological, and petrological constraints. Results highlight how more realistic crustal mechanical conditions alter the way stress and strain are partitioned in the subsurface. This impacts inferred source locations and magmatic pressures, and demonstrates how generic models may produce misleading interpretations due to their simplified assumptions. Further model results are used to infer quantitative and qualitative estimates of magma supply rate and mechanism, respectively. The simultaneous inclusion of gravity data alongside deformation measurements may additionally allow the magmatic or non-magmatic nature of the source to be characterised. Together, these results highlight how models with more realistic, and geophysically consistent, components can improve our understanding of the mechanical processes affecting volcanic unrest and geodetic eruption precursors, to aid eruption forecasting, hazard assessment and risk mitigation.

Extended Spanish Abstract

La deformación volcánica, caracterizada por pequeños cambios medibles en la morfología del volcán, a menudo, pero no siempre, precede a una erupción volcánica. Esta cuestión, sin embargo, se complica por el hecho de que los eventos de deformación (por ejemplo, inflación o deflacción) pueden ser el resultado de una fuente magmática, no magmática o de fuentes mixtas/híbridas. Utilizando tanto la amplitud y patrones espacio-temporales de datos de deformación volcánica registrados, así como la utilización de modelos matemáticos, es posible inferir la ubicación y la fuerza de la fuente impulsora subyacente. Estos métodos pueden proporcionar información esencial para la evaluación y mitigación de riesgos, así como para el pronóstico de erupciones volcánicas. Sin embargo, la mayoría de los modelos genéricos son insatisfactorios en su representación de las condiciones de la corteza en las que reside la fuente. En este trabajo presentamos una selección de estudios que emplean modelos numéricos avanzados para la interpretación de datos de deformación y gravedad. Dichos datos incorporan la heterogeneidad de la corteza, la topografía, la reología inelástica y los efectos termo-mecánicos para constreñir los parámetros asociados a la fuente de perturbación en cuatro sistemas volcánicos. Las complejidades de estos modelos están justificadas por limitaciones geofísicas, geológicas y petrológicas. El estudio realizado en el volcán Uturuncu, localizado en Bolivia, destaca la importancia de la estructura sub-superficial y de los procesos dependientes de tiempo en la fuente para explicar los patrones de deformación espacial-temporal. La combinación de dichos resultados indica un ascenso del magma de tipo diapírico. En el volcán Cotopaxi, localizado en Ecuador, los nuevos modelos de inversión que emplean el Análisis por Elementos Finitos esclarecen la ubicación y el volumen de una intrusión magmática durante un episodio de actividad asísmico y no eruptivo con una baja tasa de suministro de magma. Estos modelos también proporcionan señales observables que podrían estar asociadas con futura actividad volcánica intrusiva o eruptiva. El análisis de la deformación intra-eruptiva en el volcán Soufrière Hills, en Montserrat, mostró cómo la inflación registrada podría deberse a una serie de reservorios magmáticos apilados o un depósito alargado verticalmente, debido a respuestas termo-mecánicas de la corteza similares. Las condiciones de falla derivadas para las tasas de suministro de magma y reservorio son consistentes con las restricciones térmicas y mecánicas independientes. Utilizando datos gravimétricos y la curiosa falta de deformación asociada, el estudio en el volcán Teide, en Tenerife, se identifican, de forma separada, las contribuciones de fuentes poco profundas como de aquellas con más profundidad. La interpretación de ambos elementos sugiere que una intrusión magmática profunda activó un sistema hidrotermal superficial localizado encima de ésta, y por lo tanto demuestra un efecto causal de un origen mixto. Estos casos de estudio representan una contribución significativa para la comprensión de los procesos volcánicos durante los periodos de actividad intra-eruptiva y no eruptiva. La combinación de estos resultados enfatiza cómo una mejor parametrización de condiciones mecánicas de la corteza puede alterar fundamentalmente la forma en que el estrés y la tensión se reparten en la sub-superficie. Del mismo modo, una topografía compleja, como es el caso en los estratovolcanes con laderas empinadas, puede afectar la partición de la deformación superficial. Estos dos efectos impactan la inferencia de la localización de las fuentes y presiones magmáticas previo al fallo y la erupción, e indican cómo los modelos genéricos pueden conducir a interpretaciones engañosas debido a la simplificación de inferencias acerca de la corteza y a espacios planos y a medias. Resultados adicionales de la modelización son utilizados para inferir estimaciones cuantitativas y cualitativas de la tasa de suministro de magma y el mecanismo, respectivamente, contribuyendo así al entendimiento de la dinámica del transporte del magma. Además, la inclusión simultánea de datos de gravedad junto a mediciones de deformación, permiten la caracterización de la naturaleza magmática o no magmática de la fuente. Juntos, estos estudios destacan cómo los modelos que cuentan con componentes más plausibles y geofísicamente consistentes, pueden mejorar nuestro entendimiento de los procesos mecánicos que afectan la reactivación volcánica y de precursores geodésicos de erupciones. Estos también proporcionan un marco para ayudarnos a avanzar en el pronóstico de una erupción, así como en la evaluación y mitigación de los riesgos, proporcionando datos cuantitativos derivados de la modelización de mecanismos físicos adecuados y robustos.

Keywords

Volcano deformation · Gravity · Modelling · Crustal mechanics · Geodesy

Palabras clave

deformación volcánica · gravedad · modelización · mecánica de la corteza · geodesia

Introduction

Volcano deformation is a key observable during periods of volcanic unrest, and one of the main tools used to monitor developing crises (Sparks et al. 2012). Non-magmatic causes of volcanic deformation during an unrest period do not involve movement of new magma and are most commonly related to active hydrothermal systems (e.g., Fournier and Chardot 2012; Rouwet et al. 2014, and references therein). Magmatic causes reflect the active migration and accumulation of new magma, with usually deeper origins and wider deformation footprints. Hybrid mechanisms have also been proposed, where contributions from magmatic and hydrothermal systems combine to produce a single complex deformation pattern (Gottsmann et al. 2006a).

A variety of both ground and satellite based geodetic monitoring techniques are used to assess the spatial and temporal evolution of volcanic related surface deformation. The two most common techniques employed today are the Global Positioning System (GPS) and Interferometric Synthetic Aperture Radar (InSAR), which track changes in position due to the ground deforming and complement each other with advantages in their temporal and spatial coverage, respectively. Volcano gravimetry studies, the monitoring and interpretation of continuous or time-lapse spatio-temporal gravity variations, can also provide additional information on any density changes associated with the driving mechanism (Battaglia et al. 2008), and thus enable an estimate of its nature (e.g., magmatic or hydrothermal), which is particularly important in cases where there is no significant (observable) surface deformation.

The magnitude, spatial pattern, and temporal evolution of volcanic deformation can be used to infer the location and 'strength' of a causative subsurface source. These source parameters have important implications for volcanic hazards and risk mitigation. Geodetic data are also a key component in eruption forecasting efforts (e.g.,

Sparks et al. 2012), and constraints on source parameters (e.g., pressure or volume changes) from previous deformation episodes can help to quantitatively improve these forecasts. However, making the transition from surface deformation observations to subsurface source processes requires the use of geodetic models, which demand assumptions about crustal mechanics.

Generic analytical models of volcanic deformation (e.g., Mogi 1958) are often oversimplified and do not capture the intrinsic complexities of subsurface systems, which are essential for reliable assessment of causative processes, and thus eruption forecasting and hazard assessment. They usually represent the Earth's crust as a homogeneous, isotropic, elastic, half-space with a flat, free surface. These assumptions can cause misleading interpretations when compared to numerical models that allow for more realistic crustal conditions (Masterlark 2007). Numerical models of volcanic deformation are thus becoming increasingly popular due to this ability to estimate source parameters in crustal conditions that are beyond the analytical realm. These improvements are synonymous with the recent advances in geodetic monitoring, which deserve a more in-depth analysis than generic analytical models can offer.

The extra complexities that numerical models can account for relate to both the source itself, as well as the crustal rocks in which it resides. The deformation source does not have to be represented as a point or cavity, but can contain its own material properties (e.g., Hickey et al. 2013; Gottsmann and Odbert 2014). Moreover, numerical models can incorporate inferences from other geophysical, geological and petrological observables and thus maintain a higher level of consistency by relaxing some of the assumptions that restrict generic analytical models. For example, this enables the inclusion of subsurface heterogeneity, topography, viscoelastic rheology, and temperature-dependent mechanics (e.g., Hickey et al. 2016, and references therein). Consequently, more robust source parameters and magma transport processes can

Table 1 Summary of geodetic modelling

Volcano	Unrest period	FWM	IVM	Data	Topo	CMX	TMX	Model	Ref
Uturuncu	1992–2006	✓	–	InSAR	–	✓	–	2D	Hickey et al. (2013)
Cotopaxi	2001–2002	✓	✓	EDM	✓	✓	✓	3D	Hickey et al. (2015)
Soufrière hills	2003–2005	✓	–	cGPS	✓	✓	✓	2D	Gottsmann and Odbert (2014)
Central volcanic complex	2004–2005	–	✓	GPS & gravity	✓	–	–	3D	Prutkin et al. (2014)

FWM forward modelling, *IVM* inverse modelling, *Topo* topography, *CMX* crustal mechanics, *TMX* thermomechanics, *2D* two-dimensional axisymmetric, *3D* three-dimensional, *Ref* reference

be inferred, thereby improving the understanding of the links between deformation and eruption.

In this chapter we summarise the key findings from investigations of a variety of unrest episodes at a selection of the VUELCO target volcanoes; Cotopaxi (Ecuador), Soufrière Hills (Montserrat, British West indies) and the Central Volcanic Complex on Tenerife (Spain), as well as at Uturuncu volcano in Bolivia (Table 1). These examples highlight not only different timescales of unrest and spatial patterns of volcano deformation, but also the influence of crustal mechanical heterogeneity, thermal effects, and topography, on observed signals. The unrest episodes are investigated using both forward and inverse numerical modelling procedures and demonstrate the process of interpreting different geodetic data sets to constrain realistic source parameters. Thus, the purpose of this chapter is not to provide a detailed treatment of the mathematical and numerical approaches. Instead, it is intended to provide the reader with a general overview of the use of advanced numerical models to infer volcanic unrest driving mechanisms with realistic source characteristics.

Implementing Complex Crustal Mechanics

One of the most important aspects of a deformation model is the representation of crustal mechanics. This has a fundamental control on the way in which stress and strain is distributed and transferred through the Earth's crust, from the source to the surface. In this regard, as briefly mentioned above, generic analytical models are limited by their necessary assumptions of homogeneous and elastic conditions throughout the entire model domain. In reality, the Earth's crust is known to be layered, and volcanic regions in particular can have wide-ranging regions of stiff (high Young's Modulus) and soft (low Young's Modulus) rocks relating to the type of volcanic deposit that formed them, e.g., lava flows (stiff) compared to tuffs (soft) (Gudmundsson 2011). This is what we call subsurface heterogeneity. Where stiff and soft regions are adjacent in the crust, complex subsurface partitioning alters the way stress and strain are transferred to the surface. The outcome is a different surface deformation pattern compared to one that would be seen if the crust was in fact homogeneous. Consequently, generic analytical models restricted to homogeneous crustal mechanics will not adequately represent likely subsurface conditions, and the inferred source parameters from these models can be misleading.

Seismic studies are capable of delineating areas of relatively high and low seismic velocities. They can be used to estimate the dynamic Young's Modulus, E_D, Poisson's Ratio, v, and density, ρ, of the crust (Brocher 2005):

$$v = 0.5 \times \left[\left(\frac{V_P}{V_S} \right)^2 - 2 \right] \Big/ \left[\left(\frac{V_P}{V_S} \right)^2 - 1 \right] \quad (1)$$

$$\rho = 1.6612V_P - 0.4721V_P^2 + 0.067V_P^3 \\ - 0.0043V_P^4 + 0.000106V_P^5 \qquad (2)$$

$$E_D = \frac{V_P^2 \rho (1+v)(1-2v)}{(1-v)} \qquad (3)$$

where V_P and V_S are the primary and shear seismic velocities, respectively. The static Young's Modulus, E, is usually a factor of 2–9 smaller than E_D, and is more appropriate for deformation studies as the dominant processes taking place are significantly slower than the propagation of seismic waves (e.g., Gudmundsson 2011, and references therein). Numerical approaches can incorporate these spatially-variable calculations of Young's Modulus and Poisson's Ratio by allowing the mechanical parameterisation of the crust to vary within the model (Hickey and Gottsmann 2014). This can be achieved in one-dimension, where the values are only changed with depth, or in three-dimensions, where the mechanical representation additionally changes in the two horizontal axes. The Finite Element Method is the most common technique used when incorporating subsurface heterogeneity. With this approach, a heterogeneous model domain can be built using multiple different sized homogeneous 'blocks' of varying material properties (Hickey et al. 2013), with continuous interpolations as a function of depth or distance (Gottsmann and Odbert 2014), or with a three-dimensional coordinate interpolation (Hickey et al. 2016).

A further advantage of a numerical approach using the Finite Element Method is the ability to include a spatially variable inelastic rheology that is dependent on an estimated temperature distribution. Rocks do not always behave in an elastic manner (Ranalli 1995). Instead, where crustal conditions are hotter than the brittle-ductile transition, or subject to long-term loading, rocks can deform like very high viscosity fluids, and so viscoelastic effects are more likely (Ranalli 1995). This is especially important when a hot magmatic component is assumed to be present, or where elevated geothermal gradients are observed. Also, when models are restricted to elastic mechanical behaviour it can be difficult to constrain realistic source processes, particularly if there is a complex temporal deformation pattern that can not be reproduced with the instantaneous elastic stress-strain relationship, and may be better represented by a non-instantaneous viscous stress response. A model that incorporates temperature-dependent mechanics can overcome these difficulties. This is achieved using the Finite Element Method with a two-step procedure (Hickey and Gottsmann 2014). First, a spatially-variable, stationary, temperature distribution is derived, primarily dependent on the local geothermal gradient(s) and an assumed (magmatic) source temperature. This temperature distribution, T, is then used to define a crustal viscosity, η, for example through an Arrhenius relationship, where:

$$\eta = A_d \exp\left(\frac{H}{RT}\right) \qquad (4)$$

and A_d is the Dorn Parameter, H is the activation energy, and R is the universal gas constant. The viscosity is used in a viscoelastic material representation (Hickey and Gottsmann 2014), so where the viscosity is high the material response is dominated by elastic behaviour, and when the viscosity is lower the rocks show an increasingly higher tendency for viscous behaviour. To demonstrate the effect of temperature-dependent mechanics, and subsurface heterogeneity, the following case studies include examples of both.

Case Studies

Uturuncu

Uturuncu volcano is located in southern Bolivia, within the Altiplano-Puna Volcanic Complex. It has been steadily inflating since at least 1992 (Pritchard and Simons 2002), and combined with shallow seismicity and near-summit active fumaroles represents a volcanic system showing significant signs of unrest (Sparks et al. 2008).

Hickey et al. (2013) focused on the mechanism driving the 70 km wide region of ground uplift between 1992 and 2006. The aim was to constrain first-order source parameters that explain both the observed uplift rate of 1–2 cm/year and the large spatial deformation footprint (Pritchard and Simons 2002). Stress and strain from pressurised finite sources were solved numerically using Finite Element Analysis, accounting for both homogeneous and heterogeneous subsurface structure in elastic and viscoelastic rheologies. Crustal heterogeneity was constrained from seismic velocity data, which indicates a pervasive large low-velocity zone ~ 17 km below the surface. This is deduced to represent one of the world's largest known regions of partial-melt: the Altiplano-Puna Magma Body (APMB) (Sparks et al. 2008).

The comparison between crustal heterogeneity and homogeneity highlights the significant effect of a mechanically weak source-depth layer (Fig. 1). The weak layer, with a lower Young's Modulus, alters surface deformation patterns by accommodating more of the subsurface strain than its surrounding layers, thereby acting as a mechanical buffer. Continuous and regular time-dependent deformation, the long-lived nature of the source, and an anomalously high regional crustal heat-flux break the assumption of elastic conditions (e.g., Ranalli 1995), so a viscoelastic crustal rheology was tested, using the standard linear solid representation (e.g. Hickey and Gottsmann 2014). The elastic models could also only account for the spatial component of the observed uplift so their results were used solely to guide the parameters tested in the viscoelastic models. A range of possible source geometries were assessed, but spherical and oblate shapes were rejected on the grounds of their depth below the APMB and likely unsustainable pressurisation given the expected crustal mechanics. This left a prolate shaped source, whose minimum size was determined using maximum laboratory values for host-rock tensile strength. The final preferred model suggests that temporally-continuous pressurisation of a magma source protruding from the top of the APMB is

causing the observed spatial and temporal surface uplift, whereas the previous models could only infer that a simple-geometry source was being pressurised somewhere within the APMB (Pritchard and Simons 2002). Hence, this also demonstrates how a pressure-time function plays a first-order role in explaining time-dependent deformation.

Simultaneous work on an extended InSAR data set shows how the central uplift region at Uturuncu is surrounded by a 'moat' of subsidence (Fialko and Pearse 2012). To explain this observation they also constrained a model where magma rises out and up from the APMB with a diapiric-type ascent mechanism. Further evidence for this magmatic process is available through a complementary gravimetric study (del Potro et al. 2013). Therefore, this highlights how combinations of geodetic data and numerical models can not only constrain more plausible deformation source parameters, but can also infer magma transport dynamics.

Cotopaxi

Cotopaxi is a large, glacier-clad stratovolcano situated in the Eastern Cordillera of the Ecuadorian Andes. The 1 km^3 glacier presents a substantial lahar risk to people in the surrounding areas, and particularly to the 100,000 inhabitants that reside in the path of the 1877 lahar which descended the Inter-Andean Valley (Pistolesi et al. 2013).

Unrest was detected at Cotopaxi in 2001 and 2002. There was a significant increase in the amount of volcanic seismicity in the NE quadrant of the volcano, and this was originally interpreted to represent a dyke intrusion with subsequent gas-release and resonance of the crack (Molina et al. 2008). Hickey et al. (2015) revisited this unrest period, but approached it from a volcano deformation viewpoint. Analysis of an electronic distance meter (EDM) network over the 2001–2002 period also indicated an asymmetric inflation of the edifice that accompanied the recorded seismicity (Fig. 2). However, the irregular

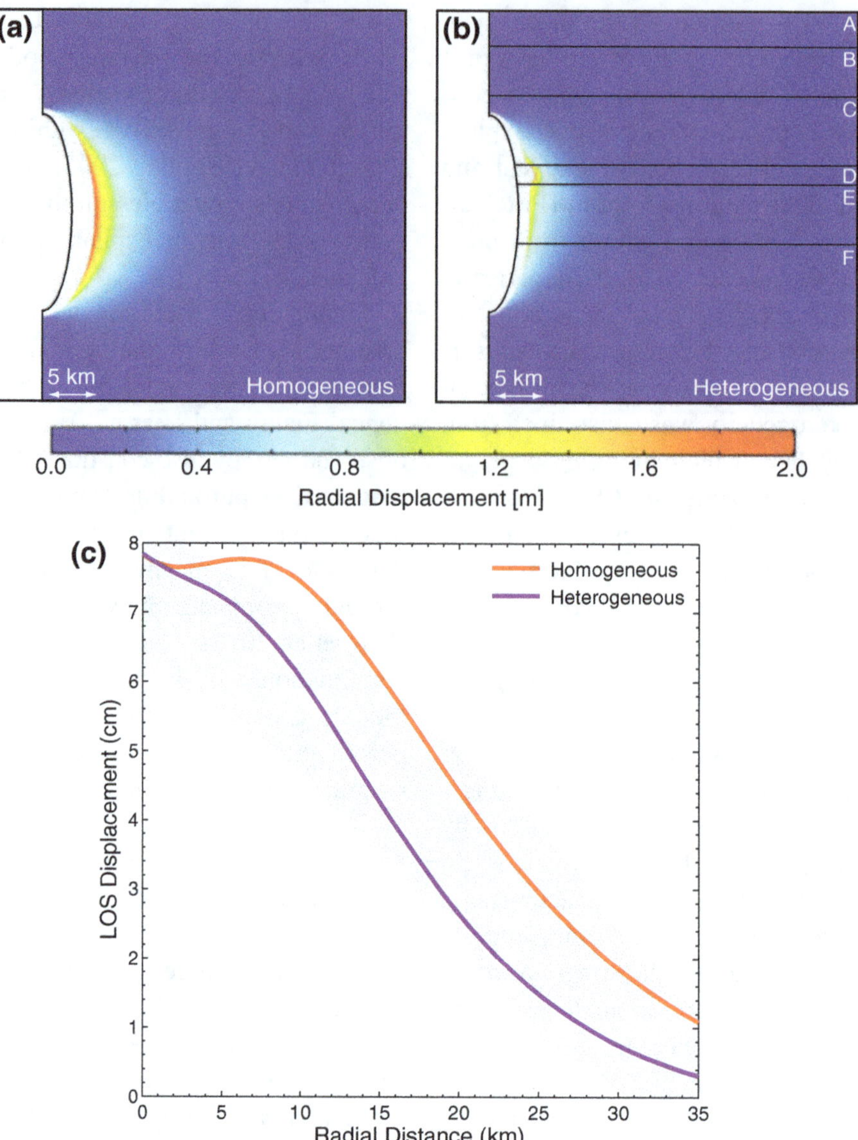

Fig. 1 A comparison of the effect crustal heterogeneity for the same prolate source geometry and depth. **a, b** The effect of a source depth soft layer on the subsurface deformation of a source. *Both panels* show the same source, embedded in a homogeneous (**a**) or heterogeneous domain (**b**). *Colours* relate to the radial displacement, and the *white shape* shows the exaggerated outline of the deformed source after the pressure is applied. In the heterogeneous model the source preferentially deforms into the softer layer (D), compared to the homogeneous medium, which exhibits a concentric deformation pattern. **c** Modelled surface displacement profiles from the homogeneous (**a**) and heterogeneous (**b**) models. The subsurface layering in **b** alters the displacement pattern produced at the surface as the soft layer modifies the subsurface strain partitioning. The *blue shaded area* represents the observed InSAR data and its estimated error bounds

acquisition in time of the EDM data prevented any systematic comparison between the two data sets. To solve for the optimum deformation source parameters, Hickey et al. (2015) implemented a novel numerical inversion procedure using Finite Element models. This is the first volcanic deformation inversion study to explicitly account for both subsurface heterogeneity and surface topography while searching for a best-fit solution with a range of source shapes. The method works by solving for the predicted EDM deformation with an initial model and source configuration. It then continually changes the source location and/or overpressure, within some predefined parameter limits, to minimise the misfit to the recorded EDM data (Fig. 2). After each inversion, the parameter limits (e.g., X and Y coordinates) are reduced around the previous solution to produce a set of decreasing size 'Russian-doll-like' parameter constraint grids that ensure a robust solution. Within this workflow, the Finite Element model geometry and mesh are automatically rebuilt, removing the need for repeated manual editing.

The inversion models converge on a shallow source beneath the SW flank. The individual best-fit model is inferred to represent a small oblate-shaped magmatic reservoir, approximately 4–5 km beneath the summit, with a volume increase of roughly 20×10^6 m³. A deformation source location in the SW is substantially different to the NE location proposed by Molina et al. (2008) when explaining the recorded seismicity. Despite this, when the deformation source was restricted to the NE quadrant the predicted

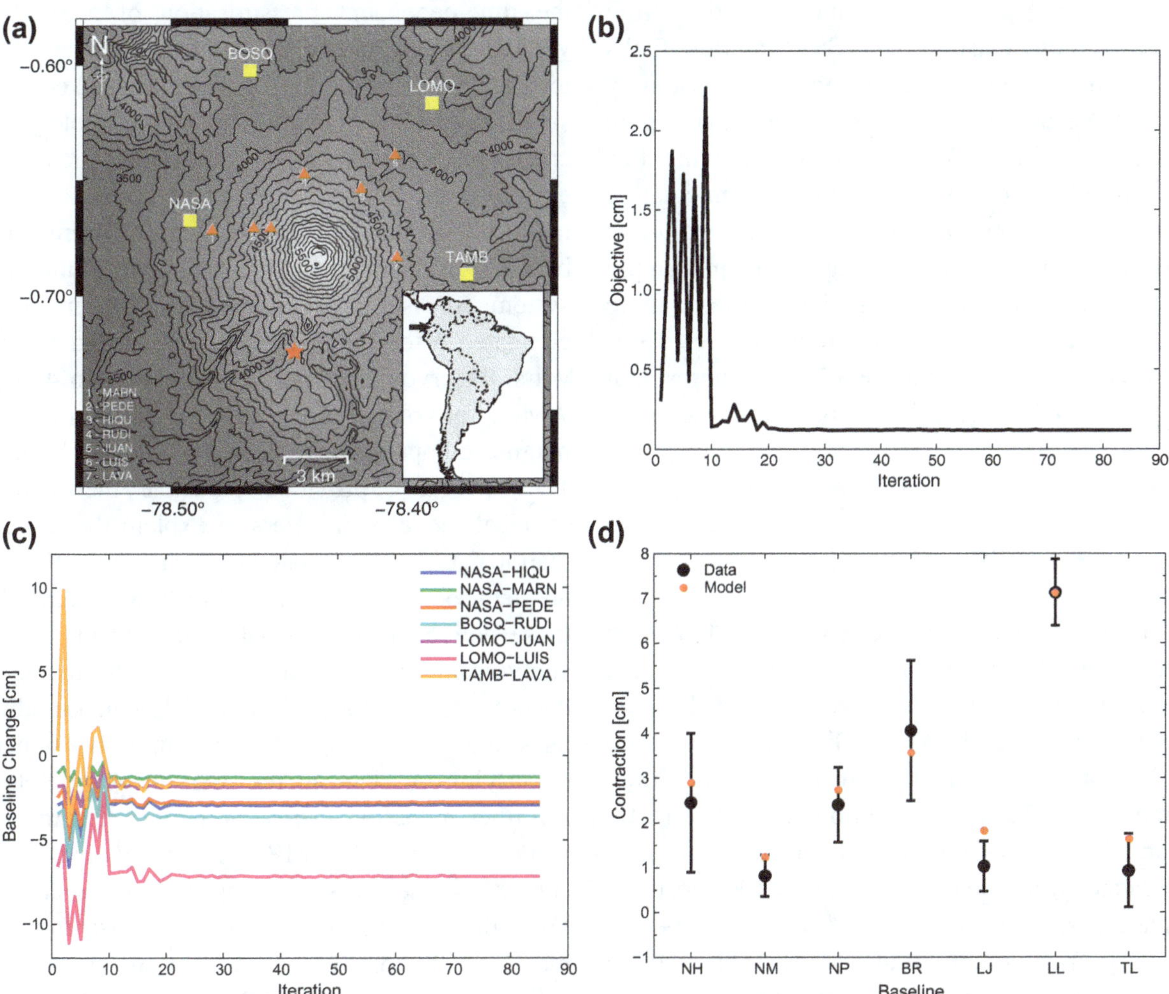

Fig. 2 Deformation and modelling results from Cotopaxi volcano. **a** Map of the EDM network operational between 2001 and 2002 around the summit of the volcano, with the best-fit source from the Finite Element inversions indicated by a *red star*. The *yellow squares* represent the EDM base-stations and the *orange triangles* are the reflecting prisms. The *inset* map and *arrow* shows the location of Cotopaxi within Ecuador, South America. **b** The variation, and eventual reduction, of the misfit objective function with each iteration from the final, best-fit Finite Element inversion model. **c** The modelled values from the final inversion for the EDM baseline changes converge to their near-absolute values after approximately 25 iterations. **d** The best-fitting model (*red circles*) fits six out of seven of the EDM observations (*black circles with error bounds*)

EDM measurements had a very poor fit to the observed data.

To clarify the difference in source location between the seismic and geodetic studies, Hickey et al. (2015) applied the best-fit source parameters from the elastic inversion models in a suite of temperature-dependent viscoelastic forward models to assess the rheology of the host-rock for a range of thermal parameters. The results indicated the most likely subsurface conditions would have promoted a large component of viscous deformation and that the deformation source in the SW would have consequently been pressurised aseismically. Fluid migration from the SW along existing NNE-SSW trending faults could have then caused the observed seismicity in the NE due to mass transport and excess pore pressures. The lack of eruption following this 2001–2002 unrest period and the aseismic nature of the event suggests that the magma supply rate for this period was low. A higher magma supply rate during a future unrest period would be more likely to produce seismicity around the reservoir, and could possibly indicate a level of unrest that signifies an increased likelihood of a forthcoming eruption.

Soufrière Hills

Starting its latest eruptive episode in 1995 and displaying a remarkable diversity of eruptive activity, Soufrière Hills volcano (SHV) on Montserrat (British West Indies) is one volcano where a correlation between observed deformation and subsequent eruption can be directly established. Ground deformation data at SHV indicate cyclic behaviour of the andesitic magmatic system from periods of several hours to a few years (Odbert et al. 2014). This section focuses on the analysis of intra-eruptive unrest associated with island-wide ground uplift observed via a network of continuous GPS receivers between 28/07/2003 and 01/08/2005 (Odbert et al. 2014). After a major lava dome collapse in 2003, which marked the end of the second phase of dome extrusion at SHV, this period preceded a restart of eruptive activity and renewed dome extrusion in August 2005.

Gottsmann and Odbert (2014) developed numerical models to test for the influence of temperature- and time-dependent stress evolution in a mechanically heterogeneous crust to explain the deformation data. Full details on the model setup and parameter derivation are given in Gottsmann and Odbert (2014) and not repeated here. They implemented two types of magma reservoir model. The first explored a series of pressurising, vertically-stacked reservoirs as proposed by Hautmann et al. (2010), while the second explored the time-dependent pressurisation of a single, vertically-elongated reservoir. Due to a similar temperature distribution from both the stacked and single reservoirs, and similar resultant rheological crustal properties, both suites of models provided equally good fits to the observed ground deformation. The study could hence not discriminate between pressurisation in a magmatic plumbing system consisting of either a single vertically elongated reservoir or a series of stacked reservoirs. Reservoir pressure changes between 4 and 7 MPa, for volumes between 60 and 100 km^3 and magma compressibility between 4×10^{-11} and 1×10^{-9} Pa^{-1}, provided plausible thermomechanical model parameters to explain the deformation data. The associated magma volume fluxes are between 0.015 and 0.021 km^3/year and match those derived from thermal modelling of active sub-volcanic systems (Annen 2009). Introducing a deep-crustal hot zone in the model, which modulates the partitioning of strain into the hotter underlying crust beneath the reservoir(s), promotes a further reduction in reservoir overpressures to values of around 1–2 MPa upon reservoir failure (Fig. 3). These pressure changes are significantly lower than those derived from models assuming a mechanically homogeneous and elastic crust. The deduced overpressures match those for sudden and rapid transcrustal reservoir activation prior to explosions at SHV from the analysis of volumetric strain data (Hautmann et al. 2014). The emerging eruption model at SHV hence involves the periodic failure of a compressible magma mush column beneath the volcano.

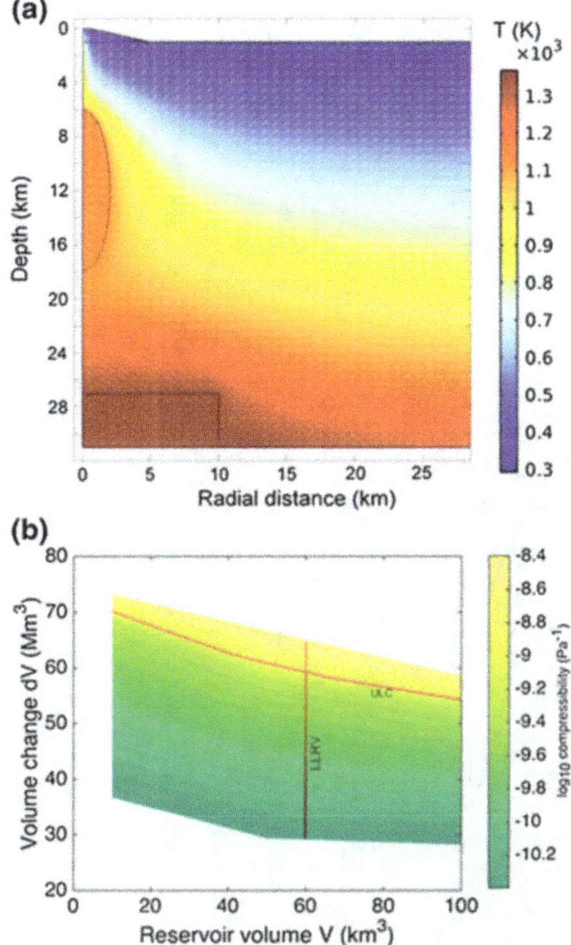

Fig. 3 **a** The temperature distribution (in K) of the 2-D axisymmetric model domains, caused by a combination of a basal heat flux, the heat from the plumbing system (magma reservoir plus feeder pipe), and a deep-crustal hot zone. Background thermal distribution in the far field is caused exclusively by a basal heat flux of 0.09 W/m^2. *Arrows* indicate deformation of the medium by reservoir pressurisation. Substantial deformation is accommodated by hot and ductile parts of the mid and lower crust. **b** The model parameter space for reservoir volume, volume change, and magma compressibility (shown in log units on the *colour bar*) from the best fit model solutions to the 2003–2005 intra-eruptive deformation time series at Soufrière Hills Volcano. The overall fit quality decreases substantially toward higher compressibility (order 10^{-9} Pa^{-1}) and toward smaller reservoir volumes (<50 km^3). The preferred parameter space is encompassed between the mapped upper limit of compressibility (*ULC*) and the lower limit of the magma reservoir volume (*LLRV*)

Las Cañadas

The Las Cañadas caldera (LCC) hosts the central volcanic complex (CVC) on Tenerife (Canary Islands), which includes the twin-volcanoes of Pico Viejo (PV) and Pico Teide (PT). Volcanic unrest on Tenerife started in April 2004 and ended a period of quiescence after the most recent magmatic eruption on the island in 1909. The unrest was geophysically characterised by anomalous seismic activity including a number of felt earthquakes and significant changes in the acceleration of gravity, yet, an absence of significant ground deformation (Gottsmann et al. 2006b). The initial interpretation of the gravity changes by Gottsmann et al. (2006b) pointed towards shallow (~ 2 km depth) fluid migration as the main source of the unrest, possibly related to a magma intrusion at greater depth. As part of the VUELCO project, Prutkin et al. (2014) revisited the microgravity data recorded between May 2004 and July 2005 and inverted the spatio-temporal residual gravity changes for the gravitational attraction of three-dimensional line segments.

The line segments are defined by 7 parameters: two end point coordinates (X, Y, Z) and a line strength (see Table 2). The latter is a proxy for the amplitude of sub-surface mass change, whereby a "weak" line segment represents a small amount of added mass compared to a "strong" line segment of greater added mass. An initial non-linear inversion yielded three line segments of different strengths at depths between 1 km a.s.l. and 2 km b.s.l. (i.e. between 1 and 4 km beneath the surface). Two of the segments were located to the NW of the PV-PT complex, while a third segment was located below the SW rim of the LCC (sources marked Sh 1–3 in Fig. 4). The locations of the segments correspond broadly to zones of heightened seismic activity and the unrest source location identified earlier by Gottsmann et al. (2006b).

Table 2 UTM coordinates X (Easting in m), Y (Northing in m), and Z (elevation in m with respect to sea level) of the end points of the modelled shallow (Sh) and deep (Dp) sources represented by line segments (see Fig. 4)

Line segment	X1	Y1	Z1	X2	Y2	Z2	ΔM
Sh1	327,500	3,137,930	1280	330,330	3,137,260	600	0.31
Sh2	336,560	3,132,050	590	334,210	3,134,040	420	0.83
Sh3	335,460	3,120,730	1380	332,100	3,121,210	380	0.53
Dp1	334,890	3,133,910	−5440	334,630	3,134,200	−5680	7.25
Dp2	334,310	3,134,590	−5840	334,630	3,134,200	−5680	8.24

Mass additions (ΔM) are given in 10^{10} kg. Data Prutkin et al. (2014)

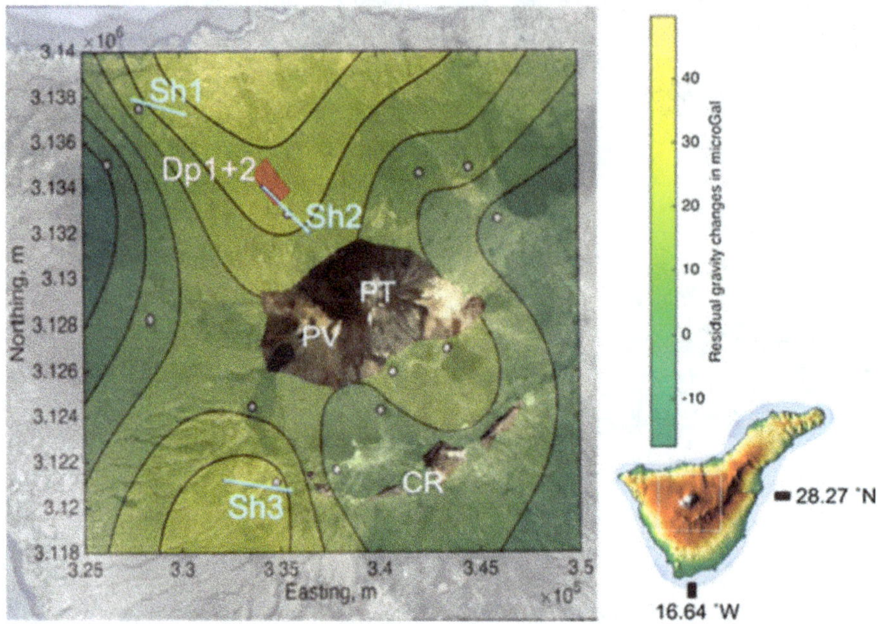

Fig. 4 Location of line segments from gravity data inversion. The figure shows the surface projections of three shallow sources (line segments Sh1–3 in *turquoise*) and the two deep sources (line segments Dp1 and 2 in *red*) superimposed over a Google Earth image of the Central Volcanic complex of Tenerife island (Spain). The source locations are derived by a nonlinear inversion of spatiotemporal residual gravity changes shown by coloured contours (in µGal) observed between 2004 and 2005 (see Prutkin et al. (2014) for details on inversion routine). The thickness of a line segment is indicative of the "strength" of the deduced mass change; i.e., the amount of added mass. The mass added to the deep sources is more than 10 times higher than to the shallow sources. Line segments Sh1–3 are interpreted to represent near-surface sources in the NW and SW part of the Pico Teide (PT; 3718 m a.s.l.) and Pico Viejo (PV; 3135 m a.s.l.) volcanic complex, associated with fluid migration as a result of an intrusion of magma at around 5.8 km b.s.l. (Dp sources). See Table 2 for details on segments and associated mass changes. The southern caldera rim (*CR*) is shown for reference and the *inset* shows a digital elevation model of Tenerife, with the study area identified by a *white rectangle*

However, closer inspection of the initial inversion results revealed that the line segments represent the superposition of deep and shallow seated sources. The decomposition of the gravity change data into shallow and deep fields (see Prutkin et al. 2014 for details) provided the basis for the separate inversions of the two fields using the three-dimensional line segment approximation. The deep field inversion constrained two connected and strong line segments at a depth of about 5.8 km b.s.l. (marked Dp1 and Dp2 in Fig. 4), while the inversion of the shallow field identified three similarly weak line segments situated at near-surface depths (<2 km from ground surface; see Fig. 4 and Table 2).

The most plausible interpretation of the inversion results is that the weak line segments represent sources dominated by hydrothermal fluids. In contrast, the deeper-seated sources can be interpreted as parts of an intrusion of new magma. This intrusion may have released fluids which consequently migrated towards shallower depths where they excited a shallow hydrothermal system. The geophysical signals resulting from these coupled magmatic-hydrothermal processes point towards a hybrid source nature for the unrest on Tenerife in 2004–2005. The identified link between deep and shallow unrest sources suggests the presence of permeable pathways for shallow fluid migration at the CVC.

Discussion

The Effect of Crustal Mechanics on Stress, Strain and Pressure

The presented investigations at Uturuncu, Cotopaxi and SHV all incorporate subsurface heterogeneity, however the effects are somewhat different. At Cotopaxi the final inferred deformation source is shallow, at a level where the model does not incorporate substantial heterogeneity due to the limited amount of available seismic data. Hence, with this model configuration, the heterogeneity has not played a significant role in altering the location of the source. It is likely, however, that in reality the volcanic edifice and shallow subsurface does have a certain level of heterogeneity. It is therefore possible that had this been taken into account the inferred deformation source might be located more centrally beneath the edifice. Conversely, at Uturuncu and SHV subsurface heterogeneity played a crucial role in determining the deformation source locations. In both cases, the inclusion of vertical layering in the Young's Modulus distribution altered the inferred depth of the source; the heterogeneous models predict deeper sources than the generic homogeneous analytical models.

From this, it also follows that a horizontal variation in mechanical properties would influence the horizontal location of a deformation source (e.g., Hickey et al. 2016).

A second effect of subsurface heterogeneity relates to source pressure requirements. It is common with homogeneous crustal mechanics to require unrealistically high source pressurisation (>100's MPa) when attempting to fit an observed deformation signal, yet there is no petrological or mechanical evidence for such conditions. Theoretical work from a mechanical viewpoint suggests that the maximum overpressure a source can sustain without failing is equal to the tensile strength of the host rock (e.g., Gudmundsson 2011). Above this limit, a magmatic reservoir would trigger a dyke intrusion. Work at Uturuncu and SHV shows that elastic heterogeneous deformation models bring source over-pressure requirements more in line with both in situ and laboratory values of tensile strength. This is due to a relative reduction in the average Young's Modulus above the source, compared to a higher homogeneous Young's Modulus through the entirety of the crust, and thus a greater amount of deformation for a given pressure increment. Furthermore, with a maximum value for the source over-pressure, constraints can be placed on the minimum size of a deformation source, given the two are directly linked. This allows, for example, an estimate to be placed on the volume of a magma reservoir. When multiple magmatic sources might be present numerical models can additionally account for stress interactions between the two, something not considered by generic analytical models (Pascal et al. 2013).

As a next step to subsurface heterogeneity, the case studies at Cotopaxi and SHV incorporated temperature-dependent mechanics to evaluate the effect of a viscoelastic rheology. At Uturuncu, investigations were also carried out using a viscoelastic rheology without temperature-dependence. In all three cases, a viscoelastic medium reduced the over-pressure requirements further when compared to the purely elastic models. This is due to the viscous expansion that follows an initial elastic inflation. The effect of thermomechanics is greater than just modifying source over-pressure requirements, however. Prolonged magma emplacement over thousands to millions of years in active volcanic areas builds up a significant thermal legacy within the crust. This results in elevated geothermal gradients, and in the case of continued active magmatism, deep crustal hot-zones (Annen 2009). These thermal perturbations are significant for the transfer of stress and strain. For example, at SHV, the combined thermomechanical effects of a deep-crustal hot zone and hot encasing rocks around a mid-crustal andesitic reservoir fundamentally alter the time-dependent subsurface stress and strain partitioning upon priming of the magma reservoir. These effects substantially influence surface strains recorded by volcano geodetic monitoring.

Hybrid Unrest and Source Characterisation

An intrusion of magma commonly leads to the exsolution of fluids upon decompression, and fluid migration (and accumulation) in itself can produce measurable geodetic surface signals (e.g., Fournier and Chardot 2012; Rouwet et al. 2014, and references therein). A deep intrusion of magma, such as the 2004–2005 unrest on Tenerife, may not necessarily lead to observable surface deformation, meaning deformation data on its own can not provide any meaningful insights on the source process(es). In this case, the combination of deformation and gravity surveys allowed the characterisation of the unrest sources in much greater detail owing to their density contrasts, relative depths and mass additions. This highlights that, especially for the case of hybrid mechanisms where both magmatic and hydrothermal components are present, multi-disciplinary geodetic surveys can provide valuable information on source characterisation to help distinguish the processes driving unrest. This identification and discrimination of sources driving volcanic unrest via mathematical modelling of surface data is of vital importance for hazard and risk characterisation, given the different connotations associated with magmatic

and non-magmatic unrest, and plays a major role in probabilistic eruption forecasting (Rouwet et al. 2014).

Application to Eruption Forecasting

The primary objective of the case studies was to develop advanced geodetic models to interpret spatial and temporal deformation monitoring signals, and provide better constraints on the subsurface processes causing volcanic unrest. This has been achieved by incorporating more plausible model components, such as subsurface heterogeneity, topography and temperature-dependent mechanics, to relax the assumptions that hinder analytical models and maintain consistency with inferences from geophysics, geology and petrology. Crucially, this highlighted the importance of pressure-time functions and inelastic rheology in deciphering temporal deformation patterns. In turn, a more thorough understanding of how a volcanic system behaves through time will benefit eruption forecasting, as quantitative estimates of key parameters such as magma supply rate and mechanism can be deduced.

When only considering the spatial deformation pattern, this work has further demonstrated some of the pitfalls associated with models assuming homogeneous, elastic, half-spaces (e.g., Mogi 1958). Crustal heterogeneity significantly effects the horizontal and vertical location of a deformation source by altering the subsurface strain distribution. Surface strain partitioning by complex topography is equally important, such as at steep-sided stratovolcanoes. The Cotopaxi case highlighted complex partitioning behaviour along deep ravines and adjacent lava flow ridges. If ground-based geodetic monitoring sites are positioned at localities with steep topographic gradients, surface strain partitioning plays a first-order role and needs to be accounted for during data modelling. Consequently, estimates of source parameters from generic, analytical models which cannot account for heterogeneities or complex topography may skew eruption forecasting and risk mitigation efforts. On the other hand, more reliable deformation source locations from numerical models can be used to better estimate where and when an eruption might occur, and the associated hazard. It should also be pointed out that data limitations at some volcanoes may prevent complex numerical models being used, in which case simpler models may be the only option and their results should be carefully scrutinised and only applied cautiously to further analyses.

Conclusions

The case studies provide new interpretations of volcanic processes during intra-eruptive and non-eruptive unrest. They also provide observations for distinct magmatic settings, thus contributing to the ongoing global comparisons of deformation and unrest, and the process of pattern recognition for identification of eruption precursors. The incorporation of multi-disciplinary data into integrated geodetic models has closed the gap between observations and interpretations of volcanic deformation and gravimetric changes. This will help to improve eruption forecasting as it moves away from a qualitative approach towards incorporation of more quantitative data derived from well-constrained physical mechanisms.

Acknowledgements Work presented herein has received funding by the European Commission (FP7; Theme: ENV.2011.1.3.3-1; Grant 282759: VUELCO). We are very grateful to Alvaro Guevara and Irving Munguía Gonzalez for their help translating Spanish, and to two anonymous reviewers for their constructive comments.

Glossary Terms

Deformation the action of changing the shape of a physical object (e.g., a volcano) under the influence of some stress.

Elastic rheological behaviour in which an applied stress causes an immediate strain that is 100% recoverable when the stress is removed. See also rheology and viscoelastic.

Finite Element Method/Analysis a numerical modelling technique that subdivides an entire problem into a set of smaller 'elements'. Mathematical problems are then solved in each of the elements and combined together to calculate the response of the whole object.

Geodesy applied mathematics with the aim of measuring the geometric spatial representation of the Earth and its gravitational field in time-varying three-dimensions, as well as their orientation in space. Volcano geodesy specifically deals with recording the spatial and temporal patterns of crustal deformation and gravimetric changes in volcanic settings.

Gravimetry field of geophysics devoted to observing, processing and interpreting minute changes in the Earth's gravitational field.

Hydrothermal system an area in which heat and fluids from a partially molten magmatic body interact with a multi-phase groundwater system, causing chemical and thermal (heating) perturbations to the water.

Mechanics (crustal, thermal) the branch of physics that studies how stress can effect a physical object. Crustal mechanics relates to the values of the material parameters that describe the mechanical behaviour of the Earth's crust. Thermal mechanics relates to the variation in mechanical material properties due to changes in temperature.

Model (inverse, forward) a simulation of a process. Inverse models solve 'backwards' to determine optimal model parameters that fit a set of data. Forward models use predefined constant input parameters to calculate the expected model response.

Rheology (crustal) the behavioural response of the Earth's crust to forces that act upon or within it. See also elastic and viscoelastic.

Strain the change in dimension of an object (e.g., ΔX) relative to the original dimension of the object (e.g., X). It has no units, as the units cancel: $\Delta X / X$.

Stress a measure of force per unit area, with the unit of pascals, Pa. $1\ Pa = 1\ N/m^2$.

Viscoelastic rheological behaviour in which an applied stress causes an elastic response, followed by a delayed (viscous) response. See also rheology and elastic.

Index Terms

Magma reservoir, deformation, gravity, GPS, InSAR, Cotopaxi, Soufrière Hills, Las Cañadas, Uturuncu, unrest, thermomechanics, modelling, finite element, inversion/inverse model, crustal mechanics, hybrid unrest, Young's Modulus, Poisson's Ratio, viscosity.

References

Annen C (2009) From plutons to magma chambers: thermal constraints on the accumulation of eruptible silicic magma in the upper crust. Earth Planet Sci Lett 284(3–4):409–416

Battaglia M, Gottsmann J, Carbone D, Fernández J (2008) 4D volcano gravimetry. Geophysics 73(6):WA3–WA18

Brocher TM (2005) Empirical relations between elastic wavespeeds and density in the Earth's crust. Bull Seismol Soc Am 95(6):2081–2092

del Potro R, Dez M, Blundy J, Camacho AG, Gottsmann J (2013) Diapiric ascent of silicic magma beneath the Bolivian Altiplano. Geophys Res Lett 40(10):2044–2048

Fialko Y, Pearse J (2012) Sombrero uplift above the Altiplano-Puna magma body: evidence of a ballooning mid-crustal diapir. Science 338(6104):250–252

Fournier N, Chardot L (2012) Understanding volcano hydrothermal unrest from geodetic observations: insights from numerical modeling and application to White Island volcano, New Zealand. J Geophys Res 117(B11):B11208

Gottsmann J, Odbert HM (2014) The effects of thermo-mechanical heterogeneities in island-arc crust on time-dependent pre-eruptive stresses and the failure of an andesitic reservoir. J Geophys Res. doi:10.1002/2014JB011079

Gottsmann J, Rymer H, Berrino G (2006a) Unrest at the Campi Flegrei caldera (Italy): a critical evaluation of source parameters from geodetic data inversion. J Volcanol Geoth Res 150(1–3):132–145

Gottsmann J, Wooller L, Marti J, Fernandez J, Camacho AG, Gonzalez PJ, Garcia A, Rymer H (2006b) New evidence for the reawakening of Teide volcano. Geophys Res Lett 33(20)

Gudmundsson A (2011) Rock fractures in geological processes. Cambridge University Press, New York

Hautmann S, Gottsmann J, Sparks RSJ, Mattioli GS, Sacks IS, Strutt MH (2010) Effect of mechanical heterogeneity in arc crust on volcano deformation with application to Soufrière Hills volcano, Montserrat, West Indies. J Geophys Res 115(B9):B09203

Hautmann S, Witham F, Christopher T, Cole P, Linde AT, Sacks IS, Sparks RSJ (2014) Strain field analysis on Montserrat (W.I.) as tool for assessing permeable flow paths in the magmatic system of Soufriere Hills Volcano. Geochem Geophys Geosyst 15:2013G. doi:10.1002/C005087

Hickey J, Gottsmann J (2014) Benchmarking and developing numerical finite element models of volcanic deformation. J Volcanol Geoth Res 280:126–130

Hickey J, Gottsmann J, del Potro R (2013) The large-scale surface uplift in the Altiplano-Puna region of Bolivia: a parametric study of source characteristics and crustal rheology using finite element analysis. Geochem Geophys Geosyst 14(3):540–555

Hickey J, Gottsmann J, Mothes P (2015) Estimating volcanic deformation source parameters with a finite element inversion: the 2001–2002 unrest at Cotopaxi volcano, Ecuador. J Geophys Res Solid Earth 120 (3):1473–1486

Hickey J, Gottsmann J, Nakamichi H, Iguchi M (2016) Thermomechanical controls on magma supply and volcanic deformation: application to Aira caldera, Japan. Sci Rep 6(August):32691

Masterlark T (2007) Magma intrusion and deformation predictions: sensitivities to the Mogi assumptions. J Geophys Res 112(B06419)

Mogi K (1958) Relations between the eruptions of various volcanoes and the deformations of the ground surfaces around them. Bull Earthq Res Inst 36:99–134

Molina I, Kumagai H, Garca-Aristizábal A, Nakano M, Mothes P (2008) Source process of very-long-period events accompanying long-period signals at Cotopaxi Volcano, Ecuador. J Volcanol Geothermal Res 176 (1):119–133

Odbert HM, Stewart RC, Wadge G (2014) Cyclic Phenomena at the Soufrière Hills Volcano, Montserrat, The Eruption of Soufrière Hills Volcano, Montserrat from 2000 to 2010, vol 39. The Geological Society, London, pp 41–60. doi:10.1144/M39.2

Pascal K, Neuberg J, Rivalta E (2013) On precisely modelling surface deformation due to interacting magma chambers and dykes. Geophys J Int 196 (1):253–278

Pistolesi M, Cioni R, Rosi M, Cashman KV, Rossotti A, Aguilera E (2013) Evidence for lahar-triggering mechanisms in complex stratigraphic sequences: the post-twelfth century eruptive activity of Cotopaxi Volcano, Ecuador. Bull Volcanol 75(3):698

Pritchard ME, Simons M (2002) A satellite geodetic survey of large-scale deformation of volcanic centres in the central Andes. Nature 418(6894):167–171

Prutkin I, Vajda P, Gottsmann J (2014) The gravimetric picture of magmatic and hydrothermal sources driving hybrid unrest on Tenerife in 2004/5. J Volcanol Geoth Res 282:9–18

Ranalli G (1995) Rheology of the Earth. Chapman and Hall, London

Rouwet D, Sandri L, Marzocchi W, Gottsmann J, Selva J, Tonini R, Papale P (2014) Recognizing and tracking volcanic hazards related to non-magmatic unrest: a review. J Appl Volcanol 3:1–17

Sparks RSJ, Folkes CB, Humphreys MC, Barfod DN, Clavero J, Sunagua MC, McNutt SR, Pritchard ME (2008) Uturuncu volcano, Bolivia: volcanic unrest due to mid-crustal magma intrusion. Am J Sci 308(6):727–769

Sparks RSJ, Biggs J, Neuberg JW (2012) Monitoring volcanoes. Science 335(6074):1310–1311

Fluid Geochemistry and Volcanic Unrest: Dissolving the Haze in Time and Space

Dmitri Rouwet, Silvana Hidalgo, Erouscilla P. Joseph and Gino González-Ilama

Abstract

The heat and gas released by a degassing magma affects the overlying predominantly meteoric aquifers to form magmatic-hydrothermal systems inside the solid body of a volcano. This chapter reviews how fluid geochemical signals help to track the evolution throughout the various stages of volcanic unrest. A direct view into a degassing magma is possible at open-conduit degassing volcanoes. Nevertheless, in most cases gas is trapped (i.e. scrubbed) by abundant water, leading to the loss of the pure signal the magma ideally provides. Deciphering how magmatic gas rises through, reacts, and re-equilibrates with the liquids in the magmatic-hydrothermal system in time and space is the only way to trace back to the pure signal. The most indicative magmatic gas species (CO_2, SO_2–H_2S, HCl and HF) are released as a function of their solubility in magma. The less soluble gas species are released early from a magma at higher pressure conditions (CO_2) (deeper), whereas the more soluble species are released later, at lower pressures (SO_2, HCl and HF) (shallower depth). When these gases hit the water during their rise towards the surface, they will be more or less scrubbed. Depending on the chemical equilibria inside the magmatic-hydrothermal system (e.g. SO_2–H_2S conversion, acidity), the gas that eventually reaches the surface will carry the history of its rise from bottom to top. Tracking volcanic unrest implies a time frame; the kinetics of magma degassing throughout the liquid cocktail inside the volcano impose the maximum resolution the volcano provides and hence the monitoring time window to be adopted for

D. Rouwet (✉)
Istituto Nazionale di Geofisica e Vulcanologia, Sezione di Bologna, Via Donato Creti 12, 40128 Bologna, Italy
e-mail: dmitri.rouwet@ingv.it

S. Hidalgo
Instituto Geofisico, Escuela Politécnica Nacional, Quito, Ecuador

E.P. Joseph
Seismic Research Centre, University of the West Indies, St. Augustine, Trinidad & Tobago

G. González-Ilama
Centro de Investigaciones en Ciencias Geológicas, Universidad de Costa Rica, San José, Costa Rica

Advs in Volcanology (2019) 221–239
DOI 10.1007/11157_2017_12
© The Author(s) 2017
Published Online: 19 July 2017

each volcano. Gas-dominated systems are "faster" and require a higher monitoring frequency, water-dominated systems are slower and require a lower monitoring frequency.

Resumen

El calor y gas liberados por la desgasificación del magma afecta los acuíferos de origen predominantemente meteórico para formar sistemas magmático-hidrotermales dentro el cuerpo sólido del volcán. Este capítulo revisa como la geoquímica de fluidos puede ayudar a trazar la evolución a través de las varias etapas de "unrest" volcánico. Una visión directa dentro de un magma en desgasificación es solo posible para volcanes de conducto abierto. Sin embargo, en la mayoría de las situaciones el gas queda atrapado (i.e. "scrubbing") en el agua, que conduce a la pérdida de la señal pura que el magma idealmente puede proporcionar. Descifrar como el magma sube a través de los líquidos, y reacciona y re-equilibra con ellos dentro el sistema magmático-hidrotermal, en un marco de tiempo y espacio, es la única manera para rastrear el origen de la señal del magma. La desgasificación de magma se da por cuatro procesos: (1) durante la subida de magma, (2) por la descompresión debido al eliminar una porción del edificio volcánico, (3) debido a la convección interna dentro la cámara magmática, o (4) después de "ebullición secundaria" siguiendo el enfriamiento y consecuente cristalización. Las especies gaseosas magmáticas más indicativas (CO_2, SO_2–H_2S, HCl y HF) se liberan en función de su solubilidad en el magma. Las especies menos solubles se liberan antes del magma, bajo regímenes de presiones más altas (CO_2), mientras que las especies más solubles se liberan después, bajo regímenes de presiones más bajas (SO_2, HCl y HF). En términos espaciales, CO_2 se libera a lo largo de una área espacial más amplia (desde lo más profundo). La presencia de SO_2 es una indicación clara de un magma que sube hacia un ambiente más somero. La llegada de HCl en la superficie generalmente indica la presencia de una remesa de magma somera (cientos de metros hasta pocos kilómetros). Especialmente un aumento en la proporción CO_2/SO_2 es indicativo para elucidar un estado de "inquietud" ("unrest"). Una disminución consecutiva en el CO_2/SO_2, después de un aumento, es una indicación de que el magma está cerca de la superficie y es propenso a una erupción. Cuando estos gases alcanzan el agua durante su ascenso hacia la superficie, serán más, o menos, absorbidos. Dependiendo de los equilibrios químicos dentro el sistema magmático-hidrotermal (e.g. conversión de SO_2–H_2S, acidez), el gas que al final llega a la superficie lleva consigo la historia de su ascenso desde el fondo hasta la superficie. La desgasificación magmática es un proceso más rápido, mientras que la dinámica hidrotermal en el sistema rocoso FeO–$FeO_{1.5}$ es más lenta. Por eso, el H_2S se suele llamar un "gas hidrotermal", y el SO_2 un "gas magmático". Trazar "unrest volcánico" implica un encuadramiento de tiempo más especifico. Si la ventana de tiempo de monitoreo es más largo que el tiempo definido por la cinética de la migración de gas, los detalles de la dinámica de desgasificación se perderán inevitablemente. Contrariamente, si la ventana de tiempo de monitoreo es más corto que la ventana definida por la cinética de la migración de gas, en tal caso, resultará en una visión demasiado detallada de lo que el sistema magmático-hidrotermal puede proveer. Estudios recientes han demostrado que acuíferos extremadamente ácidos pueden "desacelerar" la señal dejada por el sistema magmático-hidrotermal dominado por el gas (e.g. fumarolas), pero pueden "acelerar" la señal dejada por el sistema magmático-hidrotermal dominado por agua (e.g. lagos cratéricos ácidos). Estos hallazgos tienen implicaciones significativas para el encuadramiento de tiempo en reconocer la desgasificación magmática, y por tanto, para la frecuencia del monitoreo.

Keywords

Fluid geochemistry · Magmatic-hydrothermal systems · Volcanic unrest · Volcano monitoring

Palabras clave

Geoquímica de fluidos · Sistemas magmático-hidrotermales · Inquietud volcánica · Monitoreo volcánico

Introduction

Fluid geochemical monitoring tracks variations in gaseous species and fluid phases released in various manners from a volcano. Fluids infiltrate, move, migrate, rise, react and re-equilibrate in the water- and vapor-filled solid body of the volcano. These processes are invisible as they occur in the subsurface, and can only be deduced from measurements at the surface. Nevertheless, gas and water are more mobile than rock and, when a volcano shifts into the gear of unrest, a change in degassing is often the first sign to be detected. As such, fluid geochemistry offers a crucial means to recognize unrest in a timely matter.

Most volcanic edifices store a large volume of water, as cold or thermal aquifers, with various hydrogeological architectures. When gas hits water in its rise towards the surface, the original signature of the gas is mostly lost. This disadvantage can however be overcome. Understanding how gas absorbs and solutes react in the liquid phase, and how it is eventually released to the atmosphere, is key to linking surface manifestations to magma dynamics.

Volcano monitoring largely focuses on how, where and when magma migrates towards the surface, as, intrinsically, every magmatic eruption is anticipated by magma rise. Nevertheless, volcanoes can become hazardous even without an eruption. A volcano is in a state of magmatic unrest if we recognize signals of a magma-on-the-move; in any other situation, given volcanic unrest, the volcano is in a state of non-magmatic unrest (hydrothermal or tectonic) (Rouwet et al. 2014a). To recognize volcanic unrest, the background behavior of a volcanic system should be tracked for a sufficiently long period, in order to know when a deviation from this background becomes a cause for concern (i.e. unrest, Phillipson et al. 2013). This background behavior and deviations from it are volcano-specific and can be monitored in several ways, besides the geochemical approach reviewed here.

The aim of this chapter is to scan through magmatic-hydrothermal systems during the process of magmatic degassing, from bottom to top, and describe how fluids behave in time and space. What are the lessons learned from fluid geochemistry throughout the evolution of volcanic quiescence, re-awakening, volcanic unrest, magmatic unrest and non-magmatic/hydrothermal unrest?

Magmatic-Hydrothermal Manifestations

One of the first signs of re-awakening after prolonged volcanic quiescence to a state that eventually causes concern is often the appearance of fumarolic exhalations from a crater. This happened, for instance, in 1994 during the reawakening of Popocatépetl (Mexico), and recently, at Cotopaxi (Ecuador), both VUELCO target volcanoes (De la Cruz-Reyna and Tilling 2008; Hall and Mothes 2008) (Fig. 1a). For open-conduit volcanoes the presence of a plume (i.e. a visible gas-vapor cloud originating from an open volcano crater) can become the prominent manifestation of degassing.

Some volcanoes are characterized by decade to century long high-temperature fumarolic degassing in a closed-conduit setting (Fig. 1b), suggesting the presence of a stable, but shallow magma chamber. This constantly high-temperature background degassing is generally no sign of unrest (e.g. Momotombo's >700 °C, Satsuma-Iwojima's >900 °C and Kudryavy's >700 °C; Menyailov et al. 1986; Shinohara et al. 1993; Taran et al. 1995), whereas the increase of fumarolic temperatures from low (boiling point of water at a given altitude, hence atmospheric pressure) to high (above boiling to magmatic temperature) can be a sign of resumed unrest (e.g. the 1980–1990s crisis at Vulcano, Italy, Capasso et al. 1997).

Magmatic-hydrothermal systems are aquifers inside a volcano or beneath a volcanic area, heated by a magma, at an unspecified depth. The origin of the water is generally meteoric (i.e. rain, snow and its melt water). How hot, and how gas-rich such magmatic-hydrothermal systems are depend on the proportion of the water volume with respect to the heat and gas provided by the magma. The latter depends on the residing depth of the magma. Boiling-temperature fumaroles are a common manifestation at magmatic-hydrothermal systems (Fig. 1c). When the water table of such systems intersects the surface, in craters or volcano flanks, boiling pools appear (Fig. 1d). Such pools can manifest bubbling degassing, and are nothing less than a water-rich fumarole (Fig. 1 e). Depending on the dominant gas they exhale, paired with water vapor, the manifestations are called solfataras (S-rich gases) or mofettes (CO_2-rich gases); depending on the temperature and vapor/water proportion they emit they are called fumaroles (boiling or above boiling steam vents), thermal springs (liquid water emission) or geysers (water + vapor jets with a cyclic behavior). Thermal springs can discharge inside active craters or on volcano flanks in a degassed state, without bubbling or boiling (Fig. 1f). Heated water can fill (parts of) craters and form volcanic lakes (Fig. 1g). Depending on the degassing state and depth of the underlying magma, degassing features (bubbling or diffuse degassing) and evaporation can occur at the lake surface (Fig. 1h).

The pictures of the degassing manifestations in Fig. 1 show a trend from gas-dominated, active plume degassing in an open-conduit setting towards more water-dominated, hydrothermal, passive degassing in a closed-conduit setting. These visual observations only give a first glance of magmatic-hydrothermal activity, and do not reflect the state of unrest of a volcano. Throughout the life-time of magmatic-hydrothermal systems (centuries to millenia) volcanoes can evolve from gas-dominated to water-dominated, and vice versa. The next sections present what we know on the theoretical level, following the laws of chemistry. Despite these classic rules, it will become clear that the range of manifestations and variations in fluid signatures is wide, and volcano-dependent.

Fig. 1 Degassing manifestations at magmatic-hydrothermal systems. **a** Open-conduit degassing at Popocatépetl. **b** High-temperature fumarolic degassing at Vulcano. **c** Boiling-temperature fumarolic degassing at the Arbol Quemado fracture in the Turrialba crater area. **d** Boiling pools inside the El Chichón crater. **e** Bubbling thermal spring at the SE flank of El Chichón. **f** Thermal flank spring at El Chichón. **g** The El Chichón crater lake. **h** Evaporative degassing from Laguna Caliente, Poás. The *cyan arrow* points towards water-dominance; the *yellow arrow* towards more gas-dominance

Magma Degassing from Bottom to Top

Magma Degassing

The degassing of a magma increases when the confining pressure in the magma decreases. Magma decompression can occur in four ways (Fig. 2): (1) magma rise, (2) decompression by uncovering a portion of the volcanic edifice, (3) internal convection in a magma chamber, or (4) secondary boiling upon cooling and consequent crystallization. The first process is often induced by the input of a deeper magma, into a shallower magma chamber. Magma rise towards the lower pressure regime results from the buoyancy difference between the stagnant magma, the rising new melt and the surrounding rocks. The second degassing process can be triggered by the mass removal from part of the volcanic edifice. This superficial process can even trigger eruptions (e.g. 1980 Mt St. Helens). Internal magma convection causes degassing of a less dense, gas-rich magma batch. Once degassed, the now denser magma batch sinks (e.g. Stromboli, Aiuppa et al. 2009). The fourth process increases the gas/melt ratio in the magma, due to the loss of crystals from the cooling magma, hence favoring degassing.

Based on gas geochemistry only it is hard to rule out which process actually occurs at depth, and thus distinguish between magmatic and non-magmatic unrest. Geophysical signals are needed. Following the definition of magmatic unrest by Rouwet et al. (2014a) (i.e. the recognition of a magma-on-the-move), only the first process is initially consistent with the requisite of magmatic unrest.

Variations in magma degassing can be detected both qualitatively and quantitatively. Detailed insights in the degassing state of a magma can only be obtained if both are measured contemporaneously. Which gas species are released, how much and when? When a magma starts to degas, by any of the above processes, the less soluble species is released first (i.e. at higher confining pressure in the magma chamber). The order in solubility of indicative magmatic gas species is $CO_2 < SO_2 < HCl < HF$; the order of release when a magma progressively degasses is "CO_2-first till HF-last" (Giggenbach 1987). Hence, tracking variations in ratios between these species gives qualitative insights into the degassing state of a magma. A consecutive increase with time in first CO_2/SO_2, then SO_2/HCl, then HCl/HF ratios reflects the evolution in degassing state from a magma moving from depth towards the surface. Especially an increase in the CO_2/SO_2 is indicative of the state of unrest, pointing to an input of poorly degassed magma at great depths. A consecutive decrease in CO_2/SO_2 ratio, after the increase, is an indication of magma moving towards the surface. The latter two ratios come into play when eruption of magma is imminent, or even ongoing: the highly soluble species HCl and HF are released from a highly degassed magma, a situation that reflects near-surface degassing (Aiuppa et al. 2002). The arrival of HCl at the surface (e.g. in

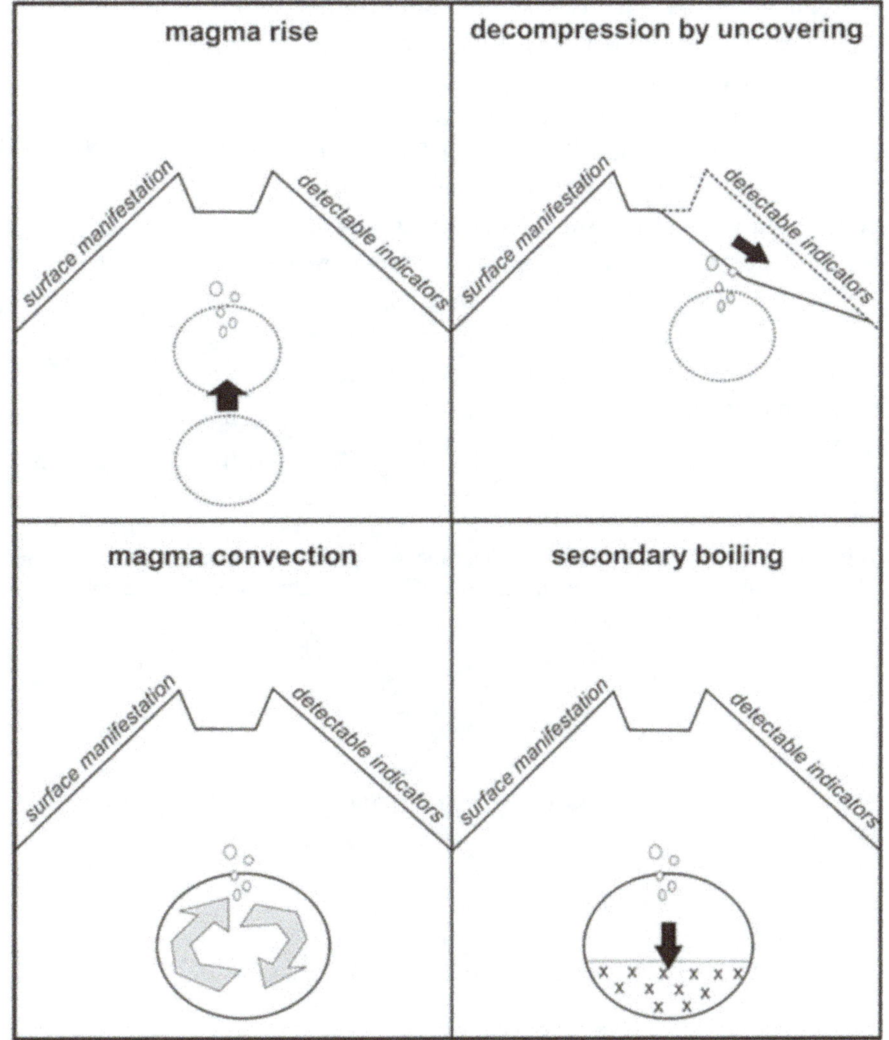

Fig. 2 Sketch of the four mechanisms that instigate magma degassing (not to scale)

fumaroles) generally indicates the presence of a shallow magma batch (hundreds of meters, or less).

The "purest" proxy of the magmatic gas is provided by direct sampling of a fumarole with a near-magmatic temperature, followed by the analysis of its chemical composition. During the past five-six decades fumarolic gases have been extensively sampled and analyzed representing a wide range of temperatures and states of volcanic activity. Table 1 presents a compilation of chemical compositions of fumarolic gases for the high- and low-temperature ranges.

Before dealing with the abundant water vapor in fumaroles, we tackle the other gas species in the "dry-gas phase". Regardless of the fumarole temperature, CO_2 is the most abundant gas species (Table 1). CO_2 is often released at the surface across a wider spatial extent, as it degasses from greater depth (Fig. 3). Old and deep magma bodies can continue to release CO_2 for tens to hundreds of millenia.

The second most abundant "dry-gas" species are the sulphur species SO_2 and H_2S. SO_2 is more soluble in magma than CO_2, and will thus be released at lower pressure. As SO_2 degasses at lower depth, the degassing tends to be more centralized along the central conduit (open or closed) of the volcano (Fig. 3). As such, SO_2 is often measured in volcanic plumes. An increase of SO_2 has often been interpreted as an indicator of magma rise into the shallower environment.

At high-temperature magmatic conditions the following reaction applies (Giggenbach 1987; Delmelle and Bernard 2015):

$$3SO_2 + 7H_2 = H_2S + 2S° + 6H_2O \quad (1)$$

This explains the presence of both SO_2 and H_2S in high-temperature fumaroles. The SO_2/H_2S ratio is sensitive to temperature and the oxidation state (i.e. the role of H_2 in Eq. 1 reflects the oxidation vs reduction state, Giggenbach 1987).

On the other hand, it is noted that the low-temperature fumaroles lack SO_2, instead, the dominant sulphur species is H_2S (Table 1). Low-temperature hydrothermal conditions (T < 300 °C) favor the reduced S-species H_2S (Table 1), equilibrated by the rock phase, following the reaction:

$$SO_2 + 6(FeO)_{rock} + H_2O = H_2S + 6(FeO_{1.5})_{rock} \quad (2)$$

or simply through reduction of SO_2 in the gaseous environment:

$$SO_2 + H_2 = H_2S + O_2 \quad (3)$$

To convert SO_2 into H_2S in a rock matrix (FeO–$FeO_{1.5}$-system), following reaction (2), a major constraint is "sufficient time". Magmatic degassing (Eq. 1) is a faster process, while hydrothermal dynamics (Eq. 2) are slower. This

Table 1 Chemical composition of fumarolic gases (concentrations in micromol/mol), for high (magmatic, 280°–1130 °C) and low (hydrothermal, 83°–160 °C) temperature conditions, expressed as the minimum and maximum measured concentrations for 12 and 59 samples, respectively

	T(°C)	H_2O	CO_2	SO_2	H_2S	HCl	HF	H_2	N_2	CH_4	CO
High-T minimum (#12)	280	311000	1200	320	4	275	21	3	40	0.1	0.2
High-T maximum (#12)	1130	993000	672000	67800	21460	14200	2500	14900	1800	7.1	4600
Low-T minimum (#59)	83	638200	2655	0	0	0	0	16	32	0.4	0.011
Low-T maximum (#59)	160	997200	355000	0	3700	0	0	1220	6800	5330	1.6

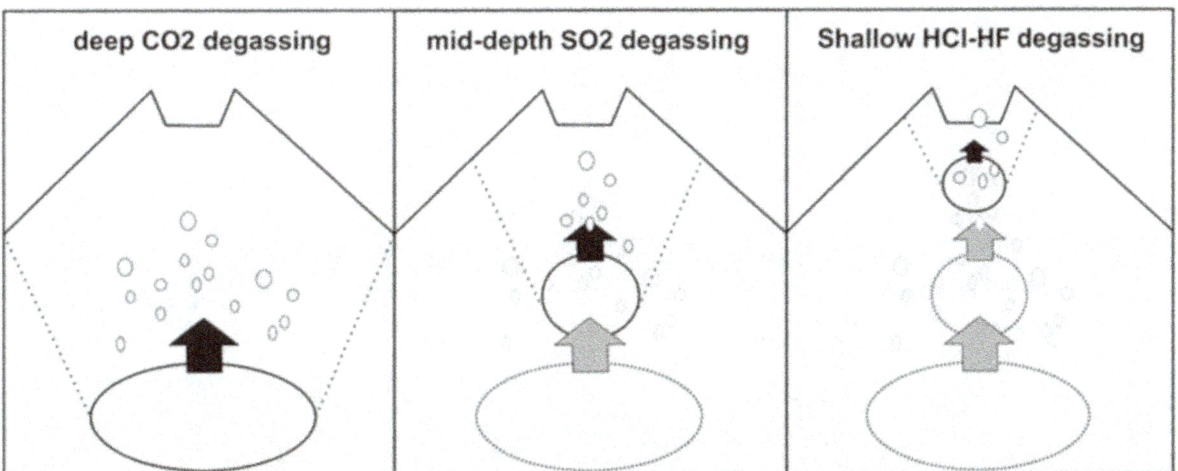

Fig. 3 Sketch of the passing "gas train" during magma degassing, resulting from the difference in solubility of the various gas species (not to scale)

is why H_2S is often called a "hydrothermal gas" and SO_2 a "magmatic gas" (Table 1).

As noted in Table 1, the major gas species in fumaroles is often water, regardless of their temperature. Even high-temperature fumaroles are water-dominated.

Knowing that water boils at 100 °C (at atmospheric pressure, at sea level), in theory, the temperature of a fumarole is buffered at 100 °C and cannot rise until water is exhausted in the underlying plumbing system. Moreover, the critical temperature of water is 374 °C (i.e. the temperature at which vapor and liquid water cannot coexist anymore), imposing a second temperature buffer. This implies that (1) water directly originates from a high-temperature magma under super-critical conditions ("andesitic water" with T > 374 °C, Taran et al. 1989), and/or (2) water is excessively present in the fumarole plumbing system with respect to the gas, and will hardly ever exhaust.

When the Gas Hits the Water

The shallow subsurface environment of the different sections of the Earth's crust hosts numerous aquifers at various depths originating from the infiltration and storage of meteoric water, or seawater in the case of low-lying islands; volcanic edifices are no different, being "small dots"

at the Earth's surface. As described before, a magma degasses and heats the space between the magma and the surface, and will inevitably heat and modify the volcanic aquifers. As the magma heats the overlying aquifers, between the magma and the surface, from bottom to top, a gas-only, vapor + gas zone, a vapor + liquid zone and a liquid-only zone can be found (Fig. 4). When a magma rises, or the heat input from the magma increases, the vapor + liquid zone or vapor + gas zone will be pushed upwards until intersecting the surface (e.g. in a volcano crater), manifested at the surface as boiling or bubbling pools and fumaroles (Fig. 4). If the distance between the magma and surface is larger, the thermal aquifer will intersect the surface and create thermal springs (Fig. 4). Facing unrest, a liquid to vapor transition at a surface manifestation reflects heating of the hydrothermal system.

When the above "gas train" (Fig. 3) consecutively reaches the liquid-only zone, gas species will be absorbed and react depending on their specific chemical properties in water. The capacity to absorb magmatic gases in the liquid phase is called "scrubbing" (Symonds et al. 2001). The CO_2 that reaches the water from greater depths during magma degassing (Fig. 3), will create CO_2-dominated bubbling thermal springs, and HCO_3-rich slightly acidic springs (pH 5–7) (Fig. 4). The second least soluble gas species that hits the water is SO_2 that will be

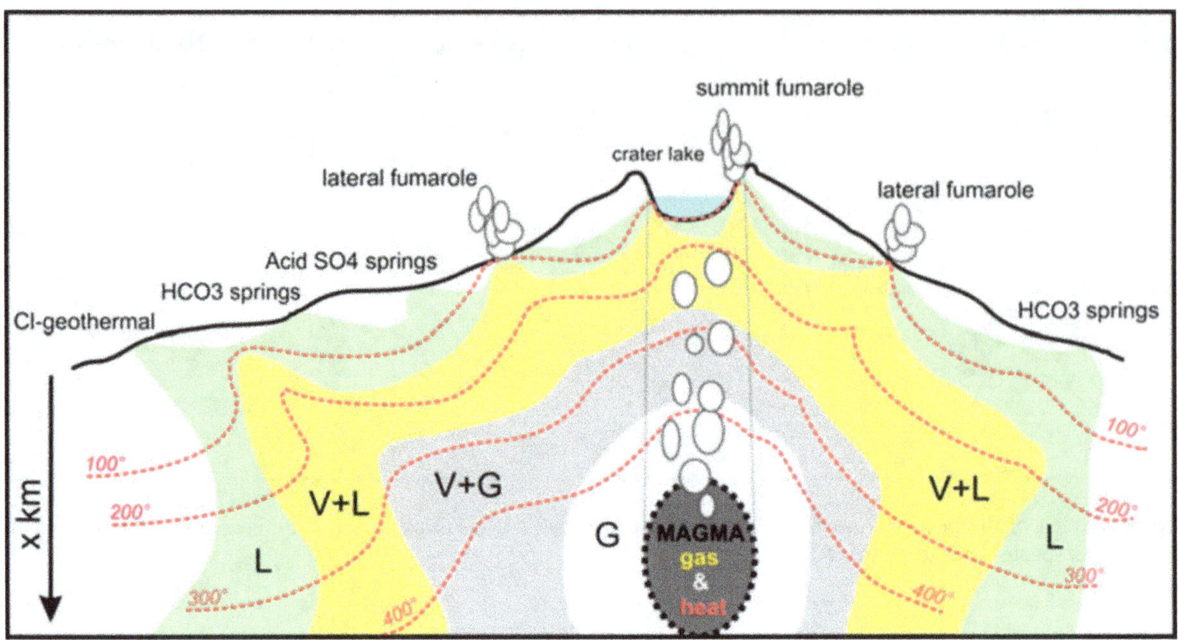

Fig. 4 Sketch of the gas that hits the water in an hypothetical "wet volcano" (not to scale). L = liquid only zone (turquoise area), V + L = vapor + liquid zone, (*yellow area*) V + G = vapor + gas zone (*grey area*), G = gas only zone (*white area* around the degassing magma, *dark grey*). *Red dotted lines* are isotherms. Cl-geothermal water are deep remnant waters, not of interest for geochemical monitoring of volcanoes for being "old and slow"

hydrolyzed as sulphuric acid (H_2SO_4) and dissociate into HSO_4^- or SO_4^{2-}, depending on the pH of the water. The dissolved H^+ creates the high acidity or low pH, following pH = $-\log$ [H^+]. Volcanic environments are renowned for being acidic; the acid is generated by the scrubbing of acidic gases into the water.

The following SO_2 disproportionation reactions occur (Kusakabe et al. 2000):

$$3SO_2 + 3H_2O = 2HSO_{4^-} + S° + 2H^+ \qquad (4)$$

or

$$4SO_2 + 4H_2O = 3HSO_{4^-} + H_2S + 3H^+ \qquad (5)$$

Reaction (4) occurs under relatively oxidizing conditions, low temperature and high total sulphur concentrations, whereas reaction (5) occurs under relatively reducing (i.e. oxygen poor) conditions, high temperature and low total sulphur concentrations. "To disproportionate SO_2" means that for each three or four moles of SO_2 that hit the water, one mole of S°, or one mole of

H_2S is given in return (reaction 4 and 5, respectively).

As SO_2 degasses at shallower depth in a magma system than CO_2, the resulting acid SO_4-rich springs are found near the central degassing conduit (e.g. inside active craters). For SO_2-dominated magmatic systems a pH near 2 or less is common; for H_2S-dominated hydrothermal systems a pH of 2–2.5 is the most acidic water can get (i.e. hydrothermal or "steam heated" waters).

The next gas species to be added to the "liquid cocktail" are HCl and HF (Fig. 3). HCl is highly volatile, but also highly hydrophile. This *contradictio in terminis* means that when HCl_{gas} reaches water it will be trapped in the liquid phase as Cl^- and H^+. Generally, Cl^- is considered "conservative" in the liquid phase, and is therefore often used as a tracer in the hydrothermal environment (see Section "Rock Leaching Upon Weathering"). Conservative means that Cl^- can be hardly lost from the solution as Cl-salts are highly soluble in acidic and hot environments, and HCl should not degas from "high pH water" (>2). Sulphate minerals in their turn can be lost from solutions by precipitation, demonstrating their non-conservative character (Rouwet and Ohba 2015). For extremely acidic environments (pH 0 or <0) the reaction

$$HCl = H^+ + Cl^- \qquad (6)$$

moves to the left due to an H^+ excess with respect to Cl^- (i.e. HCl degassing). The same counts for HF. Bearing in mind the abundance of the acid SO_2 in the magmatic-hydrothermal environment, providing a large part of the acidity, this "secondary HCl degassing from the liquid phase" is less unexpected than previously thought. This implies that HCl can degas from a low-temperature aquifer, as long as the aquifer is extremely acid (Capaccioni et al. 2016). Moreover, as hot water releases vapor, this loss enriches the remnant liquid in solutes, including H^+ (i.e. salinity and acidity increase), leading to the fact that even SO_2 tends to degas from the liquid, instead of remaining in the water phase as SO_4^{2-} or HSO_4^-.

Acid water (pH < 3.8) is completely transparent for the omni-present CO_2. This implies that CO_2, one of the "deepest signals" available for a degassing volcano, will completely outgas from bottom to top. In active craters, often underlain by acidic thermal aquifers, the release of CO_2 thus behaves as though there is no water present. This is a great advantage to monitoring volcanoes and tracking unrest, especially to detect the onset of unrest.

In conclusion, the above insights demythologize two generally accepted facts: (1) high-temperature fumaroles cannot contain water vapor (Table 1), and (2) low-temperature fumaroles cannot release acidic gases.

Within the scope of this book, tracking unrest using fluid geochemistry requires the introduction of a time frame, or a monitoring time window and frequency. Does the fumarole reflect the exact moment of degassing, or is it rather an average degassing for the longer previous period stored and steadily released from the

magmatic-hydrothermal plumbing system? What is the time delay between the moment the gas hits the water and the eventual release at the surface? The kinetics (i.e. "speed") of the gas migration from a magma towards the surface are still poorly constrained. If we can estimate the residence time of gas and water in the magmatic-hydrothermal system, we are able to define a monitoring time window, and hence adopt an adequate monitoring frequency. As explained, the acidity of the feeding aquifer plays a role. For extreme acidic conditions, less gas scrubbing occurs and the fumarolic system will react faster, and hence, shorter monitoring time windows can be adopted. If the monitoring time window is longer than the window defined by the kinetics of gas migration, details in degassing dynamics will be lost. On the contrary, the monitoring time window should not be a lot shorter than the time window defined by the kinetics of gas migration, if so, it will provide a too detailed view of what the magmatic-hydrothermal system can maximally provide.

The Other Liquid: Elemental Sulphur

Whereas water melts at 0 °C and boils at 100 °C, sulphur melts at 119 °C and boils at 444 °C (Oppenheimer and Stevenson 1989, and references therein). This physical fact on phase transitions implies that in the hydrothermal environment (boiling water) elemental sulphur is solid, and that during the initial phase of transition towards a more magmatic, high-temperature regime sulphur will become liquid. Molten sulphur in the hydrothermal plumbing system can be remobilized, clear vugs and vents and eventually be expelled as a liquid sulphur flow from fumarole mouths, or "fill-and-freeze" pores in the shallower hydrothermal system. The first process opens up degassing pathways; the second process can decrease rock porosity and permeability near the surface, thus sealing a hydrothermal system. During the evolution from low-temperature (>119 °C, unrest) towards high-temperature (occasionally magmatic unrest), the viscosity of the liquid sulphur increases

2000-fold (>160 °C) to become an extremely efficient sealer of a magmatic-hydrothermal system. Pressure build-up beneath seals in hydrothermal systems can lead to phreatic eruptions (Rouwet and Morrissey 2015; Rouwet et al. 2016). Monitoring fumarolic temperatures is thus essential, and probably one the simplest methods to apply.

Tracking Hydrothermal Unrest and Related Hazards: Methods from Case-Studies

From Quiescence to Unrest, to Phreatic Eruptions, to Magmatic Eruptions

Turrialba, Costa Rica (2001–2016)

A transition from the stable, passively degassing hydrothermal system of Turrialba volcano (Costa Rica), to hydrothermal unrest, to phreatic eruptions, to magmatic eruptions, is a recent example of an evolution from volcanic quiescence heading towards eruption.

In 2001, increased fumarolic activity (appearance of SO_2 in late 2001) was paired to seismic swarms and ground deformation (Martini et al. 2010; Vaselli et al. 2009) (i.e. unrest). In 2007, the increased SO_2/H_2S molar ratio in fumaroles (>100), combined with an increase in exhalation temperature up to 282 °C (in early 2008, Martini et al. 2010; Vaselli et al. 2009), point to more oxidized and magmatic conditions (i.e. magmatic unrest). Clear plume degassing resumed in early 2007 (Fig. 5a), and SO_2 fluxes reached 740 t/d in January 2008 (Martini et al. 2010). In late 2009, fumarolic degassing was vigorous and extended into the Arbol Quemado fracture, newly formed in 2002. The first phreatic eruption occurred on 5 January 2010, in the inner crater wall of the actively degassing SW crater. Strong jet-like degassing occurred afterwards from this new vent (Fig. 5b), while diffuse fumarolic degassing diminished in the SW crater. A second phreatic eruption occurred on 12 January 2012, from a vent inside the Arbol Quemado fracture. The day before, this eruption was preceded by liquid sulphur flowing out of the Arbol Quemado fumaroles (González et al. 2015). A third phreatic eruption episode involved the 2010 and 2012 vents simultaneously (21 May 2013). During this 3.5-year long phreatic cycle the SO_2 flux from Turrialba's plume was high: from 2500 to 4300 t/d (Campion et al. 2012; Moussallam et al. 2014). CO_2/SO_2 molar ratios in March 2013 were relatively low (2.6), hinting at a CO_2-depleted and SO_2-rich magma (Moussallam et al. 2014). From 2001 to 2013, Moussallam et al. (2014) suggest the progressive "drying-out" of the underlying hydrothermal system.

The first magmatic eruption at Turrialba since the 1864–1866 phreatomagmatic activity occurred during the night of 29 and 30 October 2014 (Mora-Amador et al. 2015). At the time of writing, the last magmatic eruptions took place in September 2016.

Despite the well-monitored and tracked evolution from volcanic quiescence to magmatic eruption it remains unclear why some volcanoes quickly evolve from quiescence to eruption, while at Turrialba it took 14 years from quiescence to magmatic eruption, passing the complete range of unrest manifestations during this relatively long time span.

Cotopaxi, Ecuador (2015–2016)

Another example of volcanic quiescence to magmatic unrest is the one of Cotopaxi volcano in 2015. As Cotopaxi is a dangerous volcano whose activity would potentially affect densely populated areas its monitoring network has been continuously improved since the late 70s. After 73 years of quiescence, the first sign of unrest at Cotopaxi was a progressive increase in the amplitude of transient seismic events in April 2015. SO_2 is measured at Cotopaxi by DOAS stations installed on the flanks of the volcano since 2008. The permanently measured SO_2 emissions showed an increase on May 20 from almost non detectable up to $\sim$3000 t/d. The fumaroles showed increased activity and a gas plume from the crater was usually observed on clear days. By early June SO_2 emissions yielded up to 5000 t/d. On July 20 a green lake was observed filling the crater of the volcano,

Fig. 5 Hydrothermal unrest at Turrialba, Poás and Irazú volcanoes. **a** Vegetation die-back due to resumed plume degassing. **b** The 2010 phreatic eruption vent (picture by S. Calabrese). **c** Lack of vegetation at Cerro Pelón (Poás) due to plume degassing and acid rain fall downwind. **d** Efficient rock mass removal through Río Sucio, 30 km downstream Irazú volcano

nevertheless no significant changes in the SO_2 emission was observed, suggesting that the lake was of high acidity and/or too small to be an efficient scrubber. The first phreato-magmatic explosions occurred on August 14 and produced ash and gas columns reaching up to 9 km above the crater. The satellite-borne instruments such as OMI and OMPS reported 16,400 and 12,500 t/d of SO_2 released to the atmosphere on that day (http://so2.gsfc.nasa.gov/pix/daily/0815/ecuador_0815z.html). Continuous ash emissions followed the initial explosions producing a permanent gas and ash plume drifting westward. SO_2 measured in this permanent plume, by mobile-DOAS traverses or by the permanent stations, reached 24,000 t/d and decreased progressively until the end of the activity in late November 2015.

Since June and more consistently since 14 August 2015, BrO was also detected in the plume (Dinger et al. 2016). Airborne Multi-GAS measurements showed that the plume had a low CO_2/SO_2 ratio, and that SO_2 was >99% of total sulfur ($SO_2 + H_2S$), indicating a shallow magmatic origin for the gas. At the time of writing (September 2016), SO_2 emissions decreased to background levels. SO_2 permanent monitoring proved to be a useful tool at Cotopaxi providing real time data contributing, together with other geophysical methods, to better evaluate volcanic unrest scenarios.

Gas Impact and Acid Rain

Persistent high-temperature fumarolic and plume degassing impact volcano flanks down-wind (Fig. 5c). This can happen when it rains through a plume, generating acid rain. Many volcanoes are subject to such long-term, non-eruptive hazard, making the surrounding ground harsh living environments. Other volcanoes suffer vegetation die-back, when activity resumes after prolonged quiescence (Fig. 5a). This was clearly visible during the increased activity at Turrialba volcano (Section "Turrialba, Costa Rica (2001–2016)", Fig. 5a, b; González et al. 2015) and downwind Poás' western flank (Cerro Pelón, Fig. 5c). To assess such hazards, a meteorological station (wind direction, speed, air humidity and rainfall), and a DOAS device to measure SO_2 fluxes are valuable tools.

Rock Leaching upon Weathering

Absorption of magmatic gases into aquifers creates acidic magmatic-hydrothermal systems that, sooner or later, will exit the volcano. Acidic water attacks the wall rock and becomes loaded with solutes (Delmelle et al. 2015). If (1) meteoric recharge is high, (2) acid input is high, (3) wall rock is fresh, and (4) (thermal) spring water discharge is high, rock leaching capacity can reach thresholds to even mechanically destabilize volcano flanks. Enhanced chemical leaching for long periods favors physical rock removal, causing rock fall, landslides, or even flank and sector collapses of volcanic edifices. Even if magmatic degassing seems absent and a volcano may be long dormant, hazards loom due to the scrubbing capacity of deeper aquifers (Delmelle et al. 2015, and references therein).

The best suited method to quantify and track rock mass removal from a volcanic edifice is by monitoring the discharge rates from thermal springs, and their dissolved solutes and solids. The "Cl-inventory" uses Cl as the conservative tracer. As mentioned earlier, this is true for less acidic magmatic-hydrothermal systems (Ingebritsen et al. 2001; Taran and Peiffer 2009; Chiodini et al. 2014; Collard et al. 2014). Measuring the Cl-release from rivers draining thermal springs, and knowing the Cl-content and Cl/solute ratios in thermal spring waters, the rock mass removal rate can be estimated by:

$$Q_r * C_r = Q_s * C_s \qquad (7)$$

where Q_r and Q_s are the discharges of rivers and thermal springs, respectively, in L/s; C_r and C_s are the concentrations of Cl of rivers and thermal springs, respectively, in mg/L. Measuring the river discharge (Q_r) and analyzing the river and spring waters for its Cl content (C_r and C_s, respectively) thus enables to estimate the spring discharge (Q_s), which would otherwise be impossible to directly measure in the field (e.g. numerous spring discharges, too irregular spring mouths). This method combines gas-water-rock interaction and hydrology of magmatic-hydrothermal systems in order to assess indirect hazard. Volcanoes with high rock mass removal rates are e.g. Irazú (Costa Rica, Fig. 5d); extremely acidic magmatic-hydrothermal rock removers are Kawah Ijen (Java), Poás (Costa Rica), and Copahue (Argentina-Chile). In the most extreme cases, the acidity and toxic metal load affects agricultural activities and human health.

Moreover, through the same Cl-inventory approach, the geothermal potential (i.e. heat output) from springs can be estimated, by multiplying the enthalpy of discharged spring waters, often based on geothermometric temperatures of the deep system, with the spring discharge rates. Such estimates were obtained for the active magmatic-hydrothermal systems of El Chichón and Tacaná (both in Chiapas, Mexico; Taran and Peiffer 2009; Collard et al. 2014), and Domuyo (Argentina; Chiodini et al. 2014), and originally of the Cascades Volcanic Range by Ingebritsen et al. (2001). Understanding the state of unrest on the long term of a specific volcano is needed to rule out if the volcano would be a feasible target for geothermal exploitation, or not.

Volcanic Lakes

Acid Peak-Activity Lakes in a State of Unrest

Volcanic lakes are the intersection of the crater surface and the underlying aquifer (Fig. 1g, h). Hence, they become "windows" into the depths of magmatic-hydrothermal systems. While fumaroles directly lose their signal from depth to the atmosphere as a "snapshot" (but see Section "When the Gas Hits the Water"), volcanic lakes preserve a past gas marker for a certain period. The time we can track back by studying the water chemistry depends on the duration the water resides in the lake (i.e. residence time). The residence time (RT) is estimated by dividing the lake volume (V in m^3) by the input or output rate (Q in m^3/s) of fluids, assuming steady-state conditions (Varekamp 2003; Rouwet et al. 2014b):

$$RT = V/Q \qquad (8)$$

Small lakes with high rates of fluid flushing offer a better time-resolution than large lakes with slow rates of fluid in- and output. The monitoring frequency (e.g. lake water sampling) should be tuned to this residence time; a higher monitoring frequency will oversample the lake chemistry, while a lower monitoring frequency will lead to the loss of information the lake potentially provides (Rouwet et al. 2014b).

Volcanic lakes are excellent gas scrubbers. Nevertheless, recent studies quantify the gas release from the lake surface of the most acidic lakes (e.g. Aso, Copahue, Poás, Kawah Ijen; Shinohara et al. 2015; Tamburello et al. 2015; Capaccioni et al. 2016; de Moor et al. 2016; Gunawan et al. 2016). Under the most extreme pH conditions (<0) HF, HCl and even SO_2 can degas freely from the lake. This means that acidic lakes are more sensitive than thought before, as acid gas flashes through the water body with only minor scrubbing. Monitoring frequency can thus increase and, hence, spectroscopic or electro-chemical sensor tools become extremely useful (DOAS to measure SO_2 fluxes from volcanic plumes, Multi-GAS to measure ratios between gas species, e.g. CO_2/SO_2).

Considering acidic lakes as "open-air" fumaroles (and fumaroles as "buried" acidic lakes) has changed the monitoring time frame, which might lead to better chances to forecast phreatic eruptions (Rouwet et al. 2014a, b; Rouwet and Morrissey 2015).

Volcanic Lake Response to External Triggers in the Absence of Magmatic Unrest

The Boiling Lake, in Dominica of the Lesser Antilles (a VUELCO target volcano), is a high-temperature volcanic crater lake, that is believed to be formed as a result of phreatic or phreatomagmatic activity (Lindsay et al. 2005). It is approximately 50 m × 60 m in size and ca. 12–15 m deep, with an estimated volume of $\sim 1.2 \times 10^4$ m^3 when filled (Fournier et al. 2009). Over the last 150 years, temperatures taken at the edge of the lake have generally ranged between 80 and 90 °C, and the pH from 4 to 6 (Joseph et al. 2011), however, the lake experiences periods of instability where water level, temperature and state of hydrothermal activity fluctuate suddenly. The long term stability of the Boiling Lake is attributed to the hypothesis that the lake is suspended above the water table by the buoyancy of rising steam bubbles from the underlying hydrothermal system (Fournier et al. 2009). Perturbations of this interaction as a result of landslides into the lake or regional seismicity is attributed to the cause of the sudden periods of instability (i.e. lake drainage and refill cycles), rather than changes in underlying magmatic activity. This is mainly because no coincident anomalous hydrothermal activity was observed in the Valley of Desolation geothermal area, which is located nearby the Boiling Lake.

The most recent episode of instability occurred during December 2004 to April 2005 (Fig. 6), when the lake water level dropped by ~ 8–10 m, and the temperature at the water's edge decreased to <30 °C. Water acidity went from the usual acidic pH of 4–6 to neutral, while

Fig. 6 The refill of an empty Boiling Lake during a period of unusual activity where water levels and geothermal activity were rapidly fluctuating. **a** An empty Boiling Lake on 7 April 2005. **b** A full Boiling Lake on 13 April 2005. Pictures by Arlington James, Forestry Officer, Forestry and Wildlife Division, Dominica (used with permission)

Cl concentration dropped from the typical 2000–6000 to 29–50 mg/L, and SO$_4$ concentration dropped from 1500–4000 to 100–270 mg/L, indicating a drastic decrease of hydrothermal fluid input into the lake (Joseph et al. 2011). Additionally, total dissolved solid content decreased from 13,400 to 4500 mg/L, suggesting strong dilution by fresh water. Measurements of temperature, pH and chemical composition taken in August 2006 indicate that the lake had returned to its normal steady-state of activity (Joseph et al. 2011). This episode is reported to have been triggered as a result of the extensional strain induced by a regional Mw 6.3 earthquake that occurred on November 21, 2004 offshore Les Saintes (Guadeloupe) (Feuillet et al. 2011), which may have contributed to diminished water inflow.

It should be noted, however, that a phreatic explosion and gas release occurred at an "empty"

Table 2 Geochemical signals and what they indicate with respect to volcanic unrest

Type of unrest	Geochemical signal	Indication
Unrest	CO_2 flux above background	Changes in deep degassing dynamics
	Increase in T of hot springs and/or fumaroles	Increased heat input
	Changes in H_2O/CO_2 ratios in fumaroles	Changes in water/gas ratio
	Appearance of new fumaroles and/or hot springs	Aerial extension of activity
Magmatic unrest	Appearance of acidic gases (SO_2, HCl, HF)	Changes in mid- to shallow magma dynamics
	T fumarole >119 °C	Remobilisation of sulphur
	SO_2 flux > X t/d	SO_2 flux above background, volcano-dependent
	Increase in CO_2/SO_2 ratio	Arrival of an undegassed magma at depth
	Extreme increase in T fumaroles (>300 °C)	Towards magmatic T
Magmatic eruption	Decreasing CO_2/SO_2 ratios after increase	More superficial magma degassing
	Increase in Cl, Br, F concentrations in hot springs/pools	Input of highly soluble acidic gases
	Decrease in H_2O/CO_2 and/or H_2S/SO_2 and/or SO_2/HCl ratios	More gas with a more magmatic signature
Hydrothermal unrest	New fumaroles	Aerial extension of activity
	Anomalous glacier defrosting	Sudden removal of water mass... lahars
	Water to vapour transition	Pushing vapour front from below
	Changes in hydrothermal features	Variations or aerial extension of activity
	Increase in B and/or NH_4 in waters	Input of vapour
	Increase in CH_4/CO_2 in fumaroles	A more hydrothermal signature in fumaroles
	Variations in phreatic level in aquifers	Pushing vapour front from below
Hydrothermal eruption	120 °C < T fumarole <200 °C	Self-sealing by a change in S viscosity
	Extension of alteration areas or fumarolic fields	Aerial extension of activity
	Appearance of muddy pools	Clearing bugs and vents, unplugging
	Boiling/bubbling of pools that previously didn't	Rising vapour front and/or extra heating and degassing

Boiling Lake on 10 December 1901 that resulted in the deaths of two individuals (Elliot 1938; Bell 1946). This suggests that hazards related to volcanic lakes such as the Boiling Lake, may occur without magmatic input.

Take-Home Ideas: Implications for Geochemical Monitoring

Over the past five-six decades, gas geochemistry at magmatic-hydrothermal systems has mainly focussed on chemical equilibria and kinetics in the subsurface environment. Over the last 10–15 years more attention has been paid to remote sensing of volcanic gas plumes (DOAS, Multi-GAS) with the obvious advantage of increased safety and frequency of data gathering. Nevertheless, the relationship between the fumarole and plume has yet to be better constrained. The best proxy of a magmatic gas remains a direct sample of a high-temperature fumarole, although such target fumaroles are often inaccessible, especially during eruptive phases. Compromises between data fidelity and safety of the operators, and the frequency of data gathering should be framed in terms of what we want and maximally can unravel. The advantage of fluid geochemistry in volcano monitoring arrises from the fact that volatiles are mobile and thus reach the surface often before physical changes manifest. Timely recognition of unrest and especially hydrothermal unrest is often possible. Table 2 summarizes geochemical signals and how they relate to the various states of volcanic unrest, useful for monitoring based on deterministic research and probabilistic modeling.

Future research should focus on better constraining degassing dynamics at the surface-atmosphere boundary. Recent studies have demonstrated that extremely acidic aquifers can "slow down" the signal released from gas-dominated magmatic-hydrothermal systems (fumaroles), but "speed up" the signal released from water-dominated systems (e.g. acidic crater lakes). These findings have strong implications for the time frame of magma degassing, and hence for the monitoring frequency.

References

Aiuppa A, Federico C, Paonita G, Valenza M (2002) S, Cl and F degassing as an indicator of volcanic dynamics: the 2001 eruption of Mount Etna. Geophys Res Lett 29(11):1559. doi:10.1029/2002GL015032

Aiuppa A, Federico C, Giudice G, Giuffrida G, Guida R, Gurrieri S, Liuzzo M, Moretti R, Papale P (2009) The 2007 eruption of Stromboli volcano: insights from real-time measurements of the volcanic gas plume CO_2/SO_2 ratio. J Volcanol Geoth Res 182(221):230. doi:10.1016/j.jvolgeores.2008.09.013

Baldoni E, Rouwet D, Mora-Amador R, Ramírez C, González-Ilama G, Lucchi F, Capaccioni B, Tranne CA, Pecoraino G (submitted) Hydrogeochemical model of the Irazú-Turrialba volcanic complex (Costa Rica) and implications for hazard assessment and volcanic surveillance. In: Caudron C, Capaccioni B, Ohba T (eds) GSL special volume geochemistry and geophysics of volcanic lakes

Bell H (1946) Glimpses of a governor's life—Dominica. Samson Low, London

Campion R, Martínez-Cruz M, Lecocq T, Caudron C, Pacheco J, Pianrdi G, Hermans C, Carn S, Bernard A (2012) Space- and ground-based measurements of sulphur dioxide emissions from Turrialba Volcano (Costa Rica). Bull Volcanol. doi:10.1007/s00445-012-0631-z

Capaccioni B, Rouwet D, Tassi F (2016) HCl degassing from extremely acidic crater lakes: empirical results from experimental determinations and implications for geochemical monitoring. In: Caudron C, Capaccioni B, Ohba T (eds) GSL special publications 437 geochemistry and geophysics of volcanic lakes. doi:10.1144/SP437.12

Capasso G, Favara R, Inguaggiato S (1997) Chemical features and isotopic composition of gaseous manifestations on Vulcano Island, Aeolian Islands, Italy: an interpretative model of fluid circulation. Geochim Cosmochim Acta 61(16):3425–3440

Chiodini G, Liccioli C, Vaselli O, Calabrese S, Tassi F, Caliro S, Caselli A, Agusto M, D'Alessandro W (2014) The Domuyo volcanic system: an enormous geothermal resource in Argentine Patagonia. J Volcanol Geoth Res 272:71–77. doi:10.1016/j.jvolgeores.2014.02.006

Dinger F, Arellano S, Battaglia J, Bobrowski N, Galle B, Hernández S, Hidalgo S, Hörmann C, Lübcke P, Platt U, Ruíz M, Warnach S, Wagner T (2016) Variations of the BrO/SO_2 molar ratios during the 2015 Cotopaxi eruption. EGU 2016-1001

Collard N, Taran Y, Peiffer L, Campion R, Jacome Paz MP (2014) Solute fluxes and geothermal potential of Tacana volcano-hydrothermal system, Mexico-Guatemala. J Volcanol Geoth Res 288:123–131. doi:10.1016/j.jvolgeores.2014.10.012

De la Cruz-Reyna S, Tilling RI (2008) Scientific and public responses to the ongoing volcanic crisis at Popocatépetl Volcano, Mexico: importance of an effective hazards-warning system. J Volcanol Geoth Res 170:121–134

Delmelle P, Bernard A (2015) The remarkable chemistry of sulfur in hyper-acid crater lakes: a scientific tribute to Bokuichiro Takano and Minoru Kusakabe. In: Rouwet D, Christenson B, Tassi F, Vandemeulebrouck J (eds) Book chapter in volcanic lakes, Springer, Heidelberg, pp 239–260. doi:10.1007/978-3-642-36833-2_10

Delmelle P, Henley RW, Opfergelt S, Detienne M (2015) Summit acid crater lakes and flank instability in composite volcanoes. In: Rouwet D, Christenson B, Tassi F, Vandemeulebrouck J (eds) Book chapter in volcanic lakes. Springer, Heidelberg, pp 289–306. doi:10.1007/978-3-642-36833-2_12

de Moor JM, Aiuppa A, Pacheco J, Avard G, Kern C, Liuzzo M, Martinez M, Giudice G, Fischer TP (2016) Short-period volcanic gas precursors to phreatic eruptions: insights from Poás Volcano, Costa Rica. Earth Planet Sci Lett 442:218–227. doi:10.1016/j.epsl.2016.02.056

Elliot EC (1938) Boiling lake—the 1900 story. In: Broken Atoms. Unpublished report presented to the Government of Dominica

Feuillet N, Beauducel F, Jacques E, Tapponnier P, Delouis B, Bazin S, Vallée M, King GCP (2011) The Mw = 6.3, November 21, 2004, Les Saintes earthquake (Guadeloupe): Tectonic setting, slip model and static stress changes. J Geophys Res 116(B10):1–25. doi:10.1029/2011JB008310

Fournier N, Withal F, Moreau-Fournier M, Bardou L (2009) The Boiling Lake of Dominica, West Indies: high temperature volcanic crater lake dynamics. J Geophys Res 114:B02203. doi:10.1029/2008JB005773

Giggenbach WF (1987) Redox processes governing the chemistry of fumarolic gas discharges from White Island, New Zealand. Appl Geochem 2:143–161

González G, Mora-Amador R, Ramirez C, Rouwet D, Alpizar Y, Picado C, Mora R (2015) Actividad historica y analisis de la amenaza del volcano Turrialba, Costa Rica. Rev Geol Am Centr 52:129–149. doi:10.15517/rgac.v0i52.19033

Gunawan H et al (2016) New insights into Kawah Ijen's volcanic system from the wet volcano workshop experiment. In: Caudron C, Capaccioni B, Ohba T (eds) GSL special volume geochemistry and geophysics of volcanic lakes

Hall M, Mothes P (2008) The rhyolitic-andesitic eruptive history of Cotopaxi volcano, Ecuador. Bull Volcanol 70:675–702. doi:10.1007/s00445-007-0161-2

Ingebritsen SE, Galloway DL, Colvard EM, Sorey ML, Mariner RH (2001) Time variation of hydrothermal discharge at selected sites in the western United States: implications for monitoring. J Volcanol Geoth Res 111:1–23

Joseph EP, Fournier N, Lindsay J, Fischer T (2011) Gas and water geochemistry of geothermal systems in Dominica, Lesser Antilles island arc. J Volcanol Geoth Res 206(1–2):1–14. doi:10.1016/j.volgeores.2011.06.007

Kusakabe M, Komoda Y, Takano B, Abiko T (2000) Sulfur isotopic effects in the disproportionation reaction of sulfur dioxide in hydrothermal fluids: implications for the d34S variations of dissolved bisulphate and elemental sulfur from active crater lakes. J Volcanol Geoth Res 97:287–307

Lindsay J, Robertson R, Shepherd J, Ali S (2005) Volcanic hazard atlas of the Lesser Antilles. In: 279. St. Augustine: Seismic Research Unit, University of the West Indies

Martini F, Tassi F, Vaselli O, Del Potro R, Martínez M, Van der Laat R, Fernandez E (2010) Geophysical, geochemical and geodetical signals of reawakening at Turrialba volcano (Costa Rica) after 150 years of quiescence. J Volcanol Geoth Res 198:416–432. doi:10.1016/j.jvolgeores.2010.09.021

Menyailov IA, Nikitina LP, Shapar VN, Pilipenko VP (1986) Temperature increase and chemical changes of fumarolic gases at Momotombo volcano, Nicaragua, in 1982–1985: are these indicators of a possible eruption? J Geophys Res 187:12199–12214

Mora-Amador R, Ramírez-Umaña C, González G, Rouwet D, Lucchi F, Forni F, Sulpizio R, Baldoni E, Alpízar-Segura Y, Tranne CA (2015) First documentation of the ongoing phreatic-strombolian eruptions of Turrialba volcano (Costa Rica). IUGG General Assembly, Prague 3746

Moussallam Y, Peters N, Ramirez C, Oppenheimer C, Aiuppa A, Giudice G (2014) Characterisation of the magmatic signature in gas emissions from Turrialba Volcano, Costa Rica. Solid Earth 5:1341–1350. doi:10.5194/se-5-1341-2014

Oppenheimer C, Stevenson D (1989) Liquid sulphur lakes at Poás volcano. Nature 342:790–793

Phillipson G, Sobradelo R, Gottsmann J (2013) Global volcanic unrest in the 21st century: an analysis of the first decade. J Volcanol Geoth Res 264:183–196

Rouwet D, Morrissey M (2015) Mechanisms of crater lake eruptions: physical and numerical modeling. In: Rouwet D, Christenson B, Tassi F, Vandemeulebrouck J (eds) Book chapter volcanic lakes. Springer, Heidelberg, pp 73–92. doi:10.1007/978-3-642-36833-2_3

Rouwet D, Ohba T (2015) Iotope fractionation and HCl partitioning during evaporative degassing from active crater lakes. In: Rouwet D, Christenson B, Tassi F, Vandemeulebrouck J (eds) Book chapter volcanic lakes. Springer, Heidelberg, pp 179–200. doi:10.1007/978-3-642-36833-2_7

Rouwet D, Sandri L, Marzocchi W, Gottsmann J, Selva J, Tonini R, Papale P (2014a) Recognizing and tracking hazards related to non-magmatic unrest: a review. J Appl Volcanol 3:17. doi:10.1186/s13617-014-0017-3

Rouwet D, Tassi F, Mora-Amador R, Sandri L, Chiarini V (2014b) Past, present and future of volcanic lake monitoring. J Volcanol Geoth Res 272:78–97. doi:10.1016/j.jvolgeores.2013.12.009

Rouwet D, Mora-Amador R, Ramírez-Umaña C, González G, Inguaggiato S (2016) Dynamic fluid recycling at Laguna Caliente (Poás, Costa Rica) before and during the 2006-ongoing phreatic eruption cycle (2005–2010). In: Caudron C, Capaccioni B, Ohba T (eds) Geochemistry and geophysics of volcanic lakes. Geological Society London Special Publications. doi:10.1144/SP437.11

Shinohara H, Giggenbach WF, Kazahaya K, Hedenquist JW (1993) Geochemistry of volcanic gases and hot springs of Satsuma-Iwojima, Japan: following Matsuo. Geochem J 27:271–285

Shinohara H, Yoshikawa S, Miyabuchi Y (2015) Degassing activity of a volcanic crater lake: volcanic plume measurements at the Yudamari crater lake, Aso volcano, Japan. In: Rouwet D, Christenson B, Tassi F, Vandemeulebrouck J (eds) Book chapter volcanic lakes. Springer, Heidelberg, pp 73–92. doi:10.1007/978-3-642-36833-2_8

Symonds RB, Gerlach TM, Reed MH (2001) Magmatic gas scrubbing: implications for volcano monitoring. J Volcanol Geoth Res 108:303–341

Tamburello G, Agusto M, Caselli A, Tassi F, Vaselli O, Calabrese S, Rouwet D, Capaccioni B, Di Napoli R, Cardellino C, Chiodini G, Bitetto M, Brusca L, Bellomo S, Aiuppa A (2015) Intense magmatic degassing through te lake of Copahue volcano, 2013–2014. J Geophys Res doi:10.1002/2015JB012160

Taran YA, Hedenquist JW, Korzhinsky M, Tkachenko SI, Shmulovich KI (1995) Geochemistry of magmatic gases from Kudryavy volcano, Iturup, Kuril Islands. Geochim Cosmochim Acta 59:1749–1761

Taran YA, Peiffer L (2009) Hydrology, hydrochemistry and geothermal potential of El Chichón volcano-hydrothermal system, Mexico. Geothermics 38:370–378

Taran YA, Pokrovsky BG, Dubik YM (1989) Isotopic composition and origin of water from andesitic magmas. Dokl Acad Sci 304:440–443

Varekamp JC (2003) Lake contamination models for evolution towards steady state. J Limnol 62(1):67–72

Vaselli O, Tassi F, Duarte E, Fernández E, Poreda RJ, Delgado Huertas A (2009) Evolution of fluid geochemistry at the Turrialba volcano (Costa Rica) from 1998 to 2008. Bull Volcanol. doi:10.1007/s00445-009-0332-4

Geophysical Footprints of Cotopaxi's Unrest and Minor Eruptions in 2015: An Opportunity to Test Scientific and Community Preparedness

Patricia A. Mothes, Mario C. Ruiz, Edwin G. Viracucha, Patricio A. Ramón, Stephen Hernández, Silvana Hidalgo, Benjamin Bernard, Elizabeth H. Gaunt, Paul Jarrín, Marco A. Yépez and Pedro A. Espín

Abstract

Cotopaxi volcano, Ecuador, experienced notable restlessness in 2015 that was a major deviation from its normal background activity. Starting in April and continuing through November 2015 strong seismic activity, infrasound registry, hikes in SO_2 degassing and flank deformation with small displacements were some of the geophysical anomalies that were registered. Obvious superficial changes, such as small hydromagmatic eruptions, emission of vapor and ash columns, thermal hotspots around the crater and in nearby orifices and exacerbated glacier melting were also observed. Our contribution provides an overview of the 2015 Cotopaxi unrest by presenting the patterns of geophysical data and the sequence of events produced by the volcano. Cotopaxi's last important VEI 4 eruption was in 1877. Then it had devastating effects because of the transit of huge lahars down 3 major drainages. Comparatively, the 2015 activity never surpassed a magnitude VEI 2 and principally produced limited hydro-magmatic explosions and semi-continuous low energy emissions and light ashfalls. Given the potential of major destruction from a large Cotopaxi eruption it is important to understand the geophysical fingerprints that characterized the 2015 episode with an eye to identifying onset of future restless periods. Overall, the monitoring activities, the data interpretation, formulation of reasonable eruptive scenarios, and finally, the preparation of a stream of constant information being relayed to concerned authorities and the public, was a real test of the IGEPN's capacity to deal with a complicated eruption situation whose outcome was not apparent at the beginning, but which concluded in a very small eruptive episode.

P.A. Mothes (✉) · M.C. Ruiz · E.G. Viracucha ·
P.A. Ramón · S. Hernández · S. Hidalgo · B. Bernard ·
E.H. Gaunt · P. Jarrín · M.A. Yépez · P.A. Espín
Instituto Geofísico, Escuela Politécnica Nacional,
Quito, Ecuador
e-mail: pmothes@igepn.edu.ec

Advs in Volcanology (2019) 241–270
DOI 10.1007/11157_2017_10

Resumen

En 2015 el volcán Cotopaxi, Ecuador experimento un notable cambio, que fue una desviación importante de su actividad normal de base. A partir de abril y hasta noviembre de 2015 fuerte actividad sísmica, registros de infrasonido, incremento en la desgasificación de SO_2 y pequeños cambios en la deformación de los flancos fueron algunas de las anomalías geofísicas registradas. Evidentes cambios superficiales también fueron observados como pequeñas erupciones hidromagmaticas, emisión de vapor, columnas de cenizas, puntos calientes alrededor del cráter y el deshielo de los glaciares. Nuestra contribución proporciona una visión general de las anomalías del Cotopaxi en el 2015, mediante la presentación de patrones de los datos geofísicos y la secuencia de eventos producidos por el volcán. La última erupción importante del Cotopaxi fue un VEI 4 en 1877. Esta tuvo efectos devastadores debido al descenso de enormes lahares por sus tres drenajes mayores. Comparativamente, la actividad del año 2015 nunca superó una magnitud VEI 2, principalmente produciendo explosiones hidromagmaticas, escasas emisiones y leves caídas de ceniza. Debido a la potencial destrucción por una eventual erupción grande del Cotopaxi es importante entender los registros geofísicos que caracterizó el episodio de 2015 para poder identificar el inicio de futuros períodos eruptivos. En general, las actividades de vigilancia, la interpretación de datos, formulación de escenarios eruptivos razonables y por último, la preparación de un flujo de información constante que llegue a las autoridades interesadas y el público, fue una verdadera prueba de la capacidad del IGEPN para hacer frente a una situación de erupción cuyo resultado no era evidente al principio, pero que finalizó como una erupción pequeña.

Keywords

Volcanic unrest · Precursory geophysical patterns · Precursory LPs and VLPs · Volcano monitoring · Cotopaxi volcano-Ecuador · State of preparedness

Introduction

Long dormant volcanoes that begin to awaken may have start and stop activity that has to be evaluated with respect to the volcano's known past. A volcano's past activity is known by study of its stratigraphy, other physical evidence and possibly historical chronicles (Tilling 1989). Many uncertainties preclude knowing the final outcome of a restless volcano (Newhall 2000; Sparks and Aspinall 2004), since a volcano may awaken for short term low-level activity, then resume repose until additional magma inputs herald an episode of further unrest (Phillipson et al. 2013). Eruptions that barely bring magma to the surface may be classified as "failed", since so little juvenile magma erupts (Moran et al. 2011).

At Cotopaxi ample geological and historical information exists with regard to its past activity (Hall and Mothes 2008; Garrison et al. 2011; Pistolesi et al. 2012). Formulation of eruptive scenarios with respect to the 2015 unrest period were based on our collective knowledge of the volcano's geology and eruptive history and published information as well as interpretation of the abundant geophysical data streams available through instrumental and observational networks operated by the Instituto Geofísico of the Escuela Politécnica Nacional (IGEPN)-Quito, Ecuador, the entity in charge of volcano and tectonic monitoring in Ecuador. The combination of these inputs allowed scientists at the IGEPN to transmit a coherent image of the evolving unrest presented during 2015 and to indicate the most likely eruption/activity scenarios. Two earlier unrest periods are known: 1975–1976, when the IGEPN had limited seismic equipment operating on the volcano and then in 2001–2002. Both periods were comprised of increased fumarolic activity both inside and outside of the crater and a hike in seismicity for the 2001 period (Molina et al. 2008). The 2015 unrest displayed important changes in seismic and deformation patterns, gas output and superficial activity, compared to background, whose level was established since around 1990. In sum, Cotopaxi's 2015 unrest displayed a progressive crescendo of geophysical signals, then minor hydromagmatic explosions, followed by overall seismic energy decrease at the end of 2015, which was accompanied by fewer superficial manifestations. Like many other volcanoes that have displayed unrest, this recent episode did not culminate in a full-fledged eruption with large volumes of juvenile pyroclastics (Phillipson et al. 2013). Moran et al. (2011) maintain that an eruption is "failed" when magma reaches but does not pass the "shallow intrusion" stage, i.e., the magma gets close to, but does not reach the surface. In the actual case, the amount of erupted material was minor, and had a dense rock equivalent volume of $\sim$0.5 Mm^3 (Bernard et al. 2016).

Cotopaxi Volcano

Cotopaxi volcano, located in central Ecuador atop the Eastern Cordillera, is a large, symmetrical stratocone with a basal diameter of 18 km and an altitude of 5897 m (Hall and Mothes 2008). Its actual glacier cap of 10.49 km^2 is rapidly diminishing due to climatic change Cáceres et al. (2016) (Fig. 1). The volcano's last important VEI = 4 eruption was on 26 June 1877. Then it generated highly erosive pyroclastic flows that melted glaciers and triggered voluminous lahars ($\sim$100 Mm^3 per drainage). Each lahar traveled hundreds of kilometers down several drainages enroute to the Amazon basin, Pacific Ocean and to the Atlantic, respectively (Mothes et al. 2004; Mothes and Vallance 2015). These past lahar routes now host sprawling suburbia, important economic activities and vital infrastructure. Ecuador's second most visited national park (Parque Nacional Cotopaxi-PNC) is centered on the volcano and draws some 200,000 tourists a year.

The volcano's five most important eruptive episodes during the historical period (since 1532) have been of andesitic composition and ranged between VEI = 3 and 4 (Pistolesi et al. 2012). Nonetheless, the volcano is bi-modal and produces VEI = 5 rhyolitic eruptions about every 2000 years (Hall and Mothes 2008). The last important rhyolitic eruption, the Peñas Blancas event, occurred about 2800 years BP (Mothes et al. 2015a). The youngest andesitic eruptive products contain intergrowths of plagioclase and pyroxene and four different populations of plagioclase crystals which indicate pervasive magma mixing (Garrison et al. 2011).

Given the high probability for the generation of long-distance lahars, wide dispersal of pyroclastic fall, and the consequential negative impact on many economic activities and the compromise of critical infrastructure should an eruptive period last for months to years, Cotopaxi is considered a "National" volcano, located in the center of Ecuador, near to Quito, the capital and other populated areas. Even a short-lived VEI 4 eruption (VEI = Volcano Explosivity Index) (Newhall and Self 1982) has the potential to gravely affect Ecuador's overall productivity and major transport lines. Lastly, the volcano is considered one of the most dangerous in Ecuador, given the possible exposure of >300,000 residents to primary lahars and ashfalls during

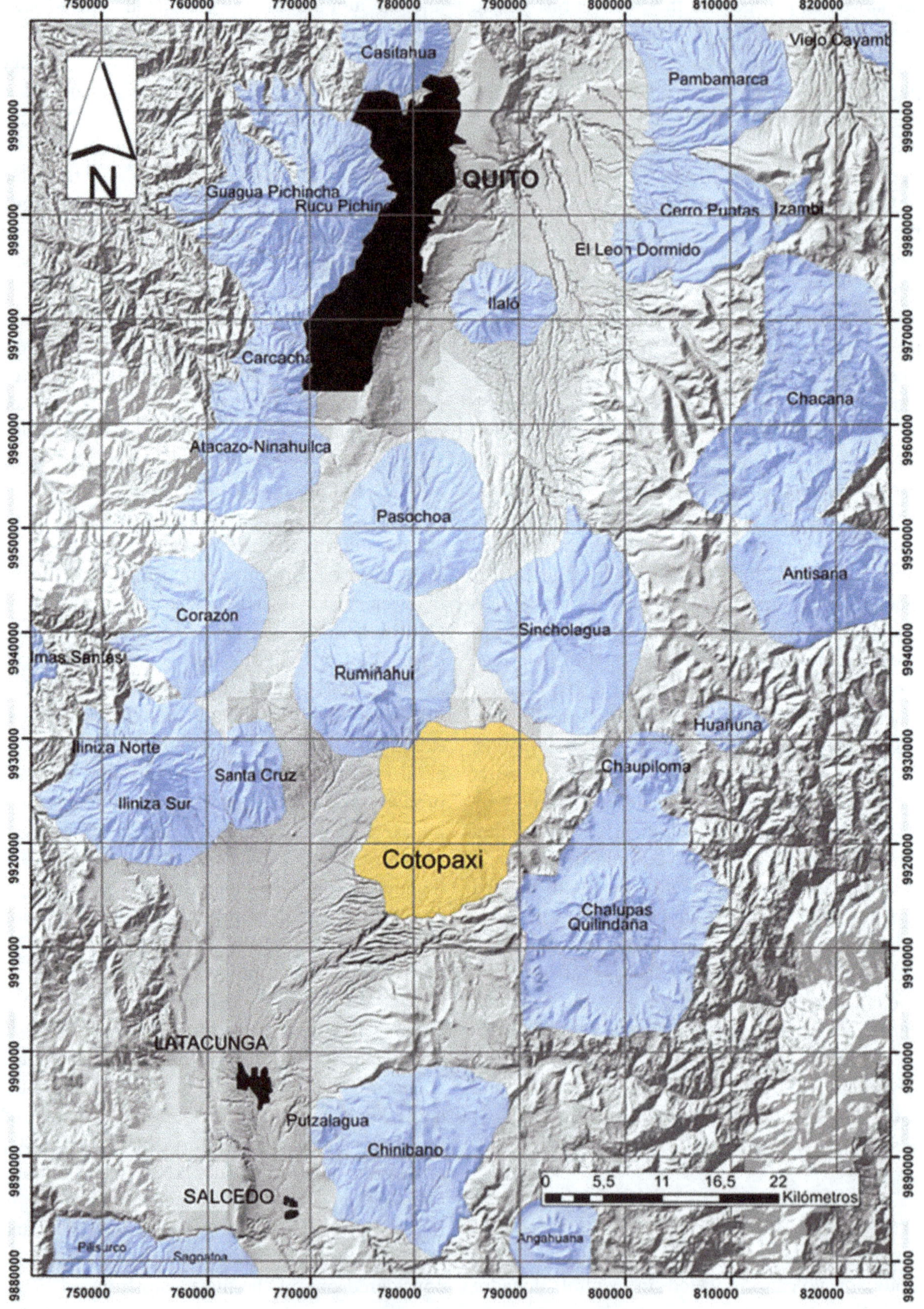

Fig. 1 Location of Cotopaxi volcano in Ecuador's Inter–Andean Valley and major cities

future VEI = 3 or greater magnitude eruptions (Mothes et al. 2016a).

Monitoring Cotopaxi

Cotopaxi has long been a producer of swarms of LP's, and Ruiz et al. (1998) hypothesized that they were related to the interaction of the hydrothermal system with heat ascending from depth. During unrest in 2001–2002 thousands of long period, volcano-tectonic and distal volcano-tectonic seismic events were registered and magma was hypothesized to have ascended to shallow levels in the center of the edifice (Molina et al. 2008). Its flanks also experienced deformation from magma input estimated at ~ 20 Mm^3 from modeling of EDM data (Hickey et al. 2015). But, observable superficial manifestations were meager, no magma erupted and the volcano returned to a relatively quiet state with only short-lived LP swarms (Lyons et al. 2012) and sporadic VLP events being registered until April, 2015 (Márquez 2012; Arias et al. 2015).

The IGEPN has monitored Cotopaxi volcano since 1986. Subsequently, over the years the monitoring coverage has become denser and more robust (Fig. 2) (Kumagai et al. 2007, 2009, 2010). Presently there are approximately 60 telemetered geophysical sensors operating on its flanks. Cotopaxi hazard maps have been published in several versions since 1976, with the newest version published in late 2016 (Mothes et al. 2016a). The IGEPN has carried out a long-term program of community education for areas that are at highest risk, although as noted by Christie et al. (2015), the attention over such a vast area ($\approx$2000 km^2), was uneven.

Cotopaxi was a VUELCO target volcano from 2013 to 2015. As part of the VUELCO project, in November 2014 a simulation exercise was carried out with the purpose of presenting a timeline of potential unrest and expected events and to test the communication between scientists, decision-makers and the public (www.vuelco.net). This present contribution is written in the spirit of holistically documenting this recent and most serious unrest of Cotopaxi to date—since 1942, when slight ash emissions and mild explosive activity were then reported (Siebert et al. 2010). Here we present the macro patterns of seismic, gas, deformation and visual observation monitoring data associated with the awakening volcano. The data are provided by the monitoring networks operated by the IGEPN. We avoid dwelling on details, as forthcoming contributions dedicated to exploiting specific datasets are in preparation. We also provide comment on selected actions in which IGEPN scientists participated to make the overall societal outcome more favorable in case Cotopaxi produced a major eruption. We impart with the philosophy, stated in Marzocchi et al. (2012) that "sound scientific management of volcanic crises is the primary tool to reduce significantly volcanic risk in the short-term". We also maintain that a constant and rapid analysis of the monitoring data is key to giving forecasts that include reasonable scenarios. Some of the IGEPN actions were guided by experiences gained in the VUELCO project, since one of the scenarios in the simulated eruptions was that the volcano would wake up, be active then return to repose.

The 14th of August, 2015 explosions and subsequent emissions pushed the first evidence of new magma to the surface, although in a limited way (Gaunt et al. 2016). Documentation of the geophysical signals and observations that we registered through late 2015 leads to the depiction of what transpired—mainly of an intrusion, which stayed deep, although the signs of intrusions that stall at depth may be very similar to those produced by intrusions that finally do erupt (e.g. Moran et al. 2011).

Synthesis of the Geophysical Fingerprints of the Unrest

Having passed 13 years with low levels of activity since cessation in 2002 of its last reactivation, in mid-April 2015 Cotopaxi began departing from background levels: higher seismic energy release, gas outputs and superficial manifestations transpired. The height of activity was

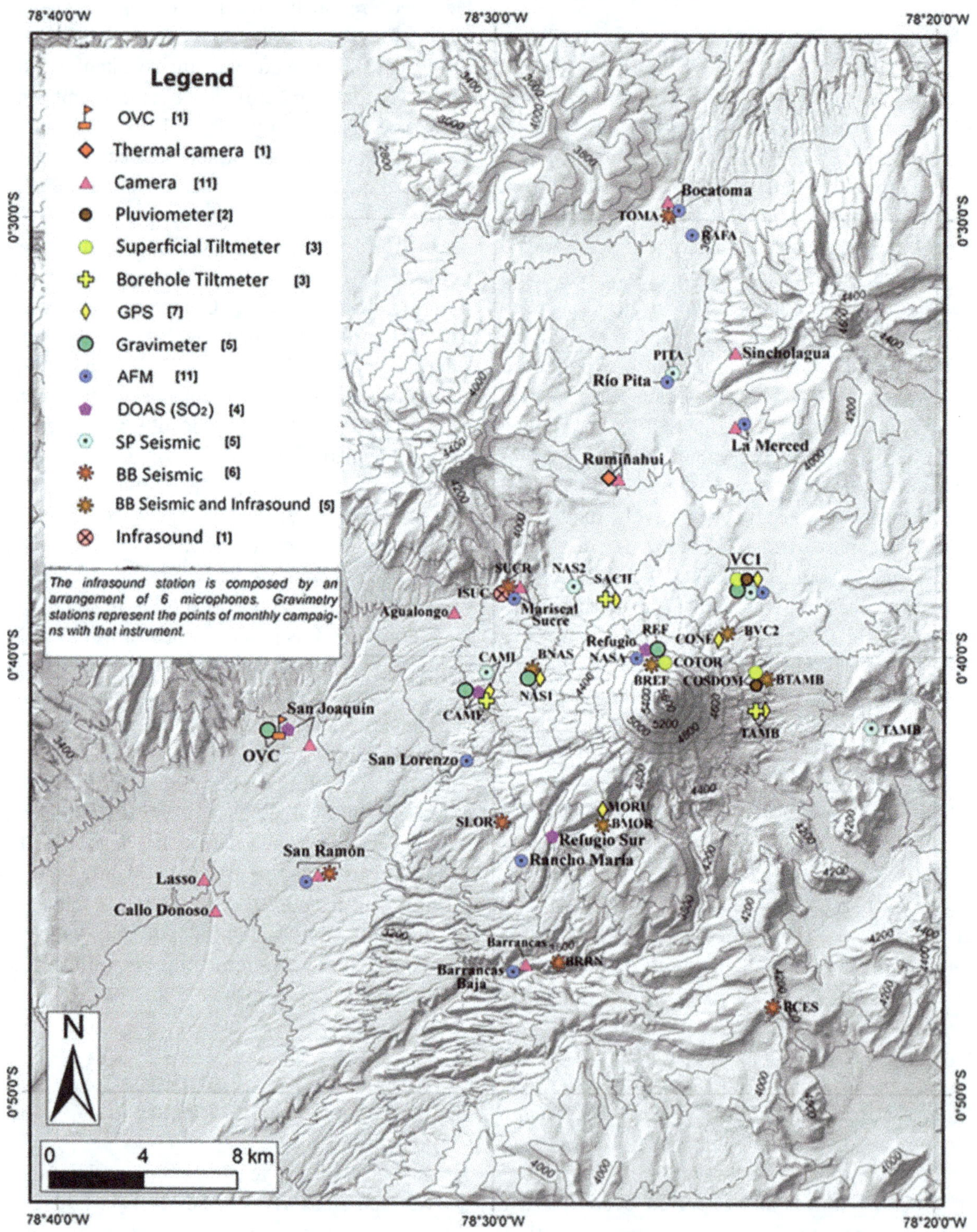

Fig. 2 Map of the instrumental monitoring network around Cotopaxi volcano, April 2016

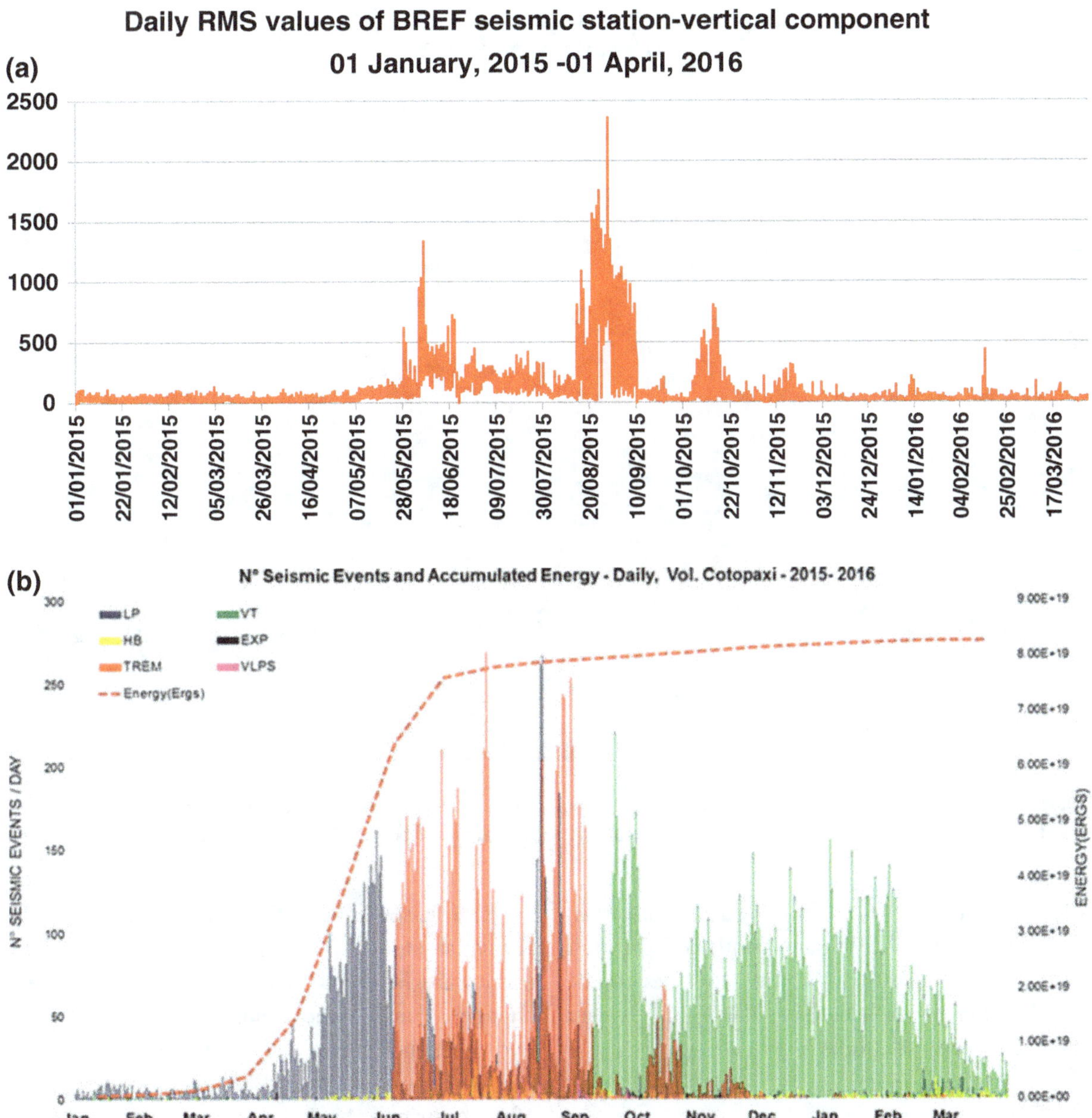

Fig. 3 **a** Plot showing RMS values of seismic data that has been segmented into 1 min windows and smoothed with a 31-point median filter. Of note is the calm period from January to April 2015, than a slight increase in seismic energy release in May. The increase in June is associated with greater emissions and strong tremor. A sharp decline in early August and later the notable increase in August and September represents the 14th of August explosions/emissions and subsequent ash emissions and emission tremor. Accumulated seismic energy values through the end of 2015 show a marked decline. **b** Number of daily seismic events versus accumulated seismic energy of these events. The acronyms for different seismic events are: *LP* Long Period; *HB* Hybrid; *VT* Volcano-tectonic; *TREM* High Frequency Emission Tremor; *EXP* Explosion; *VLPs* Very Long Period

a series of 5 explosions/energetic emissions on the 14th of August, which expulsed preexisting conduit plug material, ash and gases, but whose size did not surpass VEI = 1. By late September 2015, activity mostly died back and RSAM values showed a decline except for brief hikes in October and in November, when light ashfalls occurred. By December 2015 nearly all monitoring parameters were down to background levels (Fig. 3a), except for a protracted

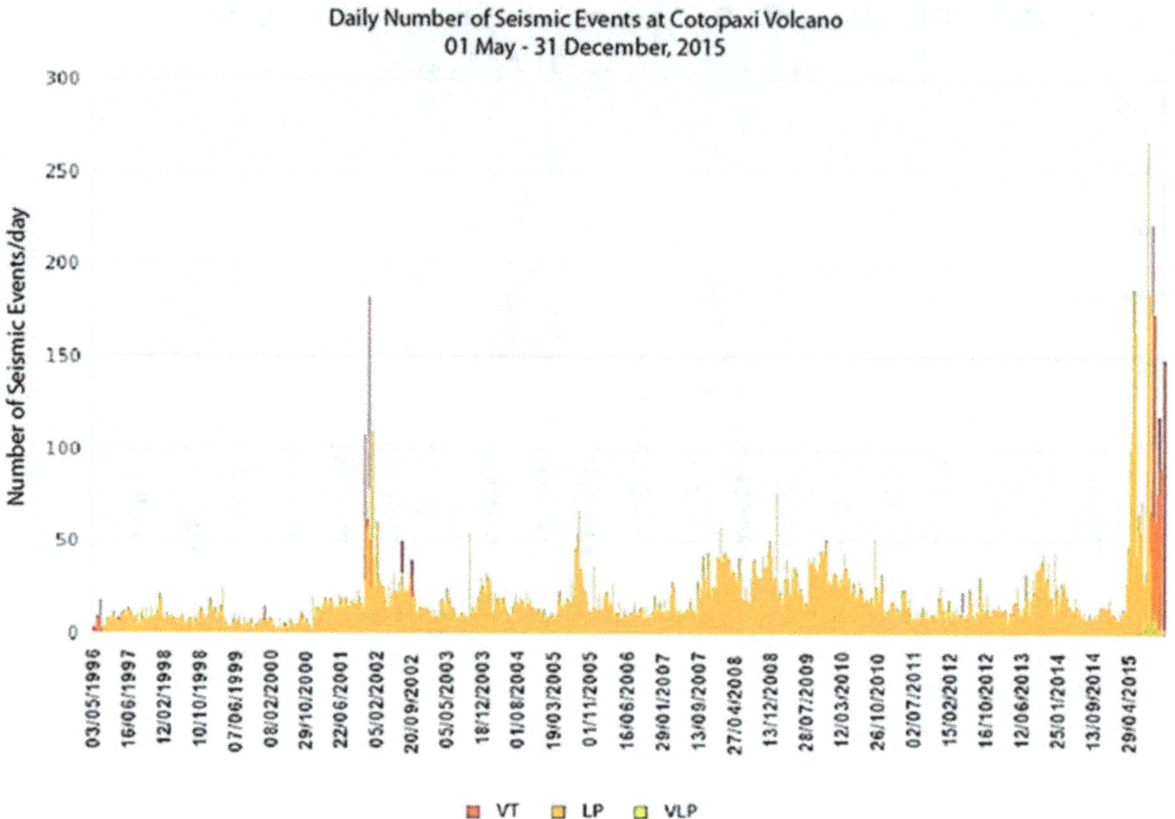

Fig. 4 Registry of VT and LP seismic events at Cotopaxi since May 1996 to 31 December, 2015. The 2001–2002 period was the only other period with a marked jump in seismic activity before the 2015 episode

volcano-tectonic (VT) seismic swarm that began on 10 September, 2015 and continued through March 2016, albeit displaying low levels of seismic energy release (Fig. 3b). This swarm produced nearly 15,000 VT events.

Geophysical Registry of Cotopaxi's Restlessness in 2015

From 2002 to April, 2015, seismic registry of mostly long period (LP) seismic events averaged around 10 events/day. In April 2015 the monthly tally was about 630 earthquakes, then rose to 3000 events in May (Fig. 4), with a jump to about 180 events/day registered on 23 May (Fig. 4).

Of significance also was the notable increase in very long period seismic events (VLPs) recorded since late May 2015. VLPs are often interpreted to signify magma movement (Zobin 2012; Jousset et al. 2013; Maeda et al. 2015; Kumagai et al. 2010; Arias et al. 2015). VLP events are believed to be generated by volume changes and movements of magmatic-hydrothermal fluids (e.g., Chouet and Matoza 2013). Between June 2006 and October 2014, 106 confirmed VLP events were identified at Cotopaxi (Márquez 2012; Arias 2015). In 2015 Cotopaxi, VLPs rarely passed 11 events/day (Fig. 5), but commonly had magnitudes of 2–3. The recent VLP events that were located under the Cotopaxi's edifice, occurred in sectors of the volcano where VLP's had been previously located by Molina et al. (2008) (Fig. 6). The greatest number of VLPs, of the 114 located events, were registered during the third week of July up to the explosions on the 14th of August. While most were between 1 and 2 magnitude, some were greater than 2.5 (Fig. 5a). The relationship between the great number of LPs which started the awakening process at Cotopaxi and

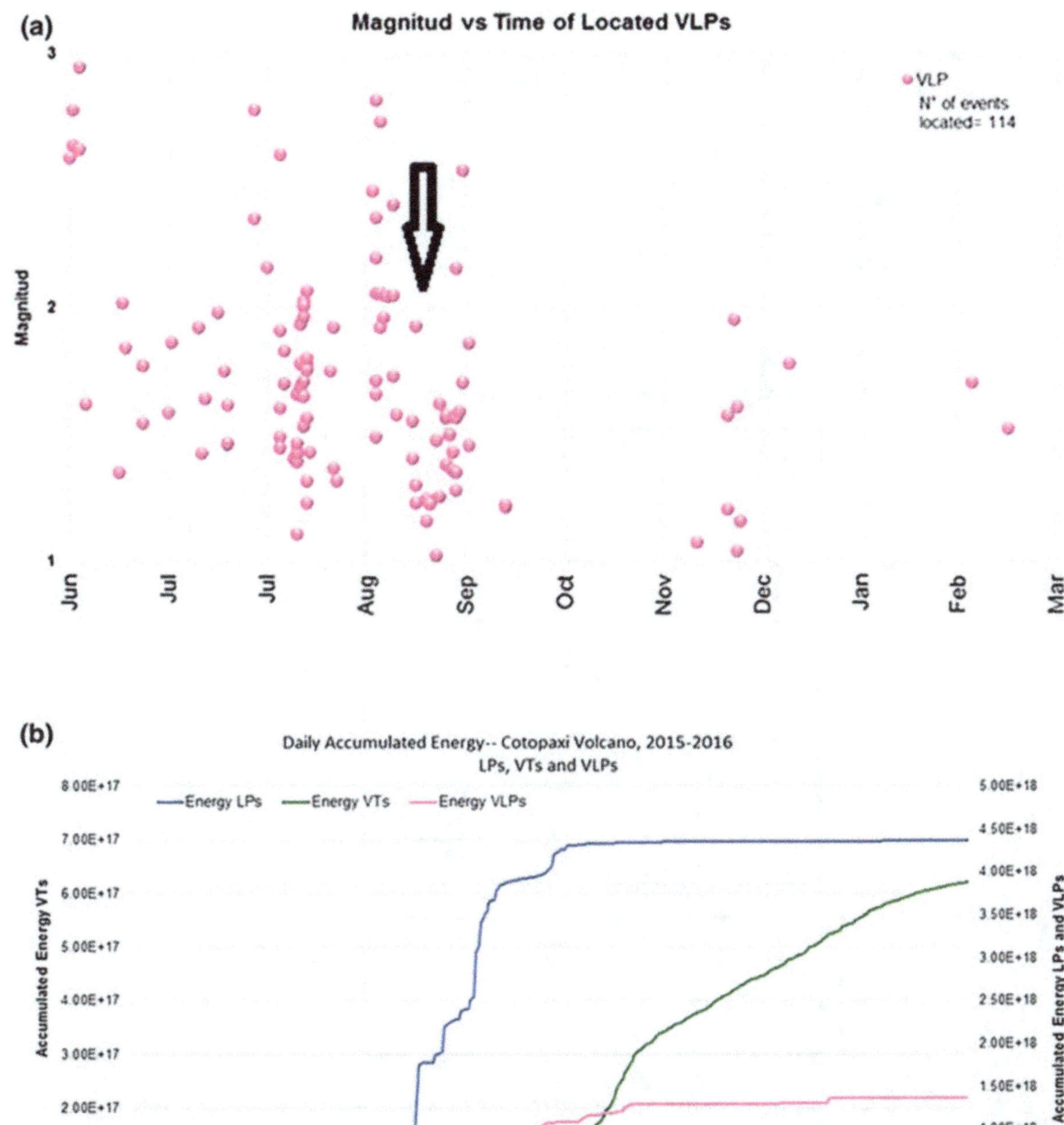

Fig. 5 **a** Occurrence of VLP events in 2015 at Cotopaxi. *Black arrow* represents 14 August hydromagmatic explosions. **b** Accumulated seismic energy from VTs, LPs and VLPs in 2015–2016 at Cotopaxi

afterwards the stalling out of these events to be followed immediately by the strong VLPs is another possible indicator of the precursory nature of this type of volcanic earthquake before the discrete eruptions on 14 August (Fig. 5b).

Most VLP events had frequencies between 0.1 and 1 Hz and had strong P and S waves, such as the example given for 04 August, 2015 which was located 3 km below summit on the NE flank of the volcano (Fig. 6).

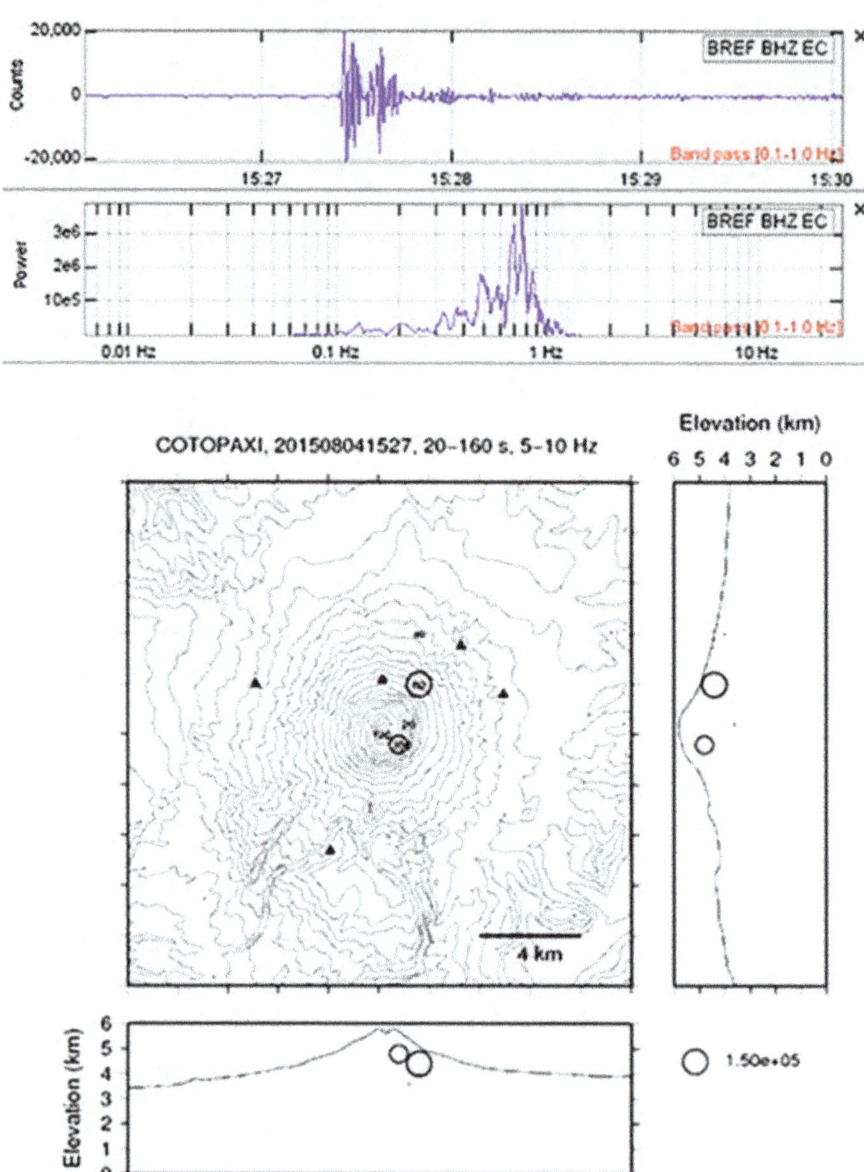

Fig. 6 Waveform of VLP registered on 04 August, 2015 15h27 GMT is one of the largest events registered at Cotopaxi (local magnitude 2.5) and was located 3 km beneath the crater on the NE flank

The locations of earthquakes (of all types) from January to December, 2015 were at two levels: at depths of about 3–5 km below the crater (Fig. 7a) and at a deeper level of 7–15 km below the crater. Most events were aligned with the conduit. However some distal VTs were registered about 15 km due north of the volcano (Fig. 7b) and were interpreted as fault slips due to stress transfer from the volcano (White and McCausland 2016). Distal VTs were also important in the reactivation of Pinatubo volcano (Harlow et al. 1996).

Overall, there was a marked increase of LP events from April to late May, followed by high frequency tremor episodes (Fig. 8) which lasted until the onset of high frequency tremor related to gas emissions and which became prominent from 04 June and lasted to the second week of August (represented by black bars), and could have been related to the boiling of the volcano's hydrothermal system, and coincided with the high water vapor and SO_2 flux then emitting from the crater (Bernard et al. 2016). In Fig. 8, the VLPs that were important especially in July

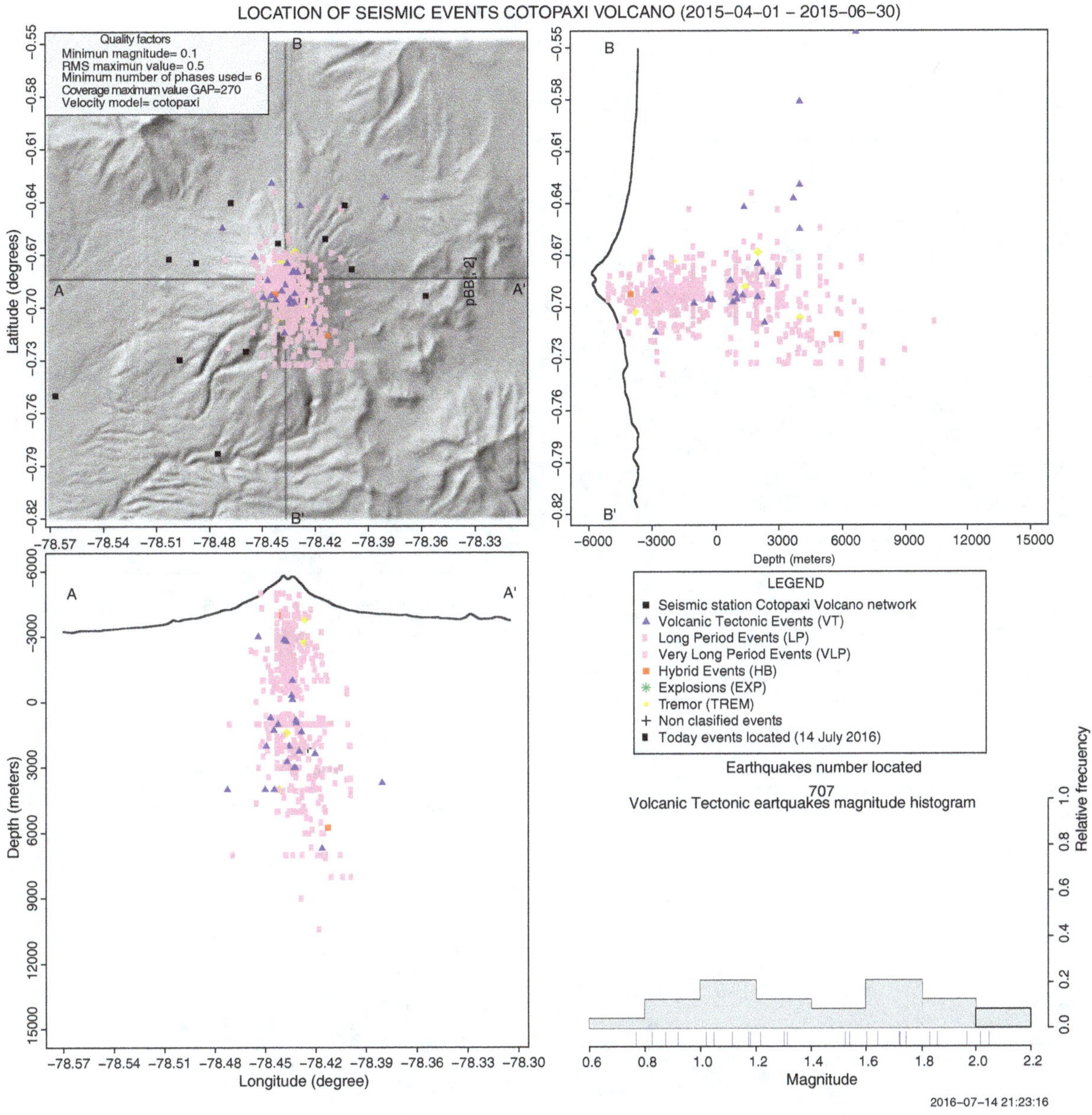

Fig. 7 Top-Locations of seismic events from a–01 April to 30 June, 2015 and Base—01 July to 31 August, 2015. Most are aligned with the conduit, however the SE flank is favored for harboring event locations

to mid-August are masked by this tremor signal, but can be observed in Fig. 3 and their accumulative energy levels are shown in Fig. 5.

The increase in SO_2 gas emissions, rose to 3000 ton/day by the end of May with a clear SO_2 signal progressively more notable through late May into June (Fig. 9) (Hidalgo et al. 2016). For example, on the 22nd–23rd of May odors of sulphur were very evident above the 5700 m level on the volcano's northern flank, as reported by Cotopaxi Park personnel.

GPS stations on the W and S flanks showed horizontal displacements of almost 16 ± 0.5 mm toward the W and SW. GPS stations on the NE and E flanks showed displacements to the N at a reduced velocity (Fig. 10). The vertical component registered a maximum uplift of 15 ± 2.3 mm. The movement to the west could have been accentuated by the volcano's morphology, as the W flank is poorly buttressed and sits upon Inter-Andean Valley volcaniclastic fill. In comparison the east and northern part of the cone

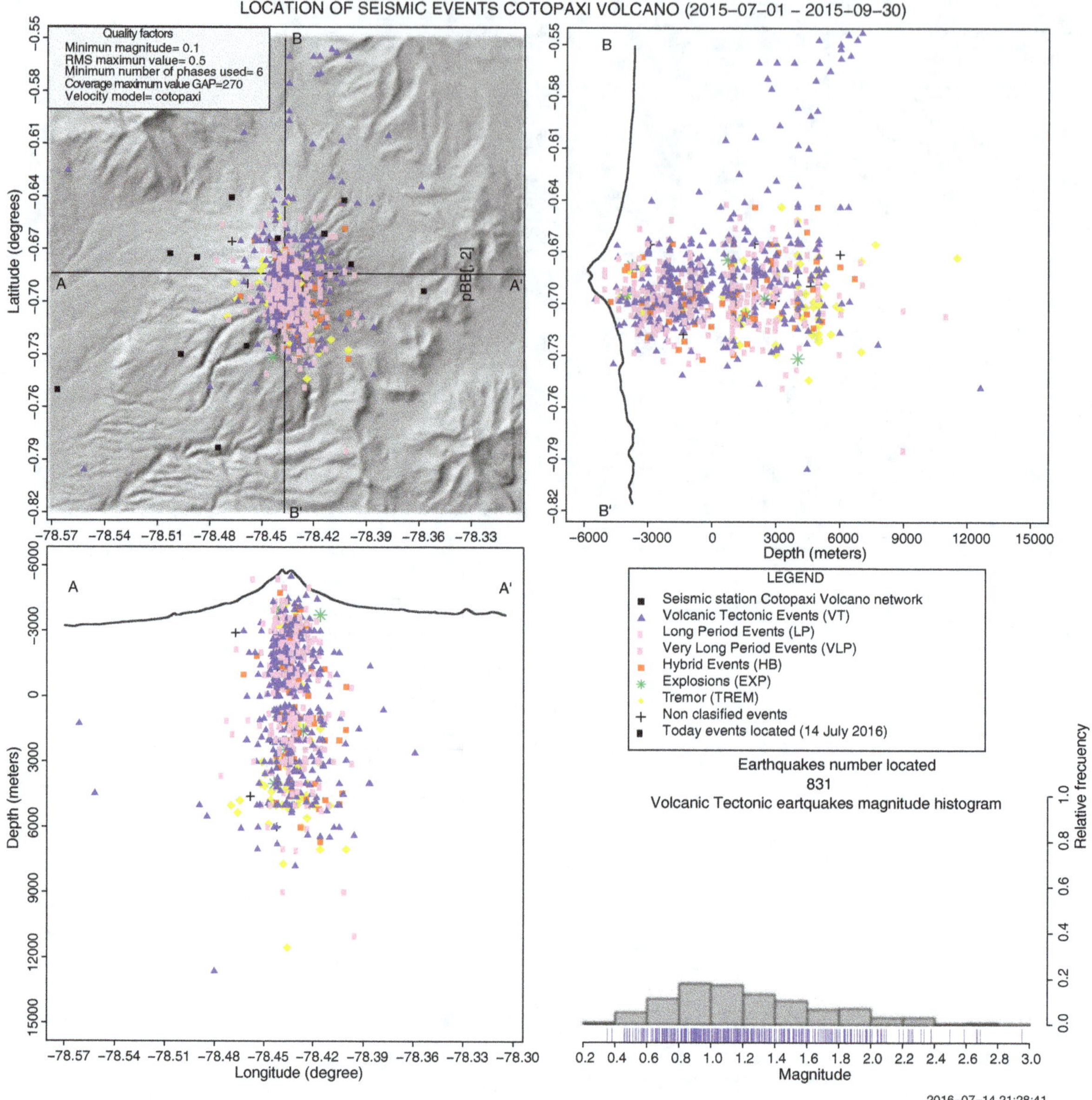

Fig. 7 (continued)

Fig. 8 a Comparative graph of LP seismic events (*orange bars*) and high frequency tremor (*black bars*) possibly related to the boiling of the hydrothermal system and gas movements from 01 April to 14 August, 2015. Ash and gas emission-related tremor (*pink bars*) abruptly began the second week of August, 2015, after the hydromagmatic explosions on the 14th. **b** Seismogram (11 June, 2015) of BREF station showing registry of spasmodic tremor related to internal fluid movements in the upper part of the edifice

lies upon a thick lava package and basement crystalline metamorphic rock and may be more resistant to lateral movement. Data processing employed the program GAMIT/GLOBK (Herring et al. 2015) and used a local reference frame with respect to fixed South America (Nocquet et al. 2014. We also defined a long-term displacement model for each GPS site by estimating a trend and annual and semi-annual components using all available data between 2008 and 2015. The transient displacements identified during the 2015 unrest period are with respect to this model. In a second step, we applied a common-mode filtering estimated from the average time series residuals for sites ∼50 km away from the volcano. Short-term repeatabilities are of the order of 1–2 mm on the

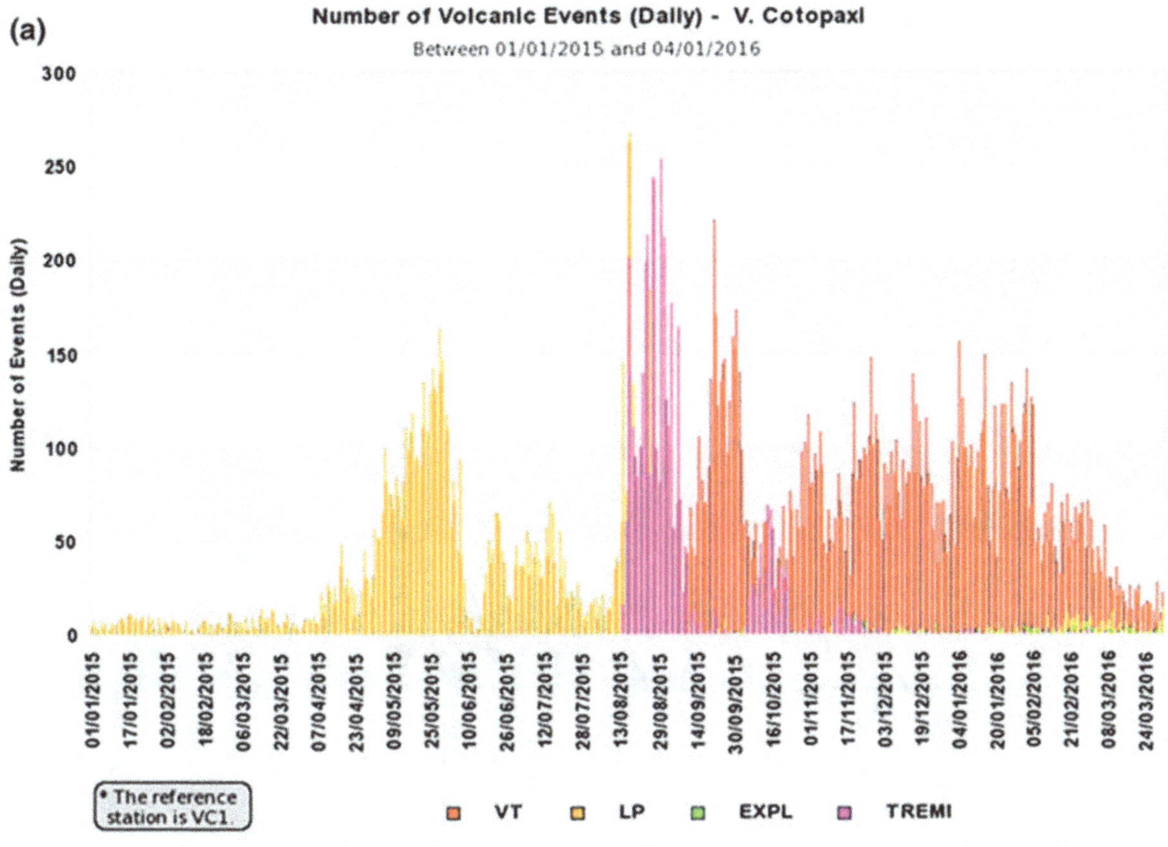

(a)
Number of Volcanic Events (Daily) - V. Cotopaxi
Between 01/01/2015 and 04/01/2016
Number of Events (Daily)
300
250
200
150
100
50
0
01/01/2015
17/01/2015
02/02/2015
18/02/2015
06/03/2015
22/03/2015
07/04/2015
23/04/2015
09/05/2015
25/05/2015
10/06/2015
26/06/2015
12/07/2015
28/07/2015
13/08/2015
29/08/2015
14/09/2015
30/09/2015
16/10/2015
01/11/2015
17/11/2015
03/12/2015
19/12/2015
04/01/2016
20/01/2016
05/02/2016
21/02/2016
08/03/2016
24/03/2016
* The reference station is VC1.
VT   LP   EXPL   TREMI

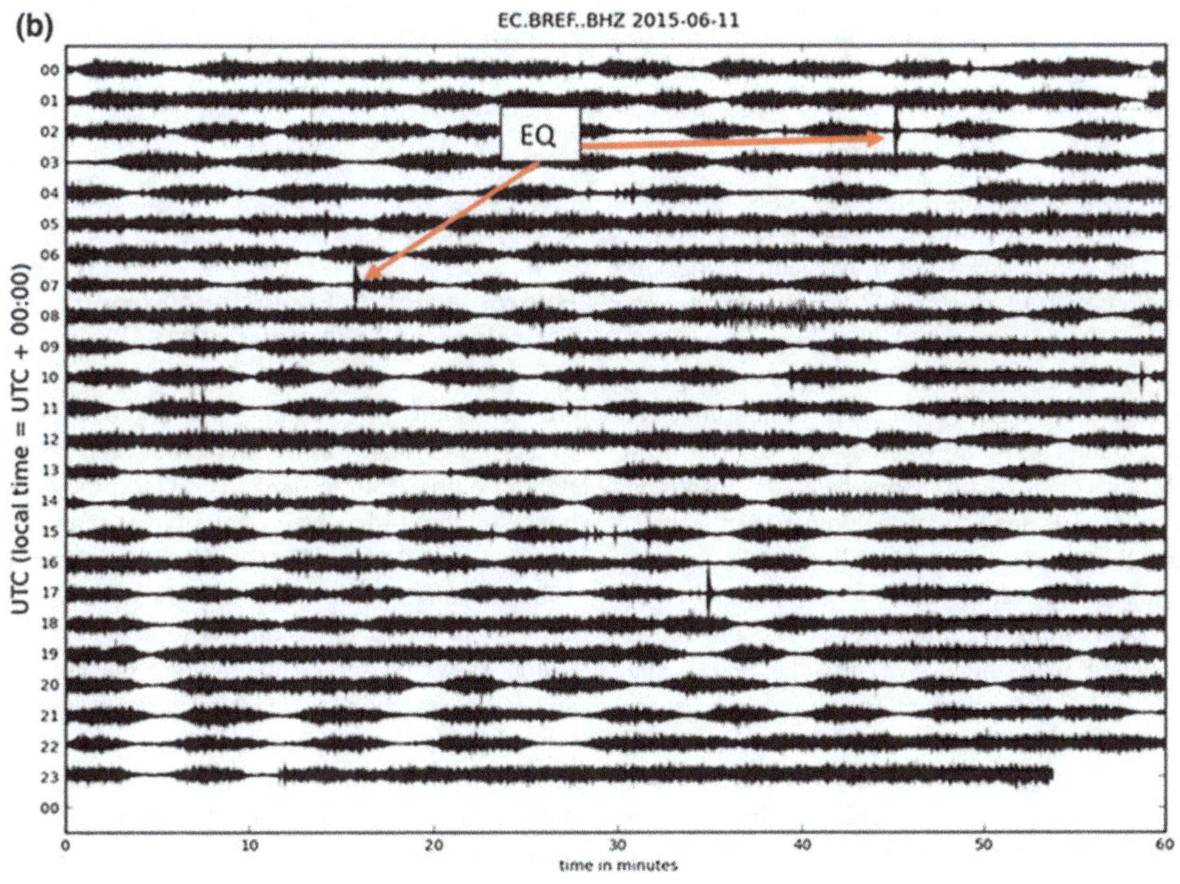

(b)
EC.BREF..BHZ 2015-06-11
UTC (local time = UTC + 00:00)
00
01
02
03
04
05
06
07
08
09
10
11
12
13
14
15
16
17
18
19
20
21
22
23
00
EQ
0   10   20   30   40   50   60
time in minutes

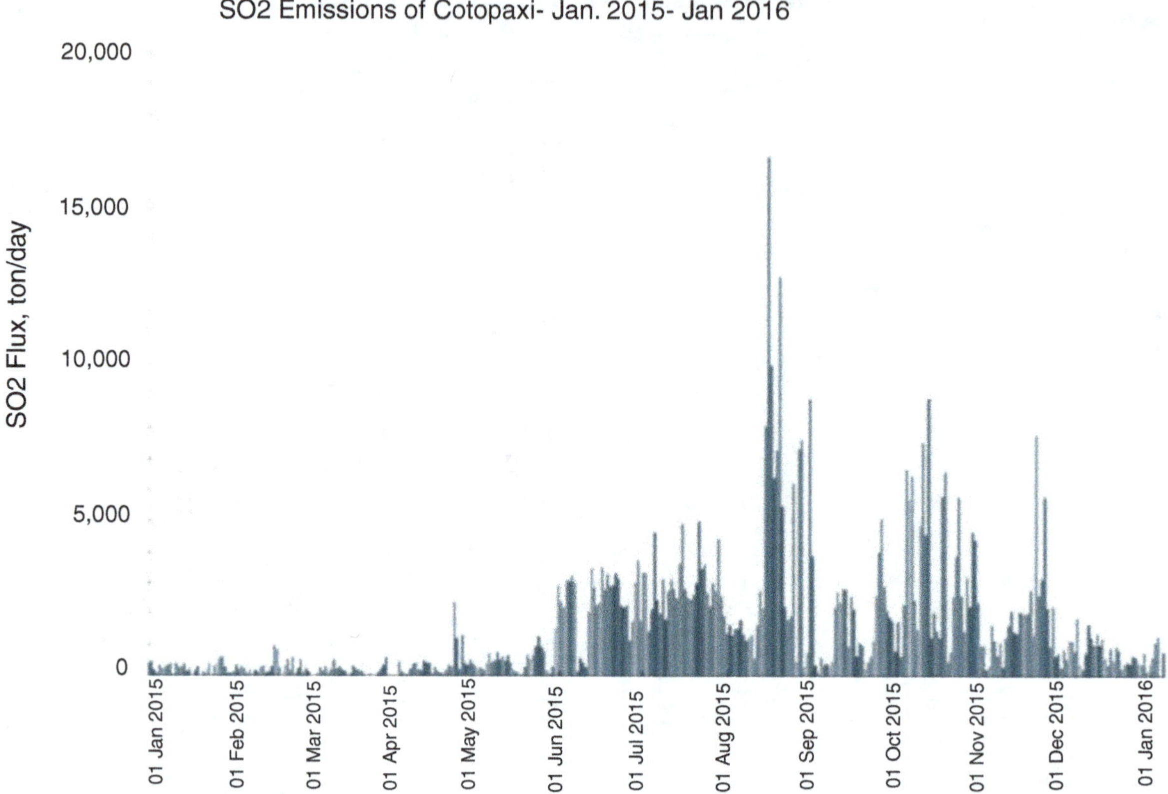

Fig. 9 Registry of SO_2 values for Cotopaxi, January 2015 until 05 January 2016, with a significant increase in SO_2 observed since early May. Data was processed daily using a single wind speed and direction obtained from the NOAA and VAAC alerts. Graph included in Cotopaxi Special Report, No. 1, 2016: http://www.igepn.edu.ec/cotopaxi/informes-cotopaxi/coto-especiales/coto-e-2016/14074-informe-especial-cotopaxi-n-01/file

horizontal components and 3 mm on the vertical component, enabling us to extract the small GPS signal observed during the unrest period.

Due to the westward movement on the GPS station PSTO, which is 22 km W of the crater (Fig. 10), we surmised that the source was deep. Subsequent modeling of the data suggested a source of about 24 km deep located under the SE flank with a volume of 42 ± 26 Mm^3 (Mothes et al. 2016b). Nonetheless, as mentioned in Sect. 3.1, analysis of the erupted ash suggests that the magma source is shallow, as least for the initial small volume that was emitted.

Data from a tiltmeter (VC1G on Fig. 2) installed in a thick lava package and located 6 km NE of the crater, showed a strong inflationary pattern that had started in April, 2015 on both axis. This tilt anomaly coincided with the notable increase in seismicity (Mothes et al. 2016b). Generally, when LP seismicity and tremor were both strong, a positive tilt signal predominated.

Hydromagmatic Explosions/Strong Emissions of 14 August, 2015

On the evening of 13 August, a swarm of VT and LP seismic events was registered between 20h03 (GMT) on the 13th to 08h55 (GMT) in the early hours of the 14th, antecedent of the explosion events (Fig. 11). At 09h02, 09h07 and later at 15h25, 18h45 and 19h29 (UT) five small explosions/strong emissions were registered at Cotopaxi which served to unblock the conduit and led to ejection of degassed altered conduit plug material and scarce juvenile components. Although infrasound from these explosions did not exceed 4 Pascals (Pa) at stations located approximately 6 km from the vent, the first two

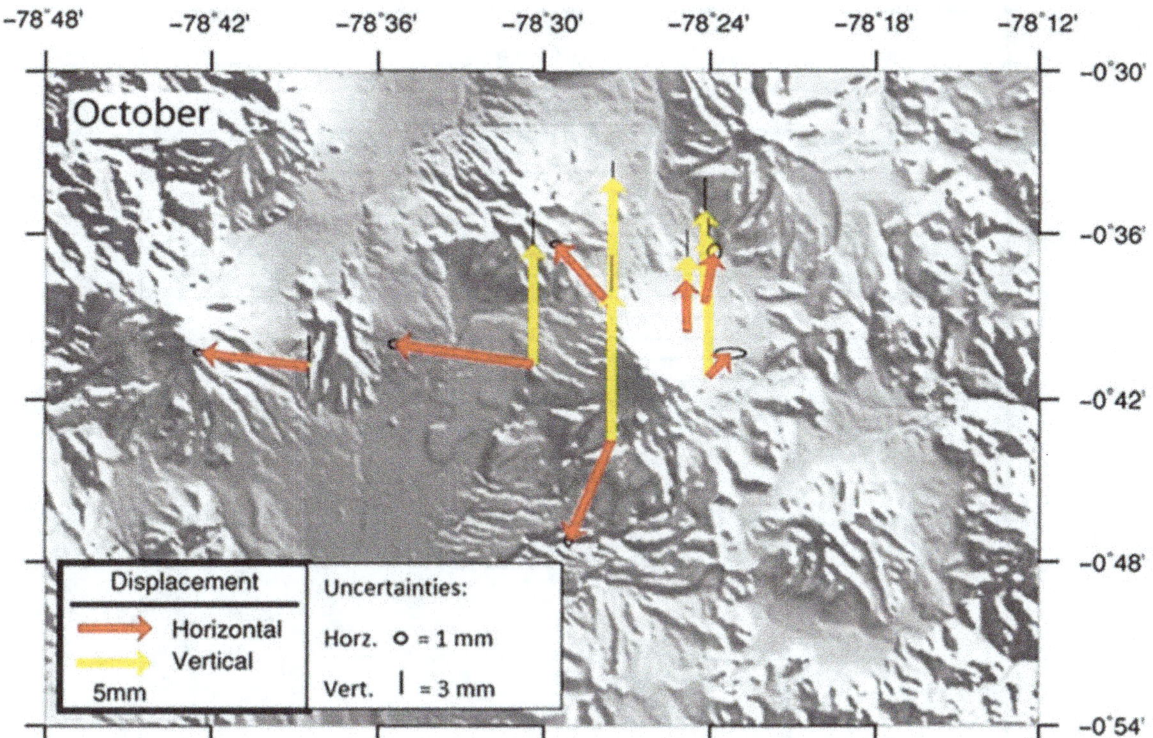

Fig. 10 GPS vectors for stations around Cotopaxi's cone and one to the west for October, 2015. Displacements are expressed with respect to the North Andean Silver and represent the comparison of GPS data collected from 01 January to 01 October, 2015 at the 7 station Cotopaxi CGPS network

explosions were heard by climbers in the Refuge on Cotopaxi's north flank, where lapilli-size fallout reached the Cotopaxi Refuge.

Two months earlier public and authorities had been forewarned in the special IGEPN reports (No. 3 and 4) that phreatic explosions would be a likely phenomenon in precursory eruptive activity (http://www.igepn.edu.ec/cotopaxi/informes-cotopaxi/coto-especiales/coto-e-2015/12990-informe-especial-cotopaxi-11-06-2015/file).

With these explosions the eruption column at 15h25 rose to 9 km above the crater rim and was clearly visible from the SW (Fig. 12a, b). Infrasound values of the explosions were less than 10 Pa at station BNAS (5 km from the crater), but the seismic source amplitudes of the tremor associated with the first two explosions were greater than those of most Cotopaxi LP events and also of some explosions registered at Tungurahua volcano (Kumagai et al. 2015). The initial explosions had evidence of water involvement. In previous weeks a small lake was observed in the crater's bottom; this was totally evacuated by the explosions. Observers also reported that the fallout had a "wet aspect" and many of the fragments were agglutinated by a fine clay-size patina. The eruption is categorized as hydromagmatic, since the rapid interaction with water caused overpressures beneath the plug, raising lithostatic pressures that overcame the capacity of the altered conduit plug rock. After these main vent-opening events the presence of hydrothermally altered material gradually waned and possible juvenile material became more prevalent (Gaunt et al. 2016).

The ash emissions from this first activity covered agricultural lands to the NW and W of the volcano with a ≤ 1 mm thick dusting of altered silt to sand lithic grit and crystals (Fig. 13a) and caused poor visibility along major highways that enter Quito from the south. This ash emission mantled over 500 km^2 with more than 80 gr/m^2 and amounted to a volume of 118,000 m^3, keeping it within the range of a VEI = 1 (Bernard et al. 2016) (Fig. 13b).

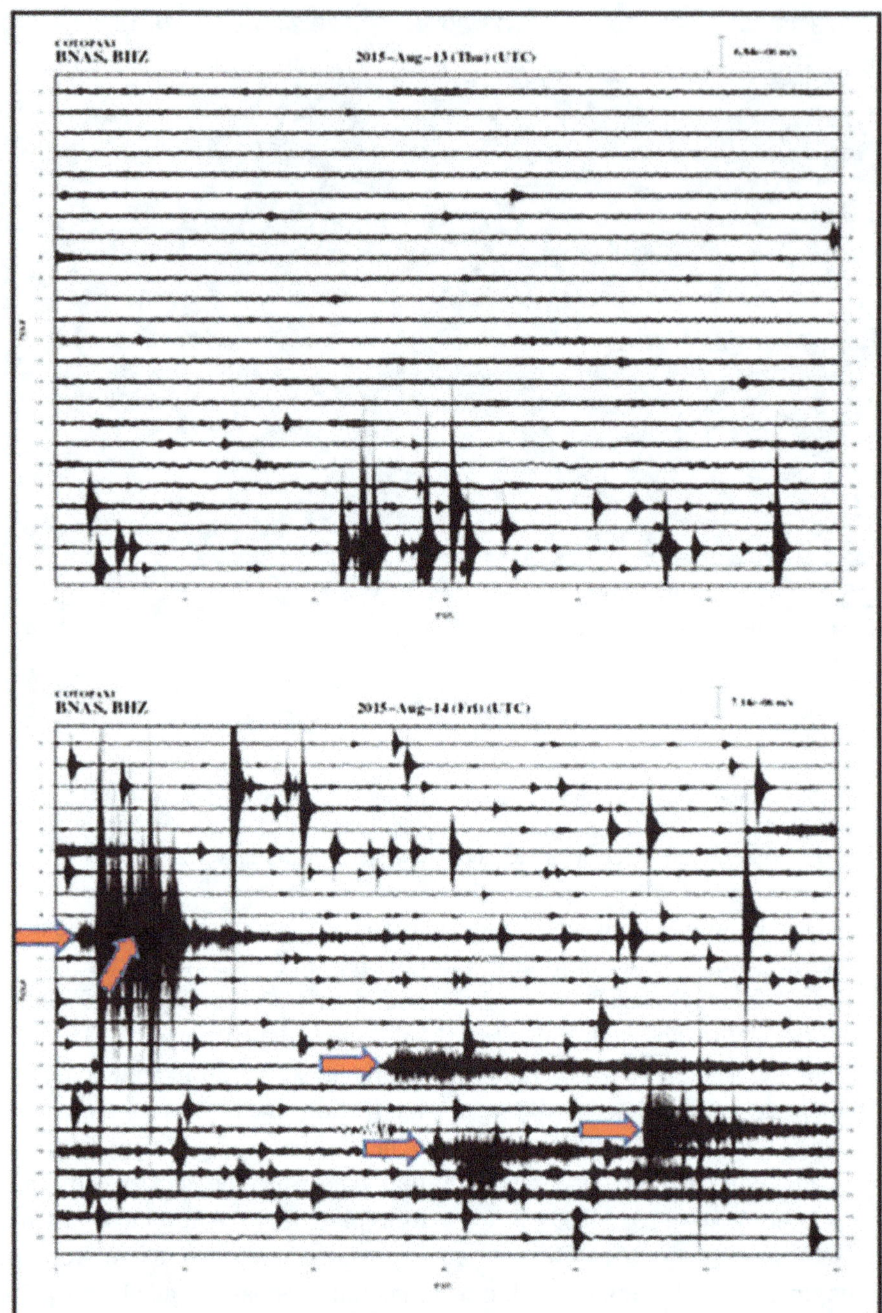

Fig. 11 Registry of VT—LP seismic swarm which begin late 13 August, 2015 and was followed by the 5 small explosions/strong emissions on 14 August, all indicated with *red arrows*. Seismograms are of the IGEPN's monitoring network

Post 14 August, 2015: Open Conduit Degassing and Ash Emissions

Ashfalls were prevalent towards the N and NW after 14 August into October and became scarce in late November (Fig. 14). A common scene was that of the ash and gas plume cascading down the W flank, with only the initial pulse rising to <1 km upon emitting from the crater (Fig. 14).

Emission tremor of varying amplitudes accompanied the ash emissions and permitted IGEPN monitoring scientists to forecast if the

Fig. 12 *Left* Cotopaxi's 14 August, continuing emission —view of the volcano from SW at 14h10UT. *Photo* E. Pinajota, IGEPN. Right- The 15h25UT strong emission produced a column that ascended 6–8 km above the summit. *Photo* by Santiago Tapia, at the Novacero company grounds, 20 SW of the volcano

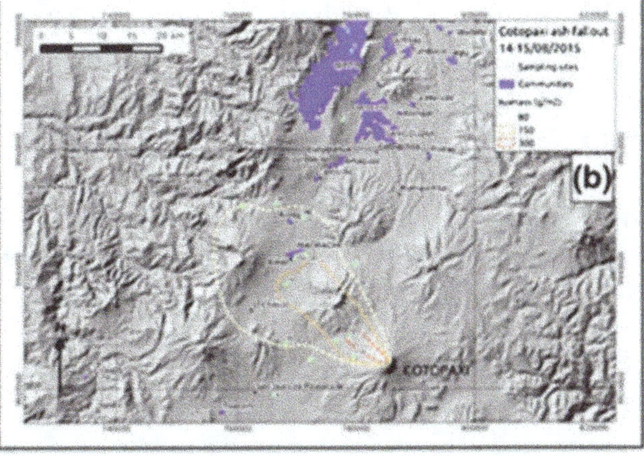

Fig. 13 **a** Ashfall from 01 September emissions accumulated in cultivated fields near El Chaupi town. **b** Ash fallout map associated with the eruptive activity of Cotopaxi on the 14th–15th of August, 2015. Map projection WGS 84, coordinates in UTMs. Values expressed in isomass of grams/m^2. *Source* Bernard et al. (2016)

rate of ash falls was increasing during frequent foggy, overcast conditions (Bernard et al. 2016). On the 14th of August, after the explosions, SO_2 levels reached 16,400 ton/day as registered by the satellite sensor OMI (http://so2.gsfc.nasa. gov/pix/daily/ixxxza/loopall3.php?yr=15&mo= 08&dy=15&bn=ecuador) (Fig. 15). Subsequently, SO_2 levels were particularly high on the 15th and 24th of August, when OMI measurements gave readings of 6500 and 6600 ton/day, respectively.

In early September ash columns still rose to over three kilometers height above the vent and carried fine ash particles to cities on the piedmont of the coastal plain, such as Santo Domingo de los Colorados, located 120 km W of the volcano,

Fig. 14 *Left* Photo with view toward south taken from Autopista Ruminahui (SE of Quito) on 20 August. *Photo-* C. Zapata- EPN. *Right* Photo taken on the 23rd of August, 2015, from the north side of Cotopaxi. A low gas and ash column trending to the NW is observed. *Photo* P. Mothes, IGEPN

essentially situated under the red swath trending W in Fig. 15.

Ashfall was still prevalent in mid-October, but had all but terminated the third week of November where it was seen W-NW of the volcano. VAAC Washington again reported suspended fine ash above Santo Domingo as well as in Los Rios province to the SW. The ash column generally rose to only 1 and 3 km above the summit and had a velocity between 6 and 10 m/s and lasted about one week. Fieldwork permitted the estimation of a mass and volume total of 3.49×10^7 kg (22,100 m^3) for this late, waning period (Bernard et al. 2016). The total ashfall dense rock equivalent (DRE) volume for the entire eruption was calculated in 0.5 Mm^3 (Bernard et al. 2016).

Ash Componentry

Analysis of ash beneath both binocular and scanning electron microscope showed clearly that there was an evolution in ash componentry from the eruption's beginning on 14 August and later. The first ash from 14 August had more hydrothermal lithics (pyrite, scoria with vesicles filled with altered material and hydrothermal quartz). As the eruptions progressed we saw an increase in more fresh magmatic components, such as free crystals, glass particles with low vesicularity and a high percentage of microlites,

which implied low magma ascent rates and stiffening of magma in the upper part of the column (Gaunt et al. 2016).

Ashes collected on the 20th of October, had a high concentration of dense microcrystalline material. Although there is evidence of few vesiculated clasts (diktytaxitic texture); about 65% of the ash is considered possibly juvenile. Gaunt et al. (2016) suggest that the origin of the ash is the top of a degassed magma column which had ascended from about 3 km below the crater.

Seismicity

For most of the post explosive period after mid-August, seismic hypocenters still remained located at the two depths mentioned above (Fig. 7). Most relevant was the sporadic occurrence of VT events with magnitudes of 3 or greater that occurred. Sometimes spasmodic tremor was registered and continued for hours, as for example, that registered on 02 September, 2015.

Starting on 10 September, a swarm of VT seismic events kicked in with a rate of approximately 100 events/day and a daily registry of coeval small internal explosions which has associated infrasound signatures, (shown in green color in Fig. 16). This swarm lasted past the New Year, but the overall seismic energy release was low (Figs. 3b and 5).

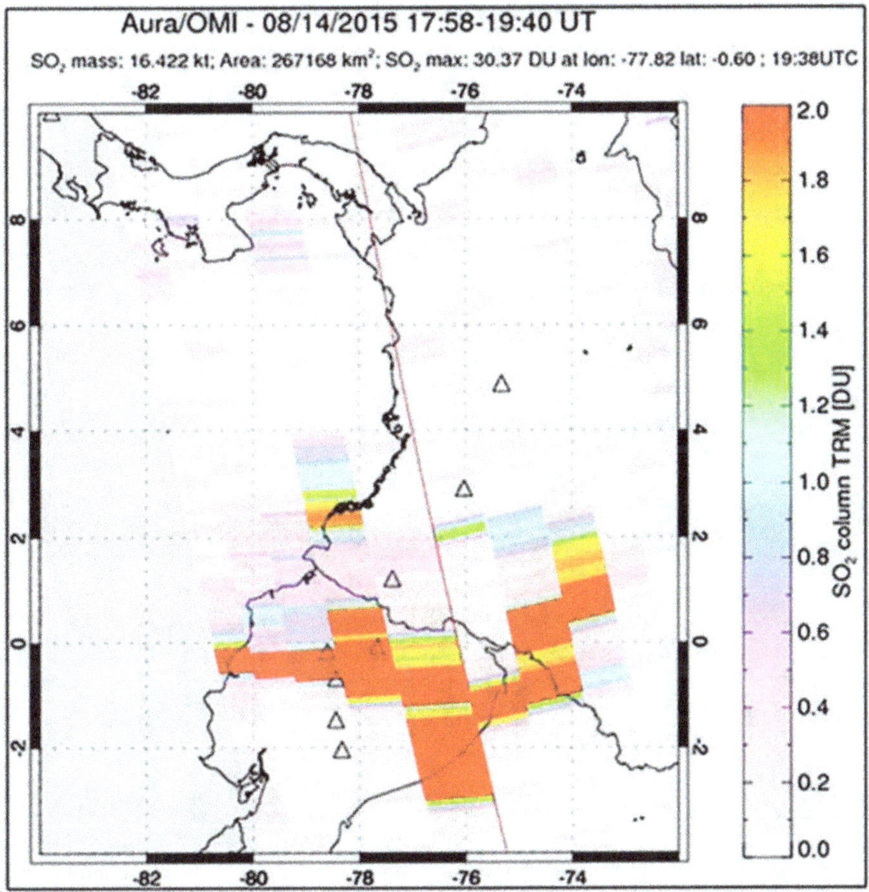

Fig. 15 Registry of SO_2 (16,400 ton/day) from satellite as detected by the OMI sensor on the 14th of August, 2015

Visual Observations and Secondary Effects

Thermal images (August/2015) showed the presence of new thermal anomalies (~ 15 °C) inside the crevices on the N side glaciers, at the same time fumarolic gases were observed coming out from those fractures. The highest temperature obtained was about 200 °C (Fig. 17) from gases ascending the crater.

On September 3, water emerging from the basal fronts on the northern glaciers was clearly observed, and countless new crevices in the majority of glacier ends and on the upper flanks were evident. All this led to the conclusion that an abnormal process was producing increased melting of the glaciers. Starting in mid September it was possible to observe the presence of small secondary lahars descending several streams and we estimated that many of them were due to increased glacier melting.

Orthophotos made on August 18 and then again on October 8, show a decrease of about 0.49 km² of the area covered by glaciers. This represents a very high rate of glacier melting, not explained exclusively by climate change (Cáceres et al. 2016).

We estimate that small volumes of magma reached surface levels in the volcano conduits causing increased circulation of hot fluids inside the edifice, apparently reaching the basal area of the glaciers and producing increased melting. It is necessary to further investigate the hazard due to instability in the melting glaciers and their eventual collapse that could lead to greater secondary lahars. Numerical modeling by

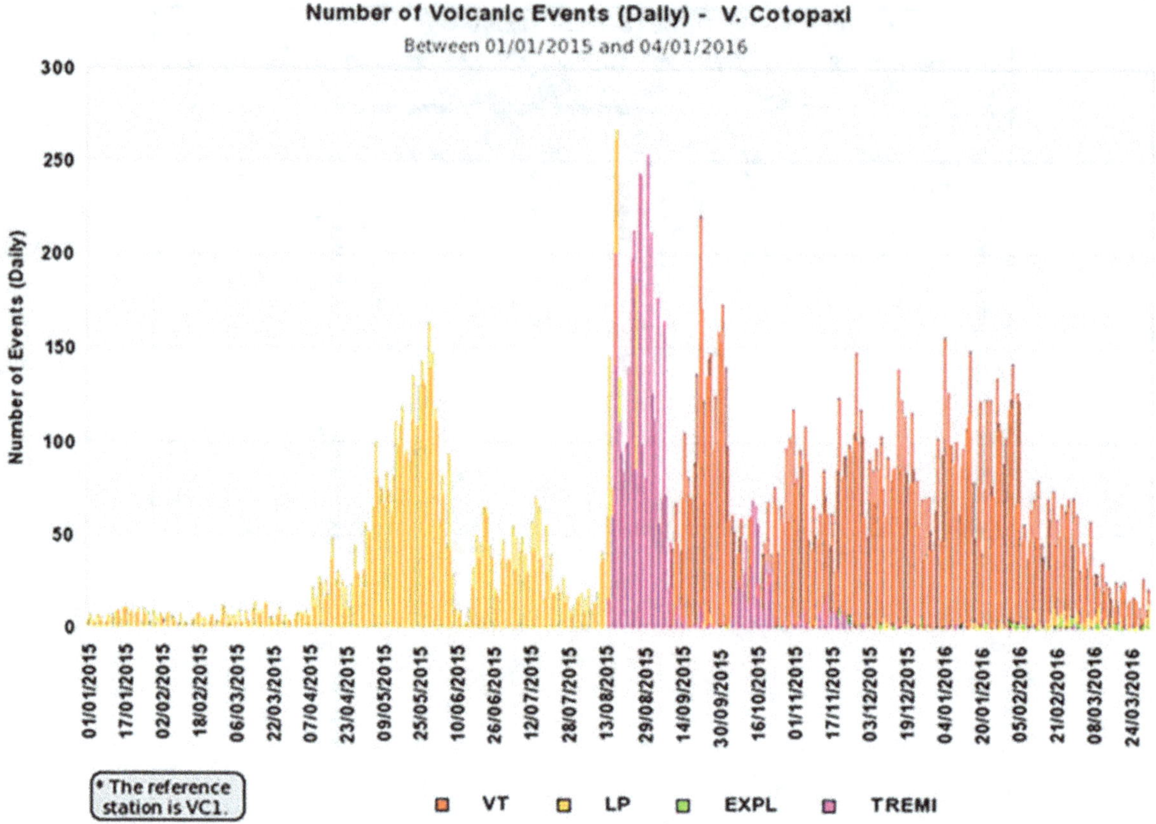

Fig. 16 Registry of overall seismicity at Cotopaxi volcano from 01 January 2015 to 01 April, 2016. Most notable is the presence of LP earthquakes in the first semester of 2015, later followed by emission tremor and finally with the advent and continuance of frequent VT seismic events with accompanying internal explosions that had no superficial manifestation, except infrasound registry

Hemmings et al. (2016) has shown the importance of hyrdrothermal perturbations at Cotopaxi in generating watery flows.

Incandescence was also occasionally observed on cold still nights with a thermal camera or by simple vision. These events were considered to have been caused by pulses of hot gases.

The glacier around the crater, on the W and N flanks, became partially covered by ash. This coating of dark ash decreased the glacier albedo and consequently increased the absorption of solar rays. Therefore expedite melting of the glacier tongues increased, leaving obvious melt water channels issuing from the glaciers' base.

As a result of the afternoon melting by insolation and perhaps also by higher temperatures of the rock beneath the glaciers, runoff increased, especially off the W flank glaciers and there were frequent small secondary lahars. Those lahars that have been especially associated with rain storms obtained the highest discharges—on the order of 10–30 m^3/sec (D. Andrade-IGEPN, Pers, Comm, 2015). The Agualongo channel, on Cotopaxi's W side was frequently flooded by lahars and on three occasions they covered partially the main road giving access to the volcano.

Interpretation and Model

As shown in Fig. 16, LP events gradually increased starting in April, 2015, beginning with small magnitudes and low energy levels (Fig. 3). We interpret the LPs to imply fluid movement occurring at 10–12 km below the crater and then up to shallower levels (Fig. 7). There were only

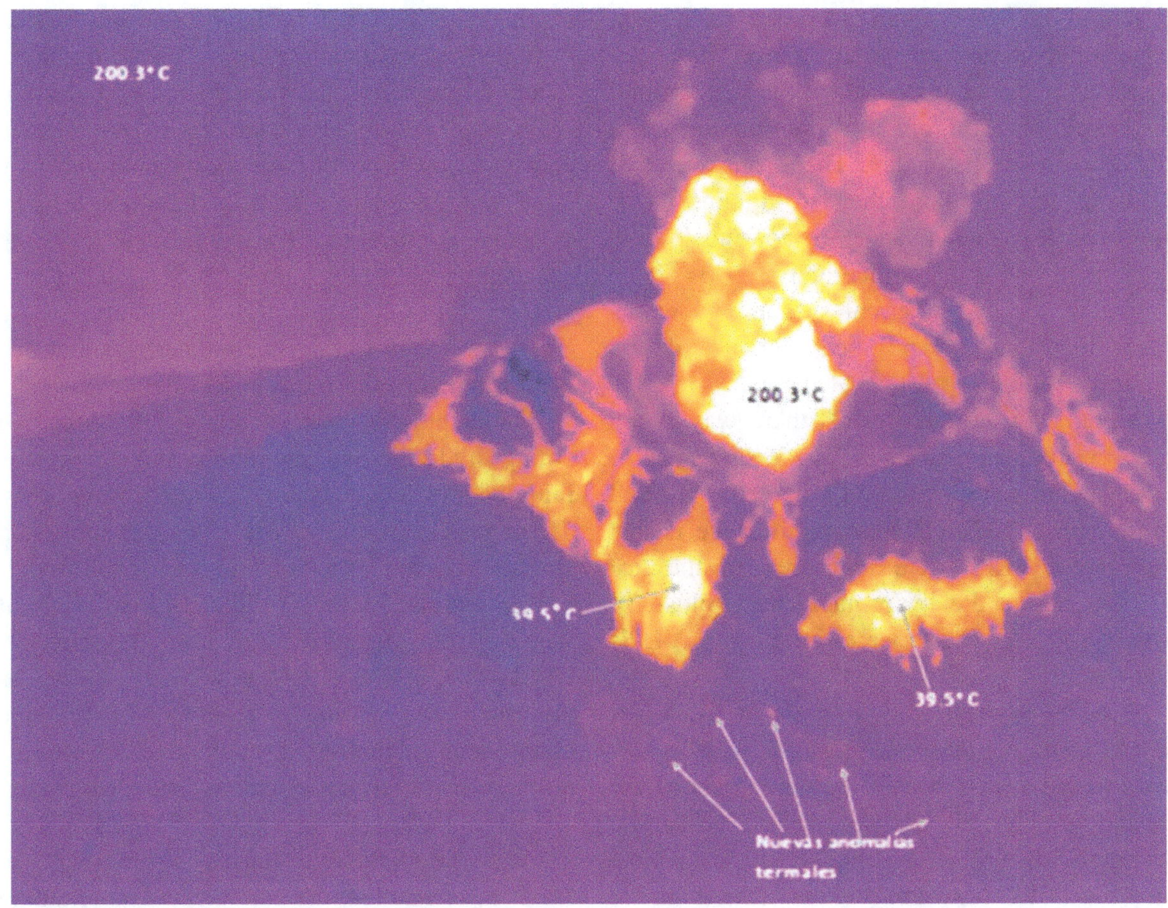

Fig. 17 Thermal image taken on 03 September, 2015 looking toward the SE sector of the upper cone. The infrared camera detected temperatures of 200 °C associated with emission from the crater and lower temperatures from fractures below the summit rim. *Photo* P. Ramón, IGEPN. *Source* http://www.igepn.edu.ec/cotopaxi/informes-cotopaxi/coto-especiales/coto-e-2015/13529-informe-especial-no-14/file

scarce VTs that occurred in concert with the LPs. In late May VLPs were registered (Fig. 5) and were interpreted to be possible mass magma transport processes as reported by Arias et al. (2015). Often VLPs are identified as eruption precursors, e.g. at Redoubt (Power et al. 2012), for example. In hindsight, vigorous VLPs were also registered in 2009–2010 at Cotopaxi and correlated with recoverable deformation patterns at borehole tilt stations (Mothes et al. 2010), but did not result in an extended seismic crisis or magmatic activity on the surface.

Nonetheless, LP and VLP events registered in April–August, 2015 were apparently responding to a slow ascent of a small magma slug and associated fluids and in April 2015 deformation recorded by tilt and GPS stations began almost synchronically with the jump in LP seismicity, implying that there may already have been magma ascent from a deeper depth to a shallower reservoir in order to show changes on the most proximal tiltmeters. The GPS stations begin to show minor displacements at the same time, particularly on the W-SW flank, where no strong evidence was detected in seismicity, but to the contrary seismicity was concentrated more on the E-SE flanks (Fig. 7). In sum, we registered both shallow and deep seismic activity. A leading hypothesis for this pattern likely was the interaction between fluids being released by a deep-seated source, say at 24 km depth as proposed in our geodetic model. These fluids ascended and perturbed a preexisting shallow-seated source, which may be the magmatic remnant that drove the 2001–2002 unrest reported by Hickey et al. (2015).

With more new magma in Cotopaxi's system, SO_2 output became prevalent in mid-May, one month after the hike in seismicity. Background SO_2 levels of <500 ton/day were surpassed and rose to over 3000 ton/day. The strong onset of bands of tremor about the 1st of June, were conjectured to be related to continual fluid movement within the edifice and perhaps to the boiling of the hydrothermal system and was a signal that more overall heat was circulating within the edifice.

Along with the rise in SO_2 there was also a trend in production of more VLP's since 114 of these events were registered between May and mid-August, 2015. Of great significance is that the largest VLPs were registered in the last 3 weeks before the hydromagmatic explosions. Afterwards too there were infrequent VLP's (Fig. 5).

With the highest energy levels of the VLPs being logged before the explosions, these events seem to have been one of the detonators of the explosions (Fig. 5b). They seemed to herald that magma/fluids were ascending. Of particular note was the VT/LP swarm of the 13th–14th of August that began 12 h before the hydromagmatic eruptions on the morning of 14 August (Fig. 11). The swarm comprised of some 40 VTs and >50 LPs was the most energetic of any seismic swarm registered at Cotopaxi since 2002 and was a warning in hindsight which presaged the subsequent explosions/strong emissions some hours later. These seismic trends and the higher SO_2 flux, would indicate that magma was working in the upper part of the system—at least in the 0 to 6 km level below the crater.

The initial explosions had a phreatic component since water was available in the small pond at the crater's base, sub-glacial melting and from pore water within the hydrothermal system. Nonetheless, Gaunt et al. (2016) argued that the most likely driving force of the initial explosions was magmatic heat interacting with the hydrothermal system providing energy to trigger hydromagmatic eruptions at Cotopaxi. Textural evidence for this process was only preserved in the deposits of the initial eruptions, but not subsequent ones. Later emissions were likely the

result of the repeated formation and destruction of a shallow magmatic plug by brittle fragmentation through mechanical stresses and decompression. Gas overpressure must have been accumulating beneath the conduit plug and may have contributed to the flank deformation, particularly as registered by the tiltmeters. In the succeeding post-explosion days SO_2 output rose to 16,000–18,000 ton/day (Hidalgo et al. 2016), a likely testimony to the accumulated gas that had been trapped in the plumbing system.

The expulsed material showed evidence of strong hydrothermal alteration and there was initially little evidence of juvenile components. There was also a low pH (3.6–5.1) and high sulfate- SO_4 concentrations (up to 13,000 mg/kg) in the expulsed ashes of the 14th to 25th of August, as detected by leachate analysis (P. Delmelle, Pers. Comm, 2015). Later, as described above, the percentage of juvenile components increased through time, to the last erupted material collected in late November, 2015.

After the explosions and strong emissions of the 14th of August, the conduit lost its retaining plug and remained open and continual fluid movement was facilitated, although pulsatile superficial activity continued, and few shallow explosions occurred. After the explosions LP events were initially high (>200 events/day), but dropped to <20 events/day by the first week of September, where they remain at this writing.

While the LPs diminished, to the contrary the VT events rose notably. Around the 1st of September 15–20 LP events/day were registered. By the third week of September these increased to >200 events/day (Fig. 16), and subsequently this value decreased to 50–100 events/day, but in all totaled nearly 15,000 events. In the last 3 months of 2015 persistent VT daily activity was registered as well as waning tilt and GPS offsets. Even with the VT swarm overall seismic energy levels decreased compared to the levels registered in May to August, which may be due to the overall successful degassing of the system, but could also be explained by closing of the conduit by a degassed magma column, thus impeding freer liberation of gases. Between

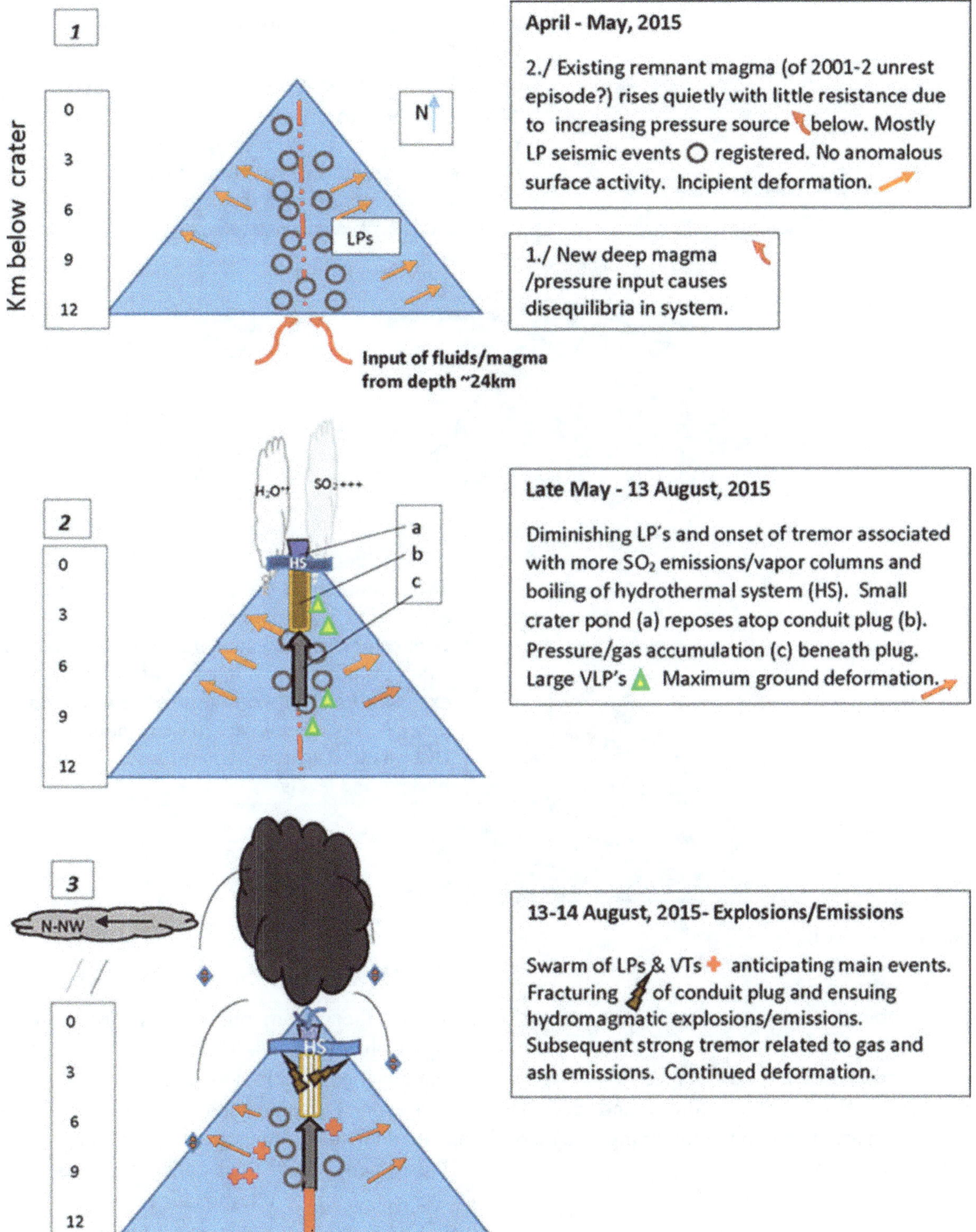

Fig. 18 Cartoons synthesizing the internal and superficial processes observed from April to December, 2015

October 2015 through April, 2016 internal explosions of deep providence were registered at a rate of 20–30 such events/month. These explosions could be interpreted as gas passing through restrictive areas within the conduit. While displaying only minimal infrasound and no detectable superficial vestiges, these explosions may be occurring due to pressurization

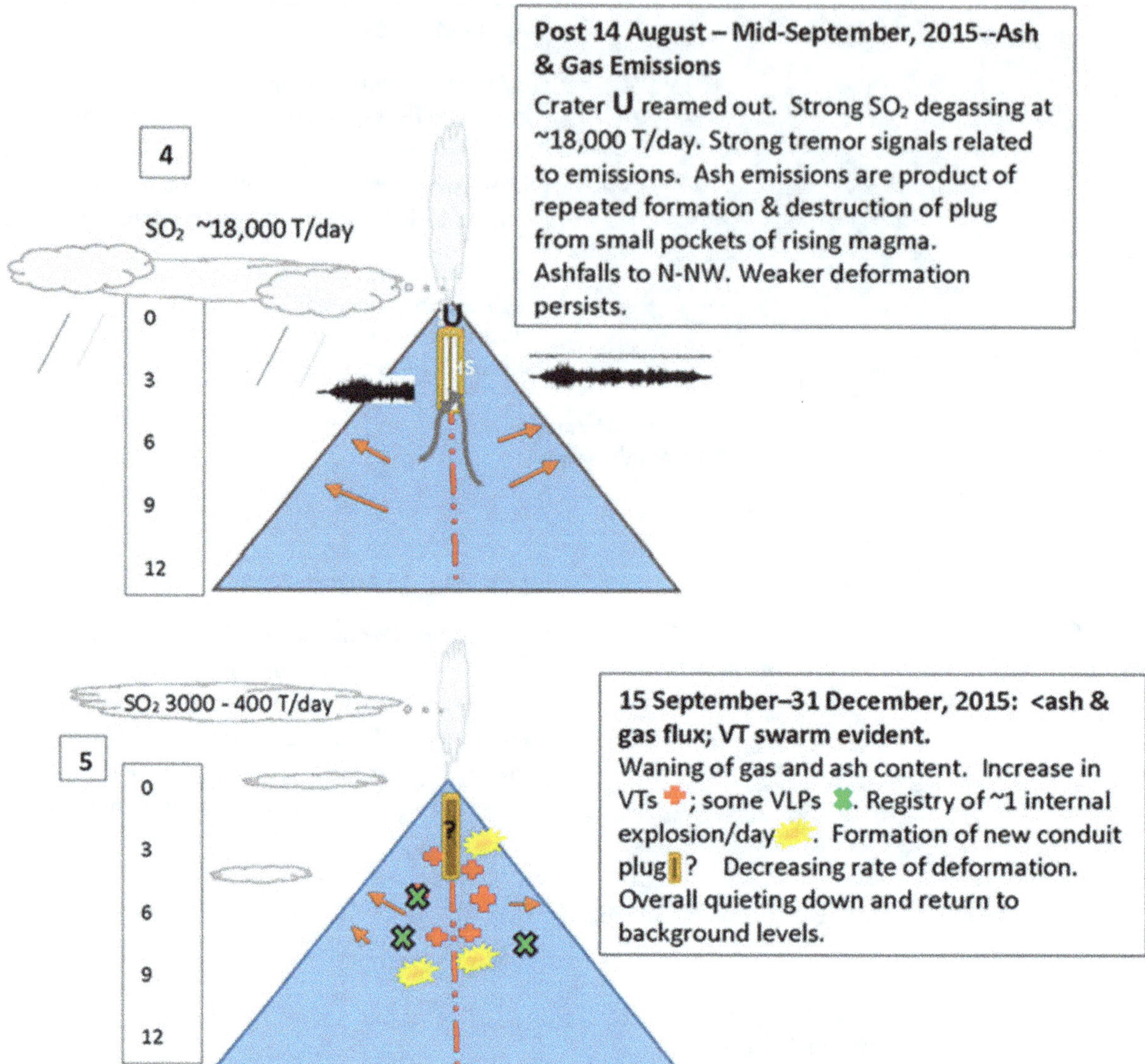

Fig. 18 (continued)

deep within the system (e.g. Valentine et al. 2014).

A synthesis of the 2015 unrest at Cotopaxi, the hypothesized driving forces and possible paths taken are shown in the following schematic cartoons (Fig. 18).

The Science-Society Interface

Strong evidence points to a magmatic component in Cotopaxi's 2015 reactivation and that this may be a precursor for future episodes, given that the bulk of the new magma that perturbed the system remained at depth. As a comparison, the 2001–2002 restless period had essentially been internal with only the weakest of superficial manifestations. For the 2015 reactivation IG scientists took the warnings very seriously and put all their collective experience to the test to make interpretations of the monitoring data and to manage successfully the great expectations of the public and authorities during the crisis period described above. Overall IGEPN volcanologists wrote 23 special reports that were disseminated via multiple media (www.igepn.edu.ec). Given the low levels of acceleration of seismic energy, small ground displacements and visual observations, eruption scenarios that the IGEPN formulated stayed within the realm of VEI 1–2 levels and clearly stated that the least likely scenarios was the generation of a paroxysmal eruption in which PDC's, voluminous ashfalls and giant lahars would be formed.

With each convincing sign that Cotopaxi was displaying stronger activity, IG scientists were proactive in improving and strengthening all monitoring systems while simultaneously helping to prepare the populations and authorities for what a major eruption of Cotopaxi could mean. Work by Christie et al. (2015) (an earlier VUELCO contribution) had shown that residents of the Chillos valley, to the east of Quito, were particularly ill-prepared to confront lahar hazards due to their recency of living in that valley. On the other hand, residents of the Latacunga valley had a clearer memory of lahar hazards, since many of their distant relatives had lived through Cotopaxi eruptions and their collective memory is better preserved. Nonetheless, social media, both beneficial and alarming, steered perceptions and actions of residents. The area of influence by the volcano, especially with respect with lahars, includes four important provinces, several counties and Quito's jurisdiction. It was particularly difficult to meet the demands and expectations and provide the personal attention of monitoring scientists to the authorities in each of these different municipalities as well as to meet with other community groups and respond to their uncertainties. There were also the constant attacks on social media of several particularly meanly-intentioned individuals who constantly tried to steer attention away from the IGEPNs scientific work by saying that it was operating with poor instrumentation or that the monitoring work at IGEPN was "carried out by amateurs".

All told, IGEPN scientists provided abundant custom guidance to local and national officials and residents with regards to volcano hazards and the proposed scenarios. A total of about 125 talks were given by IGEPN personal during the unrest period. Additionally, there was broad coordination with Ecuador's Secretary for Risk Management at all levels and participation in guiding eruption simulations. Some of the discussions with them were based on what had been reviewed in the VUELCO workshop-simulated eruption exercises carried out in late 2014 in Quito. The IGEPN also greatly benefitted by the strong collaboration and presence of members of the USGS/USAID Volcano Disaster Assistance Team who led informed discussions on the trends of the geophysical precursors and also helped to reinforce the lahar-detection network. Personal from Chalmers University of Technology (Sweden), JICA (Japan), IRD (France) and NASA (USA), DEMEX-EPN with help in SEM, LMU-Germany with grain size x-ray diffraction and UCL (Belgium) with leachates, also collaborated during the crisis.

As the eruption process waned, it was obvious that high-risk populations were tired of being constantly alert and didn't want to be perpetually attentive to volcanic processes that could threaten their livelihoods and families. This issue will have to be acknowledged and dealt with in future reactivations.

Conclusions

Seismic activity and its evolution in event types, energy release, shallowing depths and locations, elevated degassing and ash emissions and flank deformation typified the restlessness of Cotopaxi during 2015. The important accumulative energy release first of LPs than followed by registry of an important suite of large VLP's was a significant geophysical pattern indicating fluid movement, followed then by a more convincing transfer up conduit of small slugs of magma and gases to beneath the conduit plug. In the late hours of the 13th of August this plug fractured and ruptured, evidenced by the vigorous swarm of VTs and LPs before the hydromagmatic explosions that occurred early on the 14th of August 2015. Minor ground deformation, the small, limited explosions in August and subsequent ash emission suggest that the ascended magma volume was small, and indeed as calculated by Bernard et al. (2016), was only about

0.5 Mm³ DRE. This value is far inferior to the possible volume of 42 ± 26 Mm³, which is hypothesized to be at depth based on modeling of the observed GPS displacements (Mothes et al. 2016b). A second energetic magmatic pulse did not arise, and certainly not one with a sufficient volume to produce a VEI 3 or 4 eruption, which was one of the least likely scenarios, but nonetheless dreaded by the society and scientists.

Following the 14 August explosions and subsequent ash emissions we did not observe a new phase of outward GPS displacement trends in the deformation data, which could have implied a new magma input to cause another phase of deformation. The post-explosion VT seismic swarm which lasted 5 months was indicative of persistent internal perturbation but did not transpire in a new phase of deformation, thus we assumed that we were dealing with a small magma volume. The magma that tipped off the 2015 unrest may have been a remnant of that which provoked the 2001–2002 episode and was reported by Hickey et al. (2015). This residual magma could have been disturbed by the ascending heat and fluids from the new magma input at depth (~ 24 km) whose source was possibly under the SE flank, and which provoked the recorded ground deformation and the LP and VLP seismicity.

The volcano was benevolent and had awakened to only a VEI 2 level. No major damage was imparted upon the population or on livelihoods, except for temporary local economic depression, increased anxiety of the population, mild crop losses and premature selling of livestock due to fears of future losses. Overall, the volcano's manifestations served as a warning to everyone to keep attentive of Cotopaxi's capacity to cause destruction and possible severe ruin by lahar transit down major drainages which are heavily populated and host important strategic infrastructure.

An eruption process can last months to decades, and we need only to look at Tungurahua, an andesitic stratocone also in Ecuador's Cordillera Real, with ongoing eruptions for 17 years (Mothes et al. 2015a, b), or Soufrière Hills on Montserrat (Sparks and Young 2002) to suggest that the next round of Cotopaxi eruptions could last more than just several months. In the case of Tungurahua, activity started gradually in 1999 and displayed oscillating low-level behavior over the years to finally generate a rapid-onset VEI 3 eruption in 2006 (Hall et al. 2013). Such long waits test the population's resilience, but is also a time for monitoring scientists to become acquainted with the volcano's eruption style. During a reactivation period of a long dormant volcano there are many uncertainties and this demands stringent work and continual mindfulness by monitoring scientists and frequent ongoing interactive and personal communication with local communities and authorities.

During the 2015 unrest period at Cotopaxi, people living in high-risk zones (i.e. Latacunga and Valle de los Chillos) were swayed by speculation, rumors and lies concerning the status of the volcano. Some people also tended to weigh-in toward imprecise information posted on Facebook or Twitter and heed pseudo volcanologists and detractors, rather than rely on information from official channels. It was not uncommon to receive telephone calls from hysterical residents in either of these population centers inquiring if a Cotopaxi eruption was *imminent?* All told the IGEPN put out 3 reports each day about the volcano's activity and more than 24 special reports, all which are available on the IG website (www.igepn.edu.ec). The number of followers on the IGEPN's Facebook page grew to >1 million. To help stem the flow of bad information at the community level Ecuador's 911 system, in coordination with IGEPN personal, formed a pan-volcano *vigía* network comprised of volunteer observers who report via radio several times a day about their visual and audible observations of Cotopaxi or the rivers that are borne on it. This network with 55 volunteers, is in many ways a replica of the successful community-based *vigía* system that has functioned at Tungurahua volcano since 2001 (Stone et al. 2014; Mothes et al. 2015b). The information provided by the *vigía* volunteers compliments the ongoing geophysical monitoring and also serves to strengthen their capacities as community leaders and guides during volcano crisis (Espín Bedón et al. 2016).

A hypothesis for a future trend in activity weighs heavily towards hydromagmatic to Vulcanian explosions which may have a rapid onset, similar to the 14th of August episode, then evolve to sub-Plinian to Plinian eruptions of VEI 3–4 magnitude, if enough magma has accumulated at a relatively shallow depth (maybe 0–7 km below the crater) as shown in Fig. 7 for the upper level seismicity, and can make it to the surface before degassing. Vulcanian eruptions have been prominent in the volcano's historical activity (Gaunt et al. 2015). Unraveling the story will be difficult.

As the volcano is well-monitored 24 h/day/365, we anticipate that the IGEPN will provide early warnings to the public and officials before onset of important eruptive activity. This 2015 "dry run" allowed for diversification and hardening of Cotopaxi's monitoring network, frequent preparation and reappraisal of eruption scenarios and for the creation of a society-wide discussion of the possible consequences of a large Cotopaxi eruption. Some of these steps were facilitated by previous work in a VUELCO workshop. Essentially, attending to the 2015 activity was an opportunity to test the level of preparedness of the scientists and of the Ecuadorian society. All IGEPN scientists strived hard to be ready to "call it right", had the occasion arisen and a large eruption was in preparation. Since so little new magma erupted and there was no detectable subsequent shallow magmatic recharge, we consider the eruption as extremely small, and that the residual magma is in repose until a future time. Overall, the crisis was an important opportunity for learning about Cotopaxi's restlessness, with particular recognition of the increase in the VLP events and their energy levels just weeks before the mid-August explosions and the synchronous but progressive ground deformation signals, albeit small, that coincided with the increased seismicity. These two patterns more than any other geophysical signals, announced the ascent of Cotopaxi's magma, although finally only a small quantity breached the surface.

In all likelihood little or no evidence of the 2015 restless period will be preserved in the geological record. We know from written chronicles (1534–1877) that Cotopaxi often had weeks to months of ramping up before unleashing VEI 3 or 4 eruptions, i.e. there were probably several poorly preserved 2015-sized like events, and therefore unrest has been poorly documented. In this recent case, scientists had the benefit of observing and analyzing the geophysical monitoring output during the entire episode and knowing what level of activity the volcano was at. But, monitoring scientists, just like the citizens of Ecuador, experienced the anxiety of pondering what could be the volcano's next steps, i.e., the possible rapid intrusion of a new batch of volatile-rich magma or returning to calm. Fortunately, in this time around, the first scenario did not transpire. However, this training opportunity that we experienced could prove invaluable for when the next scenario is played out.

Acknowledgements This work was supported by a grant from European Commission FP 7 program (ENV.2011.1.3.3-1; grant no: 282759; VUELCO). We would like to thank the IGEPN staff for keeping all monitoring operations at Cotopaxi volcano optimally functioning. Members of the IGEPN's Volcanology group gave most of the talks and explanations of eruptions scenarios to authorities, community groups and wherever was necessary during the crisis. We acknowledge the VUELCO project for opportunities to share opinions about future eruptive crisis scenarios. Also, we are grateful for the support of the VDAP/USAID team during the crisis. The EPN, SENESCYT and SENPLADES provided funding for much of the instrumentation and daily operations. Support by JICA instrumentation and guidance is also noted. The Ecuadorian military provided overflights of Cotopaxi, while IGM provided ortophotos. The LMI project was instrumental in carrying out near real time petrologic monitoring. Thanks to Viviana Valverde for preparation of several figures. Recognition is given to the expressions of interest by STREVA project members on the societal implications of eruptive activity at Cotopaxi.

References

Arias G (2015) Estudio de las Señales Sísmicas de Muy Largo Periodo del volcán Cotopaxi. Proyecto de Titulación, Facultad de Ciencias, Escuela Politécnica Nacional

Arias G, Molina I, Ruiz M, Hernandez S, Alvarado A, Plain M, Mothes P, M Yépez, Hidalgo S, Barrington, C (2015) very long period seismicity accompanying increasing shallower activity at Cotopaxi Volcano. In Conference Paper No. S51D-2724, AGU Fall Meeting, San Francisco, 14–18 December

Bernard B, Battaglia J, Proaño A, Hidalgo S, Vásconez F, Hernandez S, Ruiz M (2016) Relationship between volcanic ash fallouts and seismic tremor: quantitative assessment of the 2015 eruptive period at Cotopaxi volcano, Ecuador. Bull Volcanol. doi:10.1007/s00445-016-1077-5

Cáceres B (2016) Dramatic reduction of Cotopaxi glaciers during the last volcano awakening 2015–2016. Conference: AGU Fall Meeting, San Francisco-Moscone Center. doi:10.13140/RG.2.2.17483.80168

Chouet BA, Matoza RS (2013) A multi-decadal view of seismic methods for detecting precursors of magma movement and eruption. J Volcanol Geoth Res 252:108–175

Christie R, Cooke O, Gottsmann J (2015) Fearing the knock on the door: critical security studies insights into limited cooperation with disaster management regimes. J Appl Volcanol 4:19. doi:10.1186/s13617-015-0037-7

Espín Bedón P, Mothes P, Montesdeoca M, Ruiz M, Alvarado A, Córdova M, Santamaria S (2016) Communication network of "Vigias" implemented at Cotopaxi Volcano, Ecuador, Abstract, Cities on Volcanoes-9, Puerto Varas, Chile, November, 2016

Garrison JM, Davidson JP, Hall M, Mothes P (2011) Geochemistry and petrology of the most recent deposits from Cotopaxi Volcano, Northern Volcanic Zone, Ecuador. J Petrol. doi:10.1093/petrology/egr023

Gaunt HE, Mothes P, Chadderton A, Lavallee Y (2015) Physical characteristics of conduit plug rocks during Vulcanian eruptions at Cotopaxi and Tungurahua volcanoes, Ecuador. Abstract, IUGG meeting, Prague-Czech Republic, June 2015.

Gaunt EH, Bernard B, Hidalgo S, Proaño A, Wright H, Mothes P, Criollo E, Kueppers U (2016) Juvenile magma recognition and eruptive dynamics inferred from the analysis of ash time series: the 2015 reawakening of Cotopaxi volcano 328 October 2016. doi:10.1016/j.jvolgeores.2016.10.013

Hall M, Mothes P (2008) The rhyolitic–andesitic eruptive history of Cotopaxi volcano Ecuador. Bull Volcanol 70(6):675–702

Hall ML, Steele A., Mothes P, Ruiz M (2013) Pyroclastic density currents (PDC) of the 16–17 August 2006 eruptions of Tungurahua volcano, Ecuador: geophysical registry and characteristics. J Volcanol Geotherm Res 265(2013):78–93. https://doi.org/10.1016/j.jvolgeores.2013.08.011

Harlow DH, Power JA, Laguerta EP, Ambubuyog G, White RA, Hoblitt RP (1996) Precursory seismicity and forecasting of the June 15, 1991, eruption of Mount Pinatubo. Fire and Mud: Eruptions and Lahars of Mount Pinatubo, Philippines. University of Washington Press, Seattle, pp 285–306

Hemmings B, Whitaker F, Gottsmann J, Hawes MC (2016) Non-eruptive ice-melt driven by internal heat at glaciated stratovolcanoes. Submitted to J Vol Geotherm Res

Herring TA, King RW, Floyd MA, McClusky SC (2015) Introduction to GAMIT/GLOBK, Release 10.6. Department of Earth, Atmospheric, and Planetary Sciences, Massachusetts Institute of Technology, PSF: http://www-gpsg.mit.edu/~simon/gtgk/Intro_GG.pdf, 50p

Hickey J, Gottsmann J, Mothes P (2015) Estimating volcanic deformation source parameters with a finite element inversion: The 2001–2002 unrest at Cotopaxi volcano, Ecuador. Journal of Geophysical Research: Solid Earth 120(3):1473–1486

Hidalgo S, Bernard B, Battaglia B, Gaunt E, et al. (2016) Cotopaxi volcano's unrest and eruptive activity in 2015: mild awakening after 73 years of quiescence. Geophys Res Abstracts Vol. 18, EGU2016-5043-1, 2016 EGU General Assembly, Viena-Austria

Jousset P, Budi-Santoso A, Jolly AD, Boichu M, Dwiyono S, Sumarti S, Hidayati S, Thierry P (2013) Signs of magma ascent in LP and VLP seismic events and link to degassing: an example from the 2010 explosive eruption at Merapi volcano, Indonesia. J Volcanol Geoth Res 261:171–192. doi:10.1016/j.jvolgeores.2013.03.014

Kumagai H, Yepes H, Vaca M, Caceres V, Nagai T, Yokoe K, Vasconez F (2007) Enhancing volcano-monitoring capabilities in Ecuador. Eos 88 (23):245–252

Kumagai H, Palacios P, Maeda T, Castillo DB, Nakano M (2009) Seismic tracking of lahars using tremor signals. J Volcanol Geoth Res 183(1):112–121

Kumagai H, Nakano M, Maeda T, Yepes H, Palacios P, Ruiz M, Yamashima T (2010) Broadband seismic monitoring of active volcanoes using deterministic and stochastic approaches. J Geophys Res: Solid Earth 115 (B8)

Kumagai H, Mothes P, Ruiz M, Maeda Y (2015) An approach to source characterization of tremor signals associated with eruptions and lahars. Earth, Planets, and Space. 67:178, 67:178, doi:10.1186/s40623-015-0349-1

Lyons J, Segovia M, Ruiz M (2012) Reporte del Enjambre Sísmico en la Zona de Volcán de Cotopaxi—Mayo 2012. Internal report. Not published. IGEPN, Quito.

Maeda Y, Kumagai H, Lacson RJ, Figueroa MS, Yamashina T, Ohkura T, Baloloy AV (2015) A phreatic explosion model inferred from a very long period seismic event at Mayon Volcano, Philippines. J Geophys Res Solid Earth 120:226–242. doi:10.1002/2014JB011440

Márquez M (2012) Caracterización de los sismos de muy largo periodo en el volcán Cotopaxi y sus implicaciones. Tesis de Grado, Carrera de Geofísica, Facultad de Ciencias Astronómicas y Geofísicas, Universidad Nacional de La Plata. Argentina

Marzocchi W, Newhall CG, Woo G (2012) The scientific management of volcanic crises. J Volcanol Geoth Res 247–248:181–189

Molina I, Kumagai H, García-Aristizábal A, Nakano M, Mothes P (2008) Source process of very-long-period events accompanying long-period signals at Cotopaxi Volcano, Ecuador. J Volcanol Geoth Res 176(1):119–133

Moran SC, Newhall C, Roman DC (2011) Failed magmatic eruptions: Late-stage cessation of magma ascent. Bull Volc 73(2):115–122. doi:10.1007/s00445-010-0444-x

Mothes P, Hall ML, Espín, PA, Vasconez F, Sierra D, Córdova M, Santamaría S, Andrade D (2016a) Mapa Regional de las Amenazas del Volcán Cotopaxi- Norte and Sur, scale 1:50,000, Instituto Geofísico and Instituto Geografico Militar, Quito http://www.igepn.edu.ec/mapas/mapas-volcan-cotopaxi.html

Mothes PA, Nocquet JM, Jarrín Tamayo P, Morales Rivera AM, Lundgren PR, Yépez YM, Viracucha EG, Gaunt EH (2016b) Geodetic signature associated with unrest at Cotopaxi in 2015–2016: modeling of GPS data, a deep magma source, synchronous seismic swarms and petrologic constraints. Abstract, Cities on Volcanoes-9, Puerto Varas, Chile, November, 2016

Mothes P, Hall ML, Andrade D, Yepes H, Pierson TC, Ruiz AG, Samaniego P (2004) Character, stratigraphy and magnitude of historical lahars of Cotopaxi volcano (Ecuador). Acta Vulcanol 16(1/2):1000–1023

Mothes P, Lisowski M, Ruiz M, Ruiz G, Palacios P (2010) Borehole Tiltmeter and CGPS Response to VLP Seismic Events under Cotopaxi Volcano. Ecuador, Fall AGU, San Francisco Abstract

Mothes PA, Vallance JW, Hall ML, Garrison JM (2015a) Cotopaxi's most recent rhyolitic eruptions, 2800 y BP. In Conference Abstract, 26th IUGG General Assembly, Prague

Mothes PA, Yepes HA, Hall ML, Ramón PA, Steele AL, Ruiz MC (2015b) The scientific–community interface over the fifteen-year eruptive episode of Tungurahua Volcano Ecuador. J Appl Volcanol 4(1):1–15

Mothes PA, Vallance JW (2015) Lahars at Cotopaxi and Tungurahua Volcanoes, Ecuador: highlights from stratigraphy and observational records and related downstream hazards. In Volcanic Hazards, Risks and Disasters. doi:10.1016/B978-0-12-396453-3.00006-X, chapter 6

Newhall CG (2000) Volcano Warnings. In: Sigurdsson et al. (Eds) Encyclopedia of Volcanoes. San Diego, CA, pp 1185–1198

Newhall, Christopher G, Self Stephen (1982) The Volcanic Explosivity Index (VEI): An Estimate of Explosive Magnitude for Historical Volcanism (PDF). J Geophys Res. 87 (C2): 1231–1238. Bibcode:1982JGR....87.1231 N. doi:10.1029/JC087iC02p01231

Nocquet JM, Villegas-Lanza JC, Chlieh M, Mothes PA, Rolandone F, Jarrin P, Martin X (2014) Motion of continental slivers and creeping subduction in the northern Andes. Nat Geosci 7(4):287–291

Phillipson G, Sobredelo R, Gottsmann J (2013) Global volcanic unrest in the 21st century: an analysis of the first decade. J Volcanol Geother Res 264:183–196. doi:10.1016/j.jvolgeores.2013.08.004

Pistolesi M, Rosi M, Cioni R, Cashman KV, Rossotti A, Aguilera E (2012) Physical volcanology of the post–twelfth-century activity at Cotopaxi volcano, Ecuador: behavior of an andesitic central volcano. Geol Soc Am Bull 123(5–6):1193–1215

Power JA, Stihler SD, Chouet BA, Haney MM, Ketner DM (2012) Seismic observations of Redoubt Volcano, Alaska—1989–2010 and a conceptual model of the Redoubt magmatic system. J Volcanol Geotherm Res 259:31–44

Ruiz M, Guillier B, Chatelain JL, Yepes H, Hall M, Ramon P (1998) Possible causes for the seismic activity observed in Cotopaxi volcano Ecuador. Geophys Res Lett 25(13):2305–2308

Siebert L, Simkin T, Kimberly P (2010) Volcanoes of the World. Univ. of California Press and Smithsonian Institute. 551 p

Sparks RSJ, Aspinall WP (2004) Volcanic activity: frontiers and challenges in forecasting, prediction and risk assessment. The State of the Planet: Frontiers and Challenges in Geophysics, Geophysical Monograph 150, IUGG Volume 19

Sparks RSJ, Young SR (2002) The eruption of Soufrière Hills Volcano, Montserrat (1995–1999): overview of scientific results Geological Society, London, Memoirs 21:45–69. doi:10.1144/GSL.MEM.2002.021.01.03

Stone J, Barclay J, Simmons P, Cole PD, Loughlin SC, Ramón P, Mothes P (2014) Risk reduction through community-based monitoring: the vigías of Tungurahua Ecuador. J Appl Volcanol Soc Volcanoes 3:11. doi:10.1186/s13617-014-0011-9

Tilling RI (1989) Volcanic hazards and their mitigation: progress and problems. U.S. Geological Survey, Menlo Park, California, p p33

Valentine G, Graettinger AH, Sonder I (2014) Explosion depths for phreatomagmatic eruptions. Geoph Res Lett doi:10.1002/2014GL060096

White R, McCausland W (2016) Volcano-tectonic earthquakes: A new tool for estimating intrusive volumes and forecasting eruptions. J Volcanol Geoth Res 309:139–155

Zobin, VM (2012) Long-Period and Very Long-Period Seismic Signals at Volcanoes. In: Zobin (Ed), VM Introduction to Volcanic Seismology (2nd ed). pp 327–354

Volcanic Unrest Simulation Exercises: Checklists and Guidance Notes

R. J. Bretton, S. Ciolli, C. Cristiani, J. Gottsmann, R. Christie and W. Aspinall

Abstract

When a volcano emerges from dormancy into a phase of unrest, the civil protection authorities charged with managing societal risks have the unenviable responsibility of making difficult decisions balancing numerous competing societal, political and economic considerations. A volcano that is threatening to erupt requires sound risk assessments incorporating trusted hazard assessments that are timely, relevant and comprehensible. Foreseeable challenges arise when the inevitable uncertainties of hazard assessment and communication meet societal and political demands for certitude. In some regions that host volcanic hazards, it would be both realistic and prudent to adopt three working assumptions. The complex legal and administrative infrastructures of risk governance will be largely untested and possibly inadequate. Many volcano observatory scientists, and probably even more risk managers and at-risk individuals/communities, will have inadequate recent experience of the challenges of hazard communication during a period of unrest. And lastly, the scientists may also have inadequate practical experience of the needs and management capacities of the risk-mitigation decision makers with whom they must communicate. "Practice doesn't make perfect. Practice reduces the imperfection." (Beta 2011). If this statement is correct, volcanic unrest simulation exercises (VUSE) have a vital role to play within the complex processes of volcanic risk governance. Consistent with the broad approach of the Sendai Framework for Risk Reduction 2015–30, this chapter argues that practical knowledge of VUSE can and should be analysed and recorded so that key lessons can be shared for the widest possible benefit. This chapter investigates five recent simulation exercises and presents six complementary checklists based upon data, insights and practice pointers derived from those exercises. The use of checklists, supported by guidance notes, is commended as a pragmatic way to create, test and develop acceptable standards of governance practice.

R. J. Bretton (✉) · J. Gottsmann · W. Aspinall
School of Earth Sciences, University of Bristol, Bristol, UK
e-mail: Richard.Bretton@bristol.ac.uk

R. J. Bretton · J. Gottsmann · R. Christie
W. Aspinall
The Cabot Institute, University of Bristol, Bristol, UK

S. Ciolli · C. Cristiani
Dipartimento della Protezione Civile (DPC), Rome, Italy

R. Christie
School of Sociology, Politics and International Studies, University of Bristol, Bristol, UK

Advs in Volcanology (2019) 271–298
DOI 10.1007/11157_2018_34

Published Online: 14 July 2018

It is argued here that well planned and executed simulation exercises are capable of informing and motivating a wide range of risk governance stakeholders. They can identify process and individual shortcomings that can be mitigated. Simulation exercises can and should play a vital role in reducing volcanic risks.

Keywords

Volcanic unrest · Risk governance · Simulations · Exercises · Training · Communication

1 Introduction

1.1 Simulation Exercises and the Sendai Framework

This chapter is about reconstructions of the evolution of past, and realistic simulations of hypothetical future, volcanic unrest events. They will be referred to as Volcanic Unrest Simulation Exercises (VUSE). A simulation is "a learning experience that occurs within an imaginary or virtual system or world" and involves role-play, which has been defined as "the importance and interactivity of roles in pre-defined scenarios" (van Ments 1999; Errington 1997, 2011; Dohaney et al. 2015). A fuller working description is set out in Sect. 3.1.

The Sendai Framework prioritises mitigation of risks before response and recovery. It also recognises the importance of "building the knowledge of, inter alia, government officials… through sharing experiences, lessons learned, good practices, and training and education on disaster risk reduction" (UN/ISDR 2015, 15). Mutual learning, dialogues and cooperation between risk governance stakeholders are encouraged. The Sendai Framework also identifies a need for quality standards.

The principal purpose of this chapter is to accept the challenge of the goals of the Sendai

Framework and to share knowledge derived from several recent simulation exercises with a wide audience.

1.2 The Managerial and Scrutiny Dimensions of Risk

1.2.1 Managerial Dimension

When a volcano emerges from dormancy into a phase of unrest, the civil authorities in charge of managing volcanic risks have to make challenging decisions (Fiske 1984; Dohaney et al. 2015). Decisions balancing safety and cost typically must made with limited information (Sparks et al. 2012a, b; Marzocchi et al. 2012; Jenkins et al. 2012) and in real time and under uncertainty, in a context of intense pressure (Marrero et al. 2015, 2). While the primary objective is to minimise the loss and damage from any volcanic event, the socio-economic losses resulting from false alarms and evacuations must also be considered" (Woo 2008; Hincks et al. 2014, 2; Donovan and Oppenheimer 2014).

Poorly handled unrest periods cause social, economic and political problems, even without an eruption. Ill-considered responses may facilitate the release of inappropriate advice and emergency declarations, and may lead to unwarranted media speculation and the premature cessation of economic activity and community services (Johnston et al. 2002, 228).

Recent crises, including the 2010 Icelandic Eyjafjallajökull eruption, have highlighted the difficulty of co-ordinating and synthesising scientific inputs from many different disciplines and institutions, and translating these into useful policy advice at very short notice (Harris et al. 2012; Dohaney et al. 2015). Effective communication, collaboration and cooperation are necessary between many expert and technical advisors, emergency management agencies and lifeline organisations (Doyle et al. 2015).

Jordan et al. (2011) have emphasised the need for scientists to have a clear role, a clear and authoritative voice and effective communication skills. The Organisation for Economic

Co-operation and Development (OECD) (2015) has identified the need for scientific advisers to have: (1) permanent authoritative structures; (2) a central contact point; (3) clear reporting procedures; (4) a pre-defined public communication strategy; and, when necessary, (5) ways to coordinate their actions internationally.

Although civil protection "authorities may have theoretical knowledge of volcanos, few have any practical experience of eruptions" (Solana et al. 2008, 312). Furthermore, the timescales of periods of volcanic unrest, especially bigger ones, do not correlate well with the tenures of political and senior management appointments (Donovan and Oppenheimer 2012; Mothes et al. 2015).

The term 'standard equivocality' relates to the "absence of commonly recognised standards capable of guiding, measuring and evolving acceptable practice" (Hood 1986; Bretton 2014; Bretton et al. 2015; Rothstein 2002). It is suggested here that there are no readily accessible standards regarding how hazard communication should be conducted during a period of volcanic unrest.

1.2.2 Scrutiny Dimension

The International Federation of Red Cross and Red Crescent Societies (IFRC) and United Nations Development Programme (2015) (IFRC/UNDP) legal checklist requires national laws to establish and promote the training of public officials and relevant professionals.

There is an emerging international law duty upon sovereign states to have substantive regulatory systems to ensure that risks from natural hazards are mitigated so that they do not endanger human lives. States must inform at-risk communities of the potential of unmitigated risks and establish sufficient co-ordination and cooperation between administrative authorities. This is an onerous duty and it is argued here that it would be prudent to assume that the general duty includes several more specific subsidiary duties. One would be to consider the merits of simulation exercises as a way of satisfying: (1) the specific requirements of national laws; (2) the general expectations of international law; (3) the education, training and knowledge-sharing goals

of the Sendai Framework; and (4) the requirements of the IFRC/UNDP legal checklist.

1.3 Academic Support for Training and Simulation Exercises

The experience and levels of expertise of observatory scientists are critical to making accurate forecasts and training is important (McGuire and Kilburn 1997). In the context of volcanic risks, Doyle et al. (2015) undertook a review of the literature on emergency management team response, decision making, mental models and situation awareness and exercising. They argue that science agencies and science advisory groups must embark on a suite of training activities to enhance their response during a disaster. These should include exercise and simulation programmes within their own organisations rather than participation solely as external players in emergency management activities. Structures, resources and time must be provided for these programmes.

It is further argued by Doyle et al. (2015) that training will enhance the future response capabilities of both scientists and risk-mitigation agencies in several ways.

2 Methodology

This chapter reviews five simulation exercises (four of which were conducted as part of the VUELCO project), which for ease of reference are summarised in Table 1. It investigates the ways in which these exercises were planned and undertaken. By design, it does not address in any detail: (1) the features of the various and varying volcanic settings in which the exercises were conducted; or (2) the nature, scope or analysis of the monitoring data that were painstakingly created for the volcanic hazard scenarios that underpinned them.

This chapter draws from ethnographic observational data (recorded in hand written field notes) collected during five simulation exercises from the perspectives of overt non-participant

274

R. J. Bretton et al.

Table 1 Brief details of the analysed five volcanic unrest simulation exercises

Name	Hosting organisations country/when	Volcanic hazard scenarios and interpretation of observables
Colima	VUELCO Mexico (Colima) 17–23/11/2012	Volcan de Colima—evidence of moderate effusive activity alternating with moderate explosive events; escalating unrest, dome growth, some explosions, a 2 km + eruptive column, ash-fall, possible evolutions included: effusive dome growth alone, a large explosive eruption after dome collapse, partial flank collapse, a mixed 'Merapi type' event and cessation of dome growth; culmination in risk sufficient to require the evacuation of a single small town—La Becerrera
Campi Flegrei	VUELCO Italy (Rome) 11–12/02/2014	Campi Flegrei—conflicting evidence of either a rise of magma from about 7 km or an increase in shallow hydrothermal activity; further evolution involving a possible 3 km sill or continued shallow fluid migration; further rapid evolution raised the possibility of phreatic explosions and small volume magmatic eruptions in the eastern sector; further dramatic evolution made explosions and eruptions likely within days or weeks in the Bagnoli-Solfatara area and small offshore eruptions were not excluded; culmination in eruption between Bagnoli and Monte Spina and a sustained eruptive column, ash fall and pyroclastic flows south and east
Mount Teide	Presidencia del Gobierno de Canarias and Instituto Geográfico Nacional (IGN), Ministerio de Fomento, Gobierno de España Tenerife, Spain (Santa Cruz de Tenerife) 25/04/2014	Teide-Pico Viejo—evidence of rumbling noises, small debris avalanches and no seismic/deformation changes; further evolution with more rumbling noises, small swarm of seismic events; further evolution with swarm of small high-frequency VT, an explosive event, new thermal events and ash emissions with LFs and tremors; further evolution with pronounced SO_2 decrease interpreted as near-surface sealing of rising magma with high probability of eruption; culmination in eruption
Cotopaxi	VUELCO Ecuador (Quito) 13/11/2014	Cotopaxi—evidence of anomalous seismic activity with returns to background levels and no physical-chemical changes; possible local tectonic activity, fluctuating gas emissions, some deformation, further evolution with many long-period events, increasing released energy, sub-vertical fault in NE flank, increasing deformation, high gas emissions, increased fumarolic activity, increased thaw in upper parts of the edifice; further evolution to sporadic vulcanian explosions, eruptive columns <5/8 km plus tephra; further evolution with escalating seismic activity, inflation and degassing changes; culmination in eruption
Dominica	VUELCO Dominica (Roseau) 14–15/05/2015	Southern volcanic region including Trois Piton—evidence of elevated seismicity, landslides and small hydrothermal changes; further evolution with many VT events and landslides, a return to hydrothermal background levels plus Boiling Lake drainage; further evolution with more VT events and hybrids followed by VT decrease with hybrids dominating, continuing geochemical background levels, no recent deformation and contraction in some areas, indications of a deeper source than in the early unrest phases, possible pressurisation of deep seated magmatic system or hydrothermal activity; culminated in the consideration of a lowering of the Alert level

observers and, to a very limited but recorded extent, of participant-observers. Four exercises relate to VUELCO volcano unrest simulations in Mexico, Italy, Ecuador and Dominica. Detailed analysis of the documents prepared before, during and after these exercises was undertaken. The challenges of avoiding researcher bias and unintended observer effects were recognised.

Additional observational and documentary data were acquired by the lead author during the Tenerife (Spain) exercise as an invited external non-participant observer commissioned to prepare a post-exercise evaluation at the request of the Presidencia del Gobierno de Canarias. No ethical agreement was signed before or during that exercise. Data from that exercise and extracts from the post-exercise report are included by kind written permission of the Presidencia del Gobierno de Canarias.

Within the main body of this chapter, no attempt is made to provide complete details of the five exercises listed in Table 1. Essential background information and additional reading sources are to be found in additional files A–E which, for ease of reference, have deposited at the 'Collaborative volcano research and risk mitigation' (VHUB) website under reference "Volcanic Unrest Simulation Exercises: checklists and guidance notes—Additional files A-E". The additional reading sources are also listed at the end of the references for this chapter.

3 Background

3.1 VUELCO Themes and Goals

As VUSE are purpose-driven learning activities, each will have tailored goals and an overall design based upon the needs of its participants. Consistent with VUELCO's stated goals, the principal purpose of each VUELCO exercise was to present a realistic simulation of the evolution of past and hypothetical future volcanic unrest events associated with a host volcano or wider volcanic setting. In the manner of a film set, a simulated volcanic event is the dynamic backdrop against which a selection of risk governance

infrastructures, policies, procedures and people is tested.

The VUELCO exercises involved the provision of raw or partially analysed monitoring data and information to geo-scientists. Their role was to undertake a scientific analysis of the data with the purpose of passing on characterisations of likely volcanic hazard scenarios, by means of hazard communications, to risk mitigation decision-makers (civil protection authorities and the representatives of at-risk individuals) and the mass media. The civil protection authorities, having considered not only societal issues but also political, economic and other values, made and communicated risk mitigation decisions. In some exercises, representatives of at-risk communities, emergency services and relevant government entities played active roles in testing emergency, law and order, rescue, medical and evacuation procedures.

3.2 Checklists and Guidance Notes

3.2.1 Checklists

The checklists within this chapter are presented in response to the perceived importance of such documents as sources of good practice in the eyes of recent commentators (e.g. Gawande 2010; Newhall 2010; IAVCEI 2015). For example, within the 'Safe Surgery Saves Lives' project, which started in 2007, the World Health Organisation (WHO) introduced a Surgical Safety Checklist developed by Dr. Atul Gawande based, inter alia, upon the success of pre-flight checklists in enhancing safety within the aviation industry. "A systematic review and meta-analysis of the effect of the WHO checklist" strongly suggests a related "reduction in post-operative complications and mortality" (Bergs et al. 2014, 150; Treadwell et al. 2014).

Drawing upon the research of Gawande (2010), this chapter's checklists aim to incorporate several critical features that are output and outcome-focussed. As far as reasonably practicable, the authors' aim has been to ensure that each checklist is: (1) concise and preferably short as well; (2) simple, precise and unambiguous;

(3) targeted by addressing only evidence-based priorities that are considered either critical or significant to risk governance; and (4) non-prescriptive and non-comprehensive. Checklists may encourage rational, systematic (i.e. both consistent and complete), routine and transparent practices whilst recognising the importance of, and encouraging careful attention to, a wide variety of constraints and expectations. However, it is not envisaged that they should be used as a regulatory device, an enforceable legal requirement, or part of a blame-avoidance strategy (Hood 2011; OECD 2015).

The checklists presented in this chapter address the general planning, funding and execution of purposeful simulation exercises. The exercises also highlighted the many challenges of hazard communication and the difficulties that may result if the needs of other stakeholders are not identified and responded to.

This chapter presents six checklists namely: (1) Planning; (2) Logistics; (3) The Volcano team; (4) The Scientific Advisory Committee (SAC); (5) The Risk Managers - Civil Protection Authorities (CPA); and (6) The Observers/ Auditors. The first two relate to exercise 'phases' and the other four relate to 'roles'. It is not suggested that these checklists will be relevant to all simulation exercises, however, they will provide an indication of some of the critical phases and major roles that must be considered.

The checklists inevitably reflect the fact that VUELCO's exercises placed particular emphasis on the scientific analysis of monitoring data, the communication of hazard characterisations to risk managers, and interactions between scientists, risk managers, the media and the general public. It is accepted that future exercises may have other goals for which more or different checklists may be helpful.

The 'Planning' checklist encourages the identification of goals and objectives, and the need for clear leadership, careful design and realistic financing.

The 'Logistics' checklist is probably the most important. The actual task of arranging the 'planned' exercise may have many benefits in its own right as it will require the careful

identification of key laws, policies, procedures and people.

The 'Volcano team' has the very difficult obligation to craft and deliver monitoring data consistent with realistic volcanic hazard scenarios. The data must be suitable to meet the hazard analysis challenges confronted by the experts within the 'Scientific Advisory Committee' (SAC). It is noted here that, in some situations, the roles of monitoring and analysis may be undertaken by the same team or with substantial personnel overlaps. This chapter retains the separation to cover circumstances where there are distinct group remits, functions and responsibilities.

The SAC receives and handles the data provided utilising, as appropriate, a variety of soft skills (including those of analysis, deliberation and communication), tools (including expert elicitation and probabilistic models) and established procedures and protocols.

The 'Risk Managers - Civil Protection Authorities' (CPA) make risk mitigation decisions based in part upon scientific communications received from the SAC or, where the existence of a SAC is not foreseen within the relevant governance system, the Volcano team. An important part of the CPA role is usually the challenging duty, not only to advise individuals and entities driven by political values, but also to interact with members of the public and respond to mass and social media demands. The press and social media "can play an important role in the dissemination of information, true or false" and social media can rapidly ferment public, anxiety, distrust or dissent (OECD 2015, 37).

Simulation exercises are learning/training exercises and, accordingly, the importance of 'Observers' should not be underestimated. If properly briefed, observers and auditors will not only contribute their own candid views about all aspects of the exercise but also encourage other participants to be reflective about their own contributions and actions.

3.2.2 Guidance Notes

To support and supplement the checklists, it is suggested here that guidance notes should be issued from time to time to provide a dynamic and helpful knowledge and innovation resource. Their aim should be: (1) to gather together, record and share the accumulated experience of other practitioners in relevant fields of expertise; and (2) to suggest ways to find optimal solutions to the most critical issues.

It is readily accepted that the observations within the guidance notes are inevitably subjective. They are not intended to be in any respect either comprehensive or prescriptive. They are presented as options to be considered along with other issues that will be found within best practice guidelines such as those identified in Doyle et al. (2015).

4 Checklists (in bold italics) and Guidance Notes (in normal font)

4.1 Planning

Leadership

Who (individual and entity) has overall responsibility for planning and delivering the exercise?

The choice of exercise leaders should be dictated by the planned goals and activities of the exercise. If a significant involvement of civil protection authorities is planned (for example, where risk mitigation decisions are to be made and/or related mitigation measures are to be tested) those authorities should probably take the lead.

The VUELCO exercises have suggested that overall responsibility for a VUSE should be assigned to just one person within the risk governance stakeholder (e.g. the civil protection authority) likely to gain most from it. That person, who will need suitable and sufficient support from a working team or steering group, should have the gravitas, personality, authority, experience and resources (both human and financial) needed to plan and handle a complex high profile project that will inevitably attract political, societal and media attention.

The MIAVITA Handbook (2012, 118) suggest that it is good practice for "a steering group to be in charge of co-ordination and leadership of the various preparatory activities".

After one VUELCO exercise it was suggested that at least "a full-time scientist" should be dedicated to preparing any exercise that focusses on scientific analysis.

Purpose and Goals

What is the overall purpose of the VUSE?

What are the short and long term goals of the exercise and its players?

A VUSE is a learning activity and the principal reasons for it must be identified and stated. It should respond to the perceived core needs of the participating stakeholders and, like any learning activity, be carefully planned with stated assumptions, aims, objectives and themes.

Before the exercise, each and every participant should know what they will learn from the exercise. Thought should also be given to how and by whom the success of the exercise will be evaluated and what will be done by whom to build upon the exercise.

The assumptions, aims, objectives and themes should be set out in the pre-exercise briefing note referred to in Sect. 4.2 of this checklist.

Scope

Which parts of the risk governance process are going to be tested?

A VUSE should be focussed, purpose-driven and planned accordingly.

No VUSE can realistically attempt to replicate all aspects and phases of, and all stakeholders interested in, a societal risk governance regime. It is possible that more regular exercises, which concentrate upon very carefully defined aspects of the system (e.g. communicating in real time with the general public and the mass media), may be more cost-effective and beneficial.

Consideration might be given to those issues and functions that require the most inter-stakeholder planning, cooperation and collaboration and accordingly excellent means of communication. A VUSE can and should be a learning exercise in communication between many key players such as:

- Geoscientists and other geoscientists.
- Geoscientists and risk governance advisers (e.g. weather forecasters, aviation and marine space managers, communication specialists etc.).
- Geoscientists and other stakeholders (e.g. individuals within civil protection authorities, interested and affected members of the public, representatives of community, religious, public utility and commercial interests, and representatives of international, national, regional, municipal and other levels of government).
- Local geoscientists (e.g. volcano observatory staff) and visiting scientists (e.g. academics contributing to a scientific advisory committee) and visiting researchers.
- Hazard analysts (e.g. local and visiting geoscientists) and experts in risk assessment and management.
- All scientific analysts/risk decision makers and (1) the general public and (2) the mass media.

In 1999 several of these interactions were addressed by IAVCEI's Subcommittee for Crisis Protocols, which issued a report entitled "Professional conduct of scientists during volcanic crises" (Newhall et al. 1999). In the context of the governance of the risks of volcanic hazards, this paper represents a rare, if not unique, example of an attempt to offer authoritative practice guidance based on past events and an extensive literature review. It was concerned principally with personal and institutional interactions during a volcanic crisis.

Consideration should also be given to whole or part of an exercise being in 'degraded mode'. The MIAVITA Handbook (2012, 118) suggests that exercises should be planned to test system level capabilities of response when some parts of the system are not fully operative. By this means, depending upon which aspects are being tested, allowance can be made for, inter alia, holidays, the malfunctioning or destruction of monitoring or telecommunication equipment, blocked escape or rescue routes, and weather conditions.

By way of illustration of needs that an exercise might seek to target, Exercise Capital Quake

identified eleven functions—"Public Information Management, Governance, National Financial System, Logistics and Other Support Co-ordination, International Assistance and Liaison, Rescue, Health, Welfare, Building Safety, Restoration of Access and Restoration of Lifelines" (NZ/MCDEM 2008).

Another possible example comes from the "Metodo Augustus", the organisation of emergency management, used at the Italian Department of Civil Protection. That organisation provided several support functions including technical-scientific, health and veterinary assistance, mass-media and information, volunteers, means and materials, transportations and mobility, telecommunications, essential services, damage assessment, operational structures, local administrations, dangerous materials, people assistance, cultural heritage and coordination (Galanti et al. 2006).

VUELCO chose a science-orientated focus for its four exercises concentrating upon the interfaces between:

- Hazard monitoring and hazard assessment:
  - Long-term monitoring
  - The host volcano's main precursors of volcanic unrest.
  - The host volcano's short-term monitoring resources (e.g. equipment, employed staff and volunteers etc.), the nature, adequacy and timing of the data output, and their capacity to respond to changing demands.
- Hazard assessment and risk assessment
  - As the period of unrest evolves: (1) real time characterisations of the host volcano's possible and most likely hazards and their temporal, physical and spatial parameters; and (2) other advice e.g. the merits/safety of further/different short-term monitoring.
- The communication of scientific analysis (with its inherent assumptions, limitations, complexities and uncertainties) to a variety of stakeholders each possibly having different requirements and expectations and related communication challenges.

If an exercise focusses predominantly upon scientific issues, it may be very difficult to engage participants from civil protections agencies and more 'risk' related functions.

A VUSE can be a good opportunity to test the processes by which scientific analysis is integrated into risk assessment agencies, such as civil protection authorities, that do not have embedded scientists.

What is the proposed active participant scope?

In a VUSE, if time and resources permit, a wide range of stakeholders can be represented and actively involved including, but not limited to:

- the host volcano as the source of scientific and visual data, which are provided sequentially in discrete pre-planned "phase" briefings/reports (hereinafter called the Volcano team);
- host volcano observatory scientists (the VOS);
- local and external scientists (collectively called the scientific advisory committee or SAC);
- risk assessors and managers possibly from a range of national, federal, state, regional and municipal tiers of government (hereinafter called civil protection authorities or CPA);
- the media;
- interested and affected members of the public.

In a more sophisticated exercise, it may be worthwhile ensuring that a "dissenting/minority view" and/or "maverick" scientists are represented to test robustly relevant democratic and communication processes.

The MIAVITA Handbook (2012, 114–116) details the benefits of informing, sharing and training involving: (1) national, regional and local authorities; (2) scientists; (3) volunteers; (4) the media; (5) pupils and students; and (6) the public.

The VUELCO Colima exercise incorporated the planned evacuation of an at-risk community. A VUSE can usefully focus upon the real time implementation of the risk mitigation strategies that result from the dynamic characterisation of an evolving unrest. However, the time and resources required to identify and engage multiple stakeholders, including members of the public, should not be underestimated.

Care must be exercised if one of the hazard scenarios will result in the need for the evacuation of representatives of at least one vulnerable community. Of course, this will have to be organised and announced well in advance of the start of the exercise. This may affect (i.e. make rather implausible) other parts of the exercise which may depend upon realistic and continuing uncertainty as to the future evolution of the initial hazard.

What is the proposed passive participant (e.g. observers and monitors) scope?

In the VUELCO exercises, very valuable contributions were made by monitors and observers. They can be from the participating organisations or entirely 'independent'. Monitors and observers (particularly with experience of previous exercises and the practices of countries other than the host country) can play a critical role in providing both hot and cold feedback whilst also gaining invaluable personal experience to assist in the planning of future exercises.

VUSE are invariably observed by local and visiting students and early career academics. With careful planning, it may be possible for less experienced scientists and civil protection authority decision makers to be involved directly (as happened in the Campi Flegrei VUSE) or indirectly in the 'shadow' SAC and CPA teams.

Reference should be made to the separate Observers/Auditors checklist in Sect. 4.6.

What is the proposed geographical scope?

The geographic scope of a VUSE should be considered carefully. A balance must be achieved between the spatial parameters that produce a risk exposure area that is realistic in the context of the host volcano and those that create an area that is too large in terms of the duration of the exercise, data production and/or the roles of the active participants.

What is the proposed administrative scope?

National laws tend to identify, authorise and fund risk governance bodies (e.g. government departments and agencies, public corporations) and public officials (e.g. individuals such as governors, mayors, prefects and village heads) within a coherent legal and administrative framework—a risk governance infrastructure.

These laws often use and build upon existing entities with existing administrative frameworks that have multi-level national, regional, district, municipal etc. political divisions and subdivisions (Bretton et al. 2015).

The MIAVITA Handbook (2012, 118) suggests that it is good practice for an exercise to "be based on the regulations and laws of the [host] country".

The administrative scope of a VUSE should be considered carefully and it is likely this scope will be related to those parts of the risk governance process selected for review and testing. It may be very difficult to involve actively all levels of governance and therefore a decision will have to be made as which levels will participate and which will merely observe.

It is likely that a good starting point will be a complete flow diagram of the existing societal risk governance infrastructure for the host volcanic region. This can then be annotated to indicate which national, regional, local bodies and individuals will participate and their respective roles (both active and passive) in the exercise.

For the SAC and CPA at least, consider setting out clearly its respective:

- legal status;
- legal remit and reporting lines; and
- the rights, responsibilities and liabilities of its members.

Duration and type

How long will the exercise last and why?

This is a very important issue with funding, resources, logistical and many other implications.

VUSE vary greatly in length as shown by the data in Table 1. Much will depend upon the critical decisions that must be made about scoping—process, participants, geographical area and administrative levels.

What type of exercise will be undertaken and why?

There are many types of management exercises including: (1) full-scale; (2) reduced; (3) orientation; (4) drill; (5) table-top/discussion; and (6) functional (see Doyle et al. 2015).

Exercises can be also announced or unannounced depending upon their objectives. It has been suggested that the use of unannounced exercises is necessary to verify the real strength of systems and levels of preparedness, when people do not expect them (MIAVITA 2012, 116–117)

How frequently should exercises be held?

The MIAVITA Handbook (2012, 116) notes that exercises are fundamental for testing existing procedures, plans, and preparedness and for maintaining the attention of stakeholders "on the spot". It also argues that exercises should be scheduled frequently with the frequency depending upon several factors including: (1) the behaviour of the host volcano; (2) the social-economic context; (3) the levels of risk perception; and (4) the democratic trend.

Finance

What are the planned budgets for all phases of the exercise including:

- **Before?**
Consider planning, field trips, printed field guides, data and scenario creation, documentation etc.

- **During?**
Consider venue, catering, data/scenario communication etc.

- **After?**
Consider feedback (hot and cold) and follow up meetings, reports, presentations, interviews, actions.

Who is funding:

- *The exercise?*
- *The participants, including invited guests, experts and invited observers/auditors (including travel, accommodation, field trips, other expenses etc.)?*

Consider whether there should be an underwriter of last resort for any unplanned deficiencies.

Do the funders have any demands, expectations etc. as to any aspects of the exercise?

If yes, these must be understood by the organiser, planned and budgeted for.

4.2 Logistics

How long will the planning stage take?

It would be easy to underestimate the time that will be needed for, and the complexity of, the planning stage upon which the whole success of the exercise will depend. It may be prudent to speak to the organisers of other VUSE to discuss this and other financial/practical issues. Reference could also be made to Dohaney et al. (2015).

The MIAVITA Handbook (2012, 118) notes that planning an announced exercise takes time and states "six to twelve months are necessary to prepare a full-scale exercise. If the promoted exercises are repeated on a fixed schedule, three can be sufficient". We consider these to be the minimum planning periods.

Will every participant/observer receive in advance a full briefing note covering:

- *Purpose*
  Refer to the Sect. 4.1 regarding the overall purpose of the training exercise.
- *Scope*
  Refer to the Sect. 4.1 regarding scoping. Issues covered may include processes, participants, geographical, administrative etc.
- *Themes*
  If some important themes have been identified for particular attention (e.g. expressions (numerical and narrative) of likelihood, expressions

of analytical/diagnostic confidence, dealing with communication network failures, using social media etc.), it might make sense to mention these in advance to encourage pre-exercise thought and preparation.

- *Models, methods and protocols to be used/made available for use*

The VUELCO exercises have proved that VUSE can be used successfully to test in simulated real time the merits of probabilistic methods and tools such as BET_EF (Bayesian Event Tree for Eruption Forecasting), HASSET (Hazard Assessment Event Tree) and QVAST (Quantifying Volcanic Susceptibility) (Sandri et al. 2009; Sobradelo and Bartolini 2014; Bartolini et al. 2013).

In advance of the exercise, there must be communication between the modellers and the Volcano team. This should ensure that all appropriate monitoring data (perhaps deliberately not all the data) is available for all relevant phases of the exercise scenario including periods between phases.

Feedback from the VUELCO exercises suggests that input from the modellers during each step of an exercise may be welcomed by other participants. Testing simultaneous integration was a specific VUELCO goal. However, if this is to be done, it must be carefully planned and sufficient time allocated within the timetable.

It is important to state in advance of the exercise, for example in the briefing documents, what role the models, tools etc. will play during the exercise. Consideration should be given to whether the results will be available in real time to the participating SACs and CPAs. If yes, how and when will the results be communicated? If no, when will feed back regarding the utility of the items tested be communicated?

VUSE can also be used to give SAC's the opportunity to use formal expert elicitation (EE) methods. EE may provide formality and direction to the deliberation of monitoring data, assist the framing of likely hazard scenarios, and facilitate the drafting of expressions of temporal certainty and analytical/diagnostic confidence.

EE were conducted in the VUELCO exercises in Colima and Campi Flegrei.

The Somma Vesuvio MESIMEX exercise undertaken in 2006 was used to conduct a before and after volcanic risk perception survey (Ricci et al. 2013).

- *Players (participants, observers etc.):*
  - *by name and organisation*
  - *by role in the exercise*

It is very likely that a large number of people will be involved from a wide range of organisations and, perhaps, several countries. It would be prudent to assume that nobody will be known to all of the other participants. Consider the use of name badges stating Name, Organisation, and Exercise Role. As found during the Campi Flegrei exercise, colour coding for teams and active/passive roles might also be helpful. The scheme adopted should be explained carefully to save time during the start of exercise when valuable time can be lost easily.

By way of illustration, more than 100 participants from over 10 European and Latin American countries attended the VUELCO Campi Flegrei exercise. The respective figures for the VUELCO Cotopaxi exercise were about 50 participants from 13 countries.

After the Dominica exercise it was suggested that, at the very start of the exercise, introductions would have been helpful "to establish who each person was and what their role was…It was challenging accepting information from a person whom we did not know or have any clue as to their background or capabilities. Also the scientist [s] must be aware of who the[y] are working with in order to frame their advice appropriately".

For numerous purposes (before, during and after the exercise) a comprehensive, up-to-date and accurate email list containing all participants will be necessary. Inadequate email lists caused difficulties during several VUELCO exercises.

- *Teams*

Consider how many teams will be required and how and when their members will be allocated. For the sake of simplicity, it may be necessary to ask representatives of several different organisations to work together. By way of illustration, this chapter refers to four teams (Volcano, SAC, CPA and Observers) but others can and should be used as dictated by the specific goals of the planned exercise.

Consideration should also be given to the use of a technical team to support the SAC's deliberations in respect to issues such as real time GPS mapping, model trials and expert elicitations. The time needed for these complex matters must be considered particularly during short exercises. In VUELCO's Campi Flegrei simulation, a technical team composed of civil protection personnel operated continuously with the aim of supporting the SAC's simulated 'real-time' deliberations. The team's results were only presented at the end of the exercise because they lagged behind the 'real time' evolution of the phases of the exercise's hazard scenarios.

- *Venue (full address, plan, transport and contact details)*
  - *Sub-venues for all aspects of the exercise (including any evening events) with a plan*
  - *Layout for the plenary sessions*
  - *Layout for any breakout sessions*

These critical issues should not be overlooked. The exercise organiser should decide in advance, in respect for each phase of the exercise, which 'team' will sit where and why, probably by reference to which other participants they will have to communicate with. For the sake of intended realism, consider whether they should be near or far apart and by what means they will communicate.

Consider whether, in a real emergency, participants would speak to each other face to face. Consider whether it would be more realistic to separate some teams physically in order to force the use of video conferences, emails and phone calls in team interactions. At the planning stage it will be necessary to allow more time for this nuance, which may highlight the challenges of long-distance communication, data sharing, data analysis, collaboration, consensus building and decision-making.

Feedback from the VUELCO exercises specifically mentioned the problems associated with having too many people in the same room and failing to plan in advance who should be where and near to whom.

It is worth considering that each refreshment break will take longer than expected unless refreshments are either provided in all rooms or available at all times. Will working lunches simulate the difficulties of a tense period of emerging unrest? For a short period of intense training, is a degree of inconvenience and/or discomfort in fact to be encouraged?

- **Timetable**
- *Dates for each part of the exercise and timings for each day addressing:*
- *Registration (including perhaps the handing out of name badges in accordance with the stated scheme)*
  This will need very careful planning and resourcing to avoid a late start on Day 1.
- *Start, finish*
  Start and finish promptly.
- *Food/drink breaks (if needed!)*

Refer to the above guidance about the need for these in an exercise that attempts to simulate real time challenges.

- *Evening events, banquets etc.*

Consider how long it will take participants to return to their accommodation before evening events. Give clear advice about the dress code, if any, and other cultural expectations (e.g. the use of cameras or the need for modest clothes and head coverings).

- *Venue tours*

It is likely that many of the participants will expect/request a tour if the exercise takes place in a major risk management/communication centre. Allow plenty of time for this and consider making these visits part of the pre-exercise field trip to avoid wasting time when all of the participants (and probably the media) are present.

- *Greetings, Introductions, Reviews, Thank you's etc.*
- *Presentations to or by representatives of central, local government, etc.*

Predicting accurate timings for these two components is notoriously difficult and careful planning in advance and strict control on the day will be necessary. Consider carefully whether these are really necessary. Consider whether it would be better for relevant entities to be given formal active or passive roles within the exercise itself and be appropriately involved in that way.

- *Dissemination of new data*

Consider when, by whom, to whom and how. During the VUELCO exercises, difficulties arose due to format incompatibilities and internet access issues.

- *Deliberation of new data*

The time necessary for deliberation may depend upon the amount/ format of the data, the need to communicate with other participants, expectations about output deliverables (e.g. written reports) etc. As a general rule, allow plenty of time within the context of the realism that the exercise seeks to simulate.

- *Dissemination of 'injects'*

Consideration should be given to the timing and drafting of 'inputs' (for example, requests for local, national and international media interviews or briefings with concerned public officials, unwelcome social media exchanges, etc.), which will be distributed to the participants at planned stages during the exercise. Participants might be asked to think about how they would respond and who would be delegated to meet these urgent requests.

- *Outputs from the SAC and other teams*

As a general rule, allow plenty of time for the dissemination of outputs within the context of the realism that the exercise seeks to simulate.

Will information about current CPA mitigation actions, risk alert levels etc. be provided at any time during the exercise?

If yes, why, what, when and to whom?

Feedback from the VUELCO exercises suggests that scientists benefit greatly from improved knowledge of civil protection authority systems and in particular better appreciation of those hazard parameters that are most relevant to risk mitigation decisions.

- *Factual background and related resources*

Consider whether a field trip is necessary before Day 1 of the exercise so that all of the participants can gain a basic grasp of the topography, geography, geology (including major structures, faults, tectonic plates, existing volcanic and other hazards, and aquifers), geo-history (a time-line of major events is often helpful), infrastructures and other essential information.

All the VUELCO exercises incorporated well-planned and informative pre-exercise field trips that were supported by printed, carefully researched field guides with relevant histories, maps, reading lists etc.

A field guide can be supplemented by more detailed essential information about the monitoring history, thresholds etc. Feedback from VUELCO exercises suggests that the briefing pack should include maps showing the positions of all relevant observatories, monitoring equipment and stations, GPS positions, cameras, sample sources etc.

In addition to a field trip, it may be helpful to have a day/half-day of short presentations about the host volcano with careful oblique references to features that will be relevant during the exercise.

- *Real/assumed legal/regulatory framework and related duty holders*

Each participant must have a clear and comprehensive understanding of their role, the roles of all other participants and all planned lines and methods of communication. To ensure a higher degree of realism, these roles should accurately reflect the governance infrastructures, roles and duties required by:

- the national laws of the host country as encouraged by the UN Sendai Framework and the IFRC legal checklist;
- any regional emergency management or response arrangements, such as those in the Caribbean involving the Caribbean Disaster Emergency Agency (CDEMA), which are set out in inter-country agreements, memoranda of understanding and protocols; and
- any applicable international law standards.

- *Information Technology (IT) including disseminating/sharing data**

This critical issue should not be underestimated based upon difficulties encountered during VUELCO exercises.

It is inevitable that most of the participants will wish to have easy access to the internet. Ensure that the Wi-Fi network has sufficient capacity and range to allow easy access.

Consideration should be given to the format requirements of the computer models and probabilistic tools that will be tested during the exercise.

- *Language arrangements**

Consider the dominant language for the exercise and its documents (particularly the pre-exercise briefing pack) and what arrangements can be made, if any, for translations/translators.

Feedback from the VUELCO exercises suggests that some participants (e.g. locally based risk managers and members of the emergency services) may become bored and disengaged if they cannot understand fully what is going on and contribute.

- *Other equipment**

Consider whether it will be necessary, for the purposes of training, publicity or planning future exercises, to use other equipment including:

- Display screens, smart boards and white boards

- Fixed and roving microphones
  A shortage of roving microphones caused irritation at several VUELCO exercises.
- Visual and audio recording equipment
- A projector for PowerPoint presentations
  Feedback sessions are enhanced by such presentations, if time allows for their preparation.
- Laser pointers
- Video conference links

- *Presentations of data, results etc.**

How, why and when will the data, results etc. be presented?

Consider imposing a strict timetable to avoid timing problems.

- *Requirements regarding pre-Day 1 reading, preparation, queries etc.*

Consider stating clearly that there will no lengthy introductions or briefings at the start of Day 1 because the participants are required and expected: (1) to read in advance the pre-exercise briefing pack; and (2) to raise any queries before Day 1 with the exercise organiser.

- *Procedures for daily and end of exercise feedback.*

If daily feedback or announcements are necessary, consider in advance what issues are likely to be covered, who will give them and allow adequate time within the timetable.

*Note

It is accepted that during a real emergency these facilities may not be readily available. Accordingly, any difficulties encountered during an exercise may provide helpful insights into the challenges that would be encountered during a fast evolving crisis.

Will role/team leaders receive in advance?

- *A full pre-exercise briefing note about the overall legal/regulatory infrastructure for the exercise?*

- *Checklists for the main aspects of the exercise?*

Day 1 (Start of exercise)

Ensure that registration does not delay a prompt start, which will thereby set the standard for all other timings during the exercise.

Who will lead the exercise and ensure that the timetable is adhered to strictly?

Consider giving this role to a person (probably supported by an assistant) who does not have any other role in the exercise and therefore can move easily from room to room and deal promptly with any difficulties that arise.

Is there a need for a short introduction? If yes, why, who will give it and how long will it last?

Is there a need for a short end of day summary? If yes, why, who will give it and how long will it last?

Day 2 etc. (Continuation of exercise)

Is there a need for a short introduction? If yes, why, who will give it and how long will it last?

Is there a need for a short end of day summary? If yes, why, who will give it and how long will it last?

Day (End of exercise)

Is there a need for a short introduction? If yes, why, who will give it and how long will it last?

Is there a need for a short end of exercise summary? If yes, why, who will give it and how long will it last?

4.3 The Volcano Team

Membership
Who will lead the team and why?
Who will be the other Volcano members?

VUELCO's exercises indicated that considerable time, exceptional skills and infinite patience are required to write any coherent and challenging hazard scenario.

Members of the Volcano team must: (1) be selected and appointed at a very early stage in the planning process; and (2) be aware of, understand and accept the needs and goals of the exercise.

The overall hazard scenario and its planned phases must be tailored to fit within the agreed timetable for the exercise. The Volcano team will need to work very closely with the exercise leader and the steering group as soon as it is formed.

Where the Volcano team is the host volcano's actual monitoring team, a chance to test the monitoring team in 'real time' action will be lost.

Preparation
How long will be needed to prepare:

- *the scenario/s?*
- *all of the briefing documents?*
- *the pre-exercise field trips?*

The exercise
What will be the time scale in months/years covered by the exercise?

VUELCO's exercise scenarios varied greatly in length—Colima (2 months), Campi Flegrei (7 years), Cotopaxi (5 years) and Dominica (2 years).

The period of time covered by the simulation, the number of phases and the length of the intervals between the phases, must take into account the needs and goals of the exercise. For example, consideration must be given to the time that will be required for data to be delivered, entered and assimilated, and for probabilistic models to be run.

How many exercise phases will be needed to cover the period of years chosen?

Each phase will involve, and must be allocated sufficient time for, the delivery, entry, analysis and modelling of the monitoring data and the drafting and delivery of any expected outputs (e.g. reports, press releases etc.).

VUELCO's exercise scenarios had between 3 and 6 phases - Colima (2 months in 4 phases over 10 h spread over 5 days), Campi Flegrei (7 years in 4 phases over 3 days), Cotopaxi (5 years in 6 phases in 1 day) and Dominica (2 years in 3 phases over 2 days).

As already stated, the precise specifications for the phases must be agreed with the exercise leader taking into account the needs of the modellers and other technical teams.

What data will be provided?

The monitoring data must be suitable and sufficient to test the whole range of geo-scientific disciplines represented in the SAC. Feedback from the VUELCO exercises identified the temptation to favour the overprovision of seismic data at the expense of adequate geochemical, geodetic, petrological and other data.

Although this might reflect real situations or even the architecture of existing monitoring networks, careful consideration should be given to the consequences of having experts from some disciplines having insufficient data to occupy them and therefore becoming disengaged from the exercise.

It may also be necessary to provide in the briefing pack related historical data and background information (such as historical thresholds) to allow the data used during an exercise to be considered in context.

The data must remain coherent, albeit deliberately unclear and uncertain, over the full duration of the scenario including periods of inactivity. Feedback after the Dominica exercise praised the quality of the data and the design of the scenario. The following expressions illustrate some of the features of a successful scenario— very realistic unsure signals creating a state of limbo, non-linearity, phreatic evidence that was not a precursor of magmatic activity, complex data and realistic missing data.

During some phases of the exercise scenario it is possible that the data will be incomplete (some critical data may be held back deliberately to await a request for more data) and/or difficult to analyse because it may be a stated objective of the exercise to test the SAC's ability to handle uncertainty, disagreement, monitoring inadequacies etc. and/or to request data from additional or different monitoring.

If data is provided for the periods between scenario phases, it is likely that the longer the inter-phase period the more complete the data will be. Time will have to be planned for data delivery and analysis. A large amount of data may also create very great difficulties for the utility of probabilistic tools such as BET.

The Volcano team should communicate with the exercise leader and modellers well before the exercise to avoid unnecessary surprises regarding the format and quantum of the monitoring data and the timing of their release. In the Colima exercise, three BET nodes were passed before Phase 1 of the exercise scenario and this degraded the utility of the BET trial.

How will the years between the periods that are covered by the data be described?

This difficult issue should not be overlooked. In the absence of data, what will the participants assume in absence of guidance? An absence of data may also create very great difficulties for the utility of probabilistic tools such as BET.

Consider issuing, along with the monitoring data for the next phase, a brief summary of the 'missing periods' in terms of volcanic activity, precursor evidence and/or the risk mitigation decisions and actions of civil protection authorities.

Will the exercise scenarios be (1) entirely fictitious; (2) based on suitably disguised real events; or (3) a mix of the two?

During the VUELCO exercises feedback suggested that:

- it may be very difficult to create realistic scenarios that are entirely fictitious because the data, under the very close expert analysis that they will receive, may appear to be inconsistent and improbable;
- a coherent and plausible chronology for the scenario is essential;
- after a period of sudden emerging unrest, a period of reducing unrest or quiescence may form the basis of a challenging scenario as proved during the Dominica exercise;
- in order to simulate reality, the Volcano team should provide monitoring data to the SAC <u>without</u> any form of analysis or interpretation, however the provision of tables comparing data sets may be helpful;
- the amount, relevance and format of the monitoring data provided should be considered carefully.

Will the exercise introduce secondary hazards (e.g. forest fires, contaminated aquifers, etc.)?

During the Mount Teide exercise, considerable attention was paid to secondary hazards. If secondary hazards are to be included, what data need to be provided before, at and after the start of the exercise?

Will plans be needed showing geographical, geological or societal features?

Will the exercise use fixed/dynamic data about weather, ground conditions etc.?

If yes, what data need to be provided before, at and after the start of the exercise? Will plans be needed showing geographical, geological or societal features?

Relevant weather/ground water data will be essential if ash dispersion and fall-out simulations are to be included within the exercise.

When, how and to whom will the monitoring data be disseminated:

- ***Before the start of the exercise?***
- ***At the start of the exercise?***
- ***During the exercise?***

This is a critical issue and important lessons were learned during the VUELCO exercises. During a VUSE, reliance on a limited Wi-Fi link may be very problematical.

In the Dominica exercise, monitoring information was provided within separate 'phase-specific' documents in password-protected pdf format and emailed before the start of exercise to all of the participants. At the beginning of each scenario phase, the relevant password was released. This solution worked very well.

During the Cotopaxi exercise, Wi-Fi and other communication difficulties were identified by the civil protection authorities and, as a result, they were better placed to improve their existing systems and resources.

Consideration should be given to the format requirements of the computer models and probabilistic tools that will be tested during the exercise.

Feedback from the VUELCO exercises suggests that when data are provided they should not duplicate data that have already been provided.

Will the SAC be able to ask for further monitoring data?

If yes, when and how can it be requested? If yes, when and how will it be provided?

Will any player other than the SAC be able to ask for monitoring data?

If yes, by whom, why, when and how can it be requested? If yes, when and how will it be provided?

Will anything other than monitoring data be provided?

Consider whether the SAC and CPA should be given, or have access to when they request it, fixed/dynamic data about weather, ground conditions etc.?

If yes, why, what, when and to whom?

4.4 The Scientific Advisory Committee (SAC)

Is a SAC necessary in every exercise?

Careful consideration should be given to this important question. It is appreciated that advice to civil protection authorities may only be provided by the head of the local volcano observatory, who is not, and would never be supported by a SAC.

Feedback from VUELCO exercises suggests that it makes good sense for exercises to simulate existing risk governance structures as far as reasonably practicable. It follows that a SAC team should be included in an exercise only: (1) where one exists, or could exist, in a period of evolving unrest or (2) where the civil protection authorities wish to test the utility of one.

If an exercise SAC is simulating the actions of a real SAC, the brief to the SAC and the briefing note for the exercise should set out clearly the SAC's constitution, mandate, responsibilities and powers.

Legal status and duties

A VUSE can be a good opportunity to consider the legal status of scientific committees and their full-time, part-time and seconded members.

Consideration should be given to the taking of appropriate steps to avoid or reduce managerial

risks by the inclusion of disclaimers and exclusion statements in hazard assessments. A VUSE may serve as, at least, a prompt for committees to seek legal advice from a competent local lawyer.

During the Dominica exercise, a member of the Scientific Advisory Committee sought legal advice from the lead author about: (1) the wording of a liability disclaimer for the benefit of visiting scientists; and (2) the way in which 'risk-related' advice could be passed to civil protection authorities without liability for subsequent risk mitigation decisions. In respect of the first, it is likely that a visiting scientist will have little or no control in respect of either the absence or adequacy of monitoring data and the selection and competence of other committee members. Based upon this advice, a disclaimer was included in the Phase 1 hazard assessment report and the Phase 2 and 3 reports were more carefully worded.

Membership

Who will be leading the SAC and why?

The SAC will require a leader who should be identified in advance.

Feedback from the VUELCO exercises suggests the role of Chairperson of the SAC may be critical to the success of an exercise.

Consider whether a deputy Chairperson should be identified in advance of the exercise.

In several of the VUELCO exercises, it became evident that the Chairperson should not be given, or retain, prime responsibility for the initial drafting of the hazard assessment report.

Who will be the other SAC members?

Consider whether it is desirable to have a range of experts to simulate the issues that might arise during a real period of emerging unrest. Consider the inclusion of: (1) local volcano observatory scientists; (2) other host country scientists; (3) host country academics (4) non-national observatory scientists; (5) non-national academics; and (6) young researchers.

The involvement of non-local and foreign scientists may represent an opportunity to share knowledge and to have access to a wider range of opinions uninfluenced by local social contexts, non-scientific values or entrenched scientific assumptions or preconceptions.

Consideration should be given to the need for rules regarding the nature and extent of the interactions between local and non-local/foreign scientists.

During the Campi Flegrei exercise a number of young researchers were co-opted in rotation onto the SAC. This worked well and gave them a unique learning experience.

Which geoscience disciplines (e.g. geophysics, geochemistry, geodetics, geo-history, and petrology) will be covered within the SAC?

If the monitoring data resources of the host volcano permit, ensure that the Volcano team provides sufficient monitoring data to keep each discipline engaged and, if possible, challenged throughout the exercise.

Will any geoscience disciplines not be covered? Why?

Consider whether this is a realistic representation of what might happen in a real period of emerging unrest.

Will the SAC have local members who are usually based near or at the host volcano?

Consider whether the briefing pack should refer to the IAVCEI protocol mentioned above.

Will the SAC have international members who are not based in the host country?

Consider whether the briefing pack should refer to Newhall et al. (1999) mentioned above.

Will the SAC be realistic in size, too big or too small?

Will any attempt be made to deal with the issue of 'maverick' scientists?

If yes, how? If no, why?

A VUSE may present a good opportunity to consider and test procedures for dealing very difficult issues such as this.

Inputs

When and in what format will the SAC get the monitoring data?

This is a critical issue. During a VUSE, reliance on a limited Wi-Fi link may be very problematical.

Will the SAC be able to ask for further monitoring data?

If yes, when and how can it be requested? If yes, when and how will it be provided?

Will the SAC get anything other than monitoring data?

Consider whether the SAC should be given, or have access to on request, fixed/dynamic data about weather, ground conditions etc.?

If yes, what data need to be provided before, at and after the start of the exercise? Will plans be needed showing geographical, geological or societal features?

Deliberations

Will the SAC's meetings/deliberations follow standardised operating procedures (SOPs) and/or use a standardised reporting template?

Feedback following the Dominica exercise advocated the use of SOPs and the keeping of a log recording and timing, inter alia, main data inputs and analytical decisions, assumptions, outputs and communications.

Careful consideration should be given to the SAC adopting working assumptions that:

- All hazard communications, including all hazard assessments, in their format, content and delivery must be focussed upon: (1) facilitating informed risk mitigation decisions that may not only be difficult but also based upon many sources of knowledge and social, political, economic and other influences; and (2) responding to the identified needs and expectations of their makers.

- Before acting as hazard communicators, hazard analysts must (by means of active and careful two-way dialogue) canvass, note and respond to the needs and expectations of the risk decision makers and the legitimate foreseeable demands of mass and social media and allow sufficient time to do this.

- The utility of hazard communications (i.e. the outputs of hazard analysis) must be judged by empirical evidence of the sentiments and actions of risk decision-makers (i.e. the outcomes of hazard analysis within wider risk mitigation processes).

- No assumptions will be made as to what risk decision-makers know about: (1) the science of volcanic hazards and in particular its complexities, uncertainties and limitations;

(2) the role of scientists and the many temporal, financial, legal and other constraints under which they operate particularly during periods of emerging volcanic unrest; and (3) the role, benefits and limitations of long-term and short-term monitoring.

- No assumptions will be made about what risk decision-makers either need or want and, accordingly, the widest possible range of reasonably practicable options for not only hazard communication but also hazard analysis should be offered and discussed.

- No assumptions will be made that risk decision-makers know what general and/or bespoke analysis, information, data, advice and guidance scientists may be able to provide if asked and given adequate resources.

Will any guidance be given as to how the data should be considered (e.g. expert elicitation, probabilistic tools etc.)?

If an expert elicitation is a possibility or is actively encouraged/required, timetable enough time for this as it is likely that a briefing will be necessary before the elicitation starts to ensure that everyone is fully aware of what will be involved. Do not underestimate the amount of time that a worthwhile elicitation will take.

Consider providing essential information in the form of an elicitation toolkit in advance. Who will prepare this toolkit?

Can the SAC take into consideration any societal factors such as high societal exposure in a particular area? If yes, what 'exposure' and 'vulnerability' data will be provided, when and how? Do not underestimate the amount of time that this will take.

Consider how and when the results of probabilistic models will be delivered to the SAC and thereafter considered. Do not underestimate the amount of time that this will take.

Feedback after VUELCO's Dominica exercise showed that scientists involved in exercises need to be briefed about: (1) the claimed benefits offered by the probabilistic models being used; (2) their limitations; (3) when and in what form their results will be presented; and (4) how it is

envisaged their results will be integrated within the overall analysis of relevant hazard scenarios.

Consider including in the pre-exercise briefing pack an additional file containing essential background information about expert elicitations and modelling.

Will the SAC have to liaise with any non-SAC players?

If yes, when and how will this be done?

During the Dominica exercise, sub-team leaders of the SAC were asked to attend early parts of the CPA's deliberations. In the post-exercise feedback, it was suggested that all of the SAC scientists would have benefitted greatly from hearing the CPA's deliberations which were often very critical of the scientific analysis upon which risk decisions had to be made.

Outputs—Hazard communications
What outputs (e.g. oral and written reports, written minutes etc.) will be required from the SAC?

The quality of the dialogue between risk management stakeholders, and the provision of information and advice, depends upon a mutual understanding by those stakeholders of their respective needs, responsibilities, functions demands and roles, and their capacity to anticipate other stakeholders' demands and decision needs (Salas et al. 1994; Ronan et al. 2008; Lipshitz et al. 2001; Paton and Jackson 2002; Doyle and Johnston 2011a, b; Doyle et al. 2014).

"Without knowing the concerns" and understandings "of the targeted audience, communication will not succeed" (Renn 2008, 147).

Careful consideration should be given to this critical issue. Based upon the VUELCO exercises, it may be very unwise to provide any guidance whatsoever as to how the outputs of the hazard analysis must be communicated to the CPA and other exercise participants. This may encourage the type of two-way dialogue referred to under the above heading "Deliberations". By providing guidance of any sort, a unique learning opportunity may be lost.

Consider briefing the observers in advance about this important issue, so that they can make an early invention if unwarranted assumptions are made about what risk decision makers know, need or want.

Will the SAC be required to communicate with teams representing:

- ***The CPA?***
- ***The public?***
- ***The media?***
- ***Representatives of local/central government?***

If yes, when and where?

A VUSE is a good opportunity to practise hazard communications and to consider the needs and expectation of risk decisions and their makers. Reference should be made to Doyle et al. (2015).

Consider how, when and by whom the needs and expectations of risk decision-makers will be sought.

4.5 The Risk Managers—Civil Protection Authorities (CPA)

Role

It is critical for the role of the CPA to be considered very carefully during the planning stage.

Feedback from the VUELCO exercises suggests that the CPA's role must be:

- clearly defined and integrated within the design of the hazard scenario;
- sufficiently interesting and demanding to keep the participants engaged; and
- known to all other participants.

The geoscientists will be engaged predominantly with the prompt analysis of monitoring data and this will lead subsequently to the delivery of hazard assessments in some form to the CPA. Depending upon the goals of the exercise, it is likely that, whilst this is being done, the CPA team must have something worthwhile to do.

Consideration might be given to a requirement within both the briefing pack and the timetable that representatives of the SAC discuss with the CPA:

- the timing, format, content etc. of hazard reports, communication channels and protocols (the preparation of a template was suggested after the Campi Flegrei exercise), and/or
- the arrangements for a possible mock joint press conference later in the exercise. A joint press conference might involve another press/media team (possibly made up of individuals from the CPA's public relations department) and/or invited observers from local and national press organisations.

Membership
Will the CPA represent a real/fictional entity?
Who will be leading the CPA and why?
Who will be the other CPA members?

Inputs
What inputs will the CPA get?
Will the CPA get monitoring data?
If yes, why?

Deliberations
Will any guidance be given as to how the CPA inputs should be considered?
Will the CPA have to liaise with any non-CPA players?
If yes, when and how will this be done?

Outputs
What outputs will be required from the CPA?
Will the CPA be required to communicate with teams representing:

- ***The SAC?***
- ***The public?***
- ***The media?***
- ***Representatives of local/central government?***

A simulated press conference, which involves both scientists and CPA officials within the same panel, can serve a number of useful purposes. In a potentially hostile environment, scientists will have to communicate effectively complex geo-scientific concepts and explain issues such as uncertainty and analytical/diagnostic confidence. They may also have to avoid being drawn into sensitive areas involving societal issues and risk decisions. Risk decision makers will have to demonstrate an understanding of complex hazard scenarios and to explain what risk mitigation decisions they have taken and the reasons for them.

Open and convincing displays of collaboration and agreement between different authorities and key individuals during real crises can build trust with representatives of the media and general public. Panel question and answers sessions involving key personnel, who will have 'communication' roles, can and should be practised during VUSE because it is likely that related training and resource needs will be identified.

The VUELCO Dominica exercise included three very realistic but successful simulated press conferences.

Feedback

The processes of hazard analysis and hazard communication should be outcome focussed and driven by the expectations and needs of risk decisions and risk decision-makers.

Consideration should be given to measuring the extent to which the exercise and its participants addressed: (1) the needs and expectations of the CPA; (2) the steps, if any, the SAC took to identify those needs and expectations; and (3) the extent to which the SAC satisfied those needs; and identified actions that might lead to improvements.

The VUELCO exercises identified the value of civil protection authorities having an internal technical-scientific structure: (1) to support the SAC in technical analysis (mapping, elicitation, models); and: (2) to improve interactions with the SAC.

4.6   Observers/Auditors

Is it necessary to have observers and/or auditors?

In the VUELCO exercises a very valuable contribution was made by monitors and observers. They can be from the participating organisations or entirely 'independent'. Monitors and observers (particularly with experience of previous exercises and the practices of countries other than the host country) can play a critical role in providing both hot and cold feedback whilst also gaining invaluable personal experience to assist in the planning of future exercises.

VUSE are invariably observed by local and visiting students and early-career geoscience academics. With careful planning, it may be possible for less experienced scientists and civil protection authority decision makers to be involved directly (as happened in the Campi Flegrei VUSE) or indirectly in 'shadow' SAC and CPA teams. The participation of young researchers in this kind of exercise represents a great chance to train future generations of advisory scientists.

What role will be played by nominated observers and auditors?

Consider whether it would be worthwhile differentiating between: (1) observers who might have responsibility for capturing and presenting immediate unstructured feedback; and (2) auditors who would compile a more formal report possibly in a format and using audit criteria agreed by the main participants in advance.

Would something be gained by having an 'independent' observer/auditor who is not part of any organisation taking part in the exercise?

Independent observers played a very useful role in the VUELCO exercises in Campi Flegrei, Cotopaxi and Dominica and the author was an invited external auditor of the Tenerife exercise.

During the Campi Flegrei exercise, it was recorded that undertaking the role of observer is particularly helpful for senior civil protection authority managers who are thinking of planning an exercise in their own locality.

Who will be nominated and/or invited to be observers/auditors?

This will depend largely upon the roles which have been chosen and whether some of the observers/auditors should be 'independent'.

Format

At the end of the exercise will there be a 'hot' feedback session?

The MIAVITA Handbook (2012, 118, 179) recommends an "on the spot debriefing straight after the exercises with the participants and another one approximately one month later with the exercise organisers, evaluators, observers, and a representative of the general public".

If yes, why, who will lead it and how long will it last?

When and how will the results be recorded, promulgated and acted upon?

Will there be 'cold' feedback meetings, surveys or questionnaires?

If yes, why and who will be responsible for their design and implementation?

When and how will the results be recorded, promulgated and acted upon?

Will a more formal audit of the exercise be undertaken?

If yes, consider whether it might address issues such as the identification of the need for better or additional:

- structures of risk governance
- information sources and resources
- long-term monitoring resources
- consideration of the role of tectonics and resulting faults and features
- geo-history data
- training
- financial, personnel and other resources
- documented procedures
- communication plans and equipment
- media response plans
- easily accessible database catalogues, which may be important and needed urgently during a crisis, of:
  - existing data (e.g. seismic, geodetic, gas, water geochemistry and geospatial) and common archiving procedures for newly acquired datasets; and
  - information (e.g. geographical information systems 'GIS' digital bases, overlays, metadata, geological maps, remote-sensing images, background levels of unrest indicators, such as seismic energy release rates, normal rates on inflation/deflation, aqueous and geo-chemistry and fluxes).
- post-exercise actions to review and revise existing procedures in particular those for any future exercises.

Will the observers give real-time feedback during the exercise?

Consideration should be given to this difficult issue. A balance must be achieved between, on the one hand, unhelpful and intrusive interventions and, on the other, allowing the exercise to proceed in ways that may waste valuable time and/or miss or diminish valuable learning opportunities.

During the Dominica exercise, the three 'independent' observers gave brief feedback to the SAC at the end of Phases 1 and 2 of the exercise. They also gave spontaneous suggestions to the CPA. The participants indicated that they found this helpful and there was clear evidence that suggestions for change were understood and put into practice immediately.

Preparation

Will instructions be given to the observers/auditors in advance?

Consider giving the observers/auditors a clear brief before the start of the exercise.

Experience from the VUELCO exercises suggests that providing observers with a list of more specific considerations, to include within their more general observations, may result in more focussed outputs.

Will instructions be given to the participants in advance of the exercise to facilitate richer feedback?

Consider asking each participant and/or each team (1) to list their main roles, responsibilities and tasks and (2) to keep a written list of needs that would help them accomplish their tasks covering for example better/different information, training, guidance, software, communication equipment, other equipment, personnel, and so on. These lists can then be used during the hot and cold feedback sessions.

Will the participants be told in advance about the planned arrangements for observing/auditing?

If yes, when? Consider whether the pre-exercise briefing pack is the best place to set out the planned arrangements and time table and what might be required from the participants in this regard.

5 Discussion

The checklists presented in this chapter are dedicated to the organisers and participants of the five exercises that this chapter has investigated. The checklists record and build upon the successes, experiences, occasional oversights, misjudgements and mistakes of a few in order to provide readily accessible sources of knowledge, learning and inspiration for others.

This chapter has identified a number of themes:

- VUSE are purpose-driven learning/training activities. Vital knowledge is acquired not only during but also before and after each exercise (see Doyle et al. 2015).
- VUSE must have clear recorded goals based upon the needs and expectations of their participants.
- VUSE are complex and require clear and firm leadership, and very careful planning, funding and execution. This finding is supported by the work of Dohaney et al. (2015), which contains very helpful guidance on design and evaluation methods.
- During the 'planning phase', the critical issues of leadership, purpose, scope, duration, type and financing should be considered.
- The mere planning of a VUSE will provide invaluable knowledge of the wider legal and administrative infrastructures in which it is framed.

- The 'logistics' phase will ensure that the execution of the exercise is not undermined by avoidable technical, communication and other related difficulties. The importance of a comprehensive and comprehensible pre-exercise briefing note and exercise plan cannot be overstated.
- A pre-VUSE field trip will serve numerous purposes including those of introducing relevant geographical, geological, cultural and governance histories and providing an opportunity for relationships to develop between the exercise participants.
- VUSE require and depend upon numerous acts of communication between a wide range of stakeholders with a variety of knowledge, expertise, experience, needs and expectations within many types of formal/informal relationships/associations (see Doyle et al. 2015).
- VUSE are not judged in terms of 'success' or 'failure' but rather in terms of whether relevant knowledge has been generated, recorded, considered and utilised.
- Whenever possible, VUSE simulate:
  - 'existing' risk governance arrangements, or those being considered for the future, with real policies, processes and people rather than contrived unrealistic governance scenarios and false role-play; and
  - volcanic hazard scenarios based upon, and significant within, the context of the host volcano region, which may experience other relevant or related natural hazards such as tectonic and weather-related hazards.
- VUSE provide unique opportunities:
  - to address many known challenges including those of inter-scientist deliberation, scientific uncertainty, analytical/diagnostic confidence, unorthodox/maverick science sources, hazard communication, and mass media and social networking relations and communications (see Doyle et al. 2015);

- to test new/prototype risk arrangements and tools, models and protocols for analysis/communication.
- Great care should be taken to ensure that comprehensive feedback and goal-related knowledge can be captured during and after each exercise and disseminated as widely as possible.

6 Conclusions

"Practice doesn't make perfect. Practice reduces the imperfection". None of the five exercises considered in this chapter was completely perfect. They involved not only well-intentioned yet imperfect policies and procedures capable of improvement but also dedicated yet not infallible people keen to seek further knowledge, training and experience. Careful analysis of these exercises shows that, as suggested by Doyle et al. (2015), with very careful planning, execution and review, worrisome periods of emerging volcanic unrest and the dynamics of real-time hazard assessments and risk decisions can be simulated for a wide variety of worthwhile purposes.

However this chapter does not attempt to dictate how future exercises must be organised. Based upon a review of the published records of four exercises and the authors' personal experiences of the five further exercises listed in Table 1, it is believed that these lists will assist the organisers of future exercises to meet the specific challenges of the volcanic hazards they face and the societal, political, economic and legal contexts in which difficult and timely risk decisions have to be made.

The authors hope that the checklists will be considered and used, and, above all, improved. Candid feedback from future exercises will ensure that the guidance notes evolve and will be supplemented by the addition of further detailed exercise studies. Will these checklists facilitate more effective future exercises and improve volcanic risk mitigation? This question can be answered only if they are used and relevant evidence is generated within future post-exercise empirical studies.

The Sendai Framework sets ambitious goals for quality standards, learning, knowledge exchange, education and training. The paramount goal of this chapter is to ensure that the pooled experiences of the few, who have had the advantages and privilege of being exercise participants, will be accessible to the widest possible audience, to encourage future exercises and thereby to improve the governance of volcanic risks.

Disclaimer

The content of this paper reflects the authors' views and not necessarily the opinion of the organisations to which they belong. The authors and their organisations are not liable for any use that may be made of the information contained therein.

References

Bartolini S, Cappello A, Martin J, Del Negro C (2013) QVAST: a new quantum GIS plug for estimating volcanic susceptibility. Nat Hazards Earth Syst Sci 13, 3013-3-34

Bergs J, Hellings J, Cleemput I, Zurel Ő, De Troyer V, Van Hiel M, Demeere J-L, Claeys D, Vandijck D (2014) Systematic review and meta-analysis of the effect of the World Health Organization surgical safety checklist on postoperative complications. BJS 101:150–158. https://doi.org/10.1002/bjs.9381

Beta T (2011) Master of Stupidity. http://www.goodreads.com/book/show/8542012-master-of-stupidity

Bretton RJ (2014) The role of science in the rule of law. http://expertsinuncertainty.net

Bretton RJ, Gottsmann J, Aspinall WP, Christie R (2015) Implications of legal scrutiny processes (including the L'Aquila trial and other recent court cases) for future volcanic risk governance. J Appl Volcanol 4:18. https://doi.org/10.1186/s13617-015-0034-x

Dohaney J, Brogt E, Kennedy B, Wilson TM, Lindsay JM (2015) Training in crisis communication and volcanic unrest forecasting: design and evaluation of an authentic role-play simulation. J Appl Volcanol 4:12. https://doi.org/10.1186/s13617-015-0030-1

Donovan A, Oppenheimer C (2012) Governing the lithosphere: insights from Eyjafjallajökull concerning the role of scientists in supporting decision-making on active volcanoes. J Geophys Res 117:B03214. https://doi.org/10.1029/2011JB0090802012

Donovan A, Oppenheimer C (2014) Science, policy and place in volcanic disasters: insights from Montserrat. Environ Sci Policy 38, 150–161

Doyle EEH, Johnston DM (2011) Science advice for critical decision-making. In: Paton D, Violanti J (eds) Working in high risk environments: developing sustained resilience. Charles C Thomas, Springfield, Illinois, USA, pp 69–92

Doyle EEH, Johnston DM, McClure J, Paton M (2011) The communication of uncertain scientific advice during natural hazard events. N Z J Psychol 40(4):39–50

Doyle EEH, McClure J, Johnston DM, Paton D (2014) Communicating likelihoods and probabilities in forecasts of volcanic eruptions. J Volcanol Geoth Res 272 (2014):1–15

Doyle EEH, Paton D, Johnston DM (2015) Enhancing scientific response in a crisis: evidence-based approaches from emergency management in New Zealand. J Appl Volcanol 4:1. https://doi.org/10.1186/s.13617-014-0020-8

Errington EP (1997) Role-play. Higher education research and development society of Australasia, Jamison Centre, ACT

Errington EP (2011) Mission possible: using near-world scenarios to prepare graduates for the professions. Int J Teach Learn High Educ 23:84–91

Fiske R (1984) Volcanologists, Journalists, and the Concerned Local Public: a tale of two crises in the Eastern Caribbean. In: Explosive volcanism: inception, evolution and hazards. Geophys. Study Committee, National Research Council, National Academy Press, Washington, pp 170–176

Galanti E, Goretti A, Foster B, Di Pasquali G (2006) Geotech Geol Earthquake Eng 369–384

Gawande A (2010) The checklist manifesto—how to get things right. Profile Books Ltd, London

Harris AJL, Gurioli L, Hughes EE, Lagreulet S (2012) Impact of Eyjafjallajökull ash cloud: a newspaper perspective. J Gepphys Res 117:35. https://doi.org/10.1029/2011JB008735

Hincks TK, Jean-Christophe Komorowski J-C, Sparks SR, Aspinall WP (2014) Retrospective analysis of uncertain eruption precursors at La Soufrière volcano, Guadeloupe, 1975–77: volcanic hazard assessment using a Bayesian Belief Network approach. J Appl Volcanol 2014(3):3. https://doi.org/10.1186/2191-5040-3-3

Hood C (1986) Administrative analysis: an introduction to rules, enforcement and organisations. Wheatsheaf Books, Sussex

Hood C (2011) The blame game—Spin, bureaucracy and self-preservation in government. Princeton University Press, Princeton, New Jersey, USA and Woodstock, Oxfordshire, UK

IAVCEI (2015) News no. 1–3. http://www.iavcei.org/

International Federation of Red Cross and Red Crescent Societies (IFRC) and United Nations Development Programme (2015) The checklist on law and disaster risk reduction, Pilot version, March 2015. IFRC, Geneva, Switzerland

Jenkins S, Spence R, Baxter P, Komorowski J-C, Sara Barsotti S, Esposti-Ongaro T, Neri A (2012) Preparation of disaster risk scenarios at La Soufriere, Guadeloupe poster at volcanic and magmatic studies group annual meeting 2013

Johnston D, Scott B, Houghton B, Paton D, Dowrick D, Villamor P, Savage J (2002) Social and economic consequences of historic caldera unrest at the Taupo volcano, New Zealand and the management of future episodes of unrest. Bull NZ Soc Earthquake Eng 35 (4):215–229

Jordan TH, Chen Y.-T, Gasparini P, Madariaga R, Main I, Marzocchi W, Papadopoulos G, Sobolev G, Yamaoka K, Zschau J (2011) Operational earthquake forecasting: state of knowledge and guidelines for implementation, findings and recommendations of the international commission on earthquake forecasting for civil protection, submitted to the Department of Civil Protection, Rome, Italy. Ann Geophys **54**, 4. https://doi.org/10.4401/ag-5350

Lipshitz R, Klein G, Orasanu J, Salas E (2001) Focus article: taking stock of naturalistic decision making. J Behav Decis Mak 14(5):331–352

Marrero JM, García A, Llinares Á, Berrocoso M, Ortiz R (2015) Legal framework and scientific responsibilities during volcanic crises: the case of the El Hierro eruption (2011–2014). J Appl Volcanol 4:13. https://doi.org/10.1186/s13617-015-0028-8

Marzocchi W, Newhall C, Woo G (2012) The scientific management of volcanic crises. J Volcanol Geoth Res. https://doi.org/10.1016/j.volgeorgres.2012.08.016

McGuire WJ, Kilburn CRJ (1997) Forecasting volcanic events: some contemporary issues Geo. Rundsch 86:439–445

MIAVITA (Mitigate and Assess risk from Volcanic Impact on Terrain and human Activities) (2012) Handbook for volcanic risk management: prevention, crisis management, Resilience, MIAVITA team, Orleans. http://miavita.brgm.fr/default.aspx

Mothes PA, Yepes HA, Hall ML, Ramón PA, Steel AL, Ruiz MC (2015) The scientific-community interface over the fifteen-year eruptive episode of Tungurahua Volcano, Ecuador. J Appl Volcanol 4:9. https://doi.org/10.1186/s13617-015-0025-y

Newhall CG (2010) A checklist for volcanic risk mitigation. Cities on volcanoes 6 presentation, earth observatory of Singapore

Newhall C, Aramaki S, Barberi F, Blong R, Calavache M, Cheminee J-L, Punongbayan R, Siebe C, Simkin T, Sparks RSJ, Tjetjep W (1999) International association of volcanology and chemistry (IAVCEI) subcommittee for crisis protocols - professional conduct of scientists during volcanic crises. Bull Volcanol 60:323–334

NZ/MCDEM (2008) Exercise Ruaumoko 2008: final exercise report. Ministry of Civil Defence & Civil Management, Wellington, NZ

OECD (2015), Scientific advice for policy making: the role and responsibility of expert bodies and individual scientists. OECD science, technology and industry policy papers, no. 21, OECD Publishing, Paris. http://dx.doi.org/10.1787/5js33l1jcpwb-en

Paton D, Jackson D (2002) Developing disaster management capability: an assessment centre approach. Disaster Prev Manage 11(2):115–122

Renn O (2008) Risk governance – coping with uncertainty in a complex world. Earthscan, London, New York

Ricci T, Nave R, Barberi F (2013) Vesuvio civil protection exercise MESIMEX: survey on volcanic risk perception. Ann Geophys 56(4):s0452. https://doi.org/10.4401/ag-6458

Ronan KR, Johnston DM, Daly M, Fairley R (2008) School children's risk perception and preparedness: a hazard education survey. Aust J Disaster Trauma Stud 1. ISSN: 1174-4707. Available at: http://www.massey.ac.uk.nz/~trauma/issues/201-1/ronan.htm

Rothstein H (2002) Neglected risk regulation: the institutional attenuation problem, centre for analysis of risk and regulation. London School of Economics and Political Science, London

Salas E, Stout RJ, Cannon-Bowers JA (1994) The role of shared mental models in developing shared situational awareness. In: Gilson RD, Garland DJ, Koonce JM (eds) Situational awareness in complex systems: proceedings of a cahfa conference. Embry-Riddle Aeronautical University Press, Daytona Beach, FL, pp 298–304

Sandri L, Guidoboni E, Marzocchi W, Selva J (2009) Bayesian event tree for eruption forecasting (BET_EF) at Vesuvius, Italy: a retrospective forward application to the 1631 eruption Bull. Volcanol 71:729–745

Sobradelo R, Bartolini S (2014) Marti J (2014) HASSET: a probability tool to evaluate future volcanic scenarios using Bayesian inference. Bull Volc 76:770

Solana MC, Kilburn CRJ, Rolandi G (2008) Communicating eruption and hazard forecasts on Vesuvius, Southern Italy. J Volcanol Geoth Res 189(2010):308–314

Sparks RSJ, Aspinall WP, Auker M, Crosweller S, Hincks T, Mahony S, Nadim F, Polley J, Syre E (2012a) Mapping and characterising volcanic risk, magmatic rifting and active volcanism conference, Addis Ababa, 12 Jan 2012

Sparks RSJ, Aspinall WP, Auker M, Crosweller S, Hincks T, Mahony S, Nadim F, Polley J, Syre E (2012b) Volcano hazard and exposure in global facility for disaster reduction and recovery priority countries and risk management measures, GVM 1st workshop 30 Apr 2012, University of Bristol and NGI Norway for the World Bank

Treadwell JR, Lucas S, Tsou AY (2014) Surgical checklists: a systematic review of impacts and implementation. BMJ Qual Saf 23:299–318. https://doi.org/10.1136/bmjqs-2012-001797

UN/ISDR (2015) Sendai framework for disaster risk reduction 2015–30

van Ments M (1999) The effective use of role-play: practical techniques for improving learning. Konan Page Ltd, London

Woo G (2008) Probabilistic criteria for volcano evacuation decisions. Nat Hazards 45:87–97. https://doi.org/10.1007/s11069-007-9171-9

Further reading sources on VHUB website

VUELCO resources

Colima Exercise Plan

Colima Debriefing Report

Ciolli S, De la Cruz-Reyna S (2013) Colima volcano exercise debriefing report. https://vhub.org/resources/2477

Campi Flegrei Simulation Plan

Campi Flegrei Simulation Debriefing Report

Ciolli S, Cristiani C, Golisciani GA (2014) VUELCO— Campi Flegrei Caldera unrest scientific simulation— debriefing report. https://vhub.org/resources/3640

Cotopaxi Debriefing Report

Dominica Exercise Plan

Dominica Exercise Debriefing Report

Ciolli S, Robertson REA (2015) VUELCO—Dominica exercise debriefing report. https://vhub.org/resources/3822

Appendix
Volcanic Unrest: Terminology and Definitions

Servando De la Cruz-Reyna, Ana Teresa Mendoza Rosas,
Joachim Gottsmann

Motivation and Background

This document provides a set of definitions of scientific and legal terminologies used in the context of volcanic unrest. The underpinning research behind this document was performed at the Universidad Nacional Autónoma de México—Geophysics Institute (Instituto de Geofísica, IGEF-UNAM) in collaboration with the National Center for Disaster Prevention (Centro Nacional de Prevención de Desastres, CENAPRED, Secretaría de Gobernación) and Colima University (Universidad de Colima, UCOL). Therefore, many definitions of the relevant legal terminology are based on Mexican law and may not be applicable *verbatim* to other jurisdictions. However, this document has been designed carefully to provide a common set of terms that are fundamental for volcanic unrest crisis management. This should provide a common platform for the interaction amongst different actors and stakeholders involved in the management of volcanic crises. The proposed definitions inherently follow on from investigations and interactions between scientists and decision-makers within the VUELCO project. It is not our intention to impose these definitions on volcanic unrest risk communication world-wide. However, we expect this compilation to instigate a dialogue and a shared platform of relevant terminologies between the different actors and stakeholders of volcanic risk governance [see also (Bretton et al. 2017)].

Analysis of past volcanic disasters shows that one of their causes were communication failures amongst stakeholders (Fiske 1984; Hall 1990; Tilling 2009; Fearnley et al. 2017). Some failures resulted from different perception of the meaning of terms used to describe or define the level of volcanic hazard or risk. This is in fact a problem related to semiotics. Definitions of scientific, social and legal concepts that are fundamental for volcanic crisis management having semantic contents that can be translated into other languages are a necessary first step to homogenize terminologies and improve risk communications. It is thus necessary to develop efficient communication protocols that univocally use the same meaning of the terminology (semantic content), integrate that meaning into a communication code that can be understood by all stakeholders (syntactic content), and translate the effects and consequences that volcanic activity may cause (pragmatic content). The following sections provide some context for a shared terminology.

S. De la Cruz-Reyna
Universidad Nacional Autónoma de México (UNAM),
Mexico City, Mexico

A. T. Mendoza Rosas
Universidad Nacional Autónoma de México (UNAM),
Mexico City, Mexico

J. Gottsmann
School of Earth Sciences, University of Bristol, Bristol, UK

Advs in Volcanology (2019) 299–313
DOI 10.1007/978-3-319-58412-6

Scientific Terminology

Term (English)	Término (Español)	Meaning	Significado	Further reading
Unrest (volcanic)	Activación; reactivación; agitación (volcánica)	A deviation from a background or baseline level behavior of a volcano which might be a prelude to an eruption, or to another hazardous event	Incremento del nivel de actividad volcánica respecto a un nivel de fondo o nivel base, el cual puede ser preludio de una erupción o de alguna otra manifestación peligrosa	Phillipson et al. (2013)
Crisis (volcanic)	Crisis (volcánica)	Situation during which a volcano shows signs of unrest, that can be interpreted as indication for potential impending eruptive activity and associated hazard. A crisis may or may not culminate in a dangerous eruption, but it may cause anxiety and socioeconomic unrest among the affected population	Situación en la que un volcán muestra signos de agitación que pueden interpretarse como precursora de actividad eruptiva inminente y de los peligros que conlleva. Una crisis volcánica puede o no culminar en una erupción peligrosa, pero puede producir ansiedad e inquietud socioeconómica en la población afectada	Tilling (1989)
Volcanic hazard (H)	Peligro volcánico (H)	Potentially damaging pre-eruptive, syn-eruptive or post-eruptive phenomena such as pyroclastic flows, windborne ash, lava flows, volcanic gases, lahars, ground deformation, major volcano-tectonic earthquakes, and landslides. In probabilistic assessments, "hazard" is sometimes referred to the probability that a specific volcanic manifestation or phenomenon will occur in a given area, within a given time frame (i.e., a threat)	Fenómenos potencialmente dañinos pre- eruptivos, co-eruptivos o post-eruptivos, como los flujos piroclásticos, ceniza transportada por el viento, flujos de lava, gases volcánicos, deformación del suelo, lahares, sismos volcano-tectónicos de magnitud considerable y deslizamientos. Para la evaluación probabilística, a veces "peligro" es la probabilidad de que una manifestación volcánica específica ocurra en una zona determinada, en un intervalo de tiempo dado	De la Cruz-Reyna and Tilling (2008)
Volcanic Risk (R)	Riesgo volcánico (R)	Probable loss (assets such as lives, property, productive capacity, etc.) in the event of the manifestation of a volcanic hazard. It is a conditional probability resulting from the product of the probability (P) that a specific volcanic hazard will occur (i.e., a threat) times the value (L) of the exposed assets times their vulnerability (V): R=P*L*V	Pérdida probable (de activos como vidas, propiedades, capacidad productiva, infraestructura, etc.) en un área sujeta a un peligro volcánico. Es la probabilidad condicional que resulta de multiplicar la probabilidad (P) de que una manifestación o fenómeno volcánico específico ocurra en un área y en un tiempo determinados (peligro), por el valor (L) de los activos por: R=P*L*V	Tilling (1989) De la Cruz-Reyna and Tilling (2008)

(continued)

Term (English)	Término (Español)	Meaning	Significado	Further reading
Acceptable risk	Riesgo aceptable	Risk that an individual or community is willing to accept, or that a public official is prepared to allow persons in his or her charge to accept. Acceptable risk is a function of the benefits of risk mitigation (safety and avoided losses) and its costs (losses of jobs and business, community disruption, the costs of transportation, housing, and food for evacuees, and costs of any structures to divert hazards). Tolerance for risk varies considerably, mainly because the benefits and costs of mitigating risks vary greatly from individual to individual and community to community	Riesgo que un individuo o comunidad están dispuestos a aceptar, o que un funcionario público puede permitir que las personas bajo su responsabilidad acepten. La aceptación del riesgo depende de los beneficios de su mitigación (seguridad y pérdidas evitadas) y su costo (pérdida de trabajos y negocios, disfunción de la comunidad, gastos de transporte, alojamiento y alimento para evacuados, y costos de estructuras para reducir peligros). La tolerancia al riesgo es muy variable, ya que los beneficios y costos de mitigación de riesgos pueden ser muy diferentes entre individuos y entre comunidades	Peterson (1988)
Vulnerability	Vulnerabilidad	The susceptibility of physical and human systems to be affected by a natural hazard. It can be expressed as a probability of damage, or as the proportion of the total exposed assets expected to be affected by a given manifestation. In probabilistic assessment "vulnerability" is the expected loss of the exposed value should a hazardous manifestation occur (i.e., the probability of loss)	La susceptibilidad de sistemas físicos y humanos que pueden ser afectados por fenómenos naturales peligrosos. Se puede expresar como una probabilidad de daño, o como la proporción del total de activos expuestos a daño por una manifestación determinada. Para evaluación probabilística, "vulnerabilidad" es el porcentaje esperado de perdida de los valores expuestos dado que una manifestación o evento volcánico ocurra (es decir, probabilidad de perdida)	Jolly and De la Cruz-Reyna (2015) De la Cruz-Reyna and Tilling (2008)
Preparedness	Preparación	A state of readiness or preparation for use or action in the case of volcanic activity or any other threat. It involves a clear understanding by the population and authorities of the natural phenomena and their destructive effects. Series of measures to reduce vulnerability	Estado de preparación y conciencia para prevenir y actuar en el caso de alguna actividad volcánica o cualquier otra amenaza. Involucra una clara comprensión de los fenómenos naturales y de sus efectos destructivos por parte de la población y las autoridades. Serie de medidas para reducir la vulnerabilidad	Jolly and De la Cruz-Reyna (2015) De la Cruz-Reyna and Tilling (2008)
Volcanic disaster	Desastre volcánico	An event or a series of events marked by a significant loss of life and/or other assets that exceed(s) the local response capacity and results from one or more volcanic hazards	Evento o serie de eventos caracterizados por una pérdida significativa de activos y del valor, que excede la capacidad de respuesta local y que resultan de uno o más peligros volcánicos	Jolly and De la Cruz-Reyna (2015)
Alert	Alerta	A condition of heightened watchfulness or preparation for	Condición de atención, vigilancia y preparación para la acción. Un	

(continued)

Term (English)	Término (Español)	Meaning	Significado	Further reading
		action. A warning method or system to make people aware of impending danger	método o sistema de aviso para que la gente perciba un peligro inminente	Jolly and De la Cruz-Reyna (2015)
Alert code	Código de Alerta	A set of rules or a system used for transmitting alert messages requiring brevity and clarity. The symbolic arrangement of instructions in a warning system	Un conjunto de reglas o un sistema utilizado para la transmisión de mensajes de alerta breves y claros. Estructura simbólica de instrucciones en un sistema de alerta.	Jolly and De la Cruz-Reyna (2015)
Consensus	Consenso	General agreement in the diagnostics and the prognosis of the majority of scientists involved in the study of an episode of volcanic activity. Consensus of the involved scientists is important for the decision making of authorities	Acuerdo general en el diagnóstico y el pronóstico de la mayoría de los científicos involucrados en el estudio de un episodio de actividad volcánica. El consenso de los científicos involucrados es muy importante para la toma de decisiones de las autoridades	Jolly and De la Cruz-Reyna (2015)
Crisis	Crisis	An unstable situation of increased danger. A crucial stage in the course of volcanic unrest, when the threat of an eruption, or the possibility of an eruption, exceeds some threshold or reference level	Una situación inestable de amenaza en aumento. Etapa crucial en el curso de una reactiviación/agitación volcánica en que la amenaza de una erupción, o la posibilidad de una erupción exceden un nivel de referencia	Jolly and De la Cruz-Reyna (2015)
Pre-event	Pre-evento	Period of time when there is significant evidence that a hazard could occur	Período de tiempo en que se tiene evidencia significativa de que un fenómeno potencialmente causante de daños puede ocurrir	Jolly and De la Cruz-Reyna (2015)
Volcanic Traffic Light Alert System (VTLAS)	Semáforo de Alerta Volcánica	A basic communications protocol that translates volcanic threat into seven levels of preparedness for the emergency-management authorities, but only three unambiguous levels of alert for the public (color coded green–yellow–red). The changing status of the volcano threat is represented as the most likely scenarios according to the opinions of an official scientific committee that analyzes all available data	Protocolo de comunicación básica que traduce las amenzas volcanicas en siete niveles de preparación para las autoridades de gestión de emergencias, pero en solo tres niveles de alerta sin ambigüedades para el público (código de colores verde-amarillo-rojo). Los cambios en el estado del volcán se representan como los escenarios de riesgo más probables de acuerdo al consenso de un Comité Científico oficial que analiza los datos disponibles	Fearnley (2013)
Seismic swarm	Enjambre sísmico	A group of many earthquakes of similar size occurring closely clustered in space and time with no dominant main shock	Grupo de múltiples sismos de tamaño similar estrechamente agrupados en el espacio y en el tiempo sin evento principal dominante	McNutt and Roman (2015)
Forecast	Pronóstico	A general description of future events, including rough estimates of time, location, and likely activity	Una descripción general de eventos futuros incluyendo estimaciones de tiempo, localización y actividad probable	Connor et al. (2015)

(continued)

Term (English)	Término (Español)	Meaning	Significado	Further reading
Precursor	Precursor	A geological event that occurs prior to an eruption and is related to the preparation processes of the forthcoming eruption	Un evento geológico que se produce antes de una erupción y se relaciona con los procesos de preparación de la próxima erupción	National Academies of Sciences, Engineering Medicine (2017)
Prediction	Predicción	A specific description of future events, including time, size, type, location, and formal errors for each	Una descripción específica de los acontecimientos futuros, incluyendo el tiempo, el tamaño, el tipo, la ubicación y los errores formales para cada uno	National Academies of Sciences, Engineering Medicine (2017)
Probabilistic	Probabilístico	A statement of the relative likelihood of an event based on study of a population of similar events that have occurred in the past	Una declaración de la probabilidad relativa de un evento basado en el estudio de una población de eventos similares que han ocurrido en el pasado	Connor et al. (2015)

Legal Terminology

Term (English)	Término (Español)	Meaning	Significado	Further reading
Accident (legal definition, Mexico)	Siniestro (definición legal, México)	Critical and harmful situation caused by the incidence of one or more hazards in a building or facility affecting the occupants and installations, with possible effects on surrounding facilities	Situación crítica y dañina generada por la incidencia de uno o más fenómenos perturbadores en un inmueble o instalación afectando a su población y equipo, con posible afectación a instalaciones circundantes	Ley General Protección Civil, 6 junio 2012, Capitulo 1, Artículo 2, Fracción LV (Mexican Civil Protection General Law, 6 June 2012, Chapter 1, 2nd Article, Section LV)
Alert (legal definition, Mexico)	Alerta (definición legal, México)	Warning of the proximity of an anthropogenic or natural hazard, or of the increase of risk associated to it	El aviso de la proximidad de un Fenómeno Antropogénico o Natural Perturbador, o el incremento del Riesgo asociado al mismo	Reglamento de la Ley General Protección Civil, 6 junio 2012, Artículo 2, Fracción I (Bylaws of the General Civil Protection Law, June 6, 2012, 2nd Article, Fraction 1)
Brigade (legal definition, Mexico)	Brigada (definición legal, México)	Basic Civil Protection Unit. A group of people organized within a facility trained in basic emergency response functions such as: first aid, fighting fires, evacuations, search and rescue; designated in the Internal Civil Protection Unit as responsible for the development and implementation of prevention, rescue and	Grupo de personas que se organizan dentro de un inmueble, capacitadas y adiestradas en funciones básicas de respuesta a emergencias tales como: primeros auxilios, combate a conatos de incendio, evacuación, búsqueda y rescate; designados en la Unidad Interna de Protección Civil como encargados del	Ley General Protección Civil, 6 junio 2012, Capitulo 1, Artículo 2, Fracción VI (Mexican Civil Protection General Law, 6 June 2012, Chapter 1, 2nd Article, Section VI)

(continued)

Term (English)	Término (Español)	Meaning	Significado	Further reading
		recovery based on the provisions of the internal civil protection program of the facility	desarrollo y ejecución de acciones de prevención, auxilio y recuperación, con base en lo estipulado en el Programa Interno de Protección Civil del inmueble	
Disaster (legal definition, Mexico)	Desastre (definición legal, México)	A consequence of the occurrence of one or more severe (and/or extreme) hazards, associated or not, with a natural origin, or caused by human activity, or from outer space that cause an amount of damage in a given time and in a determined area that exceeds the response capacity of the affected community	Resultado de la ocurrencia de uno o más agentes perturbadores severos y/o extremos, concatenados o no, de origen natural, de la actividad humana o aquellos provenientes del espacio exterior, que cuando acontecen en un tiempo y en una zona determinada, causan daños que por su magnitud exceden la capacidad de respuesta de la comunidad afectada	Ley General Protección Civil, 6 junio 2012, Capitulo 1, Artículo 2, Fracción XVI. (Mexican Civil Protection General Law, 6 June 2012, Chapter 1, 2nd Article, Section XVI)
Disaster area (legal definition, Mexico)	Zona de desastre (definición legal, México)	Specified area upon which a competent authority formally declares a temporal condition of disrupted social structure, which precludes the normal community activities. Such declaration may involve funding from FONDEN (A special federal fund that can be only used to finance the recovery of disaster areas.)	Espacio territorial determinado en el tiempo por la declaración formal de la autoridad competente, en virtud del desajuste que sufre en su estructura social, impidiéndose el cumplimiento normal de las actividades de la comunidad. Puede involucrar el ejercicio de recursos públicos a través del Fondo de Desastres	Ley General Protección Civil, 6 junio 2012, Capitulo 1, Artículo 2, Fracción LIX (Mexican Civil Protection General Law, 6 June 2012, Chapter 1, 2nd Article, Section LIX)
Disruptive natural phenomenon (legal definition, Mexico)	Fenómeno natural perturbador (definición legal, México)	Hazard of natural origin	Agente perturbador producido por la naturaleza	Ley General Protección Civil, 6 junio 2012, Capitulo 1, Artículo 2, Fracción XXII (Mexican Civil Protection General Law, 6 June 2012, Chapter 1, 2nd Article, Section XXII)
Early warning systems (legal definition, Mexico)	Sistemas de alerta temprana (definición legal, México)	A set of elements providing timely and efficient information, allowing individuals exposed to a threat to take avoidance actions aimed to prevent or reduce risk, and to be prepared for an effective response. Early Warning Systems include	El conjunto de elementos para la provisión de información oportuna y eficaz, que permiten a individuos expuestos a una amenaza tomar acciones para evitar o reducir su riesgo, así como prepararse para una respuesta efectiva. Los	Reglamento de la Ley General Protección Civil, 6 junio 2012, Artículo 2, Fracción XIII (Bylaws of the General Civil Protection Law, June 6, 2012, 2nd Article, Fraction XIII)

(continued)

Term (English)	Término (Español)	Meaning	Significado	Further reading
		hazard recognition and mapping; monitoring and forecasting of impending events; processing and dissemination of warnings or alerts understandable to authorities and population and adopting appropriate and timely response actions to such alerts	Sistemas de Alerta Temprana incluyen conocimiento y mapeo de amenazas; monitoreo y pronóstico de eventos inminentes; proceso y difusión de Alertas comprensibles a las autoridades y población; así como adopción de medidas apropiadas y oportunas en respuesta a tales Alertas	
Emergency (legal definition, Mexico)	Emergencia	Abnormal situation in which damage to society is likely, posing an excessive risk to the safety and integrity of the general population, generated or associated with the presence, imminence, or high probability of occurrence of a hazard	Situación anormal que puede causar un daño a la sociedad y propiciar un riesgo excesivo para la seguridad e integridad de la población en general, generada o asociada con la inminencia, alta probabilidad o presencia de un agente perturbador	Ley General Protección Civil, 6 junio 2012, Capitulo 1, Artículo 2, Fracción XVII (Mexican Civil Protection General Law, 6 June 2012, Chapter 1, 2nd Article, Section XVII)
Evacuee (legal definition, Mexico)	Evacuado (definición legal, México)	Person who withdraws or is removed from a place of residence in a situation of emergency or likelihood of disaster, as a preventive and temporal measure to assure his or her safety and survival	Persona que, con carácter preventivo y provisional ante la posibilidad o certeza de una emergencia o desastre, se retira o es retirado de su lugar de alojamiento usual, para garantizar su seguridad y supervivencia	Ley General Protección Civil, 6 junio 2012, Capitulo 1, Artículo 2, Fracción XIX (Mexican Civil Protection General Law, 6 June 2012, Chapter 1, 2nd Article, Section XIX)
Geological phenomenon (legal definition, Mexico)	Fenómeno geológico (definición legal, México)	Hazard caused by actions and movements of the earth's crust. This category includes earthquakes, volcanic eruptions, tsunamis, unstable slopes, flows, landslides, terrain sinking and subsidence, and ground cracks	Agente perturbador que tiene como causa directa las acciones y movimientos de la corteza terrestre. A esta categoría pertenecen los sismos, las erupciones volcánicas, los tsunamis, la inestabilidad de laderas, los flujos, los caídos o derrumbes, los hundimientos, la subsidencia y los agrietamientos	Ley General Protección Civil, 6 junio 2012, Capitulo 1, Artículo 2, Fracción XXIII (Mexican Civil Protection General Law, 6 June 2012, Chapter 1, 2nd Article, Section XXIII)
Hazard (legal definition, Mexico)	Peligro (definición legal, México)	Probability of occurrence of a potentially damaging and disruptive phenomenon with a certain intensity, in a specific region and in a given time interval	Probabilidad de ocurrencia de un agente perturbador potencialmente dañino de cierta intensidad, durante un cierto periodo y en un sitio determinado	Ley General Protección Civil, 6 junio 2012, Capitulo 1, Artículo 2, Fracción XXXVII (Mexican Civil Protection General Law, 6 June 2012, Chapter 1, 2nd Article, Section XXXVII)

(continued)

Term (English)	Término (Español)	Meaning	Significado	Further reading
Help (legal definition, Mexico)	Auxilio (definición legal, México)	Support to people at risk or victims of a mishap, emergency or disaster, as a response by official or private specialized groups, or by internal civil protection units, as well as actions to protect other exposed people and valued non-human assets	Respuesta de ayuda a las personas en riesgo o las víctimas de un siniestro, emergencia o desastre, por parte de grupos especializados públicos o privados, o por las unidades internas de protección civil, así como las acciones para salvaguardar los demás agentes afectables	Ley General Protección Civil, 6 junio 2012, Capítulo 1 Artículo 2, Fracción V (Mexican Civil Protection General Law, 6 June 2012, Chapter 1, 2nd Article, Section V)
High-risk zone (legal definition, Mexico)	Zona de riesgo grave (definición legal, México)	Human settlement located within an area of high risk caused by a hazard	Asentamiento humano que se encuentra dentro de una zona de grave riesgo, originado por un posible fenómeno perturbador	Ley General Protección Civil, 6 junio 2012, Capitulo 1, Artículo 2, Fracción LXI (Mexican Civil Protection General Law, 6 June 2012, Chapter 1, 2nd Article, Section LXI)
Imminent (impending) risk (legal definition, Mexico)	Riesgo Inminente (definición legal, México)	A risk situation that, according to a specialized technical body, requires immediate actions as there are conditions or increased probabilities of adverse effects on an endangered person or valued non-human asset	Aquel riesgo que según la opinión de una instancia técnica especializada, debe considerar la realización de acciones inmediatas en virtud de existir condiciones o altas probabilidades de que se produzcan los efectos adversos sobre un agente afectable	Ley General Protección Civil, 6 junio 2012, Capitulo 1, Artículo 2, Fracción L (Mexican Civil Protection General Law, 6 June 2012, Chapter 1, 2nd Article, Section L)
Incident (legal definition, Mexico)	Incidente (definición legal, México)	A minor event that may be considered normal and not necessarily caused by hazards, but that may generate conditions for the occurrence of a mishap, an emergency or a disaster	El suceso que sin constituir una situación anormal ni haber sido provocado por fenómenos perturbadores severos, puede crear condiciones precursoras de Siniestros, Emergencias o Desastres	Reglamento de la Ley General Protección Civil, 6 junio 2012, Artículo 2, Fracción XI (Mexican Civil Protection General Law, 6 June 2012, Chapter 1, 2nd Article, Section XXVIII)
Injured party (legal definition, Mexico)	Damnificado (definición legal, México)	Person affected by a hazard who suffered damage to their person or to their property in such a degree that requires external assistance for his subsistence, remaining is such condition as long as the emergency persists, or the pre-disaster normality is restored	Persona afectada por un agente perturbador, ya sea que haya sufrido daños en su integridad física o un perjuicio en sus bienes de tal manera que requiere asistencia externa para su subsistencia; considerándose con esa condición en tanto no se concluya la emergencia o se restablezca la situación de normalidad previa al desastre	Ley General Protección Civil, 6 junio 2012, Capitulo 1, Artículo 2, Fracción XIV (Mexican Civil Protection General Law, 6 June 2012, Chapter 1, 2nd Article, Section XIV)

(continued)

Term (English)	Término (Español)	Meaning	Significado	Further reading
Integrated management of risks (legal definition, Mexico)	Gestión integral de riesgos (definición legal, México)	Set of actions aimed to the identification, analysis, evaluation, control and reduction of risk, considering that it has a multifactorial origin, and that it is a process in continuous construction. Such actions involve all of the three levels of government, and all other sectors of society promoting the design and implementation of public policies, strategies and procedures that, within a sustained development scheme, reduce the disaster-prone structural vulnerabilities, and increase the resilience or strength of society. It involves identification of risks and their sources, and development of prevision, prevention, mitigation, preparedness, relief, recovery and reconstruction schemes	El conjunto de acciones encaminadas a la identificación, análisis, evaluación, control y reducción de los riesgos, considerándolos por su origen multifactorial y en un proceso permanente de construcción, que involucra a los tres niveles de gobierno, así como a los sectores de la sociedad, lo que facilita la realización de acciones dirigidas a la creación e implementación de políticas públicas, estrategias y procedimientos integrados al logro de pautas de desarrollo sostenible, que combatan las causas estructurales de los desastres y fortalezcan las capacidades de resiliencia o resistencia de la sociedad. Involucra las etapas de: identificación de los riesgos y/o su proceso de formación, previsión, prevención, mitigación, preparación, auxilio, recuperación y reconstrucción	Ley General Protección Civil, 6 junio 2012, Capitulo l, Artículo 2, Fracción XXVIII (Mexican Civil Protection General Law, 6 June 2012, Chapter 1, 2nd Article, Section XXVIII)
Internal Civil Protection program (legal definition, Mexico)	Programa interno de Protección Civil (definición legal, México)	Planning and operational tool circumscribed to an agency, or organization public or private and social sector, which is composed by the operating plan for the Internal Civil Protection Unit, the plan for continuity of operations and contingency plan, and aims to mitigate the previously identified risks and define preventive and response to be able to meet the event of an emergency or disaster	Es un instrumento de planeación y operación, circunscrito al ámbito de una dependencia, entidad, institución u organismo del sector público, privado o social; que se compone por el plan operativo para la Unidad Interna de Protección Civil, el plan para la continuidad de operaciones y el plan de contingencias, y tiene como propósito mitigar los riesgos previamente identificados y definir acciones preventivas y de respuesta para estar en condiciones de atender la eventualidad de alguna emergencia o desastre	Ley General Protección Civil, 6 junio 2012, Capitulo l, Artículo 2, Fracción XLI (Mexican Civil Protection General Law, 6 June 2012, Chapter 1, 2nd Article, Section XLI)

(continued)

Term (English)	Término (Español)	Meaning	Significado	Further reading
Mitigation (legal definition, Mexico)	Mitigación (definición legal, México)	Any action aimed to reduce the impact or damage by a hazard on a person or valued non-human asset	Es toda acción orientada a disminuir el impacto o daños ante la presencia de un agente perturbador sobre un agente afectable	Ley General Protección Civil, 6 junio 2012, Capitulo l, Artículo 2, Fracción XXXVI (Mexican Civil Protection General Law, 6 June 2012, Chapter 1, 2nd Article, Section XXXVI)
Monitoring systems (legal definition, Mexico)	Sistemas de monitoreo (definición legal, México)	A set of elements allowing detection, measurement, processing, prediction and study of the behavior of hazards, in order to assess them and their risks	El conjunto de elementos que permiten detectar, medir, procesar, pronosticar y estudiar el comportamiento de los agentes perturbadores, con la finalidad de evaluar Peligros y Riesgos	Reglamento de la Ley General Protección Civil, 6 junio 2012, Artículo 2, Fracción XIV (Bylaws of the General Civil Protection Law, June 6, 2012, 2nd Article, Fraction XIV)
National Risk Atlas (legal definition, Mexico)	Atlas Nacional de Riesgos (definición legal, México)	Integral geographic information system of threats and expected damages, resulting from a spatial and temporal analysis of the interaction between hazards, vulnerability and exposure of affected entities	Sistema integral de información sobre los agentes perturbadores y daños esperados, resultado de un análisis espacial y temporal sobre la interacción entre los peligros, la vulnerabilidad y el grado de exposición de los agentes afectables	Ley General Protección Civil, 6 junio 2012, Capitulo l, Artículo 2, Fracción IV (Mexican Civil Protection General Law, 6 June 2012, Chapter 1, 2nd Article, Section IV)
Prevention (legal definition, Mexico)	Prevención (definición legal, México)	Group of actions implemented in advance of the occurrence of hazards aimed to foresee and reduce the related risks, identifying, reducing or eliminating them, and to avert or reduce their destructive impact on populace, property, infrastructure, as well as prevent the social construction of risks	Conjunto de acciones y mecanismos implementados con antelación a la ocurrencia de los agentes perturbadores, con la finalidad de conocer los peligros o los riesgos, identificarlos, eliminarlos o reducirlos; evitar o mitigar su impacto destructivo sobre las personas, bienes, infraestructura, así como anticiparse a los procesos sociales de construcción de los mismos	Ley General Protección Civil, 6 junio 2012, Capitulo l, Artículo 2, Fracción XXXIX (Mexican Civil Protection General Law, 6 June 2012, Chapter 1, 2nd Article, Section XXXIX)
Prevision (legal definition, Mexico)	Previsión (definición legal, México)	To acquire awareness of the risks, and of what is needed to face them through defined steps of risk identification, prevention, mitigation, preparedness, emergency response, recovery and reconstruction	Tomar conciencia de los riesgos que pueden causarse y las necesidades para enfrentarlos a través de las etapas de identificación de riesgos, prevención, mitigación, preparación, atención de emergencias, recuperación y reconstrucción	Ley General Protección Civil, 6 junio 2012, Capitulo l, Artículo 2, Fracción XL (Mexican Civil Protection General Law, 6 June 2012, Chapter 1, 2nd Article, Section XL)

(continued)

Term (English)	Término (Español)	Meaning	Significado	Further reading
Reconstruction (legal definition, Mexico)	Reconstrucción (definición legal, México)	Transient actions aimed to recover the social and economic environment that prevailed among people before suffering the effects of a hazard in a given area or jurisdiction. This process should be aimed as much as possible to reduce existing risks, ensuring that no new risks would be created, so improving the pre-existing conditions	La acción transitoria orientada a alcanzar el entorno de normalidad social y económica que prevalecía entre la población antes de sufrir los efectos producidos por un agente perturbador en un determinado espacio o jurisdicción. Este proceso debe buscar en la medida de lo posible la reducción de los riesgos existentes, asegurando la no generación de nuevos riesgos y mejorando para ello las condiciones preexistentes	Ley General Protección Civil, 6 junio 2012, Capitulo l, Artículo 2, Fracción XLIV (Mexican Civil Protection General Law, 6 June 2012, Chapter 1, 2nd Article, Section XLIV)
Recovery (legal definition, Mexico)	Recuperación (definición legal, México)	Process starting during an emergency, consisting of actions aimed to return to the normality of the affected community	Proceso que inicia durante la emergencia, consistente en acciones encaminadas al retorno a la normalidad de la comunidad afectada	Ley General Protección Civil, 6 junio 2012, Capitulo l, Artículo 2, Fracción XLV (Mexican Civil Protection General Law, 6 June 2012, Chapter 1, 2nd Article, Section XLV)
Regulation (legal definition, Mexico, as opposite to disruption)	Agente regulador (definición legal, México)	Set of actions, tools, standards, works and more generally everything intended to protect people, property, strategic and productive infrastructure and the environment, and to reduce risk and control and prevent adverse effects from a hazard	Lo constituyen las acciones, instrumentos, normas, obras y en general todo aquello destinado a proteger a las personas, bienes, infraestructura estratégica, planta productiva y el medio ambiente, a reducir los riesgos y a controlar y prevenir los efectos adversos de un agente perturbador	Ley General Protección Civil, 6 junio 2012, Capitulo l, Artículo 2, Fracción l. (Mexican Civil Protection General Law, 6 June 2012, Chapter 1, 2nd Article, Section I)
Resilience (legal definition, Mexico)	Resiliencia (definición legal, México)	It is the ability of a hazard-exposed system, community or society to resist, assimilate, withstand and quickly and efficiently recover from damaging effects, preserving and recovering its basic and functional structures, and achieving a better protection and improved risk reduction for the future	Es la capacidad de un sistema, comunidad o sociedad potencialmente expuesta a un peligro para resistir, asimilar, adaptarse y recuperarse de sus efectos en un corto plazo y de manera eficiente, a través de la preservación y restauración de sus estructuras básicas y funcionales, logrando una mejor protección futura y mejorando las medidas de reducción de riesgos	Ley General Protección Civil, 6 junio 2012, Capitulo l, Artículo 2, Fracción XLVIII (Mexican Civil Protection General Law, 6 June 2012, Chapter 1, 2nd Article, Section XLVIII)

(continued)

Term (English)	Término (Español)	Meaning	Significado	Further reading
Risk (legal definition, Mexico)	Riesgo (definición legal, México)	Damage or probable loss of an asset, resulting from the interaction between its vulnerability and the presence of a hazard	Daños o pérdidas probables sobre un agente afectable, resultado de la interacción entre su vulnerabilidad y la presencia de un agente perturbador	Ley General Protección Civil, 6 junio 2012, Capitulo 1, Artículo 2, Fracción XLIV. (Mexican Civil Protection General Law, 6 June 2012, Chapter 1, 2nd Article, Section XLIV)
Risk identification (legal definition, Mexico)	Identificación de riesgos (definición legal, México)	Recognizing and assessing potential damage or probable loss from hazards in terms of their geographical distribution through the analysis of hazard and vulnerability	Reconocer y valorar las pérdidas o daños probables sobre los agentes afectables y su distribución geográfica, a través del análisis de los peligros y la vulnerabilidad	Ley General Protección Civil, 6 junio 2012, Capitulo 1, Artículo 2, Fracción XXXI (Mexican Civil Protection General Law, 6 June 2012, Chapter 1, 2nd Article, Section XXXI)
Risk reduction (legal definition, Mexico)	Reducción de riesgos (definición legal, México)	Preventive intervention of individuals, institutions and communities aimed to remove or reduce the adverse impact of disasters through preparedness and mitigation actions. It involves the identification of risks, and the analyses of vulnerabilities, resilience and response capabilities. Also involves developing a culture of civil protection, the public engagement and the development of an institutional framework, as well as the implementation of environmental protection measures, use of land and urban planning, protection of critical infrastructure, building partnerships for the implementation of financial and risk transfer instruments, and the development of early warning systems	Intervención preventiva de individuos, instituciones y comunidades que nos permite eliminar o reducir, mediante acciones de preparación y mitigación, el impacto adverso de los desastres. Contempla la identificación de riesgos y el análisis de vulnerabilidades, resiliencia y capacidades de respuesta, el desarrollo de una cultura de la protección civil, el compromiso público y el desarrollo de un marco institucional, la implementación de medidas de protección del medio ambiente, uso del suelo y planeación urbana, protección de la infraestructura crítica, generación de alianzas y desarrollo de instrumentos financieros y transferencia de riesgos, y el desarrollo de sistemas de alertamiento	Ley General Protección Civil, 6 junio 2012, Capitulo 1, Artículo 2, Fracción XLVI (Mexican Civil Protection General Law, 6 June 2012, Chapter 1, 2nd Article, Section XLVI)
Risk zone (legal definition, Mexico)	Zona de riesgo (definición legal, México)	Defined territorial space upon which there is a likelihood of harm caused by a hazard	Espacio territorial determinado en el que existe la probabilidad de que se produzca un daño, originado por un fenómeno perturbador	Ley General Protección Civil, 6 junio 2012, Capitulo 1, Artículo 2, Fracción LX (Mexican Civil Protection General Law, 6 June 2012, Chapter 1, 2nd Article, Section LX)

(continued)

Term (English)	Término (Español)	Meaning	Significado	Further reading
Shelter (legal definition, Mexico)	Albergue (definición legal, México)	Facility settled to provide shelter to people whose habitations have been affected by a hazard. Such people may dwell in the shelter until the recovery or reconstruction of their habitations	Instalación que se establece para brindar resguardo a las personas que se han visto afectadas en sus viviendas por los efectos de fenómenos perturbadores y en donde permanecen hasta que se da la recuperación o reconstrucción de sus viviendas	Ley General Protección Civil, 6 junio 2012, Capítulo 1 Artículo 2, Fracción III (Mexican Civil Protection General Law, 6 June 2012, Chapter 1, 2nd Article, Section III)
Sheltered (legal definition, Mexico)	Albergado (definición legal, México)	Person who is temporarily granted help, accommodation and protection against a threat, imminence or occurrence of a hazard	Persona que en forma temporal recibe asilo, amparo, alojamiento y resguardo ante la amenaza, inminencia u ocurrencia de un agente perturbador	Ley General Protección Civil, 6 junio 2012, Capítulo 1 Artículo 2, Fracción II (Mexican Civil Protection General Law, 6 June 2012, Chapter 1, 2nd Article, Section II)
Temporary shelter (legal definition, Mexico)	Refugio temporal (definición legal, México)	Physical installation enabled to temporarily provide protection and welfare to people who do not have immediate accessibility to a safe dwelling in case of impending hazard, an emergency, a major accident or a disaster	La instalación física habilitada para brindar temporalmente protección y bienestar a las personas que no tienen posibilidades inmediatas de acceso a una habitación segura en caso de un riesgo inminente, una emergencia, siniestro o desastre	Ley General Protección Civil, 6 junio 2012, Capitulo 1, Artículo 2, Fracción XLVII (Mexican Civil Protection General Law, 6 June 2012, Chapter 1, 2nd Article, Section XLVII)
Vulnerability (legal definition, Mexico)	Vulnerabilidad (definición legal, México)	Susceptibility or propensity of an asset to suffer damage or loss in the presence of a disrupting phenomenon determined by physical, social, economic and environmental factor	Susceptibilidad o propensión de un agente afectable a sufrir daños o pérdidas ante la presencia de un agente perturbador, determinado por factores físicos, sociales, económicos y ambientales	Ley General Protección Civil, 6 junio 2012, Capitulo 1, Artículo 2, Fracción LVIII. (Mexican Civil Protection General Law, 6 June 2012, Chapter 1, 2nd Article, Section LVIII)

Bibliography

Bretton RJ, Gottsmann J, Christie R (2017) The role of laws within the governance of volcanic risks advances in volcanology. Springer, Berlin, Heidelberg

Connor C, Bebbington M, Marzocchi W (2015) Chapter 51—probabilistic volcanic hazard assessment A2. In: Sigurdsson H (ed) The encyclopedia of volcanoes, 2nd edn. Academic Press, Amsterdam, pp 897–910

De la Cruz-Reyna S, Tilling RI (2008) Scientific and public responses to the ongoing volcanic crisis at Popocatépetl Volcano, Mexico: importance of an effective hazards-warning system. J Volcanol Geoth Res 170:121–134

Fearnley C, Winson AEG, Pallister J, Tilling R (2017) Volcano crisis communication: challenges and solutions in the 21st Century: Berlin. Springer, Berlin Heidelberg, Heidelberg, pp 1–19

Fearnley CJ (2013) Assigning a volcano alert level: negotiating uncertainty, risk, and complexity in decision-making processes. Environ Plan A 45:1891–1911.

Fiske RS (1984) Volcanologists, journalists, and the concerned local public: a tale of two crises in the eastern Caribbean, 110–121 p

Hall ML (1990) Chronology of the principal scientific and governmental actions leading up to the November 13, 1985 eruption of Nevado del Ruiz, Colombia. J Volcanol Geoth Res 42:101–115

Jolly G, De la Cruz-Reyna S (2015) Chapter 68—volcanic crisis management A2. In: Sigurdsson H (ed) The encyclopedia of volcanoes, 2nd edn. Academic Press, Amsterdam, pp 1187–1202

McNutt SR, Roman DC (2015) Chapter 59—volcanic seismicity A2. In: Sigurdsson H (ed) The encyclopedia of volcanoes, 2nd edn. Academic Press, Amsterdam, pp 1011–1034

National Academies of Sciences, Engineering Medicine (2017) Volcanic eruptions and their repose, unrest, precursors, and timing. The National Academies Press, Washington, DC.

Peterson DW (1988) Volcanic hazards and public response. J Geophy Res: Solid Earth 93(B5):4161–4170

Phillipson G, Sobradelo R, Gottsmann J (2013) Global volcanic unrest in the 21st century: an analysis of the first decade. J Volcanol Geoth Res 264:183–196

Tilling RI (1989) Volcanic hazards and their mitigation—progress and problems. Rev Geophy 27(2):237–269

Tilling RI (2009) El Chichón's "surprise" eruption in 1982: lessons for reducing volcano risk. Geofísica Internacional 48:3–19

Advs in Volcanology (2019) 313–313

DOI 10.1007/978-3-319-58412-6

Jin-Hua Li • Lixing Sun • Peter M. Kappeler
Editors

The Behavioral Ecology of the Tibetan Macaque

Editors
Jin-Hua Li
School of Resources
and Environmental Engineering
Anhui University
Hefei, Anhui, China

International Collaborative Research
Center for Huangshan Biodiversity
and Tibetan Macaque Behavioral Ecology
Anhui, China

School of Life Sciences
Hefei Normal University
Hefei, Anhui, China

Peter M. Kappeler
Behavioral Ecology and Sociobiology
Unit, German Primate Center
Leibniz Institute for Primate Research
Göttingen, Germany

Department of Anthropology/Sociobiology
University of Göttingen
Göttingen, Germany

Lixing Sun
Department of Biological Sciences, Primate
Behavior and Ecology Program
Central Washington University
Ellensburg, WA, USA

https://doi.org/10.1007/978-3-030-27920-2

This book is an open access publication.

Their [Tibetan macaques'] flat, broad, bearded faces provide perhaps the most humanlike countenance I have ever seen in a monkey. . . . Apart from chimpanzees, I had never seen primate males so intensely involved with each other. In chimpanzees, too, males are at the same time rivals and friends, and I would argue the same for human males.
—Frans de Waal, *The Ape and the Sushi Master* (2001)

Foreword

Mystery surrounded Tibetan macaques for a long time, even for experts. The species was not identified until the last third of the nineteenth century, and nothing more than its geographic distribution and external characters were known for the next hundred years. It was long referred to as Père David's macaque, a rather odd name in reference to the French missionary and naturalist Father Armand David, who first collected the species. Moreover, the current name, Tibetan macaque, is misleading since the species is typically found in east-central China and not within the boundaries of Tibet. This is due to the fact that Père David initially located the species at a place close to the Sino-Tibetan border of his time.

We had to wait until the 1980s to see Mount Emei and Mount Huangshan come to light on the primatology map. This is where Qikun Zhao and Ziyun Deng from the Kunming Institute of Zoology and Qishan Wang and Jinhua Li from Anhui University began to study the behavior and life history of Tibetan macaques, definitively adding a new dimension to the macaque landscape. I still have the reprints of their publications in my bibliography, some written in Chinese. The works of Hideshi Ogawa, Carol Berman, and a new generation of primatologists soon followed. Now appears this multi-authored volume entirely devoted to the Tibetan macaque. This combined effort of two dozen scientists to review 40 years of research and present new findings about a single species should be viewed as a celebration of the species. It frees Tibetan macaques from the purgatory of scientific papers scattered across various journals and collections to join the small club of primate species that are honored with this attention. Many people would consider that brown monkeys like Tibetan macaques all look similar. Although they do not have the immediate visual appeal of more brightly colored primates, brown monkeys have different but equally attractive assets. With their fiery gaze and prominent beards, Tibetan macaques are no exception, and their adaptations and behaviors attract a great deal of research interest.

As scientists we are expected to test hypotheses and theories. Some of the contributors to this book do so, addressing broad issues such as cooperative strategies, social dynamics, collective decisions, feeding ecology, and pathogen transmission. They use Tibetan macaques as a model to investigate mainstream research questions in the field of behavioral ecology and evolution. Every animal species has its singularities, however, and deserves to be studied for itself. This is why other contributors seek to identify what makes Tibetan macaques special, investigating patterns such as social play, call types, or the fascinating "bridging" interactions in which infants play a role as buffers to reduce tension between adults. Science generally values the testing of general theories more than the humble seeking of what gives a species its own touch. In the end, however, both of these approaches are necessary. Years ago, I was struggling to rank the different species of macaque according to their levels of social tolerance. Quantitative measures were available in a limited number of species and I had to rely on qualitative data for others. I remember asking Qikun Zhao about the behaviors particular to Tibetan macaques at a conference held in Japan in 1996. I was trying to guess their social style, i.e., their own touch. This resulted in a tentative scaling of macaque species which would later be amended when quantitative data became available in Tibetan macaques. It should be emphasized that the story is far from over. As discussed in the book, why and to what extent the different behavioral traits constituting social styles may covary during the evolutionary process still remains to be elucidated. I am delighted to see how the study of the particular meets the general by yielding new perspectives and hypotheses to be tested.

As the editors point out, this book should not be considered an end, but rather a beginning. This highlighting of research into Tibetan macaques has the potential to strengthen Chinese primatology and favor its development at the national and international level. It may in turn help the Tibetan macaques. Like other species of non-human primates, their populations are threatened by the loss and fragmentation of their habitat. Admiring, knowledge, and conservation should go hand in hand to save the future of this unique species.

University of Strasbourg, Strasbourg   Bernard Thierry
France

The original version of the book frontmatter was revised: For detailed information please see Correction. The correction to the book frontmatter is available at https://doi.org/10.1007/978-3-030-27920-2_15

Preface

This book is mainly based on research papers presented in a spirited international primatology symposium held in the scenic area of Mt. Huangshan, China, in the summer of 2017. The chapters were grouped into five logical parts. Part I consists of a single chapter, which offers a brief introduction to recent developments in Chinese primatology and a short history of research on the primates of China in general and the Tibetan macaque in particular.

Part II contains seven chapters (Chaps. 2–8) focusing on social behavior and social dynamics in Tibetan macaques. In Chap. 2, Jin-Hua Li and Peter M. Kappeler provide a comprehensive review of three decades of field research in Tibetan macaques at the Valley of Monkeys, highlighting the significance of this species as a model for understanding broader questions in primate behavior and evolution. Lixing Sun, Dong-Po Xia, and Jin-Hua Li follow up in Chap. 3 by introducing a new way to analyze the dynamics of macaque social hierarchy from a social mobility perspective with new insights unveiled through comparing Tibetan macaques with Japanese macaques. In Chap. 4, Dong-Po Xia, Paul A. Garber, Cédric Sueur, and Jin-Hua Li look into the internal behavioral mechanisms promoting group stability in Tibetan macaque from a behavioral exchange and biological market point of view. In Chap. 5, Xi Wang, Claudia Fichtel, Lixing Sun, and Jin-Hua Li investigate how Tibetan macaques make collective decisions during group movements. In Chap. 6, Jessica A. Mayhew, Jake A. Funkhouser, and Kaitlin R. Wright explore the significance of play behavior in the development of social cognition in juvenile Tibetan macaques. This chapter is followed by Sofia K. Blue's analysis of vocal communication in Chap. 7, which generates insights from comparing Tibetan macaques with other macaque species. In Chap. 8, Krishna N. Balasubramaniam, Hideshi Ogawa, Jin-Hua Li, Consuel Ionica, and Carol M. Berman offer a comprehensive review of Tibetan macaque's social structure, with insights from their previous work on social styles, comparative studies with other macaque species, and male–male social tolerance.

Part III contains two highly focused studies about ritualized behavior of Tibetan macaques with the implication about how culture evolves. In Chap. 9, Grant J. Clifton, Lori K. Sheeran, R. Steven Wagner, Jake A. Funkhouser, and Jin-Hua

Li examine how infants are used for the regularly observed behavior of bridging between adult females. This is further pursued in Chap. 10, where Hideshi Ogawa compares bridging behavior in two populations of *Macaca assamensis*, which are then compared with Tibetan macaques to explore the evolutionary origins of this highly ritualized behavior.

Part IV is composed of four chapters, focusing on how Tibetan macaques live with microbes, parasites, and diseases. In Chap. 11, Binghua Sun, Michael A. Huffman, and Jin-Hua Li take us into the microbial world inside the gut of Tibetan macaques and show how microbes adapt to the social behavior of the species. Then, Michael A. Huffman, Binghua Sun, and Jin-Hua Li present data in Chap. 12 to test the hypothesis that the diet of Tibetan macaques may incorporate self-medicative aspects to better survive in their environment, a proposition that has never been examined in this species before. Broadening the scope in Chap. 13, Krishna N. Balasubramaniam, Cédric Sueur, Michael A. Huffman, and Andrew J. J. MacIntosh review previous work on infectious agents at human–macaque interfaces and offer several key future directions for research in this area.

Many recent discoveries in primatology involve technological advancements in research, which is the content of Part V. In a single chapter (Chap. 14), Yong Zhu and Paul A. Garber explore the great potential of the high field MRI technology in the study of primate behavior and cognition. While promising, this new imaging technology has several obvious limitations at present.

All in all, the contributors of this volume examine a broad range of topics about the behavioral ecology of the Tibetan macaque. Although data are still far from adequate and some conclusions are tentative, we hope this volume will help remove the Tibetan macaque from the list of little known primate species. We expect that the information presented here can stimulate further comparative study of behavioral, ecological, and evolutionary questions about macaques and other primates and hope that this contribution will facilitate the integration of Chinese primatology into the mainstream field.

Hefei, Anhui, China
Ellensburg, WA, USA
Göttingen, Germany

Jin-Hua Li
Lixing Sun
Peter M. Kappeler

Acknowledgments

This volume is based mainly on the research papers presented during the 2017 International Primatological Symposium at Mt. Huangshan, China. The National Natural Science Foundation of China sponsored the meeting and also provided funding to support open access publication of this volume, as well as the publishing fund of Hefei Normal University. Hefei Normal University also provided a fund to defray the cost of publication including book purchase. We are grateful to all the contributors for sharing their work, without which the timely publication of this volume would have been impossible.

All the chapters in the volume were peer reviewed and benefited from the sharp comments and constructive suggestions by colleagues who generously donated their time and offered professional help. We are particularly thankful for the following external reviewers: Filippo Aureli, Louise Barrett, Fred Bercovitch, Marco Gamba, Andrew King, Daoying Lan, Bonaventura Majolo, Nadine Müller, Charles Nunn, Paula Pebsworth, Odile Petit, Gabriele Schino, Masaki Shimada, Wencheng Song, Bernhard Thierry, Kazuo Wada, Qi Wu, and Hongyi Yang. We also thank Rose Amrhein who offered language help for most of the chapters. We are lucky to have Srinivasan Manavalan as our in-house editor who oversaw the book project from the beginning to the end and answered all of our questions and inquiries throughout the process.

April 2019

Jin-Hua Li
Lixing Sun
Peter M. Kappeler

Contents

List of Contributors

Krishna Balasubramaniam Department of Population Health and Reproduction, School of Veterinary Medicine, University of California at Davis, Davis, CA, USA

Carol M. Berman Department of Anthropology and Graduate Program in Evolution Ecology and Behavior, State University of New York at Buffalo, Buffalo, NY, USA

Sofia K. Blue Primate Behavior and Ecology Program, Department of Anthropology and Museum Studies, Central Washington University, Ellensburg, WA, USA

Grant J. Clifton Primate Behavior and Ecology Program, Central Washington University, Ellensburg, WA, USA

Claudia Fichtel Behavioral Ecology and Sociobiology Unit, German Primate Center, Leibniz Institute for Primate Research, Göttingen, Germany

Jake A. Funkhouser Primate Behavior and Ecology Program, Central Washington University, Ellensburg, WA, USA
Department of Anthropology, Washington University in St. Louis, St. Louis, MO, USA

Paul A. Garber Department of Anthropology, Program in Ecology, Evolution, and Conservation Biology, University of Illinois, Urbana, IL, USA

Michael A. Huffman Primate Research Institute, Kyoto University, Kyoto, Japan

Consuel Ionica Biomedical Department, F. I. Rainer Anthropology Institute, Romanian Academy, Bucureşti, Romania

Peter M. Kappeler Behavioral Ecology and Sociobiology Unit, German Primate Center, Leibniz Institute for Primate Research, Göttingen, Germany
Department of Anthropology/Sociobiology, University of Göttingen, Göttingen, Germany

Jin-Hua Li School of Resources and Environmental Engineering, Anhui University, Hefei, Anhui, China
International Collaborative Research Center for Huangshan Biodiversity and Tibetan Macaque Behavioral Ecology, Anhui, China
School of Life Sciences, Hefei Normal University, Hefei, Anhui, China

Andrew J. J. MacIntosh Kyoto University Primate Research Institute, Kyoto, Japan

Jessica A. Mayhew Primate Behavior and Ecology Program, Department of Anthropology and Museum Studies, Central Washington University, Ellensburg, WA, USA

Hideshi Ogawa School of International Liberal Studies, Chukyo University, Toyota, Japan

Lori K. Sheeran Primate Behavior and Ecology Program and Department of Anthropology and Museum Studies, Central Washington University, Ellensburg, WA, USA

Cédric Sueur Université de Strasbourg, CNRS, IPHC, UMR 7178, Strasbourg, France

Binghua Sun School of Resources and Environmental Engineering, Anhui University, Hefei, China

Lixing Sun Department of Biological Sciences, Primate Behavior and Ecology Program, Central Washington University, Ellensburg, WA, USA

R. Steven Wagner Department of Biological Sciences, Central Washington University, Ellensburg, WA, USA

Xi Wang School of Resources and Environmental Engineering, Anhui University, Hefei, China
International Collaborative Research Center for Huangshan Biodiversity and Tibetan Macaque Behavioral Ecology, Anhui, China

Kaitlin R. Wright Primate Behavior and Ecology Program, Central Washington University, Ellensburg, WA, USA

Dong-Po Xia School of Life Sciences, Anhui University, Hefei, China
International Collaborative Research Center for Huangshan Biodiversity and Tibetan Macaque Behavioral Ecology, Anhui, China

Yong Zhu High Magnetic Field Laboratory, Chinese Academy of Sciences, Hefei, China
School of Life Sciences, Hefei Normal University, Hefei, Anhui, China

Part I
Introduction

Chapter 1
Recent Developments in Primatology and Their Relevance to the Study of Tibetan Macaques

Lixing Sun, Jin-Hua Li, Cédric Sueur, Paul A. Garber, Claudia Fichtel, and Peter M. Kappeler

L. Sun (✉)
Department of Biological Sciences, Primate Behavior and Ecology Program, Central Washington University, Ellensburg, WA, USA
e-mail: Lixing@cwu.edu

J.-H. Li
School of Resources and Environmental Engineering, Anhui University, Hefei, Anhui, China

International Collaborative Research Center for Huangshan Biodiversity and Tibetan Macaque Behavioral Ecology, Anhui, China

School of Life Sciences, Hefei Normal University, Hefei, Anhui, China
e-mail: jhli@ahu.edu.cn

C. Sueur
CNRS, IPHC, UMR, Université de Strasbourg, Strasbourg, France
e-mail: cedric.sueur@iphc.cnrs.fr

P. A. Garber
Department of Anthropology, Program in Ecology, Evolution, and Conservation Biology, University of Illinois, Urbana, IL, USA
e-mail: p-garber@illinois.edu

C. Fichtel
Behavioral Ecology and Sociobiology Unit, German Primate Center, Leibniz Institute for Primate Research, Göttingen, Germany
e-mail: Claudia.Fichtel@gwdg.de

P. M. Kappeler
Behavioral Ecology and Sociobiology Unit, German Primate Center, Leibniz Institute for Primate Research, Göttingen, Germany

Department of Anthropology/Sociobiology, University of Göttingen, Göttingen, Germany
e-mail: pkappel@gwdg.de

© The Author(s) 2020
J.-H. Li et al. (eds.), *The Behavioral Ecology of the Tibetan Macaque*, Fascinating Life Sciences, https://doi.org/10.1007/978-3-030-27920-2_1

1.1　Recent Trends and Developments in Primatology

Given their shared evolutionary history with humans, nonhuman primates play an exceptional role in the study of animal behavior, ecology, and evolution. This close phylogenetic relationship has led scholars from a diverse set of disciplines (e.g., biological and social sciences, notably psychology and anthropology) and theoretical perspectives (e.g., kinship theory, multilevel selection, social interactions, cultural traditions, competition, cooperation, innovation) to examine a broad range of research topics and methodologies in primatology. It is hardly an exaggeration to say that primatology is an intellectual "melting pot" in the study of animals.

The integration of different disciplines into the science of primatology has led to a major paradigm shift in the philosophy of science. Traditionally, scientists tended to assume animals had limited agency and behavioral flexibility or were incapable of engaging in complex forms of decision-making. They would be accused of committing a major scientific sin, namely, anthropomorphism, if they empathized with their research subjects or thought animals share emotions, social strategies, or cognitive abilities with humans (e.g., Masson and McCarthy 1996). Such a philosophical standpoint has become increasingly tenuous, as recent evidence in many mammals, especially nonhuman primates, have identified that they do indeed exhibit emotions, empathy, behavioral strategies, social bonding, cognitive abilities, and, in some instances, a moral sense of fairness (e.g., de Waal 1988, 2010; Kappeler and van Schaik 2006; Kappeler and Silk 2009; van Schaik 2016). In light of these findings, anthropomorphism and highlighting the behavioral and cognitive continuity between humans and other mammals may provide a more parsimonious null hypothesis than the alternative view, namely, that mammals in general and primates in particular have limited ability to respond to changes in their social and ecological environments (de Waal et al. 2006).

Primatologists have a keen and profound understanding that the sensations and emotions of researchers, as well as their subjective experiences as primates, are an essential variable in pursuing scientific questions. As in the case of quantum mechanics, where measurements and the behavior of quanta are entwined no matter what controls researchers place on the experiment, scientific objectivity in primatology is accomplished by including their own sensations and emotions as critical variables in research. We argue there exists no clear demarcation between being overtly subjective and being prudently self-reflective, and therefore in the case of primate research, we attempt to balance anthropomorphism without losing scientific objectivity.

Accompanying this philosophical shift, broad comparative studies, grounded in evolutionary, ecological, and behavioral perspectives, have led, in recent years, to new and exciting discoveries focused on social strategies, problem-solving, and cognitive abilities of nonhuman primates including complex social networks, cooperation between kin and nonkin, collective behavior, biological markets, behavioral economics, and culture (Byrne and Whiten 1988; van Schaik et al. 1999; Byrne and Bates 2007; Dufour et al. 2008; Sueur et al. 2010; Balasubramaniam et al.

2011; Sussman and Garber 2011; Pasquaretta et al. 2014; Garber 2019). Using social cognition as an example, major developments have been made in understanding primate intelligence (Machiavelli intelligence), social development, personality, empathy, theory of mind, trading, intuitive moral sense (particularly, fairness), and others (Dufour et al. 2008; Devaine et al. 2017). These discoveries would have been impossible without careful self-examination of our own behaviors, societies, and cognitive abilities. In fact, the study of our focal species in this volume, the Tibetan macaque (*Macaca thibetana*), clearly reflects such introspective, anthropomorphic thinking across a wide range of topics from social interactions and feeding ecology to gut microbe communities and self-medication. Despite the many insights anthropomorphic approaches provide, we must refrain from assuming that all similarities in social behaviors, social organizations, and cognitive abilities are derived from shared ancestry between humans and nonhuman primates because these similarities could also have arisen independently through convergent or parallel evolution. As such, distinguishing between homology and homoplasy will continue to pose a major challenge to primatologists.

1.2 Why Macaques, Especially Tibetan Macaques?

New developments in primatology in recent years have reinforced the view that nonhuman primates offer a window for us to look into human behavior, biology, and sociality from an evolutionary perspective (e.g., Kappeler and van Schaik 2006; Kappeler and Silk 2009; van Schaik 2016). Primates not only provide an instructive model to better understand human evolution, they may also provide information and insights into a range of practical issues in our society from promoting cooperation to preventing disease transmission (Romano et al. 2016). In fact, it is hard for us to underestimate how much we can learn from behavioral, ecological, and evolutionary knowledge gained from nonhuman primates.

Although nonhuman great apes represent our closest living relatives, all primate radiations can provide important model systems that unveil evolutionary patterns and processes. Currently there are over 500 species of living primates (Estrada et al. 2017). Many of these species are more abundant and evolutionarily successful (in numbers, populations, and species) than are great apes. As such they can provide a broader range of demographic and socioecological scenarios for testing evolutionary hypotheses, especially using comparative approaches (e.g., Harvey and Pagel 1991). In this sense, macaques (genus *Macaca*) may be unmatched as a model group for evolutionary studies of social behavior, social organization, and social cognition in primates. With 20–23 species, macaques are among the most successful primate radiations, with the largest geographical distribution of any taxa (Thierry et al. 2004; Fleagle 2013). The genus forms a monophyletic clade (Morales and Melnick 1998), and the evolutionary relationships among species have been mapped out with a reasonable level of certainty (Purvis 1995; Li and Zhang 2005; Jiang et al. 2016).

The availability of this critical information has paved the way for pursuing questions about the evolution of their behaviors using comparative methods.

One major challenge now for macaque behavioral ecologists is to obtain quality long-term data on behavioral variability and trade-offs between affiliative and agonistic alliances, social cohesion, and reproductive success. This level of information is missing for virtually all species, including the Tibetan macaque. In an attempt to find a general pattern of social style in macaques, for instance, Thierry (2004) classified all macaque species into four grades of social structure, from despotic (Grade 1) to egalitarian societies (Grade 4). These grades were associated with traits such as degree of social tolerance, symmetrical or asymmetrical conflict, a linear-like dominance hierarchy, and the strength of kin bonds. Clearly, there are significant challenges in attempting to quantify these variables. Based on limited information, Thierry (2004) identified Tibetan macaque as a Grade 3 species. Close examination, however, revealed that the aggressive behavior and dominance hierarchy of this species are more consistent with Grade 2 (Berman et al. 2004, 2006), which led to the revision of Thierry's classification scheme (Thierry 2011). This example illustrates the challenges of attempting to answer evolutionary questions using limited information from little known species. As such, this volume aspires to fill some glaring gaps in our knowledge of Tibetan macaques. Furthermore, several chapters attempt to directly address evolutionary questions from comparative perspectives between the Tibetan macaque and its sister species.

The Tibetan macaque is endemic to China and is listed in the IUCN Red List as near threatened (see Chap. 2). It is one of the most widespread primate species in China, distributed across 13 provinces. Its estimated population size is 20,000 individuals (Li et al. unpublished data). While rhesus macaques have been intensively studied in the field and laboratory, Tibetan macaques are far less known. As such, they provide a good comparison for understanding macaque ecology and behavior across a range of demographic, social, and ecological conditions. Also, Tibetan macaques have several unique features in the genus *Macaca*. They are the largest in body size (adult male body mass ~ 15 kg) and are found at elevations up to 2400 m (see Berman et al. 2006). They live in relatively small groups with a strict linear dominance hierarchy. Home range size is 1.62–3.62 km^2 with a pattern of habitat utilization (feeding in particular) particularly favorable for long-term observation and data collection. These features make the species well-suited for a wide range of studies for testing hypotheses related to behavior, ecology, evolution, conservation, management, human-animal interaction, and infectious disease transmission (see Chap. 14 Balasubramaniam, Sueur, and MacIntosh). Additionally, given that some populations are present in national parks or protected areas, such as those in Mt. Huangshan and Mt. Emei, they are highly accessible for educational and research purposes. (See the elaboration by Jin-Hua Li and Peter Kappeler in Chap. 2.)

1.3   A Short History of Tibetan Macaque Research

Primate research in China was largely absent until the nineteenth century, when European and American naturalists and missionaries came to China and reported their discoveries of exotic species. Following these initial descriptions, there remained no systematic studies of the behavior and ecology of Chinese primates until the 1970s, when a small number of pioneers of Chinese zoologists began to study and observe endemic species such as snub-nosed monkeys (*Rhinopithecus* spp.) and Tibetan macaques. During this period, the research was often sporadic and was published principally in Chinese journals.

The Tibetan macaque was first described to the scientific community as *M. thibetanus* by A. Milne-Edwards in 1870 based on a specimen collected by French missionary Abbé Armand David at Baoxing County in Sichuan Province (see Fooden 1983). However, in 1938, G. M. Allen believed it to be a subspecies of the stump-tailed macaque (formerly *M. speciosa* but currently *M. arctoides*) and named it *M. speciosus thibetanus*. It was not until 1983 that the Tibetan macaque regained its current species status as *M. thibetana*. This changed consensus was based on Jack Fooden's work on the anatomy of the reproductive system of the species. Recent studies using mtDNA have corroborated and validated this taxonomic distinction (Liu et al. 2006).

Despite the fact that Tibetan macaques are found in several small and isolated locations across China (a small population was recently found in eastern India) today, they were once widely distributed across a broad strip running from the foothills of southeastern Tibet to the coastal regions of East China including 13 provinces: Zhejiang, Anhui, Fujian, Jiangxi, Hubei, Hunan, Guangdong, Guangxi, Sichuan, Guizhou, Yunnan, Gansu, and Tibet (Jiang et al. 2015). Currently, four geographic subspecies are identified based on morphological characters and mtDNA (Jiang et al. 1996; Liu et al. 2006; Sun et al. 2010). Genomic studies show that the Tibetan macaque diverged from its congenic species some 0.5 Ma after a long bottleneck for the genus (Fan et al. 2014).

Field studies on Tibetan macaques have been carried out primarily at two sites, Mt. Emei in Sichuan Province and Mt. Huangshan in Anhui Province. Research on the Mt. Emei population has been led by Qikun Zhao and Ziyun Deng based at the Kunming Institute of Zoology of the Chinese Academy of Sciences. The bulk of the data from the population at this site was collected between 1987 and 1999. Research at the Mt. Huangshan site began in 1983 and led by Qishan Wang and Jin-Hua Li from Anhui University, in long-term collaboration (since 2003) with researchers from Central Washington University including Lixing Sun, Lori Sheeran, and colleagues. These research efforts have led to most of our current understanding of the species's social behavior, ecology, and population biology.

Research on ecology and population biology of Tibetan macaques has focused on morphological adaptation to ecological factors in its habitat (Xiong 1984), the current distribution of the species (Wada et al. 1987; Jiang et al. 1996), home range (Wang and Xiong 1989), adaptive relationships between body mass and elevation (Zhao and Deng 1988a, b; Zhao 1994a), effects of climate, vegetation, and slope on food resources (Zhao et al. 1989), selection of sleeping sites (Li and Wang 1994), ecological factors linked to group fission and reformation (Li et al. 1996a, b), diet (Li 1999; Zhao 1999), social organization (Deng and Zhao 1987; Li and Wang 1996), population dynamics (Wang et al. 1994), age structure and life expectancy and mortality (Li et al. 1995), and population growth in relation to population density, processes of group fission, and disease (Li et al. 1996a, b).

Several other research projects have focused on social bond formation, grooming relationships, social networks, and reproductive strategies. This has been made possible by long-term observation and monitoring of known individuals in several groups, aided by provisioning. This part of research has compared species differences in mating tactics between Tibetan and Japanese macaques (Xiong and Wang 1991; Zhao 1994c), birth timing in relation to socioecological factors such as dominance rank and altitude (Li et al. 1994; Zhao 1994c; Li et al. 2005), birth seasonality (Zhao and Deng 1988a, b; Li et al. 2005), and descriptions and evaluations of behaviors that are critical measures of social relationships such as bridging (defined as two individuals ritualistically lifting an infant accompanied by affiliative behaviors such as teeth chattering, see Zhao 1996) and grooming (Li et al. 1996a, b). These efforts have led to a complete ethogram of Tibetan macaques consisting of 32 distinct patterns of social behavior (Li 1999; Li et al. 2004). Using this ethogram, researchers can conduct in-depth analyses of behavior, social structure, and social dynamics of the species. A series of papers have been published on sociality and group stability analyzed at various levels from the individual, to dyads, social networks (cliques), and to the entire group. The topics addressed include collective decision-making and leadership in group movement (Wang et al. 2015, 2016; Fratellone et al. 2019), personality (Pritchard et al. 2014), benefit-cost analyses of grooming exchanges (Xia et al. 2012, 2013), and social networks among group members (Fratellone et al. 2019).

In recent years, new field and laboratory techniques from other disciplines have been applied to behavioral, ecological, and evolutionary studies of the species. They include the development and application of fecal DNA analysis (Zhao and Li 2008), extraction and analysis of fecal steroid hormones (Xia et al. 2015, 2018), and gut microbe analysis (Sun et al. 2016). Several of the chapters in this volume reflect these new and exciting research developments.

1.4 Tibetan Macaques at Mt. Huangshan Research Site

Many advancements in our understanding of primate behavior, biology, and evolution have been made from field studies. In this respect, long-term field studies are particularly valuable (Kappeler and Watts 2012) because they allow us to observe especially rare behaviors and biological events that may be missed based on short-term observations. For instance, long-term field studies have enabled researchers to document the cultural transmission of information within a group of Japanese macaques on Koshima Island and warfare in chimpanzees (*Pan troglodytes*). Likewise, much of our understanding of the behavior ecology and reproductive strategies of male and female Tibetan macaques reported in this volume would not be possible without long-term field studies and the dedication of primatologists working at Mt. Huangshan.

The Mt. Huangshan research site was first established in 1983, when the late Qishan Wang, then head of the Biology Department at Anhui University, founded a primate research program in collaboration with Kazuo Wada from the Primate Research Institute at Kyoto University in Japan. They jointly led a team to explore southern Anhui for a field site suitable for long-term research of the species. After some scouting, they settled on a location, now known as the Valley of Monkeys, at Yulingken, just a 15-min walk from the village of Fuxi, within the scenic area of Mt. Huangshan. This was the beginning of the first and in the meantime longest running primate research site in China, and long-term systematic observations of the Tibetan macaques have continued for more than 30 years. Today, the site is recognized as one of the eight long-term primatological study sites in the world (Kappeler and Watts 2012). Over the course of three decades, more than 150 primatologists from Japan, the United States, Germany, England, and Canada have come and conducted research in collaboration with Chinese colleagues at Mt. Huangshan.

Kazuo Wada's insights and contributions were essential, especially in getting this long-term research project started. One of the major inaugural events took place in October, 1986, when Wada presented a series of talks during the First Primatological Research Symposium of China hosted by Anhui University. The meeting was attended by only 12 participants (including Jin-Hua Li and Lixing Sun, then both beginning graduate students), representing probably fewer than ten qualified primatological researchers in the nation. At that time, some 27 primate species naturally occurred in China. In comparison, there were about 500 primatologists in Japan, working on a single endemic species, the Japanese macaque (*M. fuscata*). Wada was a dedicated field primatologist who had published extensively about the biology of macaques. Though skeptical about sociobiology, he was quite open in sharing his research ideas and experience, especially under the mild impact of Chinese whiskey. Among what he brought to Chinese primatology were two simple yet brilliant Japanese methodological perspectives that changed the face of Chinese primatological research: he recommended that primates be recognized individually and helped establish a naming system that reflected the sex, generation, and lineage information for all group members. At the time, these practices were still considered an inappropriate use of anthropomorphism by many in the West.

Since then, primatological research has thrived in China with increasingly well-trained primatologists working on virtually all of China's primate species across a range of topics from ecology, behavior, and conservation to molecular genetics and cognitive science with several highly productive and visible research groups. So, a few trickling streams of effort, initiated just a few decades ago, have grown to be a torrent of vibrant primatological research in China today, culminating in the formal launch of the Chinese Primatological Society in 2017. The inauguration conference of the society was attended by well over 200 researchers from all corners of the nation.

The chapters presented here partly illustrate the status quo of Chinese primatology from the research work conducted mainly in Tibetan macaques at Mt. Huangshan. They are also a demonstration for the fruitfulness of international collaboration by sharing research resources, ideas, and methodologies, reflecting the multidisciplinary nature of primatology. We expect that collaborations among researchers from a diverse range of academic and cultural backgrounds will continue to grow and deepen in the future.

References

Balasubramaniam KN, Berman CM, Ogawa H et al (2011) Using biological market principles to examine patterns of grooming exchange in *Macaca thibetana*. Am J Primatol 73:1269–1279

Berman CM, Ionica CS, Li J (2004) Dominance style among *Macaca thibetana* on Mt. Huangshan, China. Int J Primatol 25:1283–1312

Berman CM, Ionica CS, Dorner M et al (2006) Postconflict affiliation between former opponents in *Macaca thibetana* on Mt. Huangshan, China. Int J Primatol 27:827–854

Byrne RW, Bates LA (2007) Sociality, evolution, and cognition. Curr Biol 17:R714–R723

Byrne RW, Whiten A (1988) Machiavellian intelligence: social expertise and the evolution of intellect in monkeys, apes and humans. Clarendon Press, Oxford

De Waal FBM (1988) Chimpanzee politics: power and sex among apes. Johns Hopkins University Press, Baltimore, MD

De Waal FBM (2010) The age of empathy: nature's lessons for a kinder society. Broadway Books, New York

De Waal FBM, Wright R, Korsgaard CM et al (2006) Primates and philosophers: how morality evolved. Princeton University Press, Princeton, NJ

Deng Z, Zhao Q (1987) Social structure in a wild group of *Macaca thibetana* at Mount Emei, China. Folia Primatol 49:1–10

Devaine M, San-Galli A, Trapanese C et al (2017) Reading wild minds: a computational assay of theory of mind sophistication across seven primate species. PLoS Comput Biol 13(11): e1005833

Dufour V, Pelé M, Neumann M et al (2008) Calculated reciprocity after all: computation behind token transfers in orang-utans. Biol Lett 5(2):172–175

Estrada A, Garber PA, Rylands AB et al (2017) Impending extinction crisis of the world's primates: why primates matter. Sci Adv 3:e1600946

Fan Z, Zhao G, Li P et al (2014) Whole-genome sequencing of Tibetan macaque (*Macaca thibetana*) provides new insight into the macaque evolutionary history. Mol Biol Evol 31:1475–1489

Fleagle JG (2013) Primate adaptation and evolution, 3rd edn. Academic, San Diego, CA

Fooden J (1983) Taxonomy and evolution of the *sinica* group of macaques: 4. Species account of *Macaca thibetana*. Fieldiana Zool 11:1–20

Fratellone GP, Li J-H, Sheeran LK, Wagner RS, Wang X, Sun L (2019) Social connectivity facilitates collective decision making in wild Tibetan macaques (*Macaca thibetana*). Primates 60:183–189

Garber PA (2019) Primate cognitive ecology: challenges and solutions to locating and acquiring resources in social foragers. In: Lambert JE, Rothman JM (eds) Primate diet and nutrition: needing, finding, and using food. University of Chicago Press, Chicago, IL

Harvey PH, Pagel MD (1991) The comparative method in evolutionary biology. Oxford University Press, Oxford

Jiang X, Wang Y, Wang Q (1996) Taxonomy and distribution of Tibetan macaque (*Macaca thibetana*). Zool Res 17:361–369

Jiang Z, Ma Y, Wu Y et al (2015) China's mammal diversity and geographic distribution. Science Press, Beijing

Jiang J, Yu J, Li J et al (2016) Mitochondrial genome and nuclear markers provide new insight into the evolutionary history of macaques. PLoS One 11(5):e0154665

Kappeler PM, Silk JB (eds) (2009) Mind the gap: tracing the origins of human universals. Springer, Heidelberg

Kappeler PM, van Schaik CP (eds) (2006) Cooperation in primates and humans. Springer, Berlin

Kappeler PM, Watts DP (eds) (2012) Long-term field studies of primates. Springer, Berlin

Li J (1999) The Tibetan macaque society: a field study. Anhui University Press, Hefei

Li J, Wang Q (1994) Selection of sleeping site in Tibetan macaques in the summer. Chin J Zool 29:58

Li J, Wang Q (1996) Dominance hierarchy and its chronic changes in adult male Tibetan macaque (*Macaca thibetana*). Acta Zool Sin 42:330–334

Li Q, Zhang Y (2005) Phylogenetic relationships of the Macaques (Cercopithecidae: *Macaca*), inferred from mitochondrial DNA sequences. Biochem Genet 43:375–386

Li J, Wang Q, Li M (1994) Population ecology of Tibetan macaques II: patterns of reproduction. Acta Theriol Sin 14:255–259

Li J, Wang Q, Li M (1995) Population biology of Tibetan macaques III: age structure and life table. Acta Theriol Sin 15:31–35

Li J, Wang Q, Han D (1996a) Fission in a free-ranging Tibetan macaque troop at Huangshan Mountains, China. Chin Sci Bull 16:1377–1381

Li J, Wang Q, Li M (1996b) Migration of male Tibetan monkeys (*Macaca thibetana*) at Mt. Huangshan, Anhui Province, China. Acta Theriol Sin 16:1–6

Li J, Yin H, Zhou L et al (2004) Social behaviors and relationships among Tibetan macaques. Chin J Zool 39:40–44

Li J, Yin H, Wang Q (2005) Seasonality of reproduction and sexual activity in female Tibetan macaques *Macaca thibetana* at Huangshan, China. Acta Zool Sin 51:365–375

Liu Y, Li J, Zhao J (2006) Divergence and phylogeny of mitochondrial cytochrome B gene from Tibetan macaque and stump-tailed macaque. Ecol Sci 25:426–429

Masson JM, McCarthy S (1996) When elephants weep: emotional lives of animals. Vintage, New York

Morales JC, Melnick DJ (1998) Phylogenetic relationships of the macaques (Cercopithecidae: *Macaca*), as revealed by high resolution restriction site mapping of mitochondrial ribosomal genes. J Hum Evol 34:1–23

Pasquaretta C, Levé M, Claidiere N et al (2014) Social networks in primates: smart and tolerant species have more efficient networks. Sci Rep 4:7600

Pritchard AJ, Sheeran LK, Gabriel KI et al (2014) Behaviors that predict personality components in adult free-ranging Tibetan macaques *Macaca thibetana*. Curr Zool 60:362–372

Purvis A (1995) A composite estimate of primate phylogeny. Philos Trans R Soc B 348:405–421

Romano V, Duboscq J, Sarabian C et al (2016) Modeling infection transmission in primate networks to predict centrality-based risk. Am J Primatol 78:767–779

Sueur C, Deneubourg JL, Petit O (2010) Sequence of quorums during collective decision making in macaques. Behav Ecol Sociobiol 64:1875–1885

Sun B, Li J, Zhu Y et al (2010) Mitochondrial DNA variation in Tibetan macaque (*Macaca thibetana*). Folia Zool 59:301–307

Sun B, Wang X, Bernstein S et al (2016) Marked variation between winter and spring gut microbiota in free-ranging Tibetan macaques (*Macaca thibetana*). Sci Rep 6. https://doi.org/10.1038/srep26035

Sussman RW, Garber PA (2011) Cooperation, collective action, and competition in primate social interactions. In: Campbell CJ, Fuentes A, Mackinnon KC et al (eds) Primates in perspective, 2nd edn. Oxford University Press, New York, pp 587–599

Thierry B (2004) Social epigenesis. In: Thierry BR, Singh M, Kaumanns W (eds) Macaque societies: a model for the study of social organization. Cambridge University Press, New York, pp 267–290

Thierry B (2011) The macaques: a double-layered social organization. In: Campbell CJ, Fuentes A, MacKinnon KC et al (eds) Primates in perspective, 2nd edn. Oxford University Press, New York, pp 229–241

Thierry B, Singh M, Kaumanns W (eds) (2004) Macaque societies: a model for the study of social organization. Cambridge University Press, New York, pp 80–83

Van Schaik CP (2016) The primate origins of human nature. Wiley-Blackwell, Hoboken, NJ

Van Schaik CP, Deaner RO, Merrill MY (1999) The conditions for tool use in primates: implications for the evolution of material culture. J Hum Evol 36:719–741

Wada K, Xiong C, Wang Q (1987) On the distribution of Tibetan and rhesus monkeys in southern Anhui. Acta Theriol Sin 7:168–176

Wang Q, Xiong C (1989) Seasonal home range changes of Tibetan macaques at Yulingken. Acta Theriol Sin 9:239–246

Wang Q, Li J, Yang Z (1994) Tibetan macaques in China. Bull Biol 29:5–7

Wang X, Sun L, Li J et al (2015) Collective movement in the Tibetan macaques (*Macaca thibetana*): early joiners write the rule of the game. PLoS One 10(5):e0127459

Wang X, Sun L, Sheeran LK et al (2016) Social rank versus affiliation: which is more closely related to leadership of group movements in Tibetan macaques (*Macaca thibetana*)? Am J Primatol 78:816–824

Xia D, Li J, Garber PA et al (2012) Grooming reciprocity in female Tibetan macaques *Macaca thibetana*. Am J Primatol 74:569–579

Xia D, Li J, Garber PA et al (2013) Grooming reciprocity in male Tibetan macaques. Am J Primatol 75:1009–1020

Xia D, Li J, Sun B et al (2015) Evaluation of fecal testosterone, rank and copulatory behavior in wild male *Macaca thibetana* at Huangshan, China. Pak J Zool 47:1445–1454

Xia D, Wang X, Zhang Q et al (2018) Progesterone levels in seasonally breeding, free-ranging male *Macaca thibetana*. Mammal Res 63:99–106

Xiong C (1984) Ecological research of the Tibetan macaque. Acta Theriol Sin 4:1–9

Xiong C, Wang Q (1991) A comparative study on the male sexual behavior in Tibetan and Japanese monkeys. Acta Theriol Sin 11:13–22

Zhao Q (1994a) Birth timing shift with altitude and its ecological implication in Tibetan macaques at Mt. Emei. Oecol Mont 3:24–26

Zhao Q (1994b) Seasonal changes in body weight of *Macaca thibetana* at Mt. Emei, China. Am J Primatol 32:223–226

Zhao Q (1994c) Mating competition and intergroup transfer of males in Tibetan macaques (*Macaca thibetana*) at Mt. Emei, China. Primates 35:57–68

Zhao Q (1996) Male-infant-male interactions in Tibetan macaques. Primates 37:135–143

Zhao Q (1999) Responses to seasonal changes in nutrient quality and patchiness of food in a multigroup community of Tibetan macaques at Mt. Emei. Int J Primatol 20:511–524

Zhao Q, Deng Z (1988a) *Macaca thibetana* at Mt. Emei, China: II. birth seasonality. Am J Primatol 16:261–268

Zhao Q, Deng Z (1988b) *Macaca thibetana* at Mt. Emei, China: I. A cross-sectional study of growth and development. Am J Primatol 16:251–260

Zhao J, Li J (2008) Analysis of factors affecting DNA extracting from mammalian faecal samples. J Biol 25:5–8

Zhao Q, Xu JM, Deng Z (1989) Climate, vegetation and topography of the slope habitat of *Macaca thibetana* at Mt. Emei, China. Zool Res 10(supplement):91–99

Part II
Social Behavior and Dynamics in Tibetan Macaques

Chapter 2
Social and Life History Strategies of Tibetan Macaques at Mt. Huangshan

Jin-Hua Li and Peter M. Kappeler

2.1 Introduction

Among the more than 25 species of *Macaca*, the Tibetan macaque (*Macaca thibetana*) is relatively late to be known in primatology. There are several reasons for this. First, it had been considered as a subspecies of *Macaca speciosa* until Fooden (1983), based on his re-examination of the form and size of the glans penis and baculum and the structure of the female reproductive tract, elevated it to species status. Second, Tibetan macaques are endemic to east central China, where most of them inhabit high mountains with dense forests and cliff ledges, making field studies very challenging. Third, as argued above (Sun et al. 2019), there were few Chinese researchers interested in studying primates until the 1980s. As a result, the population status, life history, and social organization of *M. thibetana* have remained relatively poorly known.

Existing information, though still limited, indicates that the Tibetan macaque may be special in many ways. Morphologically, it resembles the stump-tailed macaque, *M. arctoides* (Delson 1980). Phylogenetically, it is close to the Assamese macaque, *M. assamensis*, as indicated by both morphological (Delson 1980) and genetic

J.-H. Li (✉)
School of Resources and Environmental Engineering, Anhui University, Hefei, Anhui, China

International Collaborative Research Center for Huangshan Biodiversity and Tibetan Macaque Behavioral Ecology, Anhui, China

School of Life Sciences, Hefei Normal University, Hefei, Anhui, China
e-mail: jhli@ahu.edu.cn

P. M. Kappeler
Behavioral Ecology and Sociobiology Unit, German Primate Center, Leibniz Institute for Primate Research, Göttingen, Germany

Department of Anthropology/Sociobiology, University of Göttingen, Göttingen, Germany
e-mail: pkappel@gwdg.de

J.-H. Li et al. (eds.), *The Behavioral Ecology of the Tibetan Macaque*, Fascinating Life Sciences, https://doi.org/10.1007/978-3-030-27920-2_2

analyses (Hoelzer et al. 1992). Ecologically and behaviorally, it is more similar to the Barbary macaque, *M. sylvanus*, because both live in montane habitats near the subtropical/temperate boundary, have a similar diet, and share intensive infant care with "bridging behavior," which involves two adults simultaneously lifting up an infant (Ogawa 2019). Yet, the four abovementioned species have been placed into different species groups within the genus *Macaca* by Fooden (1980) and Delson (1980), leading Wada et al. (1987) to suggest that the Tibetan macaque is a key species for understanding the evolution of Asian *Macaca*. These are the main reasons why *M. thibetana* has attracted much interest among primatologists lately.

Tibetan macaques have been listed as a Class II protected species in China since 1988, and they were classified as "Near Threatened" by the IUCN Red List of Threatened Species in 2017. Four subspecies of Tibetan macaque (*M. t. thibetana*, *M. t. guizhonensis*, *M. t. huangshanensis*, and *M. t. pullus*) have been identified based on morphological comparisons (Jiang et al. 1996) and mtDNA analyses (Sun et al. 2010). These subspecies represent distinct conservation units (Liu et al. 2006). Since *M. t. huangshanensis* is the most genetically distinctive and geographically isolated taxon (Sun et al. 2010), it deserves particular attention (Li et al. 2008) (Fig. 2.1).

2.2 Long-term Study of Tibetan Macaques at Mt. Huangshan

The Mt. Huangshan study site is located in southern Anhui province in East China (118.3E, 30.2 N, elevation 1841 m), about 1000 km south of Beijing. It covers an area of 154 km^2 with a south-north dimension of 40 km and an east-west extension of 30 km. This region is primarily made up of granite with many separate, sharp peaks and cliffs, which prevent people from gaining access to most parts. The weather in the mountain changes with altitude, from subtropical near the bottom, to temperate on the slopes, and to cold at the peak. Correspondingly, the vegetation changes from evergreen broad-leaved forest near the base, to deciduous and evergreen broad-leaved mixed forest on the slopes, and to montane grassland at the top. Mean temperature is 7.8 °C with the highest mean temperature in August (25.6 °C) and the lowest in January (-19.8 °C) (Li 1999). Mt. Huangshan is a well-known scenic spot and a popular tourist destination in China and is listed as a World Cultural and Natural Heritage Site, a World Geological Park, and is in the Man and the Biosphere Programme (MAB) (Fig. 2.2).

Two primate species inhabit these mountains: the Tibetan macaque and the rhesus macaque (*M. mulatta*). Tibetan macaques inhabit higher altitudes above 600 m asl with rocky cliffs, whereas the rhesus macaques live in lower areas with a relatively continuous distribution (Wada et al. 1987). Both have been strictly protected from hunting and trapping since the 1940s and have no known large predators.

Nine social groups of Tibetan macaques are currently living in Mt. Huangshan, where they maintain apparently nonoverlapping home ranges (Wada et al. 1987).

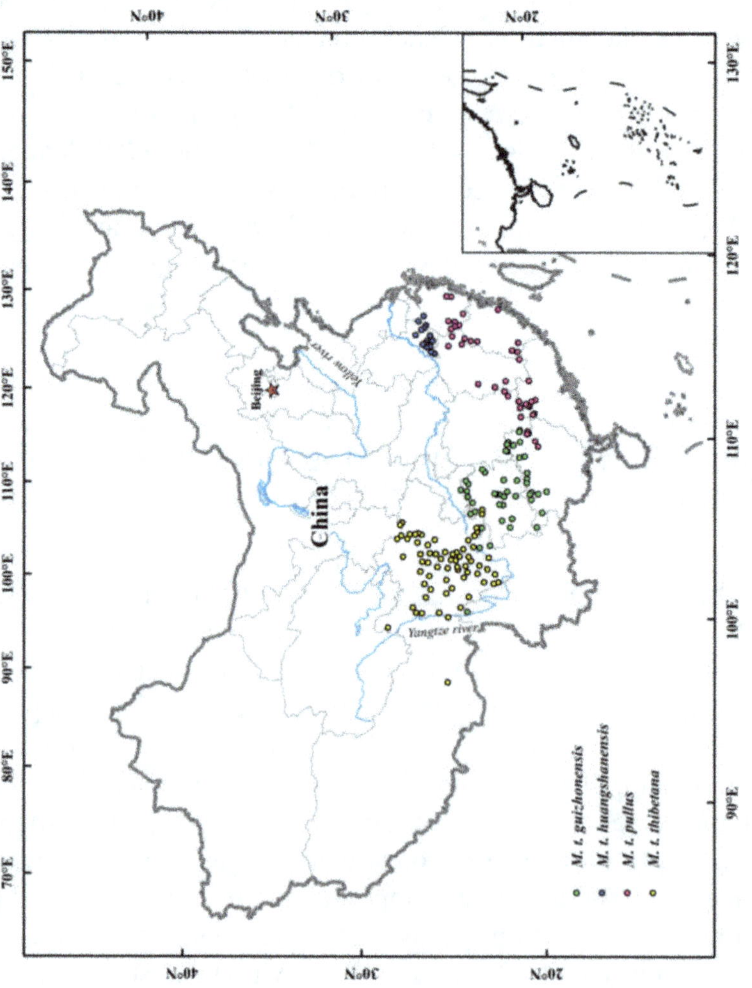

Fig. 2.1 Geographical distribution of the four subspecies of *Macaca thibetana* in China. The species sporadically inhabits in the mid-east region of China, along the Yangtze River

Fig. 2.2 Tibetan macaques inhabit high mountains with evergreen broad-leaved forest (left) and rocky cliffs (right)

They occur in two types of forest and mixed forest: the evergreen broad-forest between 500 and 1000 m asl and deciduous and evergreen mixed broad-forest between 800 and 1200 m asl. They are heavily reliant on structural plant parts for food with marked seasonality: bamboo and grass shoots in spring; fruits, nuts, acorns, and tubers in fall; and bark and mature leaves in winter, supplemented with invertebrates year-round (Xiong and Wang 1988). Geophagy has been observed in terms of ingesting yellow clay from well-established spots and licking rocks that are in contact with soil (Wang et al. 2007). In winter, Tibetan macaques huddle together and sleep on the terraces of rocky cliffs (Xiong 1984), but in summer, they spend the night in trees, which are cool and are safe from venomous snakes (Li and Wang 1994).

Our long-term study group is named Yulingken group (lately Yulingken A or YA1 following a recent group fission) at the village of Fuxi in the southern part of the mountain. The group has been monitored continuously since 1986. To facilitate observation, we have provisioned the group with dried maize to attract them to an open area by a stream. An amount of 5–6 kg of maize per day, which is about 1/3 of the daily food intake of the group, has been given by local wardens 3–4 times a day on a set schedule. When not being fed, the macaques spend most of their time in the forest near the provisioning area. Provisioning has made it possible for us to study the group in a steep terrain and has prolonged our observational time, though it may inevitably affect their natural activities to some extent (Matheson et al. 2006).

We set up a standard study protocol for long-term collection of consistent data within and among researchers at the beginning of the study in 1986. Since then, we have recognized individuals from birth on (natal individuals) or first discovery (immigrated individuals) and given them Chinese names. We have identified each group member based on individually specific physical characteristics such as body size, fur color, facial features, and body scars and taken pictures for confirmation and comparison over time. Furthermore, since Tibetan macaques have broad faces and long fur, both of which change with age, we can fairly reliably sort them into age-sex groups: infant (<1 year), juvenile (1–3 years), adolescent ($\male$3–7 years, $\female$3–5 years), young adult ($\male$7–10 years, $\female$5–10 years), middle-aged adult (10–15 years), and old adult (>15 years). Our records show that a male can live up to 28 years (Gaoshan) and a female up to 33 years old (Hua). They both died of old age. Because all females in the group have been identified after their births, we have been able to trace matrilineal relationships among all group members since 1983. Recently, we used microsatellite DNA markers to determine paternity for most group members (Appendix I). For behavioral data collection, we have followed Altmann's methods (Altmann 1974) and recognized and defined 33 distinct social behaviors (Li 1999, see Appendix II). In order to conduct genetic, physiological, and microbiological analyses, we have developed noninvasive techniques to collect cell or hormone samples from the feces and saliva of living monkeys (Zhao et al. 2005; Simons et al. 2012).

The long-term study of Tibetan macaques at Mt. Huangshan has been ongoing for over 30 years. It is the longest research project for wild primates in China. More than 150 researchers and students from the United States, Japan, Australia, England, and Germany as well as China have visited or conducted research at our field site, and 50 theses and 12 dissertations have been completed. This project is now recognized among the eight sites in the world for the long-term study of free-living monkeys (Kappeler and Watts 2012). Recently, as an integral part of the International Research Center for Huangshan Biodiversity and Tibetan Macaque Behavioral Ecology, the project has been approved by the Anhui provincial government.

2.3 Social Life History Strategies

2.3.1 The Largest Macaca

It is difficult to study Tibetan macaques in the wild because of their elusiveness and aggression, so catching and weighing them is nearly impossible. An unexpected opportunity to obtain morphometric data presented itself in July 1988, however, when 17 individuals of our study group died within 10 days for a reason that is still unknown. We obtained morphometric data from 14 of the 17 corpses. The average weight for adult males, adult females, and neonates are 16.4 ($n = 4$), 11.0 ($n = 3$), and 0.60 kg ($n = 2$), respectively. They were all heavier than members of the same age-sex classes in other *Macaca* species (Table 2.1). Apparently, the Tibetan macaque is the heaviest species in the genus *Macaca*. This may be the reason why local people call them "bear monkeys." This result has been verified by morphometric analysis of 72 linear dental and cranial variables of 11 macaque species, showing that the two stump-tailed species (*M. thibetana* and *M. arctoides*) are the largest of the macaques (Pan et al. 1998).

Table 2.1 Body weights (kg) of some *Macaca* species

Species	Adult male	Adult female	Neonate
M. thibetana	16.4	11.0	0.60
M. fuscatta	11.7	9.1	0.50
M. sylvanus	11.2	10.0	–
M. nemestrina	10.4	7.8	0.47
M. nigra	10.4	6.6	0.49
M. maura	9.5	5.1	–
M. arctoides	9.2	8.0	0.49
M. silenus	6.8	5.0	–
M. radiata	6.6	3.7	0.40
M. sinica	6.5	3.4	–
M. mulatta	6.2	3.0	0.48
M. fascicularis	5.9	4.1	0.35

Note: the data for species other than *M. thibetana* are from Smuts et al. (1987)

2.3.2 Medium-Sized Group with Even Adult Sex Ratios

We monitored the size and composition of Yulingken group annually and found a fluctuation of 21–51 individuals with an average of 35.3 individuals ($SE = 1.6$) over the 30 years from 1987 to 2017. Whenever group size reached about 50 individuals, it fissioned into two subgroups and the smaller one of the two subgroups left the study area. During this period, we recorded four fission events happening in 1993, 1996, 2001, and 2003, respectively (Li et al. 1996; Li 1999; Berman and Li 2002), and noticed that males and females who were relatives tended to separate on these occasions (Li et al. 1996). Thus, Tibetan macaques at Mt. Huangshan appear to live in medium-sized groups, in comparison with the 70–80 and sometimes more than 100 individuals found in groups of other macaque species (Macintosh et al. 2012; Waters et al. 2015). Group size is an important component of social organization that determines how many individuals a group member can potentially interact with, which in turn contributes to social complexity (Kappeler and Watts 2012).

Like other species of *Macaca*, Tibetan macaques are organized into multi-male, multi-female groups with female philopatry and male dispersal (Li et al. 1996; Li and Wang 1996). However, male Tibetan macaques do not leave their natal groups until adulthood, later than most *Macaca* in which male transfers typically occur at puberty (Zhao 1993; Li et al. 1996). Thus, Tibetan macaque groups tend to include large proportions of natal adult males and relatively even sex ratios (0.91 ± 0.05), with an average of 8.52 ($SE = 0.67$) adult males to 9.35 ($SE = 0.51$) adult females (Fig. 2.3).

2.3.3 A Rich Repertoire of Affiliative and Ritualized Behaviors

An important feature of Tibetan macaques is their rich repertoire of affiliative and ritualized behaviors. Of the 33 social behaviors we have identified, at least

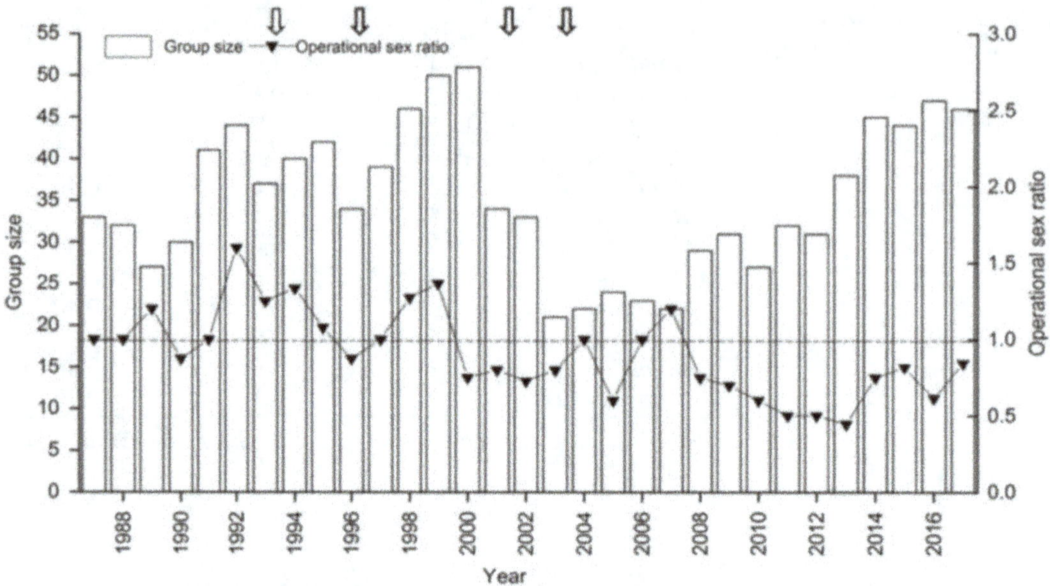

Fig. 2.3 Dynamic of group size and operational sex ratio of male to female in Yulingken A1 group during 1986 and 2017. Down arrow refers group fission events

17 (> 50%) include affiliation (Li 1999). Like other macaques, *allogrooming* (social grooming) is widespread among all group members and takes up about 20% of daily time expenditure (Wang et al. 2007). Females and juveniles are more active groomers than males in peaceful situations. Grooming is more frequent between sexual partners than between other male–female pairs. Specifically, higher-ranking males, including alpha males, groom females for a short time after copulation or during consortship. Grooming between adult males is also frequent (Xia et al. 2013). When two individuals meet, the lower-ranking individual usually *presents* to the high-ranging individual. If the two individuals are both males, they may *embrace* each other, sometimes with their hands stretching out to grasp the genitals of their partners. Additionally, the lower-ranking individual may approach a higher-ranking one, *showing its penis* to be sucked by the partner, or the higher-ranking individual *touches* a part of the body (head, back, or shoulder) of the lower-ranking individual or *mounts* the lower-ranking individual with *teeth chattering*. If the two individuals are of the opposite sex, the male may *grimace* at the female, and the female approaches the male for *genital inspection*, before copulation ensues. Also, during conflicts between two females, a third, higher-ranging female may often *approach* and *hold the bottom* of the attacking female to stop further aggression.

Tibetan macaques show an intense interest in infants. Group members, regardless of sex or age, often *hold infants*. Sometimes, an individual picks up an infant and carries it to another individual to perform the ritualized behavior of *bridging*. This behavior can occur between males, between females, and between a male and a female (Ogawa 1995a; Li 1999). The infants used in bridging are predominantly younger than 6 months, but occasionally 2–3-year-old juveniles may also be used

Fig. 2.4 Infant-holding behavior in Tibetan macaques is common. Left: three infants are attracted to two adult males, apparently waiting to be held or used in bridging. Right: a male Tibetan macaque living alone in Kowloon in Hong Kong holds an infant long-tailed macaque

(Zhang et al. 2018). Infants appear to be willing for bridging and mothers appear to be tolerant of infant handling by all group members (Fig. 2.4).

2.3.4 Despotic Dominance Style

The dominance style concept has proven useful for understanding covariation patterns in relationship qualities, particularly among macaques (Berman et al. 2004). As a member of the *sinica* lineage, Tibetan macaques are predicted to have a relaxed dominance style (Matsumura 1999; Thierry 2000). Previous studies did indicate relaxed dominance in this species. For example, males frequently engage in ritualized greetings in which they groom, mount, embrace, or touch each other's genitalia (Li 1999). In addition, both sexes engage in frequent bridging, and infants' mothers appear to be tolerant of infant handling by a wide range of group members (Ogawa 1995a).

However, a detailed study indicated that Tibetan macaques are more despotic than previously suspected. Bidirectional aggression, including counter-aggression (1.9%) and conciliatory tendency (6.4%), was consistently low in frequency across partner combinations, seasons, and locations (Berman et al. 2006). Females consistently displayed high levels of kin bias in affiliation and tolerance. Compared to other macaque species with better known dominance styles, data from Tibetan macaques generally fall within the range characteristic of despotic species and outside the range of relaxed species (Berman et al. 2004).

Despotic dominance of Tibetan macaques is consistent with our field observations. First, mating of low-ranking males was inhibited by high-ranking males. If a low-ranking male surreptitiously mated with a female in the forest but happened to be seen by a high-ranking male, the high-ranking male would immediately rush to punish him by scratching and biting. Second, when a series of conflicts happened,

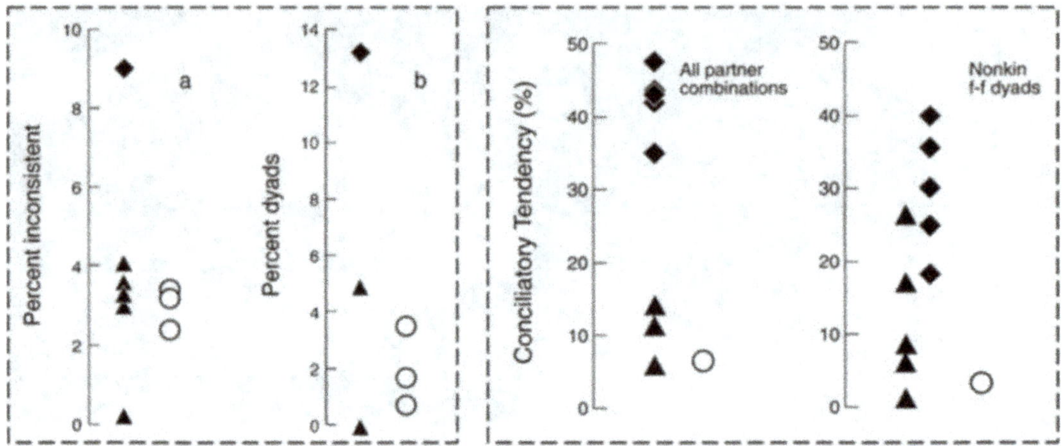

Fig. 2.5 Bidirectionality of aggression (left) and conciliatory tendencies of Tibetan macaques (right). Both a directional inconsistency index (a) and dyads-up index (percentage of dyads in which the primary direction of aggression is up the hierarchy of three data points representing three study periods for Tibetan macaques (circles)) fall within the range of despotic macaques (rhesus, long-tailed and Japanese macaques) (triangles) and are considerably lower than the one relatively relaxed species (stump-tailed macaques) (diamonds). Conciliatory tendencies of Tibetan macaque in (1) all partner combinations and degrees of relatedness combined, and (2) unrelated female–female partners (circles) are relatively low, even compared to values for despotic macaques (triangles). Data from Berman et al. (2004)

male Tibetan macaques often formed a "power coalition," in which several high-ranking males support one another to defeat low-ranking ones (Li 1999). Third, serious fresh wounds on Tibetan macaques were frequently seen immediately after intense aggressive interactions (Fig. 2.5).

2.3.5 *Reproductive Pattern with Year-Round Mating but Seasonal Births*

It took a long time for us to understand the reproductive pattern of Tibetan macaques. At first, based on intermittent field observations, they were thought to be nonseasonal breeders (Wada and Xiong 1996). After intensive year-round observations, we found that, although mating indeed takes place throughout the year, mating with high frequency and with ejaculation occurs only between July and December, with subsequent births mostly occurring between January and April (Li 1999; Li et al. 2005). Thus, Tibetan macaques are seasonal breeders (Fig. 2.6).

In order to explain this unusual reproductive pattern of Tibetan macaques, we investigated changes in the female sexual skin and nonreproductive matings in more detail. We found that adult females had a slight but detectable sexual skin in the perineal region, but the sexual skin lacked the regular changes characteristic of the physiological cycle of a female or the reproductive season. Females also had no typical behaviors to show their sexual motivation during estrus. This led to the

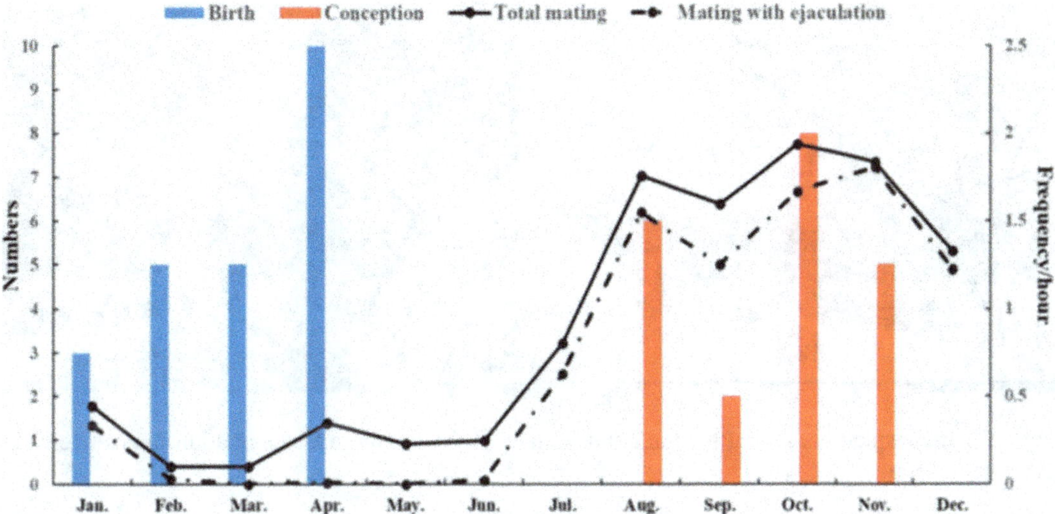

Fig. 2.6 Annual reproductive cycle of Tibetan macaques. Though Tibetan macaques mate year-round (solid line), mating with ejaculation (dotted line) is confined to a period from July to December. Therefore, the mating season is from July to December, and the birth season is from January to April

conclusion that Tibetan macaques have concealed ovulation (Li et al. 2005). Further study revealed that males could discriminate between potentially receptive females, but that they could not recognize the reproductive status of a given female (Li 1999; Zhang et al. 2010). Concealed ovulation may result in a higher mating frequency during a female's receptive period.

Compared with sexual behavior in the mating season, nonreproductive matings occurred at a lower frequency, with less frequent ejaculation, less harassment, shorter mount duration, and an absence of pauses with vocalization. It often took place in a situation in which non-lactating females were involved in social conflict or approached males for mating. Neither pregnant nor lactating females were observed to mate in the birth seasons. Copulation during the birth season did not affect a sexually receptive female's delivery the next year, nor was it associated with increased proximity, grooming, or agonistic aid within the mating pair. However, mating pairs spent more time co-feeding, presumably reflecting increased male tolerance (Li et al. 2007). Adolescent males, who rarely copulated in the mating season, engaged in mating activities during birth seasons as well. Therefore, even though birth season copulations have no reproductive function, they might fulfill social functions for females, such as post-aggression appeasement by males or improving access to resources. They may also offer good opportunities for adolescent males and females to develop their sexual skills (Li et al. 2007) (Fig. 2.7).

It is worth noting that an important aspect of the mating pattern of Tibetan macaques at Mt. Huangshan is a single mount ejaculation (SME), with average duration and thrust numbers of 23.2 s and 20 times, respectively (Xiong and Wang 1991). High-ranking males copulate more often with more females than do low-ranking males, and alpha males monopolize over 2/3 of all matings (Xiong

Fig. 2.7 Inconspicuous sexual skin in adult females (left) and large testes in adult males (right) in Tibetan macaque

and Wang 1991; Li et al. 2005). However, it has been reported that Tibetan macaques at Mt. Emei exhibit serial or multiple mount ejaculation with a mean duration and thrust number of mounts with 30.3 s and 43.3 times (MME), respectively (Zhao 1993). Since no data on the number of mounts in a MME have been provided, we do not know whether the difference is due to a difference in populations per se or observation methods.

Finally, females at the age of 7–12 years have the highest birth rate. Not only are birth rates of high-ranking females higher, but those of high-ranking females are also earlier than those of low-ranking ones in the birth season (Li et al. 1994).

2.4 Potential Contributions to Understanding Behavioral Mechanisms in Primate Societies

2.4.1 Bridge and Bond Role of Infant

Tibetan macaques exhibit some interesting and unique behavior patterns. Bridging behavior is one of them. Bridging involves two individuals and an infant and is a triadic interaction. Bridging behavior is common in Tibetan macaque society, and it can be performed between males, between females, between a male and a female, and even between an adult and a juvenile. Infants of both sexes are involved in the "bridge." Despite this variation in bridging partners, bridging lasts only a few seconds. Once bridging has ended, the two bridging individuals often groom each other or sit together, ignoring the infant. Obviously, the "bridge" between monkeys is not for infant care, but for social contact or communication. In other words, infants are a "social tool" used to facilitate older group members' associations and interactions (Ogawa 1995b). Zhao (1996) even compared infants in bridging behavior to a cigarette-like social facilitator in human social interactions.

Primatologists have always paid particular attention to the role of infants in nonhuman primate societies. However, most studies have focused on infant care or handling (Ogawa 1995a). For example, in capped langurs, *Presbytis pileata*, newborn infants of less than one month old spend nearly the same amount of time with alloparental females as with their own mothers (Stanford 1992). Likewise, young, female vervet monkeys, *Chlorocebus aethiops sabaeus*, without infants hold or carry newborn infants, which has been explained as alloparental investment (Fairbanks 1990). Bridging behavior has also been observed in *M. sylvanus* (Deag and Crook 1971). Clearly, infants in Tibetan and Barbary macaques play an active part in enhancing social relationships among group members. This is a relatively new and important research direction that is worthy of further exploration.

2.4.2 Male-Female Friendships in a Promiscuous Mating System

During the mating season from July to December, high-ranking Tibetan macaque males often follow sexually receptive females (usually those without unweaned infants). They move, feed, and rest together, often accompanied by mating; that is, they form a temporary consortship. The duration of such a consortship varies from a few days up to over 2 weeks (Li 1999). During a consortship, the male always keeps close with the female, and the male's activities are restricted by the female's movements. Sometimes, when a consorting male realizes that his female has disappeared, he shows signs of anxiety and searches for the female everywhere.

The role of consortship is unclear at present. On the one hand, the duration of consortship in Tibetan macaque is rather long, lasting as much as 18 continuous days (Li 1999). This does not only exceed the time needed to impregnate a female, but is also longer than the duration in other *Macaca* species (Li 1999). On the other hand, although a male follows one female at a time, usually he cannot monopolize all of her matings even though he is the alpha male. So why do high-ranking males use consortship as a mating tactic?

We already mentioned that female Tibetan macaques tend to conceal their ovulation by neither showing cyclical fluctuation of sexual skin nor estrus-related behaviors (Li et al. 2005). Although high-ranking males form consortship with females, they are able to opportunistically mate with other sexually receptive females, too. Thus, the co-occurrence of consort behavior and opportunistic mating may indicate a male mating tactic that is effective when mating is promiscuous and males cannot accurately detect ovulation in females from morphological and behavioral signs. That is, consortship may be a tactic related to the assurance of paternity. From an evolutionary perspective, consortship in Tibetan macaques appears to be a special type of sexual relationship, which has also been reported for some baboons, where they have been labeled as "friendships" (Smuts 1985).

2.4.3 Competitive and Cooperative Relationships Among Males

Male Tibetan macaques exhibit stable, linear dominance hierarchies typical of despotic species. On the one hand, aggression is frequent and can be accompanied by serious wounds in all sex/age classes (Berman et al. 2004; Li 1999). Additionally, the large body, large testes, and short tenures of alpha males (on average about 10 months) indicate intense competition among males (Li and Wang 1996). On the other hand, Tibetan macaque males frequently engage in friendly behaviors, such as grooming and sitting together, and engage in ritualized greetings involving mounting, embracing, and sucking penis. Many of these behaviors are rarely seen in other *Macaca* males. Tibetan macaque males also quickly reconcile after aggression. The conciliatory tendency is 19.7% in male–male dyads, compared to 4.2% in female–female dyads, and most male–male reconciliations occur within 1 min after aggression (Berman et al. 2004). It is evident that male Tibetan macaques show both competitive and cooperative relationships. This appears to agree with observations taken from other primate societies with relatively male-biased adult sex ratios.

2.4.4 Behavioral Mechanism Promoting Genetic Diversity in a Small Group

Like most nonhuman primates in China, Tibetan macaques live in small, isolated populations. Theoretically, a small population or group may suffer the detrimental effects of inbreeding and the loss of genetic diversity (Spielman et al. 2004). Our study group is a medium-sized one with 20–50 individuals and that ranges in a small habitat (Li et al. 1996). However, we have not found conspicuous changes in behavior during the 30-year period, except the sudden death of 17 individuals in the spring of 1988. Although we have noticed a decline in the frequency of affiliative interactions among group members and an increase in infant mortality in recent years, these appear to be a side effect of provisioning, which increases the rate of aggression among group members (Berman and Li 2002).

Additionally, we found that our study group has a relatively high haplotype diversity (Hd = 0.341) when compared to other subspecies (*M. t. pullus*, Hd = 0.222; *M. t. guizhouensis*, Hd = 0.478) (Sun et al. 2010). As such, loss of genetic diversity may not be a major concern in our study group at present. It is likely that some behavioral mechanisms in this group may play an important role in preventing or slowing down the loss of genetic diversity that is often observed in small groups. For instance, we found that the transfer of males into and out of the group occurs annually (Li et al. 1996). Also, kin recognition may prevent mating between matrilineal relatives. In fact, one of our studies demonstrates that no copulation involved mother-son dyads, and only 2.1% (7/329) of copulations were between maternal siblings. Females may be more averse to copulating with relatives than are males (Zhu et al. 2008).

2.5 Conclusions

Long-term studies of primate populations are essential for documenting important events or phases of individual life histories that are critical for our understanding of subtle and complex relationships of group members that determine the nature and dynamic of social organization and, as a result, variation in individual reproductive success. The medium-sized group of Tibetan macaques we study facilitates censuses and unambiguous identification of all group members in the field. As the largest macaque species with a relatively male-biased adult sex ratio and year-round mating but seasonal birth, this species offers a large spectrum of sociodemographic conditions conducive for testing hypotheses related to dominance style and group fission. Furthermore, a rich repertoire of affiliative and ritualized behaviors and a complex interplay of competitive and cooperative relationships, especially among males, can provide primatologists with a wealth of opportunities to pursue additional behavioral, ecological, and evolutionary research projects focused on Tibetan macaques.

Acknowledgments Our long-term field study has been supported by Anhui University and the Huangshan Garden Bureau. Our research would be impossible without their assistance in terms of permission, personnel, funding, and other logistic support. For more than three decades, many researchers and students, especially those from Anhui University, Kyoto University, State University of New York at Buffalo, University of Washington, and Central Washington University, have joined and contributed to the study. We cordially thank Qi-Shan Wang, Kazuo Wada, Cheng-Pei Xiong, Guo-Qiang Quan, He-Ling Shen, Ren-Mei Ren, Ming Li, Jiao Shao, Zu-Wang Wang, Zhi-Gang Jiang, Kunio Watanabe, Hideshi Ogawa, Frans de Wall, Noel Rowe, Carol Berman, Consuel Ionica, Lei Zhang, Toshisada Nishida, Randall Kyes, Lisa Jones-Engel, Lixing Sun, Lori Sheeran, Megan Matheson, Steven Wagner, Dong-Po Xia, Bing-Hua Sun, Yong Zhu, Xi Wang, Yang Liu, Jian-Yuan Zhao, Jessica Mayhew, and Paul Garber for their full-hearted supports and noted contributions. We also express our thanks to Fu-Wen Wei, Bao-Guo Li, Yan-Jie Su, Cheng-Ming Huang, Xue-Long Jiang, Ji-Qi Lu, and other members of China Primatology Society. This work was financially supported by the National Natural Science Foundation of China, the China Scholarship Council, the Key Teacher Program of the Ministry of Education of China, the Outstanding Youth Foundation of Anhui Province, the International Science and Technology Cooperation Plan of Anhui Province, Natural Science Foundation of Anhui Province, Inoue Scientific fund, Primate Conservation Inc., the L.S.B. Leakey Foundation, the Wenner-Gren Foundation, the ASP Conservation Small Grant, and the National Natural Science Foundation. Special thanks to Mr. Cheng's family and Fuxi villagers for their outstanding logistic support during our field observation for more than three decades.

Appendix I

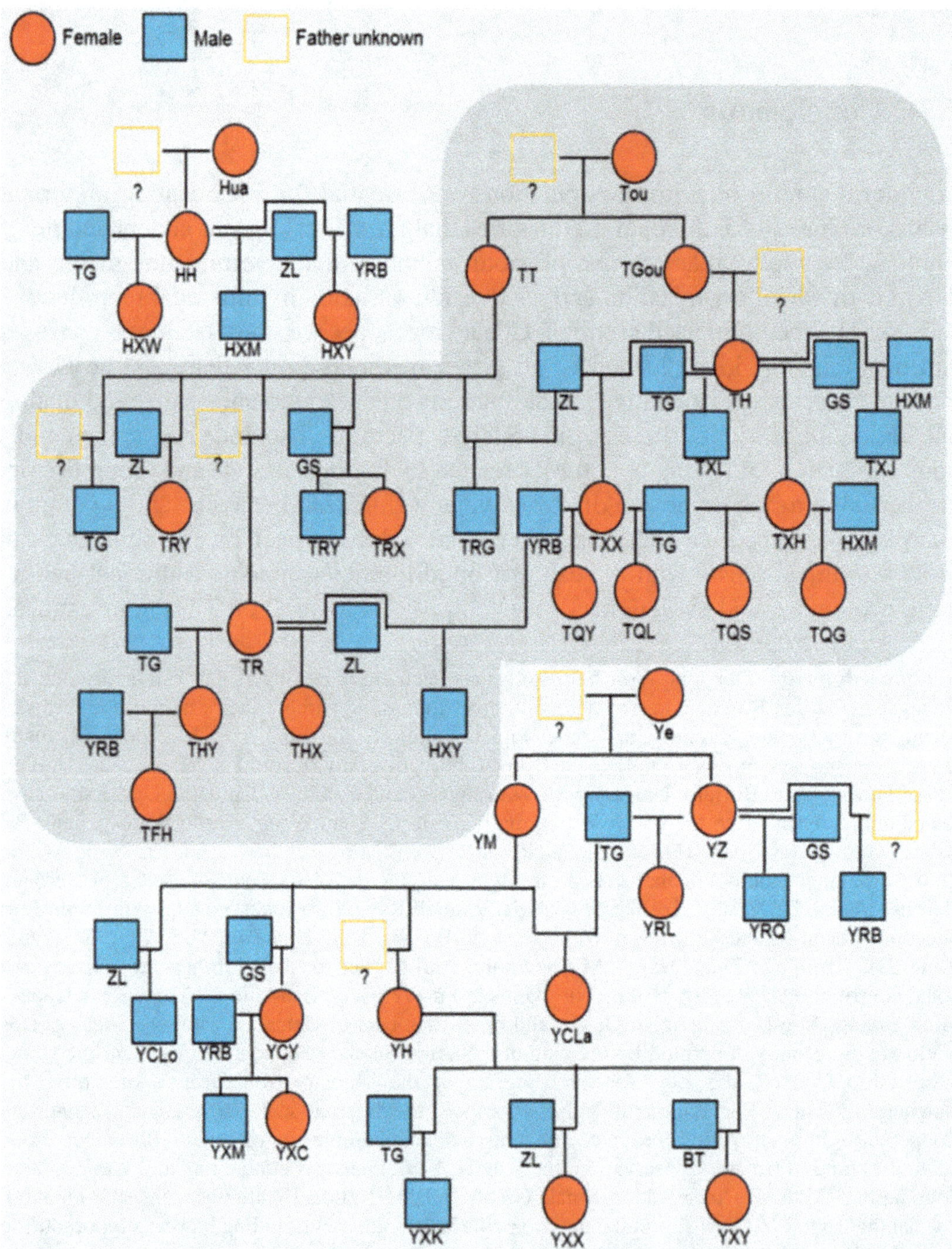

Pedigree of the YA1 social group of Tibetan macaques in Huangshan (September, 2017). Note: The pedigree was mapped based on maternal relationships of three matrilines. Males with known identities were also included. Letters under the blue squares (males) or red circles (females) represent individual names. Question marks indicate the identity of males are yet to be unknown

Appendix II

Ethogram of social behaviors in Tibetan macaque

No.	Repertoire	Definition	Picture
1	Stare	An individual looks directly at another individual with its eyes wide open and with its shoulders raised for about 3–5 s. The staring individual appears as if it is preparing to lunge or chase the recipient of the stare.	Picture 1
2	Ground slap	An individual places one hand on the ground and slaps the ground, a rock, or some grassy vegetation with the other while staring at the recipient. The ground slap may be repeated several times.	Picture 2
3	Chase	An individual stares at the recipient and rushes at him/her at great speed. The recipient typically flees.	
4	Seize	The performer grabs the body, face, neck, or ear of the recipient with one hand, shakes it, and then releases it. Sometimes the performer also comes very close (face to face) and stares at the recipient.	Picture 3
5	Bite	The performer grabs the receiver tightly, preventing him/her from fleeing, and bites the recipient vigorously.	Picture 4
6	Avoid	The performer turns its body away from the attacker as if preparing to flee while displaying a "horrified" facial expression toward the attacker.	Picture 5
7	Flee	The target of an attack will run in the opposite direction from the attacker.	
8	Scream	A vocal reaction to an attacker. The scream is high pitched and of long duration. This may be accompanied by fleeing.	Picture 6
9	Redirection	When A is attacked by B, A responds immediately by attacking C. C is an individual that is nearby and lower-ranking than both A and B. In some cases, B will join the attack against C.	Picture 7
10	Solicit support	When an individual is attacked by another individual, it intermittently scans the attacker and a higher-ranking individual nearby with screams. The higher-ranking bystander may respond by interrupting the attack on the scanner.	Picture 8
11	Hold bottom	The performer (usually the alpha female) approaches one of two other females that are engaged in a fight. She approaches from behind and holds the bottom of the female. Both females typically teeth-chatter to one another. This tends to interrupt the fight.	Picture 9
12	Approach	An individual moves directly toward another individual, coming within 1 m.	
13	Leave	An individual moves from within 1 m to more than 1 m of another individual.	Picture 10
14	Proximity	Two or more individuals are sitting or lying within 1 m or between 1 and 5 m (depending on studies).	Picture 11

(continued)

No.	Repertoire	Definition	Picture
15	Following	When one individual begins to travel from one location to another, another individual nearby immediately goes in the same direction.	Picture 12
16	Teeth chatter	The performer opens his/her mouth, places the tongue close to the teeth, and moves his/her jaw rapidly, thereby making clicking noises with his/her teeth. The eyelids are lowered, the chin is raised, and the tongue is moved rapidly back and forth across the teeth. This behavior accompanies same-sex mounting, embracing, and infant carrying. It appears to function as a friendly invitation to interact.	Picture 13
17	Play	Brief sequence of rapid, repetitive, and exaggerated movements, without clear objectives. Interacting play movements involving two or more individuals include chase, mock fight, and combinations of previous behavior.	Picture 14
18	Social grooming/ allogrooming	One individual uses his/her fingers and palms to groom the fur of another individual. The groomer may pick out small objects from the recipients fur and eat them.	Picture 15
19	Embrace	Two individuals, usually adult males, hold each other while face to face. Each partner will reach with one hand and attempt to touch the genitals of the other. Both partners typically teeth-chatter and vocalize excitedly.	Picture 16
20	Touch	A ritualistic behavior between males in which the lower-ranking individual approaches the higher-ranking individual in front. The higher-ranking male touches some part of the body of the lower-ranking individual (head, back, shoulder). Then the lower-ranking individual leaves. This may be used in a tense situation and appears to be a means by which the lower-ranking individual obtains "permission" to proceed on its pathway or with its apparent goal.	
21	Present	One individual approaches the front of the other and shows his bottom to the other. Usually the presenter is the lower-ranking of the two partners. Females also present to males.	Picture 17
22	Mount	One male (usually the lower-ranking) grabs the back hair of another male and mounts from behind, using the full ankle clasp posture. Both males teeth-chatter and scream excitedly. Then the mounter dismounts. The duration of the mount is about 3–5 s. This may be a simple friendly gesture or a post-conflict behavior.	Picture 18
23	Show penis	This is a ritualized behavior shown by juvenile males toward adult males. Usually, the lower-ranking male approaches the higher-ranking male, raises one leg, and displays his penis. The higher-ranking male puts his head on the belly of the lower-ranking male and licks or touches the penis with his hand. The juvenile may show his penis from a reclining position.	Picture 19
24	Suck penis	A young male approaches an adult male and jumps on his head. The adult male holds the younger male by the waist in such a way that his mouth can reach the young male's penis. The adult sucks the young male's penis and then the younger male leaves.	Picture 20

(continued)

No.	Repertoire	Definition	Picture
25	Hold infant	An adult male holds an infant and may carry it ventrally. Usually the infant is male. This apparently serves to invite other adult males to engage in social interaction with the adult male.	Picture 21
26	Bridge	Bridging involves three individuals, an infant or young juvenile and (a) two adult males, (b) one adult male and one subadult male, or (c) two adult females. The two older individuals hold the infant on its back, lower their heads, and lick the belly and/or genitals of the infant. They will often teeth-chatter and vocalize excitedly. The infant is usually male and his penis becomes erect immediately. Other infants, particularly females, will approach a bridging triad with excitement as if they wish to participate.	Picture 22
27	Genital inspection	One individual will touch the genitals—the vagina of a female or the anus of a male—and sniff it or lick it directly. When a male inspects a female in estrus, it is called a sexual inspection.	Picture 23
28	Sexual inspection	A genital inspection by an adult male of an estrus female.	Picture 24
29	Sexual chase	An adult male chases a female and attempts to mate with her. The sexual chase differs from the aggressive chase in that it is slower and is interrupted by pauses. In addition, although the female attempts to avoid the male, she is less likely to scream.	Picture 25
30	Sexual grimace	An individual directs a grimace toward an adult of the opposite sex that may be either nearby or far away. The male sways his body in a manner that makes him more noticeable to the female. The grimace differs from a fear grimace in that the corners of the mouth are not drawn back as far. This is a sexual solicitation.	Picture 26
31	Mating	Three types of mating are described: (1) intromission and thrusting only; (2) intromission, thrusting, and pausing; and (3) intromission, thrusting, pausing, and calling.	Picture 27
32	Harassment	Individuals approach a mating pair excitedly, touch the heads, lick, or stick their faces in the faces of the mating partners. They may also grimace or scream. This does not generally interrupt the copulation of the mating partners.	Picture 28
33	Consortship	An adult male and female form a temporary close affiliative and sexual relationship during the mating season. They will move, feed, mate, and groom together, but do not necessarily mate exclusively with one another. Their frequencies of following one another are much higher than during non-mating periods.	Picture 29

Picture 1   An individual stares another animal

Picture 2   The right adult male slaps toward his front individual

Picture 3   An adult male (middle) seizes another adult male (right)

Picture 4 An adult male bites another adult male (two monkeys in right)

Picture 5 An adult male (right) avoids adult male's (middle) aggression

Picture 6 An adult female (right) screams when she is attacked by adult male (left)

Picture 7 The middle adult male redirects aggression from right adult male to the left adult male

Picture 8 The middle adult male is getting the left male's support when he is in conflict with the right adult male

Picture 9 An adult male holds the button of a juvenile male (left)

J.-H. Li and P. M. Kappeler

Picture 10 The right adult male is leaving left adult male

Picture 11 An adult male (left) is in proximity with an adult female within 1 m

Picture 12 An adult female (left) is following an adult male

Picture 13   Teeth chatting

Picture 14   Three juveniles play together

Picture 15   Social grooming between two females

Picture 16 Embrace

Picture 17 An adult male presents to another male

Picture 18 Male–male mount

Picture 19 Show the penis

Picture 20 A juvenile (left) is sucking an infant's penis

Picture 21 An adult male holds an infant

Picture 22 Male–male bridging

Picture 23 Genital inspection

Picture 24 Sexual inspection

Picture 25 An adult male (left) chases an adult female (right) for mating

Picture 26 An adult male (right) grimace to an adult female (left)

Picture 27 An adult male (left) is mating with an adult female (right)

Picture 28   An adult male (left) is harassing a mating pair

Picture 29   An adult male (right) forms consortship with an adult female (left)

References

Altmann J (1974) Observational study of behavior: sampling methods. Behaviour 49:227–266

Berman C, Li JH (2002) Impact of translocation, provisioning and range restriction on a group of *Macaca thibetana*. Int J Primatol 23:283–397

Berman C, Ionica C, Li JH (2004) Dominance style among *Macaca thibetana* on Mt. Huangshan, China. Int J Primatol 25:1283–1312

Berman C, Ionica C, Dorner M, Li JH (2006) Postconflict affiliation between former opponents in *Macaca thibetana* on Mt. Huangshan, China. Int J Primatol 27:827–854

Deag J, Crook J (1971) Social behaviour and "agonistic buffering" in the wild Barbary macaque *Macaca sylvana* L. Folia Primatol 15:183–200

Delson E (1980) Fossil macaques, phyletic relationships and a scenario of development. In: Lindburg D (ed) The macaques: studies in ecology, behavior, and evolution. Van Nostrand Reinhold, New York, pp 31–51

Fairbanks L (1990) Reciprocal benefits of allomothering for female vervet monkeys. Anim Behav 9:425–441

Fooden J (1980) Classification and distribution of living macaques. In: Lindburg D (ed) The macaques: studies in ecology, behavior, and evolution. Van Nostrand Reinhold, New York, pp 1–9

Fooden J (1983) Taxonomy and evolution of the *sinica* group of macaques: 4. species account of *Macaca thibetana*. Fieldiana Zool 17:1–20

Hoelzer G, Hoelzer M, Melnick D (1992) The evolutionary history of the *sinica* group of macaque monkeys as revealed by mtDNA restriction site analysis. Mol Phylogenet Evol 1:215–222

Jiang XL, Wang YX, Wang QS (1996) Taxonomy and distribution of Tibetan macaque (*Macaca thibetana*). Zool Res 17:361–369

Kappeler P, Watts D (2012) Long-term field studies of primates. Springer, New York

Li JH (1999) The Tibetan macaque society: a field study. Anhui University Press, Hefei. (In Chinese)

Li JH, Wang QS (1994) Sleeping site of Tibetan macaques in summer. Chin J Zool 29(5): 58

Li JH, Wang QS (1996) Dominance hierarchy and its chronic change in adult male Tibetan macaques (*Macaca thibetana*). Acta Zool Sin 42:330–334

Li JH, Wang QS, Li M (1994) Studies on population ecology of Tibetan monkeys (*Macaca thibetana*). II. Reproductive pattern of Tibetan monkeys. Acta Therlol Sin 14(4):255–259

Li JH, Wang QS, Han DM (1996) Fission in a free-ranging Tibetan macaque group at Huangshan Mountain, China. Chin Sci Bull 41:1377–1381

Li JH, Yin HB, Wang QS (2005) Seasonality of reproduction and sexual activity in female Tibetan macaques (*Macaca thibetana*) at Huangshan, China. Acta Zool Sin 51:365–375

Li JH, Yin HB, Zhou LZ (2007) Non-reproductive copulation behavior among Tibetan macaques (*Macaca thibetana*) at Huangshan, China. Primates 48:64–72

Li DM, Fan LG, Ran JH, Yin HL, Wang HX, Wu SB, Yue BS (2008) Genetic diversity analysis of *Macaca thibetana* based on mitochondrial DNA control region sequences. DNA Seq 19 (5):446–452

Liu Y, Li JH, Zhao JY (2006) Sequence variation of mitochondrial DNA control region and population genetic diversity of Tibetan macaques *Macaca thibetana* in the Huangshan Mountain. Acta Zool Sin 52(4):724–730

Macintosh A, Huffman M, Nishiwaki K, Miyabe-Nishiwaki T (2012) Urological screening of a wild group of Japanese macaques (*Macaca fuscata yakui*): investigating trends in nutrition and health. Int J Primatol 33(2):460–478

Matheson M, Sheeran L, Li JH, Wagner R (2006) Tourist impact on Tibetan macaques. Anthrozoös 19:158–168

Matsumura S (1999) The evolution of "egalitarian" and "despotic" social systems among macaques. Primates 40:23–31

Ogawa H (1995a) Bridging behavior and other affiliative interactions among male Tibetan macaques (*Macaca thibetana*). Int J Primatol 16:707–729

Ogawa H (1995b) Recognition of social relationships in bridging behavior among Tibetan macaques (*Macaca thibetana*). Am J Primatol 35:305–310

Ogawa H (2019) Bridging behavior and male-infant interactions in *Macaca thibetana* and *M. assamensis*: insight into the evolution of social behavior in the sinica species-group of macaques. In: Li J-H, Sun L, Kappeler P (eds) The behavioral ecology of the Tibetan Macaque. Springer, Heidelberg

Pan RL, Jablonski N, Oxnard C, Freedman L (1998) Morphometric analysis of *Macaca arctoides* and *M. thibetana* in relation to Other Macaque species. Primates 39(4):519–537

Simons N, Lorenz J, Sheeran L, Li JH, Xia DP, Wagner R (2012) Noninvasive saliva collection for DNA analysis from free-ranging Tibetan macaques (*Macaca thibetana*). Am J Primatol 74:1064–1070

Smuts BB (1985) Sex and friendship in baboons. Aldine, Hawthorne, NY

Smuts B, Cheney D, Seyfarth R, Wrangham R, Struhsaker T (1987) Primate society. University of Chicago Press, Chicago, IL

Spielman D, Brook B, Frankham R (2004) Most species are not driven to extinction before genetic factors impact them. Proc Natl Acad Sci U S A 101:15261–15264

Stanford C (1992) The costs and benefits of allomothering in wild capped langurs (*Presbytis pileata*). Behav Ecol Sociobiol 30:29–34

Sun BH, Li JH, Zhu Y, Xia DP (2010) Mitochondrial DNA variation in Tibetan macaque (*Macaca thibetana*). Folia Zool 59:301–307

Sun L, Xia D-P, Li J-H (2019) Size matters in primate societies: how social mobility relates to social stability in Tibetan and Japanese macaques. In: Li J-H, Sun L, Kappeler P (eds) The behavioral ecology of the Tibetan macaque. Springer, Heidelberg

Thierry B (2000) Covariation of conflict management patterns in macaque societies. In: Aureli F, de Waal FBM (eds) Natural conflict resolution. University of California Press, Berkeley, CA, pp 106–128

Wada K, Xiong CP (1996) Population changes of Tibetan monkeys with special regard to birth interval. In: Shotake T, Wada K (eds) Variations in the Asian macaques. Tokai University Press, Tokyo, pp 133–145

Wada K, Xiong CP, Wang QS (1987) On the distribution of Tibetan and rhesus monkeys in southern Anhui Province, China. Acta Theiol Sin 7:148–176

Wang GL, Yin HB, Yu G, Wu M (2007) Time budget of adult Tibetan macaques in a day in spring. Chin J Wildlife 29:6–10

Waters S, Harrad A, Chetuan M, Amhaouch Z (2015) Barbary macaque group size and composition in Bouhachem forest, north Morocco. Afr Primates 10:53–56

Xia DP, Li JH, Garber P, Matheson M, Sun BH, Zhu Y (2013) Grooming reciprocity in male Tibetan macaques. Am J Primatol 75:1009–1020

Xiong CP (1984) Studies on the ecology of Tibetan monkeys. Acta Theriol Sin 4:1–9

Xiong CP, Wang QS (1988) Seasonal habitat used by Tibetan monkeys. Acta Theriol Sin 8:176–183

Xiong CP, Wang QS (1991) A comparative study on the male sexual behavior in Thibetan and Japanese monkeys. Acta Theriol Sin 11:13–22

Zhang M, Li JH, Zhu Y, Wang X, Wang S (2010) Male mate choice in Tibetan macaques *Macaca thibetana* at Mt. Huangshan, China. Curr Zool 56:213–221

Zhang D, Xia DP, Wang X, Zhang QX, Sun BH, Li JH (2018) Bridging may help young female Tibetan macaques *Macaca thibetana* learn to be a mother. Sci Rep 8:16102

Zhao QK (1993) Sexual behavior of Tibetan macaques at Mt. Emei, China. Primates 34(4):431–444

Zhao QK (1996) Male-infant-male interactions in Tibetan macaques. Primates 37(2):135–143

Zhao JY, Li JH, Liu Y, Yin HB (2005) Research on DNA extraction from old feces of *Macaca thibetana*. Acta Theriol Sin 25:410–413

Zhu Y, Li JH, Xia DP, Chen R, Sun BH (2008) Inbreeding avoidance by female Tibetan macaques *Macaca thibetana* at Huangshan, China. Acta Zool Sin 54:183–190

Chapter 3
Size Matters in Primate Societies: How Social Mobility Relates to Social Stability in Tibetan and Japanese Macaques

Lixing Sun, Dong-Po Xia, and Jin-Hua Li

3.1 Introduction

Social mobility refers to vertical movement of positional status of members in a society. In human societies, it refers to a wide range of upward or downward change in metrics such as income, social stature, education, and others that show some level of stratification (Lipset and Bendix 1992). As a focal issue in equality, social justice, economic development, and social stability in human societies (e.g., Wilkinson and Pickett 2009; Breen 2010; Cox 2012; Matthys 2012; Corak 2013; Clark 2014; Piketty 2014), social mobility has drawn a broad and sustained interest from social thinkers, social activists, and concerned citizens and is among the most important current research topics in social science disciplines including economics, sociology, and political science. Recently, interest in social mobility has also been on the rise in biological, psychological, and medical sciences. This is largely due to the discoveries that, along with other effects, social status can affect hormones (especially those

L. Sun (✉)
Department of Biological Sciences, Primate Behavior and Ecology Program, Central Washington University, Ellensburg, WA, USA
e-mail: Lixing@cwu.edu

D.-P. Xia
School of Life Sciences, Anhui University, Hefei, China

International Collaborative Research Center for Huangshan Biodiversity and Tibetan Macaque Behavioral Ecology, Anhui, China

J.-H. Li
School of Resources and Environmental Engineering, Anhui University, Hefei, Anhui, China

International Collaborative Research Center for Huangshan Biodiversity and Tibetan Macaque Behavioral Ecology, Anhui, China

School of Life Sciences, Hefei Normal University, Hefei, Anhui, China
e-mail: jhli@ahu.edu.cn

© The Author(s) 2020
J.-H. Li et al. (eds.), *The Behavioral Ecology of the Tibetan Macaque*, Fascinating Life Sciences, https://doi.org/10.1007/978-3-030-27920-2_3

related to stress such as glucocorticoids), immunity, and health in humans and nonhuman primates (e.g., Marmot et al. 1991; Sapolsky 2005; Seabrook and Avison 2012; Snyder-Mackler et al. 2016). As such, social mobility is among the key factors determining the well-being of primates and other social animals.

The level of social mobility is critically important for a society, especially from the long-term perspective. Both low and high social mobility are believed to weaken and disrupt social stability. A persistently low mobility will necessarily deprive low-ranking members of opportunities for social advancement, which in turn may increase the probability of revolts and revolutions. An extremely high mobility, on the other hand, will make a society constantly in flux. Therefore, a stable society is a dynamic one with the level of mobility bounded within a certain range, beyond which society may become chaotic or even collapse. Such a system dynamics view thereorized for human societies has gained some popularity recently (Acemoglu et al. 2018). New conception aside, empirical tests are hard to carry out for the obvious reason that human societies are usually too complex in structure and too diverse in culture to pursue such studies. In this respect, testing mobility-related hypotheses is more feasible in nonhuman primates because their societies are usually smaller in size and simpler in structure than human societies, including most hunter-gatherer societies (Price and Brown 1985; Hamilton et al. 2007). Furthermore, understanding how social mobility influences primate societies is interesting and important in its own right because it can give us a new tool to investigate the dynamics and evolution of primate societies.

(Note that social primates can live in groups of varying complexity in social structure, ranging from simple linear hierarchy in some macaque species (*Macaca* spp.), multilevel societies such as those in geladas (*Theropithecus gelada*) and snub-nosed monkeys (*Rhinopithecus* spp.), to highly organized human communities governed by formal or informal norms and laws. Because social mobility is a feature of an operationally definable social unit or organization regardless of the level of its organizational complexity, we here use the word "society" synonymously with "social group" for both human and nonhuman primates for the purpose of finding general and comparable patterns in social mobility.)

There are two types of social mobilities: *intragenerational* and *intergenerational*. Intragenerational mobility refers to upward and downward movement of social stature for members in a society within their lifetimes. For instance, in a Tibetan macaque society, each member has a 3-to-1 ratio of upward versus downward mobility in a year (Sun et al. 2017). Because status change tends to accumulate over time, intragenerational mobility is often measured per time unit, such as a month, a year, or a decade, to be comparable across studies. For this, it is more accurately known as the rate of social mobility (Clark 2014). Intergenerational mobility, on the other hand, refers to the status change between two consecutive generations. In American society, for instance, income mobility declined from 92% for people born in 1940 to 50% for people born in 1985 (Chetty et al. 2017). In this chapter, we will focus on intragenerational mobility.

Social mobility can be *absolute* or *relative*. In social sciences, absolute mobility refers to the total number of individuals, whereas relative mobility refers to the probability of an individual (or percentage of all individuals) moving from one social stratum to another. A society can be high in absolute mobility but low in relative mobility or vice versa. The key behind this paradox lies in the size of a society. For example, in terms of absolute mobility, two members moving up one step in rank in a group of 30 are twice as mobile as one member for the same rank change in a group of 15 during the same period of time, but the two groups are equally mobile in relative mobility, both 6.67% for this period of time. As this hypothetical example illustrates, absolute mobility does not consider the size of a society. Thus, relative mobility can better reflect the overall condition for a society and is most commonly used in characterizing upward or downward mobility (Heckman and Mosso 2014; Simandan 2018). Accordingly, social mobility, unless specified, refers to relative mobility hereafter.

The difference between absolute and relative mobility indicates that society size can play an important role in social mobility. Unfortunately, few attempts have been made to address this issue due to the reason that most mobility research has been conducted on large human societies in industrial nations. As such, society size is usually not of great interest because it has no perceivable effect on social mobility. In nonhuman primate societies, however, group size can be magnitudes smaller, typically a few dozen (up to a few hundred in some species such as baboons and snub-nosed monkeys). A group of a few dozen, such as in Tibetan macaques, *M. thibetana* (Li 1999), can be vastly different in social organization from a group of a few hundred, such as in Sichuan snub-nosed monkeys, *R. roxellana* (Zhang et al. 2006) or Japanese macaques, *M. fuscata* (Mori et al. 1989). Therefore, to understand how group size affects mobility, we compared two macaque species, the Tibetan macaque and the Japanese macaque, in this study.

Before we present our results, we briefly review some key issues related to the study of social mobility.

3.2   Social Mobility and Opportunity

In the study of animal social behavior, proximate benefits and costs for a set of alternative behaviors (such as mating versus feeding or fighting) are often estimated in terms of time and energy. However, although researchers are keenly aware of the benefit and cost in terms of opportunity intrinsic to every behavioral decision (e.g., when an animal hides in a den, it loses the opportunity to feed or mate), quantitative measurement has not been attempted due to practical difficulties in empirical studies. Measuring social mobility may provide just such a way to gauge gain or loss in opportunity for social advancement for social animals such as primates. Because mobility is an aggregate measure for a society, it can provide a baseline (the average or the expected) for opportunity in social advancement for all members in a social group (Sun et al. 2017). With this baseline as a benchmark, the opportunity consequences of many behaviors, biological identities, and social relationships of individuals (such as dispersal, conflict, cooperation, kinship, age, and sex) can be quantitatively assessed in terms of gains or losses in social opportunity relative to this benchmark. In other words, these behaviors, identities, or relationships can alter an individual's expected upward and downward mobility; that is, the benchmark. For instance, a monkey deciding to transfer from Group A to Group B must accept a complete loss in opportunity for social advancement in Group A, in exchange for a gain in opportunity in Group B. Whether this dispersal event is adaptive depends on the difference between the two opportunity measurements.

3.3 Social Mobility and Social Stability

In his classic, *Democracy in America*, the French political scientist Alexis de Tocqueville first noticed that mobility has a profound effect on the stability of human societies (de Tocqueville 1835/2002). Social scientists have since developed several explanations for why mobility is essential for stability (e.g., Lipset and Bendix 1992; Erikson and Goldthorpe 1994; Chetty et al. 2017). Studies in developed nations show that high mobility is closely related to equality and stability, and low mobility engenders inequality, which may cause a spectrum of social problems detrimental to society in general and democracy in particular (e.g., Wilkinson and Pickett 2009). It is a consensus now that a certain level of mobility is essential for the persistence of a society in the long run. Such a broadly held belief, however, hinges on the veracity of the hypothesis that mobility leads to stability. Unfortunately, this hypothesis, though logically compelling, is still short of empirical evidence (Bai and Jia 2016).

In reality, the relationship between mobility and stability is profound. On the one hand, an absolutely stable society without mobility can deprive its low-ranking members of opportunities to advance their social status, resulting ultimately in a loss in their fitness relative to that of high-ranking members (Sun 2013; Sun et al. 2017). As such, low-ranking members in a stagnant society will have a strong incentive to resort to revolt against the existing order so as to gain critical resources (such as food and mates) that are otherwise inaccessible for them. Therefore, an extremely unequal society with little mobility is evolutionarily unstable (Sun 2013). It tends to periodically experience major disruptions in the form of rebellion and revolution in humans (Bai and Jia 2016). It is in this sense that a certain level of mobility is essential for long-term social stability (Sun et al. 2017). On the other hand, mobility is the opposite of stability. That is, a higher level of mobility will necessarily decrease stability. This implies that too much mobility may also destabilize a society, which, beyond a certain point, may cascade into social upheaval (e.g., Clark 2014). However, this hypothesis, despite a broad theoretical interest, remains untested as well.

3.4 Measuring Social Mobility in Primate Societies

As mentioned earlier (Sect. 3.1), intragenerational mobility is normally calculated as the rate of status change in society within a specified time period (e.g., Lammam et al. 2012; Clark 2014). Technically, it can be measured either as the mean value in the rate of rank change or using a linear regression analysis between the ranks of all individuals in two consecutive time periods, t and $t + 1$. For the latter method, the rate of social mobility is thus $1-b$, where b is the slope of the regression line ranging from 0 to 1 (Clark 2014; Sun et al. 2017). If b is 0, there is no mobility and, in our case, no rank change between two consecutive years for all adults involved. If b is 1, there is no stability. In primates, social mobility has been measured only in Tibetan macaques, where the rate of rank mobility was calculated for all adults on a yearly basis (Sun et al. 2017).

(Note that this linear regression method belongs to a statistical analysis known as time series analysis. Although time series analysis is relatively unfamiliar to researchers in biological sciences, it is commonly used in economic and business analysis such as GDP growth from year to year and stock performance from quarter to quarter. By emphasizing changes over time, the influence of policies or the efficacy of management regimes behind these changes can be statistically analyzed. As we can see, one distinct strength of time series analysis in addressing behavioral, ecological, and evolutionary issues lies in that it allows researchers to statistically track changes over time, which in turn may lead to the unveiling of the underlying factors that have significant influences on the changes. As such, time series regression is a standard method in the study of social mobility in social sciences.)

To complicate the issue, however, dominance rank can be measured as absolute rank or relative rank. Absolute rank refers to the raw rank data regardless of group size. Relative rank, however, is standardized according to group size. (Note: absolute and relative rank is a way of measuring dominance rank for all members in a group. They have no relationship to, and should not be mistaken as, absolute and relative mobility introduced in Sect. 3.1.) As we have already discussed, the distinction is not meaningful (or even feasible to measure) when a society is very large and the hierarchy is highly nonlinear, such as an industrial society, which tends to show complex forms of strata. However, the distinction becomes important when a society is made of only a few dozen individuals typical of many macaques species. For example, we can see a huge difference between the fifth ranked individual in a group of ten versus the 25th ranked individual in a group of 50 when measured in absolute rank. However, when measured in relative rank, these two individuals are ranked the same, both in the middle of the hierarchies. Clearly, measurement of social mobility will be affected by whether we use absolute or relative rank. Yet, little is known as to how using relative or absolute rank will affect mobility measurements.

In this study, we probed into the critical issue of how group size affects intragenerational mobility by using both absolute and relative rank in Tibetan and Japanese macaques. For Tibetan macaques, data about the hierarchical relationships among all adult females come from our 29-year consecutive observation on one group (YA1) at Mt. Huangshan, Anhui, China. Details about the dataset and how the data were collected are available in Sun et al. (2017). For the Japanese macaque, we obtained data from a published paper, where 11 years of hierarchical relationships among adult females were documented in a group on Koshima Island from 1957 to 1986 (Mori et al. 1989).

Age is well known to affect dominant status in primates (e.g., de Vries 1998; Packer et al. 2000; Alberts et al. 2003; Bayly et al. 2006; Balasubramaniam et al. 2013). Thus, comparing individuals of the same age class is critically important. However, because male and female Tibetan macaques sexually mature at different ages (seven for males and five for females, see Li 1999), we only included adult females that were 7 years or older in our analysis for two reasons. First, it allowed us to compare male and female adults of the same age class in our discussion here and in other places (Sun et al. 2017). Second, the ranks of immature females from the same mother are inversed. That is, the later born are higher in dominance rank than their elder sisters (Li 1999). Therefore, the inclusion of the 5- and 6-year-old females may introduce an unnecessary confounding factor, which may in turn reduce the accuracy in measuring social mobility for adults. The data about rank and rank change in Japanese macaques were collected from adult females of 5 years or older (Mori et al. 1989). Although the research in the Japanese macaque spanned 29 consecutive years, only 11 years of rank data were published.

For the study groups of both species, we used the original rank data for all adult females as their absolute ranks. We then calculated the relative ranks by scaling all absolute rank data into values between 0 (alpha female) and 1 (omega female) using a simple conversion formula:

$$RR = (AR - 1)/(n - 1)$$

where RR and AR refer to relative and absolute rank, respectively, and n is the size of the group. We then measured mobility using the same time series regression method as that detailed in Sun et al. (2017); that is, by using linear regression between rank R (t) in year t and rank $R(t + 1)$ in year $t + 1$ to measure the annual rate of mobility for all adult group members (see Clark 2014) for both Tibetan and Japanese macaques. (Note that sample sizes about yearly rank changes were calculated as individual-years, rather than individuals in most statistics used in behavioral studies. In other words, each independent data point or sampling unit here is individual-year. This is a standard practice in time series analysis.) In our study, none of the datasets used for measuring mobility violated the normality assumption required for linear regression based on Kolmogorov-Smirnov test. As we have discussed in an earlier section, social mobility can be used to quantify the availability of social opportunities (e.g., Breen 2010; Clark 2014; Sun et al. 2017), which then can be used to quantify the benefit and cost of a behavior in terms of gain or loss in opportunity for social advancement. Here, we calculated net social mobility (subtracting the downward mobility from the upward mobility) and then used it to compare the difference between adult females of the two species. For the ease of reading, the statistical methods used are provided as they occur in the next section. As usual, all tests were two-tailed with 0.05 as the a priori level of significance for any statistical difference.

3.5 Results: Social Mobility in Tibetan and Japanese Macaques

From the adult female perspective (which may also be true if all individuals are included, if the number of adults, subadults, and infants is scaled in roughly similar proportions), the group size of Japanese macaques (33.09 ± 7.968SD; range, 21–49; $n = 11$) was more than three times larger than that of Tibetan macaques (7.62 ± 2.624 SD; range, 3–12; $n = 29$; Mann-Whitney U-test, $U = 385$; $n_1 = 11$; $n_2 = 29$; $p = 0.000$; see Fig. 3.1).

Using absolute rank, we found that group members experienced a larger rate of rank change in Japanese macaques than in Tibetan macaques. This was adequately reflected in both the mean ($U = 40{,}821$, $n_1 = 192$, $n_2 = 204$, $p = 0.013$) and variation (Levene test, $L = 65.07$, $df_1 = 1$, $df_2 = 394$, $p = 0.000$, Fig. 3.2) of annual rank change for Japanese macaques. Also, Japanese macaques had a higher rate in both upward (Chi-square test with Yates' correction, $\chi^2 = 13.187$, $df = 1$, $p < 0.001$) and downward movement ($\chi^2 = 18.705$, $df = 1$, $p = 0.000$) in hierarchy and a lower probability of staying in the same rank ($\chi^2 = 37.436$, $df = 1$, $p = 0.000$) than Tibetan macaques (Fig. 3.3). However, when we compared mobility between the two species using the regression method (rate of mobility, 0.115 for Japanese macaques and 0.144 for Tibetan macaques), the difference was not statistically significant (*t*-test for the slope, $t = 0.658$, $df = 392$, $p = 0.510$, Fig. 3.4).

For Tibetan macaques, mobility was higher using relative rank than using absolute rank, although the difference was marginally insignificant ($t = 1.706$, $df = 380$, $p = 0.089$). For Japanese macaques, the situation was the opposite. That is, mobility

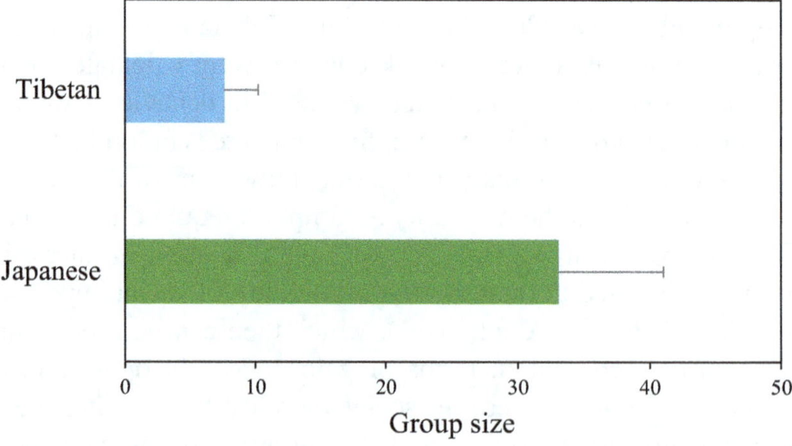

Fig. 3.1 Comparison of the mean group size (only adult females) in Tibetan and Japanese macaques. Error bars are standard deviations

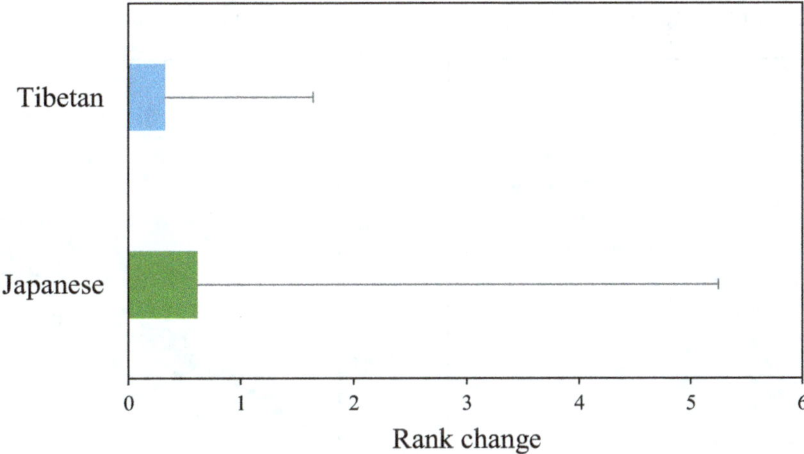

Fig. 3.2 Mean annual rate of absolute rank change in Tibetan and Japanese macaques. Error bars are standard deviations

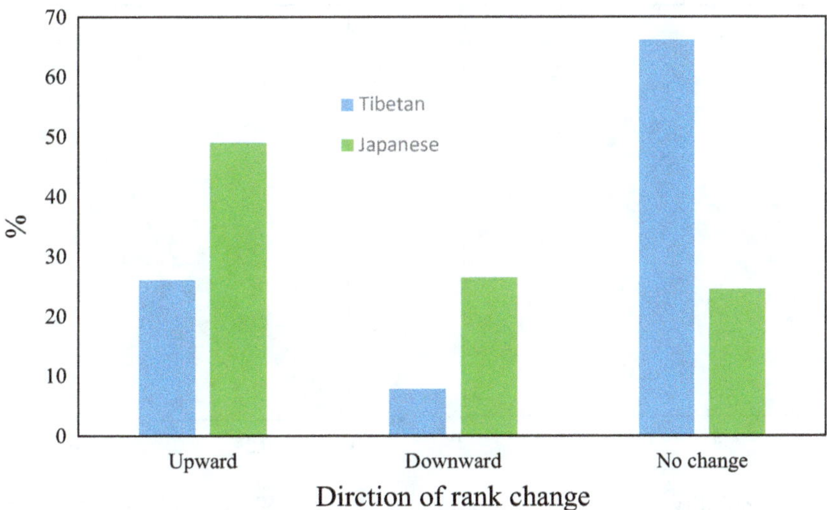

Fig. 3.3 Comparison of the frequency (%) of annual rank change (upward, downward, and no change) between Tibetan and Japanese macaques

was significantly lower when relative rank was used than when absolute rank was used ($t = 2.291$, $df = 404$, $p = 0.037$). Between-species comparison shows that there was no significant difference in absolute rank mobility, but a highly significant difference in relative rank mobility ($t = 3.901$, $df = 392$, $p = 0.0001$; see Fig. 3.5).

3.6   Discussion

This chapter presents a case study of two macaque species to illustrate how social mobility can be used as a new approach to the study of social dynamics as well as an in-depth analysis to explore the effect of group size. Our results show that social

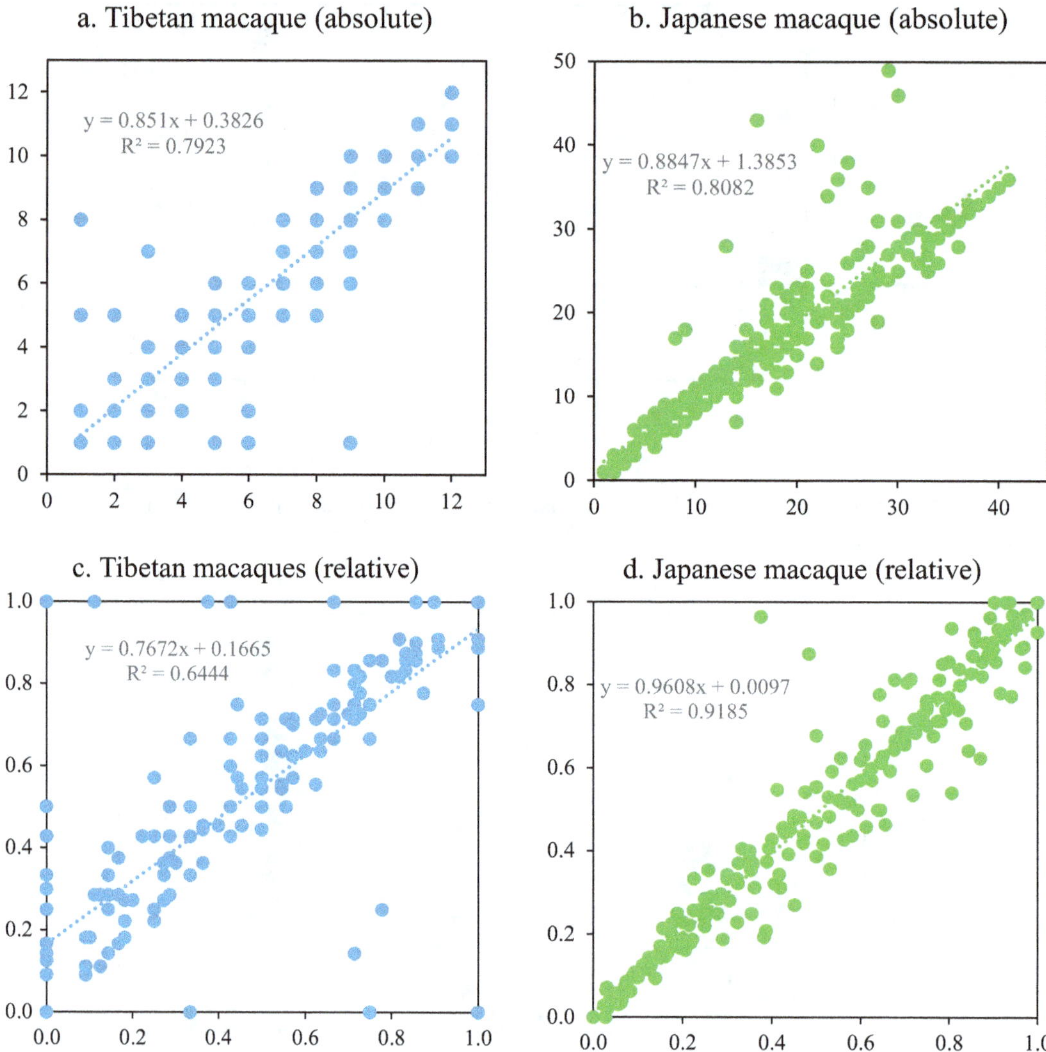

Fig. 3.4 Social mobility as measured by absolute ("absolute") and relative ("relative") rank in Tibetan and Japanese macaques using time series regression analysis. The X-axis indicates rank in year t, and the Y-axis indicates rank in the next year, year $t + 1$, for all adult individuals used in the analysis. Note that fewer data points than sample size shown in Panel (**a**) is due to data overlap

mobility can unveil new and unexpected insights into the social structure of primate societies. Our analyses show that, although Japanese macaques appeared to be more dynamic in rank changes, as shown by the higher mean and larger variation than Tibetan macaques, social mobility as measured in absolute rank was virtually the same for the two species. This result was surprising considering that Tibetan macaques and Japanese macaques are two different species with marked difference in so many aspects of their behavior and social organization (see Thierry 2011), despite some shared sociological traits such as matrilineality in social organization and strong despotism (see below).

One factor we particularly examined was group size, which was much larger in Japanese macaques than in Tibetan macaques. (Note that this difference may be

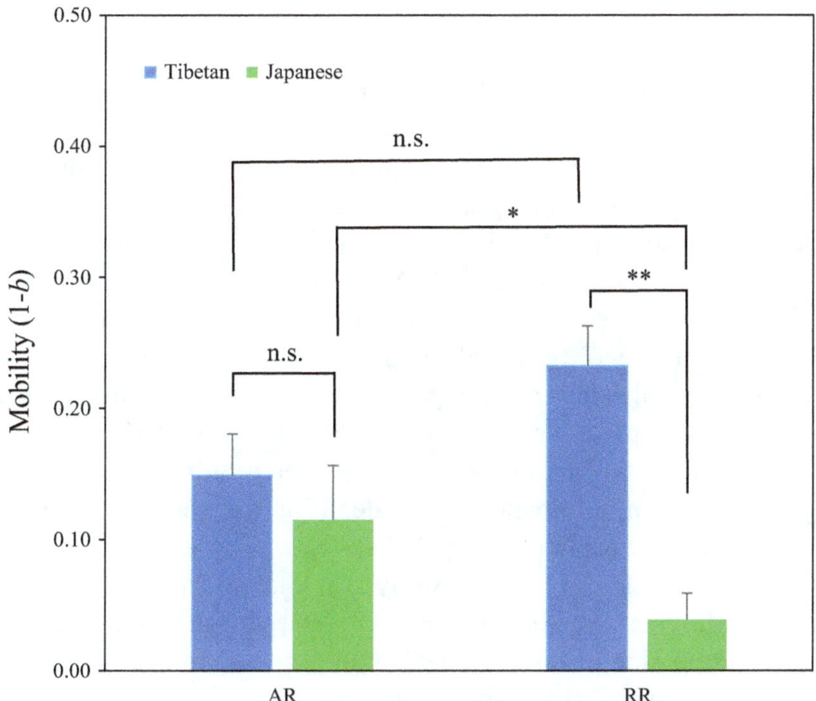

Fig. 3.5 Within- and between-species comparisons of social mobility using absolute ("AR") and relative ("RR") rank for the Tibetan and Japanese macaque. "n.s." stands for not significant, "asterisk" indicates $p < 0.05$, and "double asterisk" represents $p < 0.001$

slightly less prominent had only females of 7 years or older been included in the analysis for Japanese macaques.) This apparent discrepancy between the results from measuring the rate of rank change and from time series regression analysis might be due to two reasons. One is that, although the Japanese macaque troop studied had a larger rate of rank change, much of the significance was offset by a higher level of variation (Fig. 3.2). Also, using a nonparametric test for a non-normal dataset (Mann-Whitney U test here) might also lead to a slightly different statistical result from a parametric linear regression analysis.

Comparing the two species, we found that Tibetan macaque society appeared much more stable than that of the Japanese macaque in absolute rank change, despite no difference in mobility between the two species. Indeed, Tibetan macaques live in a small, simple, matrilineally structured group of strictly linear hierarchy (Li 1999; Berman et al. 2008) and their dominance hierarchy usually remains stable for a prolonged period of time, at least several months (Li 1999). (These behavioral conditions are particularly favorable for observing and determining social status for every adult in the hierarchy with little ambiguity. That is why Tibetan macaques are a highly desirable model species in the evolutionary study of behavior, a point compellingly addressed in the previous two chapters). Even so, social mobility in Tibetan macaques turned out to be similar to that in Japanese macaque. It appears that a larger group (society) may require a higher level of rank change to maintain the same level of mobility as a smaller group (society). Conversely, with the same level of mobility, a larger group would be more dynamic in terms of rank change. From the dynamic stability point of view, therefore, a larger society may require more rank change than a smaller society to maintain its stability in the long run, if the amount of mobility for long-term social stability is assumed to be approximately the same for all societies.

Another major reason for why the two species had similar levels of mobility may lie in the fact that they both form highly despotic societies; Grade 1 (highly despotic) for the Japanese macaque and Grade 2 (despotic) for the Tibetan macaque (Thierry 2000, 2011; Berman et al. 2004, 2006). Apparently, the two species we analyzed here may not be different enough in dominance style to show marked difference in social mobility. In other words, social mobility is likely to be related to dominance style. One logical hypothesis is that despotism may suppress social mobility. If true, this hypothesis would predict that a more relaxed or egalitarian primate society should have a higher level of mobility. This, however, requires a broader comparison with a larger diversity in dominance style, ideally in the genus *Macaca*, which has been well studied for the topic (Thierry 2011).

The most surprising finding in our study was that social mobility could vary a great deal depending on whether absolute or relative rank was used. In Tibetan macaques, the group size of adult females could be so small that rank data, once standardized to be within the range between 0 and 1, appear erratic (see in Fig 3.4c). Since relative rank can change with group size, even though there is no change in absolute rank, small changes can become prominent, which may in turn lead to increment in mobility. Obviously, the smaller the group is, the higher the mobility will become when relative rank is used. In Japanese macaques, on the contrary, larger rank changes (with the extreme records of 9 ranks upward and 27 ranks downward in a year in Japanese macaques compared with up 8 and down 7 in Tibetan macaques) became smaller after these changes were scaled to be within the range between 0 and 1. Clearly, this effect increases with group size. As a result, mobility decreases with increasing group size. These changes led to the counterintuitive result that, despite Japanese macaques appearing to be more dynamic in terms of absolute rank change, their society actually had a lower level of mobility than Tibetan macaques when relative rank was used.

This part of our results leads to an unexpected insight into social evolution. That is, group size may have a major impact on group dynamics. In both macaque species we examined, the net mobility was positive for social advancement. This may be true for most primate societies, in which social rank is more or less related to seniority and/or tenure. Consequently, as higher-ranking individuals become senile or die, younger, lower-ranking individuals can move up over time. So, even without major disruptive events, net mobility tends to be positive (upward). Overall, if everything else is equal, individuals in general should prefer smaller groups to larger groups, not only because absolute ranks go up (as the length of hierarchy becomes shorter) for most members but also because mobility goes up as well in terms of relative rank. Therefore, individuals in a larger society may have more incentive to break off to form smaller societies. This interesting and intriguing difference in mobility when measured in absolute versus relative rank may provide a novel explanation for why group fission is so common whereas group fusion is so rare in primates. It also sheds new light on the balkanization of human societies, happening in so many nations under the name of regional autonomy and sovereignty rather than the other way around. Although it is too early to claim that such events in human and nonhuman primate societies share biological roots, their uncanny similarities should warrant a close and serious examination on such possibilities from the evolutionary perspective.

Acknowledgments We are grateful to Mike Huffman, Peter Kappeler, and an anonymous reviewer for their helpful comments and constructive suggestions, which led to a considerable improvement in the clarity of our chapter. We also thank the Huangshan Garden Forest Bureau for their permission and support of this work and H.B. Cheng's family for their outstanding logistic support to our study. Rose Amrhein proofread the chapter. This work was supported in part by grants from the National Natural Science Foundation of China (No. 31772475; 31672307).

References

Acemoglu D, Egorov G, Sonin K (2018) Social mobility and stability of democracy: re-evaluating de Tocqueville. Q J Econ 133:1041–1105

Alberts SC, Watts HE, Altmann J (2003) Queuing and queue jumping: long term patterns of dominance rank and mating success in male savannah baboons. Anim Behav 65:821–840

Bai Y, Jia R (2016) Elite recruitment and political stability: the impact of the abolition of China's civil service exam system. Econometrica 84:677–733

Balasubramaniam KN, Berman CM, De Marco A et al (2013) Consistency of dominance rank order: a comparison of David's scores with I&SI and Bayesian methods in macaques. Am J Primatol 75:959–971

Bayly KL, Evans CS, Taylor A (2006) Measuring social structure: a comparison of eight dominance indices. Behav Processes 73:1–12

Berman CM, Ionica CS, Li J (2004) Dominance style among *Macaca thibetana* on Mt. Huangshan, China. Int J Primatol 25:1283–1312

Berman CM, Ionica CS, Dorner M et al (2006) Postconflict affiliation between former opponents in *Macaca thibetana* on Mt. Huangshan, China. Int J Primatol 27:827–854

Berman CM, Ogawa H, Ionica CS (2008) Variation in kin bias over time in a group of Tibetan macaques at Huangshan, China: contest competition, time constraints or risk response? Behaviour 145:863–896

Breen R (2010) Social mobility and equality of opportunity Geary lecture spring 2010. Econ Soc Rev 41:413–428

Chetty R, Grusky D, Hell M et al (2017) The fading American dream: trends in absolute income mobility since 1940. Science 356:398–406

Clark G (2014) The son also rises: surnames and the history of social mobility. Princeton University Press, Princeton NJ

Corak M (2013) Income inequality, equality of opportunity, and intergenerational mobility. J Econ Perspect 27:79–102

Cox MW (2012) Myths of rich and poor: why we're better off than we think. Basic Books, New York

De Tocqueville A (2002) Democracy in America. Regnery Publishing, Washington, DC

De Vries H (1998) Finding a dominance order most consistent with linear hierarchy: a new procedure and review. Anim Behav 55:827–843

Erikson R, Goldthorpe JH (1994) The constant flux: a study of class mobility in industrial societies. Clarendon Press, Oxford

Hamilton MJ, Milne BT, Walker RS et al (2007) The complex structure of hunter-gatherer social networks. Proc Biol Sci 274:2195–2202

Heckman JJ, Mosso S (2014) The economics of human development and social mobility. Annu Rev Econ 6:689–733

Lammam C, Karabegović A, Veldhuis N (2012) Measuring income mobility in Canada. Fraser Institute, Social Science Electronic Press, Vancouver, BC

Li J (1999) The Tibetan macaque society: a field study. Anhui University Press, Hefei

Lipset SM, Bendix R (1992) Social mobility in industrial society. Transaction Publishers, Brunswick

Marmot MG, Smith GD, Stansfeld S et al (1991) Health inequalities among British civil servants: the Whitehall II study. Lancet 337:1387–1393

Matthys M (2012) Cultural capital, identity, and social mobility. Routledge, New York

Mori A, Watanabe K, Yamaguchi N (1989) Longitudinal changes of dominance rank among the females of the Koshima group of Japanese monkeys. Primates 30:147–173

Packer C, Collins DA, Eberly LE (2000) Problems with primate sex ratios. Philos Trans R Soc Lond B Biol Sci 355:1627–1635

Piketty T (2014) Capital in the 21st century. Belknap, Cambridge, MA

Price TD, Brown JA (1985) Aspects of hunter-gatherer complexity. In: Price TD, Brown JA (eds) Prehistoric hunter-gatherers: the emergence of cultural complexity. Academic, Orlando, FL, pp 3–20

Sapolsky RM (2005) The influence of social hierarchy on primate health. Science 308:648–652

Seabrook JA, Avison WR (2012) Socioeconomic status and cumulative disadvantage processes across the life course: implications for health outcomes. Can Rev Sociol 49:50–68

Simandan D (2018) Rethinking the health consequences of social class and social mobility. Soc Sci Med 200:258–261

Snyder-Mackler N, Sanz J, Kohn JN et al (2016) Social status alters immune regulation and response to infection. Science 354:1041–1045

Sun L (2013) The fairness instinct: the Robin Hood mentality and our biological nature. Prometheus Books, Amherst, NY

Sun L et al (2017) The prospect of rising in rank is key to long-term stability in Tibetan macaque society. Sci Rep 7(1):7082

Thierry B (2000) Covariation of conflict management patterns across macaque species. In: Aureli F, de Waal FBM (eds) Natural conflict resolution. University of California Press, Berkeley, CA, pp 827–843

Thierry B (2011) The macaques: a double-layered social organization. In: Campbell CJ, Fuentes A, McKinnon KC et al (eds) Primates in perspective. Oxford University Press, New York, pp 229–241

Wilkinson R, Pickett K (2009) The spirit level: why greater equality makes societies stronger. Bloomsbury Press, New York

Zhang P, Wanatabe K, Li B et al (2006) Social organization of Sichuan snub-nosed monkeys (*Rhinopithecus roxellana*) in the Qinling Mountains, Central China. Primates 47:374–382

Chapter 4
Behavioral Exchange and Interchange as Strategies to Facilitate Social Relationships in Tibetan Macaques

Dong-Po Xia, Paul A. Garber, Cédric Sueur, and Jin-Hua Li

4.1 Introduction

Since the 1970s, sociobiologists have focused on two issues when studying social animals. One is the evolutionary processes that have favored group living, and the other is how ecological factors influence the internal structure of groups and the nature of social relationships and how changes in behavior influence the costs and benefits of group living to individuals of different age, sex, and dominance status (Wilson 1975; Barash 1977; Morse 1980; Wrangham 1980; Krebs and Davies 1984; van Schaik 1989). Studies on social animals have argued that the benefits of group living (such as social learning, cooperative hunting, collective resource and mate defense, cooperative infant caregiving, increased foraging efficiency, a reduction in stress associated with social support and reconciliatory behavior, alliance formation,

D.-P. Xia
School of Life Sciences, Anhui University, Hefei, China

International Collaborative Research Center for Huangshan Biodiversity and Tibetan Macaque Behavioral Ecology, Anhui, China

P. A. Garber
Department of Anthropology, Program in Ecology, Evolution, and Conservation Biology, University of Illinois, Urbana, IL, USA
e-mail: p-garber@illinois.edu

C. Sueur
CNRS, IPHC, UMR, Université de Strasbourg, Strasbourg, France
e-mail: cedric.sueur@iphc.cnrs.fr

J.-H. Li (✉)
School of Resources and Environmental Engineering, Anhui University, Hefei, Anhui, China

International Collaborative Research Center for Huangshan Biodiversity and Tibetan Macaque Behavioral Ecology, Anhui, China

School of Life Sciences, Hefei Normal University, Hefei, Anhui, China
e-mail: jhli@ahu.edu.cn

J.-H. Li et al. (eds.), *The Behavioral Ecology of the Tibetan Macaque*, Fascinating Life Sciences, https://doi.org/10.1007/978-3-030-27920-2_4

and partner reliability) serve to counteract the costs and offer fitness benefits to individuals living in a well-functioning and stable social unit (e.g., multilevel selection, Krause and Ruxton 2002; Sussman and Garber 2011).

According to Hinde (1976), social structures are group characteristics based on the patterns of social relationships such as affiliative, cooperative, sexual, and agonistic interactions that characterize a species or members of an established social unit. In order to maintain group cohesion, individuals are required to coordinate their daily activities, including collective movement to feeding and resting sites. Understanding how this is accomplished, whether individuals of a particular age, sex, or rank class consistently lead or direct group movement, and how "leaders" build a consensus in deciding when to leave a feeding or resting site and which location to travel to next, offers critical insights into the dynamics of animal social systems (see also Chap. 5).

A recent paper by Schino and Aureli (2017) describes two alternative processes associated with social relationships. Partner control highlights dyadic interactions between dyads in which the behavior of each subject is dependent exclusively on the previous behavior of a partner (Noë 2006). This assumes that one member of a dyad can control the behavior of its partner with no opportunity of partner switching. Partner control has been used to describe social interactions between individuals living in a social unit. Partner choice assumes that individuals can freely select and change partner preferences based on the assessed benefit each potential partner is expected to provide. For example, in studies of *Sapajus nigritus* (formerly *Cebus apella*) and *Macaca fuscata* (Schino and Aureli 2009, 2017; Schino et al. 2007, 2009), partner choice was found to influence social relationships and group stability.

Biological market theory proposes that social behaviors can be considered as valuable commodities and explains how commodities are exchanged or interchanged among group members (Noë and Hammerstein 1994, 1995; Noë 2006). This theory considers social behaviors as exchangeable commodities to obtain behavioral services as benefits (Noë 2001; Barrett and Henzi 2006; Noë & Voelkl 2013). Individuals compete over access to partners, by offering more valuable services rather than engaging in aggressive or coercive behavior. Therefore, individuals are chosen as partners depending on the services they offer, and the choice is made by comparing the offers of all potential partners depending on the services needed. As such, biological market theory provides an alternative framework to study social relationships and partner preferences and facilitates the understanding of behavioral adaptations for group stability and social cohesiveness in group-living animals.

In most nonhuman primates, behavioral exchange and interchange have been well documented (reviewed in Sánchez-Amaro and Amici 2015). For example, Seyfarth (1977) proposed that lower-ranking females compete to groom higher-ranking females who can offer them more effective aid (agonistic support) in competitive interactions with other group members. In some species such as chacma baboons, *Papio cynocephalus* (Barrett et al. 1999); red-fronted lemurs, *Eulemur fulvus rufus* (Port et al. 2009); and chimpanzees, *Pan troglodytes* (Newton-Fisher and Lee 2013; Kaburu and Newton-Fisher 2015a), grooming is reported to be exchanged for grooming (reciprocal exchange). Within a dyad consisting of individuals A and B, the amount of grooming given by A to B depends on the amount of grooming given by B to A. In some other species, studies show that behavioral services can also be interchanged for different behaviors. For example, grooming can be interchanged for food sharing to maintain social bonds and agonistic support in chimpanzees (de Waal 1997; Kaburu and Newton-Fisher 2015a). In sooty mangabeys (*Cercocebus atys*) and vervet monkeys (*Chlorocebus aethiops*), infant handling can influence dyadic relationships. In this species, grooming is interchanged for opportunities to handle infants in order to establish social relationships (Fruteau et al. 2011). In long-tailed macaques (*M. fascicularis*), grooming is interchanged for tolerance to increase spatial proximity and to reduce aggression from higher-ranking individuals (Gumert and Ho 2008). Additionally, grooming can be interchanged for copulations (Kaburu and Newton-Fisher 2015b). Mating opportunities increase for females who groom a higher-ranking male compared to females who groom a lower-ranking male (Gumert 2007a). Similarly, Gomes and Boesch (2009) found that over a 22-month period, female chimpanzees copulated more with males who shared meat with them compared to males who did not engage in food sharing. In this species, food sharing by adult males provides more mating opportunities.

In this chapter, our goal is to review the evidence for multiple behavioral exchanges and interchanges in Tibetan macaque (see Chap. 2 for more information about Tibetan macaques) groups at Mt. Huangshan, China (Fig. 4.1), and to analyze these dyadic behavioral interactions within the context of behavioral exchange or interchange. We define exchange as the reciprocation of the same behaviors (e.g., grooming for grooming) and interchange as the reciprocation between different

Fig. 4.1 A social group (known as YA1) of Tibetan macaques at Mt. Huangshan, China

behaviors of the same category (e.g., grooming for tolerance, which are different behaviors but both belong to the same category of affiliative relationships) or behaviors of different categories (e.g., agonistic support for copulation). We focus on grooming exchanges, the interchange of grooming for tolerance among female or male intrasexual dyads, the interchange of grooming for infant-handling opportunities among females, and the interchange of male-to-female agonistic support for mating opportunities among intersexual dyads.

4.2 Exchange Between the Same Behaviors

4.2.1 Grooming for Itself

Grooming has been reported to be exchanged for grooming (reciprocal exchange) in primates (such as chacma baboons, Barrett et al. 1999). In Tibetan macaques, similarly, approximately 20% of their daily activity budget is devoted to grooming, regardless of sex, social rank, age, and other factors (see Xia et al. 2012, 2013) (Fig. 4.2 shows grooming in Tibetan macaques). Xia et al. (2012, 2013) examined female and male intrasexual grooming relationships in free-ranging Tibetan macaques and hypothesized that among a broad set of fitness-maximizing strategies, grooming can be used by individuals to enhance social relationships through reciprocity and/or through the interchange of grooming for a different but equivalent service. Overall, they found that social rank played an important role in defining female and/or male social interaction and partner choice in Tibetan macaques. Among female or male intrasexual dyads, the authors found that there

Fig. 4.2 Female-to-female social grooming in Tibetan macaques

were positive correlations between the average frequency of grooming bouts received (A groom B) and reciprocated (B groom A). These patterns were consistent during the mating and non-mating seasons. Similarly, grooming effort or the time the initiator invested in grooming the recipient was positively correlated with grooming received among female and male intrasexual dyads. Furthermore, the longer the initiator groomed her partner (A groom B), the longer her partner was likely to groom her in return (B groom A). This pattern also was consistent during mating and non-mating periods.

In addition, as indicated in Fig. 4.3, in intrasexual dyads composed of individuals of equal ranks, grooming was reciprocally exchanged (equal rates and equal duration) during both the mating and non-mating seasons. This suggests that, within each sex, dyads composed of animals of equal ranks use reciprocal exchange grooming as a social tool to maintain long-term alliances and partner preferences. Grooming frequency and grooming duration were equal within female–female and male–male dyads consisting of high- and high-ranking individuals, middle- and middle-ranking individuals, and low- and low-ranking individuals during both mating and non-mating periods.

The findings that closely ranked females, regardless of their position in the hierarchy, acted as reciprocal traders are similar to the pattern reported in chacma baboons (*P. c. ursinus*, Barrett et al. 1999), Japanese macaques (*M. fuscata*, Ventura et al. 2006), and hamadryas baboons (*P. hamadryas hamadryas*, Leinfelder et al. 2001). Furthermore, the authors also documented a pattern in which the initiator (A) of a bout invested less and received greater benefit than the recipient. Thus, it appears that among female and/or male Tibetan macaques, the act of initiating a grooming bout was a behavioral investment or strategy to obtain a grooming reward or other social dividend. This effect was strongest when the initiator was higher in rank than the receiver.

4.3 Exchange Between Different Behaviors

4.3.1 Grooming for Tolerance

Grooming represents one of the most common commodities traded among primates and has long been considered a reliable indicator of both long-term and short-term social tolerance (Dunbar 2010; Schino and Aureli 2010). Seyfarth (1977) has proposed that there exists a balance between a group member's preference for particular grooming partners and competition among females for access to the most valuable grooming partners. According to Seyfarth's model, lower-ranked individuals should compete to groom higher-ranked individuals, who can offer them more effective aid in competitive interactions with other group members. This is expected to result in higher-ranked individuals receiving more grooming than lower-ranked individuals and grooming partners tending to be of similar or adjacent ranks (Henzi et al. 2003). However, Henzi and colleagues (2003) found no

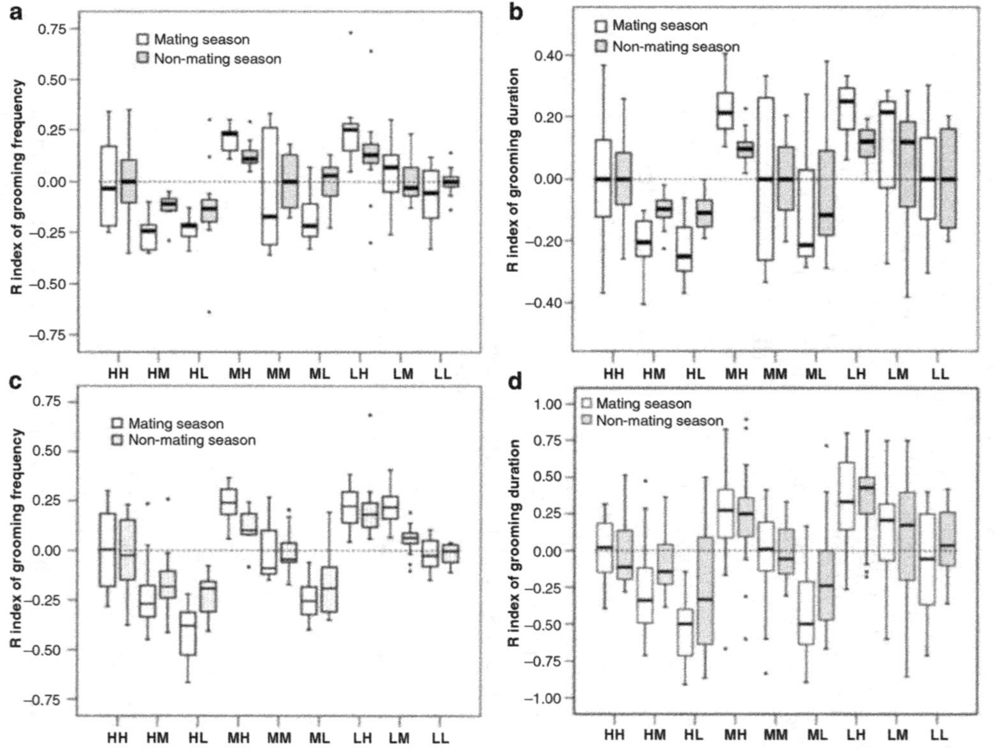

Fig. 4.3 Mean and SE (boxes) of R indices ($R = \frac{G_A - G_B}{G_A + G_B}$, in which G_A is grooming frequency or grooming duration for animal A and G_B is grooming frequency or grooming duration for animal B) within the nine female–female (**a**, grooming frequency; **b**, grooming duration) and male–male (**c**, grooming frequency; **d**,

grooming duration) categories according to rank (H, high-ranking; M, middle-ranking; L, low-ranking). The R index can range from -1 to 1, where a value of 0 represents complete reciprocity (the dashed line in each panel), negative values indicate that individual A received more grooming than it gave, and positive values indicate that individual B received more grooming than it gave. Outliers are given as dots. R indices were calculated for all grooming bouts and then were assigned each bout to one of three groups according to the level of maternal kin relationships of grooming partners. (Fig. 4.3a, b is from Xia et al. 2012, and Fig. 4.3c, d is from Xia et al. 2013.) H represents high-rankings, M represents middle-ranking, and L represents low-ranking support for Seyfarth's model in their study of chacma baboons. Rather, they argued that in these primates "grooming is unimpeded by restrictions on access to partners" and proposed that baboon females exchanged grooming for increased social tolerance.

In Tibetan macaque society, Xia et al. (2012, 2013) also found that in intrasexual dyads composed of individuals of different social ranks (e.g., high- and middle-ranking females/males, high- and low-ranking females/males, and middle- and low-ranking females/males), initiators received grooming more frequently and for a longer duration than they gave during both mating and non-mating periods. Conversely, in dyads consisting of middle- and high-ranking females, low- and high-ranking females, and low- and middle-ranking females, initiators received grooming less frequently and for a shorter duration than they gave during both mating and non-mating periods. The same results were found among male intrasexual dyads consisting of middle- and high-rankings, low- and higher-rankings, and middle- and high-rankings (see Fig. 4.3).

Xia et al. (2012, 2013) also found that the effect of grooming reciprocity among female and/or male intrasexual dyads was stronger in the non-mating season. This suggests that during periods of reduced social tension (such as the non-mating period), adult female and/or male Tibetan macaques used grooming to maintain and enhance affiliative social relationships. In contrast, during periods when the rate and the threat of male intrasexual aggression increased (such as the mating season), there was a tendency to promote social affiliation by increasing the duration of grooming. The authors found negative relationships between grooming given and aggression received both among male intrasexual dyads (e.g., as showed in Fig. 4.4) and among female intrasexual dyads. This provided supportive evidence that both female and male Tibetan macaques vary their grooming strategies based on the current rates of within-group intrasexual aggression. The results indicated that high-ranking females and/or males who received more grooming from lower-ranking males were less aggressive toward their grooming partners. In other words, lower-ranking females and/or males who devoted more time to grooming higher-ranking individuals received less aggression than did similarly ranked individuals who groomed higher-ranking females and/or males less frequently. Thus, there is evidence that, for female and/or male intrasexual dyads, Tibetan macaques interchanged grooming for social tolerance and possibly competed for access to partners of higher social rank.

Taken together, Xia et al. (2012, 2013) suggest that Tibetan macaques employed an alternative set of social strategies depending on rank relationships. This result highlights the importance of both grooming reciprocal exchange and interchange for tolerance as behavioral mechanisms that regulate female and/or male primate social relationships.

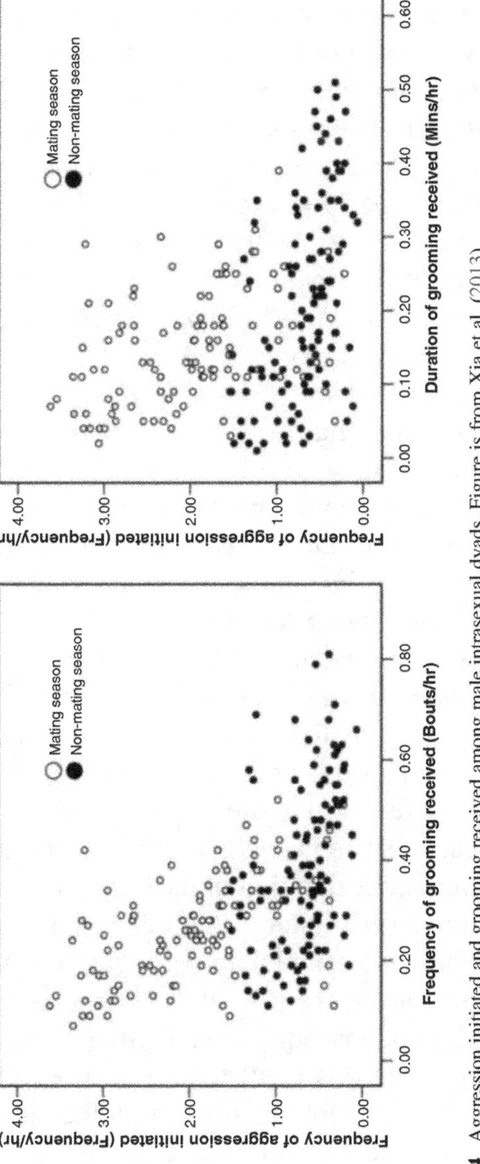

Fig. 4.4 Aggression initiated and grooming received among male intrasexual dyads. Figure is from Xia et al. (2013)

4.3.2 Grooming for Infant Handling

Infants are important components in a primate social group, and alloparental care is widespread among primates to facilitate the survival of newborn infants and increase benefits to caregivers (Maestripieri 1994). While females' attraction to infants represents a common feature of primate species, maternal response to infant handling shows a certain degree of variability (Nicolson 1987; Maestripieri 1994). In some species, such as langurs (*Presbytis pileata*) and vervet monkeys (*C. aethiops sabaeus*), mothers allow other group members to frequently hold and carry their newborn infants for long durations (Stanford 1992; Fairbanks 1990). In other species, such as baboons and macaques, although mothers are more restrictive of their infants (Nicolson 1987; Altmann 2002), group members still have opportunities to contact newborn infants (see Maestripieri 1994). It has been well documented that infants can be social tools to buffer agonistic behavior from higher-ranked individuals and facilitate dyadic social bonds (Silk and Samuels 1984 in bonnet macaques, *M. radiata*; Ogawa 1995 in Tibetan macaques). Grooming for females is an effective way to increase the opportunity for accessing *Sapajus nigritus* infants. For example, in tufted capuchin monkeys (*Sapajus apella nigritus*), Tiddi et al. (2010) proposed that potential handlers have strongly attracted to infants and grooming their mothers. Similar results have also been found in baboons, *P. anubis* (Frank and Silk 2009), long-tailed macaques (Gumert 2007b), and golden snub-nosed monkeys, *Rhinopithecus roxellana* (Wei et al. 2013). These studies imply that grooming might be interchanged for infant handling or vice versa to facilitate dyadic social bonds among females.

In Tibetan macaques, Jiang et al. (2019) found that there were bidirectional relationships between grooming and infant handling among females. Jiang and her colleagues found that there were five patterns of infant handling, including hand touching, mouth licking with teeth chattering, holding, grooming, and bridging. Female Tibetan macaques received more grooming time after giving birth than before parturition, and lactating females invested less grooming time than females without infants. For example, the duration of mothers grooming non-mothers before birth was significantly longer than after birth, whereas the duration of non-mothers grooming mothers after birth was significantly longer than that before birth. Apparently, females with infants are more attractive as grooming partners than females without infants. In a Tibetan macaque social group, adult females with infants received more grooming when this adult female allowed their newborn infants to be handled. For example, the duration of non-mother to mother post-handling-grooming was higher than the baseline values of daily grooming (see left side of Fig. 4.4) and the grooming duration when these two females were in proximity (see left side of Fig. 4.5). Moreover, with the increase in the number of infants, the amount of grooming received from females without infants decreased.

Additionally, in Tibetan macaques, the frequency of non-mother post-grooming-infant handling was higher than the baseline values of daily infant handling (see right side of Fig. 4.4) and the infant handling when these two females are in proximity (see

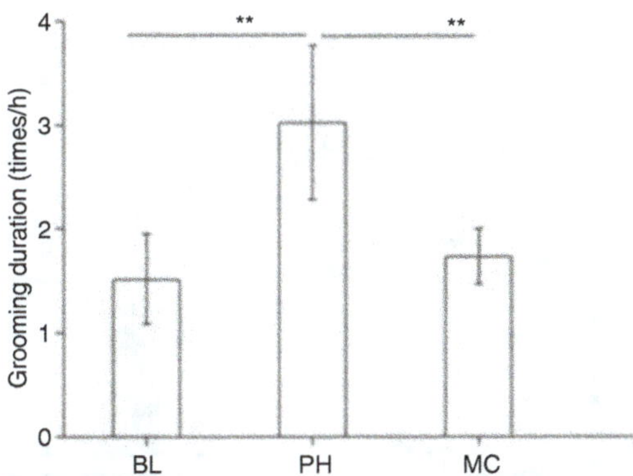
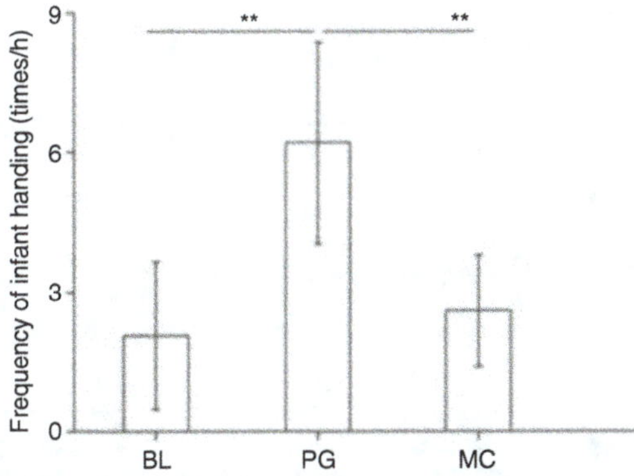

Fig. 4.5 Variation of grooming duration from non-mothers to mothers and frequency of non-mother infant handling. Baseline represents the daily activity, which is from focal sampling data. PH represents post-infant handling-grooming. PG represents post-grooming-infant handling. MC represents the mean from the data where mothers and non-mothers were in proximity. Figure is modified from Jiang et al. (2019)

right side of Fig. 4.5). These results suggested that females obtained more opportunities for infant handling when she groomed more females with infants; thus, grooming was an effective way to gain access to an infant.

4.3.3   Agonistic Support for Copulation

In multi-male–multi-female primates, mating is associated with individual's social rank, with a higher frequency for higher-ranked individuals and lower frequency for lower-ranked individuals (reviewed in Li 1999). Mating is also associated with dyadic affiliative relationships (such as grooming). For example, in a study of Barbary macaques (*M. sylvanus*), Sonnweber et al. (2015) demonstrated that males initiated grooming after copulations with ejaculation in order to keep females from mating with other males. Agonistic support is predicted to increase mating opportunity. Agonistic support has been defined as instances in which an individual joins an ongoing agonistic interaction and supports one animal by directing aggression against its opponent (Seyfarth and Cheney 1984; Hemelrijk 1994), including intrasexual and intersexual agonistic support.

Wang et al. (2013) used a 15-min focal sampling method to collect data on copulations and agonistic support as a random variable in two groups (YA1 and YA2) of Tibetan macaques in Huangshan, China. A total of 216 male-to-female agonistic support events were collected (YA1 32 and YA2 64 in the non-mating season; YA1 43 and YA2 77 in the mating season). Correlation tests showed that male-to-female agonistic support was correlated with the frequency with which the male copulated with that female in both the mating season and the non-mating season. This indicated that male-to-female agonistic support could increase the male mating opportunities with the female he supported in agonistic events.

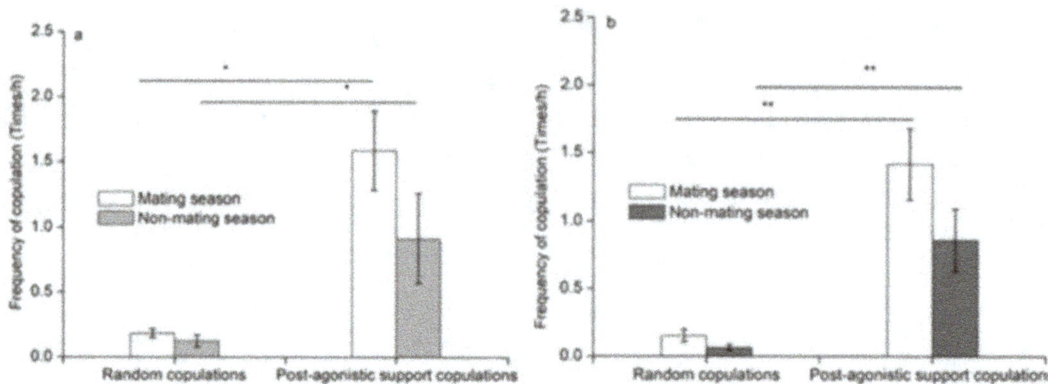

Fig. 4.6 Post-agonistic copulations and post-copulation agonistic supports among female Tibetan macaques. (**a**) from YA1 group. (**b**) from YA2 group. $*P < 0.05$. $**P < 0.01$. Figures are modified from Wang et al. (2013)

In addition, Wang et al. (2013) used a 15-min post-agonistic support focal sampling method to collect copulation data to compare with random values (the average value of copulation from the focal sampling methods). The results showed that, in the YA1 group, copulations in the post-agonistic support observation period (PO) were more frequent, but not significantly so, than random observations (RO) in the non-mating season, and post-agonistic support copulations were significantly more frequent than random observations in the mating season (see Fig. 4.6a). For the YA2 group, copulations in the post-agonistic support period were more frequent than in random observations in both mating and non-mating seasons (see Fig. 4.6b). Taken together, these results indicate that, in Tibetan macaques, male-to-female agonistic support could be interchanged for mating opportunities with females who are participating in an ongoing aggressive event.

Wang et al.'s (2013) study characterized not only the pattern of behavioral exchange in Tibetan macaque social groups but also offers new evidence for how certain mating strategies can increase mating opportunities in males. Male Tibetan macaques copulate under three different circumstances: opportunistic mating, which involves frequent copulation with many different females in the group setting (Li et al. 2015); possessive mating, which involves a single male's attempts to monopolize copulations in spite of the presence of other males (Li et al. 2015); and consortships, in which mating takes place between one male and one female, who travel apart from the rest of the group for several hours or days (Li 1999). Xia et al. (in preparation) found that males preferentially mate with their female grooming partners based on dyadic relationships. This indicates that, similar to the findings in studies in Barbary macaques (Sonnweber et al. 2015) and rhesus macaques, *M. mulatta* (Manson 1992), dyadic social relationships play a vital role in obtaining mating opportunities. Agonistic support is one of the effective interactions for males to form and maintain social relationships with females and to obtain and improve copulatory success. In Tibetan macaques, the relationships between agonistic support and copulatory behavior provide insight into understanding male–male competition and female mate choice in social primates.

4.4 Conclusions

In a Tibetan macaque social group, there are diverse exchange networks consisting of different kinds of behavioral exchanges (see Fig. 4.7). Tibetan macaques provide behavioral services for the same behavioral services in return, such as grooming in exchange for grooming. Among intra-sexual dyads (female–female and male–male dyads), grooming investment is exchanged for itself with equal value, in terms of frequency and/or duration. Secondly, Tibetan macaques can also provide behavioral services that are exchanged for different behavioral services of the same category, such as grooming for tolerance. Here, grooming and tolerance are different behaviors; however, both are friendly interactions for facilitating social relationships. Both male and female Tibetan macaques groom group members of the same sex in exchange for tolerance by decreasing aggressive interactions from higher-ranking individuals. In addition, Tibetan macaques provide behavioral services by interchanging different behavioral services with different categories, such as agonistic support for copulations.

In addition, group members can choose their partners for exchanging different types of behavioral services based on their own social status within a group. For example, grooming is regularly traded reciprocally (for grooming) among female and male intrasexual social dyads consisting of similar social ranks. Grooming can be interchanged for rank-related benefits from higher-ranking individuals, such as for tolerance from. Grooming can also be interchanged for opportunities to access newborn infants to facilitate social relationships with the infants' mothers.

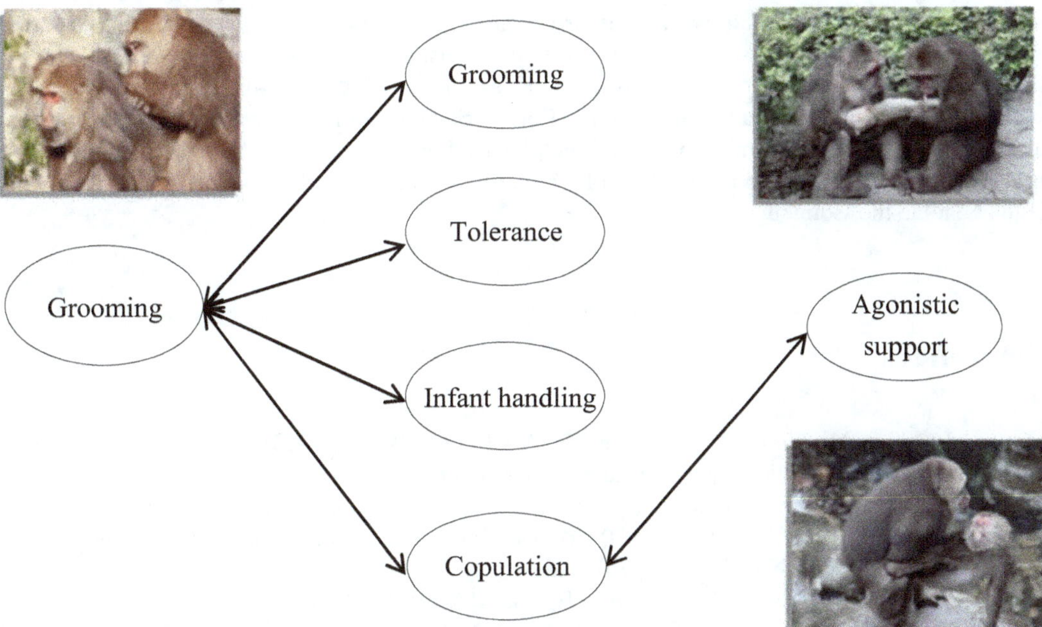

Fig. 4.7 Exchange networks in Tibetan macaque society

Finally, behavioral exchanges facilitate the establishment and maintenance of social relationships and group stability in multiple ways, by obtaining multiple behavioral services in return. For example, grooming exchanges enhance male–male and female–female social bonds. Grooming with higher-ranking group members allows lower-ranking animals to coexist peacefully with more dominant animals. Non-mothers that groom females with infants could increase their opportunity for accessing infants, and accessing newborn infants could facilitate social relationships between the females who join in infant handling. Among intersexual dyads, male Tibetan macaques support certain females against attacks by other animals and subsequently mate more frequently with that female. This implies that males can interchange copulatory behavior by supporting females to improve mating opportunity.

Although we have several case studies of behavior networks in Tibetan macaque society, we need more evidence to draw a clearer picture about behavioral exchange and interchange in this primate species. Apparently, both males and females may establish diverse forms of behavioral exchange or interchange networks with multiple group members. However, we know little about strategies for partner control or partner choice based on more complex measures of individual and social status (such as personality and dominance) to strengthen social bonds, obtain behavioral services, and maximize fitness. Future studies will need to pay closer attention to the variation of behavioral strategies in connection with group dynamics to better understand how behavioral exchange and interchange are related to the maintenance and stability of social bonds and group networks.

Acknowledgments We are very grateful to the Huangshan Garden Forest Bureau for their permission and support of this work. We also gratefully acknowledge Mr. H.B. Cheng's family for their outstanding logistic support to our study. PAG wishes to thank Chrissie, Sara, and Jenni for their support. DPX wishes to thank Randall C. Kyes, professor at the University of Washington, whom DPX worked with as a visiting scholar, Lori K. Sheeran for her encouragement and support, and Lixing Sun and Rose Amrhein for their help. This work was supported in part by grants from the National Natural Science Foundation of China (No. 31772475; 31672307; 31401981) and the China Scholarship Council.

References

Altmann J (2002) Baboon mothers and infants. Afr J Ecol 40(4):419–421

Barash DB (1977) Sociobiology and behaviour. Heinemann, London

Barrett L, Henzi SP (2006) Monkeys, markets and minds: biological markets and primate sociality. In: Kappeler PM, van Schaik CP (eds) Cooperation in primates and humans: mechanisms and evolutions. Springer, New York, pp 209–232

Barrett L, Henzi SP, Weingrill T, Lycett JE, Hill RA (1999) Market forces predict grooming reciprocity in female baboons. Proc R Soc Lond B 266:665–670

de Waal FBM (1997) The Chimpanzee's service economy: food for grooming. Evol Hum Behav 18(6):375–386

Dunbar RIM (2010) The social role of touch in humans and primates: behavioural function and neurobiological mechanisms. Neurosci Biobehav Rev 34(2):260–268

Fairbanks LA (1990) Reciprocal benefits of allomothering for female vervet monkeys. Anim Behav 9:425–441

Frank RE, Silk JB (2009) Impatient traders or contingent reciprocators? Evidence for the extended time-course of grooming exchanges in baboons. Behaviour 146:23–1135

Fruteau C, van de Waal E, van Damme E, Noë R (2011) Infant access and handling in sooty mangabeys and vervet monkeys. Anim Behav 81(1):153–161

Gomes CM, Boesch C (2009) Wild chimpanzees exchange meat for sex on a long-term basis. PLoS One 4:e5116

Gumert MD (2007a) Payment for sex in a macaque mating market. Anim Behav 74(6):1655–1667

Gumert MD (2007b) Grooming and infant handling interchange in *Macaca fascicularis*: the relationship between infant supply and grooming payment. Int J Primatol 28(5):1059–1074

Gumert MD, Ho MR (2008) The trade balance of grooming and its coordination of reciprocation and tolerance in Indonesian long-tailed macaques (*Macaca fascicularis*). Primates 49(3):176–185

Hemelrijk CK (1994) Support for being groomed in long-tailed macaques, *Macaca fascicularis*. Anim Behav 48:479–481

Henzi SP, Barrett L, Gaynor D, Greeff J, Weingrill T, Hill RA (2003) Effect of resource competition on the long-term allocation of grooming by female baboons: evaluating Seyfarth's model. Anim Behav 66(5):931–938

Hinde RA (1976) Interactions, relationships and social-structure. Man 11:1–17

Jiang Q, Xia DP, Wang X, Zhang D, Sun BH, Li JH (2019) Interchange between grooming and infant handling in female Tibetan macaques (*Macaca thibetana*). Zool Res 40:139–145

Kaburu SSK, Newton-Fisher NE (2015a) Egalitarian despots: hierarchy steepness, reciprocity and the grooming-trading model in wild chimpanzees, *Pan troglodytes*. Anim Behav 99:61–71

Kaburu SSK, Newton-Fisher NE (2015b) Trading or coercion? Variation in male mating strategies between two communities of East African chimpanzees. Behav Ecol Sociobiol 69:1039–1052

Krause J, Ruxton G (2002) Living in groups. Oxford University Press, Oxford

Krebs JR, Davies NB (1984) Behavioral ecology: an evolutionary approach. Blackwell Scientific, Oxford

Leinfelder I, de Vries H, Deleu R, Nelissen M (2001) Rank and grooming reciprocity among females in a mixed-sex group of captive hamadryas baboons. Am J Primatol 55(1):25–42

Li JH (1999) The Tibetan macaque society: a field study. Anhui University Press, Hefei. (In Chinese)

Li ZP, Li JH, Xia DP, Zhu Y, Wang X, Zhang D (2015) Mating strategies of subordinate males in Tibetan macaqus (*Macaca thibetana*) at Mt. Huangshan, China. Acta Theriologica Sinica 35 (1):29–39. (In Chinese)

Maestripieri D (1994) Social structure, infant-handling, and mother styles in group-living Old World monkeys. Int J Primatol 15:531–553

Manson JH (1992) Measuring female mate choice in Cayo Santiago rhesus macaques. Animl Behav 44:405–416

Morse DH (1980) Behavioral mechanisms in ecology. Harvard University Press, Cambridge, MA

Newton-Fisher NE, Lee PC (2013) Grooming reciprocity in wild male chimpanzees. Anim Behav 81(2):439–446

Nicolson N (1987) Infants, mothers, and other females. In: Smuts BB, Cheney DL, Seyfarth RM, Wrangham RM, Struhsaker TT (eds) Primate societies. University of Chicago Press, Chicago, IL, pp 330–342

Noë R (2001) Biological markets: partner choice as the driving force behind the evolution of mutualisms. In: Noe R, van Hooff JARAM, Hammerstein P (eds) Economics in nature: social dilemmas, mate choice and biological markets. Cambridge University Press, Cambridge, pp 93–118

Noë R (2006) Digging for the roots of trading. In: Kappeler PM, van Schaik CP (eds) Cooperation in primates and humans: mechanisms and evolution. Springer, Berlin, pp 223–251

Noë R, Hammerstein P (1994) Biological market: supply and demand determine the effect of partner choice in cooperation, mutualism and mating. Behav Ecol Sociobiol 35:1–11

Noë R, Hammerstein P (1995) Biological markets. Trends Ecol Evol 10:336–339

Noë R, Voelkl B (2013) Cooperation and biological markets: the power of partner choice. In: Sterelny K, Joyce R, Calcott B, Fraser B (eds) Cooperation and its evolution. MIT Press, Cambridge, MA, pp 131–152

Ogawa H (1995) Bridging behavior and other affiliative interactions among male Tibetan macaques (*Macaca thibetana*). Int J Primatol 16:707–729

Port M, Clough D, Kappeler PM (2009) Market effects offset the reciprocation of grooming in free-ranging redfronted lemurs, *Eulemur fulvus rufus*. Anim Behav 77(1):29–36

Sánchez-Amaro A, Amici F (2015) Are primates out of the market? Anim Behav 110:51–60

Schino G, Aureli F (2009) Reciprocal altruism in primates: partner choice, cognition, and emotions. Adv Study Behav 39:45–69

Schino G, Aureli F (2010) Primate reciprocity and its cognitive requirements. Evol Anthropol Issues News Rev 19(4):130–135

Schino G, Aureli F (2017) Reciprocity in group-living animals: partner control *versus* partner choice. Biol Rev 92:665–672

Schino G, Polizzi di Sorrentino EP, Tiddi B (2007) Grooming and coalitions in Japanese macaques (*Macaca fuscata*): partner choice and the time frame reciprocation. J Comp Psychol 121:181–188

Schino G, di Giuseppe F, Visalberghi E (2009) The time frame of partner choice in the grooming reciprocation of *Cebus paella*. Ethology 115:70–76

Seyfarth RM (1977) A model of social grooming among adult female monkeys. J Theor Biol 65:671–698

Seyfarth RM, Cheney DL (1984) Grooming, alliances and reciprocal altruism in vervet monkeys. Nature 308:541–542

Silk JB, Samuels A (1984) Triadic interactions among *Macaca radiata*: passports and buffers. Am J Primatol 6:373–376

Sonnweber RS, Massen JJM, Fitch WT (2015) Post-copulatory grooming: a conditional mating strategy? Behav Ecol Sociobiol 69:1749–1759

Stanford C (1992) The costs and benefits of allomothering in wild capped langurs (*Presbytis pileata*). Behav Ecol Sociobiol 30:29–34

Sussman RW, Garber PA (2011) Cooperation, collective action, and competition in primate social interactions. In: Campbell CJ, Fuentes A, MacKinno KC, Bearder S, Stumpf R (eds) Primates in perspective, vol 2. Oxford University Press, New York, pp 587–599

Tiddi B, Aureli F, Schino G (2010) Grooming for infant handling in tufted capuchin monkeys: a reappraisal of the primate infant market. Anim Behav 79:1115–1123

van Schaik CP (1989) The ecology of social relationships amongst female primates. In: Standen V, Foley RA (eds) Comparative socioecology: the behavioural ecology of humans and other animals. Blackwell Scientific, Oxford, pp 195–218

Ventura R, Majolo B, Koyama NF, Hardie S, Schino G (2006) Reciprocation and interchange in wild Japanese macaques: grooming, cofeeding, and agonistic support. Am J Primatol 68(12):1138–1149

Wang S, Li JH, Xia DP, Zhu Y, Sun BH, Wang X, Zhu L (2013) Male-to-female agonistic support for copulation in Tibetan macaques (*Macaca thibetana*) at Huangshan, China. Zool Res 34:139–144

Wei W, Qi XG, Garber PA, Guo ST, Zhang P, Li BG (2013) Supply and demand determine the market value of access to infants in the golden snub-nosed monkey (*Rhinopithecus roxellana*). PLoS One 8(6):e65962

Wilson EO (1975) Sociobiology. Harvard University Press, Cambridge, MA

Wrangham RW (1980) An ecological model of female-bonded primate groups. Behaviour 75:262–299

Xia DP, Li JH, Garber PA, Sun L, Zhu Y, Sun BH (2012) Grooming reciprocity in female Tibetan macaques *Macaca thibetana*. Am J Primatol 74:569–579

Xia DP, Li JH, Garber PA, Matheson MD, Sun BH, Zhu Y (2013) Grooming reciprocity in male Tibetan macaques. Am J Primatol 75:1009–1020

Chapter 5
Social Relationships Impact Collective Decision-Making in Tibetan Macaques

Xi Wang, Claudia Fichtel, Lixing Sun, and Jin-Hua Li

5.1   Introduction

Group-living offers many benefits related to survival and reproduction for animals (Bertram 1978; van Schaik 1983; Zemel and Lubin 1995; Krause and Ruxton 2002). Nevertheless, it also involves some unavoidable costs, such as mate competition or interindividual conflict (Kappeler et al. 2015). Therefore, animals need to coordinate their actions and maintain group cohesion to gain the benefits of group-living (Conradt and Roper 2003; Fichtel et al. 2011). Group coordination is often difficult to achieve because individuals may differ in their needs and interests. When these differences cannot be reconciled, cohesiveness can be at risk (Rands et al. 2003,

X. Wang
School of Resources and Environmental Engineering, Anhui University, Hefei, Anhui, China

International Collaborative Research Center for Huangshan Biodiversity and Tibetan Macaque Behavioral Ecology, Anhui, China
e-mail: xwang@ahu.edu.cn

C. Fichtel
Behavioral Ecology and Sociobiology Unit, German Primate Center, Leibniz Institute for Primate Research, Göttingen, Germany
e-mail: Claudia.Fichtel@gwdg.de

L. Sun
Department of Biological Sciences, Primate Behavior and Ecology Program, Central Washington University, Ellensburg, WA, USA
e-mail: Lixing@cwu.edu

J.-H. Li (✉)
School of Resources and Environmental Engineering, Anhui University, Hefei, Anhui, China

International Collaborative Research Center for Huangshan Biodiversity and Tibetan Macaque Behavioral Ecology, Anhui, China

School of Life Sciences, Hefei Normal University, Hefei, Anhui, China
e-mail: jhli@ahu.edu.cn

J.-H. Li et al. (eds.), *The Behavioral Ecology of the Tibetan Macaque*, Fascinating Life Sciences, https://doi.org/10.1007/978-3-030-27920-2_5

2008). How consensus is achieved and implemented at the behavioral level is often studied during natural group movements to and from specific resources and/or locations (e.g., sleeping and foraging sites). Such studies can provide an ecologically relevant context to probe into fundamental mechanisms of social coordination of collective actions (Boinski and Garber 2000; Fichtel et al. 2011).

Social interactions could affect the process of group movements. For example, in large anonymous groups, such as fish schools and bird flocks, in which members do not know each other individually, direction and action during movements are regulated by self-coordination by individuals following the simple rule of keeping a certain distance to the nearest neighbor (Parrish and Edelstein-Keshet 1999; Couzin et al. 2002; Hemelrijk 2002). In contrast, in primate groups, where members know each other individually, certain dominant or affiliated individuals may play specific roles in the context of collective behavior, such as initiating or terminating a group movement (Boinski and Garber 2000; King and Cowlishaw 2009; King and Sueur 2011).

In primate species, initiation of a group movements can be accompanied by notifying behaviors (Kummer 1968) or preliminary behaviors (Sueur and Petit 2008a) that are exhibited in the pre-departure period, directly preceeding the group movement. This recruitment process often includes visual and acoustic communication, and which can influence the recruitment success of an initiation, i.e., whether the initiator is followed and, if so, by how many group members and how quickly (Sueur and Petit 2010; Seltmann et al. 2016; Sperber et al. 2017).

A successful collective movement can be driven by an unshared decision-making mechanism, i.e., one individual leading all group movements and other members following it all the time (Conradt and Roper 2005). In this decision-making process, the highest-ranking male usually plays a major role in leadership in several species of Old World monkey (Sueur and Petit 2008a). Alternatively, shared or partially shared decision-making mechanism can also result in a collective movement. That is, all group members or a subgroup can lead the movement of the entire group on different occasions (Pyritz et al. 2011). In this case, individuals with better social connections enjoy higher rates of initiating group movements (Sueur and Petit 2008b; Strandburg-Peshkin et al. 2015; Fratellone et al. 2018).

A major focus in the study of collective decisions in primates is the joining process, which occurs once an individual has initiated a collective movement. Often, primates do not decide independently on activity changes, but, rather, base their choices on the actions of their group mates. This form of joining rule, when one individual taking an action makes it more likely for another to do so as well, has been termed mimetism (Deneubourg and Goss 1989). Mimetism can be further categorized as anonymous mimetism and selective mimetism. Anonymous mimetism refers to individuals being more likely to take the actions of other individuals irrespective of their identity, and thus group movements can simply depend on the number of individuals who have already left or performed a certain behavior (Petit et al. 2009). Selective mimetism refers to joining decisions based on some other factors, such as distance, with individuals being more likely to join a movement when in close proximity (Ramseyer et al. 2009; Ward et al. 2013), and affiliation, with individuals being more likely to follow those group members with whom they have strong social bonds (King et al. 2011; Sueur et al. 2009, 2011; Seltmann et al. 2013; Strandburg-Peshkin et al. 2015; Farine et al. 2016).

Individuals may also engage another joining process based on a quorum rule. This rule states that once a minimum number of group members joins a movement, group movement will occur all the time (Conradt and Roper 2003; Wang et al. 2015; Rowe et al. 2018). A response to a quorum is observed when the probability of members exhibiting a vote by joining a movement depends on the number of individuals already performing the voting behavior (Pratt et al. 2002; Seeley and Visscher 2004; Sumpter 2006; Ward et al. 2008).

Interestingly in macaques, social style appears to influence the organization of group movements (Sueur and Petit 2008a). Social styles of the species of *Macaca* have been divided into four grades (Thierry 2000), ranging from grade 1 (the most intolerant) to grade 4 (the most tolerant). These styles appear to influence the initiation and joining process of group movements (Sueur and Petit 2008a). In species with a more despotic dominance style, decision-making is more likely to be unshared and social rank determines leadership (Sueur and Petit 2008a, b). In contrast, in species exhibiting a more egalitarian style, decision-making is more likely to be shared and social relationships determine leadership (Sueur and Petit 2008a). For instance, in rhesus (*Macaca mulatta*) and Japanese macaques (*M. fuscata*), both of which have a despotic social style, movements are mainly initiated by dominant individuals, and joining processes are also determined by dominance order (Sueur et al. 2009; Jacobs et al. 2011). In contrast, in macaques with a more egalitarian social style, such as Tonkean (*M. tonkeana*) and Barbary macaques (*M. sylvanus*), decision-making is equally or partially shared, and joining processes are determined by affiliation (Sueur et al. 2009; Jacobs et al. 2011; Seltmann et al. 2013).

Currently, the relationship between social style and leadership and other important aspects of collective decision-making has been investigated only in a limited number of macaque species. Therefore, detailed studies of collective movement in relatively less known species, such as the Tibetan macaque (*M. thibetana*), are of special interest for a better understanding of the link between social relationships and collective decision-making. We therefore take this opportunity to review and synthesize information based largely on our studies of Tibetan macaques. We hope that some of the findings and insights (including those that have not been fully developed in our publications) will enrich our understanding of decision-making processes in collective movement in primates in general and macaques in particular.

5.2 Collective Decision-Making in Tibetan Macaques

5.2.1 A Macaque Species for Studying Decision-Making

Because *Macaca* species vary in dominance style (Thierry et al. 2004), they serve as an interesting taxon to study how social relationship influences the process of group movements (Jacobs et al. 2011). Tibetan macaques are highly gregarious and live in cohesive groups (Li 1999). Similar to other macaques, Tibetan macaques show female philopatry, male dispersal, and linear dominance hierarchies (Berman et al. 2004).

The wild group of macaques we studied (YA1, see Chap. 2 for detailed information about the history, demography, and habitat of the study group) engaged in social activities in nearby forest during most of the day without any restriction on their home range. For the convenience of viewing by tourists, they were supplied with 3–4 kg of corn daily (Berman and Li 2002; Berman et al. 2008; Xia et al. 2012). After corn feeding, they regularly switched locations from the feeding site to the nearby forest. Collective movements often occurred at the time of the switch. We therefore investigated decision-making processes during group movements in this group from August to December of 2012.

5.2.2 Decision-Making During the Initiation Process of Group Movements

It is still debated whether Tibetan macaques exhibit a despotic or a tolerant dominance style (Thierry 2000; Berman et al. 2004). In accordance with a despotic dominance style (Berman et al. 2004), Tibetan macaques should be expected to show an unshared decision-making process when initiating group movements, i.e., a single, highest-ranking individual leads most movements. In reality, however, Tibetan macaques demonstrate a shared decision-making process with affiliative individuals more often initiating group movements than less sociable group members (Wang et al. 2016). Considering the above contradiction about dominance style in Tibetan macaques, we assume that there may be a potential connection between social rank/affiliative relationship and the initiation of group movement.

We observed initiation processes when Tibetan macaques returned from the feeding site to the nearby forest. Thus, an initiator was defined as the first individual that moved more than 10 m in less than 30 s from the provisioning area to the forest. Any individual walking more than 5 m and within 45° of the direction taken by the initiator was considered as a follower (Sueur and Petit 2008a). We used the criterion of 5 min for each successive follower who joined the movement after the first mover or previous follower (Wang et al. 2016). In our study, only those movements including at least two-thirds of group members were counted as successful group movements. To quantify leadership in group movements, we standardized initiation

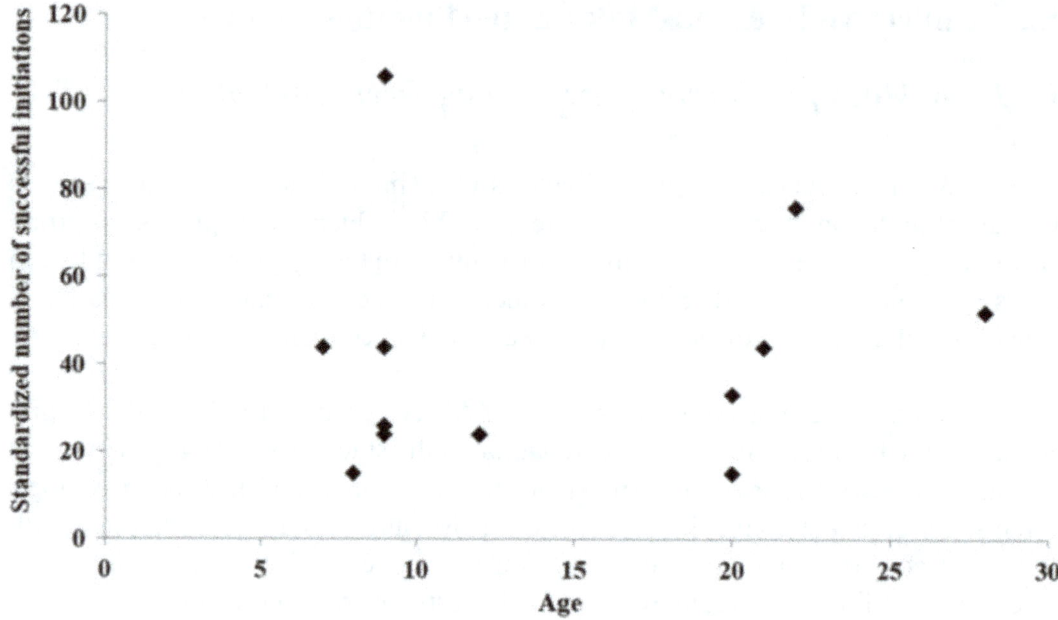

Fig. 5.1 Age of adults and their successful initiations of group movements in Tibetan macaques. There was no correlation between the two variables

data on the number of initiations of each individual by the number of times in which this individual was identified at the provisioning area (Wang et al. 2016).

During the 5 months of our observation period, we recorded more than 200 initiation attempts, all of them by adult members. Two-thirds of these initiations were considered as successful group movements. We found that all adults could initiate group movements, but that they differed significantly in the standardized number of successful initiations. This result clearly showed that decision-making during the initiation process of group movement was shared among adults.

To explore which factors might affect collective decision-making, we analyzed the relationship between several key biological/social attributes and the initiation of group movement. Interestingly, there was no significant difference in the standardized number of successful initiations between adult males and females. Second, there was neither a correlation between social rank and the standardized number of successful initiations nor with the success ratio of initiations. Also, age of adults was not correlated with the successful initiation of group movement (Fig. 5.1).

To evaluate the relationship between social affiliation of an adult and its leadership in group movements, we related the number and ratio of successful initiations of every subject in the provisioning area to its eigenvector centrality coefficient based on proximity relations among group members when they were in the forest (in comparison with the situations when they were in the feeding site). We used focal animal sampling method to collect proximity data for assessing affiliative relationships among group members (Altmann 1974; Li 1999; Berman et al. 2008). We found a positive correlation between the eigenvector centrality coefficient (based on proximity relations) and the standardized number of successful initiations

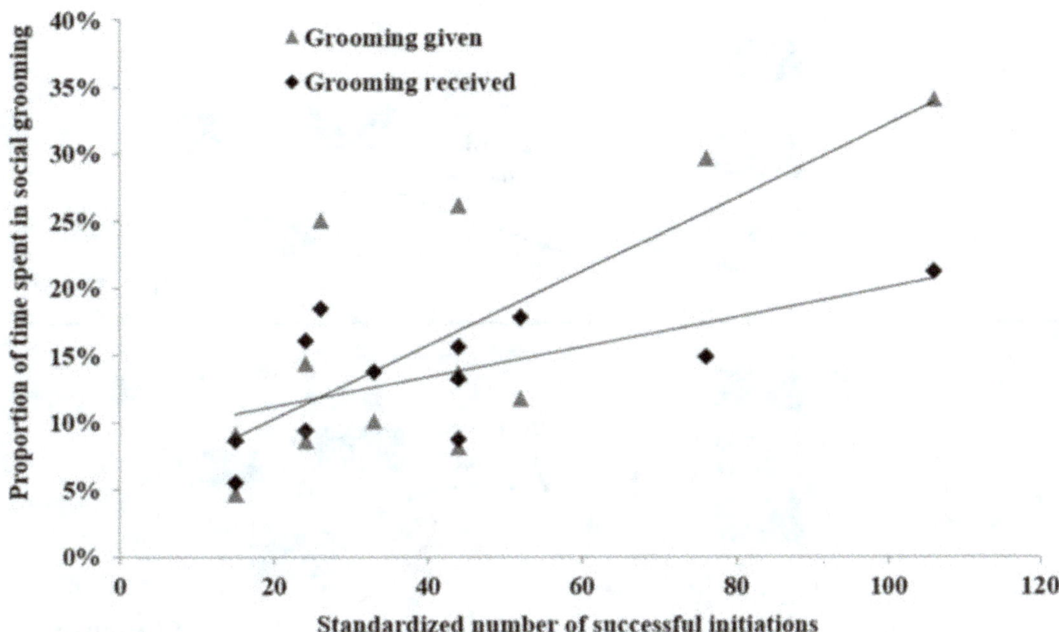

Fig. 5.2 Correlation between successful initiations of group movements and social grooming in Tibetan macaques. Proportion of time spent in social grooming indicates the duration of grooming of every subject divided by its total focal observation time. The two lines respectively represent positive correlations between grooming given/grooming received and successful initiations of group movements

across adults. Eigenvector centrality was also correlated positively with the success ratio of initiations.

Moreover, to further analyze the relationship between affiliated behavior and leadership of group movement, we correlated the initiation of group movement with social grooming among adults. Results showed that the standardized number of successful initiations was positively correlated with the duration of social grooming, including grooming given and grooming received (Fig. 5.2).

5.2.3 Decision-Making During the Joining Process of Group Movements

Joining processes were also observed when Tibetan macaques returned from the feeding site, where they were regularly provisioned, to the nearby forest. We tested whether joining occurs according to a quorum decision or mimetism as shown in a simple schematic to illustrate how the two responses would differ (Fig. 5.3).

During our preliminary observation of group YA1 (August 1–14, 2012), 5 min were used as the minimum duration of initiating a successful group movement. Therefore, an early joiner was defined as an individual that moved in the first 5 min after the initiator departed (Wang et al. 2015). According to this definition, an

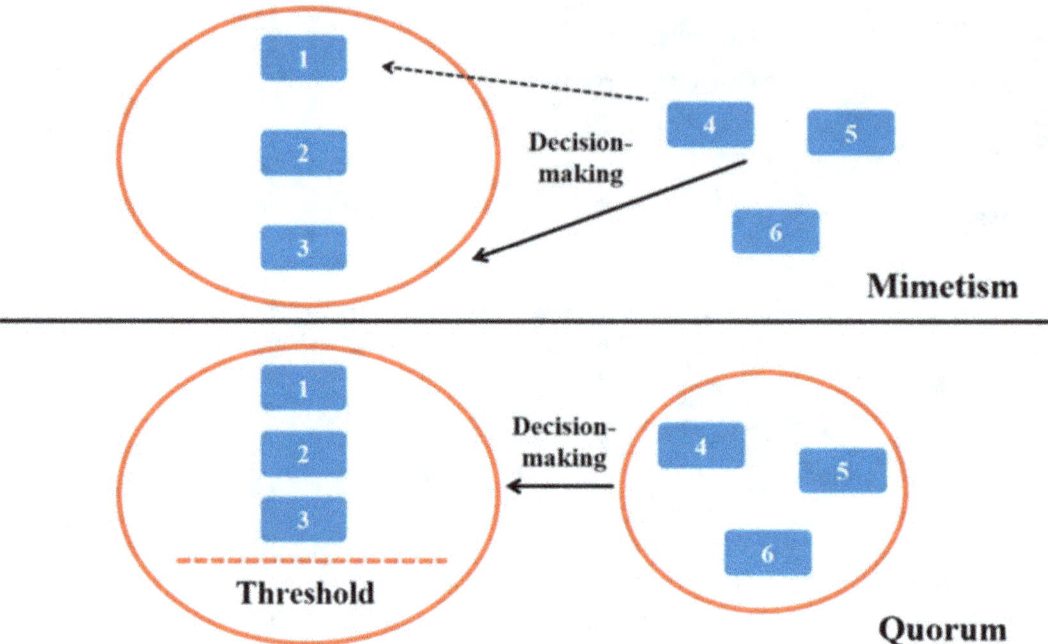

Fig. 5.3 A schematic depiction of mimetism and quorum. Nos. 1, 2, and 3 in red circles represent individuals who have joined the movement. Nos. 4, 5, and 6 are individuals waiting to join the movement. Dotted line of upper figure indicates selective mimetism: Nos. 4, 5, and 6 would join the movement based on the choice of specific members (e.g., No. 1 is the highest-ranking male). Solid line of upper figure indicates anonymous mimetism: Nos. 4, 5, and 6 would join the movement based on the number of joiners by linearly. Threshold of lower figure indicates quorum: for example, once half of the members (i.e., three individuals) have joined the movement, other individuals would follow the collective action all the time

initiator in our study was also considered as an early joiner because he/she left in the first 5 min.

We assessed the relationship between the number of adult early joiners and the probability of successful group movement. The results revealed that group movements were not successful unless three or more early joiners participated in the movement. When three to six early joiners participated, successful group movements occurred without a consistent pattern, showing some fluctuations in the probability of successful group movement. Nonetheless, once more than half of the early joiners participated in the movement, other group members followed the collective actions all the time.

To further study the role of early joiners in group movements, we performed a correlation analysis between the mean joining order and eigenvector centrality coefficient for adults in group movements (Wang et al. 2015). Results showed that the earlier an individual joined the movement, the higher its centrality was (Fig. 5.4). We then explored key attributes of early joiners in the social network of those in group movements. Our results showed that early joiners differed significantly in eigenvector centrality coefficient based on the half-weight index (HWI: co-occurrence index in group movements, Wang et al. 2015), but there was no difference between adult males and females. Also, age and eigenvector centrality

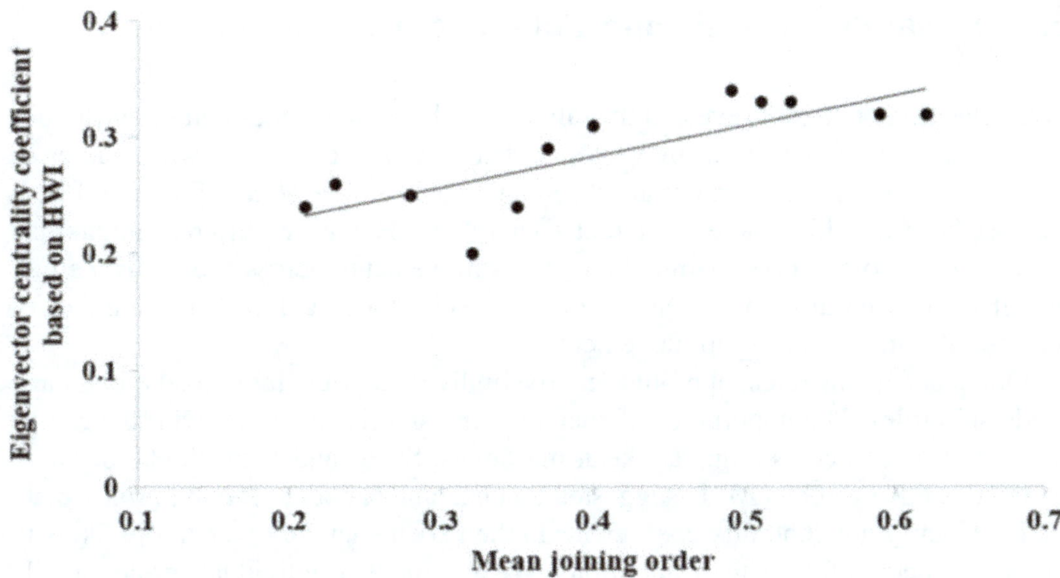

Fig. 5.4 Relationship between individuals' joining order of group movements and eigenvector centrality coefficient. HWI: co-occurrence index in group movements. The bigger the value of mean joining order was, the earlier an individual joined the movement

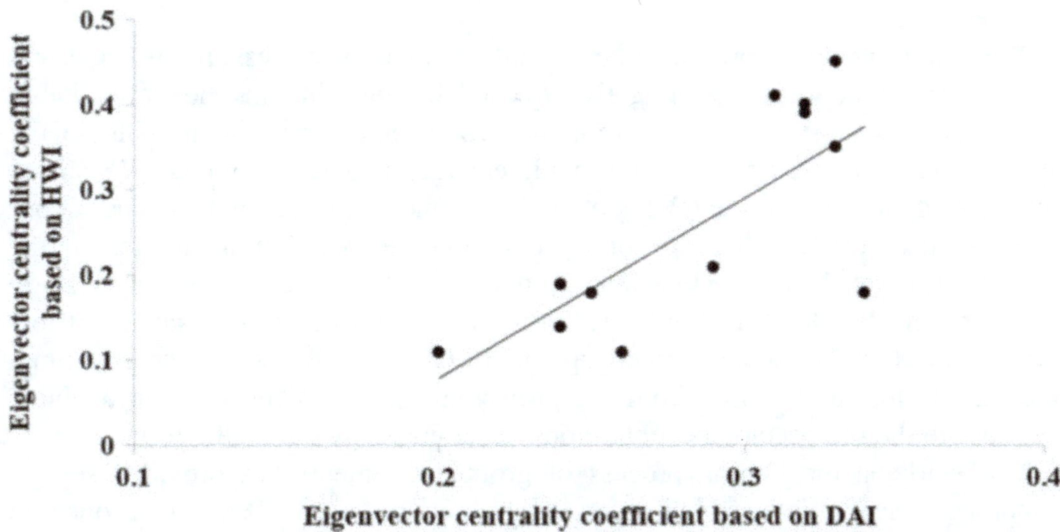

Fig. 5.5 Eigenvector centrality coefficients of individuals based on HWI and DAI. HWI: co-occurrence index in group movements. DAI: dyadic association index in proximity relations

coefficients were not correlated. However, social rank was positively correlated with eigenvector centrality coefficients in both adult males and females.

Finally, we compared two eigenvector centrality coefficients of individuals in group movements and in proximity relations (Wang et al. 2015). We found a positive correlation between the two coefficients (Fig. 5.5). This result indicates a close relation between affiliative behavior and the joining process of group movements.

5.3 Social Relationship and Collective Decision-Making

Our collection of studies revealed that all adult Tibetan macaques had opportunities to successfully initiate group movements. This result is consistent with studies in Barbary macaques (Seltmann et al. 2013) and Tonkean macaques (Sueur and Petit 2008a), both of which have a tolerant social style. However, different adults had varying times of success in initiating group movements across the study period. Eigenvector centrality coefficients were positively correlated with the number of successful initiation of group movements.

Our data demonstrate that more affiliative individuals were more likely to assume leadership roles. The importance of social relationship is consistent with the results of several other studies (e.g., Tonkean macaques: Sueur and Petit 2008a, b; Sueur et al. 2009). King et al. (2011) suggested that chacma baboons (*Papio ursinus*) with higher eigenvector centrality coefficients in their social network were more likely to attract partners to follow their initiations. We assume that individuals preferentially followed "friends" because following friends may be associated with benefits. For instance, female baboons (*P. cynocephalus*) that had closer bonds with others lived longer and their offspring also had a higher probability to survive (Silk et al. 2010), but in Tibetan macaques neither these friendships nor their potential consequences have been studied yet.

We found a positive correlation between the number of successful initiations and the amount of time spent grooming given/received, suggesting that the motivation to move may lay in staying away from the provisioning area and moving to the relatively calm forest, where animals could engage in social grooming. We carried out our study at a tourist site (Wang et al. 2016), and adult Tibetan macaques at our study site had indeed higher rates of aggression in the provisioning area than in the forest (Berman et al. 2007). Moreover, as no large predators were present at the study site in recent decades (Li 1999), the forest was supposed to be a safer and less stressful location for monkeys to engage in affiliative behaviors such as grooming. We see the location change from the provisioning area to forest as an adaptive response to the socioecological conditions experienced by our study group.

Our work on the joining process of group movements has provided several important insights into which rules might be used by Tibetan macaques in decision-making during collective movement. First, we found that the more central a group member was in the social network, the earlier it participated in a movement, as shown by the positive correlation between the mean joining order of every joiner and its eigenvector centrality coefficient (Wang et al. 2015). This result is comparable with situations in black howler monkeys (*Alouatta pigra*), where females at the front of a group movement have the highest centrality eigenvectors among the adult group members (Belle et al. 2013). Our results demonstrate the importance of early joiners in the decision-making process and indicate that the initiator was not always the only decision-maker. Other joiners in the first 5 min might also play a key role in decision-making.

Second, we found that higher-ranking early joiners tended to have higher eigenvector centrality coefficients. Because the eigenvector centrality coefficient can quantify the attraction of early joiners to other members during the joining process (Newman 2004), our data (Wang et al. 2015) showed the role of social rank in early joiners. That is, higher-ranking joiners could have more companions in group movements. A similar influence of social rank on decision-making has also been reported in other species. For instance, alpha males have been reported to be the consistent decision-makers in group movements in mountain gorillas, *Gorilla gorilla* (Watts 2000).

Third, we found that early joiners who had higher centrality coefficients in proximity activities also had higher centrality coefficients during collective movements. This means that early joiners with frequent social interactions could also attract more members during the joining process. This result is comparable to findings in chacma baboons and red-fronted lemurs, *Eulemur rufifrons* (Stueckle and Zinner 2008; King et al. 2011; Sperber et al. 2019).

Our data also showed that when the number of early joiners in a group was below seven, group members preferred to follow higher-ranking or affiliative early joiners during the joining process of group movements. By this time, selective mimetism had most likely been used as the joining rule. Apart from our study, selective mimetism has also been suggested in several other studies (Detrain et al. 1999; Camazine et al. 2001; Couzin and Krause 2003; Sumpter 2006; Gautrais et al. 2007; Sperber et al. 2019). In Tonkean macaques, for instance, how an individual decides to join a collective movement depends on whether it is strongly affiliated to departing individuals (Sueur et al. 2009).

We can explain selective mimetism in Tibetan macaques with respect to their social style. For example, Tonkean macaques exhibit an egalitarian social structure (Sueur and Petit 2008b). In this species, individuals decide when to move via a quorum. The lack of centrality for dominant or old Tonkean macaques suggests that all individuals may have equal weight in the voting process and interactions are not constrained by individual status (Sueur and Petit 2008b). However, rhesus macaques, with a despotic style, prefer to join high-ranking or related individuals during collective movements, showing selective mimetism (Sueur and Petit 2008a, b; Sueur et al. 2010). In Tibetan macaques, higher-ranking early joiners were socially connected to more individuals than lower-ranking group members, both in collective movements and during proximity activities. As our results showed that high-ranking individuals had high centrality coefficients based on the co-occurrence index in group movements, they were more attractive to other members than low-ranking individuals during the joining process.

Our data also showed that when the number of early joiners had accumulated to more than half of the adults in our study group, all group members would participate in group movement all the time. Clearly, this threshold of ">50% adult members" indicates the existence of another joining rule, the quorum rule, during collective movements in Tibetan macaques. The voting process in the group of Tibetan macaques we studied may be explained by the reduction in the risk of being left behind from the group (Wang et al. 2015). A similar joining rule has also been found in white-faced capuchins (*Cebus capucinus*), rhesus macaques, and Tonkean macaques, all of which use the threshold of four in collective movements (Petit et al. 2009; Sueur and Petit 2010).

5.4   Conclusions

In this chapter, we reviewed and synthesized studies of collective movement and decision-making in Tibetan macaques. We suggest that leadership of group movements in Tibetan macaques was distributed among adults rather than exclusively taken by a single, high-ranking individual. Different members led the group on different occasions, and social relationships were more related to leadership than social rank, age, or sex. Performing group movement presumably produced opportunities to switch locations for adults to participate in other social activities, including social grooming. Moreover, social relationship also mattered for the joining process. Tibetan macaques used selective mimetism and a quorum process in collective decision-making, and early joiners with closer affiliation played a critical role as to which rule was used. Thus, our study provided further evidence for the link between social relationships and collective decision-making in a little known macaque species. Future studies can examine whether social relationships affect the decision-making process in Tibetan macaques at the same level when group size varies. This will lead us to a better understanding as to whether a general pattern exists for group coordination and social cohesion through collective decision-making.

Acknowledgments We thank the Huangshan Monkey Management Center and the Huangshan Garden Forest Bureau for their permission for us to conduct research at the field site. We are also grateful to Haibin Cheng's family for their outstanding logistic support to our study. This work was supported by grants from the National Natural Science Foundation of China (No. 31801983; 31772475; 31672307; 31401981; 31372215), the Initial Foundation of Doctoral Scientific Research in Anhui University (J01003268), the Training Program for Excellent Young Teachers (J05011709) and China Scholarship Council.

References

Altmann J (1974) Observational study of behavior: sampling methods. Behaviour 49:227–267

Belle SV, Estrada A, Garber PA (2013) Collective group movement and leadership in wild black howler monkeys (*Alouatta pigra*). Behav Ecol Sociobiol 67:31–41

Berman CM, Li JH (2002) Impact of translocation, provisioning and range restriction on a group of *Macaca thibetana*. Int J Primatol 23:383–397

Berman CM, Ionica C, Li JH (2004) Dominance style among *Macaca thibetana* on Mt. Huangshan, China. Int J Primatol 25:1283–1312

Berman CM, Li JH, Ogawa H, Ionica C, Yin HB (2007) Primate tourism, range restriction, and infant risk among *Macaca thibetana* at Mt. Huangshan, China. Int J Primatol 28:1123–1141

Berman CM, Ogawa H, Ionica C, Yin HB, Li JH (2008) Variation in kin bias over time in a group of Tibetan macaques at Huangshan, China: contest competition, time constraints or risk response? Behaviour 145:863–896

Bertram BCR (1978) Living in groups: predators and prey. In: Krebs JR, Davies JB (eds) Behavioural ecology. Blackwell, Oxford, pp 64–96

Boinski S, Garber PA (2000) On the move: how and why animals travel in groups. University of Chicago Press, Chicago, IL

Camazine S, Deneubourg JL, Franks NR, Sneyd J, Theraulaz G, Bonabeau E (2001) Self-organization in biological systems. Princeton University Press, Princeton, NJ

Conradt L, Roper TJ (2003) Group decision-making in animals. Nature 421:155–158

Conradt L, Roper TJ (2005) Consensus decision making in animals. Trends Ecol Evol 20:449–456

Couzin ID, Krause J (2003) Self-organization and collective behavior in vertebrates. Adv Stud Behav 32:1–75

Couzin ID, Krause J, James R, Ruxton GD, Franks NR (2002) Collective memory and spatial sorting in animal groups. J Theor Biol 218:1–11

Deneubourg JL, Goss S (1989) Collective patterns and decision-making. Ital J Zool 1:295–311

Detrain C, Deneubourg JL, Pasteels JM (1999) Decision-making in foraging by social insects. In: Detrain C, Deneubourg JL, Pasteels JM (eds) Information processing in social insects. Birkhaüser Verlag, Basel, pp 331–354

Farine DR, Strandburg-Peshkin A, Berger-Wolf T, Ziebart B, Brugere I, Li J, Crofoot MC (2016) Both nearest neighbours and long-term affiliates predict individual locations during collective movement in wild baboons. Sci Rep 6:27704

Fichtel C, Pyritz L, Kappeler PM (2011) Coordination of group movements in non-human primates. In: Boos M, Kolbe M, Kappeler P, Ellwart T (eds) Coordination in human and primate groups. Springer, Heidelberg, pp 37–56

Fratellone GP, Li JH, Sheeran LK, Wagner RS, Wang X, Sun L (2018) Social connectivity among female Tibetan macaques (*Macaca thibetana*) increases the speed of collective movements. Primates 60(3):183–189. https://doi.org/10.1007/s10329-018-0691-6

Gautrais J, Michelena P, Sibbald A, Bon R, Deneubourg JL (2007) Allelomimetic synchronization in merino sheep. Anim Behav 74:1443–1454

Hemelrijk CK (2002) Understanding social behaviour with the help of complexity science. Ethology 108:655–671

Jacobs A, Watanabe K, Petit O (2011) Social structure affects initiations of group movements but not recruitment success in Japanese macaques (*Macaca fuscata*). Int J Primatol 32:1311–1324

Kappeler PM, Cremer S, Nunn CL (2015) Sociality and health: impacts of sociality on disease susceptibility and transmission in animal and human societies. Philos Trans R Soc Lond B Biol Sci 370:20140116

King AJ, Cowlishaw G (2009) Leaders, followers and group decision-making. Commun Integr Biol 2:147–150

King AJ, Sueur C (2011) Where next? Group coordination and collective decision making by primates. Int J Primatol 32:1245–1267

King AJ, Sueur C, Huchard E, Cowlishaw G (2011) A rule-of-thumb based on social affiliation explains collective movements in desert baboons. Anim Behav 82:1337–1345

Krause J, Ruxton GD (2002) Living in groups. Oxford University Press, Oxford

Kummer H (1968) Social organisation of hamadryas baboons. University of Chicago Press, Chicago, IL

Li JH (1999) The Tibetan macaque society: a field study. Anhui University Press, Hefei. (In Chinese)

Newman MEJ (2004) Analysis of weighted networks. Phys Rev E 70:056131

Parrish JK, Edelstein-Keshet L (1999) Complexity, pattern, and evolutionary trade-offs in animal aggregation. Science 284:99–101

Petit O, Gautrais J, Leca JB, Theraulaz G, Deneubourg JL (2009) Collective decision-making in white-faced capuchin monkeys. Proc R Soc B 276:3495–3503

Pratt SC, Mallon EB, Sumpter DJT, Franks NR (2002) Quorum sensing, recruitment, and collective decision-making during colony emigration by the ant *Leptothorax albipennis*. Behav Ecol Sociobiol 52:117–127

Pyritz LW, Kappeler PM, Fichtel C (2011) Coordination of group movements in wild red-fronted lemurs: processes and influence of ecological and reproductive seasonality. Int J Primatol 32:1325–1347

Ramseyer A, Boissy A, Dumont B, Thierry B (2009) Decision making in group departures of sheep is a continuous process. Anim Behav 78:71–78

Rands SA, Cowlishaw G, Pettifor RA, Rowcliffe JM, Johnstone RA (2003) Spontaneous emergence of leaders and followers in foraging pairs. Nature 423:432–434

Rands SA, Cowlishaw G, Pettifor RA, Rowcliffe JM, Johnstone RA (2008) The emergence of leaders and followers in foraging pairs when the qualities of individuals differ. BMC Evol Biol 8:51

Rowe AK, Li JH, Sun L, Sheeran LK, Wagner RS, Xia DP, Uhey DA, Chen R (2018) Collective decision making in Tibetan macaques: how followers affect the rules and speed of group movement. Anim Behav 146:51–61

Seeley TD, Visscher PK (2004) Quorum sensing during nest-site selection by honeybee swarms. Behav Ecol Sociobiol 56:594–601

Seltmann A, Majolo B, Schülke O, Ostner J (2013) The organization of collective group movements in wild Barbary macaques (*Macaca sylvanus*): dominance style drives processes of group coordination in macaques. PLoS One 8:e67285

Seltmann A, Franz M, Majolo B, Qarro M, Ostner J, Schülke O (2016) Recruitment and monitoring behaviors by leaders predict following in wild Barbary macaques (*Macaca sylvanus*). Primate Biol 3:23–31

Silk JB, Beehner JC, Bergman TJ, Crockford C, Engh AL, Moscovice LR, Wittig RM, Seyfarth RM, Cheney DL (2010) Strong and consistent social bonds enhance the longevity of female baboons. Curr Biol 20:1359–1361

Sperber AL, Werner LM, Kappeler PM, Fichtel C (2017) Grunt to go – vocal coordination of group movements in redfronted lemurs. Ethology 123:894–905

Sperber AL, Kappeler PM, Fichtel C (2019) Should I stay or should I go? Individual movement decisions during group departures in redfronted lemurs. Proc R Soc Open Sci 6(3). https://doi.org/10.1098/rsos.180991

Strandburg-Peshkin A, Farine DR, Couzin ID, Crofoot MC (2015) Group decisions. Shared decision-making drives collective movement in wild baboons. Science 348:1358–1361

Stueckle S, Zinner D (2008) To follow or not to follow: decision making and leadership during the morning departure in chacma baboons. Anim Behav 75:1995–2004

Sueur C, Petit O (2008a) Shared or unshared consensus decision in macaques? Behav Process 78:84–92

Sueur C, Petit O (2008b) Organization of group members at departure is driven by dominance style in *Macaca*. Int J Primatol 29:1085–1098

Sueur C, Petit O (2010) Signals use by leaders in *Macaca tonkeana* and *Macaca mulatta*: group-mate recruitment and behaviour monitoring. Anim Cogn 13:239–248

Sueur C, Petit O, Deneubourg JL (2009) Selective mimetism at departure in collective movements of *Macaca tonkeana*: an experimental and theoretical approach. Anim Behav 78:1087–1095

Sueur C, Deneubourg JL, Petit O (2010) Sequence of quorums during collective decision making in macaques. Behav Ecol Sociobiol 64:1875–1885

Sueur C, Deneubourg J-L, Petit O (2011) From the first intention movement to the last joiner: macaques combine mimetic rules to optimize their collective decisions. Proc R Soc B 278:1697–1704

Sumpter DJT (2006) The principles of collective animal behaviour. Philos Trans R Soc Lond B Biol Sci 361:5–22

Thierry B (2000) Covariation of conflict management patterns across macaque species. In: Aureli F, de Waal FBM (eds) Natural conflict resolution. University of California Press, Berkeley, CA, pp 106–128

Thierry B, Singh M, Kaumanns W (2004) Macaque societies: a model for the study of social organization. Cambridge University Press, Cambridge

van Schaik CP (1983) Why are diurnal primates living in groups? Behaviour 87:120–144

Wang X, Sun L, Li JH, Xia DP, Sun BH, Zhang D (2015) Collective movement in the Tibetan macaques (*Macaca thibetana*): early joiners write the rule of the game. PLoS One 10:e0127459

Wang X, Sun L, Sheeran LK, Sun BH, Zhang QX, Zhang D, Xia DP, Li JH (2016) Social rank versus affiliation: which is more closely related to leadership of group movements in Tibetan macaques (*Macaca thibetana*)? Am J Primatol 78:816–824

Ward AJW, Sumpter DJT, Couzin ID, Hart PJB, Krause J (2008) Quorum decision-making facilitates information transfer in fish shoals. Proc Natl Acad Sci U S A 105:6948–6953

Ward AJW, Herbert-Read JE, Jordan LA, James R, Krause J, Ma Q, Rubenstein DI, Sumpter DJT, Morrell LJ (2013) Initiators, leaders, and recruitment mechanisms in the collective movements of damselfish. Am Nat 181:748–760

Watts D (2000) Mountain gorilla habitat use strategies and group movements. In: Boinski S, Garber PA (eds) On the move. University of Chicago Press, Chicago, IL, pp 351–374

Xia DP, Li JH, Garber PA, Sun LX, Zhu Y, Sun BH (2012) Grooming reciprocity in female Tibetan macaques *Macaca thibetana*. Am J Primatol 74:569–579

Zemel A, Lubin Y (1995) Inter-group competition and stable group sizes. Anim Behav 50:485–488

Chapter 6
Considering Social Play in Primates: A Case Study in Juvenile Tibetan Macaques (*Macaca thibetana*)

Jessica A. Mayhew, Jake A. Funkhouser, and Kaitlin R. Wright

6.1   Introduction

The evolutionary origins and adaptive value of animal play behavior have long been contemplated. Social play in gregarious animals is a multidimensional topic that has long been debated, insufficiently investigated, and a source of enigmatic questions regarding its development, relationship to cognition, and adaptive value. Play has been characterized as the most sophisticated manifestation of communication (Bekoff 1972; Bekoff and Allen 1997; Burghardt 2005; Fagen 1981; Pellis and Pellis 2009), which is partially why it is challenging to study. Beyond its incorporation into species' activity budgets, researchers have encountered multiple logistical and theoretical stumbling blocks, including difficulty in operationalizing definitions, identifying multiple juveniles and their varied social relationships, keeping pace with changes in interaction tempo and player composition, quantifying the behaviors observed and their short- and long-term costs and benefits, and determining an individual's motivation and intention to engage with others in this specific manner. Nevertheless, social play warrants attention because of its inherent complexity,

J. A. Mayhew (✉)
Department of Anthropology and Museum Studies, Central Washington University, Ellensburg, WA, USA

Primate Behavior and Ecology Program, Central Washington University, Ellensburg, WA, USA
e-mail: MayhewJ@cwu.edu

J. A. Funkhouser
Primate Behavior and Ecology Program, Central Washington University, Ellensburg, WA, USA

Department of Anthropology, Washington University in St. Louis, St. Louis, MO, USA
e-mail: jakefunkhouser@wustl.edu

K. R. Wright
Primate Behavior and Ecology Program, Central Washington University, Ellensburg, WA, USA
e-mail: wrightkr@uw.edu

© The Author(s) 2020
J.-H. Li et al. (eds.), *The Behavioral Ecology of the Tibetan Macaque*, Fascinating Life Sciences, https://doi.org/10.1007/978-3-030-27920-2_6

perceived contributions to an individual's fitness, communicative content, and interspecies variability.

In this chapter, we provide a brief overview of play behavior, with special attention to macaques, and propose some considerations to others interested in studying play. To emphasize some of these points, we provide an example where we consider the multiple factors characterizing the social play behavior of the 2017 Yulingkeng A1 infant and juvenile Tibetan macaques (*Macaca thibetana*) at Mt. Huangshan, China. Additionally, we map their positions in a social play network and use this foundation to generate hypotheses regarding their future network positions. We encourage future studies to address how the construction of these juvenile relationships contributes to an individual's overall group position and the formation of adult relationships and the potential adaptive advantage play provides.

6.2   Play Behavior: An Overview

The diversity in the form and content of play has generated discussion about the proximate and ultimate costs/benefits and whether generalizations about this behavior can be made across taxa. There are multiple definitions of play emphasizing either structure or function (e.g., Bekoff and Allen 1997; Fagen 1981; Martin and Caro 1985), but the development of five descriptive criteria by Burghardt (2005) has provided researchers the parameters to identify play as distinct from other common behaviors in a repertoire. For a behavior to be labeled "playful," Burghardt (2005) proposes it should (1) have a limited immediate function, (2) be endogenous, (3) have structural or temporal properties that are different from "serious" behaviors, (4) be flexibly exercised and not stereotypical, and (5) be performed in a relaxed field (i.e., free of stress or social/physical pressures). Guided by these criteria, researchers now have a foundation to tackle more complex questions of how and why animals play.

Play is often broadly categorized as solitary, object, and social (Bekoff and Byers 1981; Fagen 1981), but these categories are not mutually exclusive. Solitary play typically includes intense or sustained locomotor movements performed alone, e.g., the running and gamboling of young ungulates. Object play can be solitary or social and includes the manipulation of an object for no immediate benefit. Object play is commonly observed in carnivores, in which predatory movements, such as grabbing and shaking, are performed on a non-prey item (Burghardt 2005). Primates also engage in object play from stick carrying in chimpanzees (*Pan troglodytes*) (Kahlenberg and Wrangham 2010), stone handling in Japanese macaques (*M. fuscata*, Nahallage et al. 2016; Shimada 2006, 2010), to the manipulation of one's environment in functional object substitution, sparking discussion about pretense, imagination, and theory of mind (Gómez and Martín-Andrade 2002). Social play is identified as being interactive and occurs between two or more conspecifics who may influence one another's actions (Thompson 1996). Social play often includes aggressive behaviors, such as biting and wrestling (Burghardt 2005), but tends to be reciprocal between partners (Fagen 1981).

The type and exhibition of play depends on the species observed. Typically, the occurrence of play increases throughout juvenilehood but then tapers off at sexual maturity where activity budgets and individual priorities shift toward competition and reproductive resources. In birds, orders with more altricial species tend to engage in more play than orders with mostly precocial species (Ortega and Bekoff 1986). Raptors (e.g., eagles and hawks) engage in object play (Ortega and Bekoff 1986), young herring gulls (*Larus argentatus*) have been observed performing "drop-catch" behavior with clams and nonfood objects over hard substrates (Gamble and Cristol 2002), and ravens (*Corvus corax*) slide down snowy inclines and hang upside down from tree branches (Heinrich and Smolker 1998). Members of the family Canidae, including wolves, coyotes, foxes, and jackals, engage in solitary, object, and social play. Of this group, domestic dogs (*Canis familiaris*) are the most familiar play partners to humans and even use play signals that can be readily understood and responded to (e.g., the play bow) (Bekoff 1974). Cetacean play also takes on multiple forms, including the creation of play objects in the form of bubbles (Hill et al. 2017; Jones and Kuczaj 2014). For elephants (*Loxodonta africana*), play occurs throughout the life course on land and in water, and object play can be observed by individuals of all ages (Lee and Moss 2014). Guided by Burghardt's criteria, play can also be identified in animals that are not commonly regarded as being playful, including cichlid fish (*Tropheus duboisi*, Burghardt et al. 2014), Nile softshell turtles (*Trionyx triunguis*, Burghardt et al. 1996), octopus (*Octopus dofleini*, Mather and Anderson 1999), and even spiders (Pruitt et al. 2012).

Social play intrinsically involves partner cooperation, complex communication, and learning, and these are critical variables to investigate if we are to understand the cognitive and social development in young individuals (Bekoff and Allen 1997; Palagi et al. 2007). The functional significance of this suite of "nonserious" social behaviors remains elusive yet intriguing to ethologists; nevertheless, play has been noted to nourish the physical, social, and cognitive aspects of an individual (see Palagi 2018 for review). Namely, play functions to increase and maintain physical fitness or motor performance (Byers and Walker 1995; Fontaine 1994; Martin and Caro 1985), refines social skills and increases behavioral flexibility (Baldwin and Baldwin 1974; Brown 1988; Fagen 1984), and may be an important context that enhances cognitive skills in which an individual learns to identify and respond to the intentions of others through social cues (Bekoff 1972; Bekoff and Allen 1997; Palagi et al. 2007). These types of interactions serve as an opportunity to accrue and refine adult social skills and enhance behavioral flexibility (Baldwin and Baldwin 1974; Brown 1988; Fagen 1984). The unpredictable nature of this intimate social exchange challenges the participants to literally "think on their feet" as the interaction occurs, which imposes some inherent risk, but sharpens social tactics and reactions. In general, play is expected to occur more frequently with a partner that is evenly matched in skills and ability (*M. fuscata*, Kulik et al. 2015) as well as with individuals that are likely to be frequently encountered in adulthood (e.g., Maestripieri and Ross 2004). Pellis and Pellis (1996) hypothesize that rough-and-tumble play influences the developmental of dominance relationships in postpubertal juveniles (i.e., subadult), especially in male-male play bouts, through testing a play partner's strength. Intense or more aggressive behaviors, such as bite or chase, could be used to create or maintain a competitive edge in a play bout (e.g., *Gorilla gorilla*, van Leeuwen et al. 2011) that serves as practice for future, more aggressive interactions requiring the defense or acquisition of resources. Therefore, playing at increased rates and with a diversity of partners is likely an adaptive strategy during juvenile development to enhance one's success in "serious" behaviors later in life. However, with variable social systems (between or within species), it can be expected that juvenile play rates, relationships, and partner diversity depend on the life stage (i.e., the priorities of an individual at different ages) and group composition (i.e., social partner diversity and range restrict group demography).

In addition to potentially preparing a juvenile for future social experiences (Burghardt 2005; Fagen 1981), researchers can study play in attempt to predict the context, quantity, and dyadic relationship quality of future social relationships between group members (e.g., Pellis et al. 2019). Play partner choice tends to reflect the adult social partners of cooperative, competitive, and reproductive relationships experienced in adulthood (e.g., Maestripieri and Ross 2004). At its core, play builds upon reciprocity and turn-taking between players; without reciprocity and clear communication it is difficult to maintain the interaction and playful context. Partner selection is thus important, as regularly playing against larger, stronger individuals may mean being disadvantaged more frequently. Similarly, selecting a smaller, less skilled, or younger opponent might mean you possess a more frequent advantage but may require more self-handicapping (e.g., *Cebus apella*, Lutz and Judge 2017; *Macaca mulatta*, Yanagi and Berman 2014b). Interactions with larger age disparities may require increased communication, i.e., play signaling, to maintain a playful context. For example, self-handicapping behaviors (e.g., adopting a supine position) are positively associated with play signaling in domestic dogs (*Canis lupus familiaris*, Ward et al. 2008), and gorillas often pair an open-mouth face with more intense play behaviors, such as chasing, potentially to maintain an equitable level of cooperation between play partners (van Leeuwen et al. 2011).

In the order Primates, there is marked interspecific variation in play, and even within a species, the frequency in which an individual engages in play is not uniform. Teasing apart the contributing individual- and group-level variables for these differences is no small task, and these intrinsic and external factors can be multidimensional. For example, young infants may be physically handicapped when engaged in play with larger juveniles due to their small size, limited motor coordination, and unrefined social skills. However, young infants may also be physically restricted or actively discouraged by their mothers from participating in play (e.g., the mother has a restrictive rearing style or is low-ranking in the group). Other variables, including group social organization, structure, style, and status (Ciani et al. 2012; Fagen 1981; Maestripieri 2004; Thierry 2007), an individual's opportunity (Panksepp and Beatty 1980), prior experience (Cloutier et al. 2013), personality (Lampe et al. 2017; Pellis and McKenna 1992), and brain chemistry (Siviy et al. 2011) may also influence an individual's participation in and style of play. Although it has been acknowledged that play typically declines in frequency with age, some primates continue to engage in play beyond sexual maturity, maintaining playful relationships with juveniles and even playing with other adults (Pellis and Iwaniuk 2000; e.g., *Lemur catta*, Palagi 2009; *Macaca tonkeana*, Ciani et al. 2012; *Pan paniscus*, Palagi 2006; *Pan troglodytes*, Yamanashi et al. 2018; *Theropithecus gelada*, Mancini and Palagi 2009). The motivation to continue playing into adulthood likely differs depending on a variety of factors, including the degree of social tolerance in the species (e.g., despotic vs. egalitarian society), species-specific affiliative patterns of behavior, and the evolved communicative mechanisms to maintain a playful context.

6.3 Macaque Play

Macaca spp. vary widely in their geographical distribution, but the genus shares certain foundational similarities in social organization; for example, they can be found living in female philopatric, multi-male, multi-female groups with overlapping home ranges. However, the differing geographic distribution and phylogeny of the genus has resulted in interspecific variation in patterns of affiliation, reconciliation, dominance, aggression, nepotism, and temperament (Thierry 1985, 1990; Thierry et al. 2000). From this variation, the social organization of macaques is typically regarded as a continuous, four-grade scale of dominance style: species occupying the first grade are considered hierarchical and nepotistic and those occupying the fourth grade are considered more tolerant or egalitarian (Thierry et al. 2000). Dominance style can be defined as the dominance relationships, categorized by agonistic interactions, within dyads in a social group (Thierry et al. 2000). A difference in dominance style between primate taxa can be the result of environmental variables, such as contest over food resources (Matsumura 1999), but may also be context-dependent (Funkhouser et al. 2018b). Grade one despotic species are generally marked by dominant individuals that show intense and highly asymmetrical patterns of aggression, little tolerance around resources, and infrequent reconciliation (e.g., *M. mulatta*, *M. fuscata*, and *M. cyclopis*). Species with a grade four dominance style show the opposite tendencies, with low or moderate levels of kin bias in affiliation, tolerant and supportive interactions with group members, strong group cohesion, and maternal tolerance for infant handling, for example, *M. maura*, *M. nigra*, *M. ochreata*, and *M. tonkeana*, all endemic to Sulawesi (Thierry et al. 2000). Although the above species fit easily onto this graded scale, other macaques are more difficult to categorize based on inconsistent or a lack of behavioral data. The classification of species as a grade two or three can often rely on relative comparisons: grade two macaques possess behavioral traits that are more similar to grade one species, and grade three macaques are more similar to grade four species.

Regardless of their place on this graded scale, the variation in social tolerance observed across the 23 extant macaque species provides an opportunity to directly compare different facets of behavior, including play structure and content, across multiple, differing social structures. Currently, much of the play research in macaques derives from species occupying opposite ends of the dominance scale, and the differences in observed play style, rate of play, and play signaling have been suggested to reflect this social dominance style (Petit et al. 2008; Reinhart et al. 2010; Yanagi and Berman 2014b). The social play of despotic species, such as rhesus (*M. mulatta*) or Japanese macaques (*M. fuscata*), can be characterized as competitive, whereas more tolerant species, such as Tonkean (*M. tonkeana*) or Sulawesi crested macaques (*M. nigra*), engage in a more cooperative play style (Petit et al. 2008; Reinhart et al. 2010). These differences are reflected in the targets attacked during play fighting as well as the type and frequency of play signals utilized throughout a play bout (Reinhart et al. 2010; Scopa and Palagi 2016; Yanagi and Berman 2014b). A more competitive or risky play style, one in which sensitive targets like the face are attacked, may generate miscommunication between partners and risk ending the play contact. Therefore, using play signals to indicate imminent play (Yanagi and Berman 2014a) or to reinforce playful intent (Wright et al. 2018) could be used to mitigate potential aggression and prolong the interaction (Scopa and Palagi 2016). These signals may be more specific and less interchangeable in more aggressive or competitive species to minimize the risk of misinterpretation (Scopa and Palagi 2016; Thierry et al. 2000; Yanagi and Berman 2014b) whereas in more tolerant species, these signals may be used redundantly and interchangeably to initiate/terminate play (Pellis et al. 2011; Scopa and Palagi 2016). Thus, investigating grade two and grade three macaque species, such as Tibetan macaques (*M. thibetana*), can help to determine the degree of overlap in species play patterns and play signaling.

6.4   Tibetan Macaques

Tibetan macaques are female philopatric and live in multi-male, multi-female social groups of 15–50 individuals (Berman et al. 2004; Thierry 2011; Thierry et al. 2000). This species has a strong kin bias with linear male and female dominance hierarchies, in which males generally occupy the top ranks of the hierarchy although females can outrank them (Berman et al. 2004). Despite being originally designated as having a grade three dominance style (Thierry et al. 2000), Tibetan macaques have been re-established as grade two, showing some despotism and low conciliatory tendencies, especially for female-female interactions (Berman et al. 2004). However, some female behavior is also inconsistent with a more despotic style, including that female individuals display a markedly high preference for female kin in proximity relationships and maternal tolerance for infant handling (Berman et al. 2004).

Female *M. thibetana* rank is based on matrilines with a daughter occupying the rank directly below her mother and above her older siblings (Berman et al. 2004; Thierry 2011; Zhao 1997). This hierarchy influences intergroup competition among females and generates preferential bonds between kin (Thierry 2011). Tibetan macaques groom at symmetric rates (exchange grooming for grooming received) and prefer female kin grooming partners, and females prefer to groom higher-ranking females (even if unrelated) (Xia et al. 2012). These investigations illuminate the generally despotic nature of Tibetan macaque social organization (Berman et al. 2004; Thierry et al. 2000), bias for female kin across a number of social contexts (viz., coalitionary support, grooming, and infant handling; e.g., Berman et al. 2004), and the overall value of grooming in this species (e.g., Xia et al. 2012, 2013). Tibetan macaques and other species with intermediate, graded behavioral variability may demonstrate behavioral nuances that are more likely a difference in degree rather than kind; therefore, it is important to determine the extent of these differences and how they manifest and compare across species.

Similar to other macaque species, Tibetan macaque juveniles engage in dyadic and polyadic play (Wright et al. 2018) that is fast-paced and rough and tumble. Few studies have examined social play in juvenile Tibetan macaques, primarily because few studies have been conducted on this species compared to other species of macaques. The play studies that have been performed have occurred at Valley of the Wild Monkeys in Anhui Province, Huangshan, China, specifically with the Yulingkeng A1 (YA1) group. Juveniles in this group have been noted as preferring similarly aged partners for both social play bouts and affiliation (Batts 2012). The type of play engaged in, whether solitary or social, also appears to depend on the age of the individual: infants tend to engage in more solitary play and juvenile males engage in the most social play (Batts 2012). The majority of juveniles in this group participate in play, but participation frequency, duration, and the rate of play signaling is variable (Wright et al. 2018). Nine candidate play signals, such as crouch and stare, have been observed, six of which overlap with the play signal repertoire of rhesus macaques (Yanagi and Berman 2014a). In this group, the play face is an important signal used throughout bouts but is not a reliable indicator that play will be initiated (see Wright et al. 2018 for a more detailed discussion). Having other individuals in proximity to the play bout (an audience) also seems to impact the play signaling of individuals involved, i.e., the data represent a negative parabolic trend. Play signaling in these juveniles increased as the number of audience members increased from zero to two but decreased with additional individuals beyond this threshold suggesting that the communication value of the play signal may degrade as the complexity and size of the play group increases (Wright et al. 2018). However, additional comparative research on this topic is necessary.

In the following example, we build upon previous social play studies of the YA1 juvenile Tibetan macaques of Mt. Huangshan, China, using social network analyses to supplement our current understanding of the social dynamics of this group. Although the majority of research performed on these macaques highlights the relationships between the adults, such analyses can help construct a more complete picture of group social dynamics. Additionally, this information can be used to generate future hypothesis-driven research about potential adult relationships and social position within the group as the juveniles age and become integrated into the adult social network.

6.5   Study Subjects and Data Collection

The YA1 group of Tibetan macaques resides at Valley of the Wild Monkeys in the Huangshan Scenic District, Anhui Province, China. This group has been habituated to human presence since 1986 (see Berman and Li 2002; Berman et al. 2004) and is free-ranging but regularly provisioned with corn multiple times per day in supplement to their natural diet. All group members are individually recognized, and records of the group structure, including name, sex, birth date, and matriline, are maintained by Anhui University researchers. Additional adult female grooming data, collected July 7–August 28, 2016, between 06:30 and 17:30, was used to supplement some of the following analyses. To generate maternal dominance rank, all-occurrence sampling (Altmann 1974) was used to collect agonistic data from July 14 to August 27, 2016, from 7:00 to 12:00 and 14:00 to 17:00 daily (data contributed by Lori K. Sheeran). All research herein was approved by the Central Washington University Institutional Animal Care and Use Committee (protocols: #A051602, #M061603, #A051702), and all protocols adhered to the legal requirements of the People's Republic of China and the American Society of Primatologists' Principles for the Ethical Treatment of Primates.

6.5.1   Maternal Allogrooming and Dominance Rank

In summer 2016, the YA1 group was composed of 47 individuals (19 males, 28 females): eight adult males, 13 adult females, and 26 infants and juveniles (between the ages of approximately 30 days and 6 years old) (see Table 6.1). All adult females were randomly sampled, and 10-min focal follows were conducted to collect all instances of auto- and allogrooming. This generated approximately 400 min of observation time per focal individual. Data were collected on actor/recipient identities, rank, sex, matriline, and duration. An allogrooming bout was initiated with physical touch between partners, and the bout ended when all grooming partners ceased to groom for >10 s. If a grooming bout was polyadic (more than two individuals), the identity and durations of all partners were recorded, and the interactions were treated as dyadic.

To determine maternal rank, agonistic data were collected. Agonism consisted of fear grin, scream, flee, displace, threat, lunge, chase, grab, slap, and bite (as defined by Berman et al. 2004). We coded unambiguously directed *agonistic interactions* (or "competitions") in a 1:0 dichotomous fashion, where 1 indicated the "actor" who "won" the interaction and 0 indicated the "recipient" who "lost" the interaction. For these reasons, submissive behaviors (*lack of agonism*) were reverse-coded, where the actor was said to have lost (0) to the winning (1) recipient. We then derived Elo scores for each individual using methods similar to Neumann et al. (2011) in R.

Table 6.1 Yulingkeng A1 player demographics in 2017

Name	Age (days)	Age (years)	Sex	Mother
TouQiuGuo (TQG)	30	<1	F	TouXiaHua
YeXiaYun (YXYun)	49	<1	F	YeHong
YeXiaDuo (YXD)	65	<1	F	YeChunYu
YeXiaMing (YXM)	102	<1	M	YeChunLan
HuaXiaYun (HXYun)	126	<1	F	HuaHong
TouHuaLi (THL)	399	1	F	TouRongYu
TouFuHua (TFH)	451	1	F	TouHuaYu
TouQiuYing (TQY)	474	1	F	TouXiaXue
YeXiaYue (YXYue)	794	2	F	YeHong
TouQiuSong (TQS)	802	2	M	TouXiaHua
TouHuaNan (THN)	823	2	M	TouRui
HuaXiaYue (HXYue)	852	2	F	HuaHong
YeChunHua (YCH)	1461	4	F	Unknown
YeRongLan (YRL)	1514	4	F	YeZhen
TouRongXi (TRX)	1537	4	F	TouTai
TouQiuLan (TQL)	1549	4	F	TouXiaXue
HuaXiaWei (HXW)	1579	4	F	HuaHong
YeXiaKun (YXK)	1588	4	M	YeHong
HuangYu (HY)	1643	5	M	Unknown
YeChunLan (YCL)[a]	1740	5	F	YeMai
TouXiaLong (TXL)	1817	5	M	TouHong
TouHuaXue (THX)[a,b]	1889	5	F	TouRui
TouRongYu (TRY2)	1969	5	M	TouTai
YeChunLong (YCLong)[a,b]	2557	7	M	YeMai
YeRongQiang (YRQ)[a]	2586	7	M	YeZhen
TouRongGong (TRG)[a]	2635	7	M	TouTai
HuaXiaMing (HXM)[a]	2675	7	M	HuaHong

Age was calculated from the recorded birth date to 6-21-2017, which was when Anhui University researchers prepared the demographic information for the field season. In instances where the birth date and matriline were unknown (e.g., YeChunHua and HuangYu), age was estimated based on the individual's size and sex characteristics
[a]The six individuals classified as adults during 2017, including two 5-year-old females and four 7-year-old males
[b]The two adult individuals of the six who participated in play with younger individuals (THX and YCLong)

6.5.2   *Juvenile Play Behavior*

In summer 2017, the YA1 group was composed of 46 individuals (17 males, 29 females): 10 adult males, 15 adult females, and 21 infants and juveniles (between the ages of approximately 30 days and 5 years old). Data were collected from July 6 to August 5, 2017, between 06:30 and 17:30. A complete list of player demographics, including age, sex, and matriline, are shown in Table 6.1.

Video footage of social play behavior was collected using all-occurrence sampling and a Sony Handycam camcorder. Regular scans of infant and juvenile interactions occurred until social play was initiated. If a second play bout occurred while recording the first, the video frame was widened to include both bouts. If this was not possible, the first bout was recorded to completion, and then the second bout was recorded. The juveniles were followed until play concluded and the players ceased to engage in play for >10 s. All play videos were coded using VLC media player (version 3.0.3, Vetinari) and Microsoft Excel (version 16.14) for player identity, demographic information (e.g., age, sex, and matriline), duration of the play bout, and the total number of players involved. All ages were recorded in days, and all durations were recorded in seconds.

A coding distinction was made between the longer encompassing *play bout* and the smaller component units designated here as *play periods* (see Mayhew 2013 for a discussion of play structure, including play periods and vigilance periods). Here, only play periods were examined. Guided by Burghardt (2005) and Fagen (1981), the start of a play bout was marked by the exchange of playful behavior between two or more juveniles. A play bout was considered to have concluded when (1) *a player engaged in non-playful behavior* (e.g., aggression), (2) *a non-player directed non-playful behavior toward the players* (e.g., the bout was interrupted by an adult initiating a grooming session), or (3) *player(s) withdrew from the interaction, thus dissolving the play group*. Because play dynamics change quickly, it was noted each time a player withdrew from or was added to the interaction even if play continued following this change in player composition. Each partner interchange marked the transition to a new play period; for example, players A and B are playing, and player C approaches and joins; therefore the play period between A and B terminates and a new period containing players A, B, and C begins (see example in Fig. 6.1).

6.5.3   Statistical Analyses

As indicated by its frequent appearance in recent animal behavior studies, social network analysis (SNA) has become a valuable tool for understanding the role and

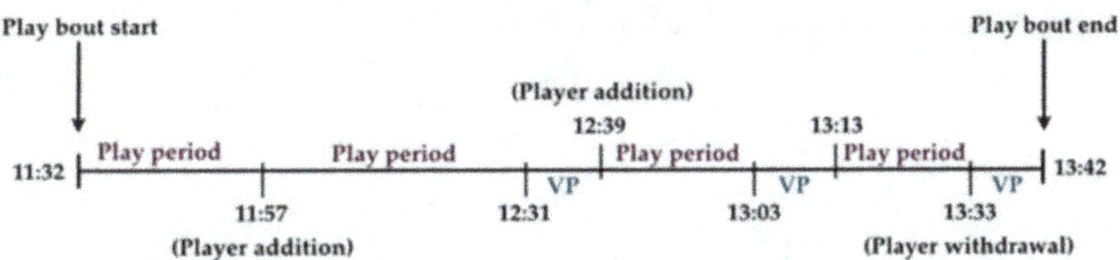

Fig. 6.1 An example of a play bout divided into its smaller components: play periods and vigilance periods (VP). A change in player composition (addition or withdrawal) is marked and indicates a transition into a new play period (adapted from Mayhew 2013)

positioning of individuals within a larger social group. SNA uses matrix-based data to analyze individual social interactions, where individuals can be depicted as nodes and network-based descriptive and statistical analyses are generated to describe the relationship between nodes (Sueur et al. 2011a; Whitehead 2008). A comprehensive overview of the utility of SNA, its terminology, methods, and analyses can be found elsewhere in comprehensive texts (see Borgatti et al. 2013 or Whitehead 2008).

SNA has been used to characterize these relationships in primate social groups, as well as identify clusters of individuals, subgroups, isolated group members, and, overall, diagram of the group's social network (e.g., Clark 2011; Farine and White-head 2015; Funkhouser et al. 2018a; Sueur et al. 2011a). For example, using SNA, researchers have examined the correlation between juvenile play network positions, ontogenetic social development, and later-life social connectedness (*P. troglodytes*, Shimada and Sueur 2014; *M. fuscata*, Shimada and Sueur 2017; *Macaca* spp., Sueur et al. 2011b). Animals that are strongly connected to one another (i.e., having strong bonds) tend to have relationships on multiple independent measures of relation; thus, it can be expected that playing preferentially with certain partners will correlate with other measures, such as grooming, copulation, and proximity (Whitehead 2008).

In this study, all statistical analyses were performed using IBM SPSS Statistics (Version 23) software ($\alpha = 0.05$). UCINET (Borgatti et al. 2002) was used to calculate network statistics, and NetDraw was used to construct sociograms. Elo scores (calculating position in the dominance hierarchy) were derived using R. To investigate correlations between independent social networks (e.g., durations of play and difference in age) we used QAP correlation analyses in UCINET; this test analyzes whole matrices against one another. The following node-by-node statistics were calculated: (1) *degree* (the sum of each node's ties with all other nodes, also known as strength), (2) *eigenvector centrality* (how well an individual is associated with others and how well the associates are associated), (3) *closeness* (how close nodes are within a network, often discussed in terms of the time it takes for information to spread from one individual to others), and (4) *betweenness* (a value used to assess a node's ability to control flow through a network). We also constructed a principal coordinate sociogram to illustrate the observed social play relationships between individuals. This sociogram plots individuals with strong associations near one another. To define the minimum edge value in this diagram, we used the mean of all directional dyadic indexes plus one standard deviation (mean + SD). Detailed information about these node-by-node statistics can be found in Borgatti et al. (2013) and Whitehead (2008).

6.6 Results

In total, 256 play bouts (12,458 s total) were observed containing 965 play periods. Twenty-three macaques (eight males, 15 females) participated in play, ranging in age from approximately 30 days (TQG) to 7 years old (YCLong) (Table 6.1). *Play bouts* ranged in duration from one to 895 s (nearly 15 min), but the mean was considerably shorter (48.99 ± 80.33 s). *Play periods* ranged in duration from 1 to 227 s with a mean of 12.91 ± 16.70 s.

6.6.1 *Player Age*

Play was observed more frequently between players of the same birth cohort (i.e., individuals born in the same year) (QAP, $r = 0.289$, $p = 0.001$) and age class (QAP, $r = 0.219$, $p < 0.001$). The mean age of a player was 867.66 ± 496.55 days old (approximately 2.4 years old), and the mean age of male players (977.75 ± 488.22 days) was slightly older than female players (736.84 ± 476.56 days). A Pearson correlation was used to determine whether the age of the player (in days) correlated with the total amount of time spent playing with any other player(s) during the study period. Player age was significantly correlated with play duration ($r = -0.246$, $p \leq 0.001$), and this negative relationship suggests that as age increased, the time spent involved in play decreased. Notably, when only dyadic play periods were examined, the ages of the players were positively correlated ($r = 0.467$, $p \leq 0.001$), indicating that as the age of Player 1 increased, the age of Player 2 also increased (Fig. 6.2).

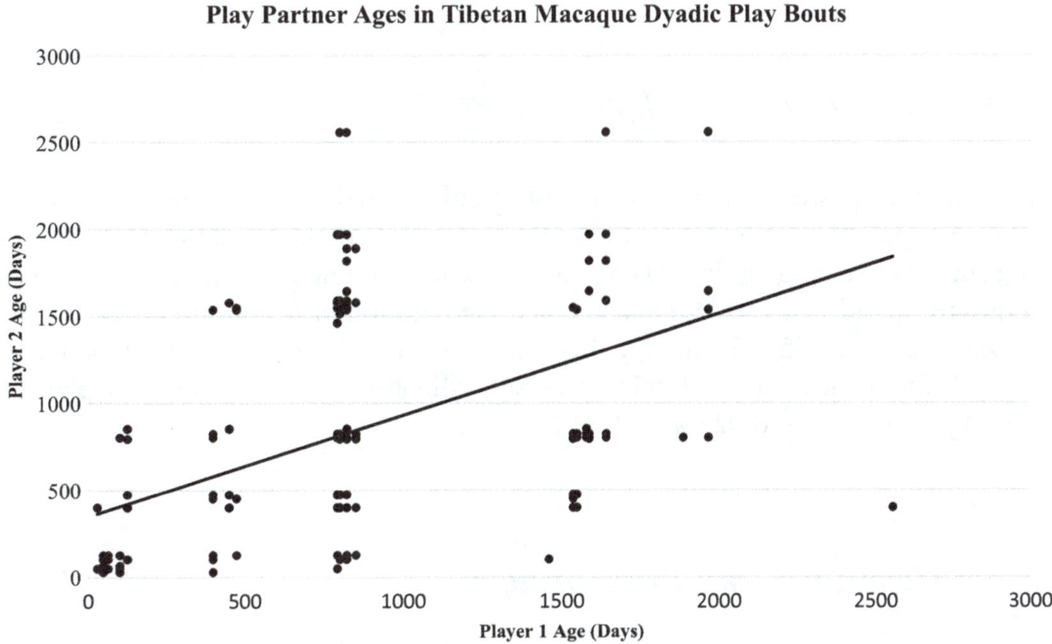

Fig. 6.2 A scatterplot of player ages in dyadic play bouts (two individuals) indicating that as the age of one partner increased, the age of the second partner also tended to increase ($r = 0.467$)

6.6.2 Number of Players

Consistent with Wright et al. (2018), polyadic play periods were observed in this group of macaques (four players, $n = 66$; five players, $n = 10$) but occurred less frequently than dyadic ($n = 628$) or triadic ($n = 261$) interactions; the mean number of players in a play period was 2.44 ± 0.67. Similarly, the total number of players in a play period was significantly correlated with the duration of a period ($r = -0.138$, $p \leq 0.001$), indicating that playing with more individuals decreased the duration of play. Additionally, the total number of players in a play period was significantly correlated with player age, indicating that as player age increased, the number of players involved decreased ($r = -0.152$, $p \leq 0.001$).

6.6.3 Player Composition

An analysis of player composition indicated that mixed sex play ($n = 533$ periods, 55.2%) represented more than half of all play periods (all-male play, $n = 273$, 28.3%; all-female play, $n = 159$, 16.5%), and this was significantly different from the expected values ($\chi^2(2) = 228.47$, $p \leq 0.001$). Therefore, juvenile males remained active participants despite the presence of more juvenile females in the group.

6.6.4 Matrilineal Relatedness and Rank

YA1 matrilines are known and documented annually by Anhui University researchers; therefore it was possible to determine and rank the degree of maternal relatedness between juveniles. There was no significant relationship between maternal difference in Elo score (dominance) and the duration of play in offspring (QAP, $r = 0.00$, $p = 0.025$). Similarly, there was no significant relationship between maternal dominance status (dominant or subordinate) and the duration of play in offspring (QAP, $r = -0.025$, $p = 0.411$).

6.6.5 Maternal Social Relationships

Adult female grooming duration data from 2016 was factored in as a proxy for maternal social dyadic relationships, but there was no significant relationship between maternal grooming (total seconds) and the duration of play in offspring (QAP, $r = 0.018$, $p = 0.274$).

6.6.6 Individual Playfulness

In addition to the above demographic variables, individualistic patterns emerged for the juveniles in this group. Predictably, some juveniles played more frequently and longer than others (see Figs. 6.3 and 6.4), and overall, TouQiuSong (TQS), TouHuaNan (THN), and YeXiaYue (YXYue) were the three most active participants (appearing in 56.58, 40.70, and 32.40% of play periods observed, respectively). All three of these players were born within 29 days of one another and belonged to the 2015 birth cohort. Interestingly, the oldest individual in this cohort, HuaXiaYue (HXYue), was born 29 days before THN, but did not participate in play nearly as much as her three peers.

Network statistics for each player were calculated, including degree, eigenvector centrality, closeness, and betweenness (Table 6.2). To better visualize these play relationships, a sociogram was constructed (minimum edge weights of mean $\pm$ SD) for all dyadic interactions (Fig. 6.5). Examining the network statistics and sociogram, TQS, THN, and YXYue appear to unite three clusters of players.

Based on a cursory examination of the play frequency and duration data as well as from personal observation, it is unsurprising that TQS, THN, and YXYue were the

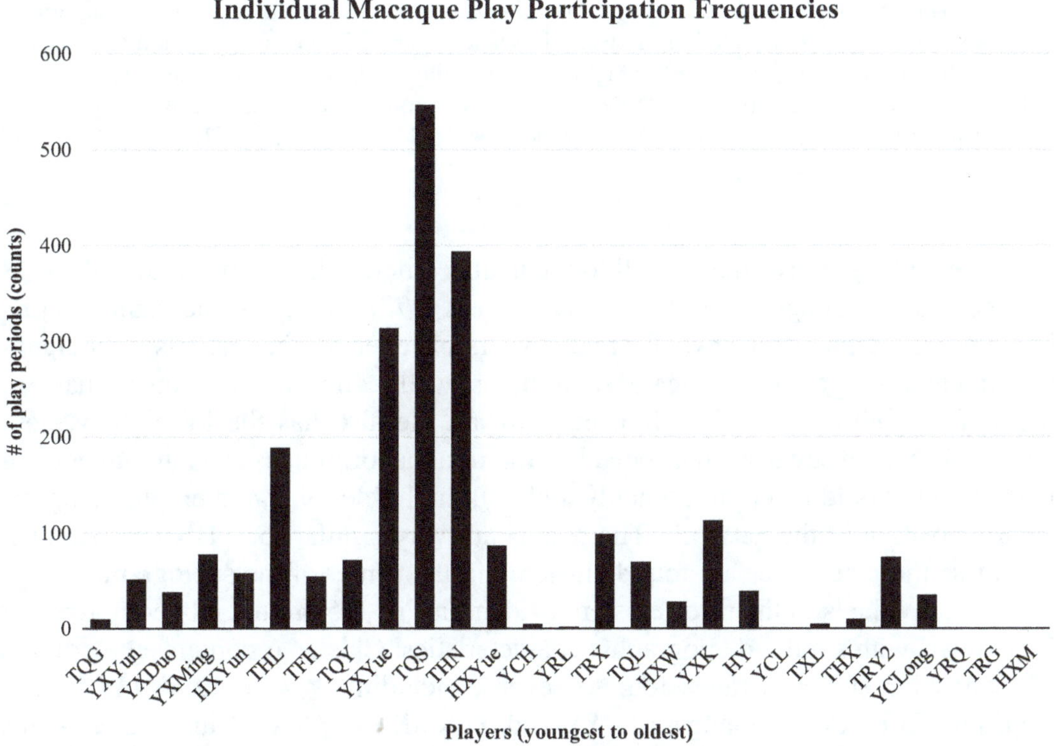

Fig. 6.3 The total individual counts of play participation for all group members ≤ 7 years old in the YA1 group during 2017. TouQiuGuo (TQG) on the left is the youngest player at 30 days old and HuaXiaMing (HXM) is the oldest at 2675 days. Some individuals were not observed to engage in play at all (e.g., HuaXiaMing), whereas others (e.g., YXYue, TQS, and THN) were frequent participants

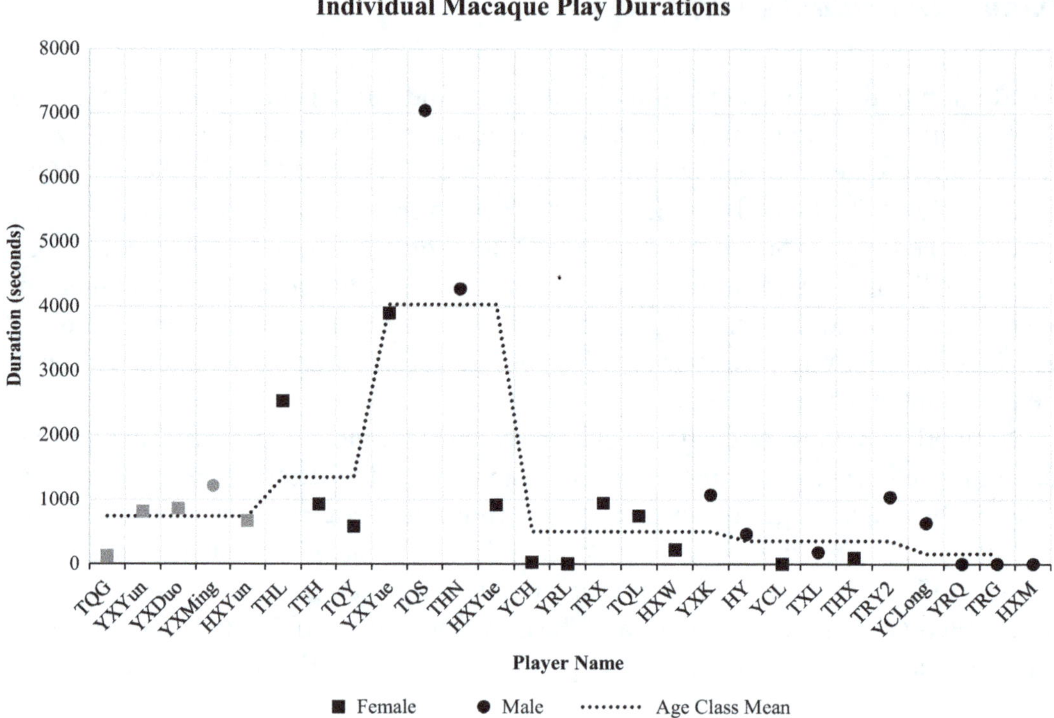

Fig. 6.4 The total individual durations of play (in seconds) for all male and female individuals ≤7 years old. Males are depicted as circles, females as squares, infants as gray, and juveniles as black. Individuals are organized left to right by age in days (TQG is the youngest, HXM is the oldest). The dotted line represents the mean play duration for each age class, including all individuals of that age class even if they were not observed playing (e.g., YCL, YRQ, TRG, and HXM)

top three ranking individuals for all four calculated network measures. Overall, TQS has the highest strength (6896) and betweenness (47.905) values, indicating that he is highly gregarious and links discrete clusters of juvenile players. His high eigen-vector centrality of 1.0 indicates that he is well connected to others that are themselves well connected within the network. He also has the lowest closeness value (24), which suggests that social information that originates at a random node in the network would reach him quickly and with high fidelity, again emphasizing his central position in the network. The results are very similar for THN and YXYue, and these three juveniles are found clustered in the center of the sociogram.

The individuals in the three clusters evident in Fig. 6.5 are notably comprised of individuals with similar demographic characteristics. Cluster A contains the group's older and larger males (ages four to seven), including YeChunLong (YCLong), TouRongYu (TRY2), HuangYu (HY), and YeXiaKun (YXK). Cluster B contains nearly all remaining juveniles (ages two to four), who are predominantly female, including TouRongXi (TRX), TouQiuLan (TQL), TouQiuYing (TQY), TouHuaLi (THL), and TouFuHua (TFH). Cluster C contains HXYue (age three) and the majority of the group's infants: YeXiaYun (YXY), YeXiaDuo (YXD), and YeXiaMing (YXM). Clusters A and C appear to be connected via cluster B,

Table 6.2 2017 Yulingkeng A1 player network measures sorted by degree

Player	Degree	Eigenvector centrality	Closeness	Betweenness
TQS	**6896**	**1**	**24**	**47.905**
THN	**4448**	**0.74**	**26**	**28.6**
YXYue	**4108**	**0.791**	**25**	**28.807**
THL	3393	0.579	30	6.211
YXMing	1351	0.062	30	13.186
TRX	1266	0.153	34	1.5
TQL	1237	0.145	35	0.25
YXK	1215	0.216	37	1.715
TRY2	1201	0.15	36	1.394
TQY	1187	0.155	32	1.313
YXDuo	1027	0.016	46	0.2
YXYun	1023	0.054	35	3.781
TFH	1005	0.124	32	1.153
HXYue	879	0.144	32	1.416
HXYun	722	0.07	32	7.197
YCLong	719	0.056	39	0.966
HY	539	0.088	36	2.276
HXW	291	0.045	34	0.862
TXL	188	0.014	43	0.2
TQG	161	0.006	40	1.069
YCH	36	0.008	40	0
YRL	4	0.001	44	0

Bold indicates the top three individuals with the highest network measures

suggesting that certain individuals or individuals of intermediate age more generally act as scaffolding for infants into the juvenile social network. One juvenile was a notable social isolate with the lowest degree, eigenvector, and betweenness values and highest closeness value: YeRongLan (YRL), who only participated in one play bout for four seconds. Similarly, other individuals had few or weak social network connections, reflected by their low degree, eigenvector, and betweenness values and high closeness scores: YeChunHua (YCH), TouQiuGuo (TQG), TouXiaLong (TXL), and HuaXiaWei (HXW). These isolated individuals were both males and females and ranged from infancy to nearly 5 years old.

6.7　Discussion

The adult macaques in the YA1 group are well known and studied, but juvenile interactions have not been well documented. The results outlined in this chapter provide a window into the social lives of these juveniles, and the results presented here build upon previous Tibetan macaque social play research performed at this site (viz., Wright et al. 2018). Importantly, the findings in this work aim to establish these

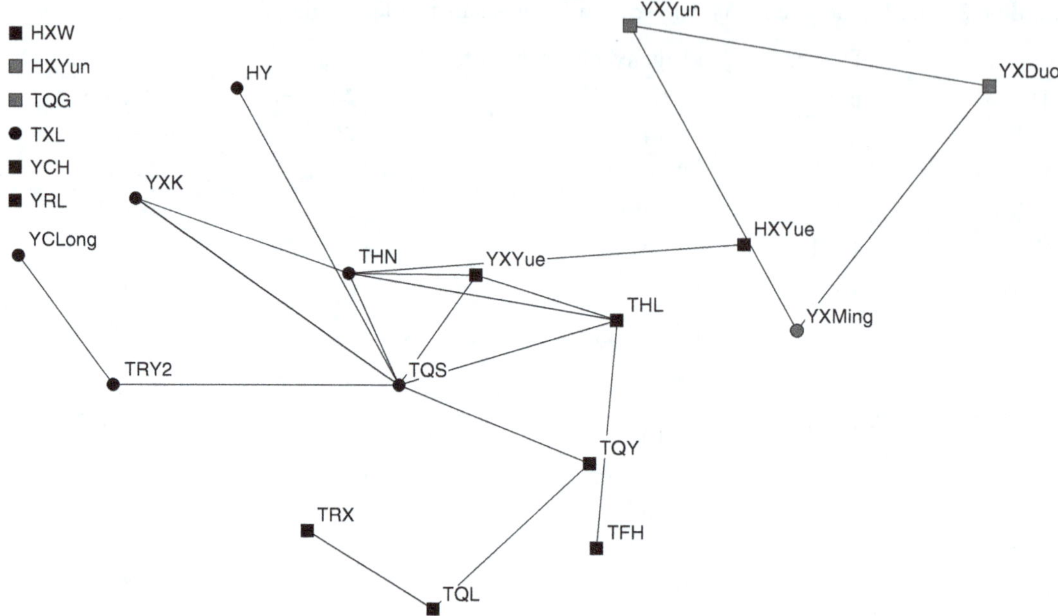

Fig. 6.5 Sociogram for the 2017 juvenile and infant play network based on duration data (mean $\pm$ SD = 242). This network highlights those juveniles who participated more frequently in social play. Males are depicted as circles, females as squares, juveniles as black, and infants ($<$1 years old) as gray. YCLong is included in this network as a juvenile but was categorized as an adult at the time of the study

juveniles as important entities within the group that have their own identities and the potential to drive the direction of group social dynamics as they age and continue to interact.

The frequency of social play in these juvenile macaques followed a similar trajectory to other primate species: play increased as age increased, peaked at approximately 2.5 years old, and then declined. Adult macaques in this group rarely participated in play, which supports previous literature indicating a general lack of adult primate play in more despotic species. Additionally, the number of players involved in a play period was fewer when the participating players were older. The QAP results indicated that juveniles had a preference for certain social partners and played more frequently with peers of their birth cohort. In dyadic play bouts, partner ages were positively correlated, which is a further indication of a preference for similarly aged partners. Taken together, partner preferences may not only be indicative of the strong social bonds that result from growing up together in the same age cohort but also reflect a preference for equity in social competition (Kulik et al. 2015; van Leeuwen et al. 2011). As juveniles age, especially males, physical strength and prowess build and are increasingly put to the test in competition with other individuals. This partner equity may be particularly important for more despotic macaque species in attempt to ensure that play occurs with an appropriately matched opponent in skills, abilities, and size in attempt to reduce the risk of play devolving into aggression by interacting with a mismatched partner. Tibetan macaques also engage in polyadic play, which can quickly become complicated for the participants as well as the researchers observing the interaction. Typically, the duration of these polyadic play events is shorter than when play occurs between dyads. This is reasonable considering that when more players become involved in an interaction, it becomes increasingly difficult for each player to monitor their own behavior as well as track the behaviors of all the other partners simultaneously. This inability to effectively monitor multiple partners may contribute to the brevity of polyadic play. Such events also open the door for future research analyzing the structure and composition of these bouts in comparison to dyadic events.

Although there may be some general conclusions to be made about play in macaques depending on their degree of social despotism or tolerance, the influence of individual identity and personality on play participation warrants further exploration. Some individuals, simply based on personality traits (e.g., boldness, tendency to explore, or sociability), may be more inclined to participate in play regardless of their partner's age, sex, or relationship. Although personality data were not collected for these juvenile macaques, and the dataset here only represents a snapshot of the relationships of this group, predictions could be made about which juveniles likely possess which traits and to what degree. This is undoubtedly a large undertaking; however, such data could generate some understanding of how personality contributes to social play participation and can aid in generating predictions about how certain individuals will behave in other realms of social behavior.

6.7.1 Does Social Position Matter?

The introduction of SNA into social play analyses is useful in that it not only literally illustrates the complex relationships shared by juveniles but also provides a method to track individual relationships and compare them across years, events, and other social networks (e.g., Shimada and Sueur 2017). Specifically, SNA allows researchers to follow an individual across time and various social networks to investigate the effects of an individual's characteristics during play as a juvenile (e.g., frequency, partner choice and diversity, and play style) on their position in adult social networks. Such a longitudinal, multilayer social network framework would provide a pointed investigation of the adaptive mechanisms present during the juvenile play period that may impact reproductive fitness later in life (e.g., dominance status, reproductive success, infant survival). The incorporation of such information into future social play studies may be of interest to researchers performing long-term studies, where data-driven predictions about the role or tenure of future juveniles within the social group is invaluable. For the YA1 group, this type of information may be helpful to future researchers interested in the development and maintenance of Tibetan macaque relationships.

In 2017, it appears that TQS, THN, and YXYue provided the foundation for this social play network in 2017, acting as social glue and connecting juveniles across multiple birth cohorts and matrilines. Interestingly, the grandmothers of THN and TQS were central in the 2016 female allogrooming network (TouTai and TouHong, respectively, unpublished data), so it is possible that well-connected maternal relatives influence the social freedom or social position of kin across years. Whether the central positioning of this trio is an artifact of their particular age (nearly 3 years old), their early cohort experience (close dates of birth), or previous sibling or kin scaffolding into the social network is currently unknown, but additional SNA data from 2015 and 2016 can help to clarify this picture.

The central nature of TQS, THN, and YXYue can also be used to make predictions about the transition of future dominance status and mating opportunities in this group. The YA1 group has a history of some resident males remaining in the group longer than expected and rising through the ranks to alpha position (e.g., YeRongBing, HuaXiaMing, and TouGui). This longer group tenure might also be expected of THN and TQS, and they may not be immediately pressured by group males to disperse. As a female, YXYue is expected to remain in YA1, and once she reaches sexual maturity, she will likely fall into place at the top of the hierarchy with the rest of her Ye female relatives, achieving this high rank namely because her mother, YeHong, is the alpha female.

Social predictions can also be made about individuals that occupy more peripheral positions in the juvenile network. For some, such as the infants of the group, this position will most likely change with age and increased social exposure to their peers or may be mediated by the infant's matriline. For others, this window of social opportunity may have closed, and this peripheral social position may persist into adulthood. One important advantage of reliably identifying unique infants and juveniles is that it allows for young individuals to be tracked across development to test for changes in the status of their social lives. Further, reliably identifying young individuals ensures control for bias in data collection, whereby ensuring that nongregarious or cryptic individuals (that might be easily overlooked because they play less) can be accurately represented in a dataset.

Although the infants in this group all had low eigenvector values (less central), they comprised their own cluster and the majority were not social isolates (peripheral). The isolation of TouQiuGuo (TQG) from the juvenile network can be explained by her very young age (a newborn). The clustering of the remaining newborn peer group may have occurred because of limited social access and ability. For infants, having access to social partners beyond their mother can be restricted because macaque mothers often carefully control their offspring's experience during their first vulnerable months. For Tibetan macaques, this can be a physical restriction, where the mother will not release the infant from her grasp, or a social restriction, where the mother actively monitors and interferes in the social interactions between her offspring and other group members for at least the initial weeks of life. Indeed, Tibetan macaque mothers often affiliate or associate with one another, thereby providing limited opportunities for their infants to socially engage while still being heavily monitored. Regardless, this mothering behavior limits the infant's ability to socially engage relative to the rate of older (juvenile) conspecifics. Aside from the influence of maternal rearing behavior, the infant itself has limited physical abilities that prohibit it from keeping up with older players. Growth and development bring physical, mental, and emotional strength, flexibility, endurance, and knowledge of the natural environment, but this is inherently a slow, prolonged process, one that requires time as well as practice and repeated experience. A newborn infant is also a novel object in the environment that others must familiarize themselves with; therefore, it is not simply that the infant must gain physical and social competence to enter into this new social world; they must also gain acceptance from members of the social group and be treated accordingly by older play partners.

A handful of juveniles were social isolates. Both YCH and YRL were rarely seen interacting with any members of the YA1 group, juvenile or otherwise. YCH is thought to be a juvenile immigrant (rare), and therefore she lacks kin relations on which to base her social behaviors; this could explain her peripheral social position. YRL's mother, YeZhen, occupied the lowest position in the social hierarchy during 2017, and this low rank could have impacted YRL's social position within the juvenile network by limiting her opportunities for social engagement. It was surprising to uncover the weak connection of TouXiaLong (TXL) to the juvenile play network because his mother, TouHong, occupied a central, well-connected position within the group's 2016 grooming network (unpublished data). Further, TXL's maternal cousin is TQS, a highly playful and central individual (see above). Although the position of TXL was wholly different from TQS, a possible explanation for his position revolves around his age: TXL was roughly 5 years old and possibly in a liminal developmental stage where he was transitioning out of the juvenile period and therefore more concerned with establishing connections to the adult social network.

6.7.2 Future Considerations

One major challenge to studying social play in primates is that there are typically a large number of participants to track. When studying a large group, especially one containing many young individuals who look quite similar, it is considerably easier to use broad age categories to ease any comparisons (e.g., infant, juvenile, subadult, and adult). However, this generalization of individuals into broad age classes can generate statistical and conceptual issues because the boundaries between age categories differ widely between sexes, species, and studies. In primates, the majority of individuals involved in play are juveniles, which often means they are between the ages of weaning and sexual maturity. This, however, is a large and variable span of developmental time, and the acquisition of physical, social, and cognitive skills may occur at different stages and paces. Not only might there be marked differences in the physical and social abilities among the age classes, there are considerable differences in abilities within age classes, for example, between a 2-year-old juvenile who is 731 days old and a 2-year-old juvenile who is 1094 days old. Therefore, it is our recommendation to use age in days to achieve a fine-grain perspective and to emphasize the individual nature and developmental trajectory of play in any group examined.

Much like any study on social play, more questions emerge and remain open for future exploration than are typically answered. In particular, the idea of social scaffolding—acting as a bridge and supporting or facilitating a connection between groups of individuals—is of particular interest. Who are the individuals that fall into this particular role? Are they kin? Do older siblings help assimilate younger siblings into the network? What is the influence of having a small familial network to begin with? These kinds of questions require longitudinal comparisons, that not only incorporate social play data but also multiple independent measures of association (Shimada and Sueur 2017), such as proximity and grooming.

Furthermore, questions of evolutionary importance surface in the study of social play. Particularly, because play is often regarded as a context for learning and practice for later serious adult interactions (Lutz et al. 2019), juvenile play is rich in opportunities to study the contributions of juvenile behavior to adult reproductive success. Do individuals who play at high frequencies when young go on to be more successful in dominance and mating interactions (relative to juveniles who rarely play)? What about those that demonstrate high play partner diversity? What is the effect of maternal or paternal social network connectedness on the juvenile's frequency or diversity of play? Do observed rates of play in different contexts (e.g., social or object) predict an individual's success in corresponding contexts later in life (e.g., mating or extractive foraging)? Researchers could test such predictions using unidimensional social network analyses (as we illustrate here) or expand on multidimensional social networks. Multiplex social network analyses are particularly useful in studying complex social systems where individuals participate in various social contexts (Smith-Aguilar et al. 2018). With multiplex statistics, researchers could ask questions of variation across multiple behaviorally defined layers (e.g., play, grooming, agonism, and mating), temporal sequences (e.g., years or developmental periods), or biologically relevant connections (e.g., kinship) to test predictions of inclusive fitness and reproductive success.

Studying play could also help to place the four grades of macaque species into clearer context: the differences between these species, especially those considered to be grades two and three, most likely reflect a nuanced degree rather than overt differences. Tibetan macaques are considered a grade two species, meaning that observers should see some evidence of despotism, hierarchical behavior, and matrilineal preference, even in juvenile play. However, researchers should expect to observe some behavioral inconsistencies when compared with grade one macaques (e.g., *M. mulatta* or *M. fuscata*). For example, contrary to the expectation for more despotic species, social play in this group is not often interrupted by an adult; instead, juveniles are fairly self-sufficient at self-regulating play bouts and choose when to participate rather than being controlled by the adults of the group (Wright et al. 2018). Additionally, neither kinship nor maternal rank appears to have a significant impact on the duration of play or partner choice (this study). These are unexpected results that warrant further inquiry and require researchers to ask themselves: just how much variation is reasonable to expect for grade two or three species?

Acknowledgments We are so grateful for the continued support and guidance of Dr. Lori K. Sheeran. Without your collaboration, this work would not be possible. Many thanks to Dr. Crickette Sanz for stimulating discussions on the utility of social networks in an evolutionary context. We are excited for the future research realized from these conversations. We are also appreciative of our collaborators at Anhui University and the helpful and positive feedback provided by two reviewers of this piece. Finally, to all of the individuals of the YA1 group, thank you for letting us spend some time in your world.

References

Altmann J (1974) Observational study of behavior: sampling methods. Behav 49(3/4):227–267

Baldwin JD, Baldwin JI (1974) Exploration and social play in squirrel monkeys (*Saimiri*). Am Zool 14(1):303–315

Batts C (2012) The impact of eco-tourism on infant and juvenile play behaviors in Tibetan macaques (*Macaca thibetana*). Thesis, Central Washington University, Ellensburg, WA

Bekoff M (1972) The development of social interaction, play and metacommunication in mammals: an ethological perspective. Q Rev Biol 47(4):412–434

Bekoff M (1974) Social play and play-soliciting by infant canids. Am Zool 14:323–340

Bekoff M, Allen C (1997) Intentional communication and social play: how and why animals negotiate and agree to play. In: Bekoff M, Byers JA (eds) Animal play: evolutionary, comparative, and ecological perspectives. Cambridge University Press, New York, pp 97–114

Bekoff M, Byers JA (1981) A critical reanalysis of the ontogeny of mammalian social and locomotor play, an ethological hornet's nest. In: Immelmann K, Barlow GW, Petrinovich L, Main M (eds) Behavioral development, the bielefeld interdisciplinary project. Cambridge University Press, New York, pp 296–337

Berman CM, Li JH (2002) The impact of translocation, provisioning and range restriction on a group of *Macaca thibetana*. Int J Primatol 23(2):383–397

Berman CM, Ionica CS, Li JH (2004) Dominance style among *Macaca thibetana* on Mt. Huangshan, China. Int J Primatol 25(6):1283–1312

Borgatti SP, Everett MG, Freeman LC (2002) UCINET for Windows: software for social network analysis. Version 6.627 [software]. 2016 Dec 14. Available from https://sites.google.com/site/ucinetsoftware/home

Borgatti SP, Everett MG, Johnson JC (2013) Analyzing social networks. Sage, Los Angeles, CA

Brown SG (1988) Play behaviour in lowland gorillas: Age differences, sex differences, and possible functions. Primates 29(2):219–228

Burghardt G (2005) The genesis of animal play: Testing the limits. MIT Press, Ann Arbor, MI

Burghardt G, Ward B, Rosscoe R (1996) Problem of reptile play: Environmental enrichment and play behavior in a captive Nile soft-shelled turtle, *Trionyx triunguis*. Zoo Biol 15:223–238

Burghardt G, Dinets V, Murphy JB (2014) Highly repetitive object play in a cichlid fish (*Tropheus duboisi*). Ethology 121:38–44

Byers JA, Walker C (1995) Refining the motor training hypothesis for the evolution of play. Am Nat 146(1):25–40

Ciani F, Dall'Olio S, Stanyon R et al (2012) Social tolerance and adult play in macaque societies: a comparison with different human cultures. Anim Behav 84:1313–1322

Clark FE (2011) Great ape cognition and captive care: Can cognitive challenges enhance well-being? Appl Anim Behav Sci 135:1–12

Cloutier S, Baker C, Wahl K et al (2013) Playful handling as social enrichment for individually- and group-housed laboratory rats. Appl Anim Behav Sci 143:85–95

Fagen R (1981) Animal play behavior. Oxford University Press, New York

Fagen R (1984) Play and behavioural flexibility. In: Smith PK (ed) Play in animals and humans. Basil Blackwell, Oxford, pp 159–173

Farine DR, Whitehead H (2015) Constructing, conducting and interpreting animal social network analysis. J Anim Ecol 84:1144–1163

Fontaine RP (1994) Play as physical flexibility training in five ceboid primates. J Comp Psychol 108 (3):203–212

Funkhouser JA, Mayhew JA, Mulcahy JB (2018a) Social network and dominance hierarchy analyses at chimpanzee sanctuary northwest. PLoS One 13(2):e0191898

Funkhouser JA, Mayhew JA, Sheeran LK, Mulcahy JB, Li JH (2018b) Comparative investigations of social context-dependent dominance in captive chimpanzees (*Pan troglodytes*) and wild Tibetan macaques (*Macaca thibetana*). Sci Rep 8(1):e13909

Gamble JR, Cristol DA (2002) Drop-catch behaviour is play in herring gulls, *Larus argentatus*. Anim Behav 63:339–345

Gómez JC, Martín-Andrade B (2002) Possible precursors of pretend play in nonpretend actions of captive gorillas (*Gorilla gorilla*). In: Mitchell R (ed) Pretending and imagination in animals and children. Cambridge University Press, Cambridge, pp 255–268

Heinrich B, Smolker R (1998) Play in common ravens (*Corvus corax*). In: Bekoff M, Byers JA (eds) Animal play: evolutionary, comparative, and ecological perspectives. Cambridge University Press, New York, pp 27–44

Hill HM, Dietrich S, Cappiello B (2017) Learning to play: A review and theoretical investigation of the developmental mechanisms and functions of cetacean play. Learn Behav 45(4):335–354

Jones BL, Kuczaj SA (2014) Beluga (*Delphinapterus leucas*) novel bubble helix play behavior. Anim Behav Cogn 1(2):206–214

Kahlenberg SM, Wrangham RW (2010) Sex differences in chimpanzees' use of sticks as play objects resemble those of children. Curr Biol 20(24):R1067–R1068

Kulik L, Amici F, Langos D et al (2015) Sex differences in the development of social relationships in rhesus macaques (*Macaca mulatta*). Int J Primatol 36:353–376

Lampe JF, Burman O, Würbel H et al (2017) Context-dependent individual differences in playfulness in male rats. Dev Psychobiol 59:460–472

Lee PC, Moss CJ (2014) African elephant play, competence and social complexity. Anim Behav Cogn 1(2):144–156

Lutz MC, Judge PG (2017) Self-handicapping during play fighting in capuchin monkeys (*Cebus apella*). Behaviour 154:909–938

Lutz MC, Ratsimbazafy J, Judge PG (2019) Use of social network models to understand play partner choice strategies in three primate species. Primates 60(3):1–14

Maestripieri D (2004) Maternal behavior, infant handling, and socialization. In: Thierry B et al (eds) Macaque societies: a model for the study of social organization. Cambridge University Press, Cambridge, pp 231–234

Maestripieri D, Ross SR (2004) Sex differences in play among western lowland gorilla (*Gorilla gorilla gorilla*) infants: Implications for adult behavior and social structure. Am J Phys Anthropol 123:52–61

Mancini G, Palagi E (2009) Play and social dynamics in a captive herd of gelada baboons (*Theropithecus gelada*). Behav Process 82:286–292

Martin P, Caro TM (1985) On the functions of play and its role in behavioral development. In: Rosenblatt JS, Beer C, Busnel MC, Slater PJB (eds) Advances in the study of behavior, vol 15. Academic, New York, pp 59–103

Mather JA, Anderson RC (1999) Exploration, play and habituation in octopuses (*Octopus dofleini*). J Comp Psychol 113(3):333–338

Matsumura S (1999) The evolution of "egalitarian" and "despotic" social systems among macaques. Primates 40(1):23–31

Mayhew JA (2013) Attention cues in apes and their role in social play behavior of western lowland gorillas (Gorilla gorilla gorilla). PhD Thesis, The University of St Andrews, St Andrews, Scotland

Nahallage CAD, Leca JB, Huffman MA (2016) Stone handling, an object play behaviour in macaques: welfare and neurological health implications of a bio-culturally driven tradition. Behaviour 153(6–7):845–869

Neumann C, Duboscq J, Dubuc C et al (2011) Assessing dominance hierarchies: validation and advantages of progressive evaluation with Elo-rating. Anim Behav 82:911–921

Ortega JC, Bekoff M (1986) Avian play: comparative evolutionary and developmental trends. Auk 104:338–341

Palagi E (2006) Social play in bonobos (*Pan paniscus*) and chimpanzees (*Pan troglodytes*): implications for natural social systems and interindividual relationships. Am J Phys Anthropol 129:418–426

Palagi E (2009) Adult play fighting and potential role of tail signals in ringtailed lemurs (*Lemur catta*). J Comp Psychol 123(1):1–9

Palagi E (2018) Not just for fun! Social play as a springboard for adult social competence in human and non-human primates. Behav Ecol Sociobiol 72:90

Palagi E, Antonacci D, Cordoni G (2007) Fine-tuning of social play in juvenile lowland gorillas (*Gorilla gorilla gorilla*). Dev Psychobiol 49(4):443–445

Panksepp J, Beatty WW (1980) Social deprivation and play in rats. Behav Neural Biol 30 (2):197–206

Pellis SM, Iwaniuk AN (2000) Comparative analyses of the role of postnatal development on the expression of play fighting. Ethology 106(12):1083–1104

Pellis SM, McKenna M (1992) What do rats find rewarding in play fighting? An analysis using drug-induced non-playful partners. Behav Brain Res 68(1):65–73

Pellis SM, Pellis VC (1996) On knowing it's only play: the role of play signals in play fighting. Aggress Violent Behav 1(3):249–268

Pellis SM, Pellis VC (2009) The playful brain. OneWorld Publications, Oxford

Pellis SM, Pellis VC, Reinhart CJ et al (2011) The use of the bared-teeth display during play fighting in Tonkean macaques (*Macaca tonkeana*): sometimes it is all about oneself. J Comp Psychol 125(4):393–403

Pellis SM, Pellis VC, Pelletier A et al (2019) Is play a behavior system, and, if so, what kind? Behav Process 160:1–9

Petit O, Bertrand F, Thierry B (2008) Social play in crested and Japanese macaques: testing the covariation hypothesis. Dev Psychobiol 50(4):399–407

Pruitt JN, Burghardt GM, Riechert SE (2012) Non-conceptive sexual behavior in spiders: a form of play associated with body condition, personality type, and male. Ethology 118:33–40

Reinhart CJ, Pellis V, Thierry B et al (2010) Targets and tactics of play fighting: competitive versus cooperative styles of play in Japanese and Tonkean macaques. Int J Comp Psychol 23 (2):166–200

Scopa C, Palagi E (2016) Mimic me while playing! Social tolerance and rapid facial mimicry in macaques (*Macaca tonkeana* and *Macaca fuscata*). J Comp Psychol 130(2):153–161

Shimada M (2006) Social object play among young Japanese macaques (*Macaca fuscata*) in Arashiyama, Japan. Primates 47:342–349

Shimada M (2010) Social object play among juvenile Japanese macaques. In: Nakagawa N et al (eds) The Japanese macaques. Springer, New York, pp 375–385

Shimada M, Sueur C (2014) The importance of social play network for infant or juvenile wild chimpanzees at Mahale Mountains National Park, Tanzania. Am J Primatol 76(11):1025–1036

Shimada M, Sueur C (2017) Social play among juvenile wild Japanese macaques (*Macaca fuscata*) strengthens their social bonds. Am J Primatol 80(1):e22728

Siviy S, Deron L, Kasten C (2011) Serotonin, motivation, and playfulness in the juvenile rat. Dev Cogn Neurosci 1:606–616

Smith-Aguilar SE, Aureli F, Busia L et al (2018) Using multiplex networks to capture the multidimensional nature of social structure. Primates 60(3):1–19

Sueur C, Jacobs A, Amblard A et al (2011a) How can social network analysis improve the study of primate behavior? Am J Primatol 3:703–719

Sueur C, Petit O, De Marco A et al (2011b) A comparative network analysis of social style in macaques. Anim Behav 82:845–852

Thierry B (1985) Patterns of agonistic interactions in three species of macaque (*Macaca mulatta, M. fascicularis, M. tonkeana*). Aggress Behav 11(3):223–233

Thierry B (1990) Feedback loop between kinship and dominance: the macaque model. J Theor Biol 145(4):511–521

Thierry B (2007) Unity in diversity: lessons from macaque societies. Evol Anthropol 16 (6):224–238

Thierry B (2011) The macaques: a double-layered social organization. In: Campbell CJ et al (eds) Primates in perspective, 2nd edn. Oxford University Press, New York, pp 229–241

Thierry B, Iwaniuk AN, Pellis SM (2000) The influence of phylogeny on the social behaviour of macaques (Primates: Cercopithecidae, genus *Macaca*). Ethology 106(8):713–728

Thompson KV (1996) Play-partner preferences and the function of social play in infant sable antelope, *Hippotragus niger*. Anim Behav 52:1143–1155

van Leeuwen EJC, Zimmerman E, Davila Ross M (2011) Responding to inequities: gorillas try to maintain their competitive advantage during play fights. Biol Lett 7:39–42

Ward C, Bauer E, Smuts B (2008) Partner preferences and asymmetries in social play among domestic dog, *Canis lupus familiaris*, littermates. Anim Behav 76(4):1187–1199

Whitehead H (2008) Analyzing animal societies: quantitative methods for vertebrate social analyses. University of Chicago Press, Chicago, IL

Wright KR, Mayhew JA, Sheeran LK et al (2018) Playing it cool: characterizing social play, bout termination, and candidate play signals of juvenile and infant Tibetan macaques (*Macaca thibetana*). Zool Res 39(4):1–13

Xia DP, Li JH, Garber P et al (2012) Grooming reciprocity in female Tibetan macaques *Macaca thibetana*. Am J Primatol 74(6):569–579

Xia DP, Li JH, Garber P et al (2013) Grooming reciprocity in male Tibetan macaques. Am J Primatol 75:1009–1020

Yamanashi Y, Nogami E, Teramoto M et al (2018) Adult-adult social play in captive chimpanzees: is it indicative of positive animal welfare? Appl Anim Behav Sci 199:75–83

Yanagi A, Berman CM (2014a) Functions of multiple play signals in free-ranging juvenile rhesus macaques (*Macaca mulatta*). Behaviour 151:1983–2014

Yanagi A, Berman CM (2014b) Body signals during social play in free-ranging rhesus macaques (*Macaca mulatta*): a systematic analysis. Am J Primatol 76:168–179

Zhao QK (1997) Intergroup interactions in Tibetan macaques at Mt. Emei, China. Am J Phys Anthropol 104(4):459–470

Chapter 7
The Vocal Repertoire of Tibetan Macaques (*Macaca thibetana*) and Congeneric Comparisons

Sofia K. Blue

7.1 Introduction

The vocal repertoires of mammals usually consist of a fixed number of calls, some of which are closely linked to particular contexts. Though not to the same extent as humans, other mammals have been documented to emit highly modifiable calls with a cognitively rich set of meanings (Seyfarth and Cheney 2010). Fixed vocal production coupled with modifiable context emission and comprehension may have been homologous traits present in our prelinguistic ancestor, since they appear to be present in a wide array of taxonomic groups. Seyfarth and Cheney (2012) suggest that the common ancestor of Old World monkeys, apes, and humans had limited vocal production and open-ended comprehension, and by making comparisons with our closest living relatives, we can further illustrate the implications of theories concerning language evolution.

Our limitless repertoire of sound combinations and capacity for vocal learning is unmatched in the animal kingdom. To date, songbirds have been the focus of vocal learning studies in non-human animals, and evidence in mammals has been restricted to species of cetaceans, pinnipeds, elephants, and bats (Lattenkamp and Vernes 2018). Surprisingly, our closest living relatives, the non-human primates, are deficient in their ability to learn new vocalizations. This inability may be a result of a lack of neuronal potential required for vocal learning although the vocal tract is speech-ready (Fitch et al. 2016). Vocal production is, for the most part, highly constrained in non-human animals, and mammalian repertoires usually consist of a variety of grunts, threatening vocalizations, alarm calls, and screams (Seyfarth and Cheney 2012). Comparing the diverse array of vocal repertoires and communication across taxa is one way to identify the selective pressures behind vocal complexity

S. K. Blue (✉)
Primate Behavior and Ecology Program, Department of Anthropology and Museum Studies, Central Washington University, Ellensburg, WA, USA
e-mail: Sofia.Bernstein@cwu.edu

J.-H. Li et al. (eds.), *The Behavioral Ecology of the Tibetan Macaque*, Fascinating Life Sciences, https://doi.org/10.1007/978-3-030-27920-2_7

and the biological underpinnings of language-specific traits. Finding an appropriate measure of vocal complexity, however, is challenging when there are no standard methods for quantifying, classifying, or describing vocal repertoires.

7.2 Measuring Vocal Complexity

7.2.1 Vocal Repertoire Size

Although communication is undoubtedly multimodal in nature, across the modalities the number of distinct signals or signaling units can be used as a representative of communicative complexity (Peckre et al. 2019). For studies investigating the modality of acoustic communication, one possible metric of vocal complexity is repertoire size. Repertoire size is defined by the number of call types produced by a species or population (Peckre et al. 2019). Therefore, repertoire size is a strictly numeric measure unable to provide any information concerning the function or usage of the calls that constitute them. Variable data collection methods among studies constrain the ability to make exact comparisons across species and genera, further limiting the informative value of repertoire size alone. In addition, comparing repertoire size leaves out the identification of species-specific calls that may have evolved.

Hohmann (1991) carried out a comparative analysis of the vocal repertoires of four species of Old World monkey (*Macaca radiata*, *Macaca silenus*, *Presbytis johnii*, *Presbytis entellus*). For the species under Hohmann's (1991) investigation, previous reports of their vocalizations were very fragmented. To bypass this challenge and make accurate comparisons, Hohmann (1991) recorded calls from these four species and used the same methods to analyze the call recordings. Even though this study involved an intensive classification of the vocal repertoire of four different species, emphasis was placed on comparisons of the frequencies of call type emission and vocal activity among sex/age classes, rather than identifying species-specific calls or call usage. Therefore, the extent to which vocal complexity among the four species was investigated is limited.

7.2.2 Identifying Homologous and Derived Calls

Gustison et al. (2012) proposed that the identification of homologous (acoustically similar calls shared between species) and derived vocalizations (acoustically unique to a species) among closely related species can be used as a measure of vocal complexity. By focusing on the identification of acoustically homologous calls (Hohmann 1991; Gustison et al. 2012) and identifying species-specific derived calls, researchers are not constrained by the variable methods used across studies to classify vocal repertoires. Furthermore, comparing homologous and derived vocalizations among closely related species is essential in the identification of phylogenetic, social, and ecological factors influencing varying degrees of vocal complexity.

To date, few studies in non-human primates have approached investigations on vocal complexity through the identification of derived calls. Kudo (1987) compared the vocal behavior of mandrills (*Mandrillus sphinx*) to savannah baboons (*Papio* spp.) and geladas (*Theropithecus gelada*) and found that differences among the three species' vocal behavior were due to the ecological pressures associated with the attenuation of sound in forest versus savannah habitats. More recently, Gustison et al. (2012) identified the homologous and derived vocalizations in two closely related species, geladas and chacma baboons (*Papio ursinus*). In this comparison, derived vocalizations were uttered in social and reproductive contexts unique to a species, suggesting that differences in sociality and reproductive ecology were the driving factors in the evolution of these species-specific vocalizations.

7.3 Understanding the Evolution of Vocal Complexity

To understand the evolution of vocal complexity, it is necessary to make comparisons across closely related species. Although broader interspecies comparisons do exist and reveal some interesting patterns (McComb and Semple 2005), in-depth investigations among closely related species allow researchers to tease apart which factors have driven the evolution of communication (Bouchet et al. 2013). Macaques may be an ideal genus for investigations of vocal complexity because of several characteristics. Vocalizations in the *Macaca* genus have garnered a considerable amount of attention over the years, and the literature is rich enough to make detailed comparisons across many species. In addition, the genus is the most geographically widespread and behaviorally diverse genus of non-human primate because macaques display a high range of interspecific variation and inhabit the greatest range of habitats (Thierry et al. 2000). By documenting the diversity across the macaques through a comparative perspective and the identification of derived calls, I can investigate how potential selective pressures, like phylogeny, sociality, and ecology, shape which calls are conserved or vary in the vocal repertoires of the genus.

The effects of phylogeny, sociality, and ecology are essential in the explorations on the evolution of vocal complexity (Freeberg et al. 2012). Phylogeny may act as a starting point for mapping the evolutionary convergence of vocal characters. For example, complex oropendola bird songs are conserved and relatively invariant among the three genera (Price and Lanyon 2002), and primate vocalizations appear to be largely genetically predetermined (Newman and Symes 1982). It is widely accepted in the literature that sociality may be an essential driving factor in species with high degrees of vocal complexity (Blumstein and Armitage 1997; Wilkinson 2003; Freeberg 2006; Furrer and Manser 2009). In their investigation on Tonkean macaques (*Macaca tonkeana*), Masataka and Thierry (1993) concluded that sociality determines the vocal repertoire of a species as strongly as phylogenetic constraints. Lastly, since macaques inhabit a wide range of habitats, interspecific variation in vocal repertoires may be the result of these ecological differences (Masataka and Thierry 1993). It is likely, however, that any one of these factors is insufficient to explain vocal diversity and that a complex interplay of phylogenetic, social, and ecological factors influences degrees of vocal complexity.

In this contribution, I explore vocal homologs and derived calls in the *Macaca* genus through the:

1. Identification of the main categories of call production
2. Selection of call types from each category of call production to compare across the genus
3. Identification of homologs based on acoustic characteristics
4. Identification of species-specific derived calls and main differences across the genus
5. Comparisons with the Tibetan macaque vocal repertoire
6. Exploration of phylogenetic, social, and ecological factors that may influence homologous and derived calls

7.4 Methods

7.4.1 *Categories of Call Production*

Two previous studies designated categories for comparison across different species of Old World monkeys (*M. radiata, M. silenus, P. johnii, P. entellus*: Hohmann 1991; *Papio ursinus, T. gelada*, Gustison et al. 2012). I followed Gustison et al. (2012) and identified calls in macaques within three main categories and included the subcategories differentiated for each: allospecific (*alarm* and *food calls*), social (long-distance and close-range calls (competitive, distress, and contact)), and other (see Fig. 7.1). Allospecific calls are elicited by external stimuli and include *alarm* and *food calls*. The majority of vocalizations in mammals are emitted during social interactions with conspecifics (Gustison et al. 2012). Such social calls are further categorized into two subclasses: long-distance and close-range calls. For close-range calls, a further subdivision is necessary; so I classified close-range calls into three additional subcategories to cover the many different contexts in which these calls are emitted: contact, competition, and distress. The last category, other, includes calls that are not strictly emitted in a particular context or the context is unknown.

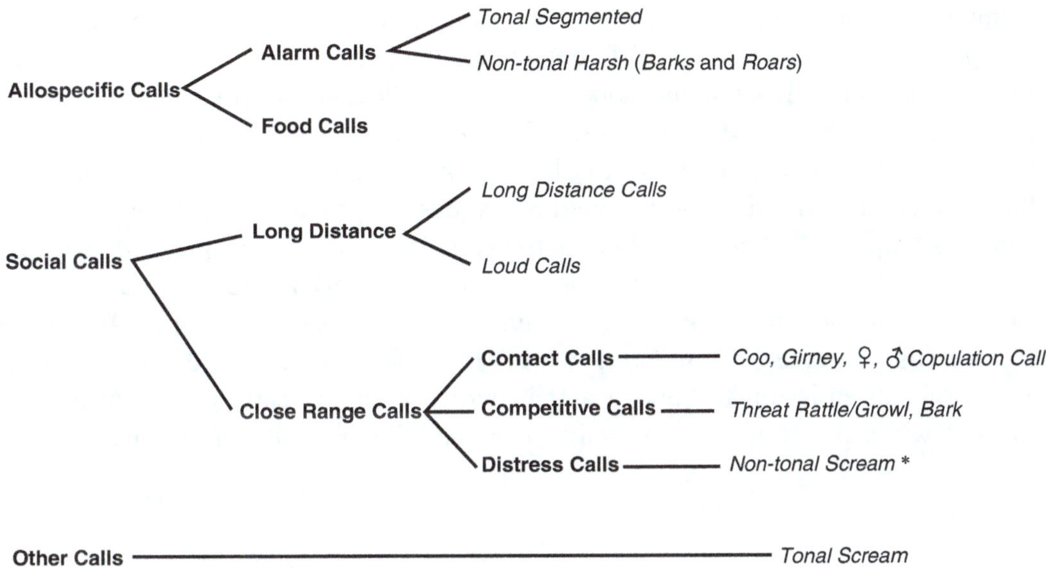

Fig. 7.1 Call categories and call types investigated in the genus *Macaca*. The main categories and subcategories are in bold, and calls in italics indicate the call types chosen for homologous and derived comparisons. (*) Non-tonal screams are not strictly distress calls since they are also emitted in feeding contexts

7.4.2   *Exploring Vocal Homologs and Derived Calls in the Genus* Macaca

The homologous call types investigated come from a review of the macaque literature and the identification of calls that were similar in acoustic structure alone, even if they were designated different names in the literature or were emitted in different contexts. Derived calls are defined as acoustically unique to a species and were also identified via visual inspection of spectrograms. Similar to my previous study (Bernstein et al. 2016), I followed Rowell and Hinde (1962) and first identified calls as either clear or harsh. Clear calls are tonal with energy concentrated in harmonic frequency bands, while harsh calls are atonal and acoustically unstructured with underlying harmonic frequency bands obscured by broadband noise and a distribution of energy across a wide frequency spectrum (Palombit 1992; Rowell and Hinde 1962). Once calls were classified as either clear or harsh, I then focused on temporal and frequency characteristics to identify homologous calls. These characteristics included the presence or absence of harmonic frequency bands, the modulation of frequency bands, spectral shifts (e.g., abrupt changes like the presence of a fast-rise transient from high pitch to low pitch), the concentration of spectral energy, whether or not calls were attached to or superimposed onto a noisy portion, and the duration of the call.

In order to make comparisons in the macaque genus, I used reports from which the entire repertoire was systematically investigated and excluded investigations that described only particular call types. This method is similar to the one adopted by McComb and Semple (2005) in their investigation of the coevolution of vocal communication and repertoire size as a measure of vocal complexity. To date there are repertoire studies from 11 different macaque species, including my study on Tibetan macaques (Tibetan macaques, *Macaca thibetana*, Bernstein et al. 2016; Formosan macaques, *M. cyclopis*, Hsu et al. 2005; Barbary macaques, *M. sylvanus*, Hammerschmidt and Fischer 1998; Tonkean macaques, *M. tonkeana*, Masataka and Thierry 1993; long-tailed macaques, *M. fascicularis*, Palombit 1992; bonnet macaques, *M. radiata*, Hohmann 1989; lion-tailed macaques, *M. silenus*, Hohmann and Herzog 1985; Sugiyama 1968; stump-tailed macaques, *M. arctoides*, Lillehei and Snowdon 1978; Japanese macaques, *M. fuscata*, Green 1975; Itani 1963; rhesus macaques, *M. mulatta*, Rowell 1962; Rowell and Hinde 1962). Comparison of call emission frequency and in which contexts calls were uttered was analyzed when the results were available across the published repertoires in the genus.

7.4.3   Comparisons with Tibetan Macaques and Phylogenetic, Social, and Ecological Factors

I included a wider array of species for comparisons with the Tibetan macaque vocal repertoire based on phylogenetic closeness, even if their repertoires have not been systematically analyzed and reported. I made this attempt to incorporate studies on specific calls for a more robust comparison with particular Tibetan macaque calls that were not included in the genus comparison. I chose interspecific call types based on their acoustic similarity and not context (for context comparisons, see Table 7.2).

After I identified the homologous and derived calls in the genus, I explored potential phylogenetic, social, and ecological factors. Depending on the source and the type of genetic study, the phylogenetic classification of macaques can vary. One of the most recent studies by Li et al. (2009) investigated the phylogeny of the macaques based on Alu elements. I used their classification for the comparisons investigated here (for each species' designation in the phylogeny, see Table 7.1). For the social component, I compared the contexts of emission and, if available, the rate at which the call types are emitted in a species' repertoire. In addition, I used reviews of the *Macaca* genus (Thierry et al. 1996, 2000; van Schaik et al. 1999; Maestripieri and Roney 2005; Pradhan et al. 2006) and the life history traits provided by Singh and Sinha (2004) to explore social and ecological factors.

Table 7.1 Shared call types in the macaque genus

Group	Species	Coo	Threat rattle/ growl	Non-tonal scream	Girney	Tonal scream	Squeak	Food call	Alarm call	♀ Cop call	♂ Cop call	Bark	Loud call
Sylvanus	*sylvanus*	x[a]	–	x	–	x	x	–	NT	x	–	x	–
Silenus	*silenus*	x	x	x	x	x	x	–	NT	x	x	–	x
	nemestrina	x	x	x	–	–	x	–	NT	x	–	x	–
	tonkeana	x	x	x	x	x	x	–	TS	x	–	x	x
Sinica	*radiata*	x	x	x	x	x	x	–	TS	x	x	x	x
	thibetana	x	x	x	–	x	x	–	TS	x	x	x	x
	arctoides	x	x	x	x	x	x	–	NT	x	x	x	x
Fascicularis	*fascicularis*	x	x	x	–	x	x	–	NT, TS	x	x	x	–
	fuscata	x	x	x	x	x	x	–	TS	x	–	–	x
	mulatta	x	x	x	x	x	x	x	TS	–	x	x	–
	cyclopis	x	x	x	x	x	x	x	TS	x	x	x	–

An x indicates that the call type is present in the repertoire of that species

NT non-tonal harsh alarm calls including barks and roars, *TS* tonal segmented alarm calls

[a]*M. sylvanus* have a limited use of coo calls in their repertoire (Bruce and Estep 1992)

7.5 Results

7.5.1 Homologous and Derived Calls in the Genus

I selected call types from the three main categories (allospecific, social, and other) of call production across the genus. I identified the following as the shared calls in the 11 macaque species investigated: *coo, threat rattle/growl, non-tonal scream, girney/ greeting call, tonal scream, squeak, food call, alarm call* (*non-tonal harsh* and *tonal segmented* types), *female* and *male copulation call, bark*, and *loud call* (Table 7.1). The main category allospecific calls includes the *tonal segmented* and *non-tonal harsh alarm calls* and the *food call. Threat rattle/growl, bark, non-tonal scream, coo, girney, female* and *male copulation call, long distance*, and *loud call* are in the social category, and the *tonal screams* are in the other category since they are emitted in feeding and distress contexts (for subcategories, see Fig. 7.1).

For the 11 species investigated, I identified 9 derived calls based solely on acoustic structure: the *krahoo* (*M. fascicularis*); *food yell, atonal greeting*, and *tonal girney* (*M. cyclopis*); *warble, harmonic arch, chirp*, and *male copulation scream* (*M. mulatta*); *male copulation grunt* (*M. arctoides*); *tonal estrus call* (*M. fuscata*); and *female copulation call* (*M. thibetana*).

The main differences found in the homologous calls concerned the frequency of emission and the range of call types within a specific context. For example, *M. arctoides* have a limited use of *coo* calls with considerably more harsh, noisy calls in a purely graded signal system (Bruce and Estep 1992). *Macaca tonkeana* and *M. cyclopis* macaques have a wider range of agonistic and *girney/greeting* calls, respectively, that is unparalleled in the rest of the genus (Masataka and Thierry 1993; Hsu et al. 2005).

7.5.2 Comparisons with the Vocal Repertoire of Tibetan Macaques

The Tibetan macaque vocal repertoire consists of five clear calls (*coo, squawk, leap coo, weeping, modulated tonal scream*) and seven harsh calls (*squeal, noisy scream, growl, bark, compound squeak, pant, female copulation call*) (for a quantitative analysis of the vocal repertoire, see Bernstein et al. (2016); see Fig. 7.2). In order to make congeneric comparisons with the Tibetan macaque vocal repertoire, I included a wider array of species based on phylogenetic closeness, even if their repertoires have not been systematically analyzed and reported. I made this attempt to incorporate studies on specific calls for a more robust comparison with particular Tibetan macaque calls that were not included in the genus comparison above (i.e., *squawk, squeal, leap coo, weeping*, and *pant*). I chose interspecific call types based on their acoustic similarity and not context (for context comparisons see Table 7.2).

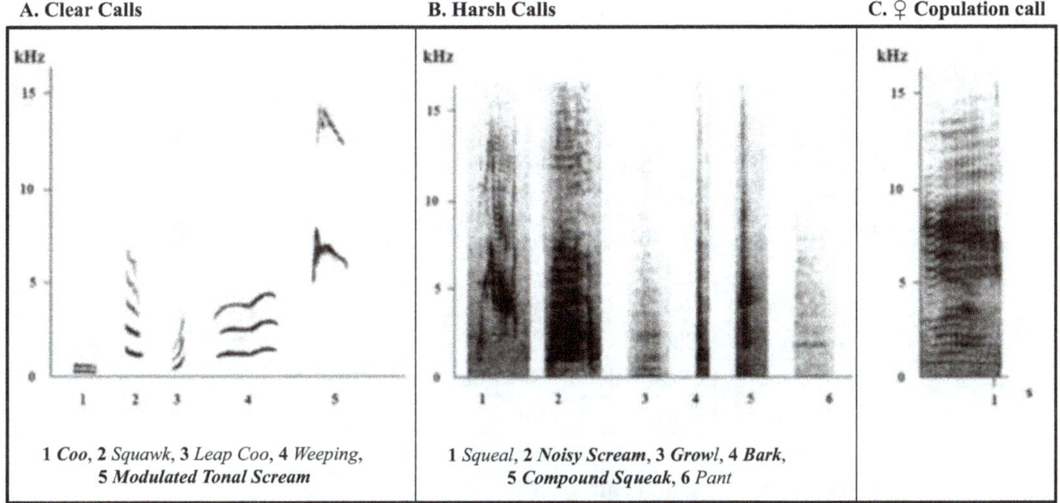

Fig. 7.2 Adapted from Bernstein et al. (2016). Calls in **bold** are homologous calls shared with the rest of the genus. (**a**) Representative spectrograms of the clear calls in the repertoire (**b**) Representative spectrograms of the harsh calls in the repertoire (**c**) Representative spectrogram of the female copulation call

Overall, the acoustic structure and context of emission was similar among the species investigated. A few main differences in context were a result of the group composition and various species-specific behaviors present in the congeneric species, but not in Tibetan macaques. For example, all of the infants died during my study on the vocal repertoire of Tibetan macaques, and therefore information about the calls associated with mother-infant interactions are limited (Bernstein et al. 2016). Also, an absence of calls from infants leaves out one vital age class that is necessary for a more complete comparison across species. Additionally, Tibetan macaques are one of two macaque species that show bridging behavior where two adults simultaneously lift an infant (Ogawa 1995, this volume). Bridging may be another context associated with infants where individuals emit calls that I was unable to observe.

Also, within the context of mating, female Tibetan macaques do not show any behaviors indicative of estrus (Li et al. 2005) and do not emit an estrous call. My study took place in the Valley of the Wild Monkeys, China, a site where the provisioned macaque groups that inhabit the mountainous sparsely covered terrain have been protected from hunting and trapping since the 1940s (Wada et al. 1987). During my observations, the Tibetan macaques did not emit an *alarm* or *food call*. Although Tibetan macaques were observed to emit *loud calls* that propagated over a large distance, these calls were clustered in the noisy scream category.

The female Tibetan macaque *copulation call* was the only call in my previous study that was a derived vocalization distinct from what has been reported in the rest of the genus (Bernstein et al. 2016). Usually, female macaques emit soft grunts during copulations. The Tibetan macaque females also emit quiet calls, but their acoustic structure does not involve biphasic "inhale-exhale" components, and they are instead shrill, undulating calls that are not always given in phrases. In

Table 7.2 Tibetan macaque vocal repertoire and congeneric comparisons

	Species	Call type	Context
Coo			*Foraging, provisioning, group movements, cohesion, dispersal*
	M. radiata	Contact whoo	Similar
	M. cyclopis	Contact coo	Similar
	M. sinica	Hum	Similar
	M. fuscata	Class II coo	Similar
	M. mulatta	Clear call	**Shut out**[a]
	M. fascicularis	Coo	Similar
	M. arctoides	Coo	Similar
	M. silenus	Whoo call	Similar
	M. sylvanus	Coo	**Limited use**[a]
Squawk			*♂ copulation call, dispersal, food begging, arrival of tourists*
	M. fuscata	Class IV high squawk	**Defensive submissive**[a]
	M. cyclopis	Squawk	**Defensive submissive**[a]
Squeal			*Submissive agonistic context, displacement by humans*
	M. radiata	Squeal	**Rejection of infants by mother, juvenile mobbing of adults**[a]
	M. cyclopis	Squeal	**Infant separation**[a]
	M. fuscata	Class VI squeal	**Estrus solicitation, infant separation**[a]
	M. silenus	Shriek	Submissive agonistic interactions
Noisy scream			*Agonistic, submissive intragroup/interspecies, provisioning, dispersal*
	M. cyclopis	Non-tonal scream	Similar
	M. radiata	Non-tonal scream	Intergroup interactions, **infant weaning**[a]
	M. mulatta	Non-tonal scream	Similar
	M. nemestrina	Agonistic scream	Similar
	M. cyclopis	Food yell	Similar
	M. sinica	Food yell	**Foraging discovery new food source**[a]
	M. fascicularis	Wraagh	Similar

(continued)

Table 7.2 (continued)

	Species	Call type	Context
Growl			*Agonistic, submissive intragroup/interspecies, provisioning, male-male mounts provisioning/ foraging, male-male mounts*
	M. cyclopis	Threat rattle	Similar
	M. radiata	Threat rattle	**Group movement**[a]
	M. silenus	Rattle	Similar
	M. fascicularis	Harr	**Juveniles threatened, during play**[a]
	M. fuscata	Class X gruff	**During threats and standoffs**[a]
Bark			*Intragroup/interspecies agonistic, provisioning, movements, weaning*
	M. radiata	Bark	Similar
	M. nemestrina	Bark	Similar
	M. tonkeana	Bark	Similar
	M. mulatta	Bark	Similar
	M. arctoides	Bark	Similar
	M. cyclopis	Bark	Similar
	M. fuscata	Bark	**Alarm response to dogs, snakes, raptors**[a]
Compound squeak			*Intragroup/interspecies agonistic context, weaning*
	M. cyclopis	Compound squeak	Similar
Leap Coo			Adults, *provisioning, foraging, group movements, arrival of tourists*; juveniles, *weaning*
	M. nemestrina	Leap coo	**During juvenile excitement**[a]
Weeping			*Rejection and weaning of juveniles*
	M. cyclopis	Weeping	Similar
	M. silenus	Weeping	Similar
	M. radiata	Pulse whoo	Similar
	M. sinica	Infant separation call	Similar
Modulated tonal scream			*Harassment of copulations, intragroup/interspecies agonistic context, provisioning, rejection and weaning of juveniles*
	M. sylvanus	Mod. tonal scream	**Forced separation from mother**[a]
	M. cyclopis	Tonal squeak	Similar
	M. radiata	Squeak	Similar
	M. fuscata	Squeak	♀ **homosexual solicitations**[a]

(continued)

Table 7.2 (continued)

	Species	Call type	Context
Pant			*Harassment of copulations, intragroup/interspecies agonistic context*
	M. mulatta	Pant threat	Similar

Context of emission for Tibetan macaques in *italics*

Comparisons made from acoustic spectra and context of emission provided in previous reports of macaque vocal repertoires (Tibetan macaques, *Macaca thibetana*, Bernstein et al. 2016; Formosan macaques, *Macaca cyclopis*, Hsu et al. 2005; Barbary macaques, *M. sylvanus*, Hammerschmidt and Fischer 1998; Tonkean macaques, *M. tonkeana*, Masataka and Thierry 1993; long-tailed macaques, *M. fascicularis*, Palombit 1992; bonnet macaques, *M. radiata*, Hohmann 1989; lion-tailed macaques, *M. silenus*, Hohmann and Herzog 1985; Sugiyama 1968; stump-tailed macaques, *M. arctoides*, Lillehei and Snowdon 1978; Japanese macaques, *M. fuscata*, Green 1975; Itani 1963; rhesus macaques, *M. mulatta*, Rowell 1962; Rowell and Hinde 1962)

[a]Tibetan macaques do not emit the homologous call in the **bold** context reported in the comparison species

comparison to the rest of the macaque genus, *female copulation call* rate was low and resembled that of *M. tonkeana* and *M. fuscata* (Masataka and Thierry 1993; Hohmann 1989; Oda and Masataka 1992; see Table 7.3). Similar to *M. radiata*, a closely related species in the same lineage, females do not emit an *estrous call* and *copulation calls* are the only auditory cue females emit in a copulation context. Males did emit *copulation calls* as well, but at a much higher frequency. Harassment of copulatory dyads is evident in Tibetan macaques, involves a wider range of age/sex classes than what is reported for most other members of the genus, and appears to be associated with the number of females in the audience.

7.5.3   *Potential Effects of Phylogeny, Sociality, and Ecology*

Phylogenetic Factors In the macaque species whose repertoires have been systematically analyzed and reported, there are two main types of acoustically structured *alarm calls*: *non-tonal barks* and *roars* (NT) or *tonal segmented* (TS) calls. Hohmann (1989) proposed that the type of *alarm call* (NT or TS) present in the repertoire of a macaque species is mostly conserved within phylogenetic groups (NT, *sylvanus* and *silenus*; TS, *sinica* and *fascicularis*). This pattern holds true for the *silenus*, *sinica*, and *fascicularis* groups (see Table 7.1). Indeed, species in the *silenus* group (*M. nemestrina* and *M. silenus*) have *non-tonal alarm calls* (Caldecott 1986; Hohmann and Herzog 1985). The only two terrestrial species of macaque with a purely graded signal system, *M. sylvanus* of the *sylvanus* lineage and *M. arctoides* of the *sinica* group, also have *non-tonal alarm calls* (Fischer and Hammerschmidt 2002; Chevalier-Skolnikoff 1974). The only exception in the *silenus* group, *M. tonkeana*, emits *tonal segmented alarm calls*. This pattern, where the acoustic structure of an *alarm call* is conserved within phylogenetic groups, is found in

Table 7.3 Copulation calls in *Macaca* and life history traits

Phylogenetic group	Species	♂ Copulation call	♀ Copulation call	Traits						
				Mount pattern	Seasonality	Number of males	♀♂ Ratio	Sexual dimorphism	Sex skin
Sylvanus	*M. sylvanus*	–	x	SME	2	4.7	1.1	0.37	Yes
Silenus	*M. silenus*	x	x	MME	1	1.6	9.9	0.38	Yes
	M. nemestrina	–	x	MME	0–1	2.4	6.3	0.54	Yes
	M. tonkeana	–	x	MME	0–1	2	3.4	0.5	Yes
Sinica	*M. radiata*	x	x	SME	2	7.3	1.7	0.54	No
	M thibetana	x	**Shrill**	SME	2	5.5	1.9	0.36	Little
	M. arctoides	**Grunt**	x	SME	1	?	1.32	0.37	No
Fascicularis	*M. fascicularis*	x	x	SME	1	4.7	4.8	0.41	Little
	M. fuscata	–	x, **Tonal estrus**	MME	2	3.4	1.3	0.32	Little
	M. mulatta	**Scream**	–	MME	2	4.5	2.9	0.22	Little
	M. cyclopis	x	x	MME	2	1.8	1	0.2	Little

Compilation from van Schaik et al. (1999), Maestripieri and Roney (2005), and Pradhan et al. (2006)

An x indicates the call is present in the repertoire of the species

Calls in **bold** are species-specific derived calls

Breeding seasonality: 0, 33% of births in 3-month period; 1, between 33% and 67%; 2, 67% in 3-month period (van Schaik et al. 1999)

sympatric species from different phylogenetic groups that may be prey to the same predators (*M. silenus* and *M. radiata*, Hohmann 1989).

Fooden (1976) sorted the macaque lineages based on the reproductive anatomy of males and females. Although his study relied only on behavioral and anatomical features, his classification is still mostly accurate even with more sophisticated genetic analyses. The evolution of different reproductive anatomy likely stemmed from each lineage going through a genetic bottleneck (Tosi et al. 2003). The three species that emit derived *copulation calls* (*M. thibetana, M. mulatta, M. arctoides*) are more closely related than they initially appear. Recent studies in the whole genome sequencing of Tibetan macaques have found that they are most closely related to Chinese rhesus macaques despite having been reported in different lineages (Fan et al. 2014). Analyses of mitochondrial loci in the macaque genus show stump-tailed macaques to potentially be a hybrid of proto-*M. thibetana* (Tosi et al. 2003).

Social Factors Most of the derived calls in the macaque genus are related to the context of mating, in particular, *female* and *male copulation calls*. Throughout the genus, males typically produce squeak-like *copulation calls*, while females emit grunts characterized by an "inhale-exhale element" (Pradhan et al. 2006). However, a few species stray from these genus characteristic calls and emit acoustically distinct calls associated with copulations. Two species emit acoustically distinct *male copulation calls*: *M. mulatta copulation screams* (Hauser 1993) and *M. arctoides copulation grunts* (Bauers 1993). Japanese macaques are the only macaque species that have studies reporting a distinct *estrous call* from the *female copulation call* (Oda and Masataka 1992), and my previous study reports an acoustically distinct shrill *female copulation call* in Tibetan macaques (Bernstein et al. 2016). All of these species that emit acoustically distinct calls in the context of copulations are seasonal breeders, have promiscuous mating systems, all exhibit the interruption of copulations by conspecifics, and have similar adult female-to-male ratios (1.3:2.9, Singh and Sinha 2004; Pradhan et al. 2006).

A large number of call types in the context of agonistic interactions is a characteristic of *M. tonkeana*. This species has high degrees of tolerance, small rank differences, a high frequency of conciliatory patterns, small interindividual distances, and a high rate of bidirectional agonistic interactions (Masataka and Thierry 1993). Context-dependent differences were also found in the derived *girney/greeting calls* of *M. cyclopis*. All of the *girney/greeting calls* were paired with particular contexts and age/sex classes, with acoustic structure changing according to the recipient of the *girney/greeting call* (Hsu et al. 2005).

Ecological Factors Overall, *loud calls* adhere to the bioacoustic requirements for long-distance propagation of sound (e.g., the repetition of units and phrases, no intergradations to other vocal patterns, a stereotyped structure, a large range of frequencies, and a concentration of energy on the lower frequencies; Hohmann and Herzog 1985). *Loud calls* are harsh calls where the fundamental frequency is in the lower frequency range. However, only some species show a large range of frequencies and utter *loud calls* that are composed of both tonal and non-tonal harsh units. Three macaque species utter *loud calls* that are acoustically distinct from the rest of the genus and are composed of tonal and harsh units in a wider range of frequencies (*M. silenus, loud call*, Hohmann and Herzog 1985; *M. fascicularis, krahoo*, Palombit 1992; *M. tonkeana, loud call*, Masataka and Thierry 1993). Also, for all three species, only males have been reported to emit these *loud calls*. The *loud calls* of *M. silenus* are mixed units with a noise-like inhalation phase followed by a tonal exhalation (Hohmann and Herzog 1985), the *krahoo* of *M. fascicularis* is distinct for its harsh "*kra*" syllable at the beginning of the call (Palombit 1992), and *M. tonkeana*'s *loud call* is characterized by a distinct flag and mast with a sharp upward and downward frequency modulation (Masataka and Thierry 1993). All three species live in rainforest habitats and are mostly arboreal.

Food calls are common in the genus, but so far only two species have been reported to give acoustically distinct calls strictly emitted in food-related contexts. *M. mulatta warbles, harmonic arches*, and *chirps* are acoustically distinct calls made exclusively in a food context (Hauser and Marler 1993). *M. cyclopis* emit *food yells* similar to the *M. sinica food call*, but are acoustically distinct (Hsu et al. 2005). Reports from these two species are from provisioned groups living in human habitations or captivity.

7.6 Discussion

Mammalian vocalizations consist mainly of grunts, harsh non-tonal threatening vocalizations, and the sometimes tonal alarm calls and screams (Seyfarth and Cheney 2012). Indeed, the macaque genus shared these basic vocalizations (e.g., grunts, *girneys, estrus calls, female copulation calls*; harsh non-tonal calls, *barks, growls*; *tonal alarm calls*; screams, *shrieks, tonal, non-tonal screams*) but also emitted additional tonal calls (e.g., *coos, squawks*, and *male copulation calls*). Overall, macaque repertoires can be described as a graded signal system with intergradations between calls, and the contexts of emission for shared calls were consistent.

Although primate vocalizations appear to be largely genetically predetermined (Newman and Symes 1982), there are differences in the flexibility and conservation of the call types investigated. The flexibility observed in the derived vocalizations seems to be a by-product of the characteristics of the contexts in which calls are emitted or species-specific features. The most salient difference found in my congeneric comparison were the calls emitted in the copulation context. Three species emit derived *copulation calls* (*M. thibetana female copulation call, M. mulatta male copulation scream, M. arctoides male copulation grunt*). These species that emit *copulation calls* uncharacteristic of the genus have species-specific features in their copulation styles.

7.6.1   The Tibetan Macaque Vocal Repertoire

In sum, the Tibetan macaque vocal repertoire was generally comparable to that of other macaque species. The context of emission was also similar across species, but with a few key differences. For example, my previous study could not investigate the vocalizations of infants because of a 100% mortality rate during the 2014 mating season (Bernstein et al. 2016). Future studies on the vocalizations of infants and infant-related contexts may find that Tibetan macaques do emit calls not previously described in the analysis of the repertoire. The main differences found between Tibetan macaques and the rest of the genus may not only be a result of a missing age class but also the method used to quantify the repertoire and species-specific aspects of their social behavior.

An absence of a *food call* or *alarm call* may be the result of site management. These call types may not be absent. It is possible that I simply did not observe them over the course of my study or that provisioning and predators being hunted out of the area have removed these contexts that would elicit calls they may have present in their repertoire. Although Tibetan macaques do emit calls that propagate over a large distance, they were not a distinct call type like the *loud calls* described for other species. Most of the species that emit *loud calls* live in rainforest habitats where certain acoustic requirements are necessary for sound propagation. Tibetan macaques inhabit mountainous terrain with sparse tree coverage, where the propagation of their calls is not limited by a densely covered forest (Xia et al. 2010). My quantitative method of relying solely on acoustic characters to define distinct call types may have clustered calls identified separately in other studies into more overarching categories. The same could be true for *greeting calls*. In some cases, variants of the *coo* call are considered as a separate category, but for my study, individual classification of *coo* call variants was not warranted.

The unique copulatory behavior of Tibetan macaques may be one reason an acoustically distinct female call was selected for. Also within the context of mating, female Tibetan macaques do not show any behaviors indicative of estrus (Li et al. 2005), and therefore, it is not surprising that their concealed ovulation would yield the absence of an *estrous call*. The similarity between *M. thibetana, M. tonkeana,* and *M. fuscata* in terms of a low *female copulation call* rate may imply a cost-benefit strategy that differs between the sexes. All of these species are seasonal breeders, living in multi-male, multi-female groups, and exhibit high rates of copulation interruptions (see Table 7.3). *Macaca tonkeana* and *M. fuscata copulation call* emission rate may also be low because females additionally produce *estrous calls* to solicit males. These species may have evolved an alternate strategy of mating promiscuously and a low call rate to reduce female competition. My preliminary investigation on the association between *female copulation call* rate and harassment of copulatory dyads does indicate that harassment increases with the number of females in the audience.

However, the presence of both *male* and *female copulation calls* might indicate that part of their function is to synchronize male and female mating behavior, especially in seasonally breeding species. Additionally, for males, it might not be so costly to call as it is for females. Adult females harass copulations, sneak copulate with lower-ranking males, and mate promiscuously with males outside of their consortship. In primates, there are very few studies that have investigated a cost-benefit analysis of signaling in the mating context (Hauser 1993). Future studies should investigate the costs and benefits of producing *copulation calls* along with emission rate and how intra- and intersexual selection plays a role in the mating strategies and auditory sexual signals of males and females.

7.6.2   *Phylogenetic, Social, and Ecological Factors Influencing Macaque Vocal Repertoires*

The type of *alarm call* was conserved within phylogenetic groups (*non-tonal, sylvanus* and *silenus*; *tonal segmented, sinica* and *fascicularis*) and followed the designation described by Hohmann (1989) with a few exceptions. *Macaca sylvanus* and *M. arctoides*, the exceptions in the *sinica* group, would be expected to have *non-tonal harsh alarm calls* since their repertoires consist of mostly harsh calls and their placement in the phylogeny of macaques is contested (*M. arctoides*, Li et al. 2009). *Macaca tonkeana*, the only exception in the *tonkeana* group, emit *tonal segmented alarm calls*, which may be the result of this species having a wide array of tonal calls in their repertoire (Masataka and Thierry 1993). In conclusion *alarm calls* are one salient example of a vocalization type that is mediated by phylogeny and a repertoire's acoustic structure, which is largely genetically predetermined (Newman and Symes 1982).

The species that emit derived species-specific *copulation calls* are all closely related. Phylogenetic and social factors associated with reproduction may be the selective factors under which derived calls have evolved in these three species. Future studies are needed to investigate the *copulation calls* of females in more detail, to understand their effect as an auditory sexual signal, and to make more in-depth comparisons of their acoustic structure with the rest of the genus.

Certain aspects of *M. tonkeana* social behavior may drive the need for a larger number of agonistic vocalizations (Masataka and Thierry 1993). The high degree of variability in the frequency of the acoustic structure of *girney/greeting calls* in *M. cyclopis* could also be a result of sociality since they were heavily context dependent. Some of these acoustically distinct *girney/greeting calls* are different from what has been reported in the rest of the genus and could be considered as derived calls.

Factors that influence call production and the extent to which a sound travels can sometimes be enhanced or restricted based on the type of environment that a species inhabits. The environment can affect the context in which a call is given, or change the motivational threshold to call (Green 1981). Macaques inhabit the widest range of habitats; therefore, environmental factors may drive the differences seen among the repertoires of various macaque species. The *loud calls* of *M. silenus* and *M. tonkeana*, and the *krahoo* of *M. fascicularis*, are acoustically distinct from the rest of the genus. Their *loud calls* are made up of tonal and harsh units in a wider range of frequencies, and their arboreal lifestyle and rainforest habitat may have selected for a call with an acoustic structure that enables the propagation of sound in a dense forest.

The acoustically distinct *warbles*, *harmonic arches*, and *chirps* of *M. mulatta* and the *food yells* of *M. cyclopis* and *M. sinica* are allospecific *food calls* emitted by species from provisioned groups living in human habitations or captivity. These allospecific calls are related to the quantity or quality of a food source (Gustison et al. 2012), and provisioning heightens these characteristics in a given environment. *Food calls* and *alarm calls* are special in that they both are emitted in response to non-conspecifics and they combine call elements to procure new meanings. Therefore, it is possible that other species that do not emit a food-specific call still have the basic acoustic requirements needed to flexibly alter the acoustic structure of a call to convey a food context-specific meaning.

7.7 Conclusions

The main differences in the genus are in the calls associated with copulation. Macaques are the most behaviorally diverse and widespread primate genus, yet reproductive features appear to be the most important discriminating factor among species. This may explain the flexibility of derived call types observed in the *copulation calls* of males and females. However, particularly for this genus, it is difficult to tease apart the effects of phylogeny and behavior on reproduction. Instead, a complex interplay of phylogenetic and social features of a species' reproduction drives the evolution of derived calls in the context of copulations. In conclusion, it is not likely that any particular factor is mutually exclusive. Instead, a complex interplay of phylogenetic, social, and ecological factors may shape the development of derived calls and the preservation of homologous calls across the macaque genus.

Acknowledgments I would like to thank the Forestry Bureau of Anhui Province, China and the Huangshan Garden Forest Bureau for permitting my research in the Valley of the Wild Monkeys, Mt. Huangshan, China. Thank you to the staff and Dr. Dongpo Xia and Dr. Binghua Sun's cooperation and help throughout the study. I am also grateful to Drs. Michael Huffman, David Hill, Fred Bercovitch, Peter Kappeler, Hideshi Ogawa, and Hiroki Koda for their extremely helpful comments in the preparation of this manuscript. I was funded by the Primatology and Wildlife Science Leading Graduate Program (PWS), and by funding allocated to Dr. Michael Huffman from PWS (U04-JSPS). Dr. Hiroki Koda's funding through the Japanese Society of the Promotion of Science (JSPS KAKENHI [Grant Number: 15K00203, 25285199 to HK as PI or co-PI]) made this study possible.

References

Bauers KA (1993) A functional analysis of staccato grunt vocalizations in the stumptailed macaque (*Macaca arctoides*). Ethology 94:147–161

Bernstein SK, Sheeran LK, Wagner RS, Li J, Koda H (2016) The vocal repertoire of Tibetan macaques (*Macaca thibetana*): a quantitative classification. Am J Primatol 78:937–949

Blumstein DT, Armitage KB (1997) Does sociality drive the evolution of communicative complexity? A comparative test with ground-dwelling sciurid alarm calls. Am Nat 150:179–200

Bouchet H, Blois-Heulin C, Lemasson A (2013) Social complexity parallels vocal complexity: a comparison of three non-human primate species. Front Psychol 4:390

Bruce KE, Estep DQ (1992) Interruption of and harassment during copulation by stumptail macaques, *Macaca arctoides*. Anim Behav 44:1029–1044

Caldecott JO (1986) Mating patterns, societies and the ecogeography of macaques. Anim Behav 34:208–220

Chevalier-Skolnikoff S (1974) Male-female, female-female, and male-male sexual behavior in the Stumptail Monkey, with special attention to the female orgasm. Arch Sex Behav 3:95–116

Fan Z, Zhao G, Li P et al (2014) Whole-genome sequencing of Tibetan macaque (*Macaca thibetana*) provides new insight into the macaque evolutionary history. Mol Biol Evol 31:1475–1489

Fischer J, Hammerschmidt K (2002) An overview of the Barbary macaque, *Macaca sylvanus*, vocal repertoire. Folia Primatol 73:32–45

Fitch WT, de Boer B, Mathur N, Ghazanfar AA (2016) Monkey vocal tracts are speech-ready. Sci Adv 2:1–7

Fooden J (1976) Provisional classification and key to living species of macaques (Primates: *Macaca*). Folia Primatol 25:225–236

Freeberg TM (2006) Social complexity can drive vocal complexity: group size influences vocal information in Carolina chickadees. Psychol Sci 17:557–561

Freeberg TM, Dunbar RIM, Ord TJ (2012) Social complexity as a proximate and ultimate factor in communicative complexity. Philos Trans R Soc Lond Ser B Biol Sci 367:1785–1801

Furrer RD, Manser MB (2009) The evolution of urgency-based and functionally referential alarm calls in ground-dwelling species. Am Nat 173:400–410

Green S (1975) Variation in vocal pattern with social situation in the Japanese monkey (*Macaca fuscata*): a field study. In: Rosenblum A (ed) Primate behavior: developments in field and laboratory research. Academic, New York, pp 1–102

Green S (1981) Sex differences and age graduations in vocalizations of Japanese and liontailed monkeys (*Macaca fuscata* and *Macaca silenus*). Am Zool 21:165–183

Gustison ML, le Roux A, Bergman TJ (2012) Derived vocalizations of geladas (*Theropithecus gelada*) and the evolution of vocal complexity in primates. Philos Trans R Soc Lond B Biol Sci 367:1847–1859

Hammerschmidt K, Fischer J (1998) The vocal repertoire of Barbary macaques: A quantitative analysis of a graded signal system. Ethology 104:203–216

Hauser MD (1993) Rhesus monkey copulation calls: honest signals for female choice? Proc Biol Sci 254(1340):93–96

Hauser MD, Marler P (1993) Food-associated calls in rhesus macaques (*Macaca mulatta*): I. Socioecological factors. Behav Ecol 4:194–205

Hohmann GM (1989) Vocal communication of wild bonnet macaques (*Macaca radiata*). Primates 30:325–345

Hohmann GM (1991) Comparative analyses of age-specific and sex-specific patterns of vocal behavior in 4 species of old-world monkeys. Folia Primatol 56:133–156

Hohmann GM, Herzog MO (1985) Vocal communication in lion-tailed macaques (*Macaca silenus*). Folia Primatol 45:148–178

Hsu MJ, Chen LM, Agoramoorthy G (2005) The vocal repertoire of Formosan macaques, *Macaca cyclopis*: acoustic structure and behavioral context. Zool Stud 44:275–294

Itani J (1963) Vocal communication of the wild Japanese monkey. Primates 4:11–67

Kudo H (1987) The study of vocal communication of wild mandrills in Cameroon in relation to their social structure. Primates 28:289–308

Lattenkamp EZ, Vernes SC (2018) Vocal learning: a language-relevant trait in need of a broad cross-species approach. Curr Opin Behav Sci 21:209–215

Li JH, Yin HB, Wang QS (2005) Seasonality of reproduction and sexual activity in female Tibetan macaques *Macaca thibetana* at Huangshan, China. Acta Zool Sin 51:365–375

Li J, Han K, Xing J et al (2009) Phylogeny of the macaques (Cercopithecidae: *Macaca*) based on *Alu* elements. Gene 448:242–249

Lillehei RA, Snowdon CT (1978) Individual and situational differences in the vocalizations of young stumptail macaques. Behaviour 65:270–281

Maestripieri DM, Roney JR (2005) Primate copulation calls and postcopulatory female choice. Behav Ecol 16:106–113

Masataka N, Thierry B (1993) Vocal communication of Tonkean macaques in confined environments. Primates 34:169–180

McComb K, Semple S (2005) Coevolution of vocal communication and sociality in primates. Biol Lett 1:381–385

Newman JD, Symes D (1982) Inheritance and experience in the acquisition of primate acoustic behavior. In: Snowdon CT, Brown CH, Petersen MR (eds) Primate communication. Cambridge University Press, New York, pp 59–70

Oda R, Masataka N (1992) Functional significance of female Japanese macaque copulatory calls. Folia Primatol 58:146–149

Ogawa H (1995) Bridging behavior and other affiliative interactions among male Tibetan macaques (*Macaca thibetana*). Int J Primatol 16:707–729

Palombit RA (1992) A preliminary study of vocal communication in wild long-tailed macaques (*Macaca fascicularis*). I. Vocal repertoire and call emission. Int J Primatol 13:143–182

Peckre L, Kappeler PM, Fichtel C (2019) Clarifying and expanding the social complexity hypothesis for communicative complexity. Behav Ecol Sociobiol 73(1):11

Pradhan GR, Engelhardt A, van Schaik CP, Maestripieri D (2006) The evolution of female copulation calls in primates: a review and a new model. Behav Ecol Sociobiol 59:333–343

Price JJ, Lanyon SM (2002) Patterns of song evolution and sexual selection in the oropendolas and caciques. Behav Ecol 15:485–497

Rowell TE (1962) Agonistic noises of the rhesus monkey (*Macaca mulatta*). Symp Zool Soc Lond 8:91–96

Rowell TE, Hinde RA (1962) Vocal communication by rhesus monkey (*Macaca mulatta*). Proc Zool Soc Lond 128:279–294

Seyfarth RM, Cheney DL (2010) Production, usage, and comprehension in animal vocalizations. Brain Lang 115:92–100

Seyfarth RM, Cheney DL (2012) Primate social cognition as a precursor to language. In: Gibson K, Tallerman M (eds) Oxford handbook of language evolution. Oxford University Press, Oxford, pp 59–70

Singh M, Sinha A (2004) Life history traits: ecological adaptations or phylogenetic relics? In: Thierry B, Singh M, Kaumanns W (eds) Macaque societies: a model for the study of social organization. Cambridge University Press, Cambridge, pp 80–83

Sugiyama Y (1968) The ecology of the liontailed macaque (*Macaca silenus* Linnaeus): a pilot study. J Bombay Nat Hist Soc 65:283–293

Thierry B, Heistermann M, Aujard F et al (1996) Long-term data on basic reproductive parameters and evaluation of endocrine, morphological, and behavioral measures of monitoring reproductive status in a group of semi-free ranging Tonkean macaques (*Macaca tonkeana*). Am J Primatol 39:47–62

Thierry B, Iwaniuk AN, Pellis SM (2000) The influence of phylogeny on the social behaviour of macaques (primates: Cercopithecidae, genus *Macaca*). Ethology 106:713–728

Tosi AJ, Morales JC, Melnick DJ (2003) Paternal, maternal, and biparental molecular markers provide unique windows onto the evolutionary history of macaque monkeys. Evolution 57:1419–1435

van Schaik CP, van Noordwijk M, Nunn CL (1999) Sex and social evolution in primates. In: Lee PC (ed) Comparative primate socioecology. Cambridge University Press, Cambridge, pp 204–240

Wada K, Xiong CP, Wang QS (1987) On the distribution of Tibetan and rhesus monkeys in Southern Anhui province, China. Acta Theriol Sin 7:148–176

Wilkinson GS (2003) Social and vocal complexity in bats. In: de Waal FBM, Tyack PL (eds) Animal social complexity: Intelligence, culture, and individualized societies. Harvard University Press, Cambridge, MA, pp 322–341

Xia DP, Li JH, Zhu Y, Sun BH, Sheeran LK, Matheson MD (2010) Seasonal variation and synchronization of sexual behaviors in free-ranging male Tibetan macaques (*Macaca thibetana*) at Huangshan, China. Zool Res 5:509–515

Chapter 8
Tibetan Macaque Social Style: Covariant and Quasi-independent Evolution

Krishna N. Balasubramaniam, Hideshi Ogawa, Jin-Hua Li, Consuel Ionica, and Carol M. Berman

8.1 Introduction: Primate Sociality and Social Structure

In animals, including primates, group living and the social tendencies that support it evolve when the benefits of living in groups (e.g., predator avoidance, cooperative resource sharing) outweigh the costs (e.g., competition for resources, disease risk) (Kappeler and Van Schaik 2002). Primate species share close evolutionary histories and are biologically, socially, and physiologically similar to humans (Cobb 1976; Kappeler et al. 2015; Suomi 2011). As such, they make excellent comparative models for understanding the evolution of aspects of human group living, an endeavor that is fundamentally important given that social connectedness is

K. N. Balasubramaniam (✉)
Department of Population Health and Reproduction, School of Veterinary Medicine, University of California at Davis, Davis, CA, USA

H. Ogawa
School of International Liberal Studies, Chukyo University, Toyota, Aichi, Japan
e-mail: hogawa@lets.chukyo-u.ac.jp

J.-H. Li
School of Resources and Environmental Engineering, Anhui University, Hefei, Anhui, China

International Collaborative Research Center for Huangshan Biodiversity and Tibetan Macaque Behavioral Ecology, Anhui, China

School of Life Sciences, Hefei Normal University, Hefei, Anhui, China
e-mail: jhli@ahu.edu.cn

C. Ionica
Biomedical Department, F. I. Rainer Anthropology Institute, Romanian Academy, Bucureşti, Romania

C. M. Berman
Department of Anthropology and Graduate Program in Evolution Ecology and Behavior, State University of New York at Buffalo, Buffalo, NY, USA
e-mail: cberman@buffalo.edu

© The Author(s) 2020
J.-H. Li et al. (eds.), *The Behavioral Ecology of the Tibetan Macaque*, Fascinating Life Sciences, https://doi.org/10.1007/978-3-030-27920-2_8

associated with many benefits, including physical and emotional well-being, coping with environmental stressors, and enhancing survival, both in humans (e.g., Cohen et al. 2015; Holt-Lunstad et al. 2010; House et al. 1988) and in nonhuman primates (e.g., Ostner and Schülke 2014; Sapolsky 2005; Schülke et al. 2010; Silk et al. 2003, 2010; Young et al. 2014).

The complexity of primate social systems may be captured by studying three interrelated aspects of social life—social organization, mating systems, and social structure (Clutton-Brock and Harvey 1977; Hinde 1976; Kappeler and Van Schaik 2002; Koenig et al. 2013; Thierry et al. 2004). Social organization refers to the distribution and composition of individuals within social groups, specifically group sizes, sex ratios, and sex-typical philopatry versus dispersal strategies (Clutton-Brock and Harvey 1977). Social organization in turn gives rise to the other two aspects of social life. The first concerns mating systems which can be seen as the outcome of individual reproductive strategies adopted by resident and dispersing adults in a group (Greenwood 1980). The second is social structure, which arises from the patterning and distribution of agonistic, affiliative, cooperative, and/or conflict-managing interactions among group members (Hinde 1976; Kappeler and Van Schaik 2002; Sterck et al. 1997; van Schaik 1989, 1996). Social structure is of particular interest to behavioral ecologists because it provides both opportunities and constraints on individuals as they strive to maximize the benefits of group living while minimizing its costs. As such, social structure provides a rich window for testing hypotheses about adaptive and evolutionary strategies that individuals may use to gain fitness.

Among primates, social structure varies widely. Over the years, several conceptual frameworks have been proposed to explain how ecological factors might explain this variation. Wrangham (1980) was one of the earliest socioecologists to point out that, in several cases, (1) species with similar social organizations display sharply different social structures and to (2) argue that the distribution and abundance of resources needed by females to reproduce influenced their tendencies to form groups and to disperse or not. He also introduced the idea that these factors would in turn influence within- and between-group agonistic and affiliative social structure. This approach was extended by van Schaik (1989) and by Sterck et al. (1997) to include predation pressure and infanticidal risk as influential ecological factors. Sterck et al. (1997) described variation in the social structure of diurnal primates (1) emphasizing the influence of competitive regimes within groups on the nature and quality of female relationships (despotic vs. tolerant vs. egalitarian) and (2) pointing out that dominance structures were often associated with other aspects of social structure, e.g., kin bias and coalitionary support (Sterck et al. 1997). Since then, researchers have continued to emphasize power relationships as central to understanding variation in social structure.

In contrast to socioecological explanations, other schools of thought argue that aspects of social structure are more strongly influenced by inherent or intrinsic characteristics (Thierry 2004, 2007). Dominance hierarchies are once again central to this framework (de Waal and Luttrell 1989), but rather than adaptive responses to current ecological conditions, they are hypothesized to be either due to ecological adaptation in the distant past (Chan 1996; Kamilar and Cooper 2015; Matsumura 1999) or to emergent properties of self-organizational processes (Hemelrijk 1999,

2005; Puga and Sueur 2017). In either case, variation in social structure is posited to be strongly influenced by species' ancestral or phylogenetic histories (Balasubramaniam et al. 2012a; Kamilar and Cooper 2015; Thierry et al. 2008, 2000) due to tendencies for dominance structures to be structurally linked to entire suites of social traits and hence for these traits to covary with one another (Thierry et al. 2008; Thierry 2000). Such contrasting explanatory models have complicated our understanding of variation for dominance and other aspects of primate social structure.

In this chapter, we review our previous studies to illustrate the complexity of a few aspects of Tibetan macaque (*Macaca thibetana*) social structure, focusing on indicators of social style. The concept of social style posits that a suite of characteristics related to dominance and social tolerance covary (and may have coevolved) with one another and that species can be placed on a continuum ranging from extremely despotic to extremely tolerant (de Waal and Luttrell 1989; Thierry 2000, 2004; Thierry et al. 2008). Early studies on Tibetan macaques (Deng 1993; Ogawa 1995; Zhao 1996) detected a suite of moderately tolerant traits that led to their categorization as a "grade 3" species (Thierry 2000). Yet later work has revealed a mixture of characteristics that are associated with both despotism and tolerance, prompting a change in classification. Here we review this more recent line of evidence to describe how quasi-independent evolution, i.e., the apparent adaptive responses by animals to variation in current sociodemographic factors and ecological conditions, may have also influenced the evolution of such a mixture of social style traits in this species. We suggest that such effects may potentially mask underlying tendencies for the covariation of traits. Similarly, we describe comparative research across the macaque genus that suggests that aspects of social style may have been influenced by species' phylogenetic histories, by the covariation of traits to each other, and by adaptive responses to variation in current conditions. In doing so, we suggest that aspects of Tibetan social styles may represent flexible responses to different types of selection pressures that are nevertheless limited to a given species-typical range of possible responses, i.e., "social reaction norms" (Berman and Thierry 2010). We end by highlighting some avenues for future research that address some crucial gaps in our current understanding of the evolution of Tibetan macaque social styles and macaque social structure in general.

8.2   The Macaques and the Study of Variation in Social Structure

To date, comparative studies of variation in primate social structure have largely focused on members of the family Cercopithecinae, specifically baboons (genus: *Papio, Theropithecus*) and macaques (genus: *Macaca*) (Cords 2013). This is because these genera are represented by many species that are among the most geographically widespread and ecologically diverse of all primates. Moreover, they show marked variation in several aspects of social structure both within and across species (Cords 2013). Members of the genus *Macaca* range from North Africa in the West to Japan in the East and from China in the North to Sulawesi, Indonesia, in the South (Abegg and Thierry 2002; Thierry 2013; Thierry et al. 2004). All macaque species show a broadly similar social organization (Thierry 2013; Thierry et al. 2004). Barring a few exceptions (Sinha et al. 2005), they live in multi-male, multi-female social groups with varying sex ratios (although usually skewed toward females). Females are philopatric and males disperse from their natal groups when they reach maturity (Cords 2013; Paul and Kuester 1987). Yet macaque species vary broadly in several aspects of social structure related to aggression intensity, dominance relationships, maternal style, tendencies to reconcile following conflicts, and tendencies to show preferences for kin in affiliative and cooperative exchanges (Thierry 2000, 2007, 2013).

Early observations found these various tendencies appeared to be clustered, particularly among some better-studied species. This led to the theoretical concept of macaque "dominance styles" (de Waal and Luttrell 1989) and shortly thereafter renamed "social styles" (Thierry 2000), a concept with dominance relationships at its core. The social style concept posits that aspects of macaque social structure covary with one another and that species can be placed on a continuum (later simplified into a four-grade scale) that ranges from extremely despotic to extremely tolerant (Thierry 2000, 2007; Thierry et al. 2004). Specifically, species categorized as "grade 1" [extremely despotic, e.g., rhesus macaques (*Macaca mulatta*) and Japanese macaques (*M. fuscata*)] and "grade 2" [moderately despotic, e.g., long-tailed macaques (*M. fascicularis*) and pigtailed macaques (*M. nemestrina*)] may be expected to show intense aggression, steep hierarchies and highly asymmetric dominance relationships (Balasubramaniam et al. 2012a), low tendencies to reconcile after conflicts (Thierry et al. 2008), strong preferences toward kin (Thierry et al. 2008), protective maternal styles (Thierry et al. 2000), and community formation or substructuring in their affiliative relationships (Sueur et al. 2011). In contrast, "grade 3" [moderately tolerant, e.g., Barbary macaques (*M. sylvanus*) and bonnet macaques (*M. radiata*)] and "grade 4" [extremely tolerant, e.g., crested macaques (*M. nigra*) and Tonkean macaques (*M. tonkeana*)] species are hypothesized to display mild aggression, shallow hierarchies, frequent counter-aggression, strong tendencies to reconcile, l'ow degrees of kin bias, relaxed maternal styles, and dense well-connected grooming social networks.

Although the concept of social style is widely employed in the primate literature, certain issues remain. First, the extent to which one can speak of single species-typical social styles is not clear. In most Cercopithecine primates, female-female relationships define the core of the group's social structure (Cords 2013; Koenig et al. 2013; Sterck et al. 1997; van Schaik 1989). As such, most designations of social style among macaques are based on female-female relationships, particularly in captive groups where the number of males available to study is often limited (Balasubramaniam et al. 2012a, b; Thierry 2000; Thierry et al. 2004, 2008). However, a number of studies have found differences in social styles among males and females of the same species (Cooper and Bernstein 2008; Preuschoft et al. 1998; Richter et al. 2009; Tyrrell et al. 2018).

It is also unclear how consistent social style indicators are over time or among groups of the same species. Although several studies have found consistent styles among different captive groups (Demaria and Thierry 2001; Petit et al. 1997; Sueur et al. 2011; Thierry 1985), and among wild, free-ranging and captive groups of the same species (Balasubramaniam et al. 2012a, b), others have found variation in social style indicators (e.g., grooming kin bias, hierarchical steepness) as groups have grown and fissioned (Balasubramaniam et al. 2011; Berman and Thierry 2010; Zhang and Watanabe 2014). Such inconsistencies in the literature have been partly responsible for ongoing debates over the origins of social style and other aspects of social structure. Specifically, the extent to which variation in social structure may be linked to inherited characteristics [e.g., genetic polymorphisms, phylogenetic history (Blomberg et al. 2003; Thierry 2007, 2013; Thierry et al. 2000)] versus outcomes of adaptive responses of individuals to current socioecological factors (Clutton-Brock and Janson 2012; Isbell 2017; Koenig et al. 2013; Sterck et al. 1997) remains unclear and hotly debated.

Some proponents of the phylogenetic argument posit that aspects of macaque social style are structurally linked and hence covary with each other across species (Petit et al. 1997; Thierry 2000; Thierry et al. 2008). In an early extreme form, this idea, called the systematic variation hypothesis (Castles et al. 1996; Petit et al. 1997), posited that species that show a single indicator of a particular social style (e.g., steep dominance gradients in grade 1 species) may be expected to display all other aspects of that style. In other words, it predicts that all indicators for species showing extreme despotism or extreme tolerance should be in the extreme range and that all indicators for species showing intermediate styles should display intermediate indicators. However, there is now a consensus that this extent of covariation is not always detectable and may be different for different traits and species. For instance, Thierry et al. (2008) found that across the range of macaques with different social styles, post-conflict affiliation covaried systematically with explicit forms of contact, but not with grooming kin bias. Further, although most studies thus far have found the predicted clusters of social style indicators and marked differences between species at the extreme ends of the social style scale (e.g., Demaria and Thierry 2001; Sueur et al. 2011; Thierry 1985), the positions of species in intermediate grades remain unclear and do not always concur with the covariation hypothesis (summarized in Balasubramaniam et al. 2012b).

Many proponents of socioecological explanations for variation in social style hypothesize that current conditions such as the distribution and abundance of resources, predation pressure, and the risk of infanticide critically influence the number and spatial distribution of females in groups. In female philopatric groups, those factors in turn influence not only the number and distribution of males but also the social structure of the group. While social style indicators are expected to cluster to some extent, this framework (named the Ecological Model of Female Social Relationships or EMFSR: Koenig et al. 2013; Sterck et al. 1997) does not necessarily predict structural linkages or tight covariation among them.

In the next sections, we summarize our research findings on aspects of Tibetan macaque social structure for both males and females. We describe indicators associated with both despotic and tolerant social styles and evaluate the extent to which these different indicators seem to covary with one another, or are labile, independent responses by individuals to external factors, within this species. Then, we review our comparative studies that examine the extent to which social style indicators covary and display phylogenetic signals and/or are influenced by variation in current sociodemographic factors, across macaque groups and species belonging to all four social style grades.

8.3   Tibetan Macaques and the YA1 Group

Tibetan macaques are the largest macaque species (Fooden 1986). Preferring montane habitats, they are found in the subtropical and temperate forests that range from South Central to Western China as far as Tibet (Li 1999; Zhao 1996). Taxonomically, they are in the *sinica-arctoides* phylogenetic clade. They are most closely related to Assamese (*M. assamensis*) and Arunachal (*M. munzala*) macaques (Balasubramaniam et al. 2012a; Chakraborty et al. 2007; Purvis 1995; Tosi et al. 2003), although they morphologically resemble stump-tailed macaques (*M. arctoides*). Like all macaques, they live in multi-male, multi-female social groups that show female philopatry, male dispersal, and linear dominance hierarchies (de Vries 1998; Zhao 1996).

Work on the ecology and social structure of Tibetan macaques began with pioneering studies on Mt. Emei, Sichuan Province, China (Deng 1993; Li 1999; Zhao 1996; Zhao et al. 1991), and at Mt. Huangshan, Anhui Province, China (118.3°E, 30.2°N) (Li 1999; Wada et al. 1987). The study at Huangshan continued by focusing on the semi-provisioned Yulingkeng A1 or YA1 group (Li 1999). Huangshan is a popular tourist destination in East-Central China and home to several groups of Tibetan macaques with (apparently) nonoverlapping home ranges. The montane vegetation consists of mixed deciduous and evergreen forests at lower elevations and sparsely covered peaks at higher elevations (Li 1999; Zhao 1996; Zhao et al. 1991). The YA1 group has been monitored, and maternal kin relationships of adults have been recorded since 1986 (Li 1999). The data we review here were collected on the YA1 group over six observation periods spread across 12 years. During each of the six observation periods, the group's hierarchy was stable, and we used similar observation methods (focal animal sampling and all occurrences sampling). H. Ogawa collected data between 1991 and 1992, and C. Berman and colleagues collected data between 2000 and 2002. During this time, the group size varied from 39 to 52 individuals. Although wild, the macaques were managed for tourism for a period of time that overlapped with data collection—inconsistently during the first three periods, but consistently so for the second three periods. Park officials herded the animals from the forest into a viewing area where tourists could observe them from a pavilion, as the monkeys were provisioned with corn. Between feedings, the group spent most of their time in the surrounding forest.

8.4 Evidence of Female Despotism Contradicts Earlier Studies

When Thierry published his classic analysis of social style and proposed a four-grade social style scale (Thierry 2000; Thierry et al. 2000), comprehensive analyses of core social style indicators were available for a limited number of macaque species. Hence, the placement of many species, including Tibetan macaques, was tentative and based on their phylogenetic closeness to better-studied species (Thierry et al. 2000), or on indirect behavioral indicators. Thierry (2000) placed Tibetan macaques in grade 3 (moderately tolerant) based on their membership in the *sinica-arctoides* lineage (which contains other tolerant macaques) and on early studies that described evidence for multiple traits that were associated with a mildly tolerant social style: frequent ritualized affiliation among males, bidirectionality in silent bare-teeth displays, triadic infant handling interactions particularly among males, and tolerant responses to infant handling by mothers (Deng 1993; Ogawa 1995; Zhao 1996). It was not until 2004 that Berman and colleagues evaluated multiple other core indicators of social style (Berman et al. 2004). Using data collected on the YA1 group between 2000 and 2002, this study expected to confirm the tentative designation of Tibetan macaques as "tolerant" or "grade 3" species. They examined (1) the degree of aggressive asymmetry and counter-aggression in dominance relationships, (2) conciliatory tendencies, and (3) kin bias in a variety of affiliative and tolerant behaviors separately for males and females.

The results were surprising for both sexes: they supported a despotic rather than a tolerant social style. First, both males and females displayed high degrees of asymmetry in their dominance relationships. Three indices were used to evaluate the extent to which aggression was bidirectional, specifically the (1) directional inconsistency index (or DII), (2) the dyads-up index, and (3) percentage of counter-aggression, and all revealed extreme aggressive asymmetry typical of extremely despotic species (Table 8.1). Similarly, submissive signaling in the form of silent bare-teeth displays was almost always given in the same direction between dyads and in the same direction as other fearful and submissive interactions. Second, conciliatory tendencies were low. These were examined using the post-conflict matched-control pairs, or the PC-MC method (de Waal and Yoshihara 1983), which examined the timing of affiliative interaction during a 5-min window of time following an act of aggression (the PC or post-conflict sample), and compared it with the timing during a comparable period of time without aggression (the MC or matched sample). The method thus evaluated whether affiliation occurred earlier, later, or at the same time in the PC as the MC. The comparisons were used to calculate a corrected conciliatory tendency (CCT) that varies from 0 to 100% and

Table 8.1 Bidirectionality of aggression during each of three data analysis periods[a] for a single group of Tibetan macaques

	8/1/00–1/28/01	2/27/01–5/29/01	12/9/01–7/25/02
A. Directional inconsistency index			
Male-male dyads	3/583 (5.2%)	2/50 (4.0%)	2/108 (1.9%)
Female-female dyads	6/160 (3.8%)	3/166 (1.8%)	4/243 (1.6%)
B. Dyads-up index			
Male-male dyads	1/36 (2.8%)	2/36 (5.6%)	0/28 (0.0%)
Female-female dyads	3/78 (3.8%)	1/78 (1.3%)	0/45 (0.0%)
C. Percent counter-aggression			
Male-male dyads	3/37 (8.1%)	0/33 (0.0%)	0/73 (0.0%)
Female-female dyads	0/97 (0.0%)	4/114 (3.5%)	6/157 (3.8%)

All three measures of bidirectionality indicate a high degree of asymmetry in dominance relationships among both males and females, typical of a despotic species
A. The percentage of total aggressive interactions that were directed in the less frequent direction within dyads
B. The percentage of dyads for which the main direction of aggression was up the dominance hierarchy
C. The percentage of instances of aggression of any kind to which the target responded with aggression of any kind
[a]Adapted from Berman et al. (2004)

Table 8.2 Kin bias among Tibetan macaques, indicated by Partial Kr coefficients between interaction rates and degree of relatedness, controlling for rank distance[a]

Partner combination	% Time w/in 5 m	Grooming	Sit near	Approaches	Co-feed
♀-♀	0.16*	0.22**	0.25**	0.24*	0.18*
♂-♂	−0.14	0.13	−0.17(*)	−0.17	−0.14

The results suggest a significant degree of kin bias among females in affiliation and tolerance
**$p < 0.01$
*$p < 0.05$
(*)$p < 0.1$
[a]Adapted from Berman et al. (2004)

estimates the degree to which dyads are more likely to reconcile (i.e., engage in affiliation soon after a conflict than at other times) (Veenema et al. 1994). For female-female dyads, the conciliatory tendency was extremely low (4.2%), a figure that is low even among grade 1 macaque species. For males, it was a moderate 19.7% (see also Berman et al. 2006), but still within the range of despotic male macaques (unrelated male Japanese macaques: Petit et al. 1997).

Finally, females, but not natal males, consistently displayed high levels of affiliative kin bias and social tolerance. Mean rates of female affiliation (grooming, sitting near, co-feeding) with maternal kin were significantly greater than affiliation rates with nonkin (Table 8.2). Indeed, the intensity of grooming kin bias, measured as the observed-to-expected ratio of grooming bouts between kin, suggested that females groomed kin more than three times the value expected by chance. All the results were consistent across seasons and locations (provisioning area vs. forest), strongly supporting a despotic social style for Tibetan macaques. Although several core indicators were similar to those of grade 1 macaques, findings from this study led to a reclassification of Tibetan macaques as grade 2 rather than grade 1 or grade 3 species (Matsumura 1999; Thierry 2007) based on previous findings of ritualized affiliative behavior among males and of maternal tolerance for infant handling among females.

Although these findings greatly expand our understanding of social style among Tibetan macaques, they pose complications for possible explanations for the origins of such despotism. On the one hand, a phylogenetic explanation is difficult since Tibetan macaques are members of the *sinica-arctoides* lineage which is composed of predominantly tolerant macaque species. Yet, they display heightened levels of despotism that are characteristic of members of the *fascicularis* lineage (Matsumura 1999; Thierry et al. 2000) like pigtailed macaques (Castles et al. 1996) and Assamese macaques (Cooper and Bernstein 2002). Socioecological explanations based on the distribution of natural resources and predation pressure are also unlikely given that Tibetan macaques have a largely folivorous diet (Zhao et al. 1991) and experience (historically) moderate levels of predation pressure (Xiong 1984), factors that socioecological frameworks associate with tolerant or egalitarian societies (Sterck et al. 1997).

Berman and colleagues (Berman et al. 2004) speculated that indicators of a despotic social style in this group may have risen recently due to human activity; activities associated with tourism, particularly range restriction, herding, and food provisioning, likely elevated levels of intragroup aggression and/or competition (Berman and Li 2002; Hill 1999; Judge 2000; Marechal et al. 2011; Ram et al. 2003) and thereby generated atypical despotism in a species that would otherwise show greater social tolerance. Together, these results demonstrate that, rather than all aspects of female social style showing moderate despotism, some are in the range of extremely despotic species, and others are in the range of moderately tolerant species. Such a mixture of both despotic and tolerant social style traits in the same species suggests that structural linkage and potential covariation of traits may be offset, at least in part, by potentially adaptive individual responses to contemporary selection pressures posed by human management.

8.5 Males Exhibit Social Tolerance Despite Evidence for Despotism

In general, less research has focused on male-male relationships than female-female relationships among Cercopithecine primates and other primates that display female philopatry and male dispersal. When males disperse, they typically join groups that lack close kin, particularly in species that disperse repeatedly (as do Tibetan macaque males, whose average tenure in a group is about 5–6 years: Berman et al., unpublished data). As such, opportunities for long-term affiliative and cooperative relationships among males are limited. More fundamentally, affiliation and cooperation would not be expected among males, because males primarily compete for resources (i.e., conceptions with fertile females) that cannot be shared in the same way as food patches or other resources (Schülke and Ostner 2008, 2013; Sterck et al. 1997). As such, theoretical considerations do not generally predict high levels of affiliation and cooperation among males.

Socioecological models predict that the social strategies of primate males are driven by the availability and distribution of fertile females (Kappeler and Van Schaik 2002; Sterck et al. 1997; van Schaik 1989). When several females form a cohesive group, single males are not likely to be able to monopolize them, particularly when they breed seasonally. This leads to the formation of multi-male, multi-female societies in which males may coexist but attempt to outcompete one another through various forms of scramble or contest competition, particularly over access to females (Kappeler 2000; van Schaik 1996; Van Schaik et al. 2004). High levels of competition should theoretically preclude strong male-male affiliative relationships in these groups. However, some macaque males regularly engage in affiliative interactions that lead to strong social bonds and that enhance their fitness (Ostner and Schülke 2014; Schülke et al. 2010; Young et al. 2014).

Indeed, contrary to predictions, particularly high levels of male-male affiliation and tolerance have been reported among groups and species with relatively even sex ratios versus those with highly female-biased sex ratios. In these cases, male-male affiliation and tolerance have been hypothesized to function as coping mechanisms against intense conflict and competition both from other group males (e.g., access to females, revolutionary coalitions) and from external threats (e.g., predators, intergroup encounters, dispersing males: Cooper and Bernstein 2002; Ogawa 1995; Preuschoft and Paul 2000). The reasoning is that cooperation with some males against others may enhance a male's competitive abilities within his group and that group action may increase probabilities of success against external threats.

Previous research on Tibetan macaques suggests that males compete vigorously with one another for rank and access to fertile females, consistent with a despotic social style (Li and Wang 1996). Nevertheless, as mentioned earlier, they also show several forms of affiliative interaction with one another, consistent with their tendencies toward relatively even sex ratios (Li 1999). These include moderate conciliatory tendencies, ritualized greetings, and bridging behavior in which infants are used to facilitate friendly contact between males and to enhance the chances of forming affiliative social bonds (Ogawa 1995).

To further examine the ways in which male Tibetan macaques may engage in tolerant and cooperative interactions with kin and nonkin, despite strong indications of a despotic social style, Berman and colleagues examined cooperative strategies among males, focusing particularly on agonistic coalitions (Berman et al. 2007). They examined affiliation, tolerance, and agonistic support to test the hypothesis that increased tolerance in otherwise despotic males may occur when high-ranking males require support from other males to maintain their positions. In this group, conservative coalitions, in which two higher-ranking males aggressively target a lower-ranking male, are the most common, and these serve to reinforce the current hierarchy. Nevertheless, revolutionary coalitions in which two lower-ranking males challenge a higher-ranking male pose a threat particularly to alpha males. Although agonistic support is unrelated to kinship and rates of grooming, high-ranking males display tolerance in the form of co-feeding toward lower-ranking males that support them, and alpha males show the most cooperation with the males that targeted them in revolutionary coalitions. Berman and colleagues suggested that high-ranking males discourage revolutionary alliances by using two cooperative strategies. They primarily rely on conservative alliances, but also offer tolerance to potential rivals in cases in which conservative coalitions are less effective.

We suggest that, contrary to socioecological theory (van Hooff and van Schaik 1994; van Schaik 1996; Van Schaik et al. 2004), Tibetan macaque males may be "tolerant despots" (*cf* Kaburu and Newton-Fisher 2015). Just as Kaburu and Newton-Fisher use the term "egalitarian despots" to describe chimpanzees that display signs of social tolerance which emerge strategically from a more despotic dominance style, we suggest that social tolerance in male Tibetan macaques is not shaped by external factors like increased between-group competition or predation pressure which were low or absent in this population. Rather, we suggest that social tolerance may be an adaptive outgrowth of within-group despotism, in that it allows despotic males to enhance their competitive abilities by selectively displaying tolerance to other males. Like the findings for female social styles, findings for male-male relationships also suggest a mixture of both despotic and tolerant social style indicators. Such evidence for unlinked traits that may be independently influenced by different, sometimes opposing, types of selection pressures may mask or offset potential underlying evidence for the structural linkage or covariation of traits.

8.6 Comparative Studies Provide Evidence for Both Covariation and Quasi-independent Evolution

Comparative research among primate species provides powerful opportunities to discern the extent to which variation in a characteristic is likely to be related to variation in ecological factors and/or evolutionary trends (Nunn 2011). This is particularly the case for comparative studies of macaque social structure, given their similar social organizations but varying social structures (Thierry 2007). To date, comparative studies of macaque social structure have largely focused on evolutionary explanations, specifically on the influence of phylogenetic relatedness. However, they have found somewhat mixed evidence for the phylogenetic model. An early study found that 7 out of a set of 22 macaque social behavioral traits analyzed appeared to be strongly influenced by phylogenetic relatedness (Thierry et al. 2000). Later work found that aspects of macaque social structure related to post-conflict affiliation and explicit contact, but not grooming kin bias, were more similar among more closely related species and that they appeared to covary continuously across species in predicted directions after accounting for phylogeny (Thierry et al. 2008). Most of these studies focused on captive groups in which it was relatively easy to restrict the analysis to behavior outside of the feeding context and hence to avoid the possibility that behavior related to social style might be masked by direct feeding competition.

Building on these pioneering studies, Balasubramaniam et al. (2012a, b) conducted three comparative studies of variation in other aspects of macaque social structure, specifically dominance hierarchies and grooming social network structure, which until then had not been tested in a comparative framework that involved multiple macaque groups and species. These studies included both captive and free-living groups of macaques, including Tibetan macaques at Huangshan, and specifically examined the potential influence of living condition on social style traits. We review the major findings from these three studies below, focusing on their implications for both the characterization of Tibetan macaque social style and for the evolutionary origins of macaque social structure in general.

In the first of two comparative studies of macaque dominance social structure that tested the predictions of phylogenetic models (Balasubramaniam et al. 2012a), Balasubramaniam and colleagues examined whether the steepness of hierarchies (de Vries et al. 2006) and degrees of dominance asymmetry or counter-aggression showed strong phylogenetic signals, i.e., whether they were more similar among more closely related species (Blomberg et al. 2003; Kamilar and Cooper 2015; Matsumura 1999; Thierry et al. 2008). They assembled a behavioral dataset on dyadic aggressive interactions and submissive displacements among adult females, for each of 14 groups of macaques representing 9 species. For each group, they calculated the steepness of the dominance hierarchy as the absolute slopes of plots between normalized David's scores (NDS: de Vries et al. 2006) and ordinal ranks of individuals attributed by David's scores. David's scores are calculated for each individual as the differences between its wins and losses in dyadic agonistic interactions, which are weighted by the relative wins and losses of all of the other individuals in the group (Gammell et al. 2003; de Vries et al. 2006). To date, hierarchical steepness represents the most comprehensive measure of the "dominance gradient" and hence the degree of despotism in the dominance hierarchy of a social group. In addition to steepness, they also calculated the degree of "dominance asymmetry," by estimating the proportion of aggressive interactions that involved counter-aggression from the initial recipient.

To test predictions of phylogenetic signals, they used multiple macaque Bayesian and maximum likelihood phylogenetic trees, both self-reconstructed and extracted from the online database for primate phylogenies *10kTrees* (Arnold et al. 2010). The results were consistent with the phylogenetic model in that both hierarchical steepness and counter-aggression showed strong, significant phylogenetic signals. Moreover, phylogenetic signals were consistent and strong when they examined a subset dataset of just free-living groups representing seven species of macaques.

In a second comparative study that used the same dataset, Balasubramaniam et al. (2012b) tested the covariation hypothesis and found mixed support for it. They asked whether hierarchical steepness and counter-aggression both covaried with the hypothesized placement of species on Thierry's four-grade social style scale and whether evolutionary shifts in social style grade corresponded consistently to covariant changes in hierarchical steepness and counter-aggression. As predicted by the covariation hypothesis, steepness and counter-aggression were strongly correlated with the placement of species on macaque social style scales in the predicted directions. Yet the nature of these relationships differed for each trait. On the one hand, hierarchical steepness appeared to covary continuously with the scale. Macaques in grades 1 and 2, including Tibetan macaques, showed the steepest hierarchies, grade 3 species showed moderately steep hierarchies, and grade 4 species showed the shallowest hierarchies. On the other hand, counter-aggression appeared highly dichotomous. Species from grades 1–3 all showed similar and markedly low levels of counter-aggression (range, 0.6–6.4% of all aggressive interactions) in comparison with the two grade 4 species (crested macaques, 50.8% of aggressive interactions; Tonkean macaques, 60.4%). This suggested that the covariation between dominance traits and social scale were driven primarily by species at the extreme ends of the scale (grade 1 vs. grade 4 macaques).

In comparison, the characteristics of species in intermediate grades 2 and 3 were inconsistent. For instance, grade 1 rhesus and Japanese macaques showed both steep dominance hierarchies (range, 0.68–0.99) and low levels of counter-aggression (range, 0–7.1% of all aggressive interactions), while grade 4 Sulawesi crested and Tonkean macaques showed the opposite characteristics (steepness range, 0.27–0.53; counter-aggression range, 50–61% of all aggressive interactions). However, grade 2 species, rather than showing more shallow hierarchies than grade 1, fell within the range for grade 1 species (steepness range, 0.85–0.98). In fact, two grade 2 species—Tibetan and long-tailed macaques—showed the steepest hierarchies of all species examined in the study, i.e., exceeding the values for grade 1 Japanese and rhesus macaques. Further, Tibetan macaques showed the lowest levels of counter-aggression (mean = 0.64%; range, 0.53–0.76% of all aggressive interactions), even lower than grade 1 macaques.

The case of grade 3 macaques was even less consistent. Although hierarchical steepness of Barbary macaques was intermediate to those for grades 2 and 4 as predicted for a moderately tolerant social style, low levels of counter-aggression seen in this same group were more indicative of a more despotic style (but see Thierry and Aureli 2006), indicating possible evidence for quasi-independent evolution. Moreover, grade 3 bonnet macaques showed both steeper hierarchies and low levels of counter-aggression, within the range of despotic species.

To test the covariation hypothesis statistically, Balasubramaniam et al. (2012b) used phylogenetic independent contrast approaches of ancestral trait reconstruction to determine whether major evolutionary shifts from despotic to tolerant social style grades corresponded to shifts in hierarchical steepness and counter-aggression in the evolutionary history of macaques. The findings were telling. For both steepness and counter-aggression, the magnitudes of independent contrasts were greatest at the nodes in the phylogenetic tree that split grade 4 macaques from all other grades. This was consistent with the finding that there were indeed marked differences between macaques at the extreme ends of the scale. However, contrary to predictions, evolutionary shifts from higher to lower social style scales, or from tolerant to despotic societies, did not result in consistent increases in steepness contrasts or decreases in counter-aggression contrasts. Moreover, the magnitude of steepness and counter-aggression contrasts were no greater at nodes in the phylogeny where there were evolutionary shifts in social style, compared to "neutral" nodes in which there were no shifts in style. In summary, the authors found strong evidence for phylogenetic signals in dominance traits and mixed evidence for the covariation hypothesis. Specifically, covariation seemed to more strongly influence the underlying differences in dominance hierarchies between some species and phyletic lineages (macaques across grades 1 and 4), but less so others (species in grades 2 and 3).

In a third study, Balasubramaniam and colleagues (2018a) examined interspecific covariation in higher-order aspects of both dominance hierarchies and social grooming relationships across an even wider range of macaque populations. To capture higher-order aspects of social structure, they used social network analysis (SNA: Farine and Whitehead 2015; Lusseau 2003). Beyond using only direct interactions among dyads, network approaches incorporate both the direct social connections and indirect pathways of interactions among individuals (Brent et al. 2011; Farine and Whitehead 2015; Kasper and Voelkl 2009; Sueur et al. 2011). Such approaches may identify consistent patterns or "motifs" of interactions that in turn describe group social structure. The focus on social grooming, a core aspect of affiliative and cooperative social structure, expanded on previous studies that had predominantly focused on aspects of post-conflict affiliation, affiliative kin bias, and dominance hierarchies (but see the inclusion of macaque grooming traits by Schino and Aureli (2008) in their interspecies analyses).

In comparison to these latter traits which may be strongly influenced by intrinsic factors such as personality (Capitanio 1999; Krause et al. 2010), matrilineal inheritance (Cords 2013; Kapsalis 2004; Sade 1972), and phylogenetic histories (Thierry 2007; Thierry et al. 2008, 2000), the patterning and distribution of grooming relationships has been assumed to be more responsive to environmental variability, such as group sizes (Berman and Thierry 2010; Kasper and Voelkl 2009; Majolo et al. 2009), the presence of infants (Fruteau et al. 2011; Gumert 2007; Henzi and Barrett 2002; Tiddi et al. 2010), and the distribution of resources (Balasubramaniam and Berman 2017; Carne et al. 2011; Ventura et al. 2006). Nevertheless, across the macaque genus, the structure of grooming social networks is also likely to be strongly influenced by varying tendencies for species to affiliate with kin, i.e., grooming kin bias (Berman 2011; Berman and Kapsalis 2009; Berman and Thierry 2010; Kapsalis 2004; Thierry et al. 2008).

Under the covariation framework, more despotic, nepotistic species that show steeper dominance hierarchies, and in which individuals prefer to interact more with close kin compared to distant kin, may be expected to show more modular or substructured grooming networks (Sueur et al. 2011). On the other hand, more tolerant species in which hierarchies are shallower, and in which individuals form affiliative relationships with a wider range of conspecifics, may be expected to show more dense, well-connected social networks that are less modular (Sueur et al. 2011). In support of this covariation framework, a study by Sueur and colleagues found that more intolerant or nepotistic grade 1 species like rhesus and Japanese macaques showed less dense but more centralized, modular, or substructured social networks than more socially tolerant, less nepotistic grade 4 species like Tonkean and Sulawesi crested macaques (Sueur et al. 2011). Balasubramaniam and colleagues used a wider range of macaques to ask whether (1) aspects of dominance and grooming social networks showed strong phylogenetic signals and (2) they covaried with each other independently of the effects of phylogeny and/or extrinsic sociodemographic factors like group size, sex ratio, and living condition.

The authors assembled a broad comparative dataset—38 dominance matrices and 34 datasets of social grooming—from captive, free-ranging, and wild groups representing 10 species of macaques. From these, they estimated two dominance social network traits, specifically triangle transitivity (Shizuka and Mcdonald 2012) and group-wide dominance certainty (Fujii et al. 2013). From the grooming datasets, they estimated three aspects of grooming social network structure: the (1) centrality coefficient, i.e., the extent to which dominant individuals are also highly central in their grooming network (Handcock et al. 2006); (2) Newman's modularity, i.e., the degree to which networks were substructured into communities of closely interacting individuals (Newman and Girvan 2004); and (3) clustering coefficient (Csardi and Nepusz 2006), which is an indicator of the localized density of a group's social connections. As in previous efforts, they replicated their comparative analyses across multiple phylogenetic trees to account for phylogenetic branch length uncertainty.

As in the previous comparative studies, the results showed mixed evidence for the phylogenetic model. Consistent with findings from the previous study on dominance steepness and counter-aggression (Balasubramaniam et al. 2012a, b), dominance transitivity and certainty both showed strong phylogenetic signals (Fig. 8.1). On the other hand, aspects of grooming social networks showed only moderate or weak phylogenetic signals (Fig. 8.1). Moreover, contrary to the covariation hypothesis, grooming network measures did not covary with dominance traits across species (Fig. 8.2). Instead, two aspects of grooming social networks, modularity and clustering coefficient, were more strongly predicted by group size, a sociodemographic factor, than by species' social styles or dominance traits. Specifically, larger groups of macaques showed more modular, less densely clustered grooming networks than smaller groups, with the results being independent of groups' living condition and variation in sampling effort.

Taken together, findings from these interspecies comparative studies suggest that different aspects of social structure may have been subject to different evolutionary and ecological selection pressures. Specifically, aspects of dominance social structure seem to be strongly influenced by phylogenetic relationships and to show tendencies to covary across some (but not all) species and lineages. In comparison to dominance, grooming social networks seem weakly influenced by phylogeny. Rather than covarying with dominance traits across groups and species, grooming networks appear to be more responsive to variation in current sociodemographic factors like group sizes and living condition. The detection of more clear-cut differences in dominance traits between the grade 4 Sulawesi macaques and members of grades 1 and 2 in the *fascicularis* lineage suggests that the influence of

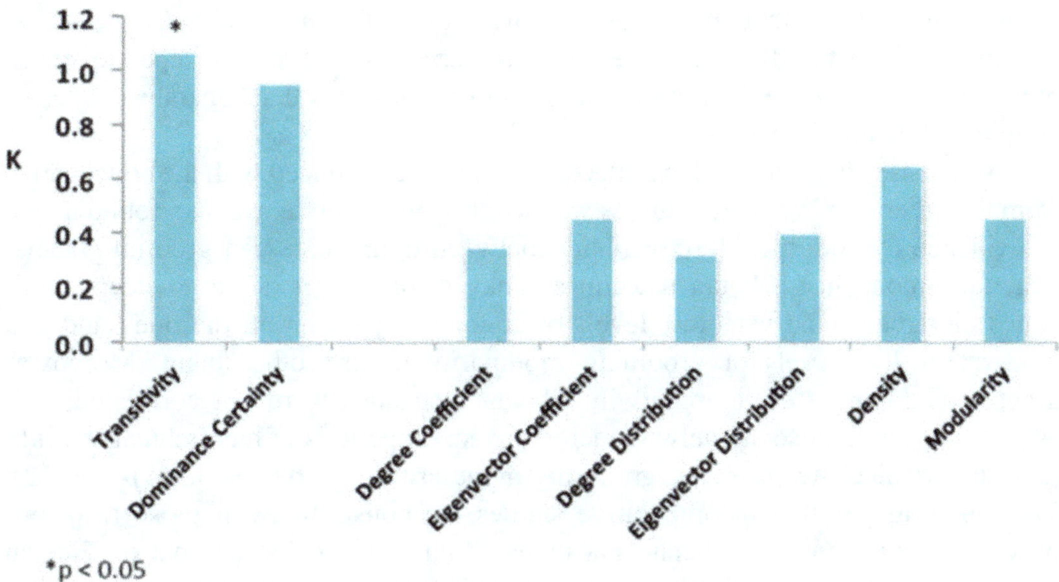

Fig. 8.1 Bar plot of the *K*-statistic calculated to determine the strength of phylogenetic signals among aspects of dominance and grooming social network structure for ten species of macaques. On the one hand, values of *K* approaching or greater than 1 indicate strong phylogenetic signals in the dominance traits. On the other hand, values of $K < 1$ indicate moderate-to-weak signals in the grooming traits

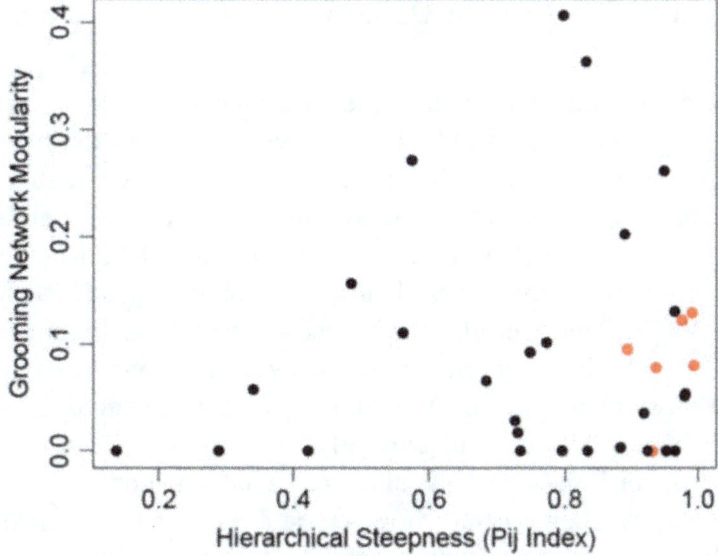

Fig. 8.2 Scatterplot showing the lack of a relationship between directional consistency index (DCI) of dominance hierarchies and Newman's community modularity of grooming social networks for 34 groups of macaques across 10 species. Red dots represent scores for the YA1 Tibetan macaque group at each of six study periods

phylogeny and the structural linkage of traits may be visible across higher organizational scales (e.g., across some species and lineages), but perhaps less so at lower organizational scales (e.g., across species within a lineage, across groups). In other words, they suggest that the structural covariation of aspects of macaque social structure has occurred over longer evolutionary timescales that separate entire phylogenetic lineages, rather than across shorter timescales that separate species or populations within lineages.

Consistent with findings at the interspecific level, and indeed with the conclusions from the other within-species studies we have reviewed above, we also found a lack of evidence for intraspecific covariation between dominance and grooming social structure among just Tibetan macaques. They displayed moderate-to-steep dominance hierarchies and very low levels of counter-aggression on the one hand and moderate-to-low levels of grooming modularity on the other hand. Moreover, aspects of dominance hierarchies in Tibetan macaques were not correlated with aspects of grooming social network across six study periods of hierarchical stability (e.g., hierarchical steepness vs. grooming modularity: $n = 6$, $r = 0.26$, $p = 0.62$). Thus, findings from our comparative studies are consistent with those from our previous studies on just Tibetan macaques. Both suggest that aspects of Tibetan macaque social style, while having potentially been shaped by phylogenetic relatedness and structural linkage in the evolutionary past, have been subject to somewhat unlinked, independent trajectories influenced by current socioecological and demographic conditions more recently in their evolutionary history.

8.7 Discussion and Future Directions

Aspects of social structure represent higher-order phenomena, i.e., phenotypically visible, group level outcomes of individuals striving to access the benefits of group living while minimizing its costs. Given the importance of social structure for individual fitness, primatologists are interested in the evolutionary underpinnings of variation in social structure (Clutton-Brock and Janson 2012; Koenig et al. 2013; Thierry 2004). Theoretical models based on species' phylogenetic history (Matsumura 1999; Thierry 2007; Thierry et al. 2000), inherent self-organization by individuals (Hemelrijk 1999, 2005; Puga and Sueur 2017), and adaptations to variation in current ecological conditions (Kappeler and Van Schaik 2002; Sterck et al. 1997; van Schaik 1989) have all been proposed, but currently there is no consensus regarding the relative extents to which these factors influence social structure.

In this chapter, we demonstrate how research on the genus *Macaca*, and in particular on Tibetan macaques, has provided useful empirical tests of some of these conflicting frameworks. Among macaque researchers, some proponents of phylogenetic arguments argued that aspects of social structure, specifically those that constitute social style, are structurally linked and covariant with each other (Petit et al. 1997; Thierry 2000; Thierry et al. 2008). Citing the detection of strong phylogenetic signals in some core aspects of macaque social style, they predicted that evolutionary change in one trait would co-occur with evolutionary change in a suite of other traits, with such covariation being more perceptible between species rather than within groups of the same species (Petit et al. 1997; Thierry 2000; Thierry et al. 2008).

Our findings on Tibetan macaques, and from comparative studies across a range of macaques, reveal incomplete evidence for such covariation across species and social style traits. We suggest that the pattern of covariation (and the lack thereof) that we found is consistent with the notion suggesting that covariation was likely present in the evolutionary past of the genus but that it has been subsequently partially offset by more contemporary, quasi-independent responses by macaques to variation in current conditions.

First, we found a mixture of both despotic and tolerant social style traits in Tibetan macaques. While aspects of the dominance structure in Tibetan macaques are within the range of despotic macaques, the structure of affiliative grooming social networks, male-male ritualized affiliation, and cooperative exchange are all characteristic of more tolerant societies. Covariation by itself, i.e., that has not been influenced by adaptive responses, would have been suggested if all these characteristics had showed moderate despotism or moderate tolerance. Second, our comparative studies revealed evidence for covariation between dominance traits and social style grade across some macaques in different phylogenetic lineages, but not across species within the same lineage. Third, the finding that grooming social network structure was more strongly influenced by group size than by phylogeny and dominance traits suggests that covariation was apparently offset by adaptive responses by macaques more recently in their evolutionary history. In other words, different aspects of social style and social structure in general likely evolved along more independent evolutionary trajectories, with some traits more strongly influenced by past ancestry and structural linkage than others, and others apparently more labile to current conditions.

This view is consistent with the concept of "social reaction norms," or the idea that species may respond similarly to major ecological changes, but may have inherently different ranges of responses to the same conditions, i.e., ranges that are related at least in part to their ancestry (Berman and Thierry 2010). The concept of social reaction norms can be seen as a group-level analog of the concept of interindividual variation in "behavioral syndromes," in which suites of correlated behaviors reflect individual variation in behavioral plasticity and in individual's capacities to show adaptive responses across multiple environmental situations and contexts (Sih et al. 2004, 2012).

Having said that, there remain many gaps in our understanding of social style and social structure in general in Tibetan macaques and in the macaque genus as a whole. Social style indicators have yet to be examined in many species, and a number of additional hypotheses about their origins remain to be explored in detail. We end this chapter by proposing and elaborating on five potentially important avenues of future research on social styles.

Paternal Kin Relatedness In animal social groups, the relatedness of individuals is strongly associated with the form, frequency, and distribution of cooperative and competitive social interactions (Hamilton 1964). In most Cercopithecine primates, females are philopatric, and there is ample evidence that matrilineal kin bias structures many aspects of social behavior within groups, more so in despotic species than tolerant species (Berman 2011; Berman and Thierry 2010; Thierry 2007; Thierry et al. 2008). As such, maternal kin bias has been considered one of the core indicators of a group's social style (Thierry 2000). In 1979, Altmann suggested that in addition to matrilineal ties, the structure of within-group relatedness may also be influenced by mating patterns, specifically by the degree to which individual males are able to monopolize fertile females (Altmann 1979). Specifically, the frequent replacement of highly successful males (i.e., males with high reproductive skew) may lead to age-structured paternal sibships within groups. If this occurs, and if individuals are able to recognize their paternal kin, inherent tendencies to favor kin could influence the patterning and distribution of cooperative and competitive interactions in a group and hence aspects of social style (Widdig 2013).

Recently, Schülke and Ostner hypothesized that the evolution of social tolerance in some macaque species may in fact be explained by high male reproductive skew (Schülke and Ostner 2008, 2013). High mean levels of paternal relatedness might explain why females in tolerant species distribute their affiliative interactions across a wide range of partners. As evidence for this hypothesis, they found that species of macaques classified as less nepotistic and more tolerant also showed higher male reproductive skew (Schülke and Ostner 2008). However, their analyses were limited by a small sample size and to comparisons of intolerant but seasonally breeding macaques with tolerant but nonseasonal macaques and hence were unable to control for the effects of seasonality on reproductive skew. In a second study, they found that paternal kin bias accounted for a small but significant proportion of the variation in the strength of grooming social bonds in a captive group of rhesus macaques (Schülke et al. 2013). Whether the reproductive skew argument can help explain some or all of the indicators of tolerance in Tibetan macaques is uncertain. Although male mating success is highly skewed toward dominants and alpha males have tenures of only a year or two, whether or not this translates into reproductive skew awaits genetic confirmation of paternity. Also needed are broader comparative studies that assess possible links between male reproductive skew, paternal relatedness, and multiple aspects of female social structure.

Disease Risk as a Selection Pressure Traditional socioecological models have speculated that multiple socioecological factors—natural resource abundance, clumped, human provisioned food, and changes in predation pressure over evolutionary time—may have all impacted variation in primate social structure (Kappeler and Van Schaik 2002; Sterck et al. 1997; van Schaik 1989). More recently, the influence of another ecological factor—disease risk—has also been proposed (Kappeler et al. 2015; Nunn 2012), but it has remained relatively understudied. Disease risk may influence primate group living and social life in at least two ways. First, strong parasite prevalence may encourage more spatially widespread social groups and subgrouping because it minimizes parasite transmission through social contact (Kappeler et al. 2015; Nunn 2012; Nunn et al. 2011). Indeed, comparative studies across a range of primates have found strong, positive correlations between parasite prevalence or diversity and community substructuring or modularity of affiliative social networks (Griffin and Nunn 2012; Nunn et al. 2015). This relationship is consistent with the hypothesis that increased parasite risk may have led to the evolution of increased levels of despotism in some primate groups and species, especially given that modular social networks have already been shown to be a characteristic of such societies (Sueur et al. 2011).

On the other hand, increased parasite risk may also cause animals to seek out more social partners, because strong and diverse social ties may enhance their social capital and lower stress levels associated with decreased disease resistance (Cohen et al. 2007; Kaplan et al. 1991; Sapolsky 2005). In accordance with this "social buffering hypothesis," increased disease risk could lead to the evolution of greater social tolerance (Silk et al. 2003, 2010; Young et al. 2014).

Evidence from macaque species, particularly wild Japanese macaques (MacIntosh et al. 2012; Romano et al. 2016) and captive rhesus macaques (Balasubramaniam et al. 2016, 2018b), suggests the transmission of gastrointestinal nematode parasites and gut pathogenic bacteria, respectively, maybe strongly influenced by the structure of affiliative grooming or huddling social networks. However, at this time, no comparable data are available for Tibetan macaques. Nevertheless, given the now well-established impact that parasite risk has on aspects of primate sociality and health (Kappeler et al. 2015), future research should focus on further unraveling such links.

Intraspecific Variation In general, comparative studies of variation in primate social organization and social structure have tended to focus more on interspecific than intraspecific variation, with the latter often overlooked or not appropriately accounted for (Clutton-Brock and Janson 2012; Nunn 2011). Given this, a long-term goal of Tibetan macaque research should be to establish the ecological and evolutionary underpinnings of intraspecific variation in social structure. To date, much of our understanding of intraspecific variation in Tibetan macaque social structure stems from longitudinal data collected on a single group at Mt. Huangshan. In one study, Berman and colleagues found that the extent of grooming kin bias among females in YA1 varied across seven different data collection periods between December 1991 and November 2004 (Berman et al. 2008). This variation was positively correlated to group size, but unaffected by other socioecological factors (e.g., within-group competition, human presence), leading to speculation that increased time constraints faced by the macaques when the group size is larger may influence their social styles (Berman et al. 2008). Later, Balasubramaniam and colleagues revealed that the steepness of the dominance hierarchy of YA1 females also varied significantly across six of these periods (Balasubramaniam et al. 2011). Yet contrary to expectation, steepness was negatively correlated to group size, leading to the speculation that the macaques responded to human presence by forming within-group alliances that may in turn have impacted their dominance style. So whether and how socioecological factors may impact one or more aspects of Tibetan macaque social structure remains unclear.

Given that the social groups studied at both Mt. Huangshan and Mt. Emei were subjected to varying degrees of tourism, provisioning, and range restriction (Berman and Li 2002), comparisons of their social styles with each other and with those of other wild populations are crucial. Evidence for intergroup differences in aspects of social structure is available for other free-living macaque species, but somewhat mixed. For instance, Majolo et al. (2009) found stark intergroup differences in the distribution of grooming and post-conflict reconciliation between two wild Japanese macaque groups of different sizes within the same population. Further, Duboscq et al. (2013) found between-group differences in rates of approaches, silent bare-teeth displays, and percentages of counter-aggression by females in two groups of wild crested macaques at Tangkoko Nature Reserve. On the other hand, a study on free-ranging rhesus macaques at Cayo Santiago revealed no differences in female dominance hierarchies, aggression intensities, and post-conflict affiliation rates across three groups of different sizes (Balasubramaniam et al. 2014). Such findings warrant establishing potential links (or the lack thereof) between socioecological factors and intraspecific variation in social structure across groups and populations of Tibetan macaques as well as other species.

Comparative Studies of Male Social Styles Most of what we know about the concept of macaque social styles has emerged from studies of female-female social relationships in captive groups. In comparison, the social relationships of males have been less intensely studied, with studies on free-ranging and wild populations of some species having just emerged during the last decade. In the light of this, we suggest that the time is right for conducting comparative studies of the evolution of male social styles.

Male relationships among the female philopatric Cercopithecine primates are expected to be predominantly intolerant or antagonistic. Coalitions are expected to be primarily (1) conservative to maintain dominance rank and (2) based on partner availability rather than the establishment of long-term affiliative social bonds among unrelated males (Cords 2013; Kappeler 2000; Preuschoft and Paul 2000). Yet, there is also evidence for overt affiliation, social bond investment, and male-male social tolerance in some species (Ostner and Schülke 2014; Schülke et al. 2010), including Tibetan macaques (see above). Evolutionary explanations for such evidence remain somewhat unclear. One possible explanation, proposed above for Tibetan macaques, involves the emergence of male "tolerant despots," i.e., the idea that male social tolerance may be an emergent outcome of a group- or species-typical tendency to exhibit a generally despotic social style. This may occur when selective tolerance enhances males' abilities to compete in a despotic society. Such indicators of social tolerance maybe a consequence of males' counter-strategies to deal with generally heightened levels of within-group resource competition (Ostner and Schülke 2014; Schülke et al. 2010).

Findings from Assamese macaques also seem to support this explanation. For instance, both male and female Assamese macaques show steep, linear dominance hierarchies that are indicative of high within-group competition and a despotic social style (Balasubramaniam et al. 2012a, b; Cooper 1999; Cooper and Bernstein 2002, 2008). Yet, males also show moderate levels of post-conflict affiliation (Cooper 1999; Cooper and Bernstein 2002, 2008) and form long-term social bonds based on partner preference rather than partner availability (Kalbitz et al. 2016; Ostner and Schülke 2014; Schülke et al. 2010). Findings from other species suggest that male social tolerance, rather than being emergent outcomes of within-group despotism, may have socioecological underpinnings. For instance, grooming and alliance formation among male Japanese macaques appear to be more strongly influenced by between-group mating competition than by within-group competition (Horiuchi 2007). In bonnet macaques, the distribution and abundance of natural resources may strongly influence male-to-female sex ratios and, thereby, male social strategies (Ram et al. 2003; Sinha et al. 2005).

When abundant and widespread resources give rise to large, multi-male-multi-female groups with near-even adult sex ratios, reduced within-group competition may give rise to unusual levels of social tolerance, ritualized greetings, huddling, and social grooming among males (Adiseshan et al. 2011; Silk 1994, 1999). When resources are clumped or seasonal, intense female-female competition may lead to group-fissioning, resulting in the emergence of atypical unimale troops in several species (Ram et al. 2003; Sinha et al. 2005). In these contexts, males actively herd females, defend them against immigrating males by engaging in intergroup encounters, and remain intolerant of resident juvenile and subadult males (Ram et al. 2003; Sinha et al. 2005). Such stark differences and explanations for male social strategies warrant more comparative studies across macaque species that examine the relative effect(s) of intrinsic characteristics (e.g., female social styles, male despotism) and extrinsic factors (e.g., resource abundance, intergroup competition, predation) on male social styles.

Phylogeographic Approaches Finally, comparative studies have now established that several (but not all) aspects of macaque social structure (Balasubramaniam et al. 2012a, 2018a; Thierry et al. 2008), and more broadly primate sociality (Kamilar and Cooper 2015), are associated at least in part with species' ancestral relationships. What is unclear is whether such phylogenetic signals are reflections of true genetic linkage or coevolution versus artifacts of adaptive responses by ancestral species to major environmental changes (Balasubramaniam et al. 2018a). To address this gap, we recommend that future research should focus on establishing links between ancestral state reconstructions of primate social structure (Pagel et al. 2004; Revell 2012) and potential changes in historic phylogeographic ranges (Abegg and Thierry 2002; Lemey et al. 2009; Ree and Smith 2008). Such approaches would help establish whether, for instance, some aspects of social structure coevolved with major dispersal events or climatic changes toward more resource abundant (or sparser) environments in the evolutionary past.

Acknowledgments We thank the Huangshan Monkey Management Center and the Huangshan Garden Forest Bureau for granting permission to collect field data on the YA1 Tibetan macaque group. We thank the following organizations: the Leakey Foundation, the Wenner-Gren Foundation, and the National Geographic Society (C.M.B.); the National Natural Science Foundation of China, Key Teacher Program of the Ministry of Education of China, and the Excellent Youth Foundation of Anhui (J.L.) for providing funding to carry out fieldwork at Huangshan. We are grateful to May Lee Gong, Krista Jones, Ming Li, Stephan Menu, Stephanie Pieddesaux, Justin Sloan, and Lei Zhang for their role as field assistants at Huangshan. We give special thanks to Xinming Chen and his family for housing and accommodation in China. For the comparative studies, we thank several past and current colleagues in primatology and their research teams, specifically Marina Butovskaya, Matthew Cooper, Arianna De Marco, Julie Duboscq, Sabina

Koirala, Bonaventura Majolo, Andrew MacIntosh, Richard MacFarland, Sandra Molesti, Odile Petit, Gabriele Schino, Sebastian Sosa, Cedric Sueur, Bernard Thierry, and Frans de Waal for contributing their datasets and playing a role in manuscript publication. Further, we also thank Katharina Dittmar, Brenda McCowan, and Brianne Beisner for providing much-needed input in phylogenetic and social network analyses, respectively. We thank the National Science Foundation (NSF), USA, for partially funding the comparative studies. Finally, we are grateful to Bernard Thierry and Bonaventura Majolo for their reviews and feedback while constructing this chapter.

References

Abegg C, Thierry B (2002) Macaque evolution and dispersal in insular south-east Asia. Biol J Linn Soc 75(4):555–576

Adiseshan A, Adiseshan T, Isbell LA (2011) Affiliative relationships and reciprocity among adult male bonnet macaques (*Macaca radiata*) at Arunachala Hill, India. Am J Primatol 73:1107–1113

Altmann J (1979) Age cohorts as paternal sibships. Behav Ecol Sociobiol 6:161–164

Arnold C, Matthews LJ, Nunn CL (2010) The *10kTrees* website: a new online resource for primate phylogeny. Evol Anthropol 19:114–118

Balasubramaniam KN, Berman CM (2017) Grooming exchange for resource tolerance: biological markets principles in a group of free-ranging rhesus macaques. Behaviour 154(11):1145–1176

Balasubramaniam KN, Berman CM, Ogawa H, Li J (2011) Using biological markets principles to examine patterns of grooming exchange in *Macaca thibetana*. Am J Primatol 73 (12):1269–1279

Balasubramaniam KN, Dittmar K, Berman CM, Butovskaya M, Cooper MA, Majolo B, Ogawa H, Schino G, Thierry B, de Waal FBM (2012a) Hierarchical steepness and phylogenetic models: phylogenetic signals in *Macaca*. Anim Behav 83:1207–1218

Balasubramaniam KN, Dittmar K, Berman CM, Butovskaya M, Cooper MA, Majolo B, Ogawa H, Schino G, Thierry B, de Waal FBM (2012b) Hierarchical steepness, counter-aggression, and macaque social style scale. Am J Primatol 74:915–925

Balasubramaniam KN, Dunayer ES, Gilhooly LJ, Rosenfield KA, Berman CM (2014) Group size, contest competition, and social structure in Cayo Santiago rhesus macaques. Behaviour 151:1759–1798

Balasubramaniam KN, Beisner BA, Vandeleest J, Atwill ER, McCowan B (2016) Social buffering and contact transmission: network connections have beneficial and detrimental effects on Shigella infection risk among captive rhesus macaques. PeerJ 4:e2630

Balasubramaniam KN, Beisner BA, Berman CM, De Marco A, Duboscq J, Koirala S, Majolo B, MacIntosh AJJ, McFarland R, Molesti S, Ogawa H, Petit O, Schino G, Sosa S, Sueur C, Thierry B, de Waal FBM, McCowan BJ (2018a) The influence of phylogeny, social style, and sociodemographic factors on macaque social network structure. Am J Primatol 80(1): e227227

Balasubramaniam KN, Beisner BA, Guan J, Vandeleest J, Fushing H, Atwill ER, McCowan B (2018b) Social network community structure is associated with the sharing of commensal *E. coli* among captive rhesus macaques (*Macaca mulatta*). PeerJ 6:e4271

Berman CM (2011) Kinship: family ties and social behavior. In: Campbell CJ, Fuentes A, MacKinnon KC, Panger M, Bearder SK (eds) Primates in perspective, 2nd edn. Oxford University Press, New York, pp 576–587

Berman CM, Kapsalis E (2009) Variation over time in grooming kin bias among female rhesus macaques on Cayo Santiago supports the time constraints hypothesis. Am J Phys Anthropol 48:89–90

Berman CM, Li J (2002) Impact of translocation, provisioning and range restriction on a group of *Macaca thibetana*. Int J Primatol 23:287–293

Berman CM, Thierry B (2010) Variation in kin bias: species differences and time constraints in macaques. Behaviour 147(13):1863–1887

Berman CM, Ionica CS, Li J (2004) Dominance style among *Macaca thibetana* on Mt. Huangshan, China. Int J Primatol 25:1283–1312

Berman CM, Ionica CS, Dorner M, Li JH (2006) Post-conflict affiliation between former opponents in *Macaca thibetana*: for males only? Int J Primatol 27:827–854

Berman CM, Ionica CS, Li J (2007) Supportive and tolerant relationships among male Tibetan macaques at Huangshan, China. Behaviour 144:631–661

Berman CM, Ogawa H, Ionica CS, Yin H, Li J (2008) Variation in kin bias over time in a group of Tibetan macaques at Huangshan: contest competition, time constraints or stress response? Behaviour 145:863–896

Blomberg SP, Garland T, Ives A (2003) Testing for phylogenetic signal in comparative data: behavioral traits are more labile. Evolution 57:717–745

Brent LJN, Lehmann J, Ramos-Fernandez G (2011) Social network analysis in the study of nonhuman primates: a historical perspective. Am J Primatol 73(8):720–730

Capitanio JP (1999) Personality dimensions in adult male rhesus macaques: prediction of behaviors across time and situation. Am J Primatol 47:299–320

Carne C, Viper S, Semple S (2011) Reciprocation and interchange of grooming, agonistic support, feeding tolerance, and aggression in semi-free-ranging Barbary macaques. Am J Primatol 73:1127–1133

Castles DL, Aureli F, de Waal FBM (1996) Variation in conciliatory tendency and relationship quality across groups of pigtail macaques. Anim Behav 52:389–403

Chakraborty D, Ramakrishnan U, Panor J, Mishra C, Sinha A (2007) Phylogenetic relationships and morphometric affinities of the Arunachal macaque *Macaca munzala*, a newly described primate from Arunachal Pradesh, Northeastern India. Mol Phylogenetics Evol 44:838–849

Chan LKW (1996) Phylogenetic interpretations of primate socioecology: with special reference to social and ecological diversity in *Macaca*. In: Martins EP (ed) Phylogenies and the comparative method in animal behaviour. Oxford University Press, Oxford, pp 324–360

Clutton-Brock TH, Harvey P (1977) Primate ecology and social organization. J Zool 183:1–39

Clutton-Brock TH, Janson CH (2012) Primate socioecology at the crossroads: past, present, and future. Evol Anthropol 21:136–150

Cobb S (1976) Social support as a moderator of life stress. Psychosom Med 38:300–314

Cohen S, Janicki-Deverts D, Miller GE (2007) Psychological stress and disease. JAMA 298:1685–1687

Cohen S, Janicki-Deverts D, Turner RB, Doyle WJ (2015) Does hugging provide stress-buffering social support? A study of susceptibility to upper respiratory infection and illness. Psychol Sci 26:135–147

Cooper MA (1999) Social tolerance in Assamese macaques (*Macaca assamensis*). Ph.D. Dissertation. University of Georgia, Athens, GA

Cooper MA, Bernstein IS (2002) Counter aggression and reconciliation in Assamese macaques (*Macaca assamensis*). Am J Primatol 56:215–230

Cooper MA, Bernstein IS (2008) Evaluating dominance styles in Assamese and rhesus macaques. Int J Primatol 29:225–243

Cords M (2013) The behavior, ecology, and social evolution of Cercopithecine monkeys. In: Mitani JC, Call J, Kappeler PM, Palombit RA, Silk JB (eds) The evolution of primate societies. University of Chicago Press, Chicago, pp 91–112

Csardi G, Nepusz T (2006) The igraph software package for complex network research. InterJ. Complex Syst. 1695

de Vries H (1998) Finding a dominance order most consistent with a linear hierarchy: a new procedure and review. Anim Behav 55:827–843 de Vries H, Stevens JMG, Vervaecke H (2006) Measuring and testing the steepness of dominance hierarchies. Anim Behav 71:585–592

de Waal FBM, Luttrell LM (1989) Towards a comparative ecology of the genus *Macaca*: different dominance styles in rhesus and stumptailed macaques. Am J Primatol 19:83–109

de Waal FBM, Yoshihara D (1983) Reconciliation and redirected affection in rhesus monkeys. Behaviour 85:224–241

Demaria C, Thierry B (2001) A comparative study of reconciliation in rhesus and Tonkean macaques. Behaviour 138:397–410

Deng ZY (1993) Social development of infants of *Macaca thibetana* at Mt. Emei, China. Folia Primatol 60:28–35

Duboscq J, Micheletta J, Agil M, Hodges K, Thierry B, Engelhardt A (2013) Social tolerance in wild female crested macaques (*Macaca nigra*) in Tangkoko-Batuangus Nature Reserve, Sulawesi, Indonesia. Am J Primatol 75:361–375

Farine DR, Whitehead H (2015) Constructing, conducting and interpreting animal social network analysis. J Anim Ecol 84:1144–1163

Fooden J (1986) Taxonomy and evolution of the *sinica* group of macaques: 5. Overview of natural history. Fieldiana Zool 29:1–22

Fruteau C, van de Waal E, Van Damme E, Noe R (2011) Infant access and handling in sooty mangabeys and vervet monkeys. Anim Behav 81:153–161

Fujii K, Fushing H, Beisner BA, McCowan B (2013) Computing power structures in directed biosocial networks: flow percolation and imputed conductance. Technical Report, Department of Statistics, UC Davis

Gammell MP, de Vries H, Jennings DJ, Carlin CM, Hayden TJ (2003) David's score: a more appropriate dominance ranking method than Clutton-Brock et al.'s index. Anim Behav 66:601–605

Greenwood PJ (1980) Mating systems, philopatry and dispersal in birds and mammals. Anim Behav 28:1140–1162

Griffin RH, Nunn CL (2012) Community structure and the spread of infectious disease in primate social networks. Evol Ecol 26(4):779–800

Gumert MD (2007) Grooming and infant handling interchange in *Macaca fascicularis*: the relationship between infant supply and grooming payment. Int J Primatol 28:1059–1074

Hamilton WD (1964) The genetical evolution of social behaviour I/II. J Theor Biol 7:1–52

Handcock M, Hunter D, Butts C, Goodreau S, Morris M (2006) Statnet: an R package for the statistical analysis and simulation of social networks. University of Washington. http://www.csde.washington.edu/statnet

Hemelrijk C (1999) An individual-orientated model of the emergence of despotic and egalitarian societies. Proc R Soc Lond B 266:361–369

Hemelrijk C (2005) Self-organisation and evolution of biological and social systems. Cambridge University Press, Cambridge

Henzi SP, Barrett L (2002) Infants as a commodity in a baboon market. Anim Behav 63:915–921

Hill DA (1999) Effects of provisioning on the social behaviour of Japanese and rhesus macaques: implications for socioecology. Primates 40:187–198

Hinde RA (1976) Interactions, relationships and social structure. Man 11:1–17

Holt-Lunstad J, Smith TB, Layton JR (2010) Social relationships and mortality risk: a meta-analytic review. PLoS Med 7:e1000316

Horiuchi S (2007) Social relationships of male Japanese macaques (*Macaca fuscata*) in different habitats: a comparison between Yakushima island and Shimokita peninsula populations. Anthropol Sci 115:63–65

House JS, Landis KR, Umberson D (1988) Social relationships and health. Science 241:50–54

Isbell LA (2017) Socioecological model. In: Bezanson M, MacKinnon KC, Riley E, Campbell CJ, Nekaris KAI, Estrada A, Di Fiore AF, Ross S, Jones-Engel LE, Thierry B, Sussman RW, Sanz C, Loudon J, Elton S, Fuentes (eds) The international encyclopedia of primatology. Wiley Online Library

Judge P (2000) Coping with crowded conditions. In: Aureli F, de Waal FBM (eds) Natural conflict resolution. University of California Press, Berkeley, pp 129–154

Kaburu SSK, Newton-Fisher NE (2015) Egalitarian despots: hierarchy steepness, reciprocity and the grooming-trade model in wild chimpanzees, *Pan troglodytes*. Anim Behav 99:61–71

Kalbitz J, Ostner J, Schulke O (2016) Strong, equitable and long-term social bonds in the dispersing sex in Assamese macaques. Anim Behav 113:13–22

Kamilar J, Cooper N (2015) Phylogenetic signal in primate behaviour, ecology and life history. Philos Trans R Soc B 368:20120341

Kaplan JR, Heise ER, Manuck SB, Shively CA, Cohen S, Rabin BS, Kasprowicz AL (1991) The relationship of agonistic and affiliative behavior patterns to cellular immune function among cynomolgus monkeys Macaca- fascicularis living in unstable social groups. Am J Primatol 25 (3):157–174

Kappeler PM (2000) Primate males: causes and consequences of variation in group composition. Cambridge University Press, Cambridge

Kappeler PM, Van Schaik CP (2002) Evolution of primate social systems. Int J Primatol 23:707–740

Kappeler PM, Cremer S, Nunn CL (2015) Sociality and health: impacts of sociality on disease susceptibility and transmission in animal and human societies. Philos Trans R Soc B 370:20140116

Kapsalis E (2004) Matrilineal kinship and primate behavior. In: Chapais B, Berman CM (eds) Kinship and behavior in primates. Oxford University Press, New York, pp 153–176

Kasper C, Voelkl B (2009) A social network analysis of primate groups. Primates 50:343–256

Koenig A, Scarry CJ, Wheeler BC, Borries C (2013) Variation in grouping patterns, mating systems and social structure: what socio-ecological models attempt to explain. Philos Trans R Soc B 368:20120348

Krause J, James R, Croft DP (2010) Personality in the context of social networks. Philos Trans R Soc B 365:4099–4106

Lemey P, Rambaut A, Drummond AJ, Suchard MA (2009) Bayesian phylogeography finds its roots. PLoS Comput Biol 5:e1000520

Li J (1999) The Tibetan macaque society: a field study. Anhui University Press, Hefei

Li J, Wang Q (1996) Dominance hierarchy and its chronic changes in adult male Tibetan macaques (*Macaca thibetana*). Acta Zool Sinica 42:330–334

Lusseau D (2003) The emergent properties of a dolphin social network. Proc R Soc Lond B 270: S186–S188

MacIntosh AJJ, Jacobs A, Garcia C, Shimizu K, Mouri K, Huffman MA, Hernandez AD (2012) Monkeys in the middle: parasite transmission through a social network of a wild primate. PLoS One 7:e51144

Majolo B, Ventura R, Koyama NF, Hardie SM, Jones BM, Knapp LA, Schino G (2009) Analyzing the effects of group size and food competition on Japanese macaque social relationships. Behaviour 146:113–137

Marechal L, Semple S, Majolo B, Qarro M, Heistermann M, MacLarnon A (2011) Impacts of tourism on anxiety and physiological stress levels in wild male Barbary macaques. Biol Conserv 144(9):2188–2193

Matsumura S (1999) The evolution of "egalitarian" and "despotic" social systems among macaques. Primates 40:23–31

Newman MEJ, Girvan M (2004) Finding and evaluating community structure in networks. Phys Rev E 69:1–15

Nunn CL (2011) The comparative approach in evolutionary anthropology and biology. University of Chicago Press, Chicago

Nunn CL (2012) Primate disease ecology in comparative and theoretical perspective. Am J Primatol 74(6):497–509

Nunn CL, Thrall PH, Leendertz FH, Boesch C (2011) The spread of fecally transmitted parasites in socially structured populations. PLoS One 6:e21677

Nunn CL, Jordan F, McCabe CM, Verdolin JL, Fewell JH (2015) Infectious disease and group size: more than just a numbers game. Philos Trans R Soc B 370:20140111

Ogawa H (1995) Bridging behavior and other affiliative interactions among male Tibetan macaques (*Macaca thibetana*). Int J Primatol 16:707–729

Ostner J, Schülke O (2014) The evolution of social bonds in primate males. Behaviour 151:871–906

Pagel M, Meade A, Barker D (2004) Bayesian estimation of ancestral character states on phylogenies. Syst Biol 53(5):673–684

Paul A, Kuester J (1987) Dominance, kinship and reproductive value in female Barbary macaques (*Macaca sylvanus*) at Affenberg Salem. Behav Ecol Sociobiol 21:323–331

Petit O, Abegg C, Thierry B (1997) A comparative study of aggression and conciliation in three cercopithecine monkeys (*Macaca fuscata*, *Macaca nigra* and *Papio papio*). Behaviour 134:415–432

Preuschoft S, Paul A (2000) Dominance, egalitarianism, and stalemate: an experimental approach to male-male competition in Barbary macaques. In: Kappeler PM (ed) Primate males: causes and consequences of variation in group composition. Cambridge University Press, Cambridge, pp 205–216

Preuschoft S, Paul A, Kuester J (1998) Dominance styles of female and male Barbary macaques (*Macaca sylvanus*). Behaviour 135:731–755

Puga I, Sueur C (2017) Emergence of complex social networks from spatial structure and rules of thumb: a modelling approach. Ecol Complex 31:189–200

Purvis A (1995) A composite estimate of primate phylogeny. Philos Trans R Soc B 348:405–421

Ram S, Venkatachalam S, Sinha A (2003) Changing social strategies of wild female bonnet macaques during natural foraging and on provisioning. Curr Sci 84:780–790

Ree RH, Smith SA (2008) Maximum likelihood inference of geographic range evolution by dispersal, local extinction, and cladogenesis. Syst Biol 51:4–14

Revell LJ (2012) phytools: an R package for phylogenetic comparative biology (and other things). Methods Ecol Evol 3:217–233

Richter C, Mevis L, Malaivijitnond S, Schülke O, Ostner J (2009) Social relationships in free-ranging male *Macaca arctoides*. Int J Primatol 30(4):625–642

Romano V, Duboscq J, Sarabian C, Thomas E, Sueur C, MacIntosh AJJ (2016) Modeling infection transmission in primate networks to predict centrality-based risk. Am J Primatol 78:767–779

Sade DS (1972) A longitudinal study of social behavior of rhesus monkeys. In: Tuttle R (ed) The functional and evolutionary biology of primates. Aldine-Atherton, Chicago, pp 378–398

Sapolsky RM (2005) The influence of social hierarchy on primate health. Science 308:648–652

Schino G, Aureli F (2008) Tradeoffs in primate grooming reciprocation: testing behavioral flexibility and correlated evolution. Biol J Linn Soc 95:439–446

Schülke O, Ostner J (2008) Male reproductive skew, paternal relatedness, and female social relationships. Am J Primatol 70:695–698

Schülke O, Ostner J (2013) Ecological and social influences on sociality. In: Mitani JC, Call J, Kappeler PM, Palombit RA, Silk JB (eds) Evolution of primate societies. University of Chicago Press, Chicago, pp 195–219

Schülke O, Bhagavatula J, Vigilant L, Ostner J (2010) Social bonds enhance reproductive success in male macaques. Curr Biol 20:2207–2210

Schülke O, Wenzel S, Ostner J (2013) Paternal relatedness predicts the strength of social bonds among female rhesus macaques. PLoS One 8:e59789

Shizuka D, Mcdonald DB (2012) A social network perspective on measurements of dominance hierarchies. Anim Behav 83:925–934

Sih A, Bell A, Chadwick Johnson J (2004) Behavioral syndromes: an ecological and evolutionary overview. Trends Ecol Evol 19:372–378

Sih A, Cote J, Evans M, Fogarty S, Pruitt J (2012) Ecological implications of behavioural syndromes. Ecol Lett 15:278–289

Silk JB (1994) Social relationships of male bonnet macaques: male bonding in a matrilineal society. Behaviour 130:271–291

Silk JB (1999) Male bonnet macaques use information about third-party rank relationships to recruit allies. Anim Behav 58:45–51

Silk JB, Alberts SC, Altmann J (2003) Social bonds of female baboons enhance infant survival. Science 302:1231–1234

Silk JB, Beehner JC, Bergman C, Crockford AL, Engh LR, Moscovice RM, Wittig RM, Seyfarth RM, Cheney DL (2010) Strong and consistent social bonds enhance the longevity of female baboons. Curr Biol 20:1359–1361

Sinha A, Mukhopadhyay K, Datta-Roy A, Ram S (2005) Ecology proposes, behaviour disposes: ecological variability in social organization and male behavioural strategies among wild bonnet macaques. Curr Sci 89(7):1166–1179

Sterck EHM, Watts DP, van Schaik CP (1997) The evolution of female social relationships in nonhuman primates. Behav Ecol Sociobiol 41(5):291–309

Sueur C, Petit O, De Marco A, Jacobs AT, Watanabe K, Thierry B (2011) A comparative network analysis of social style in macaques. Anim Behav 82(4):845–852

Suomi SJ (2011) Risk, resilience, and gene-environment interplay in primates. J Can Acad Child Adolesc Psychiatry 20:289–297

Thierry B (1985) Patterns of agonistic interactions in three species of macaque (*Macaca mulatta*, *M. fascicularis*, *M. tonkeana*). Aggress Behav 11:223–233

Thierry B (2000) Covariation of conflict management patterns across macaque species. In: Aureli F, de Waal FBM (eds) Natural conflict resolution. University of California Press, Berkley, CA, pp 106–128

Thierry B (2004) Social epigenesis. In: Thierry B, Singh M, Kaumanns W (eds) Macaque societies: a model for the study of social organization. Cambridge University Press, Cambridge, pp 267–290

Thierry B (2007) Unity in diversity: lessons from macaque societies. Evol Anthropol 16:224–238

Thierry B (2013) The macaques: a double-layered social organization. In: Campbell CJ, Fuentes A, MacKinnon KC, Bearder SK, Stumpf RM (eds) Primates in perspective. Oxford University Press, Oxford, pp 229–240

Thierry B, Aureli F (2006) Barbary but not barbarian: social relations in a tolerant macaque. In: Hodges JK (ed) Biology and behaviour of Barbary macaques. Nottingham University Press, Nottingham, pp 1–18

Thierry B, Iwaniuk AN, Pellis SM (2000) The influence of phylogeny on the social behaviour of macaques. Ethology 106:713–728

Thierry B, Singh M, Kaumanns W (2004) Macaque societies: a model for the study of social organization. Cambridge University Press, Cambridge

Thierry B, Aureli F, Nunn CL, Petit O, Abegg C, de Waal FBM (2008) A Comparative study of conflict resolution in macaques: insights into the nature of covariation. Anim Behav 75:847–860

Tiddi B, Aureli F, Schino G (2010) Grooming for infant handling in tufted capuchin monkeys: a reappraisal of the primate infant market. Anim Behav 79:1115–1123

Tosi AJ, Morales JC, Melnick DJ (2003) Paternal, maternal, and biparental molecular markers provide unique windows onto the evolutionary history of macaque monkeys. Evolution 57:1419–1435

Tyrrell M, Berman CM, Agil M, Sutrisno T, Engelhardt A (2018) Social style among wild male crested macaques (*Macaca nigra*) in Tangkoko Nature Reserve, Sulawesi, Indonesia. In Proceedings of the American Society of Primatologists, San Antonio, TX

van Hooff JARAM, van Schaik C (1994) Male bonds: affiliative relationships among nonhuman primates. Behaviour 130:309–337

van Schaik CP (1989) The ecology of social relationships amongst female primates. In: Standen V, Foley RA (eds) Comparative socio-ecology: the behavioral ecology of humans and other animals. Blackwell, Oxford, pp 195–218 van Schaik CP (1996) Social evolution in primates: the role of ecological factors and male behaviour. Proc Br Acad 88:9–31

van Schaik CP, Pandit AP, Vogel ER (2004) A model for within-group coalitionary aggression among males. Behav Ecol Sociobiol 57:101–109

Veenema HC, Das M, Aureli F (1994) Methodological improvements for the study of reconciliation. Behav Process 31:29–38

Ventura R, Majolo B, Koyama NF, Hardie SM, Schino G (2006) Reciprocation and interchange in wild Japanese macaques: grooming, cofeeding and agonistic support. Am J Primatol 68:1138–1149

Wada KC, Xiong C, Wang Q (1987) On the distribution of Tibetan and rhesus monkeys in Southern Anhui, China. Acta Theriol Sinica 7:148–176

Widdig A (2013) The impact of male reproductive skew on kin structure and sociality in multi-male groups. Evol Anthropol 22:239–250

Wrangham RW (1980) An ecological model of female-bonded primate groups. Behaviour 75:262–300

Xiong CP (1984) Ecological studies of the stump-tailed macaque. Acta Theriol Sinica 4:1–9

Young C, Majolo B, Heistermann M, Schulke O, Ostner J (2014) Responses to social and environmental stress are attenuated by strong male bonds in wild macaques. PNAS 111:18195–18200

Zhang P, Watanabe K (2014) Intraspecies variation in dominance style of *Macaca fuscata*. Primates 55:69–79

Zhao QK (1996) Etho-ecology of Tibetan macaques at Mount Emei, China. In: Fa JE, Lindburg DG (eds) Evolution and ecology of macaque societies. Cambridge University Press, Cambridge, pp 263–289

Zhao QK, Deng ZY, Xu J (1991) Natural foods and their ecological implications for *Macaca thibetana* at Mount Emei, China. Folia Primatol 57:1–15

Part III

Evolution of Rituals: Insights from Bridging Behavior

Chapter 9
Preliminary Observations of Female-Female Bridging Behavior in Tibetan Macaques (*Macaca thibetana*) at Mt. Huangshan, China

Grant J. Clifton, Lori K. Sheeran, R. Steven Wagner, Jake A. Funkhouser, and Jin-Hua Li

9.1   Introduction

Several species of the genus *Macaca* engage in an affiliative behavior commonly referred to as "bridging." Bridging is a triadic behavior in which two older individuals lift and hold an infant or juvenile between them while teeth chattering and/or licking the infant/juvenile's genitals (Ogawa 1995a). Bridging and similar affiliative

G. J. Clifton (✉)
Primate Behavior and Ecology Program, Central Washington University, Ellensburg, WA, USA

L. K. Sheeran
Primate Behavior and Ecology Program, Central Washington University, Ellensburg, WA, USA

Department of Anthropology and Museum Studies, Central Washington University, Ellensburg, WA, USA
e-mail: SheeranL@cwu.edu

R. S. Wagner
Department of Biological Sciences, Central Washington University, Ellensburg, WA, USA
e-mail: WagnerS@cwu.edu

J. A. Funkhouser
Primate Behavior and Ecology Program, Central Washington University, Ellensburg, WA, USA

Department of Anthropology, Washington University in St. Louis, St. Louis, MO, USA
e-mail: jakefunkhouser@wustl.edu

J.-H. Li
School of Resources and Environmental Engineering, Anhui University, Hefei, Anhui, China

International Collaborative Research Center for Huangshan Biodiversity and Tibetan Macaque Behavioral Ecology, Anhui, China

School of Life Sciences, Hefei Normal University, Hefei, Anhui, China
e-mail: jhli@ahu.edu.cn

© The Author(s) 2020
J.-H. Li et al. (eds.), *The Behavioral Ecology of the Tibetan Macaque*, Fascinating Life Sciences, https://doi.org/10.1007/978-3-030-27920-2_9

triadic behaviors have been recorded in Barbary (*M. sylvanus*; Deag and Crook 1971; Taub 1984; Paul et al. 1996; Kubenova et al. 2017), stump-tailed (*M. arctoides*; Estrada and Sandoval 1977; Estrada and Estrada 1984), Assamese (*M. assamensis*; Kubenova et al. 2017), and Tibetan macaques (*M. thibetana*; Ogawa 1995a, b, c; Zhao 1996; Bauer et al. 2013). Adult males' bridging tends to occur in non-agonistic contexts and is often followed by other affiliative behaviors such as grooming (Ogawa 1995a, b; Deag and Crook 1971). Adult males of all species use younger male infants as the bridge more often than female infants and juveniles of both sexes (Ogawa 1995a; Deag 1980). In *M. sylvanus*, subordinate, lower-ranked males often bring an infant to a dominant individual to engage in affiliative behaviors (Deag and Crook 1971; Deag 1980). Deag and Crook (1971) hypothesized that subordinate males use infants as an "agonistic buffer" when initiating interactions with dominant males in order to reduce the probability of aggression. Alternatively, Taub (1984) proposed that these interactions were related to paternal caretaking and instead suggested the "enforced babysitting" hypothesis. Rather than using the infant as a buffer, Taub suggested that natal males in *M. sylvanus* use infants that are matrilineally related to them in order to develop bonds with infants and to inform others of their relatedness to the individual (1984).

Paul, Kuester, and Arnemann (1996) tested both the agonistic buffer and enforced babysitting hypotheses in dyadic male-infant interactions and triadic male-infant interactions (i.e., bridging) in *M. sylvanus*. DNA evidence conflicted with the enforced babysitting hypothesis by demonstrating that males did not interact with infants related to them and did not gain any additional reproductive benefits from interacting with particular females' infants. Conversely, frequency of male-infant interactions increased during intervals of high male-male tension, which supported the agonistic buffering hypothesis in *M. sylvanus*.

Ogawa (1995a, b, c) investigated male-male bridging behavior in Tibetan macaques (*M. thibetana*) and tested the agonistic buffering and enforced babysitting hypotheses. Ogawa found that bridging almost always occurred in non-agonistic contexts and that the frequency of bridging between males positively correlated with rates of grooming (Ogawa 1995a). Subordinate males initiated bridges with dominant males more often than vice versa, and initiating males were more likely to use the infants that were preferred in social interactions by dominant males. Aggression never occurred in male-male dyadic interactions in which a bridge took place but occasionally happened in non-bridging dyadic interactions. Ogawa (1995a) suggested that in macaque species with a higher than average socionomic sex ratio, such as Barbary and Tibetan macaques, males may use bridging to reduce social tensions caused by increased male-male competition and to reduce the possibility of aggression in future interactions. While Ogawa's (1995a) findings did not support the enforced babysitting hypothesis, natal group males did not prefer to bridge with infants related to them, suggesting that male-male bridging is not related to kinship. Adult males preferred to use male infants rather than female infants, and individual males had particular immature males that they used in bridging interactions more often than they used others. However, there was no evidence to suggest that infants gained any direct benefits from bridging: male and female infants experienced equal mortality rates despite the sex bias toward male infants in male-male bridging.

Although all age and both sex classes engage in bridging, there has been little mention of bridging interactions involving females. Ogawa (1995c) studied bridging interactions between adult males and adult females. In contrast to the trend in male-male bridging, higher-ranked males were more likely to initiate male-female bridging interactions than were lower-ranked males. Additionally, males bridged more females they were in consort with. Ogawa (1995c) concluded that male-female bridging likely facilitates mating bonds.

Female-female bridging has been reported in *M. thibetana* (Ogawa 2006; Bauer et al. 2013) and *M. arctoides* (Estrada and Estrada 1984), and similar triadic female-female-infant interactions have been described in *M. sylvanus* (Deag and Crook 1971), but no studies have investigated female-female bridging in detail. When considering the existence of maternal relatedness and strong matrilines among females as well as the differing forms of within-sex competition between males and females, it is possible that female-female bridging does not serve the same purpose as male-male bridging. Moreover, maternity and females' matrilineal relationships may influence how they bridge with one another.

This study describes and analyzes female-female bridging behavior in a group of habituated Tibetan macaques and compares female-female bridging behavior to previous studies on male-male bridging behavior. We predicted female-female bridging would differ from male-male bridging and that these differences might aid in development of further hypotheses to explore female-female bridging in this species.

9.2   Methods

We conducted observations from 01 August to 21 September 2014 at the Valley of Wild Monkeys tourist site near Mt. Huangshan, Anhui Province, China (30°29′N, 118°11′W). One group of Tibetan macaques known as Yulingkeng A1 (YA1) ranges in the forest area surrounding platforms built to facilitate tourism. YA1 monkeys are free-ranging but are habituated to human presence and are provisioned with corn several times a day by park guards. Researchers have tracked and recorded matrilineal kinship, births, immigrations, and deaths since 1985 (Wada and Xiong 1996). At the study's start, the group consisted of 41 individuals: 9 adult males, 8 adult females, 4 subadult males, 4 subadult females, 6 juvenile males, 7 juvenile females, 2 male infants, and 1 female infant; however, the female infant died on 15 August 2014 (Table 9.1). We classified females as adult if they had given birth to least one offspring and were ≥ 6 years old and as subadults if nulliparous and 4–5 years old. We classified infants as under 1 year of age, with the two male infants being roughly 2 months of age and the sole female infant being roughly 5 months of age at the start of the study. To establish a dominance hierarchy for use in this investigation, Lori K. Sheeran (LKS) collected dominance interaction behavioral data from 14 July to 27 August 2014 (36 days) from 7:00 to 12:00 and 14:00 to 17:00 daily. LKS utilized all-occurrence sampling to collect dominance data (Altmann 1974). Agonistic data

Table 9.1 Interaction matrix of bridging events by initiator and receiver

Bridge initiator	Bridge receiver												
ID	YH	YM	YCY	THY	YXX	HH	TH	TXX	TR	TRY	TT	YZ	Total
YH	–	0	0	0	2	1	11	1	0	0	0	1	16
YM	2	–	0	0	0	0	0	0	0	0	0	0	2
YCY	6	0	–	0	3	0	24	6	4	0	0	1	44
THY	1	0	0	–	1	0	1	0	0	0	0	1	4
YXX	20	0	2	1	–	1	1	4	0	0	0	0	29
HH	0	0	0	0	1	–	0	0	0	0	0	0	1
TH	9	0	1	0	0	0	–	0	0	0	0	0	10
TXX	6	0	3	0	0	0	2	–	0	0	0	0	11
TR	0	0	0	0	0	0	0	1	–	0	0	1	1
TRY	0	0	0	0	0	0	0	1	0	–	0	0	1
TT	0	0	0	0	0	0	0	0	0	0	–	0	0
YZ	0	0	0	0	0	0	0	0	0	0	0	–	0
Total	44	0	6	1	7	2	39	13	4	0	0	3	

Full names of individuals are displayed in Table 9.2

Table 9.2 Select results and demographics per individual

Female individual	Date of birth	Age group	Elo score[a]	Focal time (hour)	Bridge initiations/h	Bridge receptions/h
YeXiaXue (YXX)	2010–05	Subadult	1316	2.83	2.826	0.354
HuaHong (HH)	2003–??	Adult	1173	2.00	0.498	0
YeChunYu (YCY)	2009–03	Subadult	1172	3.13	2.874	0
YeHong (YH)	2003–??	Adult	1136	3.53	1.134	1.416
YeMai (YM)	1990–04	Adult	1096	2.23	0.45	0
TouRui (TR)	2004–??	Adult	1057	2.53	0	0.396
TouHong (TH)	2003–??	Adult	987	2.72	1.464	2.196
TouXiaXue (TXX)	2008–03	Adult	885	3.13	0.318	0.96
TouRongYu (TRY)	2009–03	Subadult	838	1.00	0	0
TouTai (TT)	1991–04	Adult	766	2.23	0	0
YeZhen (YZ)	1992–01	Adult	729	2.97	0	0.336
TouHuaYu (THY)	2009–04	Subadult	656	2.67	0.378	0
Mean		–	–	2.58	0.828	0.474
Median		–	–	2.70	0.414	0.168

[a]We used Elo-rating scores as a metric of dominance status. Larger Elo scores indicate higher dominance status

consisted of fear-grin, scream, flee, displace, threat, lunge, chase, grab, slap, and bite as defined by Berman et al. (2004). Winners of these interactions were defined by either the actor in directional agonism or the recipient of a submissive behavior. Losers of these interactions were defined by those receiving directed agonism or the actors of submission. To derive a ranking order and dominance score for each individual, we analyzed these data using Elo-rating procedures (Neumann et al. 2011) in R (R Core Team 2016).

We established interobserver agreement of 100% for all adult male and female, subadult female, and infant identities on 08 August 2014. We identified juveniles by the presence of their mother or their age and sex rather than individual identities due to the difficulty of learning individual juvenile identities within the limited timeframe of our study. We collected data from 8:00 to 12:00 and 13:00 to 18:00. To record bridging events, Grant J. Clifton (GJC) used 2-min focal samples (Altmann 1974) of adult and subadult females. We selected focal order randomly each day using a random number generator. Within focal samples, we recorded behaviors from an ethogram (Ogawa 1995a). Proximity refers touching or within arm's length (Sheeran et al. 2010). GJC also used all-occurrence sampling (Altmann 1974) to record bridges occurring outside of focal samples. If a bridge occurred or was suspected to occur between non-focal individuals, then the current 2-min focal

Fig. 9.1 Successful (left) and failed (right) bridge initiations (photo credit: Anne Salow)

was suspended, and the interacting dyad was observed for 2 min or until one individual left proximity. If a bridge occurred during a focal, we extended the focal for two additional minutes after the bridge to record context, but did not include the additional focal time in the focal data.

We used an ethogram modified from Ogawa (1995a) to record affiliative behaviors that often occurred before and/or during bridging. A bridge was defined as two individuals holding an infant or juvenile between one another with at least one individual teeth chattering or genital licking the infant. The bridge initiator was considered the individual who first teeth chattered or genital licked while lifting the infant or juvenile. A bridge began when both the initiator and receiver held up an infant. The bridge ended either when one of the individuals put down the infant or when no affiliative behaviors continued (other than grooming). Bridges that occurred within short succession of one another were considered separate events as long as at least one individual stopped holding the infant in between events.

We considered a bridge successful when the receiver held the infant or juvenile in the bridging position after receiving an initiation (Fig. 9.1). If a receiver teeth chattered or genital licked but turned away while the initiator was attempting a bridge, then it was considered failed (Fig. 9.1). Following Ogawa (1995a), we classified as Type I bridges in which the initiator brings an infant to a receiver or as Type II if the initiator approaches a receiver who is holding an infant (Fig. 9.2). Ogawa also observed bridges in which both males simultaneously approached and grabbed a single infant. In his study, he classified them as Type III bridges; we, on the other hand, did not observe Type III occur among females, so they were left out of our analysis.

We analyzed bridging behavior by initiator, receiver, and dyad. For individual and group comparisons, we used bridges recorded from focal animal sampling and converted bridges to rates to account for the uneven distribution of focal times. To calculate the rate of bridge initiations and bridge receptions per individual, we divided the number of bridge initiations and receptions from within the individual's focal samples by the total focal time for that individual. To calculate the rate of bridging in dyadic interactions, we calculated the total number of initiations by each individual and divided by the sum of the total focal follows for both individuals. Since focal and all-occurrence bridging dyads were significantly correlated (Kendall's tau; $\tau = 0.65$, $n = 22$, $p < 0.01$) and since the proportions of Type I

Fig. 9.2 Type I (left) and Type II (right) bridge initiations

and Type II bridges were not significantly different between focal and all-occurrence sampling (Fisher's exact test; $p > 0.05$, $df = 2$), we pooled both sample sets for categorical analyses of bridge initiation type. However, we only included bridges recorded via focal animal sampling in our analysis of success rates. Because of our limited size and aim (to describe and categorize female-female bridging), we mainly used descriptive and simple parametric statistics. We used Pearson's correlation coefficient to analyze the relationship between dominance status (Elo scores) and bridge initiation/reception rates, paired-sample t tests to examine differences in initiations rate by subordinate or dominant individual within each dyad, chi-square tests of independence, Fisher's exact, and binomial tests to investigate differences in the distribution of nonparametric frequencies of bridging event types.

9.3   Results

We recorded a total of 31 h of focal data and 119 bridging events (27 Type I; 92 Type II; 76 successful and 43 unsuccessful; Table 9.1). All four subadults and seven of eight adults were observed involved in at least one bridge (Table 9.1). Overall, we recorded 119 bridging events, 46 (38.7%) of which were recorded via focal sampling.

Of all observed bridging events, we recorded 27 (22.7%) that occurred via Type I initiation and 90 occurred via Type II initiation, of which 76 (63.9%) were successful bridges and 43 (36.1%) were failed bridges. In focal animal sampling, we recorded 7 (13.2%) via Type I initiations and 39 (84.8%) via Type II initiations, of which 21 (45.7%) were successful bridges and 25 (54.3%) were failed bridges. Table 9.2 displays all bridging initiations and receptions for each possible dyad and focal times and bridging rates for adult and subadult females.

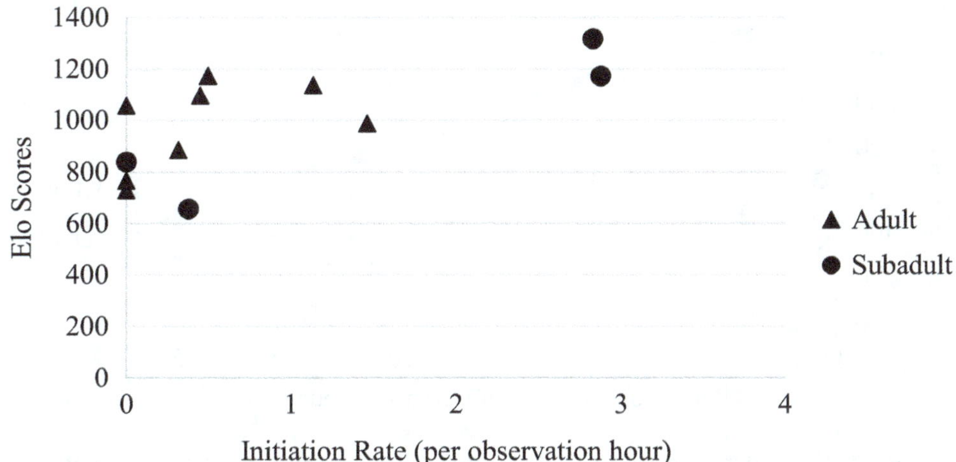

Fig. 9.3 Scatterplot of dominance status (Elo score) vs. bridge initiations (per hour of observation)

Female-female bridges were often subtle and did not involve audible vocalizations. Like male-male bridges, at least one female would lick the infant's genitals during bridges and sometimes touched the infant's genitals. Females often approached mothers who were holding infants and would peer at the infant before initiating the bridge by grabbing it. Successful bridges were never observed to be followed by aggression during our study, but receivers were occasionally observed committing aggression toward failed initiators.

The median bridge initiation rate for adults was 0.378 initiations per hour, whereas the median bridge initiation rate for subadults was 1.596 initiations per hour, but the sample size of the subadult group was too small to accurately test for significance. However, two subadult females, YCY and YXX, initiated far more often than any other adult or subadult (Table 9.1). The median bridge reception rate for adults was 0.366 receptions per hour, whereas no subadults were observed to be recipients of any bridging events. Like initiations, there were large differences in the number of bridges received by each individual within groups, with the two mothers of infants, YH and TH, receiving a large majority of bridges (Table 9.1). Individuals who initiated bridges were not more likely to have their bridges reciprocated by those with whom they initiated ($r_s = -0.45$, $n = 17$, $p > 0.05$).

We analyzed the relationship between dominance status and bridge initiations using a Pearson's rank correlation. Social rank was positively correlated with bridge initiations ($r(12) = 0.686$, $p = 0.014$; Fig. 9.3), but social rank and bridge receptions were not significantly correlated ($r(12) = 0.06$, $p = 0.85$; Fig. 9.4).

To test for significance differences between the mean rates of bridge initiations by either dominant or subordinate individuals within each dyad, we used a paired-samples t test. In contrast to what has been reported for males, within each dyad, subordinate individuals did not initiate bridges significantly more than that dominant individuals. We found no significant difference between the mean rate of bridge initiation between dominant or subordinate individuals ($t(65) = 1.398$, $p = 0.167$, Fig. 9.5). While this difference is not significant, descriptively, we found that

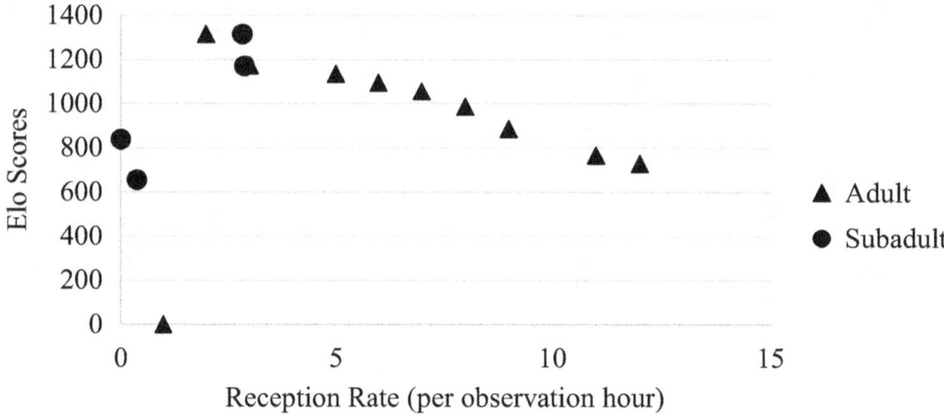

Fig. 9.4 Scatterplot of dominance status (Elo score) vs. bridge receptions (per hour of observation)

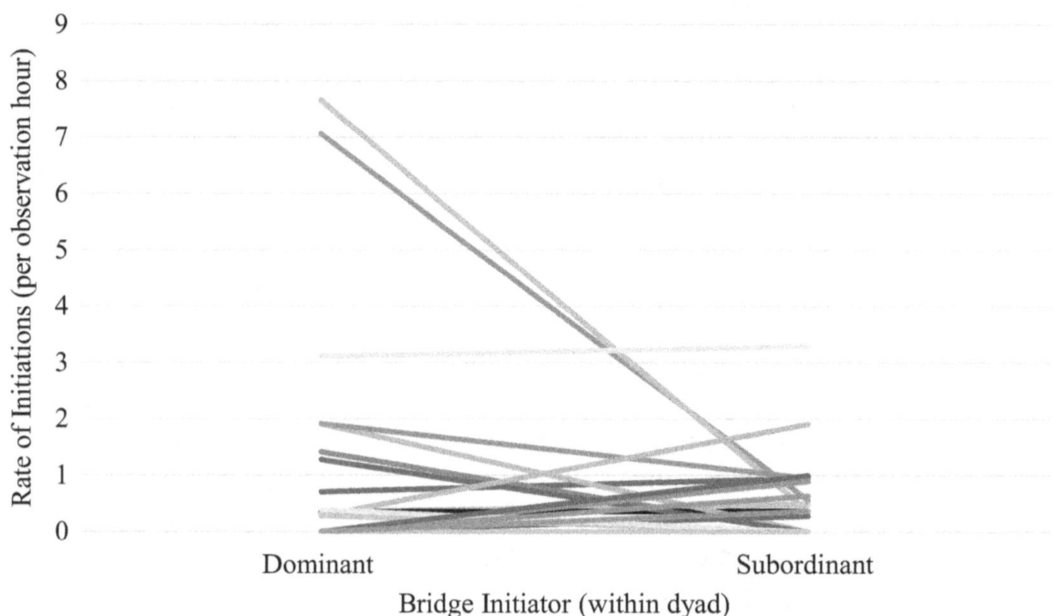

Fig. 9.5 Bridge initiation rates for the dominant vs. subordinate individual within each dyad. Each line represents a single dyad for a total of 22 dyads

dominant individuals were found to initiate bridges more (0.415 ± 1.35 initiations per observation hour) than subordinate individuals (0.195 ± 0.52 initiations per observation hour).

A majority of bridges occurred via Type II interactions. In bridges observed via focal sampling, all Type I bridges occurred by adults. Subadult females were observed initiating Type I interactions via all-occurrence sampling, but Type I interactions were still more likely to be initiated by an adult than a subadult (Fisher's exact test; $p < 0.05$, $df = 1$; Fig. 9.6). Of all but two Type II bridges, the receiver of the bridge was the mother of the infant or juvenile being used, and in the two exceptions, the same subadult female, YXX, was the receiver. Similarly, in all but two Type I interactions initiated by adults, the initiator was holding her own

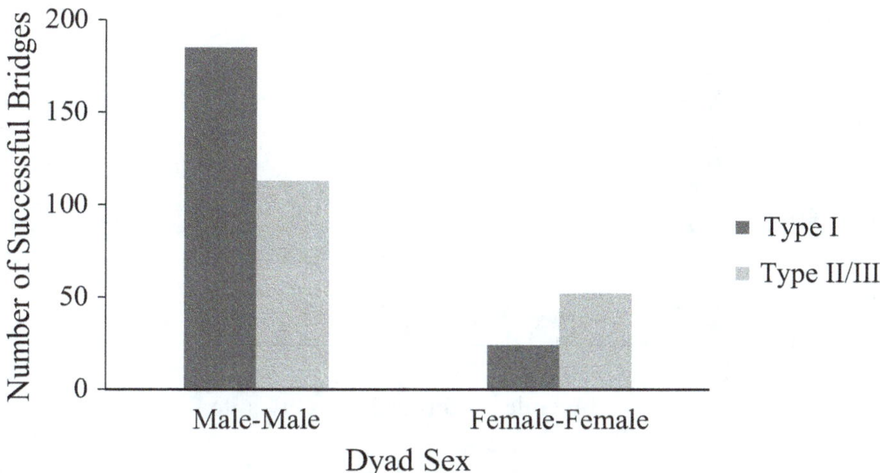

Fig. 9.6 Bar graph of successful Type I and Type II/III initiations between male and male

offspring. These two exceptions were initiated by the same individual, TXX, using TH's infant in both cases.

In Ogawa's (1995a) study of male-male bridging, 185 (62.1%) bridges occurred via Type I initiations, 70 (23.5%) occurred with Type II initiations, and 43 (14.4%) occurred via Type III interactions. We did not observe female-female bridging via Type III initiations, so to compare males and females, we pooled Type II and Type III male bridges together for a total of 185 (62.1%) Type I and 113 (37.9%) Type II/III interactions. Since Ogawa (1995a) did not mention the occurrence of failed male-male bridge initiations, we only used successful female-female bridge initiations to compare differences between male-male and female-female bridges to ensure validity between our dataset and Ogawa's. Of successful female-female bridges, 24 (31.6%) occurred via Type I initiations, and 52 (68.4%) occurred via Type II initiations. Even after pooling the male Type II and Type III initiations together, which creates a more conservative estimate than if Type III initiations were excluded, females were still significantly more likely than males to use Type II bridge initiations over Type I initiations (chi square test for independence; $\chi^2 = 28.85$, $df = 1$, $p < 0.01$; Fig. 9.6).

Bridges initiated by adults were significantly more likely to be successful than bridges initiated by subadults ($\varphi = 0.46$, $df = 1$, $p < 0.01$). In focal samples, all Type I bridges were successful, but we recorded two failed Type I bridge initiations via all-occurrence sampling. However, these two failed initiations were from a subadult. Receivers showed no significant preference in accepting bridges that utilized either infants or juveniles, and we found no significant difference in success rates of bridges that were preceded by grooming vs. those that were not preceded by grooming, although successful bridges were significantly more likely than unsuccessful bridges to be followed by grooming ($\varphi = 0.45$, $df = 1$, $p < 0.01$).

The two surviving infants in the group were used as a bridge in 88 out of 103 bridging cases, which is significantly more often than expected based on the proportions of infants and juveniles in the group (exact binomial test, $df = 1$,

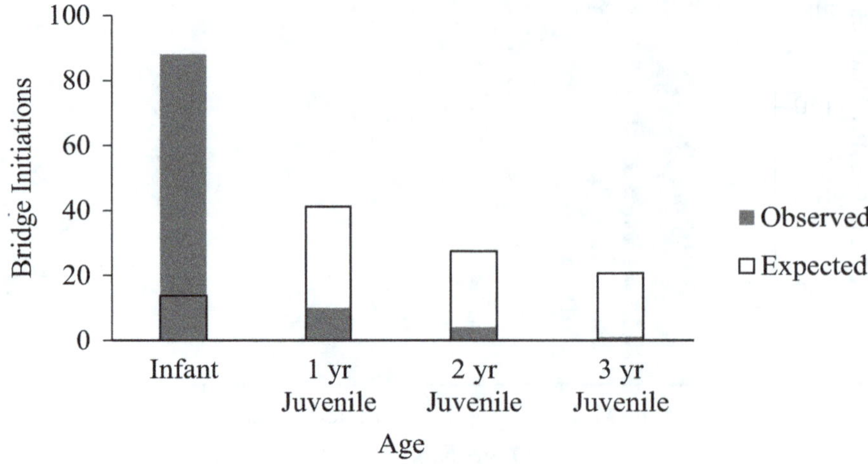

Fig. 9.7 Observed vs. expected use of infants and juveniles in bridge initiations after the death of TR's infant ($n = 103$); expected values are calculated based on the proportion of individuals in each age group

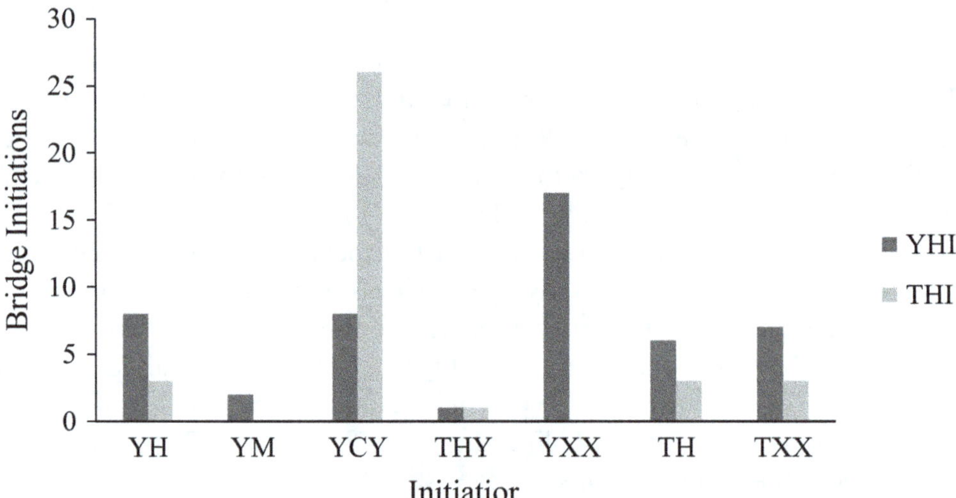

Fig. 9.8 Bar graph showing the number of bridge initiations using TH's infant (THI) and YH's infant (YHI) for each individual observed bridging with an infant at least once

$p < 0.01$, Fig. 9.7). Of individuals who used infants to bridge, some individuals demonstrated a clear preference toward a specific infant over the other (Fisher's exact test; $df = 6$, $p < 0.01$, Fig. 9.8), but our small sample size did not allow us to test the significance of the pairwise comparisons.

9.4 Discussion

The agonistic buffering hypothesis was not supported among female-female bridges within our dataset for this group of Tibetan macaques. In contrast to what has been reported in males (Ogawa 1995a; Zhao 1996; Bauer et al. 2013), social rank was positively correlated with bridge initiations. Initiators were also not more likely to be subordinate to those who received their bridge attempts, which is evidence in contrast to the idea that subordinate females initiate bridges with dominant individuals to reduce the probability of future aggression (Ogawa 1995a). Further departing from the trends seen in males, females were not more likely to initiate bridges with those from which they received bridges. These data suggest the females in this group do not bridge in order to avoid conflicts, as has been observed for males.

Although there were no significant differences in bridging initiations between adult and subadult groups in this population, two of the subadult individuals, YCY and YXX, initiated far more bridges than any other individual. In contrast, the other two subadults, THY and TRY, rarely bridged. Interestingly, YCY and YXX were both much larger and more physically mature than either THY or TRY, who were both more juvenile-like in appearance and behavior. The mother that bridged the most out of the mothers without infant offspring, TXX, was relatively young for a mother and would normally be considered a subadult if not for the fact that she previously gave birth. This individual was also the only mother to initiate Type I bridges utilizing another mother's offspring and to initiate a Type II bridge toward a receiver holding her offspring. These bridges occurred toward subadult YCY.

As a group, adults received most of the recorded bridges, though statistically there was no difference between the rate at which adults and subadults were recipients. In particular, the infant mothers TH and YH each received far more bridges than any other individual. Out of bridges that involved adults, all except for two involved the mother of the infant. In addition, female bridges were more likely to occur by Type II initiation than by Type I. Ogawa (1995a) reported far more Type I initiations between males than Type II initiations. The comparison of female-female bridges and male-male bridges from Ogawa (1995a) demonstrates that this difference in Type I vs. Type II initiations between intra-sexed bridging events is significant, meaning that females are more likely to approach and initiate a bridge with a receiver who is holding an infant, whereas males are more likely to initiate while holding the infant. Of female-female Type II bridges, a large majority involved the mother as the receiver, and all of those that involved the mother as the receiver used her infant. Type I bridges were also more likely to be successful than Type II bridges, which may suggest an importance of the infant in the bridge rather than the relationship between the other individuals. In addition, individuals were not more likely to reciprocate bridges, which further suggests that the relationship between the older members of the bridging dyad is not as important as the infant. These trends support motherhood and interest in infants as factors in how female-female bridging is both initiated and received.

Female-female bridging in our dataset did show some consistencies with male-male bridging, however. Similar to males, successful bridges never occurred in aggressive contexts and were not followed by aggression. Females also showed a bias toward initiating bridges with their own infants over other juveniles. Although initiators preferred to use infants, receivers did not appear to show a preference, as bridges initiated with juveniles were equally likely to be successful, though we should caution that the few observed juvenile bridges limits the power of our analysis. With that being said, it is nonetheless consistent with bridging trends between males observed in this same study group (Bauer et al. 2013). Moreover, among individuals that often bridged with infants, most showed a preference for one infant over the other. Interestingly, mothers of multiple offspring only initiated and received bridges utilizing their youngest offspring, further supporting the hypothesis that initiators are biased toward using younger individuals.

Unfortunately, the death of TR's infant, the only female infant, prevented us from investigating whether there are sex biases in the infants used in female-female bridges. YCY initiated four bridges using TR's infant, demonstrating that female-female bridges do occur with female infants, but we did not observe enough occurrences during this time for any sex biases to be tested. Of juvenile 1-year-olds, only females were used; however, there were five female 1-year-old juveniles and only one male 1-year-old juvenile. Moreover, TH was the mother of the single male 1-year-old juvenile, and she had a younger infant that she most often used in bridges. Without a more equally distributed sex ratio of infants and juveniles, it is impossible to state with these preliminary data whether or not females prefer to bridge with infants of a specific sex.

The preliminary results we present here can be used to generate future hypotheses regarding the function of bridging in females. In our dataset, the few individuals in the group make cross-comparisons between age classes difficult and limit a majority of the statistical analysis to less powerful, nonparametric, and descriptive procedures. Our study spanned a short time period, which limited the amount of focal data for each individual. This makes it unclear as to whether certain individuals never bridge or just do so less frequently than others. However, our study was conducted on the same group under similar conditions as past studies on bridging in male Tibetan macaques (Ogawa 1995a, b, c; Bauer et al. 2013; see also Zhao 1996 studying this species at Mt. Emei), which enabled us to directly compare male-male and female-female bridging behavior in Tibetan macaques within this particular group.

Despite the limitations, this study is the first to investigate bridging behavior between females in any species of macaque. Our results suggest that female bridging behavior may not follow the same pattern as male bridging and indicate that this behavior is related to female interest in infants than it is to the agonistic buffering hypothesis. Younger, higher-ranking female individuals may bridge more often because it allows them to gain alloparental experience from interacting with infants. Alternatively, female bridging may be a by-product of interest in mothers and infants, similar to what has been supported in female-infant interactions in bonnet macaques, *M. radiata* (Silk 1999). Future studies should follow the format of Ogawa

([1995a](#)) by observing bridges in relation to all female-female and female-infant interactions over the course of the entire year to enable a more definitive test of the agonistic buffering hypothesis.

Acknowledgments We thank Dao Zhang and the Chen family for logistical support while we were in China. We thank the members of the Huangshan Scenic Bureau for allowing us to conduct research at the site. We thank Anne Salow, Gregory Fratellone, and Sofi Bernstein for help with the data collection in the field and Dr. Dominic Klyve for assistance with statistical analysis. Grant J. Clifton particularly thanks Dr. Audrey Huerta and Dr. Linda Raubeson of CWU's Science Honors Research Program for their support. This research was funded by CWU's Science Honors Program and the National Science Foundation (OISE-1065589). Our research was reviewed and approved CWU's Institutional Animal Care and Use Committee (A011401).

References

Altmann J (1974) Observational study of behavior: sampling methods. Behaviour 49:277–266

Bauer B, Sheeran LK, Matheson MD, Li J, Wagner RS (2013) Male Tibetan macaques' (*Macaca thibetana*) choice of infant bridging partners. Zool Res 35(3):222–230

Berman CM, Ionica CS, Li JH (2004) Dominance style among *Macaca thibetana* on Mt. Huangshan, China. Int J Primatol 25(6):1283–1312

Deag JM (1980) Interactions between males and unweaned Barbary macaques: testing the agonistic buffering hypothesis. Behaviour 75:54–81

Deag JM, Crook JH (1971) Social behavior and "agonistic buffering" in the wild Barbary macaque *Macaca sylvana* L. Folia Primatol 15:183–200

Estrada A, Estrada R (1984) Female-infant interactions among free-ranging stumptail macaques (*Macaca arctoides*). Primates 25(1):48–61

Estrada A, Sandoval JM (1977) Social relations in a free-ranging troop of stumptail macaques (*Macaca arctoides*): male-care behavior I. Primates 18(4):793–813

Kubenova B, Konecna M, Majolo B, Smilauer P, Ostner J, Schulke O (2017) Triadic awareness predicts partner choice in male–infant–male interactions in Barbary macaques. Anim Cogn 20:221–232. https://doi.org/10.1007/s10071-016-1041-y

Neumann C, Duboscq J, Dubuc C, Ginting A, Irwan AM, Agil M, Widdig A, Engelhardt A (2011) Assessing dominance hierarchies: validation and advantages of progressive evaluation with Elo-rating. Anim Behav 82(4):911–921. https://doi.org/10.1016/j.anbehav.2011.07.016

Ogawa H (1995a) Bridging behavior and other affiliative interactions among male Tibetan macaques. Int J Primatol 16(5):707–729

Ogawa H (1995b) Recognition of social relationships in bridging behavior among Tibetan macaques (*Macaca thibetana*). Am J Primatol 35:305–310

Ogawa H (1995c) Triadic male-female-infant relationships and bridging behavior among Tibetan macaques (*Macaca thibetana*). Folia Primatol 64:153–157

Ogawa H (2006) Wiley monkeys: social intelligence of Tibetan macaques. Kyoto University Press, Kyoto

Paul A, Kuester J, Arnemann J (1996) The sociobiology of male-infant interactions in Barbary macaques, *Macaca sylvanus*. Anim Behav 51:155–170

R Core Team (2016) R: a language and environment for statistical computing. R Foundation for Statistical Computing, Vienna

Sheeran LK, Matheson MD, Li JH, Wagner RS (2010) A preliminary analysis of aging and potential social partners in Tibetan macaques (*Macaca thibetana*). In: Banak SD (ed) Collected readings in biological anthropology in honor of Professor LS Penrose and Dr. Sahrah B. Holt. Unas Letras Industria Editorial, Merida, pp 349–358

Silk J (1999) Why are infants so attractive to others? The form and function of infant handling in bonnet macaques. Anim Behav 57:1021–1032

Taub DM (1984) Male caretaking behavior among wild Barbary macaques (*Macaca sylvanus*). In: Taub DM (ed) Primate paternalism. Van Nostrand Reinhold, New York, pp 20–55

Wada K, Xiong C (1996) Population changes of Tibetan monkeys with special regard to birth interval. In: Shotake T, Wada K (eds) Variations in the Asian macaques. Tokai University Press, Tokyo, pp 133–145

Zhao QK (1996) Male-infant-male interactions in Tibetan macaques. Primates 37(2):135–143

Chapter 10
Bridging Behavior and Male-Infant Interactions in *Macaca thibetana* and *M. assamensis*: Insight into the Evolution of Social Behavior in the *sinica* Species-Group of Macaques

Hideshi Ogawa

10.1 Introduction

Species in the genus *Macaca* have many common features: forming multi-male multi-female social groups with female philopatry, male dispersal, and a linear dominance hierarchy (Thierry et al. 2004). However, there are variations in their "dominance style" and the degree of kin-biased social relationships between females (Berman and Thierry 2010; de Waal and Luttrell 1989). Recently, the variations in social relationships between females have been systematically studied (Balasubramaniam et al. 2012). In addition to social relationships between females, there are many inter-species differences in social relationships between males. Compared to females, however, there have been fewer comparative studies on social interactions between males. For example, frequent triadic male-infant interactions were reported in some macaque species, but not reported in most of other macaque species (Deag and Crook 1971; Kalbitz et al. 2017; Ogawa 1995, 2006). Therefore, in this paper, I focus on social interactions between adult males and infant handling by adult males in macaques.

Macaques are divided into several species-groups. Though there are several categorizations, one is the *sinica* species-group comprised of toque macaques (*Macaca sinica*) in Sri Lanka, Bonnet macaques (*M. radiata*) in India, Assamese macaques (*M. assamensis*) from India to Southeast Asia, and Tibetan macaques (*M. thibetana*) in China. Delson (1980) included stump-tailed macaques (*M. arctoides*), which are distributed in India eastward to Southeast Asia, in the *sinica* species-group. Arunachal macaques (*M. munzala*) in the Arunachal area of India and white-cheeked macaques (*M. leucogenys*) in Tibet are recently listed as new species in the *sinica* species-group (Li et al. 2015; Sinha et al. 2005).

H. Ogawa (✉)
School of International Liberal Studies, Chukyo University, Toyota, Aichi, Japan
e-mail: hogawa@lets.chukyo-u.ac.jp

J.-H. Li et al. (eds.), *The Behavioral Ecology of the Tibetan Macaque*, Fascinating Life Sciences, https://doi.org/10.1007/978-3-030-27920-2_10

Assamese macaques have been traditionally divided into two subspecies, based on morphological traits (Fooden 1982). Eastern Assamese macaques (*M. assamensis assamensis*), which inhabit areas east of the Brahmaputra River, have shorter tails than western Assamese macaques (*M. assamensis pelops*), which inhabit areas west of the Brahmaputra River. Though Assamese macaques are widely distributed in Asia, direct observations of wild Assamese macaques are difficult, because they are patchy distributed in mountainous areas (Chalise et al. 2013; Wada 2005). There are only a few direct observations of wild Assamese macaques (Kalbitz et al. 2017). However, free-ranging Assamese macaques in several social groups are provisioned and habituated to humans at several locations. I observed free-ranging provisioned Assamese macaques in Thailand, India, and Nepal and compared these populations' social behaviors to those of Tibetan macaques in China.

As may occur with regard to genetic and morphological traits, inter-species and inter-subspecies differences in social behaviors could be affected by evolutionary divergence among populations. My aim in this study is to reconstruct the evolutionary processes of macaques' social behaviors by comparing the male-male and male-infant interactions of Tibetan, eastern Assamese, and western Assamese macaques at various sites.

10.2 Methods

10.2.1 Study Sites and Study Periods

Tibetan macaques are distributed in China, and Assamese macaques are distributed in Nepal, India, Bhutan, Bangladesh, China, Myanmar, Laos, Thailand, and Vietnam (Groves 2001; Fooden 1982; Wada 2005). Among these countries, I conducted field researches at several sites in Nepal, India, China, and Thailand (Fig. 10.1).

10.2.1.1 Western Assamese Macaque (*M. a. pelops*)

Site 1 Ramdi Village (27°54′N, 83°38′E, 437 m). I stayed at Ramdi Village, Pyuthan District, Nepal, from 20 to 23 March 2016 with Pavan Paudel. There were two provisioned groups of Assamese macaques around the village.

Site 2 Nagarjun (27°44′N, 85°17′E, 1350–2093 m). Shivapuri-Nagarjun National Park, Kathmandu District, Nepal, has two separate areas, the Shivapuri area and the Nagarjun area. I stayed at Nagarjun for a total of 113 days between 2011 and 2017. Wild Assamese macaques in several social groups inhabit Nagarjun, as well as wild rhesus macaques (*M. mulatta*) (Chalise et al. 2013; Wada 2005). There is a royal palace and an active army camp in the national park, and soldiers live in the army camp. The social group of Assamese macaque was unintentionally provisioned by food waste of the army camp that was available for monkeys. Though I did not

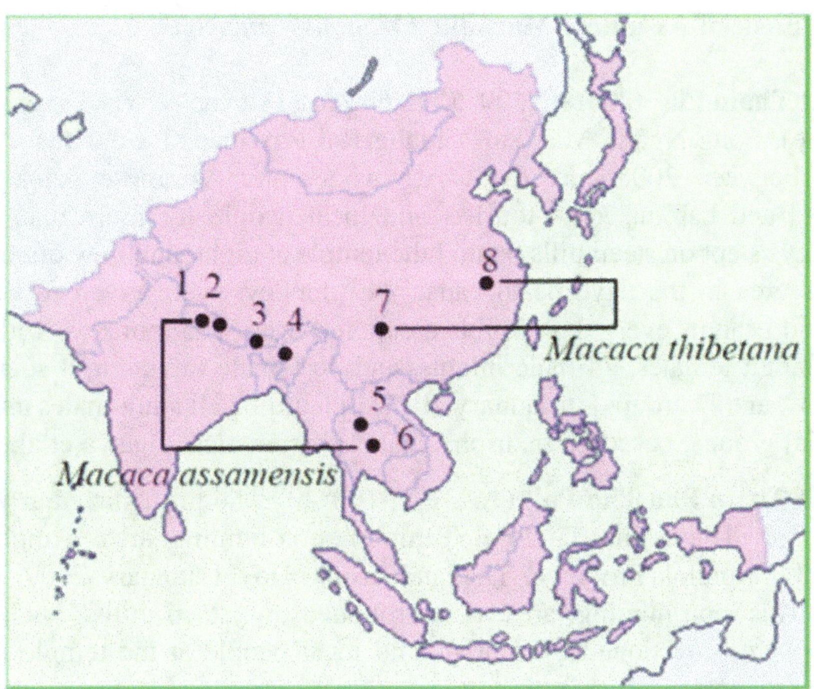

Fig. 10.1 Location of the study sites. 1, Ramdi Village (27°54′N, 83°38′E); 2, Nagarjun (27°44′N, 85°17′E); 3, Siliguri (26°54′N, 88°28′E); 4, Tukeswari Temple (26°03′N, 90°38′E); 5, Wat Tham Pla (20°19′N, 99°51′); 6, Wat Tham Pha Tha Pol (16°31′N, 100°40′E); 7, Mt. Emei (29°33′N, 103°20′E); 8, Huangshan (30°29′N, 118°11′E)

recommend it, when soldiers finished lunch and dinner, they usually disposed of their leftovers at a dumping site in the army camp. Monkeys consumed the garbage and became habituated to humans (Koirala et al. 2017). I identified all adult males and females in the AA group, based on their physical characteristics, and observed those monkeys with Sabina Koirala. The AA group was composed of 56 monkeys including 7 adult males, 13 adult females, 4 male infants (<1 year old), and 1 female infant in January 2015.

Site 3 Siliguri (26°54′N, 88°28′E, 199 m). I visited Coronation Bridge at Siliguri, West Bengal State, India, on 2 and 5 March 2009. Assamese macaques were provisioned and stayed along the road near Coronation Bridge.

10.2.1.2 Assamese macaque (*M. a. assamensis* or *M. a. pelops*)

Site 4 Tukreswari Temple (26°03′N, 90°38′E, 47 m). I stayed near Tukreswari (Tukeswari) Temple, Assam State, India, from 22 to 25 February 2009 with Mayur Bawri. Assamese macaques in two social groups were provisioned and habituated to humans on a hill surrounded by cultivated fields (Cooper and Bernstein 2008).

10.2.1.3　Eastern Assamese Macaque (*M. a. assamensis*)

Site 5 Wat Tham Pla　(20°19′N, 99°51′E, 843 m). I stayed at Wat Tham Pla (Tham Pla Temple), Pong Ngam, Mae Sai, Chiang Rai Province, Thailand, for a total of 110 days between 2008 and 2012. At this temple, Assamese macaques were provisioned and habituated to tourists and local people for more than 20 years. The monkeys slept on steep hills behind the temple at night, and they often stayed in the temple area in the daytime, because the monkeys were given foods such as bananas and peanuts every day. In this area, there were 193 monkeys with 26 adult males, 54 adult females, 24 male infants, and 16 female infants in 4 social groups (the A, B, C, and D groups) in January 2012. I identified all adult males and females in the social groups, based on their physical characteristics (Ogawa et al. 2009).

Site 6 Wat Tham Pha Tha Pol　(16°31′N, 100°40′E, 54 m). I stayed at Wat Tham Pha Tha Pol (Tham Pha Tha Pol Temple), non-hunting area, Amphoe Noen Maprang, Phitsanulok Province, Thailand, from 9 to 11 January 2009 with Eishi Hirasaki. This non-hunting area is surrounded by steep hills, but Assamese macaques were provisioned by tourists and local people at the temple outside of the non-hunting area.

10.2.1.4　Tibetan Macaque (*M. thibetana*)

Site 7 Mt. Emei　(29°33′N, 103°20′E, 1260–2100 m). I visited Mt. Emei (Emeishan), Sichuan Province, China, from 28 May to 1 June 1990 with Ming Li. Tibetan macaques were provisioned and habituated to tourists in this area (Zhao and Deng 1988). We made a round trip along the walking routes and observed monkeys in six social groups.

Site 8 Huangshan　(30°29′N, 118°11′E, 700–800 m). I stayed at Huangshan (Mt. Huang), Anhui Province, China, for a total of 382 days between 1989 and 1993. Wild Tibetan macaques in several social groups inhabited this area, as well as wild rhesus macaques (Wada et al. 1987). Since 1986, one social group, the Yulingkeng Group, has been provisioned for observations. Unlike now, no tourists visited there. The Yulingkeng Group had 42 monkeys with 7 adult males, 9 adult females, 3 male infants, and 3 female infants in November 1992. I identified all monkeys, based on their physical characteristics, and observed them sometimes with Kazuo Wada, Chenpei Xiong, Jinhua Li, Ming Li, and Qishan Wang (Ogawa 2006).

10.2.2　*Sampling Methods*

I used all occurrence behavior sampling in all of the study sites. In addition, I used focal animal sampling on adult males and adult females for 10 h each in both the mating and birth seasons, at Nagarjun in Nepal, Wat Tham Pla in Thailand, and Huangshan in China. For the focal sampling, I observed Assamese macaques at Nagarjun in Shivapuri-Nagarjun National Park from 30 July to 11 September 2014 in the birth season. S. Koirala and I observed them from 13 November 2014 to 15 January 2015 in the mating season and combined the data. I observed Assamese macaques at Wat Tham Pla in Thailand from 25 July to 7 September 2008 in the birth season and from 23 December 2010 to 8 January 2011 and from 31 December 2011 to 13 January 2012 in the mating season. I observed Tibetan macaques at Huangshan from 16 February to 26 April 1992 in the birth season and from 16 September to 4 November 1992 in the mating season.

The numbers of focal animals were 7 males and 7 females in the birth season and 7 males and 7 females in the mating season at Nagarjun; 10 males (5 in the A Group, 2 in the C Group, and 3 in the D Group) and 5 females (2 in the A Group, 1 in the C Group, and 2 in the D Group) in the birth season and 10 males (4 in the A Group, 2 in the C Group, and 4 in the D Group) and 5 females (3 in the A Group, 1 in the C Group, and 1 in the D Group) in the mating season at Wat Tham Pla; and 5 males and 9 females in the birth season and 6 males and 9 females in the mating season at Huangshan.

I calculated the frequency of behaviors, based on focal animal sampling. I combined the data by all occurrence behavior sampling and focal animal sampling to examine the presence/absence of behaviors and the rate of behaviors in which male and female infants were used in each study site.

10.2.3   Definition of Behavior

Bridging behavior is defined as "two individuals simultaneously lift up an infant" (Ogawa 1995, 2006, p. 52). If the two individuals are male, it is also called a triadic male-infant interaction. The typical interaction has the following sequence: (1) One adult holds a male or female infant ventrally. (2) The adult monkey carries the infant to another individual and presents it to the recipient. Another individual sometimes approaches the monkey who is holding the infant. (3) The two adults sat facing each other, one adult pulls up the infant's shoulder, the other pulls up its hip, the infant lays on its back, and the two adults lift up the infant together. As a result, the infant forms a bridge between them. While lifting up the infant, one or both adults often suck and/or touch the infant's genitalia with the expression of teeth chattering. Infants are handled gently and rarely show resistance or give signs of distress. After bridging, the two adults frequently stay in close proximity and groom each other, so this behavior may reduce social tension between the individuals and promote and maintain an affiliative relationship between males or between males and females.

Table 10.1 Presence/absence of social behavior in each site

Species	Assamese macaque						Tibetan macaque	
	Western Assamese macaque			?	Eastern Assamese macaque			
Country	Nepal		India		Thailand		China	
Site	Ramdi	Nagarjun	Siliguri	Tukreswari	Tham Pla	Tham Pha Tha Pol	Emei	Huangsham
Presence of behavior								
Bridging between males	No	No	No[#]	Yes	Yes	Yes	Yes	Yes
Sucking of an infant genitalia by males	No	No	No[#]	No[#]	Yes	Yes	Yes	Yes
Penis sucking between adult males	No	No	No[#]	No[#]	No	No[#]	Yes	Yes

Yes: the behavior was recorded. No: the behavior was not recorded. No[#]: Behavior was not recorded, possibly due to the short observational period at the study site. The study sites are arranged from west (left) to east (right). ?: It was not sure whether macaques here were Western Assamese macaques or Eastern Assamese macaques

10.3 Results

10.3.1 Bridging Behavior

Bridging behavior was recorded in Tibetan macaques at Huangshan and Mt. Emei in China and in Assamese macaques at Wat Tham Pha Tha Pol and Wat Tham Pla in Thailand and Tukreswari Temple in India (Table 10.1). However, no bridging behavior was recorded among any adults in Assamese macaques at Siliguri in India, Nagarjun in Nepal, and Ramdi Village in Nepal (Table 10.1). Although male Assamese macaques in these sites sometimes held an infant in front of another male, they never lifted the infant together with another male.

Figure 10.2 shows the frequency of bridging behavior at Nagarjun in Nepal, Wat Tham Pla in Thailand, and Huangshan in China, in which I observed 5–10 adult males and 5–9 adult females for 10 h each in the birth and mating seasons, respectively. Adult males of Tibetan macaque performed bridging behavior at the

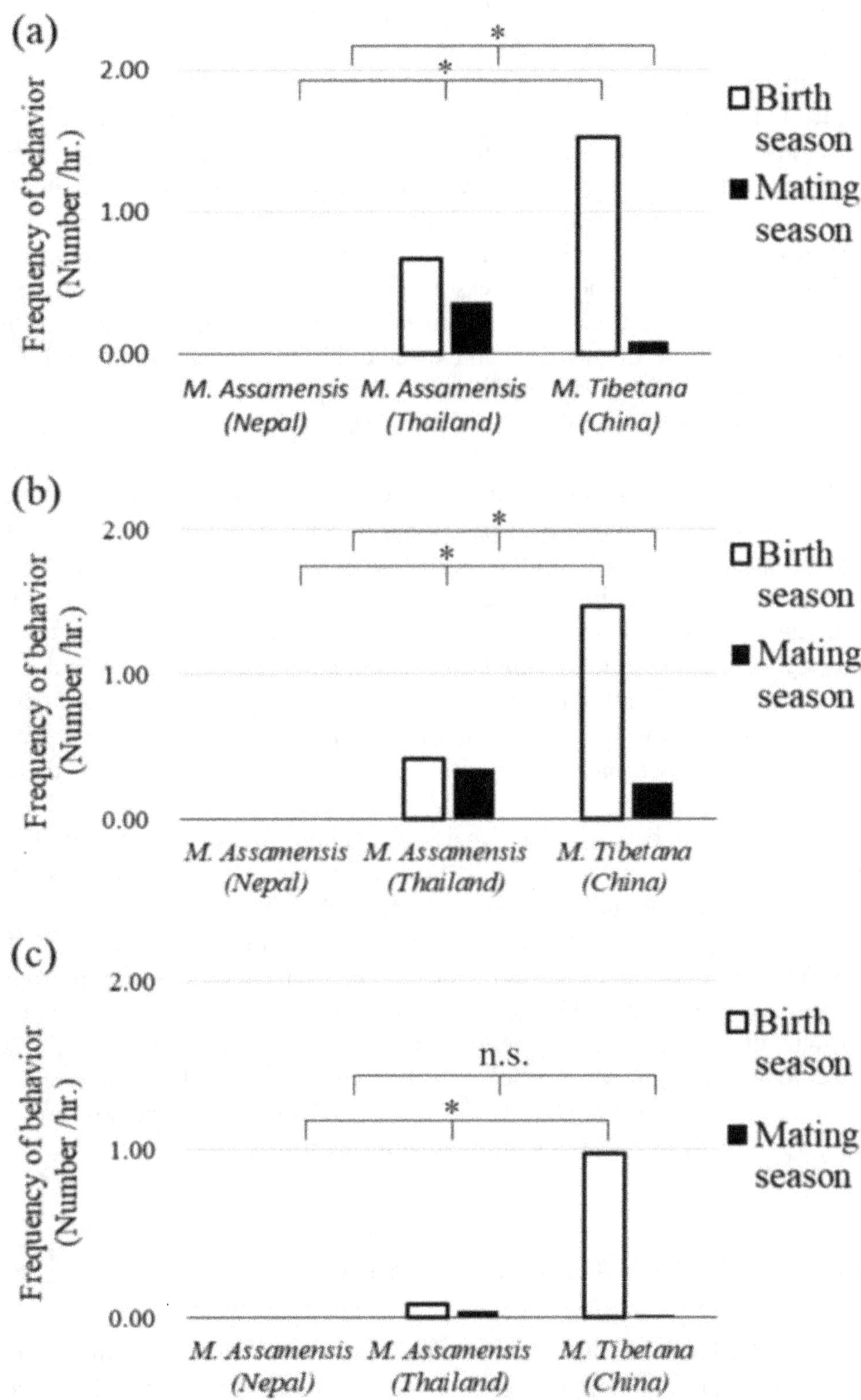

Fig. 10.2 Frequency of bridging behavior in the three study sites. Nepal: at Nagarjun, Shivapuri-Nagarjun National Park, Kathmandu District; Thailand: at Wat Tham Pla, Chiang Rai Province; China: at Huangshan, Anhui Province. $^*p < 0.05$. *n.s.* not significant. (**a**) Bridging between males, (**b**) bridging between females, (**c**) bridging between a male and a female

frequency of 1.52 (average number of behaviors of each individual per hour) in the birth season and 0.08 in the mating season. Adult males of eastern Assamese macaque in Thailand performed bridging behavior at the frequency of 0.67 in the birth season and 0.36 in the mating season, whereas adult males of western Assamese macaques in Nepal did not perform bridging behavior at all. The frequencies of bridging behavior between adult males were significantly higher in Tibetan and eastern Assamese macaques than that in western Assamese macaques who did not perform bridging behavior (Kruskal-Wallis test, $n_1 = 5$, $n_2 = 10$, $n_3 = 7$, $H = 12.964$, $p < 0.05$ in the birth season; $n_1 = 6$, $n_2 = 5$, $n_3 = 7$, $H = 13.112$, $p < 0.05$ in the mating season). Tibetan and eastern Assamese macaques used a male infant in bridging more frequently than expected, if the number of male and female infants in the study group were considered (chi-squared test, $df = 1$, $\chi^2 = 70.2$, $p < 0.05$ for Tibetan macaques; $df = 1$, $\chi^2 = 19.5$, $p < 0.05$ for eastern Assamese macaques).

Adult females of Tibetan macaque performed bridging behavior at the frequency of 1.47 in the birth season and 0.41 in the mating season. Adult females of eastern Assamese macaque performed bridging behavior at the frequency of 0.24 in the birth season and 0.34 in the mating season, whereas adult females of western Assamese macaque did not perform bridging behavior at all. The frequencies of bridging behavior between adult females were also significantly higher in Tibetan and eastern Assamese macaques than that in western Assamese macaques who did not perform bridging behavior (Kruskal-Wallis test, $n_1 = 9$, $n_2 = 10$, $n_3 = 7$, $H = 12.341$, $p < 0.05$ in the birth season; $n_1 = 9$, $n_2 = 5$, $n_3 = 7$, $H = 10.189$, $p < 0.05$ in the mating season).

Bridging behavior between adult males and adult females occurred in Tibetan macaques at the frequency of 0.98 in the birth season and 0.01 in the mating season and in eastern Assamese macaques at the frequency of 0.09 in the birth season and 0.04 in the mating season, whereas western Assamese macaques did not perform bridging behavior between males and females. The frequencies of bridging behavior were significantly higher in Tibetan and eastern Assamese macaques in the birth season than that in western Assamese macaques who did not perform bridging behavior, though it was not significant in the mating season (Kruskal-Wallis test, $n_1 = 14$, $n_2 = 20$, $n_3 = 14$, $H = 12.741$, $p < 0.05$ in the birth season; $n_1 = 15$, $n_2 = 10$, $n_3 = 14$, $H = 2.785$, n.s. in the mating season).

10.3.2   *Dyadic Male-Infant Interactions*

Adult males of Tibetan and Assamese macaques groomed and held an infant in their social group. Figure 10.3 shows the frequency of social grooming and holding an infant by adult males at Huangshan in China, Wat Tham Pla in Thailand, and Nagarjun in Nepal. There was no significant difference in the frequency of social grooming among the three sites (Kruskal-Wallis test, $n_1 = 5$, $n_2 = 10$, $n_3 = 7$, $H = 0.863$, n.s. in the birth season; $n_1 = 6$, $n_2 = 5$, $n_3 = 7$, $H = 1.380$, n.s. in the

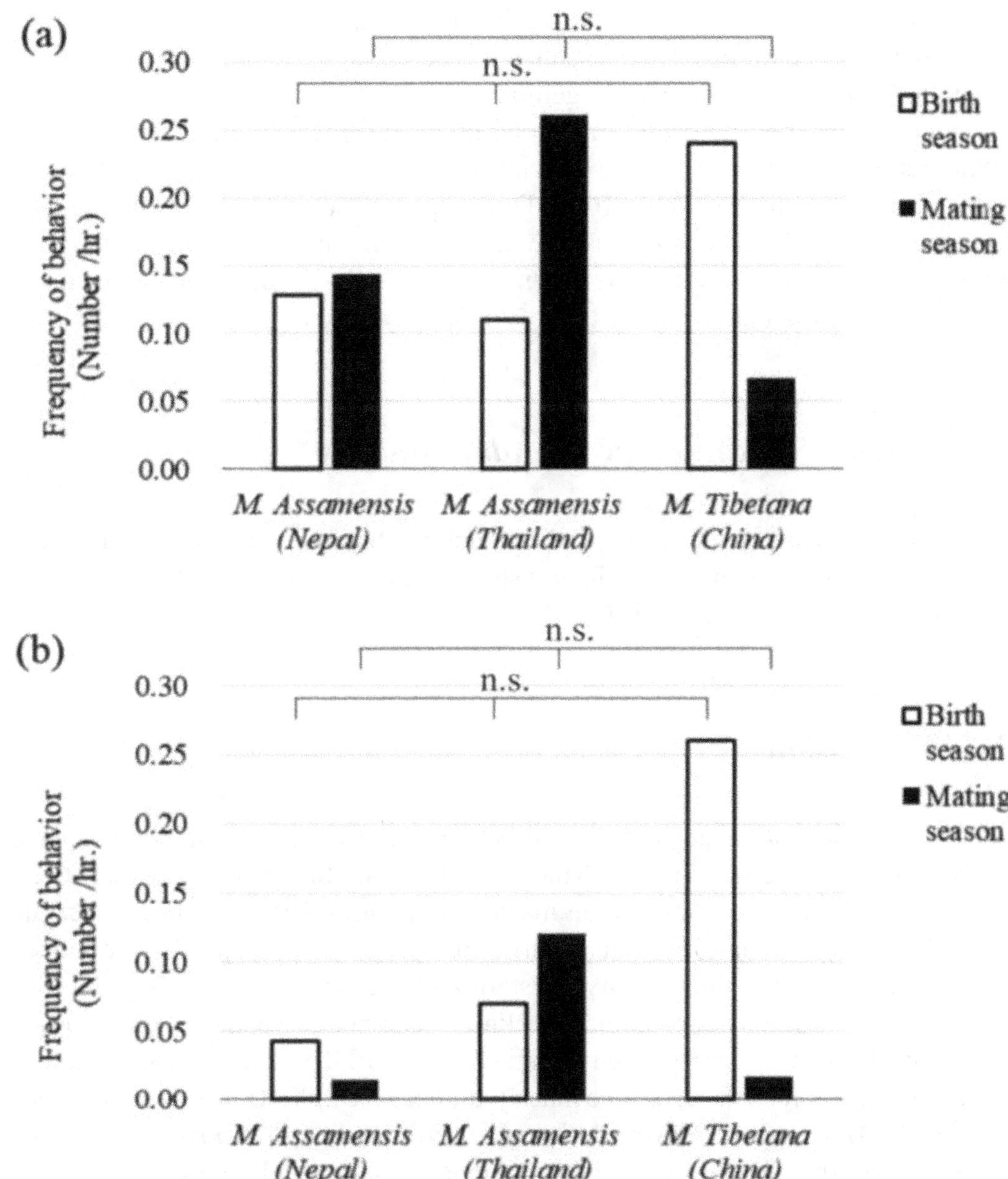

Fig. 10.3 Frequency of male-infant interactions in the three study sites. Nepal: at Nagarjun, Shivapuri-Nagarjun National Park, Kathmandu District; Thailand: at Wat Tham Pla, Chiang Rai Province; China: at Huangshan, Anhui Province. *n.s.* not significant. (**a**) Grooming an infant by adult males, (**b**) holding an infant by adult males

mating season). There was no significant difference in the frequency of holding among the three sites (Kruskal-Wallis test, $n_1 = 5$, $n_2 = 10$, $n_3 = 7$, $H = 1.022$, n.s. in the birth season; $n_1 = 6$, $n_2 = 5$, $n_3 = 7$, $H = 1.504$, n.s. in the mating season).

Adult males of Tibetan and Assamese macaques touched the genitalia of an infant, though this behavior was not confirmed at the sites in which the observational periods were short.

In addition to genital manipulation, Tibetan macaques in China and Assamese macaques at Wat Tham Pha Tha Pol and Wat Tham Pla in Thailand and Tukreswari Temple in India sucked an infant's genitalia during bridging behavior and during dyadic interactions between adult males and infants (Table 10.1). However, Assamese macaques at Siliguri in India, Nagarjun in Nepal, and Ramdi in Nepal did not suck the infant's genitalia (Table 10.1). During penis sucking behavior by an adult male on an infant, the adult male usually lifted up the infant and sucked his penis, sometimes turning the infant upside down to do so. Male infants sometimes jumped onto an adult male's face, in order to have his penis sucked.

10.3.3 Penis Sucking Between Adult Males

Only Tibetan macaques performed penis sucking behavior between adult males (Table 10.1). Assamese macaques in any study sites never performed penis sucking behavior between adult males (Table 10.1).

10.4 Discussion

The current comparative study on social behavior revealed that Tibetan macaques and eastern Assamese macaques in Thailand performed bridging behavior between males, between females, and between males and females. Males of these macaques also sucked the genitalia of an infant during bridging behavior and dyadic interactions with an infant. On the contrary, western Assamese macaques in Nepal did not perform either bridging or genital sucking behavior. Assamese macaques at Tukreswari Temple also performed bridging behavior. The macaque species there can be regarded as eastern Assamese macaques, because this site is east (on the south bank) of the Brahmaputra River. Although they have been introduced as western Assamese macaque (Biswas et al. 2011; Cooper and Bernstein 2008), their phylogenetic status remains in question. Therefore, the facts that Tibetan and eastern Assamese macaques have bridging and genital sucking behaviors are generally consistent with the genetic and morphological evidence that Tibetan and eastern Assamese macaques shared a close phylogenetic relationship (Biswas et al. 2011; Chakraborty 2007; Sukmak et al. 2014). The fact that western Assamese macaques in Nepal did not show bridging behavior indicates that they are different from Assamese macaques in other populations. Though Assamese macaques in Nepal have been regarded as the same species with Assamese macaques in other populations, the classification in the *sinica* species-group is still controversial (Khanal et al. 2018). Further studies are needed to understand phylogenetic relationships among the species in the *sinica* species-group including Arunachal and white-cheeked macaques, which were recently recorded as new species (Li et al. 2015; Sinha et al. 2005; but see Biswas et al. 2011).

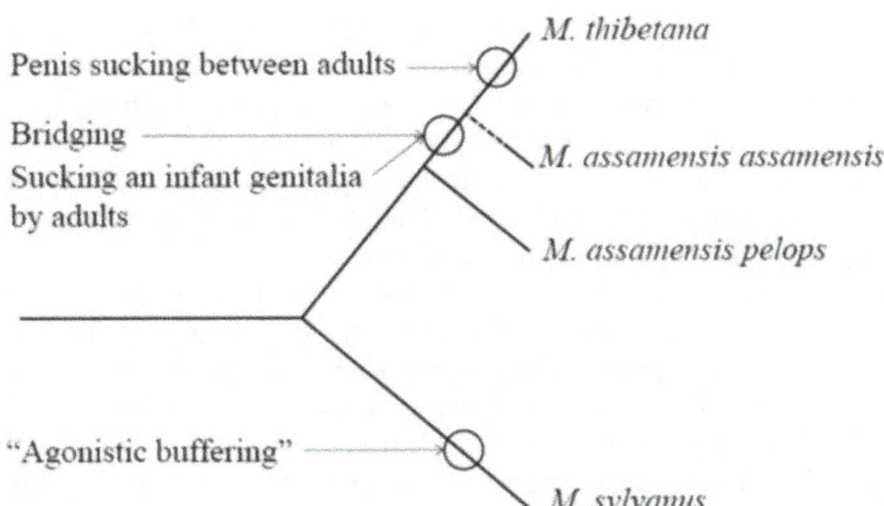

Fig. 10.4 Evolutionary scheme of bridging and penis sucking behaviors based on this comparative study, which shows that bridging and genital sucking behaviors evolved in the clade of *Macaca thibetana* and *M. assamensis assamensis*. In this figure, "agonistic buffering" in *M. sylvanus* evolved due to convergent evolution

Thus, the information on genetic and behavioral data is not complete. However, here I propose the hypothesis on the evolution of bridging and penis sucking behavior in macaques, based on the results of this study. Although it is not clear whether these social behaviors are genetically programmed or socially learned by cultural transmission, the difference of social behaviors by both of the two mechanisms is affected by the divergent evolutionary processes occurring in species, subspecies, and populations. I hypothesize that bridging and genital sucking behaviors evolved in the clade of Tibetan and eastern Assamese macaques in the *sinica* species-group, assuming that the Tibetan macaque was split from the eastern Assamese macaque (Fig. 10.4). As well as bridging behavior, many interactions between males and infants in eastern Assamese macaques in Thailand were similar to those in Tibetan macaques, though there might be some differences in frequencies, social contexts, functions, and the body form in the behavior (Ogawa, personal observation). In addition, eastern Assamese macaques in Thailand had various affiliative behaviors with body contact and sexual behaviors, most of which were similar to those of Tibetan macaques: social mount, hugging (embracing), genital showing, genital inspection, mating, and sexual harassment (Ogawa, personal observation).

It is also possible that the western Assamese macaques in Nepal lost the bridging behavior at some point in its evolutionary history. For example, similar to Tibetan and eastern Assamese macaques, stump-tailed macaques also frequently handled a newborn baby and performed bridging behavior between females one of which was the infant's mother (Estrada 1984; Maruhashi et al. 2018). Bridging behavior might occur between a mother and another female in the common ancestor of the *sinica* species-group including stump-tailed macaque, with some species engaging in frequent bridging between males and some species losing the bridging behavior.

However, it is more parsimonious to argue that bridging behavior evolved in the clade of Tibetan and eastern Assamese macaque, rather than it was lost in western Assamese macaques, because there is no report that toque and bonnet macaques have bridging behavior, except for one study reporting triadic male-infant interactions in bonnet macaques (Silk and Samuels 1984). I predict that Arunachal and white-cheeked macaques will not have bridging behavior, either, if they are phylogenetically closer to western Assamese macaques.

If bridging behavior evolved in the *sinica* species-group, especially in the clade of Tibetan and eastern Assamese macaques, this explains why similar behavior has not been reported in other species-groups of macaques, though there is one report of triadic male-infant interactions in long-tailed macaque (*M. fascicularis*) (de Waal et al. 1976). The only exception is that Barbary macaques (*M. sylvabus*) have a behavior called "agonistic buffering" (Deag and Crook 1971). The behavioral sequence, social context, and functions of "agonistic buffering" have some similarities with bridging behavior in Tibetan and eastern Assamese macaques (Deag and Crook 1971; Paul 1999; Taub 1980). Indeed, two males of Barbary macaques handle an infant together and manipulate and/or suck the genitalia of the infant; however, they seem not to lift up the infant together in the same way that Tibetan and eastern Assamese macaques do. The "agonistic buffering" in Barbary macaques and bridging behavior in the *sinica* species-group might be evolutionarily different, if the origin of these behaviors was not very old in the *sylvanus* and *sinica* species-groups.

Besides, several species in the genus *Papio* also show triadic male-infant interactions (Packer 1980; Strum 1984). However, these interactions of baboons are quite different from bridging and "agonistic buffering" in macaques, because, unlike macaques, one of the two male baboons handles an infant in agonistic interactions. Bridging behavior in the *sinica* species-group, "agonistic buffering" in Barbary macaques, and triadic male-infant interactions in baboons might occur due to convergent evolution.

Among macaques, only male Tibetan macaques directly sucked the penis of another adult male, though bridging behavior was frequently accompanied with sucking the genitalia of an infant across the taxa I studied. Penis sucking behavior between adult males evolved only in the clade to which Tibetan macaques belong. Generally, monkeys form a social relationship with another individual by dyadic interactions, and they sometimes use a third individual to regulate the social relationship. Ogawa (2006, p. 66) discussed that "during bridging behavior, a monkey may be using the penis of an infant as a substitute for his own penis because the use of the infant's penis is more effective." However, male Tibetan and eastern Assamese macaques performed bridging behavior and genital sucking of an infant. Tibetan macaques might expand penis sucking behavior between males and infants toward penis sucking behavior between adult males. The socionomic sex ratios (number of adult males/number of adult females) in the study groups were 0.78 in Tibetan macaques, 0.54 in western Assamese macaques, and 0.48 in eastern Assamese macaques during the study periods. The high socionomic sex ratio in Tibetan macaques might promote penis sucking and other affiliative behaviors between adult males in this species (Ogawa 1995).

In order to examine these hypotheses and predictions and to understand the evolution of the *sinica* species-group and the genus *Macaca*, further behavioral, genetic, and morphological studies are needed in Tibetan macaques, western and eastern Assamese macaques, and other macaque species in various areas.

10.5 Summary

1. Tibetan macaques and eastern Assamese macaques in Thailand performed bridging behavior, but western Assamese macaques in Nepal did not, though male western Assamese macaques in Nepal held an infant in front of another male.
2. Tibetan macaques and eastern Assamese macaques in Thailand sucked the genitalia of an infant, but western Assamese macaques in Nepal did not.
3. Only in Tibetan macaques did adult males perform penis sucking behavior with another adult male.

Acknowledgments I am grateful to the following people and organizations: in China, Dr. K. Wada at Kyoto University; Prof. Q. Wang, Mr. C. Xiong, Dr. J. Li, and Dr. M. Li at Anhui University; Mr. S. Zheng at Huangshan; Takashima Fund of Primate Society of Japan; JICA (Japan International Cooperation Agency); and JSPS KAKENHI Grant Numbers J1740480, 13740498; in Thailand, Dr. S. Malaivijitnond and Prof. S. Hannongbua at Chulalongkorn University; Dr. Y. Hamada at Kyoto University; Mr. L. Por, Mr. S. Win, Mr. S. Tirapanyo, and Mr. I. Ito, chief monks of Wat Tham Pla; National Research Council of Thailand (NRCT); and JSPS KAKENHI Grant Number 20255006; in Nepal, Dr. M. Chalise, Ms. S. Koirala, and Mr. B. Pandey at Tribhuvan University; Dr. Y. Kawamoto at Kyoto University; park rangers and staff of Shivapuri-Nagarjun National Park; Nepal Biodiversity Research Society; and JSPS KAKENHI Grant Numbers JP25440253 and JP16K07539; and in India, Mr. M. Bawri at Gauhati University; Dr. A. Sinha and Mr. N. Sharma at National Institute of Advanced Studies, Bangalore; Dr. M. Cooper at the University of Tennessee; Mr. G. S. S. Sarma at Tukreswari; and HOPE Project.

References

Balasubramaniam KN, Dittmar K, Berman CM, Butovskaya M, Cooper MA, Majolo B, Ogawa H, Schino G, Thierry B, de Waal FBM (2012) Hierarchical steepness and phylogenetic models: phylogenetic signals in *Macaca*. Anim Behav 83:1207–1218

Berman CM, Thierry B (2010) Variation in kin bias: species differences and time constraints in macaques. Behaviour 147:1863–1887

Biswas J, Borah DK, Das A, Das J, Bhattacharjee PC, Mohnot SM, Horwich RH (2011) The enigmatic Arunachal macaque: its biogeography, biology and taxonomy in northeastern India. Am J Primatol 73:1–16

Chakraborty D, Ramakrishnan U, Panor J, Mishra C, Sinha A (2007) Phylogenetic relationships and morphometric affinities of the Arunachal macaque, *Macaca munzala*, a newly described primate from Arunachal Pradesh, northeastern India. Mol Phylogenet Evol 44:838–849

Chalise MK, Ogawa H, Pandey B (2013) Assamese monkeys in Nagarjun forest of Shivapuri Nagarjun National Park, Nepal. Tribhuvan Univ J 18(1–2):181–190

Cooper MA, Bernstein IS (2008) Evaluating dominance styles in Assamese and rhesus macaques. Int J Primatol 29:225–243

De Waal FBM, Luttrell LM (1989) Toward a comparative socioecology of the genus *Macaca*: different dominance styles in rhesus and stumptail monkeys. Am J Primatol 19:83–109

De Waal FBM, van Hooff JA, Netto WJ (1976) An ethological analysis of types of agonistic interaction in a captive group of Java-monkeys (*Macaca fascicularis*). Primates 17:257–290

Deag JM, Crook JH (1971) Social behaviour and "agonistic buffering" in the wild Barbary macaques, *Macaca sylvana* L. Folia Primatol 15:183–200

Delson E (1980) Fossil macaques, phyletic relationships and a scenario of deployment. In: Lindburg D (ed) The macaques: studies in ecology, behavior and evolution. Van Nostrand Reinhold, New York, pp 10–30

Estrada A (1984) Male-infant interactions among free-raging stumptail macaques. In: Taub DM (ed) Primate paternalism. Van Nostrand Reinhold, New York, pp 56–87

Fooden J (1982) Taxonomy and evolution of the *sinica* group of macaques: 3. Species and subspecies accounts of *Macaca assamensis*. Fieldiana. Zoology (New Series) 10:1–52

Groves CP (2001) Primate taxonomy: Smithsonian series in comparative evolutionary biology. Smithsonian Institution Press, Washington, DC

Kalbitz J, Schülke O, Ostner J (2017) Triadic male-infant-male interaction serves in bond maintenance in male Assamese macaques. PLoS One 12(10):e0183981

Khanal L, Chalise MK, Hei K, Acharya BK, Kawamoto Y, Jiang X (2018) Mitochondrial DNA analyses and ecological niche modeling reveal post-LGM expansion of the Assam macaque (*Macaca assamensis*) in the foothills of Nepal Himalaya. Am J Primatol 80(3):e22748

Koirala S, Chalise MK, Katuwal H, Gaire R, Pandey B, Ogawa H (2017) Diet and activity of *Macaca assamensis* in wild and semi-provisioned groups in Shivapuri Nagarjun National Park, Nepal. Folia Primatol 88:57–74

Li C, Chao Z, Fan P (2015) Whit-cheeked macaques (*Macaca leucogenys*): a new macaque species from Modog, Southeastern Tibet. Am J Primatol 77(2):753–766

Maruhashi T, Toyoda A, Malaivijitnond S (2018) Timing and characters of Touch Baby Genital (TBG) behaviours of *Macaca arctoides* inhabiting the Khao Krapuk Khao Taomo Non-hunting Area, Petchaburi, Thailand. Abstract. Satellite international symposium on Asian primates. Kathmandu, Nepal, pp 26–27

Ogawa H (1995) Bridging behavior and other affiliative interactions among male Tibetan macaques (*Macaca thibetana*). Int J Primatol 16:707–729

Ogawa H (2006) Wily monkeys: social intelligence of Tibetan macaques. Kyoto University Press and Trans Pacific Press, Kyoto

Ogawa H, Malaivijitnond S, Hamada Y (2009) Social interactions among Assamese macaques at Wat Tham Pla, Thailand. Proceedings. The 3rd international congress on the future of animal research, Bangkok, p 29

Packer C (1980) Male care and exploitation of infants in *Papio anubis*. Anim Behav 28:512–520

Paul A (1999) The sociobiology of infant handling in primates: is the current model convincing? Primates 40:33–46

Silk JB, Samuels A (1984) Triadic interactions among *Macaca radiata*: passports and buffers. Am J Primatol 6:373–376

Sinha A, Datta A, Madhusudan MD, Mishra C (2005) *Macaca munzala*: a new species from western Arunachal Pradesh, northeastern India. Int J Primatol 26(4):977–989

Strum SC (1984) Why males use infants. In: Taub DM (ed) *Primate paternalism*. Van Nostrand Reinhold, New York, pp 20–55

Sukmak M, Malaivijitnond S, Schülke O, Ostner J, Hamada Y, Wajjwalku W (2014) Preliminary study of the genetic diversity of eastern Assamese macaques (*Macaca assamensis assamensis*) in Thailand based on mitochondrial DNA and microsatellite markers. Primates 55:189–197

Taub DM (1980) Testing the "agonistic buffering" hypothesis. Behav Ecol Sociobiol 6:187–197

Thierry B, Singh M, Kaumanns W (eds) (2004) Macaque societies: a model for the study of social organization. Cambridge University Press, Cambridge

Wada K (2005) The distribution pattern of rhesus and Assamese monkeys in Nepal. Primates 46:115–119

Wada K, Xiong C, Wang Q (1987) On the distribution of Tibetan and rhesus monkeys in southern Anhui, China. Acta Theriol Sinica 7:148–176

Zhao Q, Deng Z (1988) *Macaca thibetana* at Mt. Emei, China: I. A cross-sectional study of growth and development. Am J Primatol 16:251–260

Part IV
Living with Microbes, Parasites, and
Diseases

Chapter 11
The Gut Microbiome of Tibetan Macaques: Composition, Influencing Factors and Function in Feeding Ecology

Binghua Sun, Michael A. Huffman, and Jin-Hua Li

11.1 Introduction

Microbes dominated the earth for at least 2.5 billion years before multicellular life appeared in the biosphere (Hooper and Gordon 2001; Ley et al. 2008). In fact, animals have been living and evolving in a microbial world (Margaret et al. 2013). Microbes inevitably colonized animals and coevolved through the process of interactions with their hosts (Clayton et al. 2018). Microorganisms can inhabit multiple parts of a host's body, such as the skin, oral cavity, sex organs, and the gastrointestinal (GI) tract. They are usually recognized as the microbiome of a particular host. As an essential part of the host's body, the microbiome plays an important role in host physiology by influencing nutritional intake, metabolic activity, and immune homeostasis (Turnbaugh et al. 2006; Greenblum et al. 2012; Hooper et al. 2012). Thus, the complex relationship between hosts and their microbiomes provides a unique opportunity for understanding mammalian adaptation and evolution (Hird 2017).

As the most complex and diverse ecological system of the mammalian body, the GI tract is colonized by bacteria, archaea, fungi, and viruses (Underhill and Iliev 2014). Previous studies have revealed that the gut microbiota play an important role in immune regulation, vitamin synthesis, energy acquisition, and disease risk

B. Sun
School of Resources and Environmental Engineering, Anhui University, Hefei, Anhui, China

M. A. Huffman
Primate Research Institute, Kyoto University, Kyoto, Japan
e-mail: huffman.michael.8n@kyoto-u.ac.jp

J.-H. Li (✉)
School of Resources and Environmental Engineering, Anhui University, Hefei, Anhui, China

International Collaborative Research Center for Huangshan Biodiversity and Tibetan Macaque Behavioral Ecology, Anhui, China

School of Life Sciences, Hefei Normal University, Hefei, Anhui, China
e-mail: jhli@ahu.edu.cn

J.-H. Li et al. (eds.), *The Behavioral Ecology of the Tibetan Macaque*, Fascinating Life Sciences, https://doi.org/10.1007/978-3-030-27920-2_11

reduction of the host (Turnbaugh et al. 2006; Hooper et al. 2012; Sharon et al. 2012). For example, all vertebrates lack cellulase, the enzymes, which help them to digest the fiber in their food, and are reliant upon intestinal microbes for its production (Yokoe and Yasumasu 1964; Mackie 2002). As a result, mammals maintain rich gut microbial communities, which are beneficial for the digestion of cellulose and hemicellulose abundant in the foods they ingest (Amato et al. 2015).

Nonhuman primates (NHPs) evolved in tropical forest habitats and radially spread to woodland, savanna and montane environments (Hanya et al. 2011). In order to adapt to the resultant temporal food resource shifts and to maintain ordinary functions of the body, NHPs have evolved many anatomical and behavioral strategies (Rodman and Cant 1984). For example, some primates have evolved specialized stomachs and teeth, while many others have evolved distinctive habitat use patterns, ranging behavior and social organization (Milton and May 1976; Chivers et al. 1984; Cachel 1989; Yamada and Muroyama 2010; Hanya and Chapman 2013). Generally, NHPs prefer to eat high-quality food rich in lipids, proteins, and carbohydrate, such as fruits and young leaves. However, these foods are not always available year-round (Rodman and Cant 1984; Ungar 1995; Hanya et al. 2011; Hanya and Chapman 2013). NHPs living in temperate environments can only eat mature leaves, gum and roots during winter (Fan et al. 2009). These low-quality foods usually contain high proportions of cellulose and hemicellulose, which are difficult to digest (Coley 1983; Burrows 2010). In response to ecological pressures, many primates also use behavioral strategies such as adjusting their activity budgets and foraging patterns to help overcome periods of food scarcity (Zhou et al. 2007; Starr et al. 2012; Hanya and Chapman 2013; Campera et al. 2014). However, how wild primates digest low-quality foods high in cellulose during the winter has not been well understood (Amato et al. 2014).

Although low-quality foods, including woody plants, mature leaves, fungi, mature leaves, and plant exudates are difficult to digest, more recent research related to the above has revealed that these foods can be broken down and utilized by the intestinal microbes of NHPs. For this reason, the gut microbiota may provide a way to understand how primates adapt to ecological pressures during periods of food scarcity. Since the gut microbiome is dominated by bacteria, many studies on the feeding ecology of mammals have focused solely on the bacterial microbiome (Qin et al. 2010). Other common elements, such as gut fungi, also appear to affect the health and nutrition of animals, but this too is not yet well understood (Huffnagle and Noverr 2013; Underhill and Iliev 2014; Sokol et al. 2017). Although intestinal fungi have a much smaller number of cells compared to bacteria, the evidence suggests that gut mycobiota can also play an important role in host health by affecting gut bacterial composition (Hoffmann et al. 2013), interacting with immune cells, and assisting in the metabolic activity of the host (Hajishengallis et al. 2011; Romani 2011; Iliev and Underhill 2012; Rizzetto et al. 2014). Therefore, it is important to integrate information about the role of gut bacteria and fungi when discussing the role of the gut microbiome in the evolution of animal dietary strategies.

NHPs and humans have extensive similarities in their genetic characteristics, physiology and morphology. Very important animal model systems, NHPs are extremely important for understanding human physiology, behavior, cognition,

Fig. 11.1 A female Tibetan macaque carrying a baby is foraging for food in winter (Photo credit: Qixin Zhang)

health and evolution (McCord et al. 2014; Ren et al. 2015). Therefore, compared with other laboratory animals, NHPs have an advantage in helping us to understanding the role of the gut microbiome in human health maintenance, as well as the evolutionary relationships involving the gut microbiome, host diet and evolution.

The free-ranging Tibetan macaque (*Macaca thibetana*) group at Mt. Huangshan (described in Chap. 2) presents a good opportunity for studying this, since all members of the group can be identified individually, allowing for the collection of feces from individuals of known age and sex (Zhang et al. 2010). Our study group lives in a deciduous and evergreen-broadleaf-mixed montane forest. As the temperature fluctuates across seasons (highest in summer: 34.2 °C, lowest in winter: −13. 9 °C), the quality of their food also varies accordingly (see Fig. 11.1). For example, from winter to spring, Tibetan macaques experience a switch from low quality food resources (mature leaves, roots and plant stem) to high quality food resources (young leaves and flowers) (Xiong and Wang 1988; Zhao 1999). As a result, our study population is an ideal model for studying the role of the gut microbiome in primate feeding ecology and evolution.

In this chapter, we will summarize our research on the composition of the bacterial/fungal community and microbial diversity of Tibetan macaques and analysis of the factors influencing variation in the microbiome across age, sex and season in this species. Furthermore, we use this information to discuss the role of the Tibetan macaques' gut microbiome in relation to the evolution of their feeding ecology.

11.2   Gut Microbiome of Tibetan Macaque

11.2.1   Composition of Gut Bacteria

Understanding the composition of an animal's gut bacteria helps us to reveal its role in the host's feeding ecology, however, we know little about the Tibetan macaques' gut microbiota. Recently, using the methods of high-throughput sequencing and bioinformatics, Sun et al. (2016) detected 14 known bacterial phyla in this macaque's gut, the most dominant phyla (relative abundance > 1%) being Firmicutes, Bacteroidetes and Proteobacteria Spirochaetes (Sun et al. 2016). Comparing our results to those of the gut microbiome in other NHPs (Table 11.1), we found that the two most dominant phyla (Firmicutes and Bacteriodetes) were commonly reported in many other species as well (Ochman et al. 2010; Mckenney et al. 2015; Xu et al. 2015). At the genus level, the two most dominant genera present in the gut of Tibetan macaques were *Prevotella* and *Succinivibrio*. Although the genus *Prevotella* was also commonly reported elsewhere (Ochman et al. 2010; Mckenney et al. 2015; Xu et al. 2015), *Succinivibrio* has been seldom been detected as a dominant genus. The relative abundance of these main bacterial genera (relative abundance > 1%) are presented in Fig. 11.2a.

Table 11.1 The dominant phyla of gut bacteria detected in several non-human primate species

Monkey's name	Dominant phyla			Captive/ wild	Study
	Firmicutes	Bacteroidetes	Proteobacteria		
Lemur Catta	+	+		Captive	McKenney et al. (2015)
Varecia variegata	+	+		Captive	McKenney et al. (2015)
Propithecus coquereli	+	+		Captive	McKenney et al. (2015)
Rhinopithecus bieti	+	+	+	Wild	Xu et al. (2015)
Nycticebus pygmaeus		+		Wild	Xu et al. (2013)
Gorilla beringei		+	+	Wild	Ochman et al. (2010)
Gorilla gorilla		+	+	Wild	Ochman et al. (2010)
Pan troglodytes	+	+		Wild	Ochman et al. (2010)
Macaca thibetana	+	+		Wild	Sun et al. (2016)

Note: + means the relative abundance was one of the top two

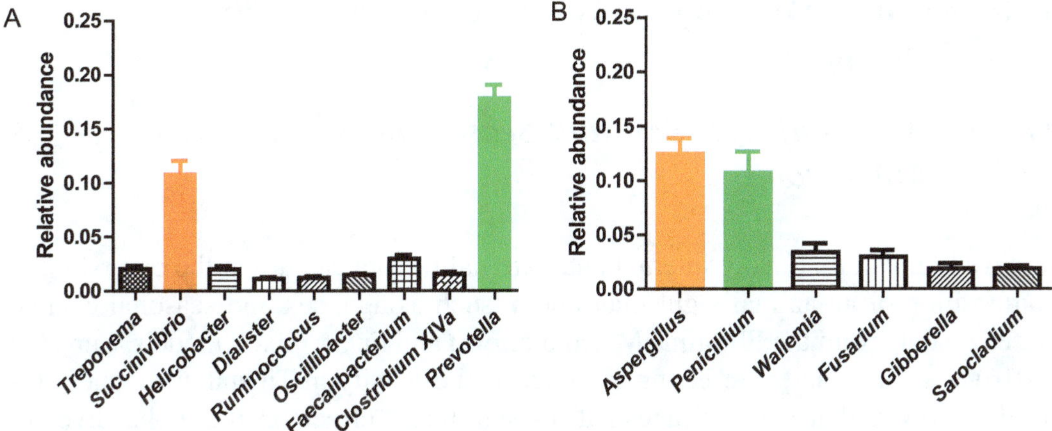

Fig. 11.2 The relative abundance of the main genera of bacteria and fungi. (a) The relative abundance of the major bacterial genera (relative abundance > 1%); data from Sun et al. (2016). (b) The relative abundance of the major fungal genera (relative abundance > 1%), data from Sun et al. (2018)

11.2.2 Composition of Gut Fungi

Previous studies on humans and mice have revealed that the mammalian gut is rich in fungi, dominated by three phyla: Ascomycota, Basidiomycota, and Zygomycota (Iliev and Underhill 2012; Qiu et al. 2015; Mar et al. 2016; Strati et al. 2016; Wheeler et al. 2016; Sokol et al. 2017). However, little is known about the gut fungi in NHPs. Using the Illlumina Miseq platform and primers of the ITS region (Bokulich and Mills 2013; Sun et al. 2018) first reported that the gut mycobiota of Tibetan macaques was dominated by two phyla Ascomycota and Basidiomycota (Sun et al. 2018), this is consistent with previous studies on mammalian gut mycobiomes. In addition, other phyla such as Zygomycota, Chytridiomycota, Glomeromycota and Rozellomycota were also detected in this macaque species. At the family level, Trichocomaceae shows the highest relative abundance (24.70%). At the genus level, the gut mycobiota of Tibetan macaques are rich in *Aspergillus* and *Penicillium* (relative abundance more than 10% respectively). Furthermore, by defining the core families and genera (both present in >90% of the samples with an average relative abundance > 0.01), Sun et al. (2018) revealed that six core taxa, namely Trichocomaceae, Nectriaceae, Davidiellaceae, *Aspergillus*, *Penicillium* and *Fusarium* can be detected in the gut of Tibetan macaques. The relative abundances of fungal genera (>1%) are presented in Fig. 11.2b. The presence of core taxa in the gut of this macaque suggests that the fungi in their guts are not random and exist as a relatively stable fungal community.

11.3 Factors Affecting the Gut Microbiome in Tibetan Macaques

11.3.1 Affects of Age, Sex, and Season on Gut Bacterial Microbiome

It is well known that many internal and external factors can affect the diversity and composition of an animal's gut microbiota, such as age, sex and seasonal change. Based on data from the Illlumina Miseq platform and linear mixed models, Sun et al. (2016) evaluated factors affecting gut microbial diversity in Tibetan macaques. The results showed that sex and age had no significant effect on the alpha diversity (including Shannon index, Chao 1, OTU richness and ACE) of the gut bacterial microbiome, as well as the beta diversity (evaluating by unweighted and weighted UniFrac distances) (Sun et al. 2016). These results differ from previous studies in humans and other vertebrates (Yatsunenko et al. 2012; Bolnick et al. 2014), but are consistent with those in chimpanzees and baboons (Degnan et al. 2012; Tung et al. 2015).

Seasonal factors on the other hand have a strong effect on the composition and diversity of the gut bacterial microbiome in Tibetan macaques (Sun et al. 2016). For example, the Shannon diversity index of the gut bacterial microbiome was significantly different between winter and spring. In addition, a significant seasonal separation of beta diversity evaluating by weighted UniFrac distances and unweighted UniFrac distances (Permanova tests, $p < 0.05$ and 0.01 respectably) were detected. Furthermore, three phyla and 20 known genera showed significant differences between the two seasons. The representative taxa rich in winter samples were Proteobacteria, Spirochaetia, *Succinivibrio*, *Clostridium* sensu stricto and *Treponema*, but Firmicutes and *Prevotella* were richest in spring samples. Differences in the composition of gut bacteria also resulted in the variations of predicted metagenomes between winter and spring. Using PICRUSt and LEfSe tests, Sun et al. (2016) found that seven KEGG pathways of Tibetan macaques' gut bacterial microbiome, including Glycan Biosynthesis and Metabolism, Amino Acid Metabolism, Signal Transduction, Cell motility, Transport and Catabolism, Neurodegenerative Diseases and Endocrine System, were significantly enriched in the winter. However six other metabolic pathways Energy Metabolism, Carbohydrate Metabolism, Cellular Processes Transcription, Signaling and Metabolism of Cofactors and Vitamins and Enzyme Families were significantly enriched in the spring (Sun et al. 2016). The potential functions of the gut microbiome in different seasons of our study group will be discussed in Sect. 11.4.

11.3.2 Gut Fungal Microbiome Affected by Age, Sex, and Season

Although previous studies in humans have revealed that age and sex can influence the composition and diversity of the gut fungal microbiome (Strati et al. 2016), little information is available for wild living primates. Using LEfSe analysis, Sun et al. (2018) first reported that only one fungal genus *Sarocladium* was enriched in the old age group. By comparing males and females, the family Mycosphaerellaceae and genus *Devriesia* were particularly enriched in females, while the phylum Ascomycota and family Tetraplosphaeriaceae were enriched in males (Sun et al. 2018). These results indicated that the effect of sex and age on the composition of gut fungi was not as strong as that found in humans. In addition, we found evidence of marked seasonal variation in the composition of this macaque species' gut mycobiota. Fifteen taxa had significant variation across the four seasons. The abundant taxa in each season were as follows, Autumn: Wallemiaceae, Hypocreaceae, *Wallemia, Trichoderma*; Spring: Sclerotiniaceae, Nectriaceae, *Ciboria, Fusarium, Gibberella, Sarocladium* and *Talaromyces*; Summer: Trichocomaceae and *Penicillium;* Winter: *Devriesia* and Teratosphaeriaceae. This result is the first report to find strong seasonal influences on primate gut mycobiota, indicating the potential relationship between gut mycobiota and seasonal changes in host diet.

Different from previous studies in humans, no evidence indicates that age or sex strongly affect the alpha diversity of Tibetan macaque gut mycobiota (Sun et al. 2018). This result is likely because even individuals living in the same environment have differences in their dietary preferences. In addition, using PCoA and PERMANOVA tests, we found that there existed significant variation in beta diversity among different age and sex classes. Interestingly, seasonal change can significantly affect alpha and beta diversity, as well as the composition of the gut fungal microbiome.

11.4 Functions of the Gut Microbiome in Tibetan Macaque Feeding Ecology

Feeding ecology plays an important role in understanding the evolution of NHPs (Lambert 2011). With advances in sequencing technology, many studies have revealed that the gut microbiome aids in the digestion of some food resources, especially those rich in cellulose and hemicellulose, substances which are difficult to digest. Therefore, this important adaptive mechanism has attracted increasing attention in recent years. Tibetan macaques living in a highly seasonal ecosystem with strongly seasonal changes in rainfall, temperature, food resources, and home range shifts (Xiong and Wang 1988; Zhao 1999). Previous studies have suggested that these environmental factors are closely reflected in the composition of the gut microbiome (Amato et al. 2015; Chevalier et al. 2015; Maurice et al. 2015; Wu et al. 2017). Therefore, our study subjects may provide an important model for understanding the functions of the gut microbiome in primate feeding ecology, especially for primates living in highly seasonal ecosystems.

11.4.1 Gut Bacterial Microbiome and the Feeding Ecology of Tibetan Macaques

It is well known that the gut bacterial microbiome can help hosts to digest food resources more efficiently, and they also can change according to variation in host diet across time and space, helping the host to meet its nutrient and energy requirements (Hooper et al. 2002; Donohoe et al. 2011; Koren et al. 2012; Amato et al. 2014; Chevalier et al. 2015). As a species living in a highly seasonal ecosystem, Tibetan macaques depend mostly on a plant diet for their subsistence, including leaves, grass, roots, fruits, and flowers, and they usually shift home ranges to adapt to seasonal changes in food availability (Xiong and Wang 1988; Zhao 1999). The diverse and responsive gut bacterial community of wild-living Tibetan macaques can be considered an important adaptation for their foraging lifestyle. The presence of the relatively abundant fiber-degrading bacteria of Firmicutes and Bacteroidetes in their gut can help them to utilize the heavily plant-based diet. Some well-known fiber-degrading bacteria genera detected in the gut of Tibetan macaques, such as *Succinivibrio* and *Clostridium*, suggest that the gut bacterial microbiome plays an important role in the Tibetan macaques' feeding ecology.

In seasonal ecosystems, one of the most important questions is how animals adapt to seasonal changes in food composition and availability. From winter to spring, Tibetan macaques experience a switch from low quality food resources (mature leaves, roots and plant stem) to high quality food resources (young leaves and flowers). Generally, low-quality food resources are rich in cellulose and hemicellulose, and high-quality food resources are rich in pectin, carbohydrates and simple sugars. Based on a comparative study of the Tibetan macaque gut bacterial microbiome between winter and spring (Sun et al. 2016), the gut bacterial adaptive mechanism for seasonal food type's changes in the Tibetan macaque diet is summarized in Fig. 11.3.

During winter, the representative gut bacterial taxa Proteobacteria, *Succinivibrio* and *Clostridium* significantly increased, and the predicted metagenomes of glycan biosynthesis and metabolic pathways significantly increased in winter samples (Sun et al. 2016). The genus *Succinivibrio*, which usually is detected in rumen microbial ecosystems, ferment glucose efficiently by producing acetic acid and succinic acid (de Menezes et al. 2011), as well as being beneficial to the metabolism of different types of fatty acids (Van Dyke and McCarthy 2002). Another genus *Clostridium* contains many organisms which produce cellulose and hemicellulose-digestive enzymes (Van Dyke and McCarthy 2002; Zhu et al. 2011). The significant increase

Fig. 11.3 Food type change and adaptive shifts of the gut bacterial microbiome in Tibetan macaques. The predicted metagenomes of KEGG pathway and bacterial taxa of the Tibetan macaque gut microbiome, which significantly increased in winter, are beneficial for the digestion of cellulose and hemicellulose rich foods in winter, and the other bacteria significantly increased in spring, which are beneficial for the digestion of pectin, carbohydrate and simple sugar rich foods which are abundant in spring

of the genus *Clostridium* in winter samples is very beneficial for the digestion of cellulose and dietary fiber. In addition, evidence from black howler monkeys, humans and mice reveals that Proteobacteria are highly correlated with energy acquisition (Bryant and Small 1956; Koren et al. 2012; Amato et al. 2014; Chevalier et al. 2015). In response to cold weather, primates experience an increase in energy loss (Tsuji et al. 2013), and Tibetan macaques are no exception. The increase of Proteobacteria is beneficial for coping with cold weather. In conclusion, the gut bacterial microbiome in winter increases the efficiency of dietary fiber digestion in Tibetan macaques, which helps them to meet their energy requirements.

In addition, Sun et al. (2016) found that Firmicutes and *Prevotella* were significantly enriched in the Tibetan macaque gut during spring. During this time, the diet consists mainly of young leaves, flowers and bamboo shoots, foods that are richer in digestible pectin, carbohydrates and simple sugars, compared to mature leaves (Xiong and Wang 1988; Zhao 1999; You et al. 2013). The genus *Prevotella* is associated with the digestion of carbohydrates simple sugars, pectin and hemicellulose (Amato et al. 2014). The significant increase of genes related to carbohydrate metabolism and energy metabolism pathways in spring samples also indicates that the gut bacterial community helps break down and make accessible these high-quality food resources, beneficial for macaques to recover from energy loss experienced during the cold conditions of winter.

11.4.2 Gut Fungal Microbiome and Feeding Ecology of Tibetan Macaques

Compared with the gut bacterial microbiome, there is very little data on the primate gut fungal microbiome, and the role of gut fungi in primate foraging ecology is unknown. Sun et al. (2018) first reported the diversity and composition of Tibetan macaque gut mycobiota. Three genera *Aspergillus*, *Penicillium* and *Fusarium* were detected as normal inhabitants in the gut. The two most dominant genera, were *Aspergillus* (12.46%) and *Penicillium* (10.72%). It has been reported that the anaerobic species of these two genera can produce cellulolytic and hemicellulolytic enzymes, which are beneficial to cellulosic biomass degradation (Boots et al. 2013; Liao et al. 2014; Shuji et al. 2014; Solomon et al. 2016; Trinci et al. 1994). Similarly, as robust cellulose and hemicellulose degrader, the genus *Fusarium* was reported in a recent study (Huang et al. 2015). At certain times of the year, Tibetan macaques depend on a plant diet, high in cellulose and hemicellulose (Xiong and Wang 1988; You et al. 2013; Campbell et al. 2001). Our previous studies indicated that the dominant genera of Tibetan macaque gut mycrobiota may play an important role in the digestion of these and other plant items.

In addition, the gut fungal microbiome of Tibetan macaques changed with the seasonal change in dietary food content. Fifteen taxa were detected to be significantly enriched in one of the four seasons (Sun et al. 2018). This result indicates that the gut mycrobiota can response to dietary shifts. It has been proposed that the gut mycobiota response to seasonal dietary shift is beneficial to the hosts' nutritional and reproductive needs (Noma et al. 1998; Tsuji et al. 2013). However, knowledge about the functions of the particular fungi in the mammalian gut is limited (Milton and May 1976), as are genomic databases and metabolic maps of gut fungi (Huffnagle and Noverr 2013), making it difficult to explain the relationship between mycobiota composition and primate feeding ecology. Based on the available information about the genus *Penicillium* (Shuji et al. 2014), we hypothesize that this summer season enriched genus may aid in the digestion of mature leaves, which contain a higher proportion of cellulose and hemicellulose. However, a possible explanation as to why this genus was not enriched during winter could be that the ingestion of fallen nuts in winter caused the amount of *Penicillium* to decrease due to some chemical interaction (Maria et al. 2014). Secondly, it has been reported that high diversity of gut mycobiota can improve the utilization efficiency of plant fiber consumption (Bauchop 1981; Akin et al. 1983; Denman and Mcsweeney 2010). Thus, the gut mycobiota with high alpha diversity during the winter, reported in our previous study, may be beneficial for Tibetan macaques to digest dietary fiber. In conclusion, the gut fungal mycobiota detected in Tibetan macaque provides evidence for the role of gut fungi in primate foraging ecology. More attention needs to be paid to this in future studies.

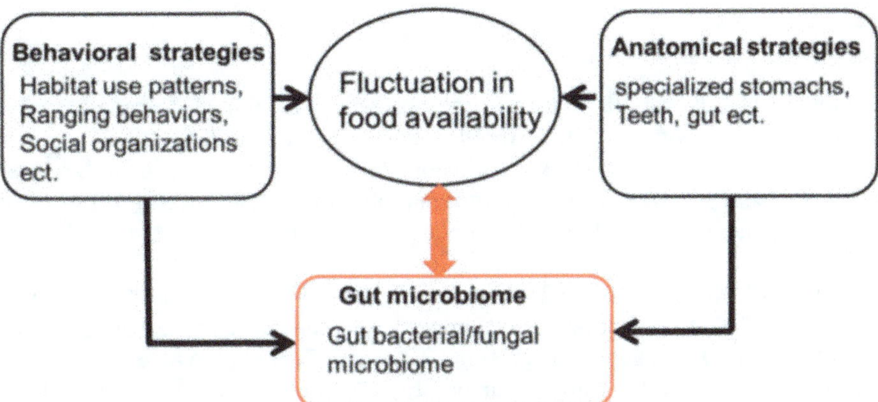

Fig. 11.4 Adaptive mechanisms of wild primates in response to fluctuations in food availability and food quality

11.5   Conclusions and Future Directions

Our studies of the Tibetan macaque gut microbiome reviewed in this chapter add important baseline data for the Tibetan macaque to our knowledge of the mammalian gut microbiome regarding difference in host age, sex and its role in feeding ecology, as well as the role of the gut microbiome in the adaptive radiation of primates. Together with anatomical and behavioral strategies, the gut microbiome is believed to be important factor in the successful evolution of NHPs, particularly as an adaption to the temporal fluctuation in food resource types and their availability. The present data on the wild Tibetan macaque gut microbiome leads us to hypothesize that gut microbiota played an important role in the adaptive radiation of primates, and other animals, which can help them to solve foraging dilemmas in seasonally fluctuating environments (Fig. 11.4). Further research is necessary to test this hypothesis on more mammalian species. In addition, mammals living in a microbial environment, and the microorganisms in the soil and plant foods can also change with seasonal fluctuations. The link between the gut microbiome and environmental microbiomes remain to be elucidated.

Acknowledgments Many thanks to the Huangshan Garden Forest Bureau for their permission and support of this work. We also gratefully acknowledge Paul A. Garber, Lori K. Sheeran and all members of our research group for their excellent work on the gut microbiome of Tibetan macaques. This chapter was supported by grants from the National Natural Science Foundation of China (No. 31870371, 31400330, 31172106), the Special Foundation for Excellent Young Talents in University of Anhui Province, China (No. 2012SQRL018ZD) and the Initial Funding for Doctoral Research in Anhui University (No. J01003229).

References

Akin DE, Gordon GL, Hogan JP (1983) Rumen bacterial and fungal degradation of *Digitaria pentzii* grown with or without sulfur. Appl Environ Microbiol 46:738–748

Amato KR, Leigh SR, Angela K, Mackie RI, Yeoman CJ, Stumpf RM, Wilson BA, Nelson KE, White BA, Garber PA (2014) The role of gut microbes in satisfying the nutritional demands of adult and juvenile wild, black howler monkeys (*Alouatta pigra*). Am J Phys Anthropol 155:652–664

Amato KR, Leigh SR, Kent A, Mackie RI, Yeoman CJ, Stumpf RM, Wilson BA, Nelson KE, White BA, Garber PA (2015) The gut microbiota appears to compensate for seasonal diet variation in the wild black howler monkey (*Alouatta pigra*). Microb Ecol 69:434–443

Bauchop T (1981) The anaerobic fungi in rumen fibre digestion. Agric Environ 6:339–348

Bokulich NA, Mills DA (2013) Improved selection of internal transcribed spacer-specific primers enables quantitative, ultra-high-throughput profiling of fungal communities. Appl Environ Microbiol 79(8):2519–2526

Bolnick DI, Snowberg LK, Hirsch PE, Lauber CL, Parks B, Lusis AJ, Knight R, Caporaso JG, Svanbäck R (2014) Individual diet has sex-dependent effects on vertebrate gut microbiota. Nat Commun 5:4500

Boots B, Lillis L, Clipson N, Petrie K, Kenny DA, Boland TM, Doyle E (2013) Responses of anaerobic rumen fungal diversity (phylum Neocallimastigomycota) to changes in bovine diet. J Appl Microbiol 114:626–635

Bryant MP, Small NJ (1956) Characteristics of two new genera of anaerobic curved rods isolated from the rumen of cattle. J Bacteriol 72:22–26

Burrows AM (2010) The evolution of exudativory in primates. Springer, New York

Cachel S (1989) Primate adaptation and evolution. Int J Primatol 10:487–490

Campbell JL, Glenn KM, Grossi B, Eisemann JH (2001) Use of local North Carolina browse species to supplement the diet of a captive colony of folivorous primates (*Propithecus* sp.). Zoo Biol 20:447–461

Campera M, Serra V, Balestri M, Barresi M, Ravaolahy M, Randriatafika F, Donati G (2014) Effects of habitat quality and seasonality on ranging patterns of collared brown lemur (*Eulemur collaris*) in littoral forest fragments. Int J Primatol 35:957–975

Chevalier C, Stojanović O, Colin DJ, Suarez-Zamorano N, Tarallo V, Veyrat-Durebex C, Rigo D, Fabbiano S, Stevanović A, Hagemann S (2015) Gut microbiota orchestrates energy homeostasis during Cold. Cell 163:1360–1374

Chivers DJ, Wood BA, Bilsborough A (1984) Food acquisition and processing in primates: concluding discussion. In: Chivers DJ, Wood BA, Alan B (eds) Food acquisition and processing in primates. Springer, New York, pp 545–556

Clayton JB, Gomez A, Amato K, Knights D, Travis DA, Blekhman R, Knight R, Leigh S, Stumpf R, Wolf T (2018) The gut microbiome of nonhuman primates: lessons in ecology and evolution. Am J Primatol 80:e22867

Coley PD (1983) Herbivory and defensive characteristics of tree species in a lowland tropical forest. Ecol Monogr 53:209–229

de Menezes AB, Lewis E, O'Donovan M, O'Neill BF, Clipson N, Doyle EM (2011) Microbiome analysis of dairy cows fed pasture or total mixed ration diets. FEMS Microbiol Ecol 78:256–265

Degnan PH, Pusey AE, Lonsdorf EV, Goodall J, Wroblewski EE, Wilson ML, Rudicell RS, Hahn BH, Ochman H (2012) Factors associated with the diversification of the gut microbial communities within chimpanzees from Gombe National Park. Proc Natl Acad Sci U S A 109:13034–13039

Denman S, Mcsweeney C (2010) Development of a real-time PCR assay for monitoring anaerobic fungal and cellulolytic bacterial populations within the rumen. FEMS Microbiol Ecol 58:572–582

Donohoe DR, Nikhil G, Xinxin Z, Wei S, O'Connell TM, Bunger MK, Bultman SJ (2011) The microbiome and butyrate regulate energy metabolism and autophagy in the mammalian colon. Cell Metab 13:517–526

Fan P, Ni Q, Sun G, Bei H, Jiang X (2009) Gibbons under seasonal stress: the diet of the black crested gibbon (*Nomascus concolor*) on Mt. Wuliang, Central Yunnan, China. Primates 50:37–44

Greenblum S, Turnbaugh PJ, Borenstein E (2012) Metagenomic systems biology of the human gut microbiome reveals topological shifts associated with obesity and inflammatory bowel disease. PNAS 109:594–599

Hajishengallis G, Liang S, Payne MA, Hashim A, Jotwani R, Eskan MA, Mcintosh ML, Alsam A, Kirkwood KL, Lambris JD (2011) Low-abundance biofilm species orchestrates inflammatory periodontal disease through the commensal microbiota and complement. Cell Host Microbe 10 (5):497–506

Hanya G, Chapman CA (2013) Linking feeding ecology and population abundance: a review of food resource limitation on primates. Ecol Res 28:183–190

Hanya G, Qarro M, Tattou MI, Fuse M, Vallet D, Yamada A, Go M, Takafumi H, Tsujino R, Agetsuma N (2011) Dietary adaptations of temperate primates: comparisons of Japanese and Barbary macaques. Primates 52:187–198

Hird SM (2017) Evolutionary biology needs wild microbiomes. Front Microbiol 8:725

Hoffmann C, Dollive S, Grunberg S, Chen J, Li H, Wu GD, Lewis JD, Bushman FD (2013) Archaea and fungi of the human gut microbiome: correlations with diet and bacterial residents. PLoS One 8:e66019

Hooper LV, Gordon JI (2001) Commensal host-bacterial relationships in the gut. Science 292 (5519):1115–1118

Hooper LV, Midtvedt T, Gordon JI (2002) How host-microbial interactions shape the nutrient environment of the mammalian intestine. Annu Rev Nutr 22:283–307

Hooper LV, Littman DR, Macpherson AJ (2012) Interactions between the microbiota and the immune system. Science 336:1268–1273

Huang Y, Busk PK, Lange L (2015) Cellulose and hemicellulose-degrading enzymes in *Fusarium* commune transcriptome and functional characterization of three identified xylanases. Enzyme Microb Technol 73–74:9–19

Huffnagle GB, Noverr MC (2013) The emerging world of the fungal microbiome. Trends Microbiol 21:334–341

Iliev ID, Underhill DM (2012) Interactions between commensal fungi and the C-type lectin receptor Dectin-1 influence colitis. Science 336:1314–1317

Koren O, Goodrich J, Cullender T, Spor A, Laitinen K, Bäckhed HK, Gonzalez A, Werner J, Angenent L, Knight R (2012) Host remodeling of the gut microbiome and metabolic changes during pregnancy. Cell 150:470–480

Lambert JE (2011) Primate nutritional ecology: feeding biology and diet at ecological and evolutionary scales. In: Campbell C, Fuentes A, MacKinnon KC, Panger M, Bearder S, Stumpf R (eds) Primates in perspective, 2nd edn. Oxford University Press, Oxford, pp 512–522

Ley RE, Micah H, Catherine L, Turnbaugh PJ, Rob Roy R, J Stephen B, Schlegel ML, Tucker TA, Schrenzel MD, Rob K (2008) Evolution of mammals and their gut microbes. Science 322 (5905):1188

Liao H, Li S, Wei Z, Shen Q, Xu Y (2014) Insights into high-efficiency lignocellulolytic enzyme production by Penicillium oxalicum GZ-2 induced by a complex substrate. Biotechnol Biofuels 7(1):162

Mackie RI (2002) Mutualistic fermentative digestion in the gastrointestinal tract: diversity and evolution. Integr Comp Biol 42:319–326

Mar RM, Pérez D, Javier CF, Esteve E, Maringarcia P, Xifra G, Vendrell J, Jové M, Pamplona R, Ricart W (2016) Obesity changes the human gut mycobiome. Sci Rep 6:21679

Margaret MFN, Hadfield MG, Bosch TCG, Carey HV, Tomislav DLO, Douglas AE, Nicole D, Gerard E, Tadashi F, Gilbert SF (2013) Animals in a bacterial world, a new imperative for the life sciences. PNAS 110:3229–3236

Maria U, Xiaoyu W, Baer DJ, Novotny JA, Marlene F, Volker M (2014) Effects of almond and pistachio consumption on gut microbiota composition in a randomised cross-over human feeding study. Br J Nutr 111:2146–2152

Maurice CF, Cl KS, Ladau J, Pollard KS, Fenton A, Pedersen AB, Turnbaugh PJ (2015) Marked seasonal variation in the wild mouse gut microbiota. ISME J 9:2423–2434

McCord AI, Chapman CA, Weny G, Tumukunde A, Hyeroba D, Klotz K, Koblings AS, Mbora DN, Cregger M, White BA (2014) Fecal microbiomes of non-human primates in Western Uganda reveal species-specific communities largely resistant to habitat perturbation. Am J Primatol 76:347–354

Mckenney EA, Melissa A, Lambert JE, Vivek F (2015) Fecal microbial diversity and putative function in captive western lowland gorillas (*Gorilla gorilla gorilla*), common chimpanzees (*Pan troglodytes*), Hamadryas baboons (*Papio hamadryas*) and binturongs (*Arctictis binturong*). Integr Zool 9:557–569

Milton K, May ML (1976) Body weight, diet and home range area in primates. Nature 259:459–462

Noma N, Suzuki S, Izawa K (1998) Inter-annual variation of reproductive parameters and fruit availability in two populations of Japanese macaques. Primates 39:313–324

Ochman H, Worobey M, Kuo CH, Ndjango JBN, Peeters M, Hahn BH, Hugenholtz P (2010) Evolutionary relationships of wild hominids recapitulated by gut microbial communities. PLoS Biol 8:8

Qin J, Li R, Raes J, Arumugam M, Burgdorf KS, Manichanh C, Nielsen T, Pons N, Levenez F, Yamada T (2010) A human gut microbial gene catalogue established by metagenomic sequencing. Nature 464(7285):59–65

Qiu X, Zhang F, Yang X, Wu N, Jiang W, Li X, Li X, Liu Y (2015) Changes in the composition of intestinal fungi and their role in mice with dextran sulfate sodium-induced colitis. Sci Rep 5:10416

Ren T, Grieneisen LE, Alberts SC, Archie EA, Wu M (2015) Development, diet and dynamism: longitudinal and cross-sectional predictors of gut microbial communities in wild baboons. Environ Microbiol 18(5):1312–1325

Rizzetto L, De FC, Cavalieri D (2014) Richness and diversity of mammalian fungal communities shape innate and adaptive immunity in health and disease. Eur J Immunol 44:3166–3181

Rodman PS, Cant JGH (1984) Anatomy and behavior. (Book reviews: adaptations for foraging in nonhuman primates). Science 226:1187–1188

Romani L (2011) Immunity to fungal infections. Nat Rev Immunol 11:275–288

Sharon G, Turnbaugh PJ, Elhanan B (2012) Metagenomic systems biology of the human gut microbiome reveals topological shifts associated with obesity and inflammatory bowel disease. PNAS 109:594–599

Shuji T, Takashi K, Tetsuo K (2014) Complex regulation of hydrolytic enzyme genes for cellulosic biomass degradation in filamentous fungi. Appl Microbiol Biotechnol 98:4829–4837

Sokol H, Leducq V, Aschard H, Pham HP, Jegou S, Landman C, Cohen D, Liguori G, Bourrier A, Nionlarmurier I (2017) Fungal microbiota dysbiosis in IBD. Gut 66:1039–1048

Solomon KV, Haitjema CH, Henske JK, Gilmore SP, Borges-Rivera D, Lipzen A, Brewer HM, Purvine SO, Wright AT, Theodorou MK (2016) Early-branching gut fungi possess a large, comprehensive array of biomass-degrading enzymes. Science 351:1192–1195

Starr C, Nekaris KAI, Leung L (2012) Hiding from the moonlight: luminosity and temperature affect activity of Asian nocturnal primates in a highly seasonal forest. PLoS One 7:e36396.1–e36396.8

Strati F, Paola MD, Stefanini I, Albanese D, Rizzetto L, Lionetti P, Calabrò A, Jousson O, Donati C, Cavalieri D (2016) Age and gender affect the composition of fungal population of the human gastrointestinal tract. Front Microbiol 7:1227

Sun B, Xi W, Bernstein S, Huffman MA, Xia DP, Gu Z, Rui C, Sheeran LK, Wagner RS, Li J (2016) Marked variation between winter and spring gut microbiota in free-ranging Tibetan Macaques (*Macaca thibetana*). Sci Rep 6:26035

Sun B, Gu Z, Wang X, Huffman MA, Garber PA, Sheeran LK, Zhang D, Zhu Y, Xia DP, Li JH (2018) Season, age, and sex affect the fecal mycobiota of free-ranging Tibetan macaques (*Macaca thibetana*). Am J Primatol 80(7):e22880

Trinci APJ, Davies DR, Gull K, Lawrence MI, Nielsen BB, Rickers A, Theodorou MK (1994) Anaerobic fungi in herbivorous animals. Mycol Res 98:129–152

Tsuji Y, Hanya G, Grueter CC (2013) Feeding strategies of primates in temperate and alpine forests: comparison of Asian macaques and colobines. Primates 54:201–215

Tung J, Barreiro LB, Burns MB, Grenier JC, Lynch J, Grieneisen LE, Altmann J, Alberts SC, Blekhman R, Archie EA (2015) Social networks predict gut microbiome composition in wild baboons. elife 4:e05224

Turnbaugh PJ, Ley RE, Mahowald MA, Magrini V, Mardis ER, Gordon JI (2006) An obesity-associated gut microbiome with increased capacity for energy harvest. Nature 444:1027–1031

Underhill DM, Iliev ID (2014) The mycobiota: interactions between commensal fungi and the host immune system. Nat Rev Immunol 14:405–416

Ungar PS (1995) Fruit preferences of four sympatric primate species at Ketambe, Northern Sumatra, Indonesia. Int J Primatol 16:221–245

Van Dyke M, McCarthy A (2002) Molecular biological detection and characterization of Clostridium populations in municipal landfill sites. Appl Environ Microbiol 68:2049–2053

Wheeler ML, Limon JJ, Bar AS, Leal CA, Gargus M, Tang J, Brown J, Funari VA, Wang HL, Crother TR (2016) Immunological consequences of intestinal fungal dysbiosis. Cell Host Microbe 19(6):865–873

Wu Q, Wang X, Ding Y, Hu Y, Nie Y, Wei W, Ma S, Yan L, Zhu L, Wei F (2017) Seasonal variation in nutrient utilization shapes gut microbiome structure and function in wild giant pandas. Proc Biol Sci 284:20170955

Xiong C, Wang Q (1988) Seasonal habitat used by Thibetan Monkeys. Acta Theriol Sinica 8:176–183

Xu B, Xu W, Yang F, Li J, Yang Y, Tang X, Mu Y, Zhou J, Huang Z (2013) Metagenomic analysis of the pygmy loris fecal microbiome reveals unique functional capacity related to metabolism of aromatic compounds. PLoS One 8(2):e56565

Xu B, Xu W, Li J, Dai L, Xiong C, Tang X, Yang Y, Mu Y, Zhou J, Ding J (2015) Metagenomic analysis of the *Rhinopithecus bieti* fecal microbiome reveals a broad diversity of bacterial and glycoside hydrolase profiles related to lignocellulose degradation. BMC Genomics 16:174

Yamada A, Muroyama Y (2010) Effects of vegetation type on habitat use by crop-raiding Japanese macaques during a food-scarce season. Primates 51:159

Yatsunenko T, Rey FE, Manary MJ, Trehan I, Dominguez-Bello MG, Contreras M, Magris M, Hidalgo G, Baldassano RN, Anokhin AP (2012) Human gut microbiome viewed across age and geography. Nature 486(7402):222–227

Yokoe Y, Yasumasu I (1964) The distribution of cellulase in invertebrates. Comp Biochem Physiol 13:323–338

You SY, Yin HB, Zhang SZ, Tian-Ying JI, Feng XM (2013) Food habits of *Macaca thibetana* at Mt. Huangshan,China. J Biol 30(5):64–67

Zhang M, Li J, Zhu Y, Wang X, Wang S (2010) Male mate choice in Tibetan macaques Macaca thibetana at Mt. Huangshan, China. Curr Zool 56:213–221

Zhao Q-K (1999) Responses to seasonal changes in nutrient quality and patchiness of food in a multigroup community of Tibetan macaques at Mt. Emei. Int J Primatol 20:511–524

Zhou Q, Wei F, Huang C, Li M, Ren B, Luo B (2007) Seasonal variation in the activity patterns and time budgets of *Trachypithecus francoisi* in the Nonggang Nature Reserve, China. Int J Primatol 28:657–671

Zhu L, Wu Q, Dai J, Zhang S, Wei F (2011) Evidence of cellulose metabolism by the giant panda gut microbiome. PNAS 108:17714–17719

Medicinal Properties in the Diet of Tibetan Macaques at Mt. Huangshan: A Case for Self-Medication

Michael A. Huffman, Bing-Hua Sun, and Jin-Hua Li

12.1 Introduction

Life history strategies include growth, maintenance, and reproduction (Gadgil and Bossert 1970), all of which are dependent upon a proper diet for metabolic functions. Animal feeding strategies are based on finding and consuming a balance of the most essential nutritional elements, carbohydrates, fats, proteins, trace elements, and vitamins, while at the same time avoiding the negative impacts of secondary metabolites in plants (Lambert 2011; Simpson et al. 2004). These secondary metabolites protect plants from predation by an array of insect and vertebrate herbivores that prey upon them by reducing palatability and digestibility (Freeland and Janzen 1974; Glander 1982; Rosenthal and Berenbaum 1992). Nonetheless, this does not always inhibit animals from ingesting such plants in tolerable amounts for purposes other than nutrition.

The idea that animals may ingest plants for their medicinal value was first suggested by Janzen (1978), based on a variety of anecdotal reports from the wild. The study of primate self-medication then began in earnest as a scientific discipline in the mid-to-late 1980s with observations of chimpanzees in the wild (see Huffman 2015). Nonetheless, it is widely documented that humans have traditionally seen animals as a source of knowledge about the use of plants for their medicinal value

M. A. Huffman (✉)
Primate Research Institute, Kyoto University, Kyoto, Japan
e-mail: huffman.michael.8n@kyoto-u.ac.jp

B.-H. Sun
School of Resources and Environmental Engineering, Anhui University, Hefei, Anhui, China

J.-H. Li
School of Resources and Environmental Engineering, Anhui University, Hefei, Anhui, China

International Collaborative Research Center for Huangshan Biodiversity and Tibetan Macaque Behavioral Ecology, Anhui, China

School of Life Sciences, Hefei Normal University, Hefei, Anhui, China
e-mail: jhli@ahu.edu.cn

J.-H. Li et al. (eds.), *The Behavioral Ecology of the Tibetan Macaque*, Fascinating Life Sciences, https://doi.org/10.1007/978-3-030-27920-2_12

(Engel 2002; Huffman 1997, 2002, 2007). Humans can learn from watching sick wild primates, because we share the same evolutionary history, possess a common physiology, and have lived together under similar environmental conditions for much of our species' history. It has been argued that we have inherited many of the same ways to combat common diseases in the environment (Huffman 2016). Indeed, recent archeological and biochemical evidence supports the idea by showing that one of our closest extinct ancestors, *Homo neanderthalensis*, also used medicinal plants still widely in use today by modern humans (Hardy et al. 2012, 2013; Huffman 2016).

Self-medication research focuses on understanding how animals respond to illness and how these behaviors can be transmitted across generations (Huffman 1997). It has also been argued to be a bio-rational for the exploration and exploitation of novel secondary plant compounds and new insights into how they can be used for the management of health in humans and livestock (Huffman et al. 1998; Krief et al. 2005; Petroni et al. 2016). At the proximate level, self-medication may be driven by the necessity to maintain physiological homeostasis to stay in relatively good condition (Foitova et al. 2009; Forbey et al. 2009).

Currently, the majority of evidence for self-medication in animals comes from the study of how they deal with parasite or pathogen-induced illness (see Huffman 1997, 2011). While some parasitic infections likely go unnoticed, when homeostasis is disrupted or threatened, it is expected to be in the best interest of the host to actively respond in ways to alleviate discomfort. However, there is no reason why self-medication should be limited to parasitosis, since animals are faced with a wide variety of health homeostatic challenges brought upon by such factors as reproductive events, climatic extremes, or other seasonal events (Carrai et al. 2003; Huffman 1997, 2011; Ndagurwa 2012). The ability of a species to defend itself against life-threatening conditions provides a significant adaptive advantage and thus should be present throughout the entire animal kingdom.

In 1997, the concept of "medicinal foods" was formally introduced to primatology, adding the extra element of passive prevention of disease based on the presence of plants in the diet that contain noticeable bioactive, physiology-modifying properties, from which the animals ingesting them could potentially benefit (Huffman 1997). This term was borrowed from the human ethnopharmacological literature (Etkin 1996). For example, among the Hausa of Nigeria, 30% of the wild plant food species they ingest are also used as medicine. Interestingly, of the species used by these people to treat symptoms of malaria, 89% are also eaten as food (Etkin and Ross 1983). Many food items eaten by primates and other mammals have also been shown to contain a variety of secondary metabolites with medicinal properties (roughly 15–25% of any population's food plant species list), suggesting that animals may benefit from the periodic ingestion, in small amounts of these plants (sifaka *Propithecus verreauxi verreauxi*, Carrai et al. 2003; gorillas *Gorilla gorilla* and *G. beringei*, Cousins and Huffman 2002; chimpanzees *Pan troglodytes*, Huffman 1997, 2003; Japanese macaques *Macaca fuscata*, Huffman and MacIntosh 2012; MacIntosh and Huffman 2010; various ungulate species, Mukherjee et al. 2011; lemurs *Eulemur fulvus*, Negre et al. 2006; wooly spider monkeys *Brachyteles arachnoides*, Petroni et al. 2016). The secondary compound rich content of some foods in the diet may play a significant role in the maintenance of health. Two examples associated with risks to parasite infection illustrate this point.

The medicinal diet of chimpanzees in the Mahale M group of Tanzania was examined (Huffman 1998) by conducting a database search using the African ethnomedicine literature. From 172 chimpanzee food species, 43 (22%) items were found to be used to treat parasitic- or gastrointestinal-related illnesses by humans. It was also common for some species to have multiple ethnomedicinal uses. While not all 43 species may have been ingested by chimpanzees in such a way as to benefit from these potential medicinal properties, 33% (20/63) of the plant parts ingested (leaf and stem = 75%, bark 15%, seed = 5%, fruit = 5%) from 16 of these species corresponded to the parts utilized by humans specifically for the treatment of intestinal parasites and gastrointestinal illness.

The medicinal diet and parasite richness of ten Japanese macaque troops were examined (MacIntosh and Huffman 2010). The study troops were selected to represent the species entire distribution, ranging from the extreme cold temperate zone of Shimokita peninsula down to the subtropical island of Yakushima. A total of 1664 plant part items (range, 56–408) from 694 species were the target of an extensive literature search for potential antiparasitic activity in the diet. Of all these ingested items, 198 (from 135 species) were found to have reported antiparasitic properties. The proportion of these antiparasitic items ranged from 12 to 18% across these ten troops. A further 167 plant items (133 species) exhibited medicinal properties not related to parasitic infection or gastrointestinal symptoms. Because nematode species richness is negatively associated with latitude among Japanese macaques (Gotoh 2000), it was predicted that the proportion of antiparasitic items in their diets would also follow the same pattern (MacIntosh and Huffman 2010). A tendency was noted for the proportion of antiparasitic food items to decrease with increasing latitude, and a strong positive statistically significant relationship between the proportions of antiparasitic items ingested and nematode species richness was found. The proportion of medicinal items unrelated to parasite activity showed no such relationship with either latitude or with nematode species richness, supporting the hypothesis that the medicinal diet was somehow influenced by the degree of parasite pressure, relative species richness, and potentially parasite load.

From these and other studies noted above, a pattern is emerging, by which primates and other mammals incorporate certain food items with medicinal properties into their diet. In primates the information thus far shows a strong connection between medicinal food consumption and parasite infection and or reproductive events. Both factors present significant challenges to the survival and fitness of an individual. Knowing what immediate homeostasis challenges that impact individuals of a group can help to better understand how the medicinal components of their diet may work in an animal's favor.

Currently, limited details are available about health and diseases affecting Tibetan macaques (*Macaca thibetana*). In a case study of the troop fission event of YA and YB, the sudden mass death of 17 individuals over a 1-week period was mentioned (Li et al. 1996). Disease was suggested to be responsible, but no diagnosis was given. The speed and widespread effect could be suggestive of an aggressive viral infection, whose virulence was perhaps exacerbated by high stress levels.

Zhu et al. (2012) report the presence of nine intestinal nematode species in Tibetan macaques, with special note of the zoonotic *Gongylonema pulchrum*, having the highest prevalence rate of 31.58%, followed by *Trichuris trichiura* (25.00%), *Oesophagostomum apiostomum* (23.68%), *Ancylostoma duodenale* (hookworm) (14.47%), *Trichostrongylus* sp. (13.16%), and other species of lower prevalence. These parasites are known to be responsible for mild to severe pathogenesis (Brack 1987). Among them, *O. apiostomum* is noted to be responsible for perhaps the severest pathogenesis, in particular among previously infected (pre-immunized) individuals. In these cases, encysted larvae are trapped in the intestinal mucosa by elevated immune response, causing the larvae to die inside the cyst, leading to inflammation, necrosis, and hemorrhaging, and in severe cases leading to secondary bacterial infections, necrosis of the intestinal mucosa, weight loss, weakness, and mortality (Brack 2008). *O. stephanostomum*, a sister species infecting chimpanzees, and other great apes, is equally pathogenic and is associated with self-medicative behaviors used in the therapeutic treatment with *Vernonia amygdalina* by chimpanzees with high-level infections during the rainy season (Huffman 1997; Huffman and Caton 2001). Furthermore, evidence for the possible role of medicinal foods for controlling *O. stephanostomum* and other infections has also been suggested (Huffman 1997).

While more work is clearly needed to understand the different factors affecting the homeostasis of Tibetan macaques (*Macaca thibetana*) possibly leading to self-medication, we take this opportunity to evaluate their diet for its potential medicinal value. We predict there is a proportion of the diet containing plants with medicinal value, and it is our goal to highlight that potential for future research, in order to better understand the role of diet in health maintenance and self-medication in Tibetan macaques.

12.2 Materials and Methods

The subjects of this investigation are Tibetan macaques living in the Valley of the Monkeys (118°11′ W, 30°29′ N), Mt. Huangshan, Anhui Province, China. The site is situated at an elevation of approximately 1840 m above sea level, and the year is divided into four seasons: winter (December to February), spring (March to May), summer (June to August), and autumn (September to November). Average temperatures range from around 0 °C in mid-winter to around 25 °C in summer (Fig. 12.1). Peaks in rainfall occur during the months of May through July, tapering off in autumn and winter. (Further details of the study site are presented in other chapters of this book.)

Analysis of the diet of Yulinkeng 1 (YA1) troop was based on a plant food list previously published by You et al. (2013). At the time of that study, the troop consisted of 28 individuals: 12 adults (4 males, 8 females), 10 juveniles (6 males,

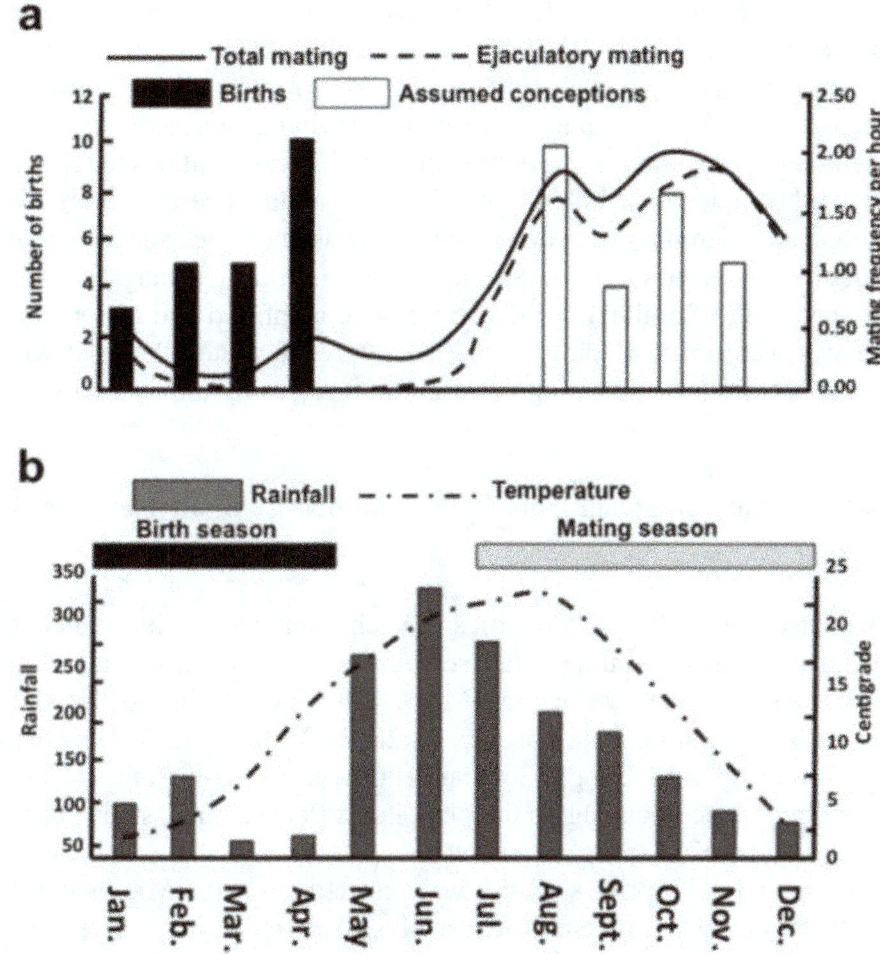

Fig. 12.1 Annual distribution of mating, births, and conceptions of Tibetan macaques (**a**) and monthly rainfall and temperature (**b**) at Mt. Huangshan. These figures were redrawn, modified, and combined from previously published material for illustrative purposes (Li et al. 2005). The meteorological data presented here is the average of 10 years (1983–1992)

4 females), and 6 infants (5 males, 1 female). Direct observations of the troop's feeding behavior in the rugged terrain of Mt. Huangshan are challenging, and to overcome these difficulties, the investigators produced a food list based on the analysis of 81 fecal samples collected over 10-day blocks, one block each during winter, spring, summer, and autumn between November 2011 and October 2012 (You et al. 2013). While the study covered just 1 year, it provides an important basis for an evaluation of the medicinal potential of food items ingested by the troop during that period.

The method used to quantify the relative contribution of each plant species (part) took a microhistopathological approach to identify ingested food remains in the feces. The method is a validated, well-established, procedure developed by Sparks and Malechek (1968) and has been used to quantify the botanical composition of diets in a variety of free-ranging, sometimes difficult to directly monitor, domestic and wild animals in their natural habitats (e.g., mule deer *Odocoileus hemionus*,

Anthony and Smith 1974; musk deer *Moschus leucogaster*, Green 1987; domestic goats *Capra aegagrus hircus*, Mellado et al. 1991; cattle *Bos taurus*, sheep *Ovis aries*, Angora goats, Alipayo et al. 1992; wild turkeys *Meleagris gallopavo*, Rumble and Anderson 1993; Yunnan snub-nosed monkeys *Rhinopithecus bieti*).

The method requires the assembly of histopathological plant tissue reference key slides prepared from identified plant species in the habitat. These keys are then used to microscopically identify plant parts in the feces based on each species' unique cell tissue structure characteristics (see Ahmed et al. 2015). At Huangshan, a total of 224 plant species (73 families) in the habitat were identified and histopathological reference keys were made (You et al. 2013). They calculated the relative density (RD) of each species (and family) present in the feces using the following formula, where:

$$RD = \text{the density of a plant particle/the total density of all plant particles} \times 100.$$

We evaluated the medicinal properties of each plant species in the resulting list for medicinal value using online database sources [Traditional Chinese Medicinal Plants (Duke and Ayensu 1985); Find Me A Cure (2018); Herbpathy Data Base (2018); Plants For A Future Data Base (1996–2012); World Agroforestry Data Base (2018)], followed up with Google Scholar article searches by plant species and/or active compound name, focusing on the plant items (leave, fruits, stems, bark, roots, flowers) ingested by YA1 troop members.

Data sorting and descriptive statistics were carried out using Microsoft® Excel® for Mac 2011 (Ver. 14.7.2.). Statistical analyses for chi-square were carried out using an online calculator: https://www.socscistatistics.com/tests/chisquare2/Default2.aspx. Statistical significance was set at 0.05.

12.3 Results and Discussion

12.3.1 Plant Food Species and Their Relative Density (RD) Values

To put the medicinal foods ingested by members of the troop into perspective, we first describe the overall trends of their feeding habits as revealed in the dietary analysis. The plant food species consumed by YA1 troop members are listed in descending order of RD by family and species in Table 12.1. The species RD values are summarized seasonally, with each season totaling 100%. Fifty species (61 different items, 26 families) from across the entire study period were analyzed here. Of the 61 items ingested, leaves accounted for 78%, by far the largest proportion of plant parts found, followed by fruits 11%, stems 3%, buds 2%, seeds 2%, young shoots 2%, and flowers 2%.

Table 12.1 Plant food species, part(s) eaten, and seasonal variation of use in the YA1 troop of Tibetan macaques at Mt. Huangshan

Family and species	Parts Consumed	Relative density (RD)			
		Winter	Spring	Summer	Autumn
Fagaceae					
*Castanpopsis eyre*i (Champion ex Bentham) Tutcher	Leaf, fruit	14.78%	9.17%	10.49%	14.48%
Lithocarpus glaber (Thunb.) Nakai	Leaf, fruit	6.60%	6.62%	4.27%	6.46%
Quercus myrsinaefolia Blume	Leaf, fruit	6.06%	7.38%	7.41%	9.37%
Quercus glauca Thunb.	Leaf, fruit	2.50%	3.12%	5.05%	7.69%
Quercus glandulifera Blume	Leaf, fruit	0	1.22%	5.41%	2.08%
Quercus aliena Blume	Leaf, fruit	0	0	0	0.60%
Lauraceae					
Litsea coreana H. Léveillé	Leaf	12.09%	8.08%	9.55%	4.66%
Machilus leptophylla Handel-Mazzetti	Leaf	4.53%	2.53%	6.84%	7.11%
Phoebe sheareri (Hemsley) Gamble in Sargent	Leaf	4.04%	2.88%	5.12%	5.55%
Machilus thunbergii Siebold & Zuccarini	Leaf, bud	0.66%	13.50%	2.20%	1.68%
Lindera aggregata (Sims) Kostermans	Leaf	0	0.17%	0.30%	0.07%
Poaceae					
Zea mays L. (corn)	Seed	8.94%	7.13%	6.76%	4.06%
Carex tristachya Thunberg in Murray	Leaf	2.08%	2.82%	4.78%	1.37%
bamboo	Young shoots	3.47%	2.88%	2.65%	2.16%
Ericaceae					
Rhododendron ovatum (Lindley) Planchon ex Maximowicz	Leaf	3.37%	3.97%	1.54%	2.65%
Vaccinium bracteatum Thunberg in Murray	Leaf	2.05%	1.34%	2%	1.06%
Rhododendron sp.	Leaf, flower	1.27%	0.87%	2.07%	1.22%
Vaccinium mandarinorum Diels	Leaf		1.40%	1.30%	0.68%
Hamamelidaceae					
Loropetalum chinense (R. Brown) Oliver	Leaf, stem	5.18%	4.09%	3.51%	2.81%
Distylium myricoides Hemsley	Leaf, stem	1.31%	0.35%	1.26%	1.61%
Liquidambar formosana Hance	Leaf	0	0.23%	0	0
Theaceae					
Camellia cuspidata (Kochs) H. J. Veitch	Leaf	5.18%	4.76%	2.84%	3.81%
Eurya alata Kobuski	Leaf	1.09%	0.64%	1.44%	0.37%
Eurya muricata Dunn	Leaf	0	0.58%	0	1.45%
Eurya nitida Korthals	Leaf	0	0.23%	3.58%	2.08%

(continued)

Table 12.1 (continued)

Family and species	Parts Consumed	Relative density (RD)			
		Winter	Spring	Summer	Autumn
Leguminosae					
Millettia dielsiana Harms. ex Diels.	Leaf	2.36%	2.53%	1.88%	1.84%
Lespedeza bicolor Turczaninow	Leaf	0	0.35%	1.44%	0
Saxifragaceae					
Itea omeiensis C. K. Schneider in Sargent	Leaf	2.09%	1.34%	1.09%	0.22%
Taxaceae					
Torreya grandis Fortune ex Lindley	Leaf	1.37%	0.64%	0	0
Aquifoliaceae					
Ilex purpurea Hassk.	Leaf	0.52%	0.29%		0.68%
Myrtaceae					
Syzygium buxifolium Hooker & Arnott	Leaf	0.49%	0.06%	0.75%	0.15%
Cephalotaxaceae					
Cephalotaxus fortunei Hooker	Leaf	0.40%	0	0	0
Taxodiaceae					
Cunninghamia lanceolata (Lambert) Hooker	Leaf, fruit	7.60%	0.11%	0.91%	2.08%
Ranunculaceae					
Thalictrum aquilegifolium L.	Leaf	0	0.11%	0.53%	0
Anacardiaceae					
Toxicodenddron sylvestre (Siebold & Zucc.) Kuntze	Leaf	0	0.17%	0.83%	0.30%
Araliaceae					
Hedera nepalensis var. *sinensis* (Tobler) Rehder	Leaf	0	0.29%	0	1.76%
Styracaceae					
Pterostyrax corymbosus Siebold & Zuccarini	Leaf	0	0.23%	0	0
Tiliaceae					
Tilia oliveri Szyszyłowicz	Leaf	0	0.52%	1.13%	0
Berberidaceae					
Epimedium davidii Franchet	Leaf	0	0.23%	0	0
Gesneriaceae					
Conandron ramondioides Siebold & Zuccarini	Leaf	0	0	0	0.30%

(continued)

Table 12.1 (continued)

Family and species	Parts Consumed	Relative density (RD)			
		Winter	Spring	Summer	Autumn
Araceae		0	1.33%	0	0
Pinellia cordata Dunn.	Leaf	0	0.11%	0	0
Grassleaf Sweetfalg	Leaf	0	1.22%	0	0
Lardizabalaceae		0	0.46%	0	2.05%
Akebia trifoliata (Thunberg) Koidzumi	Leaf	0	0.06%	0	0.83%
Akebia quinata (Houttuyn) Decaisne	Leaf	0	0.23%	0	1.22%
Sargentodoxa cuneata (Oliver) Rehder & E. H. Wilson in Sargent	Leaf	0	0.17%	0	0
Rosaceae					
Rhaphiolepis indica (Linnaeus) Lindley	Leaf	0	2.88%	1.07%	1.92%
Photinia davidsoniae Hook.	Leaf	0	0.58%	0	0.15%
Cucurbitaceae					
Thladiantha nudiflora Hemsley	Leaf	0	0.46%	0	0
Actinidiaceae					
Actinidia lanceolata Dunn.	Leaf	0	0	0	3.47%
Menispermaceae					
Menispermum dauricum Candolle	Leaf	0	1.40%	0	0
Food species consumed		18	35	25	30
Medicinal food species consumed		7	11	6	7

List is modified from You et al. (2013). RD percentages are summarized per season. $n = 50$ species
Medicinal food species consumed are highlighted in grey

Seasonal variation was noted in the number of plant species ingested from their total repertoire (Fig. 12.2), but these differences were not statistically significant ($\chi^2 = 6.93$, $df = 3$, $p > 0.05$). The largest number of species consumed was in spring ($n = 46$), followed by autumn ($n = 37$), summer ($n = 31$), and winter ($n = 25$). With the exception of one species, *Cunninghamia lanceolata* (Taxodiaceae) whose leaves and fruit were both consumed, all ingested fruit items (nuts, acorns) come from the Fagaceae family. Combined, use of leaf and fruit items in this family had the highest RD values of all plant food species identified, and none of these were classified as medicinal foods. The highest of these RD values were recorded in autumn and summer and involved the consumption of ripening and fallen ripe fruits, the main food items sought after in these two seasons. The second largest RD value family was Lauraceae, with five species, whose leaves and buds were consumed. Only one of the five species in this family, *Litsea coreana,* was classified as a medicinal food.

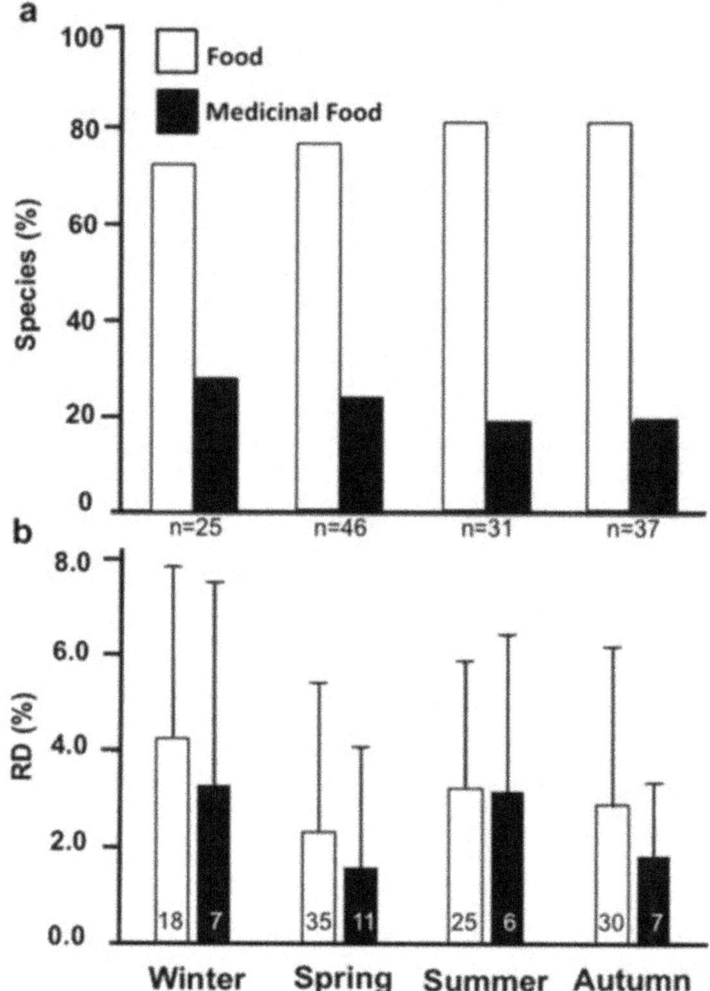

Fig. 12.2 Seasonal variation in the ingestion of food and medicinal food species (**a**) and average RD values of food and medicinal food species by YA1 troop. (**b**) Mean RD values calculated from individual species RD values per season presented in Table 12.1. Data derived from You et al. (2013)

12.3.2   Medicinal Foods in the Diet

Across the entire study period, 12 species (13 items: 12 leaves, 1 stem) in the diet had notable medicinal activity in the plant part ingested by members of YA1 troop (Table 12.1). Our literature search revealed an array of pharmacological properties of high medicinal value including antiparasitic, antiprotozoal, antibacterial, antifungal, antiviral, anti-dysentery, anti-enteritis, anticancer (antitumor), anti-inflammation, antirheumatic, antidiabetic, cardiovascular protective, neuroprotective, osteoprotective, reproductive stimulant, antidepressant, immuno-suppressant, and diaphoretic; treatment for ulcers, wounds, skin disease, weakness, and dizziness; and other health-protecting and health-promoting activities. Associated with these activities are a variety of physiologically active secondary plant metabolites (Appendix).

The following are detailed descriptions of the 12 plant species' seasonality of use, life form, pharmacological properties, and prescribed uses by humans.

***Cephalotaxus fortunei* (Leaf, Used in Winter) Evergreen Tree** Abietane diterpenoids synthesized by suspension-cultured cells displayed a wide spectrum of biological activities including antiparasitic, antibacterial, antifungal, and antiviral properties. These compounds were found to be effective against pathogens such as *Mycobacterium tuberculosis* and *Staphylococcus aureus*, including methicillin-resistant (MRSA) strains and biofilm infection of *S. aureus* (Neto et al. 2015). Abietane diterpenes isolated from aerial parts of *Plectranthus barbatus* showed remarkable activity with acceptable selectivity against the blood parasites *Plasmodium falciparum* (malaria) and *Trypanosoma brucei* (sleeping sickness in animals). Nonspecific antiprotozoal activity was also detected and is likely due to the compound's high cytotoxicity (Mothana et al. 2014). This plant also demonstrates significant anticancer activity (Duke and Ayensu 1985).

***Epimedium davidii* (Leaf Used in Spring) Perennial Herb** Compounds isolated from the leaves include Icariside II (Baohuoside I) and Baohuoside II, III, V, and VI (Ma et al. 2011). These compounds possess therapeutic activities such as osteoprotective effect, neuroprotective effect, cardiovascular protective effect, anticancer effect, anti-inflammation effect, immunoprotective effect, and enhancement of reproductive function (sexual health for males and females) (Li et al. 2015). Treatment with the compound icariin is reported to significantly increase epididymal sperm counts and testosterone levels of male rats (Chen et al. 2014) and significantly increases erectile function in castrated Wistar rats by increasing the percentage of smooth muscle and inducible nitric oxide synthase in the corpus cavernosum (Liu et al. 2005). Active ingredients administered at a dose of 5 g/kg^{-1} with an ig volume of 5 mL kg^{-1} alleviated the impact of high-intensity exercise on serum testosterone, maintaining it at normal physiological levels. It can also promote protein synthesis, inhibit degradation of amino acid and protein, and increase hemoglobin and glycogen reserves in rats receiving exercise training (Zhou et al. 2013).

A series of tests on rats and mice using icariin have also reported significant stress reduction and antidepressant properties via downregulation of glucocorticoid receptor activity and regulation of hippocampal neuroinflammation are associated with this plant species (Pan et al. 2010; Wu et al. 2011; Gong et al. 2016). The leaves of this species are also noted for their anticancer activity (Frohne and Pfänder 1984), immunosuppressive action, and inhibitory properties of lymphocyte activation (Ma et al. 2004).

***Hedera nepalensis* (Leaf Used in Spring and Autumn), Evergreen Climber or Creeper Vine** The leaves and berries are reported to have cathartic, diaphoretic, and stimulant properties. A decoction is used to treat skin diseases. Significant cancer chemo-preventative and cytotoxic properties have been demonstrated (Jafri et al. 2016). It is also reported to be an important folk medicine for the treatment of diabetes. No specific mention of *H. nepalensis* is given, but a closely related species *Hedera helix* is relevant to this discussion, as it has a similar chemical makeup and has been noted to be poisonous in large doses, but the leaves are eaten without observed side effects in wild mammals. Leaves contain hederagenin, a saponic glycoside, which can cause gastrointestinal nervous system disturbances, breathing difficulties, and coma if eaten in large amounts (Plants For A Future.com).

Ilex purpurea **(Leaf Used in Winter, Spring, and Autumn) Evergreen Tree** The plant is considered to be one of the 50 fundamental herbs in Chinese medicine. It is reported to have antitumor properties, and an extract of the leaves is made into a solution and used for treating burns and ulcers in the lower extremities. The ashes of burnt leaves are used as a dressing for skin ailments and infected wounds (Duke and Ayensu 1985).

Lespedeza bicolor **(Leaf Used in Spring and Summer) Deciduous Shrub** Leaves contain flavonoids, alkaloids, terpenes, organic acids, and stigmasterols. Potential for antioxidant, anticancer, and antibacterial activity is also noted (Ullah 2017) and is traditionally used for coughs and colds, kidney and urethra problems, fever, headache, weakness, and dizziness (Chang et al. 2017).

Liquidambar formosana **(Leaf Used in Spring) Deciduous Tree** The leaves are used in the treatment of cancerous growths (Duke and Ayensu 1985).

Litsea coreana **(Leaf Used Year Round) Evergreen Tree** Exhibits notable bioactivities, such as hepatoprotection, hyperglycemia, anti-inflammation, antioxidation, and antibacterial properties through multiple molecular mechanisms. These compounds augment immunoglobulin M and G values and show significant inhibitory effects on the pathogens *Bacillus anthracis*, *Proteusbacillus vulgaris*, *Staphylococcus aureus*, and *Bacillus subtilis*. Anti-HSV-1 activity and anti-gastric carcinoma and anti-colon carcinoma HT-29 activity have been reported. Leaves contain polysaccharides, polyphenols, essential oils, and numerous flavonoids (Jia et al. 2017).

Loropetalum chinense **(Leaf, Stem Used Year Round) Evergreen Woody Shrub** The leaves are crushed and pulverized for external application on wounds (Find Me A Cure.com), while a decoction of the whole plant is used to treat coughing in tuberculosis patients and is a treatment against dysentery and enteritis. This species also promotes wound healing and possesses antibacterial, anti-inflammatory, antioxidant, and antitumoral activity, as well as adjusts fat metabolism, and protects from cardiovascular disease (Zhou et al. 2014).

Millettia dielsiana **(Leaf Used Year Round) Deciduous Shrub** A decoction or tincture of leaves is used to treat hookworm, roundworm (nematodes), and filarial infections and is a treatment for amenorrhea, metrorrhagia, anemia, traumatic injuries, and rheumatoid arthritis (Anonymous 1977; Yeung 1985). Stem extracts exhibit anti-enterovirus activity and is effective against Coxsackie virus B3, Coxsackie virus B5, Poliovirus I, Echovirus 9, and Echovirus 29 (Guo et al. 2006). It also aids in the prevention of cardiovascular and cerebrovascular disease. It possesses high antioxidant activity (Gan et al. 2010). Seven isoflavones have been isolated from the stem and identified as 6-ethoxyca lpogonium isoflavone A, durmillone, ichthynone, jamaicin, toxicaro l isoflavone, barbigerone, and genistein (Gong et al. 2007).

Sargentodoxa cuneata **(Leaf Used in Spring) Deciduous Climbing Shrub** The stem is anthelmintic, antibacterial, antirheumatic, carminative, diuretic, and tonic (Usher 1974; Anonymous 1977; Yeung 1985; Duke and Ayensu 1985). The mashed leaves are plastered onto sores (Duke and Ayensu 1985). A decoction or tincture is used in the treatment of anemia, traumatic injuries, rheumatoid arthritis, hookworm disease, roundworm, and filariasis (Anonymous 1977; Yeung 1985). Stem extracts exhibited anti-enterovirus activity against anti-Coxsackie virus B3, Coxsackie virus B5, Polio virus I, Echovirus 9, and Echovirus 29 (Guo et al. 2006).

***Syzygium buxifolium* (Leaf Used Year Round) Evergreen Shrub** The juice of macerated leaves are taken to reduce fever (Duke and Ayensu 1985). Leaf powder is rubbed on the skin of smallpox patients for its cooling effect (World Agroforestry. org).

***Vaccinium bracteatum* (Leaf Used Year Round) Evergreen Shrub** Leaves contain isoorientin, orientin, vitexin, isovitexin, isoquercitrin, quercetin-3-O-α-L-rhamnoside, and chrysoeriol-7-O-β-D-glucopyranoside. Radical scavenging activity and protection against KBrO3-mediated kidney damage have been demonstrated (Zhang et al. 2014a, b). Anticancer and anti-inflammatory activity was demonstrated (Landa et al. 2014).

12.3.3 Seasonality of Medicinal Food Ingestion

In total, 76% of the 50 species (61 food items from 26 families) reported in YA1 troop's diet showed no evidence of significant toxicity or pharmacological activity, supporting the assumption that most plants in the diet are indeed selected for their nutritional value. However, the remaining 24% ($n = 12$) of the species are considered medicinal foods, with potential health-promoting properties.

There was no statistically significant seasonal difference in the number of medicinal foods in the diet ($\chi^2 = 0.9371$, $p > 0.05$, $df = 3$; Fig. 12.2). However, more medicinal food species were consumed in winter (28%) and spring (24%) and then summer (19%) or autumn (19%). All the consumed medicinal plant items were leaves. Seven of these species are evergreen trees, shrubs, or creepers, so leaves of these species are available year-round. The remaining five species were deciduous trees or shrubs, with leaves only in spring and summer months. These results suggest a broad based potential for access to medicinal food across the year. While the number of potential medicinal species ingested is not significantly different across seasons, the particular species ingested varied between seasons in some cases and present some interesting patterns related to their possible seasonal benefits. The three following categories of health concern extracted from our analysis provide some important areas for future investigation.

12.3.4 Antiparasitic Properties

Previous reports of medicinal food ingestion and therapeutic self-medication in response to parasite infections point out the relationship between seasonality of reinfection and plant ingestion as evidence for the context of plant use (Huffman et al. 1997, 1998). While no clear-cut seasonal trend for medicinal food ingestion was apparent in the YA1 troop diet, the relationship between medicinal properties of five specific medicinal food species and their seasonality of use provides information for future detailed investigation. *C. fortunei, L. bicolor, L. coreana, L. chinese,* and *S. cuneata* are reported to possess significant broad-spectrum antiparasitic, antibacterial, and antiviral activity. One of the highest RD values of all medicinal foods was assigned to *L. coreana* (RD = 12.09%) consumption in winter. However, two of the most intriguing medicinal food candidates with wide-spectrum antiparasitic properties are *C. fortunei* and *S. cuneata*, even though both had low RD values (0.40 and 0.17, respectively). Ingestion was restricted to winter in the former and spring in the later.

Winter and spring, the mating and birth seasons, respectively, could be potentially key seasons for investigating the parasite infection status of individuals before and after the ingestion of these plant species. One of the three zoonotic parasites identified in YA1 troop is *O. apiostomum* (Zhu et al. 2012). Infection in Japanese macaques by *O. apiostomum* occurs more frequently during winter months (MacIntosh et al. 2010), and self-medication in response to a sister species, *O. stephanostomum*, occurs in chimpanzees (Huffman 1997). The reduction of worm burden and temporary relief from related gastrointestinal upset has been linked to the ingestion of the bitter pith of *V. amygdalina* (Huffman et al. 1993, 1996a). In vitro pharmacological assays of *V. amygdalina* demonstrate a broad range of antiparasitic activities (e.g., Ohigashi et al. 1994; Oyeyemi et al. 2018).

For respiratory viruses, seasonal changes in humidity improve viral survival and increase opportunities for infection (Altizer et al. 2006). A general survey of disease prevalence in YA1 troop is necessary to provide further insights about the ecology of infection dynamics that could lead to important links with their medicinal diet. *L. bicolor* is associated with possible respiratory health, and a number of other species could inhibit respiratory viruses from establishment (Chang et al. 2017).

12.3.5 *Reproductive Modulation*

Throughout history, humans have utilized a number of plant hormones to suppress or enhance their reproductive and sexual activity (Lewis and Elvin-Lewis 1977). The reproductive behavior of male and female Tibetan macaques has been studied from a variety of perspectives including endocrinology, behavior, and seasonal variation (e.g., Li et al. 2005; Xia et al. 2018; Zhao 1993). Tibetan macaques at Mt. Huangshan exhibit high levels of sexuality inside and sometimes outside of the mating season (Li et al. 2007, see Fig. 12.2 above; Xia et al. 2010). Could there be something in the diet that stimulates or enhances Tibetan macaque sexual behavior?

One medicinal food consumed by YA1 troop in particular deserves attention in this respect. Used as an aphrodisiac in Chinese traditional medicine, *E. davidii* is known as horny goat weed or rowdy lamb herb (Ma et al. 2011), suggesting an origin for the use of this plant from watching the behavior of animals. Experimental studies have demonstrated significant enhancement of reproductive function, including increased sperm count, testosterone levels, and enhanced erectile function in rats (Chen et al. 2014; Li et al. 2015; Liu et al. 2005), but ingestion of *E. davidii* by macaques at Mt. Huangshan is limited to spring, the birth season of this troop. The seasonal timing of the consumption of this plant is preceded by a decline in ejaculatory mating frequency at the end of the mating season in late winter. For a few months in spring, non-ejaculatory mating continues at low levels with a slight peak around April (Fig. 12.1). Could the ingestion of this plant be having some effect on their reproductive activity? There is precedence in the literature to believe there might be.

Ingestion of plant hormones by animals has been found to have a number of other influences on reproductive behavior (e.g., Berger et al. 1977; Starker 1976; Sadlier 1969). Wasserman et al. (2012) report the seasonal influence of estrogenic plant consumption on hormonal and behavioral fluctuations in red colobus monkeys (*Procolobus rufomitratus*) in Uganda. Peaks in the consumption of young leaves (*Millettia dura*) with high estrogen levels coinciding with both fecal estradiol and cortisol levels in the feces. This was associated with increased levels of both copulation and aggressive interactions. In a study by Whitten (1983), the timing of onset, duration, and ending of seasonal mating behavior in female vervet monkeys (*Cercopithecus aethiops*) were closely correlated with the availability and ingestion of the flowers of *Acacia elatior* (Mimosaceae). Later Garey et al. (1992) analyzed the flowers of this species and found them to be estrogenic. Garey and colleagues determined that the amount of flowers consumed by vervet monkeys could provide adequate exogenous estrogen to stimulate the onset of mating activity. *Sargentodoxa cuneata* is only ingested during the spring birth season and has been found to assist in menstrual regulation among people, as a prophylactic against amenorrhea or metrorrhagia (Anonymous 1977; Yeung 1985). Therefore, including this species in the diet could have some sex steroid-like properties that have a role in modifying female reproductive status after birth.

Phytoestrogens present in the diet of many primates have been proposed to affect birth spacing, influence the sex of offspring, and regulate fertility. Glander (1980) proposed that inter-annual variation in birth spacing of howler monkeys (*Alouatta palliata*) was due to inter-annual variation in food quality. That is births were concentrated seasonally in years when secondary compound concentration in plant foods were low (high food quality) and spread across the year when concentrations were high (low food quality).

An in-depth examination of sex and reproduction in the Gombe chimpanzees by Wallis (1995, 1997) noted significant seasonal patterns, and multiple reproductive parameters including conception, anogenital swelling, infant mortality, and fertility. Wallis proposed that intensive foraging on seasonally available plant foods containing phytoestrogens mediate these fertility factor (Wallis 1992, 1994, 1997).

Hence, both male and female Tibetan macaque reproductive biology and behavior might be influenced by the inclusion in the diet of plants that have been shown in other species to influence reproductive hormones.

12.3.6 Stress Reduction

Stress disrupts health homeostasis, affecting reproductive function, overall health, and well-being of animals. Stress can be induced by both environmental and social factors, leading to physiological and behavioral imbalances (e.g., Takeshita et al. 2013, 2014; Wooddell et al. 2016). For example, primates living in seasonally cold habitats have a number of behavioral means for ameliorating cold stress, such as staying warm by sleeping and resting site selection, huddling, and, in one unique case, taking therapeutic hot spring baths (e.g., Hori et al. 1977; Zhang and Watanabe 2007; Kelley et al. 2016; Takeshita et al. 2018).

For stress induced by social interactions relating to social instability, dominance interactions, intergroup encounters, competition for food or mates, etc., affiliative behaviors such as grooming, reconciliation, consolation, and nonreproductive sexual behavior have been reported to be mechanisms of physiological reduction of stress in socially living species. Primates in particular have received wide attention (e.g., Aureli et al. 2002; Berry and Kaufer 2015; Carter et al. 2008; Fraser et al. 2008).

Tibetan macaques are classified as having a despotic, strongly linear, dominance style (Berman et al. 2004). They are well known for their kin-biased affiliation, tolerance, post-conflict reconciliation, nonreproductive mating, and the use of infants as a buffer to reduce tension between adults: "bridging" behavior (e.g., Bauer et al. 2014; Berman et al. 2004, 2007; Li et al. 2007; Ogawa 1995a, b). This suite of behaviors is linked to stress reduction through conflict buffering, suggestive of an undercurrent of social stress in their daily lives. Schenepel (2015) recorded high levels of agonistic and submissive behaviors around the provisioning area of YA1 troop compared to non-provision areas in the forest. Within the context of our study, we pose the question: "Do Tibetan macaques also have a dietary choice that aids in stress reduction?"

E. davidii is a prime candidate. Experimental evidence from several studies on hormonal and behavioral stress amelioration has been reported in relation to the administration of icariin, a major flavonoid isolated from *E. davidii*, in stress-induced rats and mice (Pan et al. 2010; Wu et al. 2011; Liu et al. 2015; Gong et al. 2016). Wu et al. (2011) demonstrated that icariin markedly decreased stress-induced downregulation of glucocorticoid receptors in mice subjected to "social defeat" by conspecifics. The compound also displays antidepressant activity and is used in Chinese traditional medicine for this purpose.

To the best of our knowledge, a dietary strategy for stress reduction has not yet received attention in the animal self-medication literature. Further investigation of the context of stress and the ingestion of plants like *E. davidii* may allow us to expand our knowledge of the role of diet in this area as well (Table 12.2).

Table 12.2 Medicinal properties of the 12 candidate "medicinal foods" in the diet of YA1 troop of Tibetan macaques at Mt. Huangshan

Species (part ingested) [season of use] form	Medicinal properties (see text for references and further details)
Cephalotaxus fortunei (leaf) [winter] evergreen, tree	Antiparasitic antibacterial, antifungal, and antiviral properties. Effective against pathogenic *Mycobacterium tuberculosis* and *Staphylococcus aureus*, including methicillin-resistant (MRSA) strains and biofilm infection of *S. aureus*. Contains abietane diterpenes showing remarkable activity against *Plasmodium falciparum* (malaria), *Trypanosoma brucei* (sleeping sickness in animals). Nonspecific antiprotozoal activity likely due to high cytotoxicity. Cancer prevention properties
Epimedium davidii (leaf) [spring] herbaceous, perennial	Enhancement of reproductive function (erectile, sperm count, testosterone levels). Stress reduction and antidepressant properties via downregulation of glucocorticoid receptor activity and regulation of hippocampal neuroinflammation. Anticancer activity, immunosuppressive action, and inhibition of lymphocyte activation. Osteoprotective effect, neuroprotective effect, cardiovascular protective effect, anti-inflammation effect, and immunoprotective effect
Hedera nepalensis (leaf) [spring, autumn] evergreen climber or creeper vine	Purgative action, sweat-inducing and stimulant properties. Used to treat skin diseases. Cancer preventative and cytotoxic properties. Important folk medicine for the treatment of diabetes. Contains saponins and is toxic. Ingestion induces gastrointestinal nervous system disturbances
Ilex purpurea (leaf) [winter, spring, autumn] evergreen tree	One of the 50 fundamental herbs in Chinese medicine. Antitumor properties. Used for treating burns, ulcers in the lower extremities
Lespedeza bicolor (leaf) [spring, summer] deciduous shrub	Antioxidant, anticancer, and bactericidal activity. Traditionally used for coughs and colds, kidney and urethra problems, fever, headache, weakness, and dizziness. Contains flavonoids, alkaloids, terpenes, organic acids, and stigmasterols
Liquidambar formosana (leaf) [spring] deciduous tree	Treatment of cancerous growths
Litsea coreana (leaf) [year-round] evergreen tree	Augments immunoglobulin M and G values and shows significant inhibitory effects on pathogenic *Bacillus anthracis, Proteusbacillus vulgaris, Staphylococcus aureus*, and *Bacillus subtilis*. Anti-HSV-1 activity, anti-gastric carcinoma and colon carcinoma HT-29 activity. Exhibits antibacterial, hepatoprotective, hyperglycemic, anti-inflammatory, and antioxidation activity. Contains polysaccharides, polyphenols, essential oils, and numerous flavonoids
Loropetalum chinense (leaf, stem) [year-round] evergreen woody shrub	An external application to wounds. A treatment for coughing in patients with tuberculosis, dysentery, and enteritis. Promotes wound healing, possesses antibacterial activity. Anti-inflammatory and antioxidant activity. Adjusts fat metabolism. Contains antitumoral activity and protects from cardiovascular disease

(continued)

Table 12.2 (continued)

Species (part ingested) [season of use] form	Medicinal properties (see text for references and further details)
Millettia dielsiana (leaf) [year-round] deciduous shrub	Used for prevention of cardiovascular and cerebrovascular disease. Significant antioxidant activity
Sargentodoxa cuneata (leaf) [spring] deciduous climbing shrub	Anthelmintic (hookworm, roundworm, filariasis) and antibacterial activity. Stem extracts exhibit anti-enterovirus activity against anti-Coxsackie virus B3, Coxsackie virus B5, Poliovirus I, Echovirus 9, and Echovirus 29. Also possess antirheumatic, carminative, diuretic, and tonic properties. Treatment for sores, amenorrhea, metrorrhagia (irregular uterine bleeding), traumatic injuries, rheumatoid arthritis, and anemia
Syzygium buxifolium (leaf) [year-round] evergreen shrub	Taken as a febrifuge (reduce fever). Cooling effect when rubbed on the bodies of smallpox patients
Vaccinium bracteatum (leaf) [year-round] evergreen shrub	Radical scavenging activity and protection against KBrO3-mediated kidney damage. Anticancer and anti-inflammatory activity. Contains isoorientin, orientin, vitexin, isovitexin, isoquercitrin, quercetin-3-O-α-L-rhamnoside, and chrysoeriol-7-O-β-D-glucopyranoside

12.4 Future Research

This study was designed to evaluate the potential for self-medication in Tibetan macaques. At present we cannot completely rule out the possibility that macaques consumed some or all of these plants described above only to meet some micronutrient deficiency, or that the amounts consumed were insufficient to bring about physiological change. However, the evidence presented in this chapter is compelling enough to warrant further research.

Where do we go from here? In order to demonstrate therapeutic self-medication, there are four basic requirements: (1) identify the disease or symptom(s) being treated, (2) distinguish the use of a therapeutic agent from that of everyday food items, (3) demonstrate a positive change in health condition following self-medicative behavior, and (4) provide evidence for plant activity and or direct pharmacological analysis of compounds extracted from these therapeutic agents (Huffman 2010). The pharmacological activity reported here for 12 species is the first stage of fulfilling requirement (4). This suggests that ingesting these plants may elicit significant physiological benefit if ingested in sufficient amounts.

The broad spectrum of confirmed pharmacological activity reported here suggests several avenues of research to pursue in the future and reason to believe that Tibetan macaques self-medicate. Depending on the pharmacological activities of the plant in question, future work needs to attempt to directly link the context of use with the health status of the individual. This requires longitudinal investigation with attention to (1) seasonal influences (e.g., birth, mating, infection seasonality), (2) presence or absence of large-scale disruptive social influences likely to induce psycho-physiological stress (e.g., troop fission or the death of a leader or principle care-giver), (3) reproductive state (e.g., pregnancy, estrus, reproductive history), and (4) age and health status. Monitoring of identified individuals representative of all age-sex classes, recording reinfection seasonality, infection intensity and behavioral indicators of poor health, weakened body condition, poor appetite, plant food selection patterns, etc. are required. These data will help to provide the necessary context of medicinal plant use to strengthen the case for self-medication in Tibetan macaques (e.g., see Huffman et al. 1996a, b; Huffman and Caton 2001; Alados and Huffman 2000; MacIntosh et al. 2011; Burgunder et al. 2017). Only recently has the effect of seasonal dietary change on micro- and mycobiota composition of Tibetan macaques been investigated (Sun et al. 2016, 2018).

In closing, it should be noted that there are still other medicinal properties in the medicinal diet that have not been discussed in detail. They need to be looked at more closely in the future. These include the possible roles of anti-inflammation, immunoprotection, antibiotic, antibacterial, and antiviral properties, in the passive protection or treatment of seasonal afflictions brought on by cold-damp or hot-humid weather conditions. Do troop members ingest these items more in some seasons than others? Noteworthy too about these understudied properties of the diet are the widespread anticancer (antitumoral), osteoprotective, cardiovascular protective, and neuroprotective effects. Do older members of the troop ingest items with these properties more often than younger ones?

In the aging Western society today, such diseases form the core of many of our health problems, and it has been argued that the cause of this is due to the change in our diets, shifting towards more processed foods and away from more natural food sources (Johns 1990). The properties of the Tibetan macaque diet may provide us with important insights into the long-term dietary strategy of primates occurring at the interface of food and medicine.

Acknowledgments MAH is grateful to the organizers of the International Symposium "Nonhuman Primates: Insights into Human Behavior and Society" held in Mt. Huangshan from July 21 to the 25, 2017, for being able to attend and exchange valuable scientific ideas and cultivate friendships and collaborations. This chapter is one result of that opportunity. MAH also thanks Yamato Tsuji and Massimo Bardi for their statistical advice and comments on the manuscript. BH SUN is very grateful to the Huangshan Garden Forest Bureau for their permission and support to his study. BH SUN also wishes to thank Xiaojuan Xu and Jayue Sun for their support during the writing of this manuscript.

Appendix: Plant Secondary Metabolites in Plant Items Ingested by Tibetan Macaques at Mt. Huangshan

Abietane (diterpenoid)
Alkaloids
Essential oils
Flavonoids
Icariin (prenylated flavonol glycoside)
Icariside II (Baohuoside I) and Baohuoside II, III, V, VI
Isoflavones
i6-ethoxyca lpogonium isoflavone A, durmillone, ichthynone, jamaicin, toxicaro l
Isoflavone, barbigerone, genistein
Organic acids
Polysaccharides
Polyphenols
Saponic glycosides (hederagenin)
Stigmasterols
Terpenes

References

Ahmed T, Khan A, Chandan P (2015) Photographic key for the identification of some plant of Indian Trans-Himalaya. Not Sci Biol 7:171–176

Alados CL, Huffman MA (2000) Fractal long-range correlations in behavioural sequences of wild chimpanzees: a non-invasive analytical tool for the evaluation of health. Ethology 106 (2):105–116

Alipayo D, Valdez R, Holechek JL et al (1992) Evaluation of microhistological analysis for determining ruminant diet botanical composition. J Range Manag 45:148–152

Altizer S, Dobson A, Hosseini P, Pascual M, Rohani P (2006) Seasonality and the dynamics of infectious diseases. Ecol Lett 9:467–484

Anonymous (1977) A barefoot doctors manual. Running Press, Philadelphia

Anthony RG, Smith NS (1974) Comparison of rumen and fecal analysis to describe deer diets. J Wildl Manag 38:535–540

Aureli F, Cords M, van Schaik CP (2002) Conflict resolution following aggression in gregarious animals: a predictive framework. Anim Behav 64:325–343

Bauer B, Sheeran LK, Matheson MD, Li J-H, Wagner RS (2014) Male Tibetan macaques' (*Macaca thibetana*) choice of infant bridging partners. Zool Res 35(3):222–230

Berger PJ, Sanders EH, Gardner PD, Negus NC (1977) Phenolic plant compounds functioning as reproductive inhibitors in *Microtus montanus*. Science 195:575–577

Berman CM, Ionica CS, Li J-H (2004) Dominance style among *Macaca thibetana* on Mt. Huangshan, China. Int J Primatol 25(6):1283–1312

Berman CM, Ionica CS, Li J-H (2007) Supportive and tolerant relationships among male Tibetan macaques at Huangshan, China. Behaviour 144:631–661

Berry AK, Kaufer D (2015) Stress, social behavior, and resilience: insights from rodents. Neurobiol Stress 1:116–127

Brack M (1987) Agents transmissible from simians to man. Springer, Berlin

Brack M (2008) Oesophagostomiasis. EAZWV transmissible disease fact sheet no 116. https://c. ymcdn.com/sites/www.eazwv.org/resource/resmgr/Files/Transmissible_Diseases_Handbook/ Fact_Sheets/116_Oesophagostomiasis.pdf. Accessed 19 Jun 2018

Burgunder J, Pafco B, Petrelkova KJ, Modry D, Hashimoto C, MacIntosh AJJ (2017) Complexity in behavioral organization and strongylid infection among wild chimpanzees. Anim Behav 129:257–268

Carrai V, Borgognini-Tarli SM, Huffman MA et al (2003) Increase in tannin consumption by sifaka (*Propithecus verreauxi verreauxi*) females during the birth season: a case for self-medication in prosimians? Primates 44(1):61–66

Carter CS, Grippo AJ, Pournajafi-Nazarloo H, Ruscio MG, Porges SW (2008) Oxytocin, vasopressin and sociality. Prog Brain Res 170:331–336

Chang N, Luo Z, Li D et al (2017) Indigenous uses and pharmacological activity of traditional medicinal plants in Mount Taibai, China. Evid Based Complement Alternat Med 2017:8329817. https://doi.org/10.1155/2017/8329817

Chen M, Hao J, Yang Q et al (2014) Effects of icariin on reproductive functions in male rats. Molecules 19(7):09502–09514

Cousins D, Huffman MA (2002) Medicinal properties in the diet of gorillas – an ethnopharmacological evaluation. Afr Study Monogr 23:65–89

Duke JA, Ayensu ES (1985) Medicinal plants of China. Reference Publications, Algonac. isbn:0-917256-20-4

Engel C (2002) Wild health. Houghton Mifflin, Boston

Etkin NL (1996) Medicinal cuisines: diet and ethnopharmacology. Int J Pharmacog 34(5):313–326

Etkin NL, Ross PJ (1983) Malaria, medicine, and meals: plant use among the Hausa and its impact on disease. In: Romanucci-Ross L, Moerman DE, Tancredi LR (eds) The anthropology of medicine: from culture to method. Praeger, New York, pp 231–259

Find Me A Cure (2018). https://findmeacure.com. Accessed 30 Apr 2018

Foitova I, Huffman MA, Wisnu N et al (2009) Parasites and their effect on orangutan health. In: Wish SA, Utami SS, Setia TM et al (eds) Orangutans—ecology, evolution, behavior and conservation. Oxford University Press, Oxford, pp 157–169

Forbey J, Harvey A, Huffman MA et al (2009) Exploitation of secondary metabolites by animals: a behavioral response to homeostatic challenges. Integr Comp Biol 49(3):314–328

Fraser ON, Stahl D, Aureli F (2008) Stress reduction through consolation in chimpanzees. PNAS 105(25):8557–8562

Freeland WJ, Janzen DH (1974) Strategies in herbivory by mammals: the role of plant secondary compounds. Am Nat 108(961):269–289

Frohne D, Pfänder JA (1984) Colour atlas of poisonous plants. Timber Press, Portland

Gadgil M, Bossert WH (1970) Life historical consequences of natural selection. Am Nat 104:1–24

Gan R-Y, Xu X-R, Song F-L et al (2010) Antioxidant activity and total phenolic content of medicinal plants associated with prevention and treatment of cardiovascular and cerebrovascular diseases. J Med Plant Res 4(22):2438–2444

Garey J, Markiewicz L, Gurpide E (1992) Estrogenic flowers, a stimulus for mating activity in female vervet monkeys. In: XIVth Congress of the International Primatological Society Abstracts, Strassbourg, p 210

Glander KE (1980) Reproduction and population growth in free-ranging mantled howling monkeys. Am J Phys Anthropol 53:25–36

Glander KE (1982) The impact of plant secondary compounds on primate feeding behavior. Yrbk Phys Anthropol 25:1–18

Gong T, Wang H-Q, Chen R-Y (2007) Isoflavones from vine stem of *Millettia dielsiana*. Zhongguo Zhong Yao Za Zhi 32(20):2138–2140

Gong M-J, Han B, Wang S-M, Liang S-W (2016) Icariin reverses corticosterone- induced depression-like behavior, decrease in hippocampal brain-derived neurotrophic factor (BDNF) and metabolic network disturbances revealed by NMR-based metabonomics in rats. J Pharm Biomed Anal 123:63–73

Gotoh S (2000) Regional differences in the infection of wild Japanese macaques by gastrointestinal helminth parasites. Primates 41(3):291–298

Green MJ (1987) Diet composition and quality in Himalayan musk deer based on faecal analysis. J Wildl Manag 51:880–892

Guo J-P, Pang J, Wang X-W et al (2006) *In vitro* screening of traditionally used medicinal plants in China against enteroviruses. World J Gastroenterol 12(25):4078–4081

Hardy K, Buckley S, Collins MJ et al (2012) Neanderthal medics? Evidence for food, cooking and medicinal plants entrapped in dental calculus. Naturwissenschaften 99:617–626

Hardy K, Buckley S, Huffman MA (2013) Neanderthal self-medication in context. Antiquity 87:873–878

Herbpathy Data Base (2018). https://herbpathy.com. Accessed 27 Apr 2018

Hori T, Nakayama T, Tokura H, Hara F, Suzuki M (1977) Thermoregulation of Japanese macaque living in a snowy mountain area. Jpn J Physiol 27:305–319

Huffman MA (1997) Current evidence for self-medication in primates: a multidisciplinary perspective. Yrbk Phys Anthropol 40:171–200

Huffman MA (2002) Animal origins of herbal medicine. In: Fleurentin J, Pelt J-M, Mazars G (eds) From the sources of knowledge to the medicines of the future. IRD Editions, Paris, pp 31–42

Huffman MA (2003) Animal self-medication and ethno-medicine: exploration and exploitation of the medicinal properties of plants. Proc Nutr Soc 62:371–381

Huffman MA (2007) Animals as a source of medicinal wisdom in indigenous societies. In: Bekoff M (ed) Encyclopedia of human-animal relation, vol 2. Greenwood Publishing Group, Westport, CT, pp 434–441

Huffman MA (2010) Self-medication: passive prevention and active treatment. In: Breed MD, Moore J (eds) Encyclopedia of animal behavior, vol 3. Academic, Oxford, pp 125–131

Huffman MA (2011) Primate self-medication. In: Campbell C, Fuentes A, MacKinnon K et al (eds) Primates in perspective. University of Oxford Press, Oxford, pp 563–573

Huffman MA (2015) Chimpanzee self-medication: a historical perspective of the key findings. In: Nakamura M, Hosaka K, Itoh N, Zamma K (eds) Mahale chimpanzees—50 years of research. Cambridge University Press, Cambridge, pp 340–353

Huffman MA (2016) An ape's perspective on the origins of medicinal plant use in humans. In: Hardy K, Kubiak-Martens L (eds) Wild harvest: plants in the hominin and pre-agrarian human worlds. Oxbow Books, Oxford, pp 55–70

Huffman MA, Caton JM (2001) Self-induced increase of gut motility and the control of parasitic infections in wild chimpanzees. Int J Primatol 22:329–346

Huffman MA, MacIntosh AJJ (2012) Plant-food diet of the Arashiyama Japanese macaques and its potential medicinal value. In: Leca J-B, Huffman MA, Vasey PL (eds) The monkeys of Stormy Mountain: 60 years of primatological research on the Japanese macaques of Arashiyama. Cambridge University Press, Cambridge, pp 356–431

Huffman MA, Gotoh S, Izutsu D, Koshimizu K, Kalunde MS (1993) Further observations on the use of the medicinal plant, *Vernonia amygdalina* (Del) by a wild chimpanzee, its possible effect on parasite load, and its phytochemistry. Afr Stud Monogr 14(4):227–240

Huffman MA, Koshimizu K, Ohigashi H (1996a) Ethnobotany and zoopharmacognosy of *Vernonia amygdalina*, a medicinal plant used by humans and chimpanzees. In: Caligari PDS, Hind DJN (eds) Compositae: biology and utilization, vol 2. Royal Botanical Gardens, Kew, pp 351–360

Huffman MA, Page JE, Sukhdeo MVK et al (1996b) Leaf-swallowing by chimpanzees, a behavioral adaptation for the control of strongyle nematode infections. Int J Primatol 17(4):475–503

Huffman MA, Gotoh S, Turner LA et al (1997) Seasonal trends in intestinal nematode infection and medicinal plant use among chimpanzees in the Mahale Mountains National Park, Tanzania. Primates 38(2):111–125

Huffman MA, Ohigashi H, Kawanaka M et al (1998) African great ape self-medication: a new paradigm for treating parasite disease with natural medicines? In: Ebizuka Y (ed) Towards natural medicine research in the 21st century. Elsevier Science BV, Amsterdam, pp 113–123

Jafri L, Saleem S, Kondrytuk TP et al (2016) *Hedera nepalensis* K. Koch: a novel source of natural cancer chemopreventive and anticancerous compounds. Phytother Res 30(3):447–453. https://doi.org/10.1002/ptr.5546

Janzen DH (1978) Complications in interpreting the chemical defenses of trees against tropical arboreal plant-eating vertebrates. In: Montgomery GG (ed) The ecology of arboreal folivores. Smithsonian Institution Press, Washington, DC, pp 73–84

Jia X, Li P, Wan J, He C (2017) A review on phytochemical and pharmacological properties of *Litsea coreana*. Pharm Biol 55(1):1368–1374. https://doi.org/10.1080/13880209.2017.1302482

Johns T (1990) With bitter herbs they shall eat it. University of Arizona Press, Tucson

Kelley EA, Jablonski NG, Chaplin G, Sussman RW, Kamilar JM (2016) Behavioral thermoregulation in *Lemur catta*: the significance of sunning and huddling behaviors. Am J Primatol 78:745–754

Krief S, Hladik CM, Haxaire C (2005) Ethnomedicinal and bioactive properties of the plants ingested by wild chimpanzees in Uganda. J Ethnopharmacol 101:1–15

Lambert JE (2011) Primate nutritional ecology, feeding biology and diet at ecological and evolutionary scales. In: Campbell CJ, Fuentes A, MacKinnon KC et al (eds) Primates in perspective. Oxford University Press, Oxford, pp 512–522

Landa P, Skalova L, Bousova I et al (2014) *In vitro* anti-proliferative and anti-inflammatory activity of leaf and fruit extracts from *Vaccinium bracteatum* Thunb. Pak J Pharm Sci 27(1):103–106

Lewis WH, Elvin-Lewis MPF (1977) Medical botany. Wiley, New York

Li J-H, Wang Q, Han D (1996) Fission in a free-ranging Tibetan macaque troop at Huangshan Mountain, China. Chin Sci Bull 41(16):1377–1381

Li J-H, Yin HB, Wang QS (2005) Seasonality of reproduction and sexual activity in female Tibetan macaques (*Macaca thibetana*) at Huangshan, China. Acta Zool Sin 51(3):365–375

Li J-H, Yin HB, Zhou L-Z (2007) Non-reproductive copulation behavior among Tibetan macaques at Huangshan, China. Primates 48:64–72

Li C, Li Q, Mei Q et al (2015) Pharmacological effects and pharmacokinetic properties of icariin, the major bioactive component in Herba Epimedii. Life Sci 126:57–68

Liu WJ, Xin ZC, Xin H et al (2005) Effects of icariin on erectile function and expression of nitric oxide synthase isoforms in castrated rats. Asian J Androl 7:381–388

Ma A, Qi S, Xu D et al (2004) Baohuoside-1, a novel immunosuppressive molecule, inhibits lymphocyte activation in vitro and in vivo. Transplantation 78(6):831–838

Ma H, He X, Yang Y et al (2011) The genus *Epimedium*, an ethnopharmacolgical and phytochemical review. J Ethnopharmacol 143:519–541

MacIntosh AJJ, Huffman MA (2010) Towards understanding the role of diet in host-parasite interactions in the case of Japanese macaques. In: Nakagawa F, Nakamichi M, Sugiura H (eds) The Japanese macaques. Springer, Tokyo, pp 323–344

MacIntosh AJJ, Hernandez A, Huffman MA (2010) Host age, sex, and reproductive seasonality affect nematode parasitism in wild Japanese macaques. Primates 51:353–364

MacIntosh JJJ, Alados CI, Huffman MA (2011) Fractal analysis of behavior in a wild primate: behavioural complexity in health and disease. J R Soc Interface 8(63):1497e1509

Mellado M, Foote RH, Rodriquez A et al (1991) Botanical composition and nutrient content of diets selected by goats grazing on desert grassland in northern Mexico. Small Rumin Res 6:141–150

Mothana RA, Al-Said MS, Al-Musayeib NM et al (2014) *In vitro* antiprotozoal activity of abietane diterpenoids isolated from *Plenctranthus barbatus* Andr. Int Mol Sci 15:8360–8371. http://www.mdpi.com/1422-0067/15/5/8360. Accessed 27 Apr 2018

Mukherjee JR, Chelladurai V, Ronald J et al (2011) Do animals eat what we do? Observations on medicinal plants used by humans and animals of Mudanthurai Range, Tamil Nadu. In: Kala CP (ed) Medicinal plants and sustainable development. Nova Science, New York, pp 179–195

Ndagurwa HGT (2012) Bark stripping by chacma baboons (*Papio hamadryas ursinus*) as a possible prophylactic measure in a pine plantation in eastern Zimbabwe. Afr J Ecol 51:164–167

Negre A, Tarnaud L, Roblot JF et al (2006) Plants consumed by *Eulemur fulvus* in Comoros Islands (Mayotte) and potential effects on intestinal parasites. Int J Primatol 27(6):1495–1517

Neto I, Faustino C, Rijo P (2015) Antimicrobial abietane diterpenoids against resistant bacteria and biofilms. In: Méndez-Vilas A (ed) The battle against microbial pathogens: basic science, technological advances and educational programs, Microbiology book series. Formatex Research Center, Badajoz

Ogawa H (1995a) Triadic male-female-infant relationships and bridging behavior among Tibetan macaques. Folia Primatol 64:153–157

Ogawa H (1995b) Recognition of social relationships in bridging behavior among Tibetan macaques (*Macaca thibetana*). Am J Primatol 35:305–310

Ohigashi H, Huffman MA, Izutsu D, Koshimizu K, Kawanaka M, Sugiyama H, Kirby GC, Warhurst DC, Allen D, Wright CW, Phillipson JD, Timmon-David P, Delmas F, Elias R, Balansard G (1994) Toward the chemical ecology of medicinal plant use in chimpanzees: the case of *Vernonia amygdalina*, a plant used by wild chimpanzees possibly for parasite-related diseases. J Chem Ecol 20(3):541–553

Oyeyemi IT, Akinlabi AA, Adewumi A, Aleshinloye AO, Oyeyemi OT (2018) *Vernonia amygdlina*: a folkloric herb with anthelmintic properties. Beni-Suef Univ J Basic Appl Sci 7:43–49

Pan Y, Wang F-M, Quang L-Q, Zhang D-M, Kong L-D (2010) Icariin attenuates chronic mild stress-induced dysregulation of the LHPA stress circuit in rats. Psychoneuroendocrinology 35:272–283

Petroni LM, Huffman MA, Rodriguez E (2016) Medicinal plants in the diet of woolly spider monkeys (*Brachyteles arachnoides*, E. Geoffroy, 1806)—a bio-rational for the search of new medicines for human use? Rev Bras Farm 27(2):135–142

Plants For A Future Data Base (1996–2012). https://pfaf.org/user/Default.aspx. Accessed 27 Apr 2018

Rosenthal GA, Berenbaum MR (1992) Herbivores: their interactions with secondary plant metabolites. Academic, San Diego

Rumble MA, Anderson SH (1993) Evaluating the microscopic fecal technique for estimating hard mast in turkey diets. USDA Forest Service Research Paper RM-310. https://www.fs.fed.us/rm/pubs_rm/rm_rp310.pdf

Sadlier RM (1969) The ecology of reproduction in wild and domestic mammals. Methuer, London

Schenepel BL (2015) Provisioning and its effects on the social interactions of Tibetan Macaques (Macaca Thibetana) at Mt. Huangshan, China. All Master's Theses. 399. https://digitalcommons.cwu.edu/etd/399

Simpson SJ, Sibly RM, Lee KP, Behmer ST, Raubenheimer D (2004) Optimal foraging when regulating intake of multiple nutrients. Anim Behav 68:1299–1311

Sparks DR, Malechek JC (1968) Estimating percentage dry weight in diets using a microscope technique. J Range Manage 21:264–265

Starker LA (1976) Phytoestrogens: adverse effects on reproduction in California quail. Science 191:98–100

Sun B, Wang X, Bernstein S, Huffman MA, Dong-Po Xia D-P, Gu Z, Chen R, Sheeran LK, Wagner RS, Li J (2016) Marked variation between winter and spring gut microbiota in free-ranging Tibetan Macaques (*Macaca thibetana*). Sci Rep 6:26035. https://doi.org/10.1038/srep26035

Sun B-H, Gu Z, Wang X, Huffman MA, Garber PA, Sheeran LK, Zhang D, Zhu Y, Xia D-P, Li J-H (2018) Season, age, and sex affect the fecal mycobiota of free-ranging Tibetan macaques (*Macaca thibetana*). Am J Primatol 80:e22880. https://doi.org/10.1002/ajp.22880

Takeshita RSC, Huffman MA, Bercovitch FB, Mouri K, Shimizu K (2013) The influence of age and season on fecal dehyroepiandrosterone-sulfate (DHEAS) concentrations in Japanese macaques (*Macaca fuscata*). Gen Comp Endocrinol 19:39–43

Takeshita RSC, Bercovitch FB, Huffman MA, Mouri K, Garcia C et al (2014) Environmental, biological, and social factors influencing fecal adrenal steroid concentrations in female Japanese macaques (*Macaca fuscata*). Am J Primatol 76:1084–1093

Takeshita RSC, Bercovitch FB, Kinoshita K, Huffman MA (2018) Beneficial effects of hot springs bathing on stress levels in Japanese macaques. Primates 59(3):215–225

Ullah S (2017) Methanolic extract from Lespedeza bicolor: potential candidates for natural antioxidant and anticancer agent. J Tradit Chin Med 37(4):444–451

Usher G (1974) A dictionary of plants used by man. Macmillan, London

Wallis J (1992) Socioenvironmental effects on timing of first postpartum cycles in chimpanzees. In: Nishida T, McGrew WC, Marler P, Pickford M, de Waal F (eds) Topics in primatology, Human origins, vol 1. University of Tokyo Press, Tokyo, pp 119–130

Wallis J (1994) Socioenvironmental effects on full anogenital swellings in adolescent female chimpanzees. In: Roeder JJ, Thierry B, Anderson JR, Herrenschmidt N (eds) Current primatology, Social development, learning and behavior, vol 2. University of Louis Pasteur, Strasbourg, pp 25–32

Wallis J (1995) Seasonal influence on reproduction in chimpanzees of Gombe National Park. Int J Primatol 16:435–451

Wallis J (1997) A survey of reproductive parameters in the free-ranging chimpanzees of Gombe National Park. J Reprod Fertil 109:297–307

Wasserman MD, Chapman CA, Milton K, Gogarten JF, Wittwer DJ, Zeigler TE (2012) Estrogenic plant consumption predicts red colobus monkey hormonal state and behavior. Horm Behav 62:553–562

Whitten PL (1983) Flowers, fertility and females. Abstract. Am J Phys Anthropol 60:269–270

Wooddell LJ, Kaburu SSK, Rosenberg KL et al (2016) Matrilineal behavioral and physiological changes following the removal of a non-alpha matriarch in rhesus macaques. PLoS One 11(6): e0157108

World Agroforestry Data Base (2018). http://www.worldagroforestry.org/treedb/. Accessed 27 Apr 2018

Wu J, Du J, Xu C, Le J, Xu Y, Liu B, Dong J (2011) Icariin attenuates social defeat-induced down-regulation of glucocorticoid receptor in mice. Pharmacol Biochem Behav 98:273–278

Xia D-P, Li J-H, Zhu Y, Sun B-H, Sheeran LK, Matheson MD (2010) Seasonal variation and synchronization sexual behaviors in free-ranging male macaques (*Macaca thibetana*) at Huangshan, China. Zool Res 31(5):509–515

Xia D-P, Wang X, Zhang Q-X, Sun B-H, Sun L, Sheeran LK, Li J-H (2018) Progesterone levels in seasonally breeding, free-ranging male *Macaca thibetana*. Mamm Res 63:99–106

Yeung H-C (1985) Handbook of Chinese herbs and formulas. Institute of Chinese Medicine, Los Angeles

You S-Y, Yin H-B, Zhang S-Z et al (2013) Food habits of *Macaca thibetana*. China J Biol 30 (5):64–67

Zhang P, Watanabe K (2007) Extra-large cluster formation by Japanese macaques (*Macaca fuscata*) on Shodoshima Island, Central Japan, and related factors. Am J Primatol 69:1119–1130

Zhang J, Chu C-J, Li X-L et al (2014a) Isolation and identification of antioxidant compounds in *Vaccinium bracteatum* Thunb. by UHPLC-Q-TOF LC/MS and their kidney damage protection. J Funct Foods 11:62–10

Zhang Y, Ren F-X, Yang Y et al (2014b) Chemical constituents and biological activities of *Loropetalum chinense* and *Loropetalum chinense* var. *rubrum*: research advances. J Int Pharm Res 41(3):307–312

Zhao Q-K (1993) Sexual behavior of Tibetan macaques at Mt. Emei, China. Primates 34 (4):431–444

Zhou H-T, Cao J-M, Lin Q et al (2013) Effect of *Epimedium davidii* on testosterone content, substance metabolism an exercise capacity in rats receiving exercise training. Chin Pharm J 48 (1):25–29

Zhou J, Wang Y-S, Wu Z-G (2014) Study of medicinal value of *Loropetalum chinense*. CJTCMP 29(7):2283–2286

Zhu Y, Ji H, Li J-H, Xia D-P, Sun B-H, Xu Y-R, Kyes RC (2012) First report of the wild Tibetan macaque (*Macaca thibetana*) as a new primate host of *Gongylonema pulchrum* with high incidence in China. J Anim Vet Adv 11(24):4514–4518

Chapter 13
Primate Infectious Disease Ecology: Insights and Future Directions at the Human-Macaque Interface

Krishna N. Balasubramaniam, Cédric Sueur, Michael A. Huffman, and Andrew J. J. MacIntosh

13.1 Introduction

The expansion of human populations has increased interactions and conflict between humans and nonhuman primates (hereafter primates) throughout their range. Assessing the causal factors and thereby mitigating such conflict pose a major challenge for anthropologists, primatologists, and conservation biologists. This is because human-primate interactions are spatiotemporally variable in form and frequency (reviewed in Dickman 2012; Paterson and Wallis 2005). For instance, some of these interactions include (1) human-induced changes to primate habitat that lead to the fragmentation and decline of primate populations [e.g., Zanzibar red colobus monkeys (*Procolobus kirkii*): Siex 2005], (2) increased crop-raiding by primates leading to transactional costs on humans [e.g., Buton macaques (*Macaca ochreata*): Priston et al. 2012], (3) human-primate competition for space and resources [e.g., chimpanzees (*Pan troglodytes*): Hockings et al. 2012], (4) injuries to both humans and primates on account of direct aggression [e.g., rhesus macaques (*Macaca mulatta*): Southwick and Siddiqi 1994, 1998, 2011], and (5) primate-induced damage to human property and landscapes that generate transactional or opportunity costs to humans (Barua et al. 2013).

In comparison to such readily discernible negative effects, one outcome of conflict that is subtler and hence often goes undetected or unchecked is the

K. N. Balasubramaniam (✉)
Department of Population Health and Reproduction, School of Veterinary Medicine, University of California at Davis, Davis, CA, USA

C. Sueur
IPHC, UMR 7178, Université de Strasbourg, CNRS, Strasbourg, France
e-mail: cedric.sueur@iphc.cnrs.fr

M. A. Huffman · A. J. J. MacIntosh
Primate Research Institute, Kyoto University, Kyoto, Japan
e-mail: huffman.michael.8n@kyoto-u.ac.jp

J.-H. Li et al. (eds.), *The Behavioral Ecology of the Tibetan Macaque*, Fascinating Life Sciences, https://doi.org/10.1007/978-3-030-27920-2_13

acquisition and transmission of infectious diseases (Barua et al. 2013; Wolfe et al. 2007). Our shared evolutionary histories, along with physiological and behavioral similarities, make many primate species natural reservoirs of human parasites (Fiennes 1967; Nunn and Altizer 2006; Tutin 2000). Likewise, the acquisition of parasites from humans has led to disease outbreaks among free-living primates (Kaur and Singh 2009; Kaur et al. 2008, 2011; Nunn and Altizer 2006). From an ecological standpoint, free-living primates may acquire parasites from humans in many ways. For example, increasing epidemiological assessments continue to establish the sharing of parasites between humans and populations of socioecologically flexible primates like baboons and macaques which have become increasingly reliant on human-provisioned food or garbage in areas of overlap (Engel and Jones-Engel 2011; Engel et al. 2008; Jones-Engel et al. 2005). Humans may also indirectly influence primate exposure to parasites by altering the environment, which may potentially subdivide primate populations and change their behavioral and foraging strategies (Chapman et al. 2005, 2006a; Huffman and Chapman 2009). Third, wild primates may also sometimes be exposed to "spillovers" of parasites from international travelers during ecotourism and biological field research (Carne et al. 2017; Engel et al. 2008; Jones-Engel et al. 2005; Marechal et al. 2011; Muehlenbein and Ancrenaz 2010). Such a wide range of potential disease acquisition and transmission routes make human-primate interfaces hot spots for emerging infectious diseases (EIDs) (Nunn et al. 2008; Wolfe et al. 2007). This is especially significant in the light of the growing call for a global, transdisciplinary strategy to deal with zoonoses in both humans and animals (the One Health, hereafter OH, concept: Destoumieux-Garzon et al. 2018; Zinsstag et al. 2011, 2015). Finally, human activities like agricultural and urban land development, tourism, and provisioning, aside from directly influencing exposure as stated above, may also influence variation in the susceptibility of primates to parasites once exposed, for example, by altering levels of stress and immune function (Chapman et al. 2006b; Marechal et al. 2011, 2016; Muehlenbein and Ancrenaz 2010).

In this chapter, we focus on how human-macaque interfaces, being hot spots for the transmission of a diverse array of parasites, present opportunities for human-primate infectious disease ecology research. We first briefly outline the significance and primary objectives behind research on primate infectious disease ecology, highlighting the greater focus to date on research implementing such approaches to study wild primates in comparison to research at human-primate interfaces. We next reveal how macaques, and more broadly the variable nature of human-macaque interfaces, present opportunities to study human-primate disease transmission from a socioecological perspective (Engel and Jones-Engel 2011; Jones-Engel et al. 2005; Nahallage and Huffman 2013). We then provide a detailed account of previous studies we extracted from the online Global Mammal Parasite Database (Nunn and Altizer 2005; Stephens et al. 2017) that have detected parasites at human-macaque interfaces. Finally, we demonstrate how the implementation of novel conceptual frameworks like the Coupled Natural and Human Systems (An and Lopez-Carr 2012; Destoumieux-Garzon et al. 2018; Liu et al. 2007) and One Health concepts (Destoumieux-Garzon et al. 2018; Zinsstag et al. 2011, 2015), as well as the implementation of cutting-edge methodological approaches like Social Network Analyses (e.g., Drewe and Perkins 2015; Pasquaretta et al. 2014; Rushmore et al. 2017; VanderWaal and Ezenwa 2016) and community-level bipartite or multimodal Networks (e.g., Dormann et al. 2017; Gomez et al. 2013; Latapy et al. 2008), can address some of the critical gaps in these studies to offer key future directions for epidemiological research at these interfaces.

13.2 Primate Infectious Disease Ecology

An infectious disease is a disorder that is caused by an infectious agent, or in ecological terms a "parasite," that causes pathology in its host (MacIntosh 2016). In the ecological realm, a "parasite" is considered any organism that lives within (or on) another "host" organism, at some cost to the latter (MacIntosh 2016). For the remainder of the chapter, we deal with enteric parasites or "endoparasites" (hereafter just "parasites"), which live within the body of the host organism. These typically fall under seven major types of organisms. Five of these, specifically bacteria, viruses, rickettsia, prions, and fungi, are conventionally pathogenic microorganisms. The last two, protozoa and helminths, include both pathogenic and non-pathogenic species. All parasites typically disrupt the normal, homeostatic functioning of the body, both directly as a result of their own activity and indirectly by stimulating the host's immune system to produce a defensive response. They may do so by their sheer presence, by competing with host cells and symbiotic microbes, and, in extreme cases, by releasing toxins that increase the severity of diseases. Depending on their ecologies or life histories, parasites may enter hosts via their exposure to contaminated environmental sources such as food, water, and soil (e.g., enteric bacterial pathogens: Kilonzo et al. 2011; Sinton et al. 2007). They may also spread rapidly through host populations via mechanisms such as (1) direct host-to-host contact (e.g., respiratory viruses), (2) the sharing of common, contaminated environmental space or resources (e.g., enteric bacteria such as *Salmonella* sp., *Shigella* sp.), (3) exchange of body fluids (e.g. blood-borne pathogens like HIV and HPV), or via (4) vector-borne transmission [e.g., mosquitoes spreading malarial parasites (*Plasmodium* sp.)] (summarized in Engel and Jones-Engel 2011; Nunn and Altizer 2006).

Infectious disease ecology is a subfield that deals with the evolutionary and environmental factors that influence the exposure, acquisition, and transmission dynamics of parasites within and (more recently) between human and animal populations (Grenfell and Dobson 1995; Hudson et al. 2002). As we have now entered the Anthropocene epoch, human influence on the environment has generated an increased awareness of the importance of both public health and the conservation of natural ecosystems. So it is not surprising that over the last two decades in particular we have seen an incredible surge in research related to infectious disease ecology and evolution (reviewed in Huffman and Chapman 2009; Kappeler et al. 2015; MacIntosh and Frias 2016; Nunn 2012; Nunn and Altizer 2006), with interdisciplinary approaches drawing on theory and methods from several biological sciences including anthropology, evolutionary genetics, behavioral ecology, epidemiology, network theory, and statistics.

Nonhuman primates have served as especially useful model host systems in these endeavors (summarized in Huffman and Chapman 2009; Nunn 2012; Nunn and Altizer 2006). In addition to sharing evolutionary histories and, increasingly, ecological space with humans, primates also exhibit diverse forms of social systems, characterized by heterogeneity in group composition and size, dispersal patterns, foraging strategies, mating systems, and social structures (Hinde 1976; Kappeler and Van Schaik 2002; Sterck 1998; Thierry 2007a). For these reasons, they are physiological, ecological, and behavioral model host systems for infectious disease research (MacIntosh 2016). There is now consensus among scientists that the evolutionary, ecological, and social diversity of free-living primates is impacted by (or indeed impact) the risk of acquisition and transmission of parasites (Sueur et al. 2018).

Broadly, empirical research on primate infectious disease ecology to date has had five major foci. First, in studies related to (1) *parasite-host co-evolution*, evolutionary anthropologists have attempted to establish links between the phylogenetic relationships of parasites and their primate hosts (MacIntosh and Frias 2016; Nunn 2011; Nunn and Altizer 2006; Petrášová et al. 2011; Vallo et al. 2012). Second, studies on (2) *primate parasite socioecology*, in addition to the relative role(s) of resource abundance, predation pressure, and infanticidal risk, have also begun to examine the role of parasites in shaping the evolution of primate group sizes and social network structure (Chapman et al. 2009; Nunn et al. 2011; meta-analyses by Griffin and Nunn 2012; Nunn et al. 2015; Patterson and Ruckstuhl 2013; Rifkin et al. 2012). Conversely, the idea that group-living and social structure may also impact the diversity and prevalence of parasites in hosts (Drewe and Perkins 2015; VanderWaal and Ezenwa 2016) has led to such socioecological approaches to also focus on the identification of potential "super spreaders" or "social bottlenecks" of infection (Balasubramaniam et al. 2016, 2018; Duboscq et al. 2016; Griffin and Nunn 2012; MacIntosh et al. 2012; Romano et al. 2016). Other studies have used agent-based models to predict the prevalence and transmission of parasites through artificial primate groups and networks (Griffin and Nunn 2012; Nunn et al. 2015). Yet social life does not always equate to disease transmission or threats to homeostasis. Indeed, studies on both captive and wild primates that assess the links between (3) *infection risk and sociality, stress, and immune function* have tested the opposite paradigm, i.e., that possessing strong, diverse social connections, rather than increasing the risk of pathogenic acquisition, may function to socially buffer some primates against infection (Balasubramaniam et al. 2016; Duboscq et al. 2016; Sapolsky et al. 2000; Young et al. 2014). More research has focused on the impact of (4) *parasites in primate conservation and management*—while some deal with the implications of introduced species on the prevalence and diversity of parasites in indigenous primates (Petrášová et al. 2010, 2011), other research has attempted to quantify differences in parasite richness or diversity in primates living in varying degrees of human influence or in relation to their threatened status(es) (Bublitz et al.

2015; Chapman et al. 2006a; Gillespie et al. 2005; Goldberg et al. 2007; Kowalewski et al. 2011). Finally, emerging lines of research have focused on (5) *primate counter-strategies*, including avoidance behaviors to minimize exposure to parasites (Amoroso et al. 2017; Poirotte et al. 2017, 2019; Sarabian and MacIntosh 2015; Sarabian et al. 2017), and self-medication that removes or minimizes the impact of an infection on the host (Huffman 2016).

To date, much of the empirical work related to primate infectious disease ecology has focused on wild or red-listed primate populations (reviewed in Frias and MacIntosh 2018). Aside from habitat loss and fragmentation (Hussain et al. 2013), red-listed populations also face the risk of extinction on account of infectious diseases transmitted from humans or livestock (reviewed in Frias and MacIntosh 2018). In comparison, less research has focused on the relationship between host socioecology and transmission of parasites between humans and free-living primates at overlapping interfaces (Kaur and Singh 2009). This is despite the wide recognition that humans and primates strongly influence each other's biology, behavior, and health (Fuentes 2012; Fuentes and Hockings 2010) and that such human-primate interfaces are also potential sources of EIDs (Jones-Engel et al. 2005; Nunn et al. 2008; Wolfe et al. 2007).

13.3 Human-Macaque Interfaces

The genus *Macaca* is the most diverse, geographically widespread, and ecologically successful group of primates (Cords 2013; Thierry 2007a, b). They constitute 23 extant species, which range from North Africa in the West (Barbary macaques: *M. sylvanus*), across the Indian subcontinent [e.g., rhesus macaques (*M. mulatta*), bonnet macaques (*M. radiata*)], China [e.g., rhesus macaques (*M. mulatta*), Tibetan macaques (*M. thibetana*)], and Southeast Asia [e.g., long-tailed macaques (*M. fascicularis*), Sulawesi macaque species (e.g., *M. nigra*, *M. tonkeana*)], and up to Japan in the Far East [Japanese macaques (*M. fuscata*)] (Thierry 2007a, b). Across this range, their ecological flexibility is evidenced by the fact that macaque species, and indeed populations of the same species, inhabit a wide variety of habitats, from tropical rainforests to snowcapped mountains and from dry scrub forests to urbanized human settlements (Cords 2013; Gumert et al. 2011; Thierry 2007a, b).

In nature, all macaque species show broadly similar social organization [but see Sinha et al. (2005) for an exception]—they live in multi-male multi-female social groups in which females are philopatric and males disperse from their natal groups (Cords 2013; Thierry 2013). At the same time, they show a remarkable degree of inter- and intraspecific variation in the structure of social relationships, ranging from despotic, nepotistic societies with steep dominance hierarchies and modular, centralized, and kin-biased social networks (e.g., rhesus macaques, Japanese macaques) to tolerant or egalitarian societies characterized by shallower dominance relationships and dense, decentralized, and well-connected social networks (e.g., Sulawesi

Fig. 13.1 Macaques at human-macaque interfaces, specifically (**a**) rhesus macaques in Himachal Pradesh, Northern India; (**b**) long-tailed macaques in Kuala Lumpur, Malaysia; (**c**) toque macaques in Colombo, Sri Lanka; (**d**) bonnet macaques in Kerala, Southern India. Photo credits: (**a**), (**b**) and (**d**): K. N. Balasubramaniam; (**c**): M. A. Huffman

crested macaques: Balasubramaniam et al. 2012; Thierry et al. 2008; Sueur et al. 2011a).

More pertinently, macaques also vary in the extent to which they show adaptive or maladaptive responses to human disturbance and anthropogenic landscapes, i.e., along a spectrum of overlap at human-macaque interfaces (Priston and McLennan 2013; Radhakrishna and Sinha 2011; Radhakrishna et al. 2013). At the upper end of this spectrum lie rhesus and long-tailed macaques (Fig. 13.1a, b). Large populations of these "weed" species, categorized as "Least Concern" by IUCN since 2010 (IUCN 2019), gravitate toward and even preferentially exploit human settlements (Jaman and Huffman 2013; Southwick and Siddiqi 1994, 2011; Southwick et al. 1983). Long-tailed macaques are even listed among the IUCN Invasive Species Specialist Group's (ISSG) top 100 invasive species in the world (Lowe et al. 2000). Thus, they inhabit a variety of human-macaque interfaces: from buffer zones of ecotourism in national parks, to agricultural fields bordering rural villages, to urbanized cities like Delhi, Dhaka, and Kuala Lumpur (Fig. 13.1a, b). Other species like bonnet macaques and toque macaques (*M. sinica*) are not far behind, with both wild and semi-urban populations that inhabit the smaller town-, temple-, and university campus-interfaces of Southern India and Sri Lanka, respectively (Huffman et al. 2013a; Nahallage and Huffman 2013; Nahallage et al. 2008; Radhakrishna et al. 2013; Ram et al. 2003; Sinha et al. 2005) (Fig. 13.1c, d). Yet some of these species, like toque macaques, remain listed as "vulnerable" or "endangered" on account of the negative effects of ecotourism and habitat loss throughout their range (IUCN 2019). Finally, some less ecologically flexible species like lion-tailed macaques (*M. silenus*), Tibetan macaques (*M. thibetana*), and Sulawesi crested macaques (*M. nigra*) are still exposed to the negative impact of human activity in the form of habitat loss affecting their socioecology (Kumara et al. 2014; Singh et al. 2001), ecotourism-related stressors and mortality rates (Berman et al. 2007; Marechal et al. 2011), and hunting for bush-meat impacting mortality rates (Kyes et al. 2012; Palacios et al. 2012; Riley 2007; Riley and Fuentes 2011). Indeed, many of these species are classified as being "endangered" or "critically endangered" as a result (IUCN 2019).

The rise of *ethnoprimatology* as a subfield of biological anthropology has occurred simultaneously with the rise of primate infectious disease ecology. Specifically, ethnoprimatology is related to understanding how humans and primates impact each other's niche construction, behavioral biology, and health-related outcomes (Dore et al. 2017; Fuentes 2012; Fuentes and Hockings 2010). Unsurprisingly, human-macaque interfaces in North Africa, India, Sri Lanka, and Southeast Asia have been the primary foci of most ethnoprimatological research, with some more recent studies in Africa also having been conducted on baboons (Fehlmann et al. 2016; Kaplan et al. 2011; Hoffman and O'Riain 2012), chimpanzees (Hockings et al. 2012), and lemurs (Loudon et al. 2017). To date, this work has revealed that the nature, frequency, and severity of interactions and conflict at human-macaque interfaces vary broadly by context (Radhakrishna and Sinha 2011; Radhakrishna et al. 2013). For instance, across the Indian subcontinent, China, and Southeast Asia, conflict is heavily influenced by whether macaques also play more positive roles with resident and/or visiting human communities, e.g., monkeys as religious symbols, pets, trade commodities, or tourist attractions (Jones-Engel et al. 2004; Nahallage and Huffman 2013; Radhakrishna et al. 2013). At the same time, some intrinsic characteristics of macaques, including the age-sex class, personalities of individuals, and/or species-typical adaptive responses, have also been shown to influence interface interactions (Beisner et al. 2014; Fuentes 2006; Marechal et al. 2011; Sha et al. 2009). Such variation in human- and macaque-specific features across interfaces generates a broad variety of direct and indirect interactions, such as (1) human provisioning of macaques, (2) macaques using anthropogenic landscape features (e.g., buildings, fences, water tanks), (3) mutual contact- and non-contact aggression, (4) the exchange of body fluids like blood and saliva, (5) the fragmentation of macaque populations on account of the loss of natural habitat, (6) the hunting and consumption of macaques by humans as bush-meat, and (7) the use of macaques as pets, trade commodities, or tourist attractions (Fuentes et al. 2011; Gumert et al. 2011; Hussain et al. 2013; Jones-Engel et al. 2004; Radhakrishna et al. 2013; Riley and Fuentes 2011; Riley 2003). Naturally, the dynamic nature of such environments provides myriad mechanisms for the acquisition and transmission of parasites (Engel and Jones-Engel 2011; Nunn 2012).

13.4 Parasites at Human-Macaque Interfaces

To extract and review previous studies that report parasites among free-living macaque populations at human-macaque interfaces, we relied on the Global Mammal Parasite Database (or GMPD, Version 2.0: Stephens et al. 2017). The GMPD is a compilation of studies that report disease-causing organisms—bacteria, viruses, protozoa, helminths, and fungi—isolated from wild or free-living populations of some of the major mammalian taxa, specifically ungulates, carnivores, and primates (Nunn and Altizer 2005; Stephens et al. 2017). The database now contains 24,000 records, from over 2700 literature sources including journal articles, books and book chapters, and reports at conference proceedings. Records may be filtered on the basis of different parasite or host-specific characteristics, such as taxonomic categories, geographic location, and mode of transmission.

A search of the GMPD database filtered by host genus (macaques) and type of parasite (bacteria, viruses, helminths, and protozoa) revealed 570 records from across 80 different studies. Figure 13.2 indicates the distribution of these records by study period. Aside from the general geographic location, the GMPD does not offer more specific filtering options that aid in the classification of studies in accordance with socioecological conditions under which they were conducted. So, we manually screened for "human-macaque interface" studies as those among the above studies that reported one or more of the following: (1) the direct transmission of these agents between humans and macaques in either direction (e.g., contact

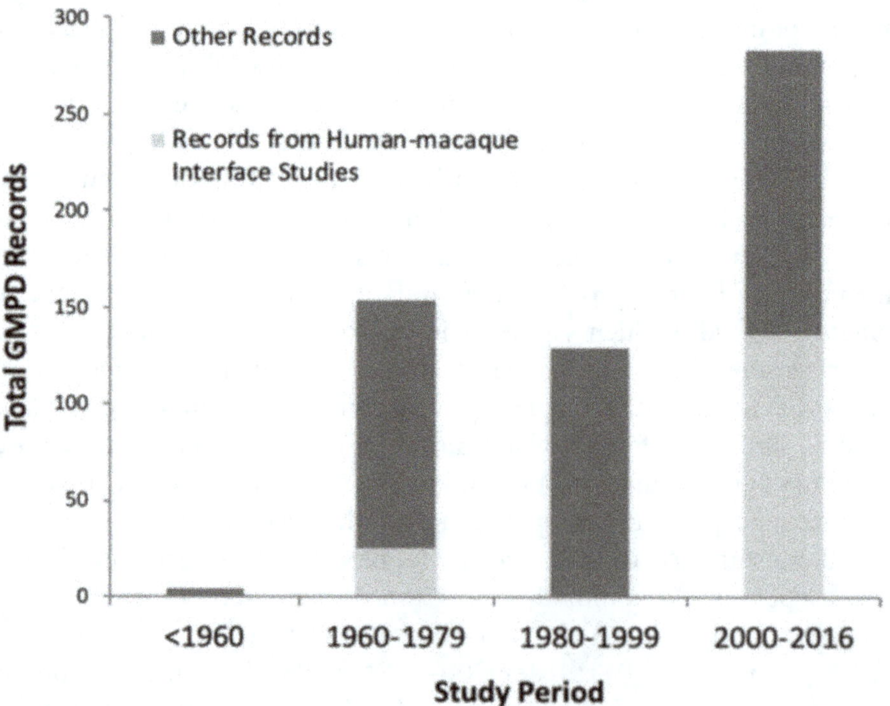

Fig. 13.2 Records of parasites reported among free-living macaque populations from studies in the Global Mammal Parasite Database (GMPD)

aggression or provisioning, pet macaques released into free-living populations), (2) the possible or potential transmission of such agents, or (3) the indirect impact of humans or anthropogenic factors (e.g., habitat fragmentation, livestock, macaque foraging on provisioned food) being identified to have influenced the acquisition of these parasites among macaques. Since this chapter primarily deals with the socioecological impact of human-macaque interfaces on disease risk, we also did not include studies on the phylogenetic co-evolutionary roots of parasites and their primate hosts.

These criteria led to the extraction of 161 (out of 570) macaque records, the vast majority of which were dated post 2000 (130 out of 161: Fig. 13.2). These records were spread across nine cited studies conducted on six different species of macaques. In Table 13.1, we summarize information from these studies, providing details on the parasites isolated, host macaque species, geographic location, prevalence and the number of individuals sampled (where the information is available), and the type of acquisition reported or speculated. Unsurprisingly, viral agents dominate this list with 88 of the 161 entries (or 55%) from 4 citations. During the last two decades, several zoonotic viruses have been described and studied in nonhuman primates in Africa and Asia from evolutionary and virulence perspectives (e.g., Ebola in great apes, reviewed in Leendertz et al. 2017; respiratory viruses in macaques, reviewed below). From a socioecological standpoint, studies on macaques have revealed strong associations between the frequency of intense human-macaque contact behaviors that involve the exchange of body fluids (e.g., aggressive bites and scratches) and the prevalence of respiratory viruses. Early work showed that wild-caught rhesus macaques in Northern India that had the highest degrees of exposure to human contact were also the most likely to show blood serum antibodies against human respiratory viruses (Shah and Southwick 1965). Later work in Nepal revealed correlations between the frequency of intense contact interactions with humans like aggressive bites and scratches and the seroprevalence of respiratory viruses such as the simian foamy virus (SFV), simian type-D virus, Cercopithecine herpesvirus-1 (CHV-1), and simian virus-40 (SV-40) (Jones-Engel et al. 2006) among rhesus macaques living in human settlements. In comparison, the Barbary macaques of Gibraltar, which engage in less intense aggression and have lower rates of contact bites in comparison to the Asian rhesus populations, showed a markedly lower seroprevalence (or were even seronegative) of these same respiratory viruses (Engel et al. 2008). More recently, humans traveling with performing "pet" rhesus macaques were found to indirectly influence the genetic structure and translocation of macaque SFV across rhesus populations in Bangladesh (Feeroz et al. 2013). Finally, other work not included in the GMPD has recorded the prevalence of macaque-borne viral pathogens like SFV and retroviruses among humans that regularly come into contact with these populations (Jones-Engel et al. 2005).

Gastrointestinal protozoa (36 entries out of 161, or 22%) and helminths (26 entries out of 161, or 16%) were the next most commonly reported parasites among the human-macaque interface studies examined. In nature, these are among the most commonly occurring parasites in wild primates (Huffman and Chapman 2009; Nunn and Altizer 2006). Yet, a few ecological assessments of human-perturbed landscapes

Table 13.1 Summary of studies extracted from the GMPD that have reported parasites at human-macaque interfaces

Citations	GMPD entries	Parasite	Parasite genus	Host species	Location	Lat.	Long.	Prev.	Type of acquisition (reported or speculated)
Huffman et al. (2013a)	7990	Bacteria	*Escherichia coli*	*Macaca sinica*	Sri Lanka	7	81	0.25	Direct: human-macaque interactions
Huffman et al. (2013a)	7992	Helminth	*Trichuris* sp.	*Macaca sinica*	Sri Lanka	7	81	0.22	Indirect: habitat fragmentation
Huffman et al. (2013a)	7991	Protozoa	*Entamoeba histolytica*	*Macaca sinica*	Sri Lanka	7	81	0.07	Direct: human-macaque interactions
Hussain et al. (2013)	7873	Helminth	*Ancylostoma* sp.	*Macaca silenus*	Southern India	10.4	77.02	0.26	Indirect: habitat fragmentation
Hussain et al. (2013)	7880	Helminth	*Diphyllobothrium* sp.	*Macaca silenus*	Southern India	10.4	77.02	0.02	Indirect: habitat fragmentation
Hussain et al. (2013)	7875	Helminth	*Strongyloides* sp.	*Macaca silenus*	Southern India	10.4	77.02	0.32	Indirect: habitat fragmentation
Hussain et al. (2013)	7877	Helminth	*Trichuris* sp.	*Macaca silenus*	Southern India	10.4	77.02	0.32	Indirect: habitat fragmentation
Hussain et al. (2013)	7881	Protozoa	*Balantidium* sp.	*Macaca silenus*	Southern India	10.4	77.02	0.06	Indirect: habitat fragmentation
Ekanayake et al. (2006)	5945–5948; 6090	Helminth	*Enterobius* sp.	*Macaca sinica*	Sri Lanka	7.9	81	0.51	Indirect: soil/livestock
Ekanayake et al. (2006)	5961–5964; 6094	Helminth	*Balantidium coli*	*Macaca sinica*	Sri Lanka	7.9	81	0.26	Indirect: soil/livestock
Ekanayake et al. (2006)	5969–5972; 6096	Protozoa	*Chilomastix* sp.	*Macaca sinica*	Sri Lanka	7.9	81	0.12	Indirect: soil/livestock
Ekanayake et al. (2006)	5937–5940	Protozoa	*Cryptosporidium* sp.	*Macaca sinica*	Sri Lanka	7.9	81	0.29	Indirect: soil/livestock
Ekanayake et al. (2006)	5965–5968; 6095	Protozoa	*Entamoeba coli*	*Macaca sinica*	Sri Lanka	7.9	81	0.30	Indirect: soil/livestock

Ekanayake et al. (2006)	5973–5976, 6097	Protozoa	*Entamoeba hartmanni*	*Macaca sinica*	Sri Lanka	7.9	81	0.27	Indirect: soil/livestock
Ekanayake et al. (2006)	5977–5980; 6098	Protozoa	*Entamoeba histolytica*	*Macaca sinica*	Sri Lanka	7.9	81	0.30	Indirect: soil/livestock
Ekanayake et al. (2006)	5957–5960; 6963	Protozoa	*Iodamoeba* sp.	*Macaca sinica*	Sri Lanka	7.9	81	0.27	Indirect: soil/livestock
Ekanayake et al. (2006)	5949–5952; 6091	Helminth	*Strongyloides* sp.	*Macaca sinica*	Sri Lanka	7.9	81	0.29	Indirect: soil/livestock
Ekanayake et al. (2006)	5953–5956; 6092	Helminth	*Trichuris* sp.	*Macaca sinica*	Sri Lanka	7.9	81	0.09	Indirect: soil/livestock
Engel et al. (2008)	7131; 7302; 7308; 7314; 7321	Virus	*Betaretrovirus Mason-Pfizer monkey virus*	*Macaca sylvanus*	Gibraltar	36	−5.6	0.00	Direct: human-macaque interactions
Engel et al. (2008)	7128; 7304; 7310; 7315; 7320	Virus	*Deltaretrovirus STLV 1*	*Macaca sylvanus*	Gibraltar	36	−5.6	0.00	Direct: human-macaque interactions
Engel et al. (2008)	7130; 7301; 7307; 7313; 7319	Virus	*Lentivirus SIV* sp.	*Macaca sylvanus*	Gibraltar	36	−5.6	0.00	Direct: human-macaque interactions
Engel et al. (2008)	7127; 7303; 7309; 7316; 7323	Virus	*Simplexvirus Herpes simplex virus 1*	*Macaca sylvanus*	Gibraltar	36	−5.6	0.00	Direct: human-macaque interactions
Engel et al. (2008)	7132; 7300; 7306; 7312; 7318	Virus	*Spumavirus Simian foamy virus*	*Macaca sylvanus*	Gibraltar	36	−5.6	0.59	Direct: human-macaque interactions
Feeroz et al. (2013)	7833–7845; 8102–8107	Virus	*Spumavirus Simian foamy virus*	*Macaca mulatta*	Bangladesh			0.92	Direct: human-macaque interactions, geographic isolation
Jones-Engel et al. (2006)	3729; 5422; 5458; 5521; 5581	Virus	*Cytomegalovirus Cercopithecine herpesvirus 8*	*Macaca mulatta*	Nepal	27.7	85.3	0.95	Indirect: habitat fragmentation, human provisioning

(continued)

K. N. Balasubramaniam et al.

Table 13.1 (continued)

Citations	GMPD entries	Parasite	Parasite genus	Host species	Location	Lat.	Long.	Prev.	Type of acquisition (reported or speculated)
Jones-Engel et al. (2006)	3730; 5423; 5459; 5522; 5582	Virus	*Polyomavirus SV-40*	*Macaca mulatta*	Nepal	27.7	85.3	0.90	Direct: human-macaque interactions
Jones-Engel et al. (2006)	3731;5424; 5560; 5523; 5583	Virus	*Simplexvirus Cercopithecine herpes-virus 1*	*Macaca mulatta*	Nepal	27.7	85.3	0.64	Direct: human-macaque interactions
Jones-Engel et al. (2006)	3732; 5425; 5461; 5524; 5584	Virus	*Spumavirus Simian foamy virus*	*Macaca mulatta*	Nepal	27.7	85.3	0.97	Direct: human-macaque interactions
Lee et al. (2011)	7157; 7164	Protozoa	*Plasmodium coatneyi*	*Macaca fascicularis*	Borneo			0.45	Indirect: anopheline vector
Lee et al. (2011)	7160; 7161	Protozoa	*Plasmodium cynomolgi*	*Macaca fascicularis*	Borneo			0.48	Indirect: anopheline vector
Lee et al. (2011)	7159; 7162	Protozoa	*Plasmodium fieldi*	*Macaca fascicularis*	Borneo			0.02	Indirect: anopheline vector
Lee et al. (2011)	7158; 7163	Protozoa	*Plasmodium inui*	*Macaca fascicularis*	Borneo			0.80	Indirect: anopheline vector
Lee et al. (2011)	7156; 7165	Protozoa	*Plasmodium knowlesi*	*Macaca fascicularis*	Borneo			0.89	Indirect: anopheline vector
Shah and Southwick (1965)	3135; 5519; 5579	Virus	*Alphavirus Chikungunya*	*Macaca mulatta*	Northern India	28	80	0.00	Direct: human-macaque interactions
Shah and Southwick (1965)	5508; 5568	Virus	*Dengue virus group Dengue 1*	*Macaca mulatta*	Northern India	28	80	0.00	Direct: human-macaque interactions
Shah and Southwick (1965)	5516; 5576	Virus	*Enterovirus poliovirus 1*	*Macaca mulatta*	Northern India	28	80	0.00	Direct: human-macaque interactions

Shah and Southwick (1965)	5517; 5577	Virus	*Enterovirus poliovirus 2*	*Macaca mulatta*	Northern India	28	80	0.13	Direct: human-macaque interactions
Shah and Southwick (1965)	5518; 5578	Virus	*Enterovirus poliovirus 3*	*Macaca mulatta*	Northern India	28	80	0.00	Direct: human-macaque interactions
Shah and Southwick (1965)	5507; 5567	Virus	*Flavivirus Japanese encephalitis*	*Macaca mulatta*	Northern India	28	80	0.00	Direct: human-macaque interactions
Shah and Southwick (1965)	5514; 5574	Virus	*Pneumovirus Human respiratory syncytial virus*	*Macaca mulatta*	Northern India	28	80	0.00	Direct: human-macaque interactions
Shah and Southwick (1965)	5510; 5570	Virus	*Polyomavirus SV-40*	*Macaca mulatta*	Northern India	28	80	0.60	Direct: human-macaque interactions
Shah and Southwick (1965)	5511; 5571	Virus	*Respirovirus Human parainfluenza virus 1*	*Macaca mulatta*	Northern India	28	80	0.05	Direct: human-macaque interactions
Shah and Southwick (1965)	5513; 5573	Virus	*Respirovirus Human parainfluenza virus 3*	*Macaca mulatta*	Northern India	28	80	0.51	Direct: human-macaque interactions
Shah and Southwick (1965)	5512; 5572	Virus	*Rubulavirus Human parainfluenza virus 2*	*Macaca mulatta*	Northern India	28	80	0.01	Direct: human-macaque interactions
Shah and Southwick (1965)	5509; 5569	Virus	*Simplexvirus Herpes simplex virus 1*	*Macaca mulatta*	Northern India	28	80	0.44	Direct: human-macaque interactions
Wenz-Mücke et al. (2013)	7862; 7866	Helminth	*Oesophagostomum* sp.	*Macaca fascicularis*	Thailand	16.2	103.1	0.07	Indirect: habitat fragmentation, human provisioning

(continued)

Table 13.1 (continued)

Citations	GMPD entries	Parasite	Parasite genus	Host species	Location	Lat.	Long.	Prev.	Type of acquisition (reported or speculated)
Wenz-Mücke et al. (2013)	7858; 7863	Helminth	*Strongyloides fuelleborni*	*Macaca fascicularis*	Thailand	16.2	103.1	0.41	Indirect: habitat fragmentation, human provisioning
Wenz-Mücke et al. (2013)	7859; 7864	Helminth	*Trichuris* sp.	*Macaca fascicularis*	Thailand	16.2	103.1	0.54	Indirect: habitat fragmentation, human provisioning

have revealed that anthropogenic factors may indirectly influence their acquisition among macaques [but see Lane et al. (2011) who report a decrease in such acquisition]. In Sri Lanka, for instance, the prevalence of both gastrointestinal protozoan parasites like *Cryptosporidium* sp., *Entamoeba* sp., and *Balantidium coli* and nematodes like *Enterobius* sp. and *Strongyloides* sp. was more common among toque macaques in more human-disturbed than pristine environments (Ekanayake et al. 2006). Further, Huffman et al. (2013a) speculate that increased human impact may in part be responsible for why the prevalence of helminth parasites was lower among toque macaques sampled at lower altitudes. Increased contact with anthropogenic landscapes and human-provisioned food was strongly linked to the prevalence of *Strongyloides fuelleborni* in wild long-tailed macaques in Thailand (Wenz-Mücke et al. 2013). Among populations of critically endangered lion-tailed macaques in Southern India, the anthropogenic fragmentation of their natural habitat was positively correlated to the diversity of gastrointestinal helminths and protozoan parasites (Hussain et al. 2013).

We extracted 10 records (6% of 161 records) of the protozoan parasite *Plasmodium* sp., the causative agent of malaria, all from a single study that surveyed wild populations of long-tailed and pig-tailed macaques (*Macaca nemestrina*) in Borneo (Lee et al. 2011). This study revealed especially high prevalence of three *Plasmodium* species—*P. knowlesi*, *P. cynomolgi*, and *P. inui*—among the macaque populations. They also revealed that *P. knowlesi*, previously hypothesized as having been transmitted to humans via anopheline vectors from overlapping macaque populations, was derived from an ancestral malarial parasite that existed before humans came to Southeast Asia. Since then, high prevalence of *P. knowlesi* has been detected among free-living macaques at vegetation mosaics and forest fragments in other parts of Southeast Asia, including Indonesia, Cambodia, Laos, and Vietnam (Huffman et al. 2013b; Zhang et al. 2016). Malaria is now widely recognized as being a threat at human-primate interfaces (Singh et al. 2004). In addition to thriving macaque populations acting as natural reservoirs for these parasites, the fragmented mosaic landscapes of Southeast Asia are also highly conducive to the proliferation of mosquito vector complexes like *Anopheles dirus* and *A. leucosphyrus*, which may transmit malaria into otherwise infection-naive macaque and human populations (Moyes et al. 2014).

The least reported type of parasite was bacterial, with only one record speculating that anthropogenic factors may be responsible for the prevalence of *Escherichia coli* in toque macaques (Huffman et al. 2013a). This was more broadly reflective of the general lack of studies that have focused on the detection of bacterial pathogens in free-living primates (Kaur and Singh 2009; Nunn and Altizer 2006). Bacterial pathogens like *Salmonella* sp., *Shigella* sp., and *E. coli* O157:H7 routinely cause acute diarrheal infection among humans and domestic livestock (Gorski et al. 2011; Rwego et al. 2008; Sinton et al. 2007; Suleyman et al. 2016). They have been previously isolated from wild primate populations in Africa that live in human-perturbed, fragmented habitats (chimpanzees: McLennan et al. 2017; lemurs of Madagascar: Bublitz et al. 2015). Since they strongly overlap and rely heavily on anthropogenic resources, urban and semi-urban macaques may be natural reservoirs of these agents, with the potential to disseminate them into overlapping human and critically endangered wildlife populations. A preliminary study at human-rhesus macaque interfaces in Northern India revealed that anthropogenic factors, such as rates of human-macaque aggression and provisioning, were positively correlated with the prevalence of enteric bacteria like *Salmonella* sp. and *E. coli* O157:H7 (Beisner et al. 2016). This finding should lead to future assessments of the relative prevalence of enteric bacteria among humans, livestock, and other overlapping macaque populations.

Our GMPD search yielded no studies on parasites in Tibetan macaques, which is somewhat surprising. There is scope for future work to focus on parasite transmission at human-Tibetan macaque interfaces. At both Mt. Emei and Mt. Huangshan, China, where they have been best studied (Zhao 1996; Li 1999), wild Tibetan macaque groups are indeed exposed to anthropogenic factors, particularly tourism. At Mt. Emei, a Buddhist community that is visited by tourists for its temples, there is no regulation of tourist-macaque interactions. Tourists regularly hand-provision the macaques, and there have been reports of tourists suffering fatal injuries from macaque attacks (Zhao 2005). Such intense and frequent contact presents scope for the transmission of parasites. On the other hand, a primate tourism program that is currently in place at Mt. Huangshan restricts the scope for macaque-tourist interactions. This program was laid down following a period between 1994 and 2004, when a group of Tibetan macaques at Mt. Huangshan was "managed" for tourist activity by restricting their home range (Berman et al. 2007). Studies on this group have revealed that intragroup aggression, attacks on infants, and infant mortality rates were all much higher during periods when the group's home range was restricted for tourist viewing than in periods prior to such activity (Berman et al. 2007). Later work in the mid-2000s that was conducted following the period of severe range restriction revealed that the macaques showed increased self-directed behaviors (e.g., self-scratching, yawning, body shake) as well as stress-coping social behaviors (e.g., allogrooming, body contact) when they were closer to tourists, in comparison to when there were no tourists present (reviewed in Matheson et al. 2013). Such tourist activity, now more controlled, may have presented or may continue to present a stressful environment to Tibetan macaques that may heighten the acquisition and transmission of parasites.

13.5 The Future of Human-Macaque Disease Ecology

Our review of studies on human-macaque interfaces reveals the detection and confirmation of a range of parasites. Yet many of these studies, based on either symptomatic or mortality-based evidence of pathogenic infection in either humans or macaques, have inferred that disease transmission has occurred without ever having established that transmission did occur (VanderWaal and Ezenwa 2016; VanderWaal et al. 2014). In other words, little or no work has assessed the precise mechanisms and pathways of parasite transmission at human-macaque interfaces. In this section, we illustrate how implementing (A) the conceptual frameworks of Coupled Natural and Human Systems and One Health, in combination with (B) cutting-edge network analytical techniques, may significantly enhance our current knowledge of infectious disease transmission at human-macaque interfaces.

(A) Unifying Conceptual Frameworks: Coupled Systems and One Health
Conflict at human-macaque interfaces may be spatiotemporally variable in form and frequency and may affect parasite transmission in dynamic and sometimes unpredictable ways. In this light, one of the biggest challenges facing research on infectious disease ecology at these, and indeed all human-wildlife interfaces, is the lack of a consensual theoretical or conceptual framework applicable across multiple types of systems.

One framework that may prove useful in this regard stems from the broader conceptualization that human interactions with nature and the environment may be viewed as dynamic, coupled systems (Liu et al. 2007). Since its proposition, the Coupled Natural and Human Systems (or CNHS) approach has presented a significant advancement in our understanding of human impact on abiotic and (more recently) biotic factors. Traditionally, studies examining the interactions between humans and natural phenomena have been largely reductionist in nature (summarized in Liu et al. 2007). They have adopted principles from biology, anthropology, geography, and environmental sciences, with an almost exclusive focus on how a single component of the human system may influence a given property of a natural system or vice versa. Further, they have tended to focus on short-term effects rather than conduct long-term assessments of the feedback effects of such interactions on both human and natural systems. Expanding significantly on these assessments, the CNHS approach explicitly acknowledges that aspects of human systems and natural systems are coupled or interlinked and must therefore be assessed as a collective whole. Multiple, dynamic components of human and natural systems are expected to influence the nature and types of interactions at interfaces, with such impact being expected to reciprocally impact long-term indicators of the overall stability, sustainability, and health of both human populations and natural components.

In its short history, studies implementing CNHS frameworks have primarily focused on the impact of humans on *inanimate, abiotic* factors (e.g., landscape ecology, climatic conditions: (Foley et al. 2005; Postel et al. 1996) and their long-term effects (e.g., via environmental degradation, natural disasters: Dilley et al. 2005) on human population dynamics and ecology (Liu et al. 2007)). In comparison, fewer studies have tackled the relationship between humans and *animate* natural systems like wildlife populations (Dickman 2010, 2012). The well-documented nature of interactions and conflict at human-macaque interfaces (reviewed above) offer opportunities to address this gap. Or conversely, the CNHS framework maybe useful to assess the mechanistic processes through which the variant nature of human-macaque interfaces may favor or inhibit the transmission of parasites across human and macaque systems.

As we allude to earlier, macaques may acquire parasites in many ways, including direct physical contact with humans or during social interactions with infected conspecifics, changes in foraging strategies induced by anthropogenic landscapes, or human-induced stressors increasing macaques' susceptibility to infection. Implementing the CNHS approach would entail examining the relative likelihood (s) of these mechanisms and indeed whether specific (suites of) attributes of the human system (e.g., community type, history of interactions with macaques, visitors versus tourists) or the macaque system (e.g., age-sex class, group size, species-typical social style) are linked with the degree to which one type of interface interaction may be expected to prevail over another in influencing parasite acquisition. More tellingly, we reckon that the CNHS framework would finally take research on human-macaque infectious disease research beyond mere descriptions of parasites at interfaces. Expanding on these findings, a CNHS approach would naturally lead to more long-term assessments of the impact of parasite diversity and distribution on indicators of macaque and human population health (e.g., symptomatic evidence for disease outbreaks, stress-induced illness), reproductive success (e.g., the number and fitness of offspring), and survival (e.g., population demographics and infant mortality rates).

Closely related to the CNHS framework is the One Health (OH) concept (derived from the "One Medicine" concept: Schwabe 1984), or the idea that addressing the challenges surrounding human health issues cannot be dissociated from environmental health (or EcoHealth) or from veterinary medical practices associated with treating wild and domestic animals (Destoumieux-Garzon et al. 2018; Zinsstag et al. 2011, 2015). The OH concept stems from the acknowledgment that the impact of human population expansion on the environment generates negative health outcomes, such as the occurrence of chronic, non-infectious diseases in humans, human and animal exposure to environmental toxins and emerging pollutants like plastics (Kannan et al. 2010; Waters et al. 2016), as well as the emergence of infectious diseases at human-wildlife interfaces (Gomez et al. 2013; Hudson et al. 2002; Nunn et al. 2008; Wolfe et al. 2007). So, it constitutes a global strategy highlighting the need for a holistic, transdisciplinary approach in dealing with the health of humans, animals, and ecosystems (the One Health Initiative).

Since its proposition more than a decade ago, proponents of OH approaches have (with varying degrees of success) proposed to deal with some of the barriers facing infectious disease research (summarized in Destoumieux-Garzon et al. 2018). We highlight three as being particularly relevant to human-macaque interfaces. The first is a resolution of the extent to which the factors that influence human health outcomes overlap with those that influence the health of natural ecosystems. Human-macaque interfaces are useful to conduct such assessments. This is because many (although not all) social and environmental factors that may potentially drive parasite transmission from humans to macaques, such as direct physical contact, contaminated food or water sources, and the exchange of body fluids, are also likely to transmit agents from macaques to humans (Engel and Jones-Engel 2011; Engel et al. 2008; Jones-Engel et al. 2005; Kaur and Singh 2009).

A second barrier is related to the promotion of interdisciplinary projects that combine veterinary medical assessments to detect and diagnose infectious diseases, with ecological and evolutionary approaches to understand the relationships between parasites and their hosts (MacIntosh and Frias 2016; Nesse et al. 2010; Nunn and Altizer 2006). Among all the primates, macaques (particularly rhesus macaques and long-tailed macaques) continue to be the most common genus used in captivity as models for biomedical research (Hannibal et al. 2017; Phillips et al. 2014). Further, as we review above, infectious disease research among free-living macaque populations have had variant foci, ranging from the detection and diagnosis of parasites (Engel and Jones-Engel 2011; Engel et al. 2008; Jones-Engel et al. 2004, 2005), through establishing co-evolutionary links between parasites and macaque hosts (Huffman et al. 2013b), to assessing the social and environmental underpinnings of parasite prevalence and transmission (Duboscq et al. 2016; MacIntosh et al. 2010, 2012; Romano et al. 2016). The OH concept, along with our above-stated argument that human-macaque interfaces are functionally interdependent, coupled systems, would provide a means to bring such diverse foci under a single, unifying framework (Destoumieux-Garzon et al. 2018).

Finally, a third direction involves placing a strong emphasis on complementing strong medical and theoretical knowledge, with current advancements in methodological and data analytical approaches. Below we elaborate on how one set of approaches—Network Analyses—may be especially significant for future research on infectious disease ecology at human-macaque interfaces.

(B) Network-Based Analytical Approaches

In the last two or three decades, network-based analytical techniques have revolutionized infectious disease epidemiology and ecology (Craft 2015; Craft and Caillaud 2011; Drewe and Perkins 2015; Godfrey 2013; Keeling 2005; Moore and Newman 2000; Newman 2002; VanderWaal and Ezenwa 2016). From a biological perspective, networks are reconstructions of entities (nodes) that are connected to each other based on one or more shared characteristics (edges) (Fig. 13.1a–c). For instance, *animal social and spatial networks* capture relationships between individuals in a social group linked together based on the frequency with which they interact or the degree of spatial overlap, respectively (reviewed in Brent et al. 2011; Croft et al. 2008; Farine and Whitehead 2015; Kasper and Voelkl 2009; Krause et al. 2007; Lusseau and Newman, 2004; Newman 2004; Sueur et al. 2011b; Wey et al. 2008) (Fig. 13.3a). *Bipartite or multimodal networks* add a level of complexity by distinguishing two or more components or layers of organization within a system, such that distinctions can be made between the edges that link nodes within the same layer to nodes across layers (Dormann et al. 2017; Kane and Alavi 2008; Latapy et al. 2008) (Fig. 13.3b, c). Third, links based on the degree of genotypic similarity of gastrointestinal microbes isolated from potentially interacting hosts within the same time frame may be used to reconstruct *microbial sharing or transmission networks* (VanderWaal and Ezenwa 2016; VanderWaal et al. 2014). Finally, in the absence or sparsity of real data, mathematical *agent-based models and artificial networks* have proven exceptionally useful in modeling the transmission of parasites (Bente et al. 2009; Griffin and Nunn 2012; Nunn 2009; Romano et al. 2016). Such heterogeneity in connectedness within and between the components of socioecological systems may strongly influence the likelihood of parasite acquisition and transmission.

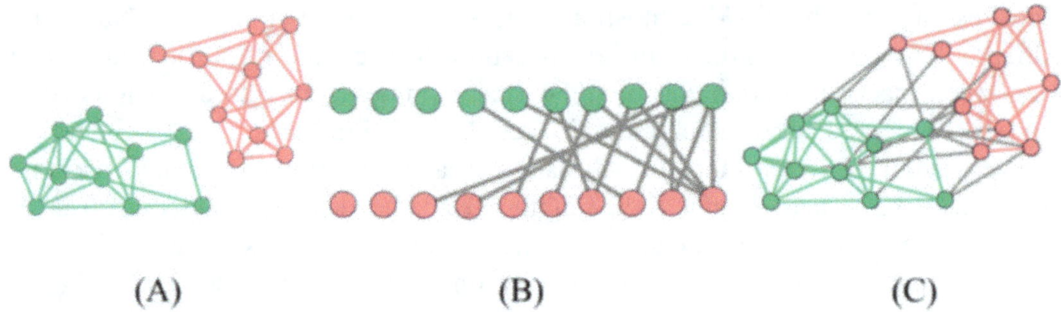

Fig. 13.3 Hypothetical social (**a**), bipartite (**b**), and multimodal (**c**) networks of the same individuals. Green nodes might represent macaques and pink ones might be humans. Green links are interactions between macaques, pink links are interactions between humans, and gray links are interactions between macaques and humans (interspecies)

Not surprisingly, network approaches, particularly social network analysis, have already found a wide range of applications in infectious disease epidemiology (Craft 2015; Craft and Caillaud 2011; Drewe and Perkins 2015; Godfrey 2013; Keeling 2005; VanderWaal and Ezenwa 2016). We briefly review these applications and related studies below. Recognizing the relative dearth in the implementation of network approaches at human-macaque interfaces, we also highlight some context (s) in which they may be implemented in human-macaque infectious disease ecological research.

Social Networks and Parasite Acquisition In humans and other animals, heterogeneity in space use overlap or contact social behavior may strongly influence the acquisition of parasites (reviewed in Drewe and Perkins 2015; Kappeler et al. 2015; Silk et al. 2017). Such heterogeneity can be modeled using social network analysis (Brent et al. 2011; Croft et al. 2008; Farine and Whitehead 2015; Krause et al. 2007; Sueur et al. 2011b). The first applications of social network approaches in the context of disease transmission were focused on humans, particularly in the spread of sexually transmitted diseases (or STDs) (Klovdahl 1985) and later following the detection of the severe acute respiratory syndrome (SARS) outbreak in 2003 (Meyers 2007). Since then, social networks across a wide range of taxa have been used to identify central or well-connected individuals, which may be potential "super spreaders" of parasites [e.g., lizards (*Egernia stokesii*): Godfrey et al. 2009; spider monkeys (*Ateles geoffroyi*): Rimbach et al. 2015; meerkats (*Suricata suricatta*): Drewe 2010; Japanese macaques: MacIntosh et al. 2012; reviewed in Drewe and Perkins 2015; VanderWaal and Ezenwa 2016]. In wild Japanese macaques, for instance, high-ranking individuals with more direct and indirect connections, or *eigenvector centrality* (Newman 2006) in their social grooming networks, were also shown to have greater species richness and infection intensities of nematode parasites (MacIntosh et al. 2012). Yet having strong and diverse social connections, rather than increasing parasite acquisition owing to contact-mediated transmission, may actually decrease the likelihood of such acquisition by mitigating stressors or enhancing immune function (e.g., Balasubramaniam et al. 2016; Cohen et al. 2015;

Hennessy et al. 2009). Consistent with this "social buffering hypothesis," work on captive rhesus macaques revealed that individuals with the strongest and most diverse social grooming and huddling connections were also the least prone to the acquisition of environmental bacterial pathogens (Balasubramaniam et al. 2016). Finally, aside from inter-individual differences in contact patterns, the higher order structure of social networks may also influence parasite transmission in contrasting ways. For instance, increased community modularity or substructuring in social networks, on account of individuals interacting more with subsets of preferred partners (Fushing et al. 2013; Newman 2006; Whitehead and Dufault 1999), may enhance parasite transmission within subgroups while presenting "social bottlenecks" to the group-wide spread of parasites (Griffin and Nunn 2012; Huang and Li 2007; Nunn et al. 2011, 2015; Romano et al. 2018; Salathe and Jones 2010). On the other hand, dense, well-connected networks, with a higher efficiency (the inverse of the number of shortest paths in the network), may facilitate rather than hinder the rapid transmission of parasites (Drewe and Perkins 2015; Griffin and Nunn 2012; Pasquaretta et al. 2014).

Such dynamic relationships between social networks and parasite acquisition suggest that broader socioecological contexts may determine the circumstances under which social life may be beneficial versus detrimental to infectious disease risk. The variant nature of the human-macaque interface may present such contexts and thereby influence parasite acquisition by altering the structure and connectedness of macaque social networks. For instance, higher frequencies of interactions with humans and/or changes to macaque movement or foraging behavior in landscapes altered by anthropogenic disturbance, by constraining the time available for macaques to engage in social interactions (Dunbar 1992; Kaburu et al. 2019; Marty et al. 2019), may lead to more modular, substructured social networks which may present bottlenecks to parasite transmission. On the other hand, such interactions or anthropogenic changes may present environmental stressors to the macaques (e.g., Barbary macaques: Carne et al. 2017; Marechal et al. 2011, 2016), in which individuals possessing strong and diverse social networks may benefit by being socially buffered against infection. In summary, future research may focus on establishing the precise mechanism(s) of super spreading, social bottlenecking, or stress-induced acquisition of parasites, through which social or contact network connectedness may influence parasite transmission dynamics at human-macaque interfaces.

Bipartite and Multimodal Networks Social networks have proven exceptionally useful to model the acquisition and transmission of parasites. Yet by themselves, they are somewhat limited in not capturing heterogeneity at higher organizational scales, for instance, across interactions between different components of a social or ecological system. This may be the reason why most epidemiological studies implementing network approaches to model heterogeneity in contact patterns have focused on either human systems or more recently wildlife systems (reviewed above), but almost never at the human-wildlife interface. Bipartite and multimodal networks, which establish connections between different interlinked components of a system, may prove especially useful in this regard (Dormann et al. 2017; Kane and Alavi 2008; Latapy et al. 2008; Finn et al. 2019). Recently, bipartite networks are beginning to feature in ecological and evolutionary research (reviewed in Bascompte and Jordano 2014; Cagnolo et al. 2011; Dormann et al. 2017), as illustrated by their being used to model marine food webs (Rezende et al. 2009), mutualistic interactions between flowers and seed-dispersing animal pollinators (Spiesman and Gratton 2016; Stang et al. 2009; Vazquez et al. 2009), and, more pertinently, host-parasitoid relationships (Laliberte and Tylianakis 2010; Poulin et al. 2013). For networks that combine links both within and across system components, some researchers have coined the term "multimodal networks," aka "multilayer" or "multislice networks" (Kane and Alavi 2008; Finn et al. 2019).

Their potential to model the connections of complex systems make bipartite and multimodal networks highly relevant tools for infectious disease research at human-wildlife interfaces. Yet to our knowledge, they have not been extensively used in this context. In the context of primates and EIDs, Gomez et al. (2013) used a combination of bipartite network construction and social network analytical tools to identify primate species that are likely to harbor EIDs. They constructed a bipartite network that linked each primate species with each parasite isolated from them. From this, they projected a unipartite "social" network in which primate species as nodes were linked to each other by edges weighted by the number of parasites they shared. They then revealed that species that were highly central in this "primate-parasite" network were also the most likely to function as "super spreaders" of EIDs to humans or at least most likely to share those EIDs with humans, suggesting potential conservation implications as well. However, only one macaque species—the toque macaque—was among the top 10 most central primates in this network. So, the extent to which macaques pose threats as transmitters of EIDs remains unclear [and more generally reflective of a knowledge gap in infectious disease research in East and Southeast Asia (Hopkins and Nunn 2007)], though this is undoubtedly likely to vary across contexts.

As reviewed in the previous section, humans and macaques engage in a variety of interactions at interfaces, some forms of which have already been linked to the acquisition and transmission of parasites (summarized in Table 13.1). To assess the mechanistic bases of such transmission, we recommend that future studies, implementing a CNHS approach, focus on constructing bipartite and multimodal networks based on intra- and interspecies interactions and spatial overlap at human-macaque interfaces. For such networks, the choice of system components may also be informed by the characteristics of the parasite studied (Craft 2015). For instance, the transmission of RNA respiratory and retroviruses between humans and macaques may require intense contact events such as bites and scratches and the exchange of body fluids like blood and saliva. Yet their fast mutation rates and short generation times may make them difficult to detect. So, multimodal networks of mild and severe contact aggressive interactions among and between humans and macaques over shorter durations of time may be useful for these purposes. On the other hand, the transmission of gastrointestinal helminths, protozoa, and enteric bacteria may require just subtle interactions such as human provisioning of macaques, acquisition from soil and anthropogenic surfaces, and macaque social grooming and contact huddling interactions, any or all of which would involve fecal-oral transmission (Balasubramaniam et al. 2016; Beisner et al. 2014, 2016; MacIntosh et al. 2012). Given that enteric bacteria may also survive longer in anthropogenic environments such as human-contaminated food, water sources and substrates (Sinton et al. 2007), moist soil (Kilonzo et al. 2011), and livestock (Craft 2015; Rwego et al. 2007; VanderWaal et al. 2014), unraveling their transmission routes may involve the construction of "multipartite" networks connecting contact patterns and spatial overlap between these biotic and abiotic factors. An even higher level of complexity may be required to detect the transmission routes of vector-borne malarial parasites like *Plasmodium knowlesi* (Abkallo et al. 2014; Huffman et al. 2013b; Lee et al. 2011). These may depend heavily on geospatial variation in the distribution and overlap of humans, reservoir macaques, and other host wildlife populations, anopheline vectors, and a host of environmental factors that may be conducive to the completion of both vector and pathogen life histories (Lee et al. 2011; Moyes et al. 2014).

Microbial Transmission Networks More recently, the phylogenetic relationships of symbiotic gut microbes isolated from animal hosts have been used to construct microbial transmission networks (VanderWaal and Ezenwa 2016). Such networks offer special advantages to detecting the potential transmission pathways of parasites that spread through the fecal-oral route (Sears et al. 1950, 1956; Tenaillon et al. 2010). Symbiotic microbes like gastrointestinal *E. coli* are present in almost every individual, have shared evolutionary histories with intestinal pathogens, and are typically acquired via fecal-oral routes (Caugant et al. 1981; Sears et al. 1950, 1956). So, if two individuals have genotypically similar or identical strains of *E. coli*, they are likely to have either shared the strain via fecal-oral transmission, which may occur either through direct social contact or through using a common environmental source (Chiyo et al. 2014; Springer et al. 2016; VanderWaal et al. 2013, 2014). Further, they rarely (if ever) alter the behavior of the host (VanderWaal et al. 2014), which allows researchers to potentially detect subtle transmission events that may signal the potential for a more devastating outbreak. Limited research to date has revealed strong links between the degree of dyadic similarity in *E. coli* and the frequency of animal space use overlap and/or social contact patterns (Balasubramaniam et al. 2018; Chiyo et al. 2014; Springer et al. 2016; VanderWaal et al. 2013, 2014). Recently, a study on captive rhesus macaques established that macaques in the same social network communities were more likely to share strains of *E. coli* among themselves than they were to macaques from other communities (Balasubramaniam et al. 2018). Previous assessments at human-primate interfaces have revealed that the population genetic structure of gut *E. coli* from great ape populations living in human-perturbed habitats was more similar to *E. coli* from humans and livestock than they were to bacteria. Such findings encourage future efforts to construct microbial transmission networks between humans and overlapping macaque populations, which may serve as models to gauge the potential transmission pathways of parasites with greater precision. However, the lack of discernible mortality or sickness behaviors associated with acquiring non-pathogenic microbes makes them models, rather than accurate forecasters, for parasite transmission.

Agent-Based Modeling When data on biological systems are either unavailable or incomplete, mathematical models offer ways of dealing with such inadequacies. In simple terms, a model may be thought of as a simplified version of a study system used to better understand it (Epstein and Axtell 1996; Minsky 1965). Thus, the complexity of a model has to be lower than that of the study system; otherwise its usefulness is lost. The main advantage of a computational model is that it can be tested infinitely by recreating or simulating a situation in the same way many (thousands of) times by adding, removing, or varying parameters and measuring emergent characteristics. These results are then usually compared with findings from empirical data to confirm or reject the tested hypotheses and accept (at least temporarily) the one for which the simulations best explain the data (Epstein and Axtell 1996; Minsky 1965). Furthermore, outcomes from these models may reveal complex or unexpected effects which may be different from, or even go undetected, based on a priori predictions. This might in turn provide a stronger basis to make other predictions, including those related to the relative role(s) of different predictive factors, in future assessments of biological systems (Epstein and Axtell 1996; Minsky 1965).

Unsurprisingly, the bottom-up approaches of simulated models have made them exceptionally useful tools to understand the acquisition and transmission of parasites. The first mathematical models of parasite transmission did not implement social network approaches: virtual individuals moved and interacted randomly in their environments (Wilensky and Stroup 1999). From individual characteristics (e.g., age, sex, hierarchical rank) and basic interaction rules (social contact, spatial proximity, conflicts, etc.), these tried to assess how the more global phenomena of parasite prevalence and outbreak potential emerge in a system (Romano et al. 2016; Rushmore et al. 2014). In these models, only the R_0, the initial number of infected agents and their interaction rates, mattered. R_0 is the basic reproduction number used to quantify the transmission potential of a parasite, defined as the number of secondary infections caused by a single infected individual introduced into a population made up entirely of susceptible individuals. However, following the acknowledgment that individuals do not move or interact randomly, social networks have been integrated into these models during the last decade or so (Griffin and Nunn 2012; Huang and Li 2007; McCabe and Nunn 2018; Nunn 2009).

Expanding beyond homogenous populations that contain only susceptible individuals that interact randomly, recent studies have integrated network approaches with a classic set of individual-based models used in epidemiology that classify individuals or "agents" into moving between susceptible infected and resistant (or SIR) classes or compartments (Bansal et al. 2007; Brauer 2008; Grimm and Railsback 2005; Kohler and Gumerman 2000). Such integrated network-based SIR models now form the basis of many epidemiological assessments in primate systems (Griffin and Nunn 2012; Kohler and Gumerman 2000; McCabe and Nunn, 2018;

Rushmore et al. 2014). For instance, the likelihood of parasite transmission from A to B may be a function of (1) whether A is already infected, (2) the likelihood of a link between A and B in their social network, and (3) the per-contact transmission probability "Beta." Simulations are run until set criteria are reached, e.g., either the extinction or saturation of infection throughout the group, at which time the average outbreak size or R_{infinity} is calculated (Diekmann et al. 1998; Keeling 1999). The implemented social network might be theoretical (Sueur et al. 2012), based on empirical data (Romano et al. 2016; Rushmore et al. 2014), or both (Griffin and Nunn 2012) depending on the aims of the study.

In a comparative evolutionary analysis that used both natural primate datasets and simulated networks, Griffin and Nunn (2012) used an SIR model to reveal how increased community modularity in the social network, despite a larger group size, negatively impacts parasite success. In a more applied example of the implementation of agent-based models, Rushmore et al. (2014) modeled the social network of a wild population of chimpanzees in order to target specific individuals to vaccinate. Based on their social centrality, they revealed that it was sufficient to target fewer, more central individuals for the same result regarding controlling the spread of infection. In this study, social network position was a model parameter, but it might also be the study object if the model includes feedback loops between the social network and parasite acquisition. Corner et al. (2003), for instance, not only showed that social networks of Australian possums (*Trichosurus vulpecula*) have an effect on the transmission rate of tuberculosis (TB) but also that the spread of TB had a feedback effect on the social network: infected possums showed higher proximity degree and betweenness centrality than non-infected possums. To our knowledge, agent-based models have not been applied to assess parasite transmission at human-macaque interfaces. In order to do so, we recommend that future research implement the SIR approach, but focus on using multimodal network models in the place of social networks which may account for an added level of complexity to model the heterogeneity between human-macaque and macaque-macaque interactions.

13.6 Conclusions

Our goals in writing this chapter were threefold. First, we wanted to convey how, despite a recent surge in primate infectious disease ecological research during the last two decades, relatively few efforts have focused on the impact of humans and anthropogenic factors on disease risk in wild primates. Second, we hope to have illustrated how our current knowledge of human-macaque interactions in particular present "starting points" from which long-term, in-depth assessments of the ecology of parasite acquisition and transmission at human-primate interfaces may be conducted. We believe that such efforts would need to integrate ethnoprimatology and infectious disease ecology in order to be successful. Finally, we hope to have convinced readers of how research on infectious disease ecology at human-macaque interfaces are also in need of the implementation of both novel conceptual frameworks (e.g., One Health, Coupled Systems) and cutting-edge analytical approaches (e.g., Network Analyses, mathematical modeling). Such approaches not only bolster the scope of epidemiological research but are also imperative for the conservation and management of both threatened and potentially problematic free-living primate (and indeed other wildlife) populations.

References

Abkallo HM, Liu W, Hokama S, Ferreira PE, Nakazawa S, Maeno Y, Quang NT, Kobayashi N, Kaneko O, Huffman MA, Kawai S, Marchand RP, Carter R, Hahn BH, Culleton R (2014) DNA from pre-erythrocytic stage malaria parasites is detectable by PCR in the faeces and blood of hosts. Int J Parasitol 14:467–473

Amoroso CR, Frink AG, Nunn CL (2017) Water choice as a counterstrategy to faecally transmitted disease: an experimental study in captive lemurs. Behaviour 154:1239–1258

An L, Lopez-Carr D (2012) Understanding human decisions in coupled natural and human systems. Ecol Model 229:1–4

Balasubramaniam KN, Dittmar K, Berman CM, Butovskaya M, Cooper MA, Majolo B, Ogawa H, Schino G, Thierry B, de Waal FBM (2012) Hierarchical steepness, counter-aggression, and macaque social style scale. Am J Primatol 74(10):915–925

Balasubramaniam KN, Beisner BA, Vandeleest J, Atwill ER, McCowan B (2016) Social buffering and contact transmission: network connections have beneficial and detrimental effects on Shigella infection risk among captive rhesus macaques. PeerJ 4:e2630

Balasubramaniam KN, Beisner BA, Guan J, Vandeleest J, Fushing H, Atwill ER, McCowan B (2018) Social network community structure is associated with the sharing of commensal *E. coli* among captive rhesus macaques (*Macaca mulatta*). PeerJ 6:e4271

Bansal S, Grenfell BT, Meyers LA (2007) When individual behaviour matters: homogeneous and network models in epidemiology. J R Soc Interface 4:879–891

Barua M, Bhagwat SA, Jadhav S (2013) The hidden dimensions of human-wildlife conflict: health impacts, opportunity and transaction costs. Biol Conserv 157:309–316

Bascompte J, Jordano P (2014) Mutualistic networks. Princeton University Press, Princeton, NJ

Beisner BA, Heagarty A, Seil S, Balasubramaniam KN, Atwill ER, Gupta BK, Tyagi PC, Chauhan NPS, Bonal BS, Sinha PR, McCowan B (2014) Human-wildlife conflict: proximate predictors of aggression between humans and rhesus macaques in India. Primates 57:459–469

Beisner BA, Balasubramaniam KN, Fernandez K, Heagerty A, Seil SK, Atwill ER, Gupta BK, Tyagi PC, Chauhan PS, Bonal BS, Sinha PR, McCowan B (2016) Prevalence of enteric bacterial parasites with respect to anthropogenic factors among commensal rhesus macaques in Dehradun, India. Primates 57:459–469

Bente D, Gren J, Strong JE, Feldmann H (2009) Disease modeling for ebola and marburg viruses. Dis Model Mech 2:12–17

Berman CM, Li J, Ogawa H, Ionica C, Yin H (2007) Primate tourism, range restriction, and infant risk among *Macaca thibetana* at Mt. Huangshan, China. Int J Primatol 28:1123–1141

Brauer F (2008) Compartmental models in epidemiology, chapter 2. In: Brauer F, van den Driessche P, Wu J (eds) Mathematical epidemiology. Springer, Berlin

Brent LJN, Lehmann J, Ramos-Fernandez G (2011) Social network analysis in the study of nonhuman primates: a historical perspective. Am J Primatol 73(8):720–730

Bublitz DC, Wright PC, Rasambainarivo FT, Arrigo-nelson SJ, Bodager JR, Gillespie TR (2015) Pathogenic enterobacteria in lemurs associated with anthropogenic disturbance. Am J Primatol 77:330–337

Cagnolo L, Salvo A, Valladares G (2011) Network topology: patterns and mechanisms in plant-herbivore and host-parasitoid food webs. J Anim Ecol 80:342–351

Carne C, Semple S, MacLarnon A, Majolo B, Marechal L (2017) Implications of tourist–macaque interactions for disease transmission. EcoHealth 14:704–717

Caugant DA, Levin BR, Selander RK (1981) Genetic diversity and temporal variation in the *E. coli* population of a human host. Genetics 98:467–490

Chapman CA, Gillespie TR, Goldberg T (2005) Primates and the ecology of their infectious diseases: how will anthropogenic change affect host-parasite interactions? Evol Anthropol 14:134–144

Chapman CA, Speirs ML, Gillespie TR, Holland T, Austad KM (2006a) Life on the edge: gastrointestinal parasites from the forest edge and interior primate groups. Am J Primatol 68:397–409

Chapman CA, Wasserman MD, Gillespie TR, Speirs ML, Lawes MJ, Saj TL et al (2006b) Do food availability, parasitism, and stress have synergistic effects on red colobus populations living in forest fragments? Am J Phys Anthropol 131:525–534

Chapman CA, Rothman JM, Hodder SAM (2009) Can parasite infection be a selective force influencing primate group size? A test with red colobus. In: Huffman MA, Chapman CA (eds) Primate parasite ecology. Cambridge University Press, Cambridge, pp 423–440

Chiyo PI, Grieneisen LE, Wittemyer G, Moss CJ, Lee PC, Douglas-hamilton I, Archie EA (2014) The influence of social structure, habitat, and host traits on the transmission of *Escherichia coli* in wild elephants. PLoS One 9:e93408

Cohen S, Janicki-Deverts D, Turner RB, Doyle WJ (2015) Does hugging provide stress-buffering social support? a study of susceptibility to upper respiratory infection and illness. Psychol Sci 26:135–147

Cords M (2013) The behavior, ecology, and social evolution of Cercopithecine monkeys. In: Mitani JC, Call J, Kappeler PM, Palombit RA, Silk JB (eds) The evolution of primate societies. University of Chicago Press, Chicago, pp 91–112

Corner LAL, Pfeiffer DU, Morris RS (2003) Social-network analysis of *Mycobacterium bovis* transmission among captive brushtail possums (*Trichosurus vulpecula*). Prev Vet Med 59:147–167

Craft ME (2015) Infectious disease transmission and contact networks in wildlife and livestock. Philos Trans R Soc Lond B Biol Sci 370:1–12

Craft ME, Caillaud D (2011) Network models: an underutilized tool in wildlife epidemiology. Interdiscip Perspect Infect Dis 2011:1–12

Croft DP, James R, Krause J (2008) Exploring animal social networks. Princeton University Press, Princeton, NJ

Destoumieux-Garzon D, Mavingui P, Boetsch G, Biossier J, Darriet F, Duboz P, Fritsch C, Giradoux P, Le Roux P, Morand S, Paillard C, Pontier D, Sueur C, Voituron Y (2018) The one health concept: 10 years old and a long road ahead. Front Vet Sci 5:14

Dickman AJ (2010) Complexities of conflict: the importance of considering social factors for effectively resolving human-wildlife conflict. Anim Conserv 13:458–466

Dickman AJ (2012) From cheetahs to chimpanzees: a comparative review of the drivers of human-carnivore conflict and human-primate conflict. Folia Primatol 83:377–387

Diekmann O, De Jong MCM, Metz JJ (1998) A deterministic epidemic model taking account of repeated contacts between the same individuals. J Appl Probab 35:448–462

Dilley M, Chen SR, Deichmann U, Lerner-Lam LA, Arnold M, Agwe J, Buys P, Kjekstad O, Lyon B, Yetman G (2005) Natural disaster hotspots: a global risk analysis. In: Bank W (ed) Disaster risk management. World Bank and Columbia University, Washington, DC

Dore KM, Riley EP, Fuentes A (2017) Ethnoprimatology: a practical guide to research on the human-nonhuman primate interface. Cambridge University Press, Cambridge

Dormann C, Frund J, Schaefer MH (2017) Identifying causes of patterns in ecological networks: opportunities and limitations. Annu Rev Ecol Evol Syst 48:559–584

Drewe JA (2010) Who infects whom? Social networks and tuberculosis transmission in wild meerkats. Proc Biol Sci 277:633–642

Drewe JA, Perkins SE (2015) Disease transmission in animal social networks. In: Krause J, James R, Franks DW, Croft DP (eds) Animal social networks. Oxford University Press, Oxford, pp 95–110

Duboscq J, Romano V, Sueur C, MacIntosh AJJ (2016) Network centrality and seasonality interact to predict lice load in a social primate. Sci Rep 6:22095

Dunbar RIM (1992) Time: a hidden constraint on the behavioural ecology of baboons. Behav Ecol Sociobiol 31:35–49

Ekanayake DK, Arulkanthan A, Horadagoda NU, Sanjeevani M, Kieft R, Gunatilake S, Dittus WP (2006) Prevalence of cryptosporidium and other enteric parasites among wild non-human primates in Polonnaruwa, Sri Lanka. Am J Trop Med Hyg 74:322–329

Engel GA, Jones-Engel L (2011) The role of *Macaca fascicularis* in infectious disease transmission. In: Gumert MD, Fuentes A, Jones-Engel L (eds) Monkeys on the edge: ecology and management of long-tailed macaques and their interface with humans. Cambridge University Press, Cambridge, pp 183–203

Engel GA, Pizarro M, Shaw E, Cortes J, Fuentes A, Barry P, Lerche N, Grant R, Cohn D, Jones-Engel L (2008) Unique pattern of enzootic primate viruses in Gibralter macaques. Emerg Infect Dis 14:1112–1115

Epstein J, Axtell R (1996) Growing artificial societies: social science from the bottom Up. MIT Press, Cambridge, MA

Farine DR, Whitehead H (2015) Constructing, conducting and interpreting animal social network analysis. J Anim Ecol 84:1144–1163

Feeroz MM, Soliven K, Small CT, Engel GA, Pacheco MA, Yee JL, Wang X, Kamrul Hasan M, Oh G, Levine KL, Rabiul Alam SM, Craig KL, Jackson DL, Lee EG, Barry PA, Lerche NW, Escalante AA, Matsen Iv FA, Linial ML, Jones-Engel L (2013) Population dynamics of rhesus macaques and associated foamy virus in Bangladesh. Emerg Microbes Infect 2:e29

Fehlmann G, O'Riain MJ, Kerr-Smith C, King AJ (2016) Adaptive space use by baboons (*Papio ursinus*) in response to management interventions in a human-changed landscape. Anim Conserv 20:101–109

Fiennes R (1967) Zoonoses of primates. Cornell University Press, Ithaca, NY

Finn KR, Silk MJ, Porter MA, Pinter-Wollman N (2019) The use of multilayer network analysis in animal behaviour. Anim Behav 149:7–22

Foley JA, DeFries R, Asner GP, Barford C, Bonan G, Carpenter SR, Chapin FS, Coe MT, Daily GC, Gibbs HK, Helkowski JH, Holloway T, Howard EA, Kucharik CJ, Monfreda C, Patz JA, Prentice IC, Ramankutty N, Snyder PK (2005) Global consequences of land use. Science 309 (5734):570–574

Frias L, MacIntosh AJJ (2018) Threatened hosts, threatened parasites? Parasite diversity and distribution in Red-Listed Primates. In: Behie AM, Teichroeb J, Malone N (eds) Primate research and conservation in the anthropocene. Cambridge University Press, Cambridge

Fuentes A (2006) Patterns and context of human-macaque interactions in Gibraltar. In: Hodges JK, Cortes J (eds) The Barbary macaque: biology, management, and conservation. Nottingham University Press, Nottingham, pp 169–184

Fuentes A (2012) Ethnoprimatology and the anthropology of the human-primate interface. Annu Rev Anthropol 41:101–117

Fuentes A, Hockings KJ (2010) The ethnoprimatological approach in primatology. Am J Primatol 72(10):841–847

Fuentes A, Rompis ALT, Arta Putra IGA, Watiniasih NL, Nyoman Suartha I, Wandia IN et al (2011) Macaque behavior at the human-monkey interface: the activity and demography of semi-free-ranging *Macaca fascicularis* at Padangtegal, Bali, Indonesia. In: Gumert MD, Fuentes A, Jones-Engel L (eds) Monkeys on the edge: ecology and management of long-tailed macaques and their interface with humans. Cambridge University Press, Cambridge, pp 159–179

Fushing H, Wang H, VanderWaal K, McCowan B, Koehl P (2013) Multi-scale clustering by building a robust and self correcting ultrametric topology on data points. PLoS One 8(2):e56259

Gillespie TR, Chapman CA, Greiner EC (2005) Effects of logging on gastrointestinal parasite infections and infection risk in African primates. J Appl Ecol 42:699–707

Godfrey SS (2013) Networks and the ecology of parasite transmission: a framework for wildlife parasitology. Int J Parasitol Parasites Wildl 2:235–245

Godfrey SS, Bull CM, James R, Murray K (2009) Network structure and parasite transmission in a group-living lizard, the gidgee skink, *Egernia stokesii*. Behav Ecol Sociobiol 63:1045–1056

Goldberg T, Gillespie TR, Rwego IB, Wheeler E, Estoff EL, Chapman CA (2007) Patterns of gastrointestinal bacterial exchange between chimpanzees and humans involved in research and tourism in western Uganda. Biol Conserv 135:511–517

Gomez JM, Nunn CL, Verdu M (2013) Centrality in primate–parasite networks reveals the potential for the transmission of emerging infectious diseases to humans. PNAS 110:7738–7741

Gorski L, Parker CT, Liang A, Cooley MB, Jay-Russell MT, Gordus AG et al (2011) Prevalence, distribution, and diversity of *Salmonella enterica* in a major produce region of California. Appl Environ Microbiol 77(8):2734–2748

Grenfell BT, Dobson AP (1995) Ecology of infectious diseases in natural populations. Cambridge University Press, Cambridge

Griffin RH, Nunn CL (2012) Community structure and the spread of infectious disease in primate social networks. Evol Ecol 26(4):779–800

Grimm V, Railsback SF (2005) Individual based modeling in ecology. Princeton University Press, Princeton, NJ

Gumert MD, Fuentes A, Jones-Engel L (eds) (2011) Monkeys on the edge: ecology and management of long-tailed macaques and their interface with humans. Cambridge University Press, Cambridge

Hannibal DL, Bliss-Moreau E, Vandeleest J, McCowan B, Capitanio J (2017) Laboratory rhesus macaque social housing and social changes: implications for research. Am J Primatol 79:1–14

Hennessy MB, Kaiser S, Sachser N (2009) Social buffering of the stress response: diversity, mechanisms, and functions. Front Neuroendocrinol 30:470–482

Hinde RA (1976) Interactions, relationships and social structure. Man 11:1–17

Hockings KJ, Anderson JR, Matsuzawa T (2012) Socioecological adaptations by chimpanzees (*Pan troglodytes*) verus inhabiting an anthropogenically impacted habitat. Anim Behav 83:801–810

Hoffman TS, O'Riain MJ (2012) Monkey management: using spatial ecology to understand the extent and severity of human-baboon conflict in the Cape Peninsula, South Africa. Ecol Soc 17:13

Hopkins ME, Nunn CL (2007) A global gap analysis of infectious agents in wild primates. Divers Distrib 13:561–572

Huang W, Li C (2007) Epidemic spreading in scale-free networks with community structure. J Stat Mech 2007:P01014

Hudson PJ, Rizzoli A, Grenfell BT, Heesterbeek H, Dobson AP (2002) The ecology of wildlife diseases. Oxford University Press, New York

Huffman MA (2016) Primate self-medication, passive prevention and active treatment—a brief review. Int J Multidiscip Stud 3:1–10

Huffman MA, Chapman CA (2009) Primate parasite ecology: the dynamics and study of host-parasite relationships. Cambridge University Press, Cambridge

Huffman MA, Nahallage CAD, Hasegawa H, Ekanayake S, De Silva LDGG, Athauda IRK (2013a) Preliminary survey of the distribution of four potentially zoonotic parasite species among primates in Sri Lanka. J Natl Sci Found 41:319–326

Huffman MA, Satou M, Kawai S, Maeno Y (2013b) New perspectives on the transmission of malaria between macaques and humans: the case of Vietnam. Folia Primatol 84:288–289

Hussain S, Ram MS, Kumar A, Shivaji S, Umapathy G (2013) Human presence increases parasitic load in endangered lion-tailed macaques (*Macaca silenus*) in its fragmented rainforest habitats in southern India. PLoS One 8:e63685

IUCN (2019) The IUCN Red list of threatened species. Version 2019-2. http://www.iucnredlist.org. Accessed 18 July 2019

Jaman MF, Huffman MA (2013) The effect of urban and rural habitats and resource type on activity budgets of commensal rhesus macaques (*Macaca mulatta*) in Bangladesh. Primates 54:49–59

Jones-Engel L, Engel GA, Schillaci MA, Kyes K, Froehlich J, Paputungan U, Kyes RC (2004) Prevalence of enteric parasites in pet macaques in Sulawesi, Indonesia. Am J Primatol 62:71–82

Jones-Engel L, Engel GA, Schillaci MA, Rompis ALT, Putra A, Suaryana KG, Fuentes A, Beer B, Hicks S, White R, Wilson B, Allan JS (2005) Primate to human retroviral transmission in Asia. Emerg Infect Dis 7:1028–1035

Jones-Engel L, Engel GA, Heidrich J, Chalise M, Poudel N, Viscidi R, Barry PA, Allan JS, Grant R, Kyes R (2006) Temple monkeys and health implications of commensalism, Kathmandu, Nepal. Emerg Infect Dis 12:900–906

Kaburu SSK, Marty PR, Beisner B, Balasubramaniam KN, Bliss-Moreau E, Kaur K, Mohan L, McCowan B (2019) Rates of human-macaque interactions affect grooming behavior among urban-dwelling rhesus macaques (*Macaca mulatta*). Am J Phys Anthropol 168:92–103

Kane GC, Alavi M (2008) Casting the net: a multimodal network perspective on user-system interactions. Inf Syst Res 19:253–272

Kannan K, Yun SH, Rudd RJ, Behr M (2010) High concentrations of persistent organic pollutants including PCBs, DDT, PBDEs and PFOS in little brown bats with white-nose syndrome in New York, USA. Chemosphere 80:613–618

Kaplan BS, O'Riain MJ, van Eeden R, King AJ (2011) A low-cost manipulation of food resources reduces spatial overlap between baboons (*Papio ursinus*) and humans in conflict. Int J Primatol 32:1397–1412

Kappeler PM, Van Schaik CP (2002) Evolution of primate social systems. Int J Primatol 23:707–740

Kappeler PM, Cremer S, Nunn CL (2015) Sociality and health: impacts of sociality on disease susceptibility and transmission in animal and human societies. Philos Trans R Soc Lond Ser B Biol Sci 370:20140116

Kasper C, Voelkl B (2009) A social network analysis of primate groups. Primates 50:343–256

Kaur T, Singh J (2009) Primate-parasitic zoonoses and anthropozoonoses: a literature review. In: Huffman MA, Chapman CA (eds) Primate parasite ecology: the dynamics and study of host-parasite relationships. Cambridge University Press, Cambridge, pp 199–230

Kaur T, Singh J, Humphrey C, Tong S, Clevenger D, Tan W, Szekely B, Wang Y, Li Y, Alex Muse E, Kiyono M, Hanamura S, Inoue E, Nakamura M, Huffman MA, Jiang B, Nishida T (2008) Descriptive epidemiology of fatal respiratory outbreaks and detection of a human-related metapneumovirus in wild chimpanzees (*Pan troglodytes*) at Mahale Mountains National Park, western Tanzania. Am J Primatol 70:755–765

Kaur T, Singh J, Huffman MA, Petrzelkova KJ, Taylor NS, Xu S, Dewhirst FE, Paster BJ, Debruyne L, Vandamme P, Fox JG (2011) *Campylobacter troglodytes* sp. nov., isolated from feces of human-habituated wild chimpanzees (*Pan troglodytes schweinfurthii*) in Tanzania. Appl Environ Microbiol 77:2366–2373

Keeling MJ (1999) The effects of local spatial structure on epidemiological invasions. Proc Biol Sci 266:859–867

Keeling MJ (2005) The implications of network structure for epidemic dynamics. Theor Popul Biol 67:1–8

Kilonzo C, Atwill ER, Mandrell R, Garrick M, Villanueva V (2011) Prevalence and molecular characterization of *Escherichia coli* O157:H7 by multiple locus variable number tandem repeat analysis and pulsed field gel electrophoresis in three sheep farming operations in California. J Food Prot 74:1413–1421

Klovdahl AS (1985) Social networks and the spread of infectious diseases: the AIDS example. Soc Sci Med 21:1203–1216

Kohler TA, Gumerman GJ (2000) Dynamics of human and primate societies: agent-based modeling of social and spatial processes. Oxford University Press, Oxford

Kowalewski MM, Salzer JS, Deutsch JC, Raño M, Kuhlenschmidt MS, Gillespie TR (2011) Black and gold howler monkeys (*Alouatta caraya*) as sentinels of ecosystem health: patterns of zoonotic protozoa infection relative to degree of human–primate contact. Am J Primatol 73:75–83

Krause J, Croft DP, James R (2007) Social network theory in the behavioural sciences: potential applications. Behav Ecol Sociobiol 62:15–27

Kumara HN, Singh M, Sharma AK, Santhosh K, Pal A (2014) Impact of forest fragment size on between-group encounters in lion-tailed macaques. Primates 55:543–548

Kyes R, Iskandar E, Onibala J, Paputungan U, Laatung S, Huettmann F (2012) Long-term population survey of the Sulawesi black macaques (*Macaca nigra*) at Tangkoko Nature Reserve, North Sulawesi, Indonesia. Am J Primatol 75:88–94

Laliberte E, Tylianakis JM (2010) Deforestation homogenizes tropical parasitoid–host networks. Ecology 91:1740–1747

Lane KE, Holley C, Hollocher H, Fuentes A (2011) The anthropogenic environment lessens the intensity and prevalence of gastrointestinal parasites in Balinese long-tailed macaques (*Macaca fascicularis*). Primates 52:117–128

Latapy M, Magnien C, Del Vecchio N (2008) Basic notions for the analysis of large two-mode networks. Soc Networks 30:31–48

Lee K, Divis PCS, Zakaria SK, Matusop A, Julin RA, Conway DJ, Cox-Singh J, Singh B (2011) *Plasmodium knowlesi*: reservoir hosts and tracking the emergence in humans and macaques. PLoS Pathog 7:1–11

Leendertz SAJ, Wich SA, Ancrenaz M, Bergl RA, Gonder MK, Humle T, Leendertz FH (2017) Ebola in great apes—current knowledge, possibilities for vaccination, and implications for conservation and human health. Mammal Rev 47:98–111

Li J (1999) The Tibetan macaque society: a field study. Anhui University Press, Hefei

Liu J, Dietz T, Carpenter SR, Alberti M, Folke C, Moran E et al (2007) Complexity of coupled human and natural systems. Science 317(5844):1513–1516

Loudon JE, Patel ER, Faulkner C, Schopler R, Kramer RA, Williams CV, Herrera JP (2017) An ethnoprimatological assessment of human impact on the parasite ecology of silky sifaka (*Propithecus candidus*). In: Fuentes A, Riley EP, Dore KM (eds) Ethnoprimatology: a practical guide to research at the human-nonhuman primate interface. Cambridge University Press, Cambridge, pp 89–110

Lowe SJ, Browne M, Boudjelas S (2000) Published by the IUCN/SSC Invasive Species Specialist Group (ISSG), Auckland

Lusseau D, Newman MEJ (2004) Identifying the role that individual animals play in their social network. Proc Biol Sci 271:S477–S481

MacIntosh AJJ (2016) Pathogen. In: Fuentes A, Bezanson M, Campbell CJ (eds) The international encyclopedia of primatology. Wiley, Hoboken, NJ

MacIntosh AJJ, Frias L (2016) Coevolution of hosts and parasites. In: Fuentes A, Bezanson M, Campbell CJ (eds) The international encyclopedia of primatology. Wiley, Hoboken, NJ

MacIntosh AJJ, Hernandez AD, Huffman MA (2010) Host age, sex and reproduction affect nematode parasitism among wild Japanese macaques. Primates 51:353–364

MacIntosh AJJ, Jacobs A, Garcia C, Shimizu K, Mouri K, Huffman MA, Hernandez AD (2012) Monkeys in the middle: parasite transmission through the social network of a wild primate. PLoS One 7:e51144

Marechal L, Semple S, Majolo B, Qarro M, Heistermann M, MacLarnon A (2011) Impacts of tourism on anxiety and physiological stress levels in wild male Barbary macaques. Biol Conserv 144:2188–2193

Marechal L, Semple S, Majolo B, MacLarnon A (2016) Assessing the effects of tourist provisioning on the health of wild Barbary macaques in Morocco. PLoS One 11:e0155920

Marty PR, Beisner B, Kaburu SSK, Balasubramaniam KN, Bliss-Moreau E, Ruppert N, Sah S, Ahmad I, Arlet ME, Atwill RA, McCowan B (2019) Time constraints imposed by anthropogenic environments alter social behaviour in long-tailed macaques. Anim Behav 150:157–165

Matheson MD, Sheeran LK, Li J, Wagner S (2013) Tourist impact on Tibetan macaques. Anthrozoos 26:158–168

McCabe CM, Nunn CL (2018) Effective network size predicted from simulations of pathogen outbreaks through social networks provides a novel measure of structure-standardized group size. Front Vet Sci 5:71

McLennan MR, Mori H, Mahittikorn A, Prasertbun R, Hagiwara K, Huffman MA (2017) Zoonotic enterobacterial pathogens detected in wild chimpanzees. EcoHealth 15:143–147

Meyers LA (2007) Contact network epidemiology: bond percolation applied to infectious disease prediction and control. Bull Am Math Soc 44:63–86

Minsky M (1965) Matter, mind, and models. In: Minsky M (ed) Semantic information processing. MIT Press, Cambridge, pp 425–432

Moore C, Newman MEJ (2000) Epidemics and percolation in small-world networks. Phys Rev E 61:5678–5682

Moyes CL, Henry AJ, Golding N, Huang Z, Singh B, Baird JK, Newton PA, Huffman MA, Duda KA, Drakeley CJ, Elyazar IRF, Anstey NM, Chen Q, Zommers Z, Bhatt S, Gething PW, Hay SI (2014) Defining the geographical range of the *Plasmodium knowlesi* reservoir. PLoS Negl Trop Dis 8:e2780

Muehlenbein MP, Ancrenaz M (2010) Minimizing pathogen transmission at primate ecotourism destinations: the need for input from travel medicine. J Travel Med 16:229–232

Nahallage CAD, Huffman MA (2013) Macaque-human interactions in past and present-day Sri Lanka. In: Radhakrishna S, Huffman MA, Sinha A (eds) The macaque connection: cooperation and conflict between humans and macaques. Springer, New York, pp 135–148

Nahallage CAD, Huffman MA, Kuruppu N, Weerasingha T (2008) Diurnal primates of Sri Lanka and people's perception of them. Primate Conserv 23:1–7

Nesse RM, Bergstrom CT, Ellison PT, Flier JS, Gluckman P, Govindaraju DR, Niethammer D, Omenn GS, Perlman RL, Schwartz MD, Thomas MG, Stearns SC, Valle D (2010) Making evolutionary biology a basic science for medicine. PNAS 107:1800–1807

Newman MEJ (2002) Spread of epidemic disease on networks. Phys Rev E66:016128

Newman MEJ (2004) Analysis of weighted networks. Phys Rev E 70:056131

Newman MEJ (2006) Finding community structure in networks using the eigenvectors of matrices. Phys Rev E 74:036104

Nunn CL (2009) Using agent-based models to investigate primate disease ecology. In: Huffman M, Chapman CA (eds) Primate parasite ecology: the dynamics and study of host-parasite relationships. Cambridge University Press, Cambridge, pp 83–110

Nunn CL (2011) The comparative approach in evolutionary anthropology and biology. University of Chicago Press, Chicago

Nunn CL (2012) Primate disease ecology in comparative and theoretical perspective. Am J Primatol 74:497–509

Nunn CL, Altizer SM (2005) The global mammal parasite database: an online resource for infectious disease records in wild primates. Evol Anthropol 14:1–2

Nunn CL, Altizer SM (2006) Infectious diseases in primates: behavior, ecology and evolution. Oxford University Press, Oxford

Nunn CL, Thrall PH, Stewart K, Harcourt AH (2008) Emerging infectious diseases and animal social systems. Evol Ecol 22:519–543

Nunn CL, Thrall PH, Leendertz FH, Boesch C (2011) The spread of fecally transmitted parasites in socially structured populations. PLoS One 6:e21677

Nunn CL, Jordan F, McCabe CM, Verdolin JL, Fewell JH (2015) Infectious disease and group size: more than just a numbers game. Philos Trans R Soc Lond Ser B Biol Sci 370:20140111

Palacios JFG, Engelhardt A, Agil M, Hodges K, Bogia R, Waltert M (2012) Status of, and conservation recommendations for, the critically endangered crested black macaque *Macaca nigra* in Tangkoko, Indonesia. Oryx 46:290–297

Pasquaretta C, Levé M, Claidière N, van de Waal E, Whiten A, MacIntosh AJJ, Pele M, Bergstrom ML, Borgeaud C, Brosnan S, Crofoot MC, Fedigan LM, Fichtel C, Hopper LM, Mareno MC,

Petit O, Schnoell AV, di Sorrentino EP, Thierry B, Tiddi B, Sueur C (2014) Social networks in primates: smart and tolerant species have more efficient networks. Sci Rep 4:7600

Paterson JD, Wallis J (2005) Commensalism and conflict: the human-primate interface. American Society of Primatologists, Norman, OK

Patterson JEH, Ruckstuhl KE (2013) Parasite infection and host group size: a meta-analytical review. Parasitology 140:803–813

Petrášová J, Petrželková KJ, Huffman MA, Mapua MI, Bobáková L, Mazoch V, Singh J, Kaur T, Petrášová KJ (2010) Gastrointestinal parasites of indigenous and introduced primate species of Rubondo Island National Park, Tanzania. Int J Primatol 31:920–936

Petrášová J, Uzliková M, Kostka M, Petrželková KJ, Huffman MA, Modrý D (2011) Diversity and host specificity of Blastocystis in syntopic primates on Rubondo Island, Tanzania. Int J Primatol 11:1113–1120

Phillips KA, Bales KL, Capitanio JP, Conley A, Czoty PW, Hart BA, Hopkins WD, Hu SL, Miller LA, Nader MA, Nathanielsz PW, Rogers J, Shively CA, Voytko ML (2014) Why primate models matter. Am J Primatol 76:801–827

Poirotte C, Massol F, Herbert A, Willaume E, Bomo PM, Kappeler PM, Charpentier MJE (2017) Mandrills use olfaction to socially avoid parasitized conspecifics. Sci Adv 3:e1601721

Poirotte C, Sarabian C, Ngoubangoye B, MacIntosh AJJ, Charpentier M (2019) Faecal avoidance differs between the sexes but not with nematode infection risk in mandrills. Anim Behav 149:97–106

Postel SL, Daily GC, Ehrlich PR (1996) Human appropriation of renewable fresh water. Science 271:785–788

Poulin R, Krasnov BR, Pilosof S, Thieltges DW (2013) Phylogeny determines the role of helminth parasites in intertidal food webs. J Anim Ecol 82:1265–1275

Priston NEC, McLennan MR (2013) Managing humans, managing macaques: human–macaque conflict in Asia and Africa. In: Radhakrishna S, Huffman MA, Sinha A (eds) The macaque connection, developments in primatology: progress and prospects. Springer, New York, pp 225–250

Priston NEC, Wyper RM, Lee PC (2012) Buton macaques (*Macaca ochreata brunnescens*): crops, conflict, and behavior on farms. Am J Primatol 74:29–36

Radhakrishna S, Sinha A (2011) Less than wild? Commensal primates and wildlife conservation. J Biosci 36:749–753

Radhakrishna S, Huffman MA, Sinha A (2013) The macaque connection. Springer, New York

Ram S, Venkatachalam S, Sinha A (2003) Changing social strategies of wild female bonnet macaques during natural foraging and on provisioning. Curr Sci 84:780–790

Rezende EL, Albert EM, Fortuna MA, Bascompte J (2009) Compartments in a marine food web associated with phylogeny, body mass, and habitat structure. Ecol Lett 12:779–788

Rifkin J, Nunn CL, Garamszegi LZ (2012) Do animals living in larger groups experience greater parasitism? A meta-analysis. Am Nat 180:70–82

Riley EP (2003) "Whose woods are these?" Ethnoprimatology and conservation in Sulawesi, Indonesia. Am J Phys Anthropol:179–179

Riley EP (2007) The human-macaque interface: conservation implications of current and future overlap and conflict in lore Lindu National Park, Sulawesi, Indonesia. Am Anthropol 109:473–484

Riley EP, Fuentes A (2011) Conserving social–ecological systems in indonesia: human–nonhuman primate interconnections in Bali and Sulawesi. Am J Primatol 73:62–74

Rimbach R, Bisanzio D, Galvis N, Link A, Di Fiore A, Gillespie TR (2015) Brown spider monkeys (*Ateles hybridus*): a model for differentiating the role of social networks and physical contact on parasite transmission dynamics. Philos Trans R Soc Lond Ser B Biol Sci 370:20140110

Romano V, Duboscq J, Sarabian C, Thomas E, Sueur C, MacIntosh AJJ (2016) Modeling infection transmission in primate networks to predict centrality-based risk. Am J Primatol 78:767–779

Romano V, Shen M, Pansanel J, MacIntosh AJJ, Sueur C (2018) Social transmission in networks: global efficiency peaks with intermediate levels of modularity. Behav Ecol Sociobiol 72:154

Rushmore J, Caillaud D, Hall RJ, Stumpf RM, Meyers LA, Altizer S (2014) Network based vaccination improves prospects for disease control in wild chimpanzees. J R Soc Interface 11:0349

Rushmore J, Bisanzio D, Gillespie TR (2017) Making new connections: insights from primate-parasite networks. Trends Parasitol 33:547–560

Rwego IB, Isabirye-basuta G, Gillespie TR, Goldberg T (2007) Gastrointestinal bacterial transmission among humans, mountain gorillas, and livestock in Bwindi Impenetrable National Park, Uganda. Conserv Biol 22:1600–1607

Rwego IB, Gillespie TR, Isabirye-basuta G, Goldberg TL (2008) High rates of *Escherichia coli* transmission between livestock and humans in rural Uganda. J Clin Microbiol 46:3187–3191

Salathe M, Jones JH (2010) Dynamics and control of diseases in networks with community structure. PLoS Comput Biol 6:e1000736

Sapolsky RM, Romero LM, Munck AU (2000) How do glucocorticoids influence stress responses? Integrating permissive, suppressive, stimulatory, and preparative actions. Endocr Rev 21:55–89

Sarabian C, MacIntosh AJJ (2015) Hygienic tendencies correlate with low geohelminth infection in free-ranging macaques. Biol Lett 11:20150757

Sarabian C, Ngoubangoye B, MacIntosh AJJ (2017) Avoidance of biological contaminants through sight, smell and touch in chimpanzees. R Soc Open Sci 4:170968

Schwabe C (1984) Medicine and human health. William & Wilkins, Baltimore

Sears HJ, Brownlee I, Uchiyama JK (1950) Persistence of individual strains of *Escherichia coli* in the intestinal tract of man. J Bacteriol 59:293–301

Sears HJ, Janes H, Saloum R, Brownlee I, Lamoreaux LF (1956) Persistence of individual strains of *Escherichia coli* in man and dog under varying conditions. J Bacteriol 71:370–372

Sha JCM, Gumert MD, Lee B, Jones-Engel L, Chan S, Fuentes A (2009) Macaque-human interactions and the societal perceptions of macaques in Singapore. Am J Primatol 71:825–839

Shah KV, Southwick CH (1965) Prevalence of antibodies to certain viruses in sera of free-living rhesus and of captive monkeys. Indian J Med Res 53:488–500

Siex KS (2005) Habitat destruction, population compression, and overbrowsing by the Zanzibar red colobus monkey (*Procolobus kirkii*). In: Paterson JD, Wallis J (eds) Commensalism and conflict: the human-primate interface. American Society of Primatologists, Norman, OK, pp 294–337

Silk MJ, Croft DP, Delahay RJ, Hodgson DJ, Boots M, Weber N, McDonald RA (2017) Using social network measures in wildlife disease ecology, epidemiology, and management. Bioscience 67:245–257

Singh M, Kumara HN, Kumar MA, Sharma AK (2001) Behavioural responses of lion-tailed macaques (*Macaca silenus*) to a changing habitat in a tropical rain forest fragment in the Western Ghats, India. Folia Primatol 72:278–291

Singh B, Kim Sung L, Matusop A, Radhakrishnan A, Shamsul SS, Cox-Singh J, Thomas A, Conway DJ (2004) A large focus of naturally acquired *Plasmodium knowlesi* infections in human beings. Lancet 363:1017–1024

Sinha A, Mukhopadhyay K, Datta-Roy A, Ram S (2005) Ecology proposes, behaviour disposes: ecological variability in social organization and male behavioural strategies among wild bonnet macaques. Curr Sci 89:1166–1179

Sinton L, Hall C, Braithwaite R (2007) Sunlight inactivation of *Campylobacter jejuni* and *Salmonella enterica*, compared with *Escherichia coli*, in seawater and river water. J Water Health 5:357–365

Southwick CH, Siddiqi MF (1994) Primate commensalism: the rhesus monkey in India. Rev Ecol 49:223–231

Southwick CH, Siddiqi MF (1998) The rhesus monkey's fall from grace. In: Ciochon RL, Nisbett RA (eds) The primate anthology. Prentice Hall, Upper Saddle River, pp 211–218

Southwick CH, Siddiqi F (2011) India's rhesus population: protection versus conservation management. In: Gumert MD, Fuentes A, Jones-Engel L (eds) Monkeys on the edge: ecology and management of long-tailed macaques and their interface with humans. Cambridge University Press, Cambridge, pp 275–292

Southwick CH, Siddiqi MF, Oppenheimer JR (1983) Twenty-year changes in rhesus macaque populations in agricultural areas of Northern India. Ecology 64:434–439

Spiesman BJ, Gratton C (2016) Flexible foraging shapes the topology of plant-pollinator interaction networks. Ecology 97:1431–1441

Springer A, Mellmann A, Fichtel C, Kappeler PM (2016) Social structure and Escherichia coli sharing in a group-living wild primate, Verreaux's sifaka. BMC Ecol 16:6

Stang M, Klinkhamer PGL, Waser NM, Stang I, van der Meijden E (2009) Size-specific interaction patterns and size matching in a plant–pollinator interaction web. Ann Bot 103:1459–1469

Stephens PR, Pappalardo P, Huang S, Byers JE, Farrell MJ, Gehman A, Ghai RR, Haas SE, Han B, Park AW, Schmidt JP, Altizer S, Ezenewa VO, Nunn CL (2017) Global mammal parasite database version 2.0. Ecology 98(5):1476

Sterck EHM (1998) Female dispersal, social organization, and infanticide in langurs: are they linked to human disturbance? Am J Primatol 44(4):235–254

Sueur C, Petit O, De Marco A, Jacobs AT, Watanabe K, Thierry B (2011a) A comparative network analysis of social style in macaques. Anim Behav 82(4):845–852

Sueur C, Jacobs A, Amblard F, Petit O, King AJ (2011b) How can social network analysis improve the study of primate behavior? Am J Primatol 73:703–719

Sueur C, Deneubourg JL, Petit O (2012) From social network (centralized vs. decentralized) to collective decision-making (unshared vs. shared consensus). PLoS One 7:e32566

Sueur C, Romano V, Sosa S, Puga-Gonzalez I (2018) Mechanisms of network evolution: a focus on socioecological factors, intermediary mechanisms, and selection pressures. Primates 60:1–15

Suleyman G, Tibbetts R, Perri MB, Vager D, Xin Y, Reyes K, Samuel L, Chami E, Starr P, Pietsch J, Zervos MJ, Alangaden G (2016) Nosocomial outbreak of a novel extended-spectrum beta-lactamase *Salmonella enterica* serotype isangi among surgical patients. Infect Control Hosp Epidemiol 37:954–961

Tenaillon O, Skurnik D, Picard B, Denamur E (2010) The population genetics of commensal *Escherichia coli*. Nat Rev Microbiol 8:207–217

Thierry B (2007a) The macaques: a double-layered social organization. In: Campbell CJ, Fuentes A, MacKinnon KC, Panger M, Bearder SK, Stumpf RM (eds) Primates in perspective. Oxford University Press, New York, pp 224–239

Thierry B (2007b) Unity in diversity: lessons from macaque societies. Evol Anthropol 16:224–238

Thierry B (2013) The macaques: a double-layered social organization. In: Campbell CJ, Fuentes A, MacKinnon KC, Bearder SK, Stumpf RM (eds) Primates in perspective. Oxford University Press, Oxford, pp 229–240

Thierry B, Aureli F, Nunn CL, Petit O, Abegg C, de Waal FBM (2008) A comparative study of conflict resolution in macaques: insights into the nature of covariation. Anim Behav 75:847–860

Tutin CEG (2000) Ecologie et organisation des primates de la foret tropicale africaine: aide a la comprehension de la transmission des retrovirus. For Trop Emerg Virales 93:157–161

Vallo P, Petrželková KJ, Profousová I, Petrášov J, Pomajbíková K, Leendertz F, Hashimoto C, Simmons N, Babweteera F, Piel A, Robbins MM, Boesch C, Sanz C, Morgan D, Sommer V, Furuichi T, Fujita S, Matsuzawa T, Kaur T, Huffman MA, Modry D (2012) Molecular diversity of entodiniomorphid ciliate Troglodytella abrassarti and its coevolution with chimpanzees. Am J Phys Anthropol 48:525–533

VanderWaal KL, Ezenwa VO (2016) Heterogeneity in pathogen transmission: mechanisms and methodology. Funct Ecol 30:1606–1622

VanderWaal KL, Atwill ER, Isbell LA, McCowan B (2013) Linking social and pathogen transmission networks using microbial genetics in giraffe (*Giraffa camelopardalis*). J Anim Ecol 83:406–414

VanderWaal KL, Atwill ER, Isbell LA, McCowan B (2014) Quantifying microbe transmission networks for wild and domestic ungulates in Kenya. Biol Conserv 169:136–146

Vazquez DP, Chacoff N, Cagnolo L (2009) Evaluating multiple determinants of the structure of plant–animal mutualistic networks. Ecology 90:2039–2046

Waters CN, Zalasiewicz J, Summerhayes C, Barnosky AD, Poirier C, Gauszka A, Cearreta A, Edgeworth M, Ellis EC, Ellis M, Jeandel C, Leinfelder R, McNeill JR, Richter DB, Steffen W, Syvitski J, Vidas D, Wagreich M, Williams M, Zhisheng A, Grinevald J, Odada E, Oreskes N, Wolfe AP (2016) The Anthropocene is functionally and stratigraphically distinct from the Holocene. Science 351:aad2622

Wenz-Mücke A, Sithithaworn P, Petney TN, Taraschewski H (2013) Human contact influences the foraging behaviour and parasite community in long-tailed macaques. Parasitology 140:709–718

Wey T, Blumstein DT, Shen W, Jordan F (2008) Social network analysis of animal behaviour: a promising tool for the study of sociality. Anim Behav 75:333–344

Whitehead H, Dufault S (1999) Techniques for analyzing vertebrate social structure using identified individuals: review and recommendations. Adv Study Behav 28:33–74

Wilensky U, Stroup W (1999) NetLogo HubNet Disease model [computer software]. Center for Connected Learning and Computer-Based Modeling, Northwestern University, Evanston, IL

Wolfe ND, Dunavan CP, Diamond J (2007) Origins of major human infectious diseases. Nature 447:279–283

Young C, Majolo B, Heistermann M, Schülke O, Ostner J (2014) Responses to social and environmental stress are attenuated by strong male bonds in wild macaques. PNAS 111:18195–18200

Zhang X, Kadir KA, Quintanilla-Zarinan LF, Villano J, Houghton P, Du H, Singh B, Glenn Smith D (2016) Distribution and prevalence of malaria parasites among long-tailed macaques (*Macaca fascicularis*) in regional populations across Southeast Asia. Malar J 15:450

Zhao QK (1996) Etho-ecology of Tibetan macaques at Mount Emei, China. In: Fa J, Lindburg DG (eds) Evolution and ecology of macaque societies. Cambridge University Press, Cambridge

Zhao QK (2005) Tibetan macaques, visitors, and local people at Mt. Emei: problems and counter-measures. In: Paterson JD, Wallis J (eds) Commensalism and conflict: the human–primate interface. American Society of Primatologists, Norman, OK, pp 376–399

Zinsstag J, Schelling E, Waltner-Toews D, Tanner M (2011) From "one medicine" to "one health" and systemic approaches to health and well-being. Prev Vet Med 101:148–156

Zinsstag J, Schelling E, Waltner-Toews D, Tanner M (2015) One health, the theory and practice of integrated one health approaches. CAB International, Oxfordshire

Part V

Emerging Technologies in Primatology

Chapter 14

MRI Technology for Behavioral and Cognitive Studies in Macaques In Vivo

Yong Zhu and Paul A. Garber

14.1   Introduction

Primate behavior, especially social behavior, might seem a strange place to begin in trying to advance our understanding of the brain or human mind (Opstal 1996; Critchley and Harrion 2013). Primate brains differ in structural details, proportion among functional areas, as well as in overall size, implying specific adaptations to the challenges posed by particular ecological and social environments across that species' evolutionary history. Given that brains require a disproportionate amount of nutrients and energy relative to other body organs (Aiello and Wheeler 1995), an understanding of the functional implications of evolutionary changes in brain organization is critical for evaluating relationships between cognition, decision-making, and behavior (Rilling 2006) (Fig. 14.1).

The recent decade has seen an explosion of interest and information about brain connectivity and functions over a wide range of spatial scales, including macroscopic, microscopic, and mesoscopic levels (Essen et al. 2016). Noninvasive imaging studies with magnetic resonance imaging (MRI) technologies have been used for over 30 years (Vanduffel 2018) and have made fundamental contributions to our understanding of animal behavior and cognition, especially in our understanding of the relationships between brain structures and brain function. For example, the brain's default mode network (DMN) consists of discrete, bilateral, and symmetrical cortical areas in the medial and lateral parietal, medial prefrontal, and medial and lateral temporal cortices of the human, nonhuman primate, cat, and rodent brains

Y. Zhu (✉)
High Magnetic Field Laboratory, Chinese Academy of Sciences, Hefei, China

School of Life Sciences, Hefei Normal University, Hefei, Anhui, China

P. A. Garber
Department of Anthropology, Program in Ecology, Evolution, and Conservation Biology, University of Illinois, Urbana, IL, USA
e-mail: p-garber@illinois.edu

J.-H. Li et al. (eds.), *The Behavioral Ecology of the Tibetan Macaque*, Fascinating Life Sciences, https://doi.org/10.1007/978-3-030-27920-2_14

287

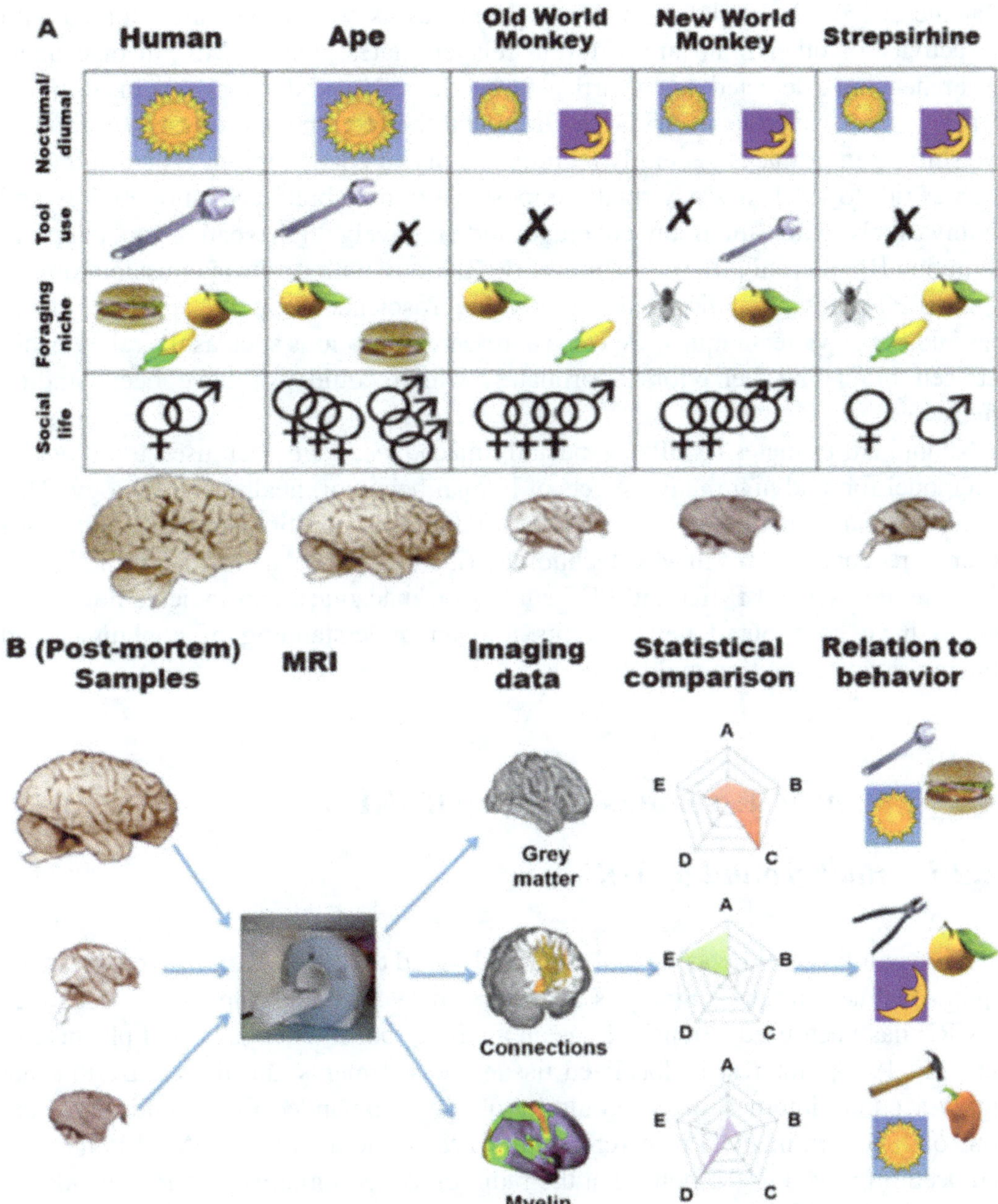

Fig. 14.1 Analyzing the relationship between brains and behavior (Mars et al. 2014). (**a**) In a multivariate comparative approach, each brain is viewed as a unique combination of variables, including whether the animal is active during night or day; whether it uses tools regularly, occasionally, or not at all; what its diet is; or how complex its social life is. By using a whole-brain and multivariate approach, it is possible for us to investigate how differences in specific aspects of brain organization are related to different ecological and social variables. (**b**) MRI of whole-brain (postmortem) samples allows a number of measures to be collected, for which comparative analysis techniques have now been developed and validated

(Raichle 2015). The DMN consistently decreases its activity when compared with the activity of other brain areas during relaxed states. Using fMRI in macaques, researchers have detected that a cortical network, activated during shifts in cognitive activity, largely overlapped with the DMN and therefore have proposed that cognitive shifting in primates generally recruits activity in DMN regions (Arsenault et al. 2018). Prior to 1991, it was virtually impossible to map brain activation rapidly and noninvasively with full brain coverage and relatively high spatial and temporal resolution (Bandettini 2009). The use of fMRI along with positron emission tomography (PET), has revolutionized cognitive neuroscience (Logothetis 2008). Using these noninvasive techniques, we can ask research questions such as the relationship between brain and behavior in primates, which could not have been studied otherwise.

Nonhuman primates (NHPs), especially macaques, have been used traditionally as a model for studying many aspects of human behavior, health, and biology. This includes social structure, social behavior, and social cognition. In this chapter, we describe recent advances in MRI technology (including the state-of-the-art high field MRI), review several fMRI and PET studies on macaques, and indicate how these studies have contributed new insights into an understanding of nonhuman and human primate cognition and behavior.

14.2 Magnetic Resonance Imaging (MRI)

14.2.1 Background of MRI

MRI is a painless, noninvasive tool commonly used to diagnose disease progression, injury, or other ailments. Since its discovery in 1945, nuclear magnetic resonance (NMRI) has been used extensively as an analytic tool in chemistry and physics. In the early 1970s, interest in localized tissue measurements and the realization that highly accurate internal images could be obtained expanded the potential applications of NMRI in medical research. By the 1980s, the quality of NMRI in humans had been improved to the point that the radiological community considered MRI as the next high technology imaging modality.

Today, we refer to "NMRI" as MRI. The use of word "nuclear" in the acronym was dropped to avoid a negative association with the potential exposure to nuclear radiation. Its working principle is based on the fact that certain atomic nuclei are able to absorb and emit radio-frequency energy when placed in an external magnetic field. In both its clinical application and in MRI research, hydrogen atoms are most often used to generate a detectable radio-frequency signal that is received by antennas in close proximity to the anatomical structure being examined. Hydrogen atoms are naturally abundant, particularly in water and fat. For this reason, most MRI scans essentially map the location of water and fat in the body. Pulses of radio waves excite the nuclear spin energy transition, and magnetic field gradients localize the signal in space. By varying the parameters of the pulse sequence, different contrasts may be generated between tissues based on the relaxation properties of the hydrogen atoms.

14.2.2 The Advantages of MRI

Compared with other medical imaging techniques such as X-ray imaging, ultrasonic imaging, and computed tomography (CT), MRI is a painless and noninvasive method to view human or animal tissue and obtain anatomical and functional diagnostic information. While the hazards of X-rays are now well-controlled in most medical contexts, MRI may still be seen as a better choice than CT. MRI scanners are designed to visualize non-bony parts or "soft tissue" areas such as muscles, ligaments, and tendons. In particular, the brain, spinal cord, and nerves are seen much more clearly with MRI than with regular X-rays and CT scans.

What are the advantages of an MRI scan? To summarize, there are several as follows:

1. High accuracy

   In clinical diagnostics and studies, MRI is used as the preferred medical examination tool owing to its high accuracy in the detection of serious ailments such as tumors and cancers. In research, it can be used in structural and functional connectivity studies. In addition, the MRI scan parameters can be adjusted for a more comprehensive view of the area of interest.

2. Less confining

   An MRI scanner has a design that makes it less confining for the patient or research subject. It is possible to assign tasks to the subject in the scanner, for example, evaluate light or sound stimulation, and this facilitates directly linking behavior and functional anatomy.

3. Noninvasive

   MRI does not involve X-rays or the use of ionizing radiation, distinguishing it from CT or CAT scans. This removes all ethical concerns regarding harm that could be done to patients or study subject. This permits the ethical use of fMRI research studies because normal subjects face no risk of harm or injury.

   With these four main advantages, MRI represents a powerful tool well suited for studying in vivo behavioral responses, especially functional studies that link neuroanatomy and cognitive behavior in primates.

14.2.3 State of the Art at High Field MRI

As far as MRI is concerned, the need to achieve the highest possible magnetic field in clinical settings has been a major motivation for scientists and engineers to improve the imaging technology. As a result, MRI scanners have evolved from the first generation (0.5 Tesla or 0.5 T) to the conventional (1.0–1.5 T) scanner and then to the high field (3 T) scanners in clinical applications. Most recently, ultra-high field MRI scanners ranging from 7 to 11.7 T have been designed. Generally speaking, the higher the magnetic field, the better the resolution of the images.

Using MRI, a higher magnetic field intensity leads to a higher signal-to-noise ratio (SNR), resulting in a higher image resolution. However, higher magnetic field intensity can lead to a smaller caliber machine (the diameter of the machine that allows the animal to be scanned) meaning that animals larger than macaques (e.g., gorilla) could not be examined. For example, the 3 T MRI scanner in clinical applications usually has a 70 cm caliber, suited for human or gorilla body scans. The 9.4 T MRI scanner with a caliber ranging from 30 to 40 cm is suitable for NHPs ranging in size from a mouse lemur to a macaque. The 9.4 T MRI scanner obtains enhanced image resolution compared to the 3 T MRI scanner. Therefore, considering the image resolution, a 9.4 T MRI scanner is the most suitable for studying mammals ranging from rodents to most species of NHPs.

There are many 9.4 T MRI scanners currently in use. For example, a 9.4 T MRI scanner with a 40 cm diameter bore (Agilent, US) is operated at the High Magnetic Field Laboratory (HMFL), Chinese Academy of Sciences, Hefei. This scanner is dedicated to the study of larger mammals (e.g., dog, sheep, and macaque). Another 9.4 T/30 cm MRI scanner (Bruker, Germany) is operated at the Shanghai Institute of Biological Sciences of the Chinese Academy of Sciences. This scanner focuses on scientific research requiring a high magnetic field. At the University of Chicago, Chicago, Illinois, a 9.4 T/30 cm MRI facility is used for basic science research within the Department of Radiology. The goal is to advance state-of-the-art animal imaging technology at the anatomical and functional level (https://mris.uchicago.edu/). Many experimental studies of NHPs that focus on research questions including brain development, brain structural changes, and functional studies of different brain regions have been conducted using other types of 9.4 T MRI scanners such as the 9.4 T/35 cm scanner in Germany (Goebel et al. 2009) and France (Même et al. 2015) and the 9.4 T/39 cm scanner at University College London Centre for Advanced Biomedical Imaging, London (Ramasawmy et al. 2016).

Some primate studies have been carried out on 1.5 T or 3 T scanners (Nelissen and Vanduffel 2017), with a sub-optimal resolution in the range of 2–3 mm. In vivo studies of structural brain imaging at 7 T have been reported in the rhesus macaque (*Macaca mulatta*), with resolution from 0.3 to 0.5 mm (Zitella et al. 2015). A much finer image resolution (under 100 µm) is needed to pick up minor changes, such as in brain aging that would be required to examine questions of ontogenetic changes in social behavior related to aging. However, only a limited number of primate studies have used 9.4 T MRI scanners, which would generate in vivo data with a resolution on the order of 0.1 mm. This level of resolution is needed to examine questions related to the substructure of the hippocampus. Because dynamic brain alterations can be detected on a much finer scale, high field MRI greatly facilitates studies integrating primate brain structure and behavioral change.

14.3   In Vivo MRI Study in Macaques

Investigations into the structure and function of the NHP brain have significantly contributed to our overall understanding of cognition.

14.3.1 Structural MRI in Brain Imaging Study

To understand the anatomical localization and functional activity in different cortical areas, it is necessary to obtain an accurate map of neural architectonic areas. In most MRI studies, primates are anesthetized when scanned. The latest research, published in the journal *NeuroImage*, has created an anatomical MRI brain template derived from 31 rhesus macaques (the macaques were juveniles and adults between 3.2 and 13.2 years old). The template also includes tissue maps, surfaces, and transformation scripts to assist in data analysis (Seidlitz et al. 2018). These data can be used to determine the variance of cortical topographies in the same individual over time and also compare cortical differences between infants, juveniles, and adults. This can be used as a framework for examining correlational relationships between changes in behavior and behavioral and brain development.

Unlike morphological studies using postmortem tissue samples, MRI allows the noninvasive, in vivo assessment of many different brain parameters including the topography and volume of gray and white matter in brain structures. At the level of the cerebral hemisphere, gray matter is mainly distributed in the periphery (cortex) while the white matter is located deep within the cortex.

Phillips and Sherwood (2008) described growth patterns in total brain volume, cortical gray and white matter volume, frontal lobe gray and white matter volume, and corpus callosum area in 29 brown capuchin monkeys (*Sapajus nigritus*, formerly *Cebus apella*) ranging in age from 4 days to 20 years. Of the total subjects, 12 were adults (≥ 5 years) and 17 were juveniles (between 4 days to 5 years). The results revealed that nonlinear age-related changes in total brain volume, cortical white matter volume, and frontal white matter volume occur from birth to 5 years of age (subadult period of development). The implications of these results is the rapid increase of frontal lobe white matter during the first few years of life corresponds with opportunities for social learning and acquiring technical skills related to object manipulation, prey search, possibly tool use, and other complex foraging behaviors (Phillips and Sherwood 2008). Similarly, Wisco et al. (2008) studied the age-related white and gray matter volume changes in eight young adult (5–12 years), six middle-aged adult (16–19 years) and eight old adult (24–30 years) rhesus macaques. The results found an overall decrease in the total forebrain (5.01%), forebrain parenchyma (5.24%), forebrain white matter (11.53%), forebrain gray matter (2.08%), caudate nucleus (11.79%), and *globus pallidus* (18.26%) with increasing age. Corresponding behavioral data for five of the younger, five of the middle-aged, and seven of the old adults on the delayed non-matching to sample (DNMS) task, the delayed recognition span task (DRST), and the cognitive impairment index (CII) found no correlation between these cognitive measures and ROI volume changes.

In a recent study, Scott et al. (2016) longitudinally assessed normative brain growth patterns using MRI in rhesus macaques. Cohort A consisted of 24 individuals (12 males, 12 females) and cohort B of 21 individuals (11 male, 10 female). They scanned the macaques at 1, 4, 8, 13, 26, 39, and 52 weeks of age. Cohort A had additional scans at 156 weeks (3 years) and 260 weeks (5 years) (Fig. 14.2). The

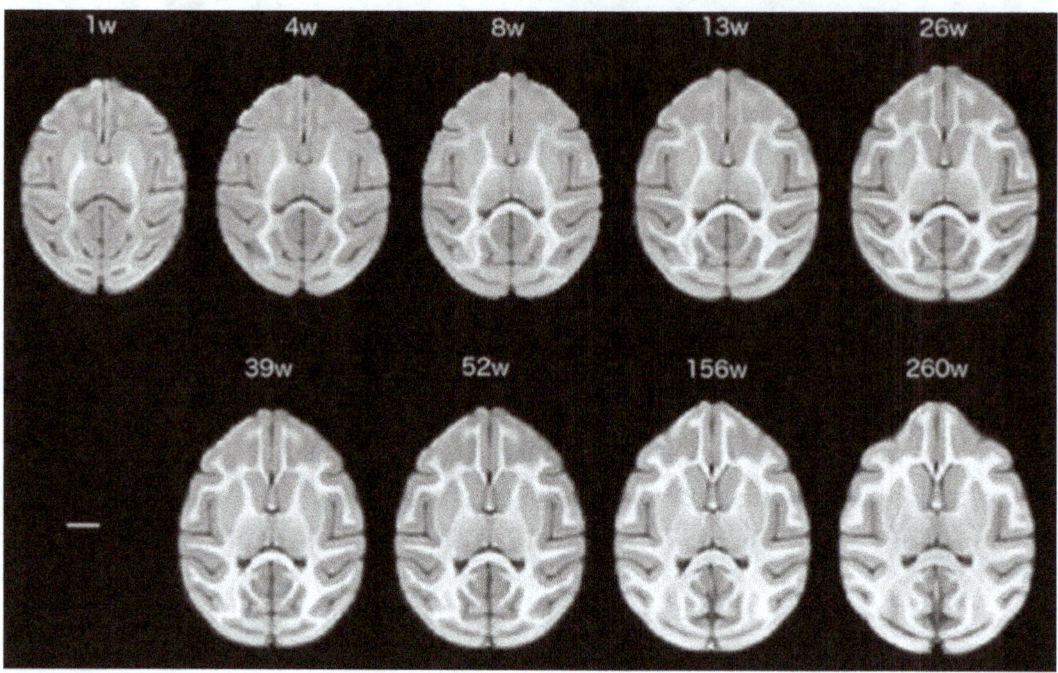

Fig. 14.2 Brain MRI at each study time point (Scott et al. 2016). Age-specific, horizontally oriented templates are shown for each study time point. Each template represents an average brain image constructed from individual subject scans. Images display enhanced signal-to-noise and optimal shape characteristics relative to individual scans (scale bar represents 1 cm)

results showed that total brain volume at 1 week was approximately 64% of that of the adult. Brain volume was larger in male rhesus macaques compared to females. While brain volume generally increased between any two imaging time points, there was a transient plateau of brain growth between 26 and 39 weeks in both males and females (Scott et al. 2016). This study serves as a starting point for more extensive analyses into the relationship between structural development of the brain and behavioral development of the rhesus macaques. The image database is available to behavioral scientists for addressing additional questions examining the relationship between changes in behavior and changes in neural development, including the emergence of sex and species typical behavior.

To evaluate hippocampal development in rhesus macaques, Hunsaker et al. (2014) obtained longitudinal structural MRI scans at 9 time points between 1 week and 260 weeks (5 years) of age in 24 rhesus macaques (12 males, 12 females) (Fig. 14.3). The results showed that the hippocampus reached 50% of its adult volume by 13 weeks of age and full adult volume by 52 weeks in both males and females. The hippocampus appears to be slightly larger at 3 years than at 5 years of age, and damage to the hippocampus deficit can result in permanent changes in behavior such as learning and memory, as well as neurocognitive function. Male rhesus macaques have a 5% larger hippocampi than females from 8 weeks of age onward. Neuroimaging studies in rhesus macaques can provide critical information about the relationships between MRI volumetric changes and behavior during individual development.

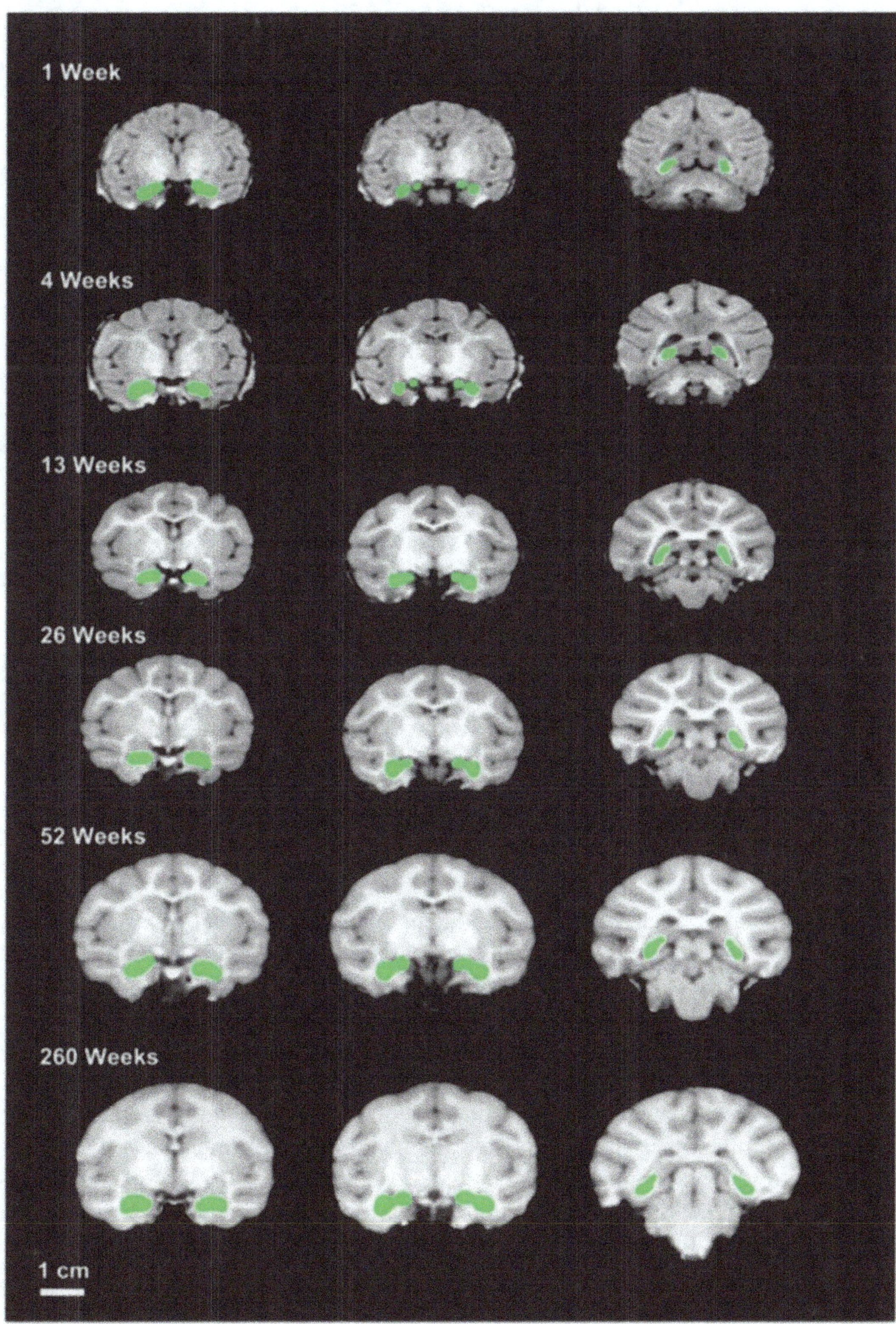

Fig. 14.3 MRI for hippocampal tracings at different ages (Hunsaker et al. 2014). Sample hippocampal tracings from a single male rhesus macaque. Shown are scans at 1 week, 4 week, 26 weeks, 39 weeks, and 260 weeks of age. Note the difference in white matter at the different ages. All scans are shown at the same scale for direct visual comparison. Note the general shape of the traced hippocampus and size relative to the rest of brain at different ages (scale bar = 1 cm)

Liu et al. (2015) examined brain development in 14 male rhesus macaques from 6 to 16 months of age. The results showed rapid growth (6.21%) in brain volume during early development between 6 and 10 months of age compared to a 2.81% growth rate between 10 and 16 months of age. Early expansion is mainly the result of a significant increase in white matter volume, while the later decline can be partly explained by a significant decrease of gray matter volume after 10 months of age. Compared with macaque brain development, human brain volume increases by 100% during the first year of life and then maintains cubic growth into adolescence. The pattern of human brain growth differs from the rhesus macaque in that humans have a less mature brain at birth (approximately 25% of adult size) and a longer period of early rapid brain growth during the first year of life (Phillips and Sherwood 2008).

14.3.2   Functional MRI in Brain Imaging Study

Functional magnetic resonance imaging (fMRI) has been used extensively in comparative sensory and cognitive experiments associated with brain activation mapping in NHPs and humans (Logothetis et al. 1999; Nakahara et al. 2002; Mantini et al. 2012). Moreover given that the subject is alert, one can study relationships between behavior, decision-making, and neural activity in real time. What exactly does fMRI tell us? We know that its signals arise from changes in local brain hemodynamics that, in turn, result in alterations in neuronal activity. However, exactly how neuronal activity, hemodynamics, fMRI signals, and decision-making are related is still unclear. It has been assumed that the fMRI signal is proportional to the local average neuronal activity, but many factors can influence this relationship (Heeger and Ress 2002).

For comparative studies of human and nonhuman primates, subjects are exposed to the same set of images as their brains are scanned using fMRI. The time course of fMRI activity during viewing is extracted from "seed" regions in human participants and correlated with the fMRI signal in NHP, or vice versa, with an adjustment made for interspecific differences in vascular hemodynamics. After statistical thresholding, which is a method of image processing, the resulting maps for each species show areas that are stimulated and potentially homologous with the targeted seed region in the other species (Wager and Yarkoni 2012) (Fig. 14.4). fMRI has been used to study the primate visual system (Logothetis et al. 1999; Nakahara et al. 2002; Russ and Leopold 2015), the auditory system (Mantini et al. 2012; Ortiz-Rios et al. 2015), and the motor system (Bauman et al. 2013), again principally in macaques. We present the results of some of these studies below.

Ortiz-Rios et al. (2015) investigated how species-specific vocalizations are represented in auditory and auditory-related regions of the brain using fMRI in rhesus macaques. The results indicated that these vocalizations preferentially activated the auditory ventral stream and in particular areas of the anterolateral belt and parabelt.

Fig. 14.4 Humans and monkeys watch the same film as their brains are scanned with fMRI (Wager and Yarjoni 2012). The time course of fMRI activity during viewing is extracted from "seed" regions in human participants and correlated with the fMRI signal in monkeys, and vice versa, with an adjustment made for interspecific differences in vascular hemodynamics

A major limitation of fMRI studies of nonhuman primates is the difficulty of training monkeys and apes to remain calm when placed in the restraining devices required to limit movement during scanning. Srihasam et al. (2010) developed a technique for holding subjects' heads motionless during scanning using a custom-fitted plastic helmet, chin strap, and mild suction supplied by a vacuum blower. This vacuum helmet method appeared to have few adverse effects on subject physical health (although the monkeys were stressed) even after repeated use for several months.

14.4   Conclusion

Studies of the brain of nonhuman primates are vital for understanding aggressive and cooperative interactions (Rilling 2014). Noninvasive imaging technologies, such as fMRI, are an important research tool that provide data that field primatologists can use to better understand how age, sex, and species differences in neural development, social experience, and neurohormones affect behavior and cognitive processes. For example, differences in patterns and rates of brain growth may help explain taxonomic differences in maternal pre- and postnatal offspring investment; differences in the manner in which species encode, store, and integrate temporal, spatial, and quantitative information in deciding where to feed; and species-specific differences in cooperative behavior. For example, oxytocin, a neurohormone synthesized in the hypothalamus, can increase cooperative behavior and reduce fear of cheaters in humans by stimulating brain regions that associate cooperative interactions with feelings of pleasure or reward (Rilling 2014). In the case of wild chimpanzees, Wittig et al. (2014) found that during periods of food sharing, levels of urinary oxytocin concentration increased significantly compared to periods when chimpanzees fed alone. Moreover this effect was evident in chimpanzees who shared food with close partners and in chimpanzees who shared food with nonpartners (Wittig et al. 2014). Similarly, Crockford et al. (2013) found that oxytocin levels in wild chimpanzees increased equally when individuals groomed kin or nonkin. Thus, there exists an important neurohormonal mechanism for promoting social cooperation between both related and nonrelated group members. In the case of common marmosets (*Callithrix jacchus*), a small-bodied New World monkey, Finkenwirth et al. (2016) examined relationships between nonmaternal infant caregiving and urinary oxytocin levels. In this species, both male and female helpers were found to exhibit significantly higher levels of urinary oxytocin when caring for offspring than during periods when not providing infant care. Finally, different species of macaques vary in dominance style from highly aggressive to highly affiliative. Although early experience may contribute to the ways in which individuals interact with others, a study by Rosenblum et al. (2002) reported that bonnet macaques (*M. radiata*), which are highly affiliative, were characterized by higher levels of oxytocin than pig-tailed macaques (*M. nemistrina*) which are highly aggressive. Combined, these studies highlight the integrated role of neurohormones, neuroreceptors, and social experience in understanding primate social interactions and behavior.

Neurodevelopmental research is expected to expand in the future, offering new insights and understanding into links between neural anatomy and behavior, both within and between species. For example, neurodevelopmental studies may allow us to understand how the brain neuronal functions organize and mature during different stages of development. This will allow researchers to address questions linking behavior, ontogeny, and social cognition in primates.

Acknowledgments We would like to thank Dr. K. Zhong, H. Y. Yang, J. Zhang, Q. J. Zhu, and H. Y. Tong at the High Magnetic Field Laboratory, the Chinese Academy of Sciences. We also thank Dr. L. X. Sun at Central Washington University for his many helpful comments and suggestions on this chapter. PAG wishes to thank Chrissie, Sara, and Jenni for their love and support. This study was supported by the National Natural Science Foundation of China (31501866) and the Science and Technology Innovation Program of the Ministry of Science and Technology of the People's Republic of China (2014BAL03B00).

References

Aiello LC, Wheeler P (1995) The expensive-tissue hypothesis: the brain and the digestive system in human and primate evolution. Curr Anthropol 36:199–221

Arsenault JT, Caspari N, Vandenberghe R, Vanduffel W (2018) Attention shifts recruit the monkey default mode network. J Neurosci 38:1202–1217

Bandettini PA (2009) Functional MRI limitations and aspirations. In: Kraft E, Gulyás B, Pöppel E (eds) Neural correlates of thinking. Springer-Verlag, Berlin, pp 15–38

Bauman MD, Losif AM, Ashwood P, Braunschweig D, Lee A, Schumann CM, Water JV, Amaral DG (2013) Maternal antibodies from mothers of children with autism alter brain growth and social behavior development in the rhesus monkey. Transl Psychiatry 3(7):e278

Critchley HD, Harrison NA (2013) Visceral influences on brain and behavior. Neuron 77:624–638

Crockford C, Wittig RM, Langergraber K, Ziegler TE, Zuberbuhler K, Deschner T (2013) Urinary oxytocin and social bonding in related and unrelated wild chimpanzees. Proc Biol Sci 280:20122765

Essen DCV, Donahue C, Dierker DL, Glasser MF (2016) Parcellations and connectivity patterns in human and macaque cerebral cortex. In: Kennedy H, Essen DCV, Christen Y (eds) Micro-, meso- and macro-connectomics of the brain. Springer International, Cham, pp 89–106

Finkenwirth C, Martins E, Deschner T, Burkart JM (2016) Oxytocin is associated with infant-care behavior and motivation in cooperatively breeding marmoset monkeys. Horm Behav 80:10–18

Goebel H, Spehl T, Paul D, Markl M, Harloff A (2009) Histopathological correlation and feasibility of atherosclerotic carotid lesion classification using T_{2*} weighted imaging at 9.4t MRI. Virchows Archiv 455:50–51

Heeger DJ, Ress D (2002) What does fMRI tell us about neuronal activity? Nat Rev Neurosci 3:142–151

Hunsaker MR, Scott JA, Bauman MD, Amaral DG (2014) Postnatal development of the hippocampus in the Rhesus macaque (*Macaca mulatta*): a longitudinal magnetic resonance imaging study. Hippocampus 24:794–807

Liu CR, Tian XG, Liu HL, Mo Y, Bai Y, Zhao XD, Ma YY, Wang JH (2015) Rhesus monkey brain development during late infancy and the effect of phencyclidine: a longitudinal MRI and DTI study. NeuroImage 107:65–75

Logothetis NK (2008) What we can do and what we cannot do with fMRI. Nature 453:869–878

Logothetis NK, Guggenberger HG, Peled S, Pauls J (1999) Functional imaging of the monkey brain. Nat Neurosci 2:555–562

Mantini D, Hasson U, Betti V, Perrucci MG, Romani GL, Corbetta M, Orban GA, Vanduffel W (2012) Inter-species activity correlations reveal functional correspondences between monkey and human brain areas. Nat Methods 9:277–282

Mars RB, Neubert FX, Verhagen L, Sallet J, Miller KL, Dunbar RIM, Barton RA (2014) Primate comparative neuroscience using magnetic resonance imaging: promises and challenges. Front Neurosci 8:298

Même S, Joudiou N, Yousfi N, Szeremeta F, Lopes-Pereira P, Beloeil JC, Herault Y, Meme W (2015) *In vivo* 9.4T MRI and ^{1}H MRS for evaluation of brain structural and metabolic changes in the Ts65Dn mouse model for Down syndrome. World J Neurosci 4:152–163

Nakahara K, Hayashi T, Konishi S, Miyashita Y (2002) Functional MRI of macaque monkeys performing a cognitive set-shifting task. Science 295:1532–1536

Nelissen K, Vanduffel W (2017) Action categorization in rhesus monkeys: discrimination of grasping from non-grasping manual motor acts. Sci Rep 7:15094

Opstal VA (1996) Dynamic patterns: the self-organization of brain and behavior. Complexity 2:253–254

Ortiz-Rios M, Kuśmierek P, Dewitt I, Archakov D, Azevedo FAC, Sams M, Jääskeläinen IP, Keliris GA, Rauschecker JP (2015) Functional MRI of the vocalization-processing network in the macaque brain. Front Neurosci 9:113

Phillips KA, Sherwood CC (2008) Cortical development in brown capuchin monkeys: a structural MRI study. NeuroImage 43:657–664

Raichle ME (2015) The brain's default mode network. Annu Rev Neurosci 38:433–447

Ramasawmy R, Johnson SP, Roberts TA, Stuckey DJ, David AL, Pedley RB, Lythgoe MF, Siow B, Walker-Samuel S (2016) Monitoring the growth of an orthotopic tumour xenograft model: multi-modal imaging assessment with benchtop MRI (1T), high-field MRI (9.4T), ultrasound and bioluminescence. PLoS One 11(5):e0156162

Rilling JK (2006) Human and nonhuman primate brains: are they allometrically scaled versions of the same design? Evol Anthropol 15:65–77

Rilling JK (2014) Comparative primate neuroimaging: insights into human brain evolution. Trends Cogn Sci 18:46–55

Rosenblum LA, Smith ELP, Altemus M, Scharf BA, Owens MJ, Nemeroff CB, Gorman JM, Coplan JD (2002) Differing concentrations of corticotropin-releasing factor and oxytocin in the cerebrospinal fluid of bonnet and pigtail macaques. Psychoneuroendocrinology 27:651–660

Russ BE, Leopold DA (2015) Functional MRI mapping of dynamic visual features during natural viewing in the macaque. NeuroImage 109:84–94

Scott JA, Grayson D, Fletcher E, Lee A, Bauman MD, Schumann CM, Buonocore MH, Amaral DG (2016) Longitudinal analysis of the developing rhesus monkey brain using magnetic resonance imaging: birth to adulthood. Brain Struct Funct 221:2847–2871

Seidlitz J, Sponheim C, Glen D, Ye FQ, Saleem KS, Leopold DA, Ungerleider L, Messinger A (2018) A population MRI brain template and analysis tools for the macaque. NeuroImage 170:121–131

Srihasam K, Sullivan K, Savage T, Livingstone MS (2010) Noninvasive functional MRI in alert monkeys. NeuroImage 51:267–273

Vanduffel W (2018) Long-term value memory in primates. PNAS 115:1956–1958

Wager TD, Yarkoni T (2012) Establishing homology between monkey and human brains. Nat Methods 9:237–239

Wisco JJ, Killiany RJ, Guttmann CR, Warfield SK, Moss MB, Rosene DL (2008) An MRI study of age-related white and gray matter volume changes in the rhesus monkey. Neurobiol Aging 29:1563–1575

Wittig RM, Crockford C, Deschner T, Langergraber KE, Ziegler TE, Zuberbuhler K (2014) Food sharing is linked to urinary oxytocin levels and bonding in related and unrelated wild chimpanzees. Proc Biol Sci 281(1778):20133096

Zitella LM, Xiao YZ, Teplitzky BA, Kastl DJ, Duchin Y, Baker KB, Vitek JL, Adriany G, Yacoub E, Harel N, Johnson MD (2015) *In vivo* 7T MRI of the non-human primate brainstem. PLoS One 10(5):e0127049

Correction to: The Behavioral Ecology of the Tibetan Macaque

Jin-Hua Li, Lixing Sun, and Peter M. Kappeler

Correction to:
J.-H. Li et al. (eds.), *The Behavioral Ecology of the Tibetan Macaque*, **Fascinating Life Sciences,**
https://doi.org/10.1007/978-3-030-27920-2

In the original version of the book,

The following sentence **"The Chinese National Natural Science Foundation sponsored the meeting and also provided funding to support open access publication of this volume."** in Acknowledgements section has been updated as follows **"The National Natural Science Foundation of China sponsored the meeting and also provided funding to support open access publication of this volume, as well as the publishing fund of Hefei Normal University. Hefei Normal University also provided a fund to defray the cost of publication including book purchase."**

The following sentence **"This book is mainly based on research papers presented in a spirited international primatology symposium held in the scenic area of Mt. Huangshan, China, in the summer of 2017. The chapters were grouped into four logical parts."** in Preface has been updated as follows **"This book is mainly based on research papers presented in a spirited international primatology symposium held in the scenic area of Mt. Huangshan, China, in the summer of 2017. The chapters were grouped into five logical parts."**

The following sentence **"Many recent discoveries in primatology involve technological advancements in research, which is the content of Part IV. In a single chapter (Chap. 14), Yong Zhu and Paul A. Garber explore the great potential of the high field MRI technology in the study of primate behavior and**

The updated online version of this book can be found at
https://doi.org/10.1007/978-3-030-27920-2

J.-H. Li et al. (eds.), *The Behavioral Ecology of the Tibetan Macaque*, Fascinating
Life Sciences, https://doi.org/10.1007/978-3-030-27920-2_15

cognition." in Preface has been updated as follows **"Many recent discoveries in primatology involve technological advancements in research, which is the content of Part V. In a single chapter (Chap. 14), Yong Zhu and Paul A. Garber explore the great potential of the high field MRI technology in the study of primate behavior and cognition."**